Handbook of
Behavioral Neurobiology

Volume 5
Motor Coordination

HANDBOOK OF BEHAVIORAL NEUROBIOLOGY

General Editor:
Frederick A. King
Yerkes Regional Primate Research Center, Emory University, Atlanta, Georgia

Volume 1   Sensory Integration
Edited by R. Bruce Masterton

Volume 2   Neuropsychology
Edited by Michael S. Gazzaniga

Volume 3   Social Behavior and Communication
Edited by Peter Marler and J. G. Vandenbergh

Volume 4   Biological Rhythms
Edited by Jürgen Aschoff

Volume 5   Motor Coordination
Edited by Arnold L. Towe and Erich S. Luschei

A Continuation Order Plan is available for this series. A continuation order will bring delivery of each new volume immediately upon publication. Volumes are billed only upon actual shipment. For further information please contact the publisher.

Handbook of
Behavioral Neurobiology

Volume 5
Motor Coordination

Edited by

Arnold L. Towe
University of Washington, School of Medicine
Seattle, Washington

and

Erich S. Luschei
University of Washington, School of Medicine
Seattle, Washington

SPRINGER SCIENCE+BUSINESS MEDIA, LLC

Library of Congress Cataloging in Publication Data

Main entry under title:

Motor coordination.

  (Handbook of behavioral neurobiology ; v. 5)
  Bibliography: p.
  Includes index.
  1. Human mechanics. 2. Animal mechanics. 3. Motor ability. 4. Sensory-motor inte-
gration. I. Towe, Arnold Lester, 1927-   . II. Luschei, 2. Movement. 3. Muscle con-
traction. 4. Neurophysiology. W1 HA511 v. 5 / WL 102 M922]
  QP301.M69   152.3'85   81-17925
  ISBN 978-1-4684-3886-4   ISBN 978-1-4684-3884-0 (eBook)   AACR2
  DOI 10.1007/978-1-4684-3884-0

Contributors

MARJORIE E. ANDERSON, *Departments of Rehabilitation Medicine and Physiology and Biophysics, University of Washington School of Medicine, Seattle, Washington*

H. PETER CLAMANN, *Department of Physiology, Medical College of Virginia, Richmond, Virginia*

EBERHARD E. FETZ, *Department of Physiology and Biophysics and Regional Primate Research Center, University of Washington, Seattle, Washington*

ALBERT F. FUCHS, *Department of Physiology and Biophysics and Regional Primate Research Center, University of Washington, Seattle, Washington*

LEON G. HOWELL, *Department of Psychology, University of Arizona, Tucson, Arizona*

DONALD M. LEWIS, *Department of Physiology, Medical School, Bristol University, Bristol, England*

RODOLFO R. LLINÁS, *Department of Physiology and Biophysics, New York University Medical Center, New York, New York*

PETER B. C. MATTHEWS, *University Laboratory of Physiology, Parks Road, Oxford, England*

LEWIS M. NASHNER, *Neurological Sciences Institute of Good Samaritan Hospital and Medical Center, Portland, Oregon*

PETER C. SCHWINDT, *Epilepsy Center, Veterans Administration Hospital, and Departments of Physiology and Biophysics and Medicine (Neurology), University of Washington School of Medicine, Seattle, Washington*

vi

CONTRIBUTORS

JOHN I. SIMPSON, *Department of Physiology and Biophysics, New York University Medical Center, New York, New York*

MARY C. WETZEL, *Department of Psychology, University of Arizona, Tucson, Arizona*

MARIO WIESENDANGER, *Department of Physiology, University of Fribourg, Fribourg, Switzerland*

Preface

The focus of this volume differs from what is suggested by the series title, for it is on muscle contraction and movement rather than on behavior. The lone overnight flight of a ruby-throated hummingbird across the Gulf of Mexico is a migratory behavior mediated through an incredibly lengthy, repetitive series of wing movements, each movement being produced by a complex sequence of muscle contractions. It is significant that these same movements may be used to mediate other behaviors, and that these same muscle contractions, in different sequence, may be used to produce other movements. The immense journey of white-bearded gnus across the Serengeti plains to suitable calving grounds is likewise a migratory behavior mediated through rather more varied, yet repetitive, limb movements, each produced by a complex sequence of muscle contractions. Again, these same movements may be used to mediate other behaviors, and again, the details of each limb movement may be varied through variations in the strength and the sequence of muscle contractions. A laboratory rat may learn to perform an escape behavior in a shuttle box, bringing its performance to a high level of efficiency by modifying its movement on successive trials. After intraperitoneal injection of pentobarbital sodium in an amount sufficient to render the animal severely incoordinated, the escape behavior is still performed, albeit through a different sequence of movements, even to "rolling" out of the compartment in response to the warning signal. The lesson is that, even though the topography of the movements making up a behavior may be much the same each time the behavior occurs, the animal is not restricted to that particular sequence of movements in carrying out the behavior. The analogy to a vocabulary *(movements)* and language *(sequencing of movements)* through which messages *(behaviors)* are expressed is not far amiss.

This book is about the physiology of muscles, of spinal reflexes and of supraspinal regulating systems. It is a journey of exploration that leads along familiar routes to the edges of the unknown, and includes several forays into the unknown. It is a guided tour by a group of seasoned explorers. The group was assembled by two physiologists who themselves have explored the dizzying heights of behavioral research and descended to *terra firma* (the more *firma,* the less *terra*) and who—perhaps tainted by their experience—share a deep conviction that behavior is more than a simple assemblage of movements. They also share the conviction that a thorough knowledge of the mechanisms through which behavior is expressed is essential to understanding behavior. Thus, this volume is an important part of the series on behavioral neurobiology. The reader will establish a firm perspective through this volume, and will find the later chapters looking in the direction of behavior. The tour through the lowlands of neurophysiology leads to the base camp of posture and locomotion, from which it is possible to discern the misty outlines of the enchanted peak of behavior. Our hope is that some young and thoughtful reader may ultimately find a path by which to scale that elusive peak. The view from its heights back down the path to fundamental physiology promises to be spectacular.

We wish to thank Dr. Remedios W. Moore and Ms. Kate Schmitt for extensive editorial advice and help with the production of finished manuscripts of several of the chapters. Dr. Moore was supported by the Department of Physiology and Biophysics at the University of Washington, and Ms. Schmitt by the Regional Primate Research Center at the University of Washington, a Center supported by NIH grant RR 00166.

ARNOLD L. TOWE
ERICH S. LUSCHEI

Contents

CHAPTER 4

Control of Motoneuron Output by Pathways Descending from the Brain
Stem 139

Peter C. Schwindt

CHAPTER 5

Cerebellar Control of Movement 231

Rodolfo R. Llinás and John I. Simpson

CHAPTER 6

Albert F. Fuchs

CHAPTER 7

Marjorie E. Anderson

CHAPTER 11

Properties and Mechanisms of Locomotion 567

Mary C. Wetzel and Leon G. Howell

The Physiology of Motor Units in Mammalian Skeletal Muscle

DONALD M. LEWIS

THE HISTORICAL CONCEPT OF THE MOTOR UNIT

The term "motor unit" was first used in a paper by Liddell and Sherrington (1925) to describe the "motoneurone axon and its adjunct muscle fibers." Later Sherrington (1925) made it clear that the term should be taken to include "together with the muscle-fibers innervated by the unit, the whole of the axon of the motoneurone from its hillock on the perikaryon down to its terminals in the muscle." This definition clearly excludes the motoneuron soma. Although this chapter will adhere to Sherrington's definition by concentrating on properties of the axons and of the muscle fibers of motor units, the relationships between these two components of the motor unit can be discussed most profitably together with a knowledge of the firing patterns of the motoneurons, which are determined largely by the properties of the cell bodies and their synaptic connections.

An understanding of the motor unit is equally important today in several lines of physiological study. Current research concentrates on differences between motor units within an animal and the mechanisms by which the properties of the neural and muscular elements of the unit are interrelated, particularly how the nervous system determines the properties of the muscles. Studies of motoneurons have revealed a wide range of properties that, to a large extent, are responsible for the range of activity exhibited by motor units; details of spinal organization may be as important as those of descending pathways in the precise control of muscle contraction. A quantitative assessment of the two factors will be an important contri-

DONALD M. LEWIS   Department of Physiology, Medical School, Bristol University, Bristol, Avon BS8 1TD, England.

bution to our understanding of motor control. In addition, however, the effect of a motoneuron discharge pattern depends on the response of the muscle part of the motor unit, and a full knowledge of both components is essential in understanding the role of a motoneuron in any particular movement. The realization that muscle properties are linked to those of the motor nerve at the level of the motor unit was an important step in understanding how regulation of motor response might develop and be modified in an individual. The link between motoneuron and muscle properties may be the most important part of these studies: the mechanism by which nerve influences muscle development, and subsequently can produce activity-related changes in muscle, may have far-reaching parallels within the central nervous system.

The concept of the motor unit seems to have been introduced by Sherrington and his colleagues in order to allow quantitative description of muscle contraction. This was essential because muscle tension was the index of motoneuron excitation in their studies of reflex activity. Sherrington (1925) made it clear that the motor unit was the smallest functional unit of muscular contraction because it responds in an all-or-none manner, in contrast to the graded response of a whole muscle. Earlier, Mines (1913) had shown that as the stimulus to the motor nerve of a frog muscle was increased continuously, the twitch responses increased in a small number of finite steps. The steps were too few to be ascribed to individual muscle fibers and represented recruitment of single axons, each of which brought in a new group of muscle fibers (Fig. 1).

The all-or-none response of the motor unit was explicitly discussed by Sherrington, but a second important property is implicit in these early papers. It is assumed that the mechanical responses of motor units summate linearly, so that the contraction of a whole muscle can be predicted by simple summation of individual motor-unit contractions. A requirement for linear summation is that no muscle fiber be innervated by more than one motoneuron. Hunt and Kuffler (1954) found nonlinear summation of twitch contractions and suggested that some polyneuronal innervation was possible in the mammal. However, linear summation of tetanic contractions was later demonstrated by Brown and Matthews (1960), who concluded that polyneuronal innervation was either not present in the mammalian

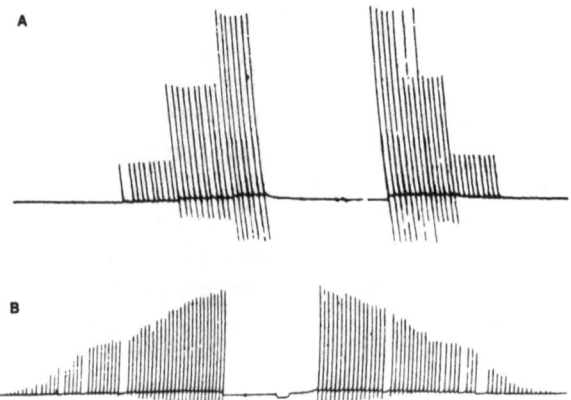

Fig. 1. An early record of motor units. Twitch contractions have been recorded from frog sartorius muscle. In (A) the nerve was stimulated at gradually increasing stimulus strength, and three motor units are recruited showing all-or-none responses. In (B) the muscle has been curarized and the stimulus applied directly to the muscle. The smaller steps of tension increase now correspond to individual muscle fibers. From Mines (1913).

limb muscle or was at a very low level. The nonlinear summation of twitch contractions was ascribed to mechanical interactions (see below), which would cause as much error as would polyneuronal innervation in the use of contraction force as an index of motoneuron activity.

TWITCH AND TONIC MUSCLES

In lower vertebrates a number of the striated muscles contain tonic fibers as well as twitch fibers (Hess, 1970; Kuffler and Vaughan Williams, 1953). Tonic muscle and twitch muscle are qualitatively similar in many details of fiber structure and both are described as striated muscle. The two types differ, however, in their motor innervation. Typically, tonic fibers have end plates at a number of places along their surface; the end plates are simpler than the single terminals of twitch muscles, not having the complexity of branching of the nerve terminals or the great infolding of the underlying muscle membrane. The twitch end plate is organized to give a large end-plate potential that triggers an all-or-none propagated muscle-fiber action potential. In tonic muscle fibers there is typically no muscle action potential either in response to physiological stimulation or to strong direct stimulation with intracellular current injection. The distributed local end-plate potentials in the muscle fibers initiate muscle excitation directly. Since the resultant muscle depolarization is graded according to the frequency of nerve stimulation, the muscle contraction is also graded according to frequency. There is no tension or shortening in response to a single stimulus. Tonic fibers are also termed *slow fibers* but this phrase should not be confused with *slow twitch fibers* (see below).

It is important here to distinguish between two terms. Multiterminal innervation indicates that a muscle fiber has more than one motor nerve ending on it. These endings can be derived from branches of just one axon. Alternatively, the endings may be supplied by more than one axon derived from different motoneurons; this type of innervation is referred to as *polyneuronal*. Tonic muscle fibers are multiterminally innervated and in most cases probably polyneuronally innervated. Many nonmammalian muscles are polyneuronally innervated even when they are of the twitch type (e.g., frog sartorius). Unfortunately, it is not always clear whether a muscle fiber with more than one end plate derives these from one or more motoneurons, because axons may branch at some distance from the muscle (Dunn, 1900), and if the experimental conditions only allow use of a short length of muscle nerve, then it will be impossible to distinguish between two axons and two branches of a single axon.

Mammalian skeletal muscle is mostly of the twitch type. Tonic fibers have been demonstrated in muscles innervated by cranial nerves: they have been found in the extraocular muscles (Browne, 1976; Hess and Pilar, 1963), the tensor tympani of the middle ear (Erulkar, Shelanski, Whitsel, and Ogle, 1969), and possibly in laryngeal muscles (Rossi and Cortesina, 1965). It is of interest that those muscles found to contain tonic fibers also have twitch fibers that show the body's most rapid twitch contractions (see below). Some intrafusal fibers of mammalian muscle spindles might be considered to be related to tonic fibers more closely than to twitch fibers, since many receive multiterminal innervation from more than one axon and do not produce propagating action potentials following motor stimulation. Apart from intrafusal fibers, mammalian skeletal muscles innervated by spinal nerves and the majority of those innervated by cranial nerves are of the twitch type, and generally the fibers receive only one axon. There is one major exception to the rule, however, in that polyneuronal innervation is common in muscles innervated by actively growing nerves. A naturally occurring example is found in normal postnatal development, during which there is clear evidence for multiterminal (Redfern, 1970) and polyneuronal (Bagust, Lewis, and Westerman, 1973, and Fig. 2) innervation. Secondly, when nerve growth and muscle reinnervation occur after axotomy, polyneuronal innervation again becomes common even in the adult mammal. Early after reinnervation, polyneuronal innervation may be extensive, but the incidence falls rapidly (Guth, 1962), although a small number of fibers may retain two nerves for long periods (Frank, Jansen, Lømo, and Westgaard, 1975) after special reinnervation procedures (Gutmann and Hanzliková, 1967).

EARLY STUDIES

ENUMERATION OF MOTOR UNITS

The first quantitative investigation of motor units was an attempt by Eccles and Sherrington in 1930 to count the number of axons supplying several cat hindlimb muscles and so deduce the mean size of a motor unit (by dividing the total muscle tension by the number of axons). In a parallel investigation Clark (1931) counted the fibers in a muscle, allowing estimates to be made of the mean number per motor unit. These authors knew that the histograms of motor axon diameters showed two clear groups (Fig. 3) in muscle nerves from which the sensory fibers had been removed by resection of the appropriate dorsal root ganglia, seven to fourteen weeks earlier. They concluded that the total number of motor axons should be used to calculate mean motor-unit sizes, since the axons of the group of small fibers passed through the ventral roots, had cell bodies within the spinal cord

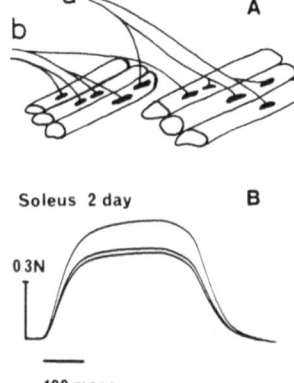

Fig. 2. Polyneuronal innervation of neonatal muscle. (A) shows two ventral roots (*a* and *b*) innervating two sets of muscle fibers. Some fibers are innervated by one axon only, others by more than one axon. (B) shows experimental records from such a muscle (kitten soleus). The two smaller myograms were elicited by tetanically stimulating root *a* or root *b*, the larger myogram by stimulating *a* and *b* simultaneously. A proportion of fibers are innervated by axons in *a* and *b*, so the total response is less than the sum of the individual responses to *a* and *b*. (B) is from Bagust *et al.* (1973).

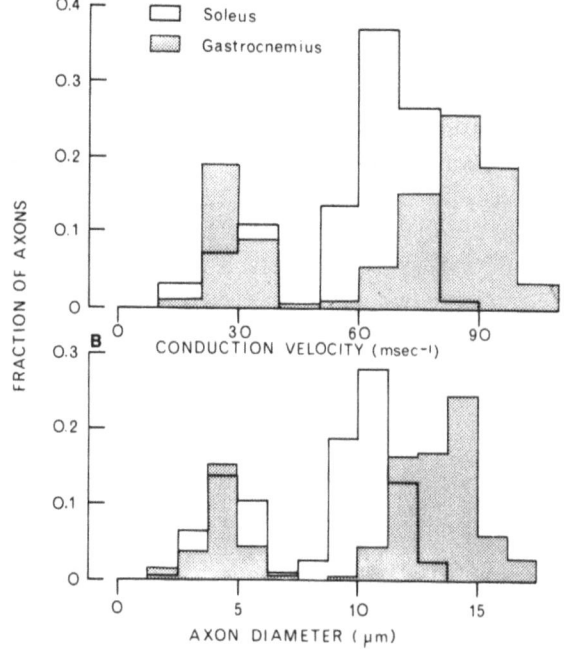

Fig. 3. Variation in motor axon properties. The histograms also show the differences between axons innervating fast-twitch muscle (gastrocnemius) and slow-twitch muscle (soleus). In (A) axonal conduction velocities are from McPhedran *et al.* (1965) and Weurker *et al.* (1965); in (B) axonal diameters from Eccles and Sherrington (1930).

(i.e., were not postganglionic automomic), and were not branches of large axons. It is now known that the small γ motoneurones supply only intrafusal muscle fibers, which contribute negligible tension to the total muscle contraction, and that the early estimates of mean motor-unit size should be corrected.

A second problem met in these investigations was the fact that motor axons branched extensively at large distances from the muscle. Eccles and Sherrington (1930) found that, for a nerve to medial gastrocnemius muscle, the number of axons increased from 393 at 60 mm from the muscle (the farthest distance that this nerve could be separated cleanly) to 482 at 6 mm from the muscle, an increase of 23%. Since branching was uncommon in the smaller axons, the increase in the number of large α-axons may have been as much as 33%. Axon branching may be a source of major error in studies of muscles with short motor nerves such as the extraocular muscles. Muscle tension and number of axons tended to covary so that mean motor-unit tension was relatively constant for any one muscle. For example, in a sample of 11 semitendinosus nerves, the number of α-axons (estimated from the total motor nerve counts) ranged from 84 to 444 and the muscle tensions from 7 to 45N*. Most muscles, however, had a much more limited range. The mean motor-unit tension, calculated as muscle tension divided by the number of α-axons, varied only between 83 and 101 mN. This consistency was not maintained between muscles; semitendinosus gave the smallest values while the mean motor-unit tension for medial gastrocnemius was 450 mN.

*The Newton (N) is the SI unit of force. It is defined as the force required to accelerate a mass of 1 kg to a velocity of 1m/sec in 1 sec (compare the dyne which is 10^{-5}N). In more practical terms, 0.981 N is the force exerted by gravity on a 100 g weight (or on a Standard Newtonian Apple: unpublished demonstration to The Physiological Society at University College, London, 1978).

More recent measurements (Olson and Swett, 1966) have provided the basis for similar estimates from cat hind-limb muscles of intermediate size: 260 mN for flexor hallucis longus, 150 mN for soleus, and 150 mN for flexor digitorum longus. No axon counts are available for small muscles of the extremities in the cat. However, as described later, it is now possible to measure motor-unit tension directly and obtain a mean value from an adequate sample of units. In three muscles (Bagust, 1974; Bagust, Knott, Lewis, Luck, and Westerman, 1973; Olson and Swett, 1966) mean unit tension has been measured both from motor axon counts and directly from a sample of motor tensions: in all three muscles the agreement between the two techniques has been good. The small superficial lumbrical muscle of the cat had its mean motor-unit tension measured directly (Appelberg and Emonet-Dénand, 1967; Emonet-Dénand, Laporte, and Proske 1971) and gave a value of 68 mN. This value of 68 mN is probably a minimum since many motoneurons supply both first and second superficial lumbricals (Emonet-Dénand *et al.,* 1971) and may well branch to other lumbricals, although no tests have been reported. Therefore, cat lumbrical motor units share what are considered to be anatomically separate muscles, making completely independent use of such muscles impossible unless the central organization exists to recruit, for discrete movements, only those units that are confined to one muscle. In man, for whom independent movement of the digits is important, there is evidence against motor units spreading outside the bounds of one anatomical muscle in the hand.

To summarize, it has been shown that the mean motor-unit tension is consistent, when examining one muscle in a number of individual animals. More remarkable is the fact that the variation in mean motor-unit tension between different muscles in the cat hind limb is only over a sevenfold range. The relative consistency of the mean motor-unit tension of a particular muscle is remarkable when the great variation of motor units within a muscle is taken into consideration (see below).

Measurements of mean motor-unit tension have been made in few muscles in man but mean motor-unit size can be characterized instead by the "innervation ratio," which is calculated as the ratio of the number of fibers in a muscle to the number of α-axons in its motor nerve. A detailed series of measurements was carried out by Feinstein, Lindegard, Nyman, and Wohlfart (1955) who also reviewed earlier work. They obtained innervation ratios in man ranging from 25 for platysma (in the neck) to 1,934 for medial gastrocnemius. Intrinsic hand muscles had small intermediate innervation ratios (107 for lumbrical, 340 for interosseus), whereas other limb muscles had higher intermediate values (brachioradialis, 410; tibialis anterior, 562). An even smaller value of 9 has been recorded for external rectus, an extrinsic eye muscle (Björkman and Wohlfart, 1936), but the problems of axon branching make determination especially uncertain in this site.

All these measurements must be considered with some reservation because of difficulties inherent in any human measurements. The authors had to work with mixed motor nerves, which requires that assumptions be made about the ratio of motor to sensory axons. This ratio has only been measured in experimental animal studies and one human case. Boyd and Davey (1968) have shown that the motor:sensory ratio could vary between 0.4 and 1.6 (and possibly up to 3.4), even between motor nerves in the hind limb of the cat. The presence of afferent fibers, some of which are in the β range, also obscures the division between α- and γ-axons, adding to the uncertainty of α-axon estimates. A further difficulty is that although some of the authors discussed the problem of axon branching near the muscles (Eccles and Sherrington, 1930), they were not always able to correct for branching. Finally, if some axons innervated anatomically separate muscles (Emonet-Dénand et al., 1971) no knowledge of this would be obtained from cadaver material.

Motor-unit size has been estimated as innervation ratio in many species but only with sufficient accuracy and controls in the cat. Clark (1931) counted fibers in soleus and extensor digitorum longus, and if his figures are corrected by later estimates of $\alpha : \gamma$ ratios in nerves to these muscles (Boyd and Davey, 1968), they show innervation ratios of 213 and 286, respectively. Cat flexor digitorum longus has a ratio of 248 muscle fibers per axon (Westerman, Lewis, Bagust, Edjtehadi, and Pallot, 1974). The innervation ratio in medial gastrocnemius may be estimated as 590 from measurements of fibers (Burke and Tsairis, 1973) and axons (Boyd and Davey, 1968) in different individuals. The first superficial lumbrical muscle has 2,000 fibers (Lewis and Edjtehadi, unpublished) and may be estimated to have 10 to 15 motor axons (Appelberg and Emonet-Dénand, 1967), giving an innervation ratio of 133 to 200. However some 30% of the motor axons innervating the first superficial lumbrical muscle also innervate other lumbrical muscles (Emonet-Dénand et al., 1971). Moreover, those axons that branched to two muscles were from motor units that developed higher tension (85 mN) than the mean (68 mN) and may have contained more muscle fibers (see below). Therefore, the real innervation ratio for the axons supplying these foot muscles will be greater than the directly calculated value of 200 derived above.

If only the limb muscles are considered (no eye-muscle estimates have been made in cat), the range of innervation ratios in man (from 107 to 1934) is greater than that in cat (180 to 590) but is not grossly dissimilar.

In spite of the difficulties of interpretation, it would seem that the mean number of muscle fibers per motor unit extends over at least a tenfold range when comparing different muscles and different species, and that the smallest motor units are found in the smallest muscles. It must be emphasized again that these are only mean values, and the distribution of motor-unit sizes within a muscle is at least as important. The precision of control must be related to the force demanded of a muscle. Delicate finger movements involve use of the intrinsic muscles of the hand as well as small contractions of those muscles in the forearm that act on the fingers. If more force is required, the forearm muscles will do most of the work and, although the motor units involved are larger, the precision of control must be considered relative to the total output of the muscle and need not be less. On a comparative basis the number of motor units in a muscle is the important estimator of potential accuracy in control.

The sensory innervation of a muscle is also important in accuracy of control and, although muscle sensory nerves are discussed elsewhere, it is worthwhile to compare briefly here sensory and motor innervation ratios. As with motor innervation, sensory innervation is denser in muscles that are used for finely controlled tasks. The similarity between motor and sensory innervation ratios may be illustrated by looking at the ratio of α-motor to Groups I and II sensory axons (which innervate muscle spindles and Golgi tendon organs). The counts in cat hind-limb muscle nerves made by Boyd and Davey (1968) yield motor:sensory ratios that range from 0.3 to 1.77 between muscles, but that showed no clear relationship to the total number of large axons.

It has been suggested not only that the spindle:muscle ratio is highest in small, finely controlled muscles but also that this ratio varies over two orders of magnitude between muscles. The spindle:muscle ratio has been calculated, however, as the number of spindles per gram of muscle, and muscle mass is not an appropriate measure for the present purpose. The mass of a muscle depends on fiber length, fiber diameter, and number of fibers (among other factors), but the first two are irrelevant to fineness of control. Two muscles could have the same number of motor and sensory axons and therefore be capable of the same relative accuracy of control. If, however, the fibers of one muscle were twice as long and had twice the diameter, that muscle would have eight times the mass of the other. Estimates of number of spindles per gram would appear to show, incorrectly, an eightfold difference in accuracy of sensory information.

VARIATION OF MOTOR UNITS WITHIN MUSCLES

So far, discussion has been restricted to comparisons of mean motor-unit sizes between muscles. Eccles and Sherrington (1930) were well aware that motor-unit properties need not be uniform within a muscle. Indeed, it was clear that the one characteristic which they could quantify, the diameter of the axon, showed great variation (see Fig. 3B). They also observed that large axons branched more frequently than small ones and deduced from this that the large axons would supply more muscle fibers and form larger motor units. This difference in branching was clear between α- and γ-axons; the mean diameter of 13 axons which were seen to branch was 15.6 μm (range 14–22 μm), significantly greater than the overall α-axon mean of 13.4 μm (range 9–20 μm).

Apart from size there were indications that the contracile properties of motor units would not be uniform. There are differences between the contraction speeds of individual muscles (Cooper and Eccles, 1930). A broad classification into fast-twitch and slow-twitch muscles can be made, for there is approximately a threefold difference in the duration of isometric twitch contractions of the two types. Isometric and isotonic contractions and biochemical properties exhibit similar differences (reviewed by Close, 1972). Examples of the isometric twitches of a fast and slow limb muscle are illustrated in Fig. 4, derived from a paper by Cooper and Eccles (1930). Their results also include the twitch of an internal rectus muscle of the eye, which is much briefer than that of the fast-twitch muscle. In one way, however, limb fast-twitch and eye muscles can be classified together, for Close and Luff (1974) have shown that the isotonic shortening velocities of an extrinsic eye muscle and a fast limb muscle are similar. The faster isometric twitch duration of the eye muscle presumably is due to a smaller release or faster re-uptake of calcium ions within the fibers following a single stimlus, rather than a different contractile protein.

Some slow-twitch muscles are anatomically distinct muscles, although they are all essentially one head of a muscle complex: soleus tendon unites with those of the gastrocnemii and plantaris to form the Achilles tendon. In some muscle groups the slow head is less distinct, for example, crureus in quadriceps. As well as the different contraction speed, slow muscle is redder in color than fast muscle. Other muscles (cat tibialis anterior) have a superficial white layer and a deeper red layer, and by artificial separation of these layers, Gordon and Phillips (1953) showed that the deep, redder layer also had a slower contraction.

Electromyographic (EMG) studies were the first to distinguish individual motor units by the use of fine needle electrodes that record from only a small number of muscle fibers. As early as 1929, Denny-Brown had shown that red muscle had a lower threshold to stretch and labyrinthine reflexes. Adrian and Bronk (1929) noted that motor-unit potentials in quadriceps muscles activated by the stretch reflex (probably recorded from the low threshold, red parts) began to discharge at about 10/sec and could increase their rate up to 25/sec. In contrast, flexor reflexes caused motor units of flexor muscles to fire at rates of up to 90/sec. Gordon and Phillips (1953) found differences in unit firing rates between superficial (50/sec) and deep layers (28/sec) of tibialis anterior during flexor reflexes. Systematic observations by Fischbach and Robbins (1969) have confirmed these earlier observations that motor units of fast muscle discharge at higher rates than those of slow muscle, but there was so much variation within muscles that the ranges overlapped. This variation is also seen in fast human muscles (Milner-Brown, Stein, and Yemm, 1973a; Tanji and Kato, 1973).

There were several early attempts to record mechanically from motor units. Denny-Brown (1929) recorded oscillations in myograms during stretch. Eccles and Sherrington (1930) recorded tendon jerk–reflex contractions of muscles, the motor nerves of which had been cut down to contain few axons, and found step increases in twitch-like contractions. Porter (1929) and Porter and Hart (1923) also recorded threshold twitch contractions from tenuissimus and tail muscles excited in flexor reflexes. Gordon, with Holbourn (1949) and with Phillips (1953), made use of the well-known fact that, during minimal contractions, dimpling could be seen on muscle surface (or heard through a stethoscope) owing to contraction of individual motor units, and they were the first researchers to show clearly the differences in twitch speed. They were unable to be sure of tensions because they were only recording from a sample of the fibers, which was true of earlier work involving nerve section, which might leave intact only a branch of an axon. Systematic mea-

Fig. 4. Isometric myograms of three types of twitch muscle. In each panel, a fused and unfused tetanus are illustrated, and the relaxation of a twitch superimposed with dashed lines. (A) Internal rectus muscle of the eye, (B) medial gastrocnemius and (C) soleus muscles of the hind leg of the cat. All myograms have the same time scale. Modified from the records of Cooper and Eccles (1930).

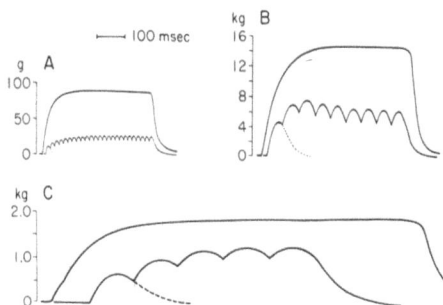

surements began to be made soon after this, but before discussing the results, it is worthwhile to consider methods that can be used.

TECHNIQUES OF STUDY OF MUSCLE AND MOTOR UNITS

The methods that are available for the study of motor units are based on, and often more restricted than, those that have been used for whole muscle. The two could come together in single fiber work, which is essential for the most fundamental understanding of muscle, for it is possible to identify the motor unit, or at least the type of muscle, from which the fiber was isolated. Studies have been made of the contractile, electrical, structural, biochemical (histochemical) and pharmacological properties of motor units. In this section, the basic muscle physiology on which motor-unit studies have been made is discussed. There is also, however, a more general consideration of those muscle properties that are of interest in the use of muscles in locomotion. For a more general treatment of muscle, the reader is referred to a monograph by Wilkie (1976); mammalian fast-and slow-twitch muscle is described by Buller (1975) and Buller and Buller (1979); muscle literature is reviewed by Close (1972).

MECHANICAL RESPONSES

ISOMETRIC AND ISOTONIC RECORDING. In whole muscle and in single fiber studies, two extreme types of the contraction (isometric and isotonic) have been commonly used in the detailed analysis of the mechanisms involved. In an ideal isometric contraction, the muscle length is held constant and measurement is made of tension developed after stimulation (Fig. 5A). A tension transducer works by converting force to a change of length that is then measured, but the device is designed so that the change in length is negligibly small. During isotonic contraction (Fig. 5B) the muscle is allowed to shorten against a constant force. In the after-loaded isotonic contraction illustrated in Fig. 6A, the constant force is applied only after the contraction begins by the simple method of supporting the weight that provides the load, so that the muscle is held at a constant length before con-

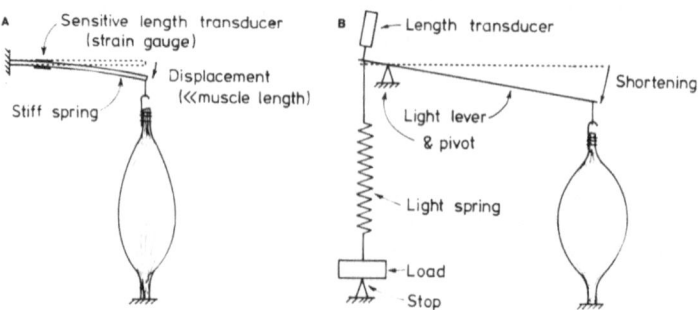

Fig. 5. Recording the mechanical response of muscle: the principles of (A) isometric and (B) isotonic myographs.

traction begins. The apparatus is complicated because, if the muscle were made to lift the load directly, inertial forces during acceleration would be added to the constant force of gravity on the mass. The lever and spring represented in Fig. 4B reduce the inertial force on the muscle. The inertia of the length transducer must also be minimized. The shortening of the muscle recorded against time after the onset of a stimulus is illustrated in Fig. 6A, and it can be seen that both the extent and speed of shortening decrease as the load on the muscle increases. When the load is sufficiently large, there is no shortening and the contraction is isometric; if it is made greater than maximum isometric tension, the muscle will be lengthened. At the top of Fig. 6A, the velocity of shortening has been derived in order to illustrate some details of the contraction. There is an initial phase without shortening, during which tension rises until it exceeds the force of the load. The duration of this isometric phase increases with load. The initial shortening shows an acceleration phase and, if the inertia of the apparatus is too high, an overshoot in velocity and even some cycles of oscillation. Finally, there is a period of constant velocity that lasts until either the muscle fibers reach a limiting length or the stimulus is removed. The relationship between the constant velocity of shortening and the load on the muscle is indicated in Fig. 6B. The relationship has been represented by Hill (1938) as a hyperbola (velocity a reciprocal function of load), or by Aubert (1956) as an exponential function. Curves derived from both equations fit experimental points equally well. Hill's equation is used more frequently, and may be written in the form

$$V = V_o \frac{(1 - P/P_o)}{(1 + P/a)}$$

where V is the velocity of shortening at load P, P_o is maximum isometric tension, and V_o and a are constants. V_o is the extrapolation of the curve to zero load and

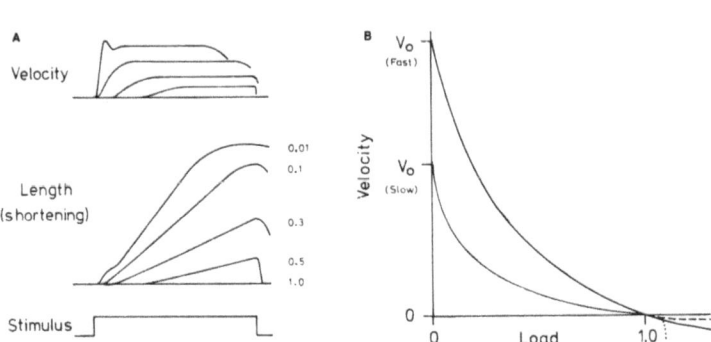

Fig. 6. Isotonic shortening. (A) Diagrammatic myograms at different loads (indicated as a fractiom of the maximum isometric force of the muscle). The lower line indicates the time of application of tetanic stimulation, the middle traces show muscle shortening, and above is the velocity of shortening derived from the length records. (B) The constant velocities have been plotted against load (expressed as the fraction of maximum isometric force), i.e., the force-velocity relationship. The upper graph is for fast-twitch muscle and the lower for slow-twitch muscle. The graph for fast-twitch muscle has been extrapolated for loads greater than maximum isometric tension. Superimposed on the extrapolation are measurements of velocities of extension at high loads (see Fig. 7).

is termed the maximum velocity of shortening. Fast muscle has a higher velocity of shortening than slow-twitch muscle (compare the two curves in Fig. 6B). The ratio a/P_o is an expression of the curvature of the force–velocity relationship, which is greater for slow-twitch muscle.

EXTENSION OF ACTIVE MUSCLE. If the after-load is made greater than maximum isometric tension, the muscle is extended (Katz, 1939). The initial length change is rapid (Fig. 7A), and if the load is large (about twice maximum isometric tension), it will extend the muscle up to the after-loading safety stop. At lighter loads the initial give is smaller and is followed by a phase of constant velocity lengthening. If the constant lengthening velocities are added to the force–velocity plot of Fig. 6B (dashed line), it can be seen that the rates of extension are several times slower than would be predicted by extrapolating the graph of shortening velocities at loads less than maximum isometric tension. A muscle is better able to resist forcible extension than would be predicted from its shortening properties. This last statement disregards the initial rapid extension. The initial lengthening velocity is greater than predicted from the shortening curve and almost independent of load. This initial give may be produced by the rapid disruption of many actin-myosin cross bridges. The faster the stretch, the more bridges broken, and the greater the fall of tension and the slower the recovery curve. Functionally, this "give" to rapid extensions may protect muscle against excessive loads and may play a part in muscle interactions in limb movements (see below). Neither Hill's nor Aubert's equations fit lengthening reactions.

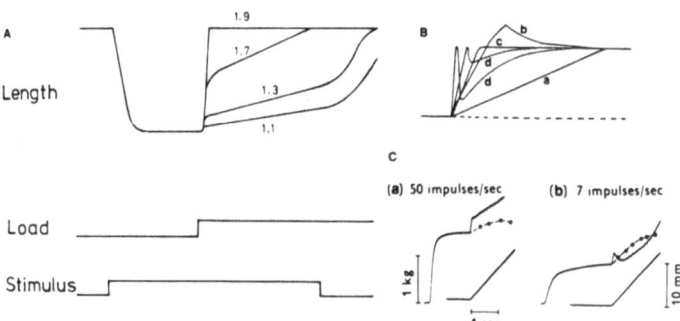

Fig. 7. Isotonic lengthening by two methods. (A) Series of loads greater than maximum isometric tension has been applied after a period of isotonic shortening. At some loads there is an initial rapid give, the extent of which is proportional to load. The velocity of the give is plotted as the dotted curve in Fig. 6. When the give is not too great, there is a second phase of constant velocity extension plotted as the dashed curve in Fig. 6. Data from Katz (1939). (B) and (C) Tension developed by a muscle when extended at constant velocity after a period of isometric contraction. (B) Diagrams by Gasser and Hill (1924) to show the effects of extension at various rates. At the lowest rate (a) tension rises smoothly from that at the shorter length to that at the longer length. At a higher rate (b) tension overshoots the final value. At the highest rate (d) there is a rapid rise of tension which equally rapidly collapses and redevelops only slowly. (C) Experimental results by Joyce et al. (1969) using distributed stimulation (see Fig. 8) in cat soleus. The muscle is extended at the same rate in the two myograms, but whereas tension overshoots when the muscle is stimulated at a high rate (a), at lower rates, with little tension fusion in individual fibers, the overshoot is transient (b). The dots in each case are the constant length (isometric) tetanic tensions corresponding to points in the extension ramps.

Effects of load on muscle movement have also been studied by measuring force in active muscle, the length of which is changed in a controlled manner, usually at constant velocity (Gasser and Hill, 1924). Experiments are illustrated in Fig. 7B. At low velocities of extension, the tension rises steadily to reach a new isometric value defined by the length-tension diagram (see Fig. 9C). At highest velocities the tension rises very rapidly but, sometimes before overshooting, falls almost as rapidly before slowly redeveloping to its final value. The rapid fall may be compared with the initial give at high loads (Fig. 7A) and may again represent the breaking of bonds involved in the contractile mechanism. Joyce, Rack, and Westbury (1969) have demonstrated in cat soleus that tension collapses at a lower rate of stretch when a muscle tetanus is elicited by distributed stimulation at a low rate, and this feature may be important in physiological contractions (Fig. 7C).

DISTRIBUTED STIMULATION. In a moderate voluntary contraction, motor units are firing at frequencies sufficiently low to produce unfused tetani (Fig. 8A), but because of asynchrony, contraction is smooth. In some experiments described above, stimuli were delivered at a high frequency in order to produce a smoothly fused tetanus. There are two objections to this procedure. First, these high frequencies do not occur naturally, except possibly in maximal voluntary effort. Second, fully fused tetani produce pressures within the muscle greater than systolic blood pressure and so cut off the blood supply to the muscle. An experimental technique has been developed to overcome these problems, and is indicated in Fig. 8 (B and C). The ventral roots supplying the muscle are divided into at least five parts, and each part is stimulated independently. The frequencies applied to all the root divisions are often identical, but because they are delayed in time relative to each other, the resulting unfused partial contractions are desynchronized, and the total contraction is as smooth as a voluntary one. (Incidentally, results obtained with this preparation are often similar to those obtained using synchronized high-frequency stimulation, but one could not have been certain of that result).

MUSCLE STIFFNESS. Rack and Westbury (1974) have used the technique of distributed stimulation described above to study the effects of physiological changes

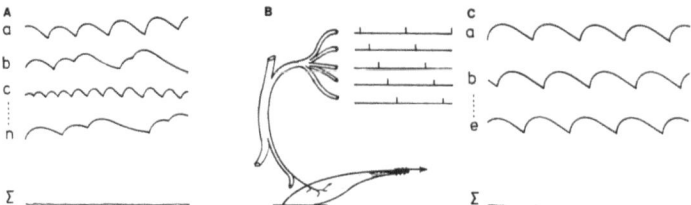

Fig. 8. Natural and distributed stimulation. (A) Normal voluntary contraction. The contractions of individual motor units (a, b, c . . . n) are indicated above. The tetani of natural responses are unfused, but because units fire at different frequencies, their oscillations are not synchronous, and therefore the total muscle tension (Σ) is smooth. (B) To mimic (A), the ventral root is divided into five (or more) parts, and each is stimulated at a rate which need not result in fusion; the five stimulus trains are delayed in time on each other. (C) Component contractions resulting from stimulation in (B) are shown as a, b . . . e, the overall muscle response (Σ) is smooth because of the desynchronization.

of length on active muscle. One of the experiments they performed was to apply regular triangular length changes to the stimulated muscle (Fig. 9A) and to measure the resultant oscillations of tension. Their results are best indicated by plotting length against tension (Fig. 9B). At small-length changes (central trace in Fig. 9B), the muscle behaved as a simple spring; tension was linearly related to length and there was no difference between lengthening and shortening. At higher amplitudes, there were two differences. First, the muscle showed hysteresis; the tension was higher at any length during lengthening than it was at the same length during shortening (see the arrows on the outer trace of Fig. 9B). Second, the slope of the curve was not constant; soon after a change of direction (either to lengthening or to shortening) the curve was steep with a slope about equal to that obtained with small amplitude movements, but toward the end of the length change, the slope became flatter. The slope of the traces of Fig. 9B represents the dynamic stiffness of the muscle (change of tension for unit change of length). The preceding description can be summarized by the statements that with small oscillatory movements the stiffness of the muscle is large and linear, but with larger oscillations, although the stiffness is initially as large as with small oscillations, it then decreases with the amplitude of the movement.

Length–Tension Diagram. It is of interest to compare the stiffness derived from such small movements with that given by the conventional length–tension diagram (Fig. 9C). It should be recalled that the change of tension with length (solid line) may be separated into two components. The first component (dotted line) is the tension developed by an unstimulated muscle (passive tension). This is present even in single muscle fibers as a result of the nonlinear stiffness of structural components such as the sarcolemma and fiber membrane. In an intact muscle, the stiffness of surrounding connective tissue is added, and passive tension is higher (in both absolute and relative terms). In many preparations used in the experiments reported here, all surplus connective tissue has been removed, but, if all the lateral attachments of a muscle are left intact, passive tension is even greater (Grill-

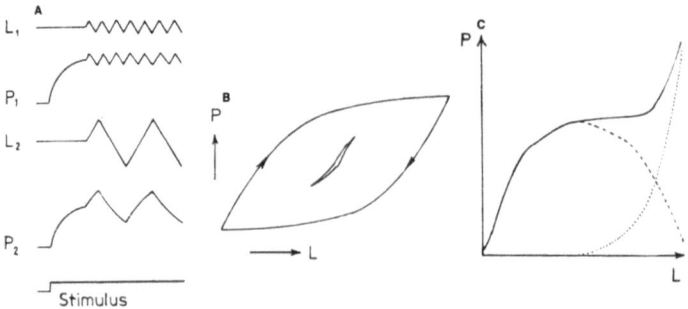

Fig. 9. Tension-length relationships of muscle. (A) Dynamic responses: oscillatory length changes (L_1 and L_2) applied to contracting muscle and the resultant muscle tension (P_1 and P_2). (B) Relation between tension and length from (A) (data from rack and Westbury, 1974). (C) Total tension developed by a muscle stimulated at a number of resting lengths (continuous curve). The initial, passive tension (dotted curve) was recorded before each stimulus, and the active tension (dashed curve) calculated as the difference between the two measurements.

ner, 1972), and this represents the state in the body for most muscles. In passing, it might be noted that connective tissue is not purely elastic; it has a higher tension during rapid extension than at the extended length. The dynamic reaction of passive tissues will reinforce the responses of active muscle described above, but is much smaller.

The second component of the muscle tension is the active tension (dashed line, Fig. 9C) derived as the difference between the total and passive tension (there is an assumption of linear summation). The shape of the active tension curve depends on the overlap between active sites on thick and thin filaments in the muscle (Gordon, Huxley, and Julian, 1966).

The steady state, or constant length, stiffness of the muscle is the slope of the length–tension curve. Rack and Westbury (1974) used muscle lengths to the left of the peak, where active tension increases with length, and where passive tension is relatively small. In another paper, Rack and Westbury (1969) have reported measurements of tension at constant length in cat soleus activated by distributed stimulation (their results are illustrated in Fig. 21). The steepest part of the length–tension curve for 10/sec stimulation has a slope of about 1.5 N/mm so that an additional load equal to 10% of maximum tetanic tension (or about 20% of the tension at the steep part of the curve) would increase muscle length by about 1.5 mm or about 3% of its initial length, and 10% of its physiological range of movement.

Dynamic Stiffness. The figure for constant length stiffness may be compared with the dynamic stiffness of the muscle to small oscillations of length. Dynamic stiffness was greatest if the oscillations were less than 1 to 1.5 mm. Maximum dynamic stiffness when movement was small was as high as 15 N/mm, some ten times higher than the constant length values.

The linear range of maximal stiffness increased as velocity increased within the tested range of 5 to 40 mm/sec. The maximum velocity of shortening of soleus in isotonic measurements is some 200 mm/sec and will be higher for its fast-twitch agonist and antagonist muscles. Recently, direct measurements of the tension and length of soleus and some agonist muscles have been made in conscious cats during normal activity (Prochazka, Westerman, and Ziccone, 1977; Walmsley, Hodgson, and Burke, 1978). In quiet walking, velocities of muscle lengthening and shortening reach a maximum of 30 mm/sec, rising to 100 mm/sec in running, and to 250 mm/sec in jumping or landing after a moderate fall. The experimentally imposed velocities of movement described above were therefore well within the physiological range of velocities.

Measurements of muscle stiffness are important since they estimate the contribution of the muscle itself to resisting unexpected imposed loads; the stiffer the muscle the more it will resist and the more faithfully it will follow a command signal from the nervous system. The muscle has built-in negative feedback that will provide servo length control. This property may be at least as important as reflex feedback since it is not subject to the delays inherent in any reflex. The muscle stiffness is greatest for small fast extensions.

Activity-Related Stiffness. An important property of the muscle stiffness is that a part of it is directly related to the force of contraction. The mechanism for this proportionality has been demonstrated in experiments on single fibers, a preparation in which the complications of parallel and serial connective tissue is minimized, and in which it is possible to measure and control accurately and rapidly changes in length and tension of a small, uniform group of sarcomeres within the fiber (Ford, Huxley, and Simmons, 1977). In actively contracting single fiber preparations, it has been found that the initial change in elastic tension in response to a very rapid stretch or release is linearly related to the initial active tension of that fiber. The initial tension was varied by holding the fiber at different lengths, which varies the overlap between thick and thin filaments, and therefore varies the number of cross bridges that are formed in a tetanus (Gordon *et al.*, 1966). The conclusion is that a part of the muscle elasticity resides in the cross bridges, so that the more cross bridges formed between thick and thin filaments, the greater will be the stiffness of the muscle. The proportion of cross bridges that are present will depend not only on the length of the muscle (as above) but also on the rate of stimulation and on the time course of stimulation; the rate of development of tension in an isometric tetanus appears to be determined by the rate of formation of cross bridges (Huxley and Simmons, 1971).

Whole Muscle Response. When whole muscle responds to length change, the cross-bridge elasticity will be modified by factors that were eliminated in the experiments described above. Most important will be the elasticity of the tendon and other connective tissue links. Rapid length changes applied to the tendon of a whole muscle are slowed by transmission along the fiber. Also, tension changes within the fiber will be attenuated at the tendon of insertion by the additional compliance and by frictional and other viscous forces. Further, the largest tension responses of single fibers are in part transient, with substantial recovery occurring within milliseconds. Despite these modifications, the stiffness of a whole muscle to a sudden change of load will be greater if the muscle is contracting forcibly. Rack and Westbury (1974) have demonstrated the relationship between stiffness and activity in the cat soleus muscle and calculated that the stiffness in small movements is increased by 1 to 1.5 N/mm for each 1 N of muscle tension. An alternative expression of this dynamic stiffness is that there will be an increase in active muscle tension of at least 50% in response to a 1% increase in fiber length. If the values of the cross-bridge stiffness measured by Ford *et al.* (1977) in isolated frog fibers are assumed to apply to cat soleus, they lead to an estimate of about 5 N/mm per 1 N of muscle tension, or about half this value if the early tension, which rapidly decays, be discounted. Therefore, the values of force-related stiffness for single fibers are greater than those for whole muscle, but it is clear that the stiffness of a whole muscle increases with the force generated by that muscle, and the force-related stiffness will have consequences for the performance of a muscle in responding to rapid length perturbations. The more force a muscle is generating the larger will be the load change it can tolerate for a given change of length, that is, the muscle can tolerate a sudden change of tension which is proportional to the tension being developed. The argument may be taken a stage further. Instead of developing the required tension, a muscle may develop a greater force, the excess being balanced by a contraction of antagonists. The resultant force about the joint will be identical in the two cases, but the stiffness of the muscles will be greater in the second case where both agonist and antagonist are active. In many tasks, balanced agonist-antagonist action is employed; palpate the biceps and triceps muscles in the upper arm when trying to resist flexion of the elbow, or consider that force is (or should be) developed in both arms on the steering wheel of a car. Even in normally walking cats, simultaneous activity can be recorded in flexors and extensors (Engberg and Lundberg, 1969), which would increase the stiffness of all muscle groups with resulting increased resistance to unpredicted loads.

MUSCLE STIFFNESS AND STRETCH RELEXES. If, as is normally the case, the muscle is connected to the nervous system, changes of length will elicit reflex changes in motoneuron activity that will modify muscle tension. For a complete analysis, it is necessary to assess the parts played by the direct muscle response, by the reflex response, and by the interaction of these two factors. Matthews (1969) has measured the tension response to small (0.2 mm) changes of length, from a number of initial lengths, in soleus of decerebrate cats in which there is an active stretch reflex. In these experiments, the reflexly controlled muscle had a stiffness of about 1.5 N/mm measured from the tension over a range of constant muscle lengths. Small oscillations of length elicited a stiffness of about 6 N/mm, and this stiffness was independent of the mean length of the muscle and therefore of the mean tension, which varied from 9 to 16 N in the range of tested lengths. It would be expected that the dynamic muscle stiffness would have increased as the mean muscle tension increased (see above). If so, the superadded dynamic reflex response might be concluded to have decreased reciprocally with length. Nichols and Houk (1976) have produced experimental evidence of a reciprocity between muscle and reflex stiffness. The reciprocal behavior that Houk proposes is one that compensates for nonlinearity in the stiffness of the muscle component seen in the difference between responses to increase and decrease of length. Matthews (1969) concluded that the contribution of muscle was much less than that of the reflex to stretch. In contrast Goodwin, Hoffman, and Luschei (1978) have found, in displacements of monkey jaw muscle, that a part of the early tension response was a direct contribution from muscle properties. The discrepancy may represent different states of the muscle and nervous system under different experimental conditions.

It is certain that the relative contributions of muscle and reflex will depend on temporal factors. Following a change of length, there is the immediate tension change due to the elastic elements. This tension change decays over tens of milliseconds to the value given by the constant length–tension relationship. The reflex stiffness will have time lags that occupy some 30 to 50 msec. If matched temporally, the muscle and reflex components could give a constant stiffness that would not be subject to gross temporal variation. However, this is speculation until precise analysis is made of the time course of the two components under identical conditions.

EFFICIENCY. The elastic behavior of muscle has importance from another point of view, that of muscle efficiency. During mammalian locomotion (except on a bicycle) the center of gravity rises and falls rhythmically, a process that involves alternate expenditure and dissipation of energy. To a certain extent, some of the energy released in the fall of the center of gravity can be stored as elastic energy in muscles and tendons—the Achilles tendon and the calf muscles being especially important in this respect. This stored energy is used in the next step cycle as the elastic elements shorten and help to lift the body, thus reducing the work that has to be done by the energy-requiring active shortening of the muscles. In jumping animals, such as the kangaroo, the Achilles tendon is long, increasing the elasticity which greatly improves the efficiency of hopping (Morgan, Proske, and Warren, 1978). Efficiency also depends on shortening velocity, being maximal at about one-third of maximum velocity or about one-third of maximal tension. The load for maximal efficiency is relatively lower for slow muscle although slow muscle is intrinsically more efficient.

PHYSIOLOGICAL CONTRACTIONS. At the beginning of this section it was stated that the classical method of studying muscles was under either isometric or isotonic conditions. Other conditions have also been used, however, in order to mimic more closely the conditions under which physiological contractions occur. In the simplest natural contraction, the muscle has been described as passing from an isometric to an approximately isotonic phase ending in a final isometric phase in which external or internal masses are maintained against gravity. In the isometric phase, tension is built up in the muscle to overcome the sum of forces opposing movement. These forces include any external load on the muscle, inertial forces of the external load or internal masses (bone and the muscle itself), and frictional or viscous forces. The latter forces are small except under water. The second phase only approximates to isotonic for a fraction of its duration since it begins with an acceleration and ends with a deceleration that, in normal animals, occurs without the overshoot of velocity shown in Fig. 6A. Undoubtedly, the smoothness of movement depends primarily on neural control. The increase and decrease of motoneuron discharge frequency are gradual and asynchronous in normal animals and people, not rectangular and correlated for all motor units, as in the artificial stimulation of Fig. 6A. Disorders of the central nervous system, particularly of the cerebellum, illustrate how lack of proper central control may produce gross errors in performing smooth movement.

The properties of muscle fibers also play a part in the production of smooth movement in that the rate constants for the contraction of particular muscles are matched to the inertial loads these muscles have to move. The matching is seen whether the potential load is compared with the velocity of isotonic shortening or the rate of isometric tension development, which is reflected in both the rise time of the isometric twitch and in the rate of rise of tetanic tension. The isometric rates have been studied extensively, and matching to load is seen whether considering differences between species or between muscles in one species. Between species, the twitch contraction time of an ankle extensor may be considered; this is very short in the mouse (9 msec) and increases progressively with body size, from rat (12 msec), to cat and rabbit (20 msec) and to man (80 msec). In man, a similar progression is observed when moving from low mass structures such as the eye, where the extrinsic muscles have twitch contraction times of about 20 msec, to the fingers (40 msec), to the arm (60 msec), and ending in the most massive part, the leg (120 msec). Such progressions ensure that delays in achieving peak isometric tension are related to the inertial loads on which the muscles operate (Hill, 1956).

RELATION OF ISOTONIC TO ISOMETRIC CONTRACTION. These principles of matching will apply to isotonic contractions since isotonic shortening velocity is inversely related to isometric contraction time (Close, 1971); this relationship is not unexpected since both properties are functions of the contractile proteins. However, the relationship need not be absolute since isometric tension development depends on the rate at which cross bridges between myosin and actin can form, whereas shortening velocity is limited by the rate of breaking of cross bridges. There is no *a priori* reason why the two rate constants should convary precisely, and certainly comparisons between species suggest no rigid link. Detailed examination reveals, for example, that cat flexor digitorum longus (FDL) has a twitch contraction time about twice that of rat extensor digitorum longus (EDL) (21 msec and 12 msec, respectively), although the former has a maximum shortening velocity which is considerably more than half that of the latter (35 and 45 μm/sec/sarcomere, respectively). Similarly, the intrinsic shortening velocities of hind-limb and extrinsic eye muscles are very close in the rat, although there are considerable differences in isometric twitch contraction times (Close and Luff, 1974). Close and Luff considered the differences between isometric twitches of eye and limb muscles to be due to differences in the activation process, presumably in the release or re-uptake of intrafiber calcium. Some authors have assumed a simple relationship that allows isotonic velocity of a motor unit to be deduced from its isometric contraction. The assumption may be true, but, if not, it will lead to serious error. The section on twitch–tetanus ratio (see below) offers some reasons to indicate that the simple relationship may not be true.

It is uncertain whether or not isotonic contractions can be recorded sensibly when only a small fraction of the muscle fibers are shortening against frictional forces and inertial loads of adjacent fibers, although in some of the earlier experiments (Porter, 1929) the technique of recording attempted to achieve isotonic conditions. Similar doubts have occurred about isometric contractions, but at least tetanic tension appears to be similar in motor unit and whole muscle contractions, as judged by the fact that the average of the tensions recorded from a number of motor units closely matches the mean motor-unit tension calculated from the muscle tetanus and the number of α-axons (Bagust, 1974; Bagust *et al.*, 1973; Burke and Tsairis, 1973; Olson and Swett, 1966). This is not true for twitch contractions of motor units (Lewis, Luck, and Knott, 1972) or even of large fractions of the muscle (Brown and Matthews, 1960). The differences between myograms of single motor units in an otherwise inactive muscle, and of the motor units in a whole muscle contraction seem largely to lie in the twitch tension rather than in contraction time (Lewis *et al.*, 1972). Both aspects of motor-unit response are extremes, rarely are just one or all the motor units active in a normal contraction, and synchrony is rare.

CONDUCTION VELOCITY. Although the differences between isotonic properties in different muscles may be less than the differences in isometric properties, there is an additional factor that, in the body, limits the acceleration of muscles. This factor is the conduction time along the muscle. Muscle fibers conduct at velocities between 3 and 6 m/sec (Buller, Lewis, and Ridge, 1965). Motor units with the shortest contraction time have the fastest conduction velocities (Knott, Lewis, and Luck, 1971). In a rat limb muscle, fibers are about 20 mm long. If end plates are situated about the middle third of the fibers, the longest conduction times would be some 2 msec. In man, leg muscle fibers may be as long as 300 mm, and fibers might not be activated along their whole length until 60 msec after the arrival of the first impulse at the end plate. In both cases, conduction time is a large fraction of the twitch contraction time. Therefore, the longest muscles, which will have to work against the highest inertial loads, will reach maximal fiber activation with the greatest delay, slowing the rate of rise of isometric tension and the acceleration toward maximum shortening velocity. The effects of temporal dispersion of active sarcomeres along a fiber have been assessed quantitatively (Lewis, 1972), but the effects were not calculated of the long length of inactive sarcomeres before the action potential is conducted all along a fiber, which will add to the series compliance. The effects of conduction along fibers can be expected to be much greater. All the considerations above will, of course, apply to the end of activity as much as to the onset.

INTERACTIONS OF MUSCLES. The experimental techniques that measure muscle lengths, forces, and electrical activity during normal movements (Prochazka *et al.*, 1977; Walmsley *et al.*, 1978) are just beginning to be applied. Already, however, they indicate complicated patterns that will need investigation under controlled conditions. It has been clear for some time that simultaneous agonist–antagonist activation is common in natural movements, and muscle records now show, for example, that stretching of active muscle occurs in hind-limb extensors during the stance phase of walking or crouching before a jump. What is not certain is the nature of the mechanism brought into play. One can speculate that if, just before a jump, the extensors are active as well as the flexors, there may be a balance of tone that allows the joint angle to be held relatively stable until the explosive release of power necessary for the jump to be achieved. It may also be that the stretch of active extensors initially produces a rapid build-up of force that will be balanced by an increase in the opposing flexor force. When the phase of extensor shortening begins, the lengthening of active flexors may elicit a rapid yield and fall of tension (see Fig. 7B) in the flexors, resulting in an explosive increase in the excess of tension of the extensors over the flexors. Then the extensors would shorten without the acceleration being reduced by delays due to muscle action potential conduction, or to high compliance of tendons (which would be extended into their stiffest ranges), or to other factors discussed above. Certainly, such a mechanism is theoretically possible, but there is insufficient information to know if conditions, such as speed of stretching, are correct to allow the muscle yield to occur.

Although it is possible to design experiments to test such hypotheses, there will remain complications and uncertainty owing to the motor-unit composition of muscle. It will be difficult to match experimental conditions to the way in which the nervous system activates motor units selectively, even when the latter process has been described in detail. The extent to which motor-unit contractile properties have been studied is very limited when compared with the information from whole muscles.

HISTOLOGICAL STUDIES OF MUSCLE

Histology is another general technique that has been applied extensively to motor units and is therefore worth more general discussion. In general, histology has been used to examine biochemical aspects of muscle fibers (for which the term *histochemistry* is used) or to examine the microstructure of fibers and their supporting tissues and innervation.

HISTOCHEMISTRY. Color and metabolic differences between muscles have been known for many years, and an early report by Ranvier (1873) showed that red muscle contracted more slowly than white muscle. Red muscle has a deeper color partly because of a higher capillary density and higher resting blood flow, but the color difference persists when the muscles are drained of blood (Kühne, 1865) and is caused by a higher content of myoglobin. Later it was shown that not all red muscles were slowly contracting (Paukul, 1904), the red masseter having a fast twitch. The red muscles (such as soleus in the lower leg and crureus in the thigh) are usually uniform and typically lie deep to a group of white muscles from which they are more (soleus) or less (crureus) separate. In contrast, white muscles often have a deep layer of fibers that are redder in color. The precise description of the nonuniformity of white muscles became possible with histochemical techniques. In the early stages, lipid and glycogen stains showed a difference between fibers, but true histochemistry depends on a reaction produced by specific enzymes in the muscle fibers. Muscles are rapidly frozen to prevent ice-crystal formation which results in subsequent artifacts, and sections are cut at about $-20°C$ to preserve the activity of enzymes. The sections are incubated with a suitable substrate in a medium that is optimal for enzyme activity. Finally, one of the products of reaction is treated with reagents that form a colored compound. If all conditions are properly adjusted, fibers react consistently with different intensities. Examples of some reactions are given in Fig. 10.

The first histochemical techniques to be applied to muscle looked at metabolic enzymes, and classifications were made based on differences in staining intensities of fibers. An early classification that has proved very useful was the one adopted by Ogata (1958) and described in detail by Stein and Padykula (1962). Stein and Padykula examined, among other enzymes and constituents the fiber content of succinic dehydrogenase (SDHase) in rat muscle. SDHase is involved in the major energy-supplying pathway of muscle in oxidative metabolism and is concentrated in mitochondria. Three fiber types were described and were called A, B, and C. Type A fibers had low SDHase activity (stained weakly with the reagents). Types B and C had high SDHase acitvity, but Type C was distinguished by having a ring of heaviest activity around the outside of the fiber, immediately subsarcolemmal. The types, defined on SDHase activity, had different fiber diameters; Type A fibers were largest and Type C were smallest, on average.

Type B and especially the Type C fibers were surrounded by greater numbers of capillaries compared, with Type A fibers, despite their smaller size (Olson and Swett, 1966).

Phosphorylase was the second type of metabolic enzyme for which methods were developed to examine muscle. Phosphorylase is responsible for the breakdown of glycogen and is important in the only major energy-producing pathway available when oxygen supply is limited, as in prolonged maximal activity, which reduces muscle blood flow. Romanul (1964) used phosphorylase in addition to SDHase and also described the three groups that he labeled I, II, and III. Fibers low in SDHase had high phosphorylase activity, and those high in SDHase either had intermediate or low phosphorylase reactions. Despite some differences in descrip-

DONALD M. LEWIS

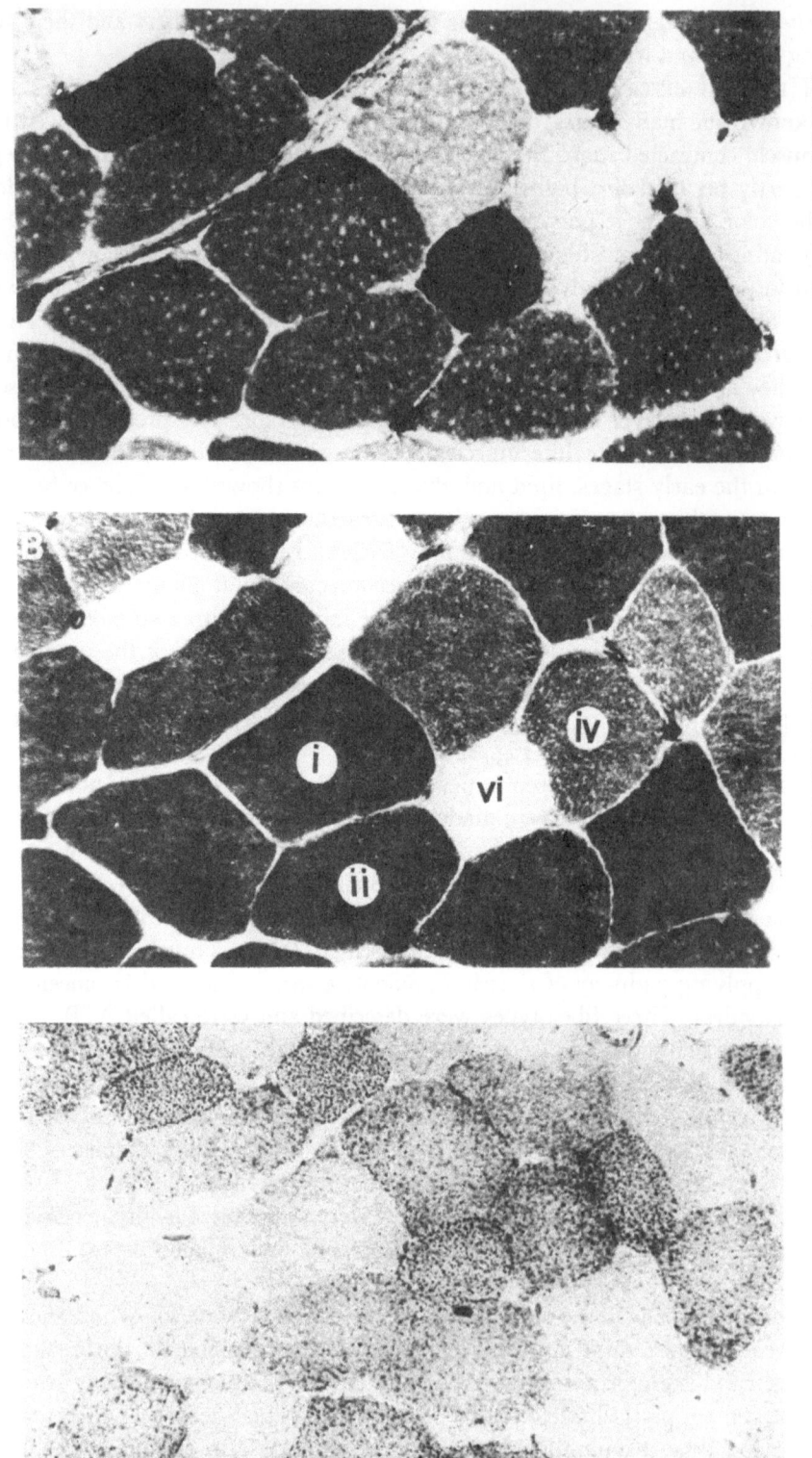

Fig. 10. Histochemistry of cat fast-twitch muscle (FDL). Three serial sections have been stained for ATPase after acid (A) or alkaline (B) pre-incubation and for SDHase (C). A Type B or I fiber is labeled as (vi), a Type C or IIa fiber is (iv), and a type A or IIb fiber is (i). Fiber (ii) has many characteristics of Type A but has high SDHase activity (see Table II). Calibration bars all 100 μm. From Edjtehadi and Lewis (1979).

tions, it is probable that Stein and Padykula's Type A corresponded to Romanul's Group I, Type B to Group III, and Type C to Group II. These two examples represent only a fraction of the number of classifications that have been produced in various studies of muscle histochemistry. Comparisons sometimes are difficult because of slight differences in histochemical techniques between studies as well as differences between muscles and species used. It is natural to assume that there should be some consistency and that there is some correspondence between the classifications produced by groups of workers. Table I is an attempt to show the correspondences where these are easily seen to exist. These comparisons, in general, will not be considered in detail in the remainder of the text.

The third type of enzyme to be used in muscle is adenosine triphosphatase (ATPase). Under carefully controlled conditions, the activity revealed is thought to be that of actomyosin, the contractile protein. It is known that myosin isoenzymes isolated from fast-twitch and slow-twitch muscle have different biochemical activities (Bárány, 1967), and presumably determine the speed of isotonic shortening and, in part, the rate of isometric contractions. It is probable that an immunochemical staining technique (Gauthier, Lowey, and Hobbs, 1978) will supplant present ATPase methods; the immunochemical method can be made as specific as the biochemical purification of the myosin isoenzymes allows (Weeds, 1978). Using ATPase activity, it has been possible to subdivide muscle fibers by virtue of the differences between the stability of the ATPases of fast and slow muscle to formaldehyde or to extremes of pH. Preincubation of sections at pH values around 4.5 or 9.5 results in differences of ATPase staining intensity between fibers. Guth,

TABLE I. FIBER TYPE CLASSIFICATIONS AND THEIR PROBABLE CORRELATIONS

Histochemical							Physiological
Stein and Padykula (1962)	Brook and Kaiser (1970)	Peter, Barnard, Edgerton, Gillespie, and Stempel (1972)	Yellin and Guth (1970)	Davies and Gunn (1972)	Prewitt and Salafsky (1970)	Edjtehadi and Lewis (1979)	Burke (1974)
A	IIb	FG	A	Ah Ph Sl	2	DIDL	FF
					(3)	DIDD	(FR)
	IIc				(2)	ILDL	(FF)
C	IIa	FOG	C	Ah Ph Sl	3	ILDD	FR
					4	LDLL	
B	I	SO	B	Al Pl Sh	1	LDLD	S
				Ah Pl Sh		(ILLD)	

Samaha, and Albers (1970) found some fibers that still reacted strongly for ATPase after alkaline preincubation ("actomyosin ATPase") and called these α. Those that had low residual activity were called β and those having intermediate activity were called $\alpha\beta$. Acid preincubation ("intermyofibrillar ATPase") also produces three grades of reaction. but the relation to the results of alkaline preincubation is complex, being species dependent (Yellin and Guth, 1970), as detailed in Table II. Table II also includes details of SDHase activity (with Stein and Padykula's classification) which was studied in parallel with ATPase by Yellin and Guth (1970). The differences between species are not present if only SDHase and acid preincubated ATPase are considered.

Another labeling system uses the numerals I and II to indicate fibers with high and low SDHase activity, respectively. Dubowitz and Pearse (1960) used this system, also studying phosphorylase, and suggested that there was a reciprocal relationship between the activity of the two enzymes. However, it is clear that some fibers may have high activity for both enzymes (Edgerton and Simpson, 1969) or even low activity for both (Prewitt and Salafsky, 1970). The I, II labeling has also been adopted by Engel (1962) and Brooke and Kaiser (1970). The latter authors used ATPase activity and subdivided the type II fibers into IIA, IIB, and IIC according to small differences of pH at which enzyme activity disappeared.

Other authors have preferred to label fiber types by indicating directly the activity of specified enzymes. Thus Davies and Gunn (1972) referred to fibers that had strong ATPase and phosphorylase reactions and weak SDHase as Ah, Ph, Sl. Edjtehadi and Lewis (1979) used four enzyme reactions (alkaline and acid preincubated ATPase, phosphorylase, and SDHase) and indicated by initials the intensity of responses as dark, intermediate, or light. Thus a fiber described as DIDL produced dark reactions with alkaline ATPase and phosphorylase, an intermediate response with acid ATPase, and a light reaction with SDHase.

Alternative schemes of labeling that depend on functional descriptions have become more popular recently. (A change perhaps precipitated by a shortage of suitable alternative alphanumeric series!) Red, white, and intermediate have been used but perhaps are unsuitable because red and white muscle are well-established terms, and they are often composed of mixtures of fiber types that may vary between species. Fast and slow have also been used, often abbreviated as F and S. A little caution is necessary here because the classification into F and S is sometimes

TABLE II. FIBER TYPES IN DIFFERENT SPECIES[a]

Species	Rat, mouse			Cat, rabbit, man, guinea pig		
Alkaline (actomyosin) ATPase	Dark (α)	Intermediate ($\alpha\beta$)	Light (β)	Dark (α)	Intermediate ($\alpha\beta$)	Light (β)
Acid (intermyofibrillar) ATPase	Light	Intermediate	Dark	Intermediate	Light	Dark
SDHase	Dark (C)	Light (A)	Intermediate (B)	Light (A)	Dark (C)	Intermediate (B)

[a]Yellin and Guth (1970).

based either on ATPase activity (a low ATPase reaction following alkaline preincubation corresponding to slow) or on direct measurements from motor units that
are subsequently identified for histochemical description. An example of the first
usage is the system used by Edgerton and his colleagues (Peter, Barnard, Edgerton,
Gillespie, and Stempel, 1972) who, in addition, used the initials O and G to indicate high oxidative and glycolytic enzyme activity, respectively, and grouped fibers
into three classes: FG, FOG, and SO. Burke (1974) used the physiological characteristics of isometric contraction speed (F and S, indicating fast and slow) and
resistance to fatigue (F and R, indicating fatiguing and resistant) to classify motor
units. Burke described three groups of motor units (FF, FR, and S), but has been
careful to point out that the intensity of myofibrillar ATPase staining should not
be overinterpreted in terms of the supposed speed of muscle fibers contraction
(Burke, Levine, Saloman, and Tsairis, 1974). However, Burke's physiological
motor-unit classification is included, for convenience, in the histological classification of Table I.

Two points of caution must be observed with the histochemical work. First,
the biochemical reactions involved typically have been stopped at a particular stage
and the extent of the reaction estimated. It is rare for true rates of reactions to be
measured as is done in most test-tube biochemical investigations. Pette, Smith,
Standte, and Vrbová (1973) have managed to measure rate constants for single
fibers, but the techniques are too complex for general use. Second, a decision
whether to divide a population of fibers into two, three, or more categories generally has been made on subjective, and possibly biased, grounds. Recently, some
authors have applied measurements of light transmission through fibers stained by
histochemical methods to justify subjective classification (Schmalbruch and
Kamieniecka, 1975) or to determine how many modes of intensity are present
(Edjtehadi and Lewis, 1979). An example of such results is given in Fig. 11, which
illustrates three histochemical reactions in a cat fast-twitch muscle. There was a

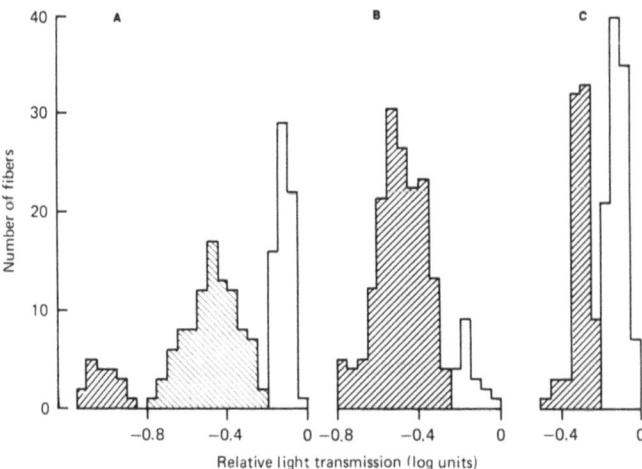

Fig. 11. Histochemistry of a fast-twitch muscle. Light transmission through fibers of the photo-
micrographs illustrated in Fig. 10. The reactions studied were acid preincubated ATPase (A), phosphorylase (B), and SDHase (C). From Edjtehadi and Lewis (1979).

clear separation of fibers in some cases, for example, those with a dark reaction for acid preincubated ATPase. In all other cases, although there was a continuous distribution of intensities, the peaks were sufficiently clear to feel justification to divide into further groups of fibers. Spurway (1978) makes quantitative measurements and uses them directly for each fiber rather than placing the fiber in a class on the basis of the measurement, a technique that allows much greater power of analysis and that he has applied to a large number of enzymes. All these quantitative techniques have restrictions because total light transmission of a fiber, or especially of an arbitrarily chosen part of a fiber, is not the best estimator of the quantity of reaction product. The techniques are also insensitive to the differences in distribution within the fiber which have been used by Stein and Padykula (1962) to distinguish between Type B and Type C fibers.

The general opinion at present is that fibers should be classified into three main types, and I shall use Stein and Padykula's (1962) nomenclature but with the wider implications listed in Table I. White muscle contains predominantly Type A fibers, but Types B and C fibers are common, especially in the deeper and darker layers of some muscles. Red muscle contains Type B fibers, alone in some species (for example the cat and guinea pig) but with a large proportion of Type C fibers in other species (rat and mouse). Some authors explicitly ignore fibers that do not fit the simple classification. Prewitt and Salafsky (1970) found four types based on only two enzyme systems (see below). Two groups of workers have suggested independence of enzyme systems, which implies that types must only be defined for individual enzyme systems. For example, Edjtehadi and Lewis (1979) agree with other workers in defining three types of fibers that showed consistent reactions when examination was restricted to three enzymes. One type gave a dark reaction for alkaline ATPase, an intermediate reaction for acid ATPase and a dark reaction for phosphorylase and was labeled as DID, the other two types were, IID and LDL. However, it was found that fibers of any one of the three types could give a light or a dark reaction for SDHase (compare Prewitt and Salafsky, 1970), resulting in a total of six types of fibers (labeled i to vi, four of which are illustrated in Fig. 10). Spurway (1978) goes further and, from his extensive range of histochemical reactions, suggests that many enzymes are independently distributed between fibers. For example, SDHase and lactic dehydrogenase do not rigidly covary in intensity. It would be important to know how far differences between enzymes in different types of fibers are due to qualitative rather than quantitative differences. Certainly, the quantitative studies suggest modes in the distributions that are clear although overlapping. The basis for ATPase histochemistry may become better understood with the use of specific immunochemical staining. Pette and Schnez (1977) have identified, by electrophoresis, myosin light chains extracted from single muscle fibers that were also typed histochemically. The light chains from IIa (or Type C) fibers were indistinguishable from those of Type IIb (Type A) fibers. Type I (B) fibers contained the same light chains whether dissected from fast- or slow-twitch muscle. The differences found histochemically between the ATPase activity of types A and C fibers must depend on a biochemical difference in constituents other than the myosin light chains. There seems to be a mixture during early postnatal development (Gauthier et al., 1978) but this is complicated by the superaddition of fetal myosin isoenzymes. Even late in life, intermediate transitional fibers may well be those in which the proportion of two isoenzymes is changing from one end of a range to the other. The position for respiratory enzymes such as SDHase is even less clear. Do the fiber divisions represent different isoenzymes (Margreth, Angelini, Valfre, and Salviati, 1970) or differences in concentrations of one enzyme type? Certainly quantitative separation of SDHase types appears to be as good as that of some ATPase types (Fig. 11).

MICROSTRUCTURE. The fibers of both fast- and slow-twitch muscle are elongate cylinders, although in cross section they show a wide range of shapes (Fig. 10). Nevertheless, it is common to express fiber size in terms of diameters. Fiber diameters range from 10 to 100 μm, in general but not universally, being larger in mammals with larger body size. There is much variability within a muscle, and Fig. 12 shows the distribution of fiber cross-sectional areas within a cat fast-twitch muscle and how the differences are related to muscle fiber type. The fibers of slow muscles tend to be more uniform in size but have approximately the same mean area as those of a fast muscle, although the slow-twitch fibers in a fast muscle are considerably smaller. The fibers are multinucleate, being formed by fusion of a large number of myotubes in fetal life. The nuclei are peripheral, under the outer sarcolemma membrane. There are only small differences between the motor nerve endings on the two types of muscle. The regular sarcomeres of the fibers have almost identical dimensions in fast and slow muscle (A1-Amood and Pope, 1972), the lengths of the thick and thin filaments being very close. The myofibril bundles are smaller in fast muscle and in fast motor units (Howells and Jordan, 1978). A

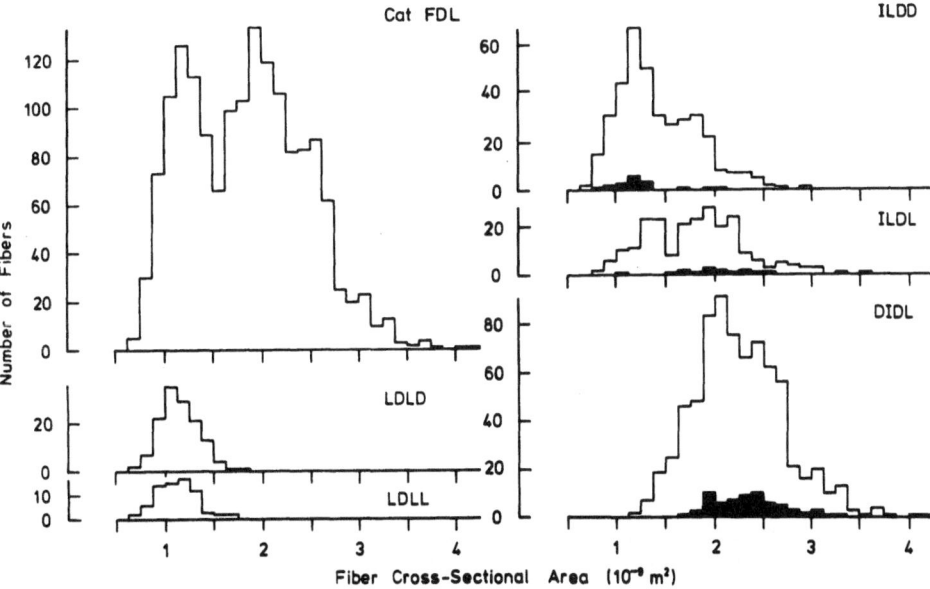

Fig. 12. Fiber areas in a fast-twitch muscle. The histogram *(top left)* presents the areas of more than 1000 fibers from one muscle. Histograms of five histochemical types of fiber are shown separately (see Table II). The solid histograms represent minor groups of fibers that differed from the six major types in only one histochemical reaction. From Edjtehadi and Lewis (1979).

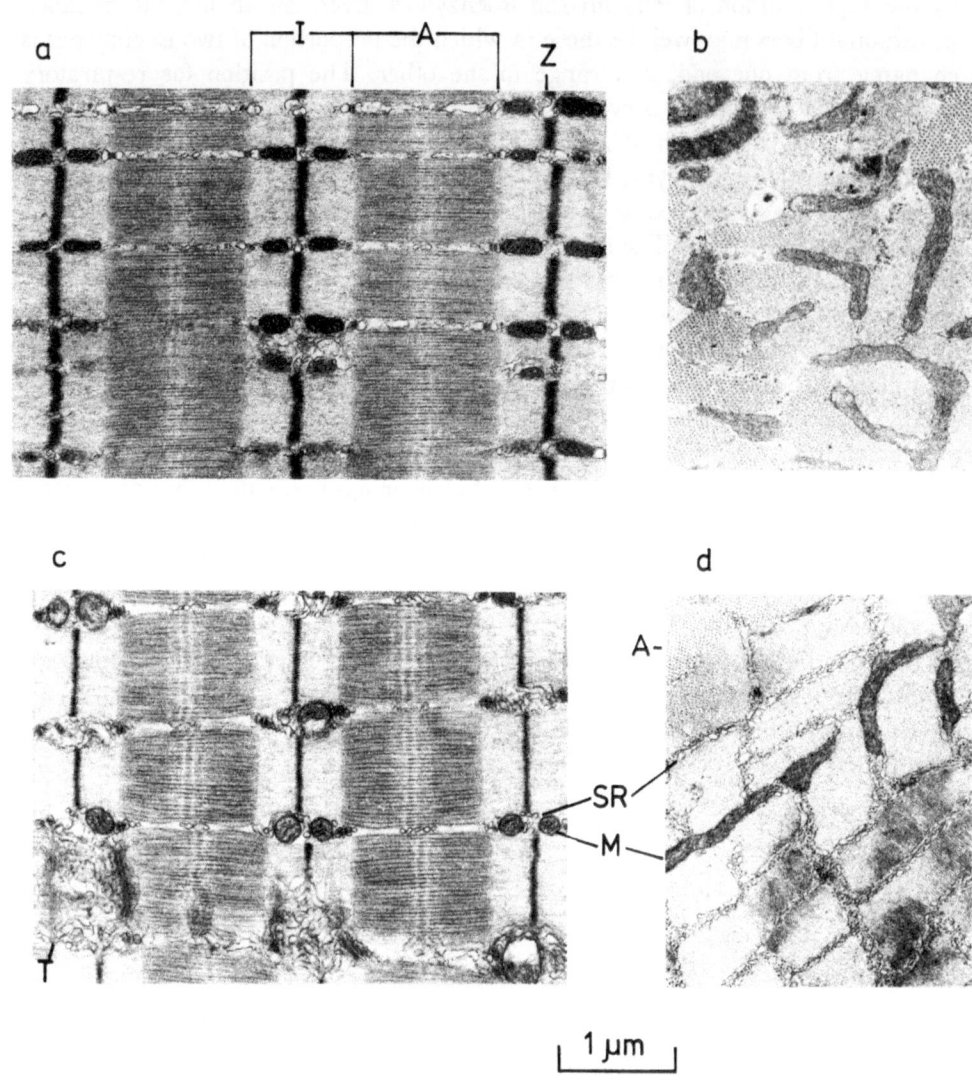

Fig. 13. Electron micrographs of mouse muscles: fast-twitch (c) and (d) medial gastrocnemius, and slow-twitch (a) and (b) soleus, in longitudinal (a) and (c) and transverse (b) and (d) section. The A-band is indicated in the longitudinal and transverse sections, in the latter thick and thin filaments can be seen. The A-band has the same width in fast- and slow-twitch muscle; the I-bands are different but the length depends on the degree of stretch of the muscle during fixation. Slow muscle has a much wider Z-disc. Note in slow muscle that the mitochondria (M) are more numerous, whereas longitudinal elements of the sarcoplasmic reticulum (SR) are more common in fast muscle. (T) indicates a transverse element of the sarcoplasmic reticulum that happens to have been cut over a long distance, but short segments of these transverse tubules are seen in both muscles at other sites near the A–I junction. There is a more regular arrangement of myofilaments (outlined by the arrays of mitochondria and SR) in the fast muscle. All sections are at similar magnifications. Photographs kindly provided by Professor L. W. Duchen, Institute of Neurology, London University.

very characteristic difference is in the Z-line, which is thicker in slow muscle. The transverse tubules and triads are similarly placed on either side of the A-bands in all mammalian striated muscle fibers. There are differences in the longitudinal elements of the sarcoplasmic reticulum, in that they are more abundant in fast contracting muscle (Engel and Stonnington, 1974) and also have different appearance when isolated (Mussini, Margreth, and Salviati, 1972). Mitochondria are more abundant in slow muscle (Engel and Stonnington, 1974). The electronmicrograph reveals most clearly the differences in microstructure between fast and slow muscle (Fig. 13), but its application to motor-unit physiology is limited in the absence of any marker corresponding to the glycogen depletion used for light microscopy.

FUNCTIONAL ISOLATION OF MOTOR UNITS

As described above, the earliest attempts to record the responses of single motor units made use of the fact that some motor units respond in a reflex at a lower threshold than others and can be studied in isolation. This technique would have been selective if the most excitable units had properties different from the remainder, and more recent experiments indicate such selectivity. Electrical stimulation of the muscle nerve at threshold was also used in early studies and recently in man (Sica and McComas, 1971a), and may also be selective, but, although larger axons have an average threshold lower than that of smaller ones, the anatomical position of the axon in a nerve trunk relative to the stimulation electrodes is of more importance in determining threshold.

Both of these procedures for motor-unit isolation have also been used after partial nerve transection, making them less selective. If, however, the remaining nerve was only one division of a motor-unit axon, which had branched central to the transection, the results will be misleading. When stimulus strength is increased above threshold, successive increments of tension are seen and the increments might be deduced to be due to motor units. This procedure works well in muscles in which motor units are fairly uniform in size (Mines, 1913), but in most muscles there are a number of difficulties apart from selectivity. First, the contractions may not sum linearly (Lewis et al., 1972). Second, more than one motor unit may be recruited at each increment of stimulus, that is, two or more nerves may have the same threshold. Third, threshold is not absolute; the probability of a response from a given axon may vary from zero to unity over a range of voltages. At one stimulus voltage, the response of a motor unit has a certain probability (between zero and unity) of occurring. Although the effect of each stimulus will be all or none (see Fig. 14), when a response occurs the amplitude of the twitch will be dependent on the time since the previous twitch; after a long delay the response amplitude will be small for a fast motor unit and large for a slow unit (Bagust, Lewis, and Luck, 1974). Small differences in tension (or EMG) amplitude or latency (utilization time varies at threshold) would give minor differences in the response when the two motor units fire together, and the differences could be interpreted as being very small motor units (Lewis, 1975). Equally, small units could be lost in the noise.

Despite these difficulties, this type of technique is one of the few that can be used to study motor units in man. McComas developed the technique (Sica and McComas, 1971a) and has reduced the problem of nonlinear summation by restricting his measurements to the EMG amplitudes of the motor units (related to tension). He has shown reduction in numbers of motor units in several diseases. The multiplicity of modifications of the technique emphasizes the difficulty of precise interpretation. Recently, the linearity of summation of EMG potentials has been questioned (Parry, Mainwood, and Chan, 1977).

If stimulation is made within the muscle through a needle electrode, it is much easier to restrict the response to one motor unit. Moreover, the method will be less selective, except that the probability of finding a motor unit is related to the number of axon branches, and therefore to the size of the motor unit. The technique has proved practical and has recently been used in man by Garnett, O'Donovan, Stephens, and Taylor (1979). Buchthal and Schalbruch (1970) introduced intramuscular stimulation earlier, although it was assumed that fascicles of muscle fibers were being stimulated. The method does assume that the action potential in the stimulated axon branch can propagate into all other branches as well as it can from the parent axon. This assumption may not be universally true because the parent axon is larger (Eccles and Sherrington, 1930), and so the safety factor will be smaller in the antidromic direction. Indeed, it is not even true that the action potential is transmitted from an axon to all the terminal branches, and there is evidence that different branches may be invaded intermittently (Thiele and Stalberg, 1974).

EMG with sufficiently small electrodes can reveal the activity of single motor units (Adrian and Bronk, 1929; Denny-Brown, 1929). In itself, the unit EMG potentials can give information about discharge patterns, but recently they have been used to study the mechanical properties of motor units (Milner-Brown *et al.,* 1973). Muscle tension is recorded, and an average is made of a large number of periods following each unitary EMG potential. The tensions due to other motor units occur at random times relative to the EMG potential triggering the averager (there are exceptions owing to synchronization of motor units, but these can be recognized and are rare except in some individuals). If a sufficient number of

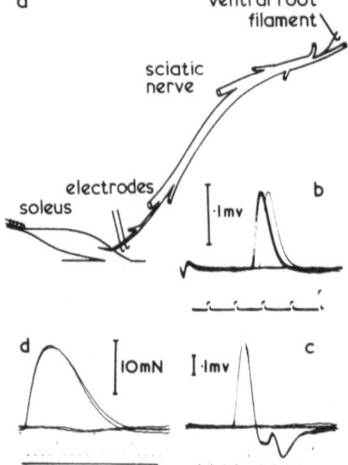

Fig. 14. Isolation of motor units by splitting ventral roots. The filament is judged to contain a single functional unit by the all-or-none behavior of the antidromic nerve action potential (b), and the muscle EMG (c) and twitch (d). From Bagust (1971).

periods is summed, the random events will average out, and only the increase in tension following the EMG potential will appear in the final record. As always, a technique applicable to man has restrictions. The method is selective since maximal contractions cannot be used, first, because individual motor-unit potentials cannot be distinguished when too many are active, and second, because motor units are discharging at higher rates at which the responses become partially fused. The fluctuations of tension in an unfused tetanus can give useful information (Milner-Brown *et al.*, 1973a) about the nature of the response to a single nerve impulse in a natural contraction. In general, the fluctuations of an unfused tetanus will be smaller and briefer than the discrete twitch, the difference becoming larger the higher the frequency (Burke, Rudomin, and Zajac, 1976; Lewis, 1972).

Finally, there are the neuron isolation techniques that have been used only in experimental animals. In one technique, functionally single motor axons are dissected from nerves with fine watchmaker's forceps. The nerves chosen for dissection have almost always been the ventral roots, which are most convenient because there the axons are not bound together with the large amount of connective tissue found in peripheral nerves. Theoretically, results could be biased by finer axons being more difficult to isolate, but comparison of the distributions of axon diameters and conduction velocities (Fig. 3) rules this out. Theoretically, branching within the spinal cord would also upset results but has not been reported histologically. Ventral root splitting was first successfully adapted for motor units by Denslow and Gutensohn (1950, 1951) but was not systematically exploited until used by Bessou, Emonet-Dénand, and Laporte (1963) and by Henneman, McPhedran, and Wuerker (1963), McPhedran, Wuerker, and Henneman (1965), Weurker, McPhedran, and Henneman (1965), and has since been most productive. Ventral root splitting is illustrated in Fig. 14, which emphasizes the care that must be taken to ensure that a motor unit is single and also illustrates that the dissected nerve filament does not consist of a single axon but may contain several dozen, all but one of which supply other muscles that are denervated in the preparation of the animal. These multiaxon strands may survive for many hours, allowing extensive examination of the motor unit. The second neuron isolation technique involves direct stimulation of single motoneurons that are impaled by an intracellular microelectrode (Fig. 15). This technique ensures that the motor unit is stimulated in isolation since much higher currents are necessary to spread the response to other motoneurons, but there is a danger of selection against small neurons. The earliest account by Devanandan, Eccles, and Westerman (1965) indeed failed to find very small motor units, but agreement with the ventral root technique has since been obtained by others (Burke, Levine, Tsairis, and Zajac, 1973).

VARIETY OF MOTOR UNITS

UNIFORMITY OF THE MOTOR UNIT

Once the properties of individual motor units had been investigated directly, it became clear that there was great variability within a muscle. Before looking at these observations, it is pertinent to ask whether the muscle fibers within a motor unit are uniform. This has been tested in two ways, and no great deviation from uniformity has been demonstrated. The first method has been to examine the histochemical properties of a motor unit. The unit is fatigued by repeated tetanic stimulation, and glycogen depleted fibers are investigated by histochemical staining. This method was first employed in the rat by Edström and Kugelberg (1968) and Kugelberg and Edström (1968) who concluded that the muscle fibers of a motor unit were nearly, but not completely, uniform. Two units were presumed to be composed of only Type C fibers (small diameter with high SDHase activity) in that no fibers were depleted. Those composed of Type B fibers (intermediate in diameter and SDHase activity) were less pure in that half of eight motor units contained about 1% of Type A fibers. Four out of seven Type A motor units (large diameter, with low SDHase activity) were pure, the others contained between 2 and 4% Type B fibers. A difficulty was that the method could not deplete the glycogen of about 10% of Type B fibers, or any of the Type C, so the proportion of atypical fibers could have been higher in the last two groups of motor-unit types. This study was limited in the number of motor units examined, but Burke and his colleagues (Burke, Levine, Tsairis, and Zojac, 1973; Burke, Levine, Zajac, Tsairis, and Engel, 1971) have made more extensive studies in cat gastrocnemius. They achieved a more complete histochemical identification by using ATPase. In Burke's studies, about 100 fibers in each of five units from each of the three histochemical types were systematically examined for uniformity, and all were found to have consistent histochemical characteristics.

The second approach to the problem of uniformity has been to compare the properties of the parts of motor units found in separate heads of the cat lumbrical muscles (Emonet-Dénand *et al.*, 1971). It was found that the isometric twitch contraction time of the fibers of a motor unit in one lumbrical muscle was within 10% of the twitch time of the fibers in a second lumbrical in 70% of the 58 motor units studied. This difference was within experimental error, but greater differences

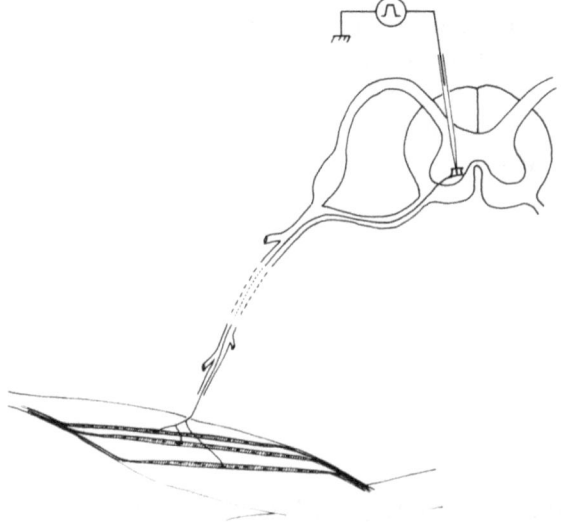

Fig. 15. Method of isolating a single motor unit by motoneuron penetration and stimulation with a microelectrode. (Thanks to Ros Owens).

amounting to 20% to 35% were seen in 7% of the motor units. The discrepancies may have been due to anatomical differences in the position of fibers in the two muscles, rather than to nonuniformity.

The problem of uniformity cannot be said to be completely settled, but the majority of motor units probably contain fibers with closely similar properties. This result is expected in those muscles in which the fibers are supplied by only one axon. It is known that the motor nerve influences the differences between fast- and slow-twitch muscles as far as a number of properties are concerned: the isometric twitch (Buller, Eccles, and Eccles, 1960a) and tetanus (Buller and Lewis, 1965a, b), isotonic shortening velocity (Close, 1969), actomyosin ATPase activity (Buller, Mommaerts, and Seraydarian, 1971), composition (Weeds, Trentham, Kean, and Buller, 1974) and other properties (Bárány and Close, 1971; Samaha, Guth, and Albers, 1970) of myosin, calcium pumping of the sarcoplasmic reticulum (Margreth, Angelini, Valfre, and Salviati, 1970; Margreth, Salviati, and Garraro, 1973), histochemistry (Guth *et al.*, 1970; Karpati and Engel, 1967; Romanul and van der Meulen, 1966), and blood flow (Hudlická, Brown, Cotter, Smith, and Vrbová, 1977) have been investigated. Buller *et al.* (1960), in their initial demonstration of this neural influence, showed that, if the nerve from a fast muscle was cut and the muscle reinnervated by a nerve that had supplied a slow-twitch muscle, the twitch of the fast muscle was converted toward that of a slow muscle. The reverse was also true. Two possible mechanisms were offered to explain the effects. One was that some aspect of the pattern of nerve impulses (frequency or total number) might control the speed of contraction. Alternatively, a hypothetical trophic chemical might be transported along the nerve and produce the changes in the muscle. The influence of impulse frequency is now well established (Salmons and Vrbová, 1969), and there is some evidence against the trophic chemical (Lømo, Westgaard, and Dahl, 1974). It is reasonable to assume that neuronal influence is operative at the motor-unit level, so that the properties of all muscle fibers of a motor unit would come under the same neural factor or factors, and so develop the same properties. This is a reasonable argument, but other possibilities are by no means improbable. For example, some fiber properties that are determined genetically may be modifiable only within certain limits. Since fibers are multinucleate, and are formed by fusion of a number of myoblast cells, this is perhaps unlikely. However, it is possible that partial or intermittent conduction block where axon branching occurs would result in different impulse patterns in two sets of terminals (Thiele and Stalberg, 1974). Such differences in activity between fibers could result in nonuniformity of motor units.

DISTRIBUTION OF MOTOR UNITS

Histochemical investigations of motor units indicate that the individual fibers typically are separated by fibers of other units, and occasional groupings of two to six are rare, probably occurring by chance 1968; (Burke and Tsairis, 1973; Edström and Kugelberg, 1968). This is not true of muscle reinnervated after nerve section (Kugelberg, Edstrom, and Abbruzzese, 1970), in which the fibers of a motor unit are closely grouped. Presumably, during regrowth of an axon into a muscle, it branches and the terminals capture fibers within a limited area. That this does not occur during normal development may be because the fibers receive end plates from a number of axons (Redfern, 1970), for each of which the packing of fibers may be close and localized as in reinnervated muscle. Later, if connections are lost at random, the grouping of fibers of motor units would also be lost. It is not clear why this does not occur following reinnervation, as polyneuronal innervation also occurs under these conditions (Guth, 1962). Similar grouping has been deduced to occur in man following some types of disease involving slow destruction of some of the motoneurons supplying a muscle. In the adult pig, groups of histochemically similar muscle fibers are seen and are found to enlarge in numbers during normal postnatal development (Davies, 1972), being absent in neonatal animals. Groups of similar fibers may be parts of motor units that have expanded during development at the expense of adjacent motor units.

Although the fibers of a motor unit are typically separate from one another, they do not normally extend throughout a muscle in the cat. The majority of fibers of some units were to be found in restricted regions, as had been clear from early observations of localized reflex contractions described above. The restriction of cat motor units was investigated by Denslow and Guthensohn (1950a,b) in the first use of ventral root filaments for motor-unit isolation. More detailed mapping has been done histochemically (Burke and Tsairis, 1973) and electrically in the cat (Knott, Lewis, and Luck, 1971) and in man (Buchthal, Guld, and Rosenfalck, 1957; Stalberg and Thiele, 1975). The degree of restriction seems to be less in limb muscles of smaller mammals, such as the rat (Edstroöm and Kugelberg, 1968), and there appears to be no localization at all in rat diaphragm (Krnjevic and Miledi, 1958). The wide distribution does not, however, prevent there being special regions within a muscle as George and Susheela (1961) report variation in proportions of fiber types in different areas of the rat diaphragm.

THE ISOMETRIC CONTRACTION

TWITCH CHARACTERISTICS

Twitch Contraction Time. The contraction time of functionally isolated single motor units was the first characteristic to receive detailed attention. Gordon with Holbourn (1949) and with Phillips (1953) recorded the twitch times of reflexly activated units. In the early 1960s more systematic studies were reported by Steg (1962), Bessou *et al.* (1963), and Anderson and Sears (1964). The work of Bessou *et al.* (1963) was a study of one of the small lumbrical muscles of the cat's foot, but systematic studies of the larger hind-limb muscles, both fast- and slow-twitch, soon followed from the laboratory of Henneman (McPhedran *et al.*, 1965; Wcurker *et al.*, 1965). A wider range of cat muscles was reported by Devanandan *et al.* (1965) using intracellular stimulation. Early investigations of human motor-unit twitch characteristics were made by Sica and McComas (1971b) and Stein, French, Mannard and Yemm (1972).

All reports agree that the range of motor-unit twitch contraction times is much greater than that of the whole muscle contractions. This is illustrated in the two top histograms of Fig. 16, in which the parent muscle contractions were recorded (shaded areas) as well as the motor units. The range was wider in the slow-twitch soleus than in the fast-twitch FDL. The wider range in slow muscle is not seen if the motor-unit time to peak is expressed as a fraction of that of the parent muscle; indeed, it may then be wider in the fast muscle. In absolute terms, other hind-limb fast muscles have wider ranges than FDL (e.g. gastrocnemius: Weurker et al., 1965). Some experiments were done at lower temperatures, which make the twitch slower and increase the twitch time to peak (Close and Hoh, 1968b). Gastrocnemius muscle does have a somewhat longer contraction time than FDL (even at the same temperature) although both are classified as fast-twitch muscles. It would seem that the variation in times to peak between different fast-twitch muscles is caused by the varying composition of motor units. There are also differences between contractions of slow muscles: crureus in the thigh is slightly faster than soleus (Buller et al., 1960a). Unfortunately, only soleus has had its motor units examined in detail (there are anatomical difficulties with the others), and it may be dangerous to generalize from soleus alone to differences between fast and slow muscles (although this chapter will).

An asymmetry of fast muscle motor units in both cat and man is seen as a skew toward long contraction times. The skewed distribution probably results from a mixture of two groups of motor units, one with short and the other with intermediate twitch contraction times. There is evidence from fatigue characteristics in favor of such a division (see below), and a very clear separation has been found in the rat by Close (1967) who examined soleus and EDL. In the rat, the fast-twitch EDL is relatively pure in that its motor-unit twitch times to peak are almost sym-

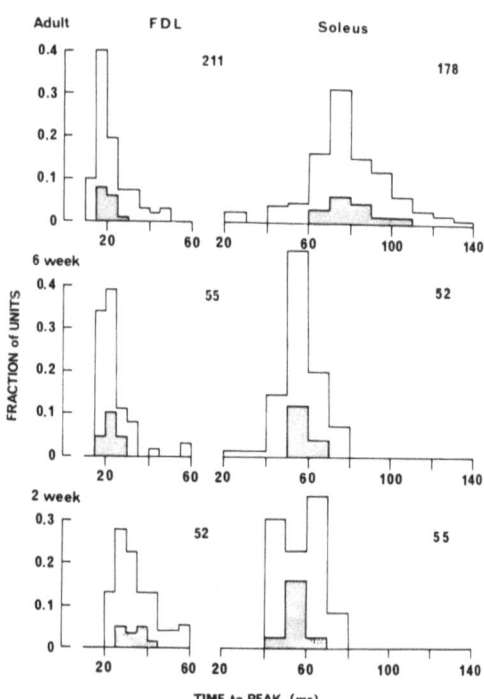

Fig. 16. Distribution of motor-unit twitch times to peak in cat fast-twitch (FDL) and slow-twitch (soleus) muscles at three ages. Shaded areas represent the responses of the parent muscles. From Bagust (1974); Bagust et al. (1973, 1974); and Finol, Lewis, and Webb (unpublished observations).

metrically and unimodally distributed. It is rat soleus that has a mixture of motor units, and its distribution histogram has two modes (one of which corresponds to the few nonfast units in EDL). Presumably, in the cat, there is more variance within the two groups in fast-twitch muscles so that, instead of the overall distribution being bimodal, it appears skewed. Variation occurs in the motor-unit twitch contraction times of nonmammalian species (Cliff and Ridge, 1973; Knott, 1971; Luff and Proske, 1977).

Twitch:Tetanus Ratio. The mechanisms underlying the differences in twitch contraction times are unknown. One possibility is that they represent differences in the intrinsic activities of contractile proteins, most probably arising from different mixtures of a small number of protein types. Close (1971) has drawn attention to the reciprocal relationship between twitch time to peak and maximum velocity of shortening if a wide range of muscles is considered (covering almost two orders of magnitude). Although Close's reciprocal relationship is a good broad description, detailed examination has shown that two muscles may have the same shortening velocity but different twitch times to peak (Close and Luff, 1974, see above). This discrepancy has also been found between FDL muscles sampled from two groups of cats with different genetic and rearing histories (Kean, Lewis, and McGarrick, 1974). In both cases a second variable, the ratio of peak active twitch tension to maximum tetanic tension (the twitch:tetanus ratio), was also found to be different. The muscles with the shorter contraction times had the smaller twitch:tetanus ratio. The simplest explanation of these differences would be that the contractile proteins of the two members of each pair of muscles were similar, but in one muscle less calcium is released following a single action potential (or calcium is removed faster) so that the twitch is both briefer and smaller.

A relationship between motor-unit twitch time to peak and twitch:tetanus ratio has been described by one group of workers (Bagust, 1974; Bagust, Knott, Lewis, Luck, and Westerman, 1973) and is illustrated in Fig. 17. This figure

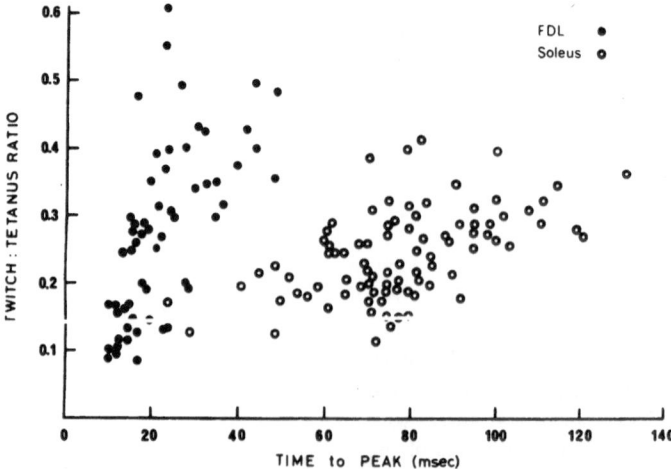

Fig. 17. Relations between twitch time to peak and twitch:tetanus ratio of motor units of a fast-twitch (FDL) and slow-twitch (soleus) muscle. Data from Bagust (1974) and Bagust *et al.* (1973).

shows clearly that, although the slower motor units of fast muscle had the same contraction times as the faster motor units of the slow muscle, these two sets of motor units had very different twitch:tetanus ratios. Other workers have failed to demonstrate the relationship between twitch contraction time and twitch:tetanus ratio (Applcberg and Emonet-Dénand, 1967; Proske and Waite, 1974). The failure may have resulted from different recording conditions. For example, if the motor-unit twitch is measured after a tetanus, a fast contracting unit will be potentiated and a slow one depressed (see below). This tetanic effect would increase the twitch:tetanus ratio for fast and decrease it for slow units and, if sufficiently large, would result in a failure to observe any regression of twitch:tetanus ratio on twitch contraction time. However, the difference between groups of workers probably also represents a difference between animals or muscles. In the rat, Close (1967) finds there to be no difference in twitch:tetanus ratio between fast, intermediate, and slow motor units. Lewis and Bagust (Bagust, 1974; Bagust *et al.*, 1973), who found the relationship between twitch:tetanus ratio and time to peak in one group of cats, found no such relationship during early postnatal development (Bagust, Lewis, and Westerman, 1974) or in another group of adult cats from a different population (unpublished).

It is tempting to draw a parallel showing the relationship between twitch contraction time and twitch:tetanus ratio of motor units with that in the pairs of muscles that have similar isotonic properties but different isometric twitch contraction times and correspondingly different twitch:tetanus ratios (see above). One conclusion would be that within a group of fast-twitch motor units, the variability in twitch contraction time is the result of differences in the calcium transients, without differences in contractile proteins. There are other, equally possible, explanations; crucial experimental tests await more precise biophysical measurements of fibers from single motor units. A beginning was made in identifying the myosin composition of single fibers by dissection, extraction, and electrophoresis (Weeds, Hall, and Spurway, 1975). Pette and Schnez (1977) have extended this work to show that fibers contain one or other of only two types of myosin light chains (see above). It is known that muscles that are fast and slow contracting under both isotonic and isometric conditions are composed of different isoenzymes of myosin (Sréter, Seidel, and Gergely, 1966). Myosin ATPase activity is also related to isotonic shortening velocity in whole muscles (Bárány and Close, 1971) but, although histochemical ATPase reactions are generally assumed to depend on the same parameter, there have been no suitable quanitative measures of biochemical activity of ATPase for single fibers.

Factors Modifying the Twitch. There are a number of conditions in which fast and slow muscle react differently. Following a short tetanus (one second or less), the twitch response of a fast muscle may double in amplitude at optimal length (see below), or more at shorter muscle length; this is termed *posttetanic potentiation* (Brown and von Euler, 1938). Similarly, there is an increase in twitch tension on increasing the rate of stimulation from 1/min to 1/sec; this is termed the *staircase* or *treppe effect.* The increase in twitch tension occurs with little change in the time course of the twitch since there is also an increase in the rates of rise and fall of tension. Posttetanic effects are completely different in slow muscle, the twitch of which is decreased in amplitude and shortened in duration. Twitch tension, contraction time, and relaxation time all decrease in the same proportion by as much as 30% in the cat. In rat soleus, the changes are smaller, possibly because of the substantial number of fast fibers mixed in the muscle.

In the whole animal, these changes may have a functional significance. During activity after a period of rest, the force available from twitches or from unfused tetani of fast muscle will increase, as will the rate at which force can develop, and therefore the maximal acceleration in movements will be increased. This is not true of the tension of fully fused tetani, but limb muscles are probably not physiologically activated at the high rates necessary for full fusion (Buller and Lewis, 1965a; Rutherford, 1887). The slow-twitch muscle might oppose the contraction of fast antagonists during rapid movement, but after a few seconds of increased activity their speed of relaxation would increase, and they would offer less opposition to rapid movements, while still adding to the total force available. The details of the use of motor units in relationship to their properties is only beginning to be understood (Milner-Brown, Stein, and Yemm, 1973b), but it is clear that the traditional view that slow-twitch muscles and slow motor units in fast-twitch muscles are concerned only in postural activity is probably incorrect. Recent *in situ* measurements of tension in cat soleus (Prochazha *et al.*, 1977; Walmsley *et al.*, 1978) show that this muscle, at least, is fully used during normal walking. At slow rates of stepping, soleus may contribute at least as much force as its synergists, the gastrocnemeii and plantaris, which are potentially much more powerful. The slower contraction speed of soleus is probably no hindrance at these slow speeds of walking. During faster movement, posttetanic depression would occur at the known rates of activation, and the resulting increase in rate of relaxation of soleus would probably improve the smoothness of the step cycle.

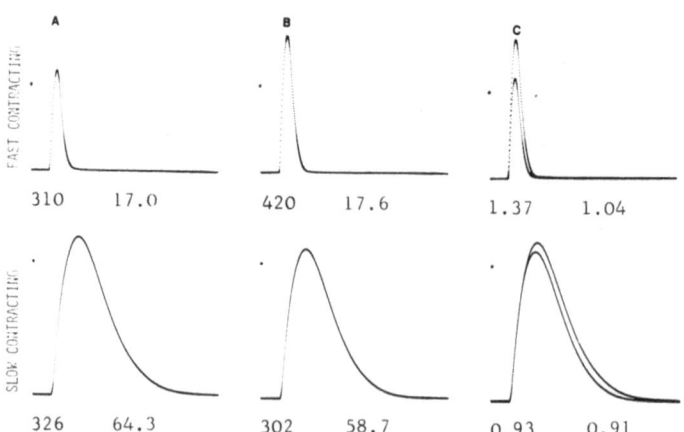

Fig. 18. Post tetanic effects on motor units of a fast-twitch muscle. In (A) are the resting twitches, the numbers below each myogram representing tension (mg, left) and time to peak (msec, right). In (B) the records were taken 10 sec after a 300 msec tetanus at 100/sec. The pre- and posttetanic myograms have been superimposed in (C), and the numbers represent the ratios of tensions and times. The fast unit is potentiated with little change in time course; the slow unit is depressed with shortening of contraction and relaxation. From Bagust *et al.* (1974).

Motor units have been examined for posttetanic effects (Bagust, Lewis, and Luck, 1974; Olson and Swett, 1971) and a continuous gradation has been found, so that those with the longest contraction times in a fast muscle show least potentiation and may even be depressed (Fig. 18).

Burke, Rudomin, and Zajac (1976) have described a second form of posttetanic potentiation. Very prolonged tetanic stimulation will potentiate slow as much as fast motor units, the potentiation being accompanied by some prolongation of contraction time and a great increase in relaxation time. The conditions necessary to elicit this second type of posttetanic potentiation are probably unphysiological but the results provide useful information about the mechanisms of muscle contraction.

Adrenaline, but not noradrenaline, has effects on the twitches of fast and slow muscle (Fig. 19). Like a preceding tetanus, adrenaline potentiates fast muscle and depresses slow. However, fast-twitch tension potentiation is not accompanied by an increase in the rate of change of tension, so that the twitch contraction and relaxation times are increased. In slow muscle, the fall of twitch tension brought about by adrenaline is much less than the shortening of contraction and relaxation times, there being an increased rate of tension development. At physiological levels of adrenaline, soleus is maximally affected. Although FDL gives a submaximal response, this may still amount to 20% potentiation. It is not known how motor units respond, but cat tibialis anterior shows no response to adrenaline; this is a mixed muscle, and if the fast units are potentiated and the slow units depressed by the drug, this would account for there being no total change.

Another difference between fast and slow muscle that requires examination in terms of motor units is the effect of temperature. An increase of temperature

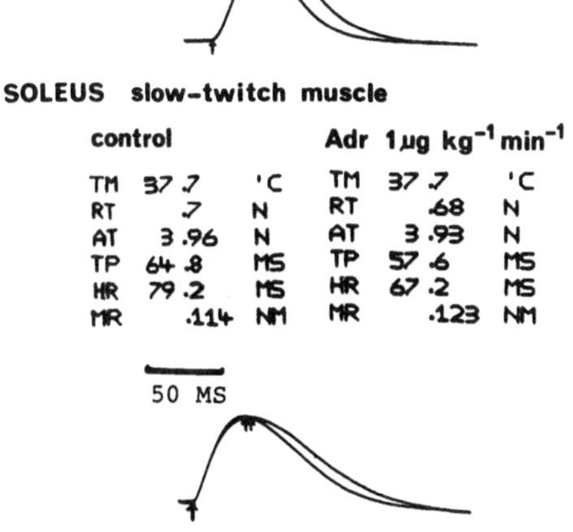

FDL   fast-twitch muscle
control   Adr 5μg kg^{-1}min^{-1}

TM	37 .4	'C	TM	37 .4	'C
RT	.59	N	RT	.48	N
AT	3 .87	N	AT	4 .52	N
TP	17 .8	MS	TP	20 .8	MS
HR	13 .8	MS	HR	16 .2	MS
MR	.341	NM	MR	.341	NM

20 MS

Fig. 19. Effect of adrenaline on twitch contractions of fast- and slow-twitch muscles. Adrenaline was perfused into a jugular vein at the rates indicated and selected for near maximal effects in each muscle. In the alphanumeric display, TM = temperature, RT = resting tension, AT = active tension, TP = time to peak (indicated by arrows in the myograms), HR = time to half relaxation, and MR = maximum rate of tension development (units = N/msec). Note that adrenaline potentiates and prolongs the fast muscle twitch, but shortens and slightly depresses the slow muscle. By Lewis and Webb (1976).

SOLEUS   slow-twitch muscle
control   Adr 1μg kg^{-1}min^{-1}

TM	37 .7	'C	TM	37 .7	'C
RT	.7	N	RT	.68	N
AT	3 .96	N	AT	3 .93	N
TP	64 .8	MS	TP	57 .6	MS
HR	79 .2	MS	HR	67 .2	MS
MR	.114	NM	MR	.123	NM

50 MS

shortens the contraction time of both types of muscle, but it depresses the twitch of fast muscle while increasing the twitch tension of slow muscle. There is relatively little effect on tetanic tension (Close and Hoh, 1968a). In a cool environment, muscle temperature increases by at least 7°C between rest and full activity in limbs, a change that will produce appreciable effects on muscle. Unlike prior activity or adrenaline, the effect of exercise on muscle temperature causes a decrease in twitch tension in fast muscle, but the increase in temperature causes an increase in the rate of development of tension and rate of shortening. The effect on rate may be the more important in correct functioning of the limb.

The preceding paragraphs have outlined several physiological factors that influence fast- and slow-twitch muscle differently. The study of these factors at the motor-unit level has been minimal. Greater detail might provide clues about the mechanisms underlying differences between motor units, and would help us to understand their role in activity of the muscle as a whole.

In human muscle, the need for data is even greater, as many of these factors have not been investigated even at the whole muscle level. Problems of extrapolation are particularly difficult in man, since the identification of slow muscle with postural activity and fast muscle with phasic responses probably does not hold. Differences between muscle twitch times seem better related to the inertial loads, where the hind-limb muscles contract more slowly than forelimb muscles, and distal muscles (moving only the digits) more rapidly than proximal ones.

GENESIS OF TETANUS. When a muscle or motor unit is stimulated repetitively at intervals that are less than the duration of the twitch (Fig. 4), the responses sum with each other to produce an unfused tetanus, in which greater tension (or more shortening) is developed. The maximum tension is related to the frequency of stimulation (Fig. 20A). Above a certain frequency, tension does not increase further,

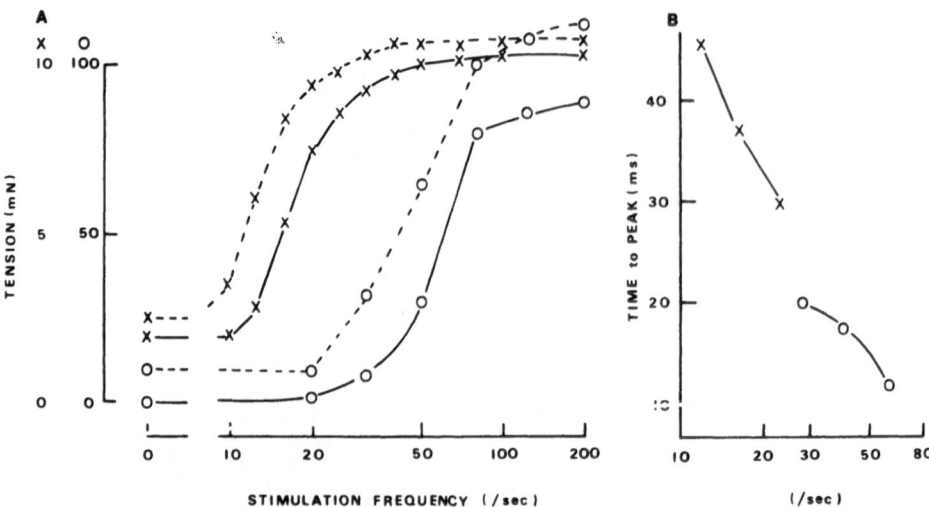

Fig. 20. (A) Effect of stimulus frequency on the tensions of two motor units (X,O) held at optimum length (continuous curve) and with the muscle extended by 4 mm (dashed curve). (B) Twitch time to peak plotted against the tetanic stimulation rate required to produce half maximal tension. The same two units as (A) but at three muscle lengths.

and the tetanus may be said to be fused. Buller and Lewis (1965a) showed small oscillations in the plateau of tension, even at rates of stimulation of 300/sec in fast muscle, confirming the observations of Rutherford in the last century (1887) who was able to hear a musical note "corresponding to the lower F of the treble clef" from muscle stimulated at 352/sec. Even though there is no increase in maximum tension at these higher rates of stimulation, the rate at which the muscle develops tension continues to increase with frequency up to 600/sec in fast muscle (Buller and Lewis, 1965a). This is well above the rate at which α-motor nerves discharge in the body, with the possible exception of those to the extraocular muscles (Robinson, 1970).

The stimulation rate necessary to produce fusion, estimated for convenience and accuracy as the frequency at which tension is half maximal in Fig. 20, has been measured for motor units by a number of authors (McPhendran *et al.*, 1965). This parameter is related fairly simply to contraction time (Fig. 20B) fusion occurring at lower frequencies with the slower motor units.

The maximum rate of rise of isometric tetanic tension is another stimulus frequency related variable, motor units with long times to peak develop tension slowly in a fused tetanus as well as in the twitch. It is difficult to be sure that optimal stimulation rates have been used in measuring rate of motor-unit tension development since the refractory period of nerve (and muscle) increases during repetitive stimulation; as a result, the frequency of action potentials in the muscle fibers will be less than that elicited at high frequencies in a ventral root filament.

Interesting information is provided by responses to double stimuli. Both fast- and slow-twitch muscle show a maximal response when the interval between stimuli lies between five and ten msec (Eccles and Sherrington, 1930). In contrast, muscle at low temperature (Hartree and Hill, 1921; Ranatunga, 1977) and denervated muscle (Lewis, 1972) produce a maximal response when the second stimulus is given near the time to peak of the muscle twitch, when the series elastic element is maximally extended in the isometric contraction. In normal mammalian muscle, the short stimulus interval gives maximal response, probably because it allows optimal summation of the excitation–contraction coupling processes for maximal calcium release. At low temperature and in denervation, activation presumably already is maximal for activation of the contractile proteins following a single stimulus. Burke and his colleagues (1976) have studied these phenomena extensively in motor units. They have shown also that if a short interval is interposed in an unfused tetanus, the tension is increased for more than one second. Thus one short interval in a low-frequency train of pulses will produce both an immediate and a long-term increase in muscle tension. Such phenomena may be important physiologically because interposed short intervals have been observed in the EMG patterns of motor units both in voluntary human contractions (Denslow, 1948) and in the stretch reflexes of decerebrate rats (Hoff and Grant, 1944). A mechanism of generation of such short intervals, by application of a ramp of current to a motoneuron, has been described by Baldissera and Campadelli (1977) who further suggest that the motor-unit and motoneuron properties are such that the rate of tension development in the muscle fibers (and this probably is applicable to rate of shortening, too) is a direct function of the rate of injection of current into the motoneuron. This linearization may simplify the organization necessary for the control of movement and certainly may make it easier for the physiologist to analyse the system.

Burke has used the phrase "catch properties" to describe the effects of a single short interval interimposed in an otherwise relatively low-frequency stimulus train. It is quite different from the catch mechanism of some molluscan muscles, which can maintain tension for hours after a short period of stimulation, with little expenditure of energy (Twarog and Munoeka, 1972).

Burke and his colleagues (1973) have used unfused tetani (stimulus interval 125% of time to peak) to distinguish between motor units. Fast motor units showed an increase in peak tension of the first four to eight oscillations of such a tetanus and then fell to a lower value (termed a "sag" in the response). Slow motor units in a fast muscle showed no sag at the test frequency.

FATIGUE. Slow muscle has a richer blood supply and the fibers have more numerous mitochondria than fast muscle (Gauthier, 1969), and is less easily fatigued (McPhedran et al., 1965). Burke (1974) has investigated resistance to fatigue by stimulating motor units tetanically, once every second (at 40/sec for 330 msec). Slow motor units in fast muscle (and most motor units in slow muscle) show little or no fatigue, even after one hour of this treatment. The fast motor units can be divided into two groups. Those of one group are relatively resistant to fatigue; tetanic tension falls by less than 25% after two minutes. In the other fast units, tension falls by more than 75% in the same time. The fatigue-resistant fast units had a longer average time to peak than the easily fatigued fast units, but there was considerable overlap of the ranges in the cat, but little in the rat (Kugelberg, 1976). Histochemically, the fibers of fatigue-resistant fast units have relatively high concentrations of oxidative enzymes, approaching those of slow motor units.

EFFECTS OF MUSCLE LENGTH. If the length of muscle is increased in steps, there is a change in all the parameters that have been discussed above. The passive tension increases approximately exponentially with length. Also discussed earlier is the active tetanic tension, which increases with length up to a certain value (the optimum length) and then decreases at longer lengths. The optimum length often corresponds to the rest position of the limb in the body. At optimum length, there is maximum overlap of the active sites on the thick and thin filaments of the muscle (Gordon et al., 1966), and therefore maximum force is generated. If a muscle is overstretched for any period, new sarcomeres are laid down at the end of the fibers (Crawford, 1961; Goldspink, Tabary, Tabary, Tardieu, and Tardieu, 1974). By this mechanism the muscle would be kept near its optimum length.

Twitch tension changes in a similar way with length but is maximal at lengths slightly longer than the tetanic optimum (by about 10% of the length of the fibers: Westerman et al., 1974). As muscle length is increased, the twitch responses of a muscle become prolonged, the effect being greater on the relaxation phase than on the contraction phase (Buller and Lewis, 1963). A stimulation rate sufficient to produce a certain degree of fusion at a short length will therefore cause more fusion at longer length (Fig. 20). The magnitude of the effect is related to the change in twitch time to peak.

The augmentation of the tension of an unfused tetanus with increase of length is sufficiently large to compensate for the decrease of tension beyond the optimum length. Rack and Westbury (1969) represented the complete response of a muscle to length by measuring a series of tension:length curves at different frequencies of stimulation (Fig. 21). The optimum length for a fully fused tetanus is shorter than that for the twitch, but both are less than the optima for unfused tetani. This is important since most physiological responses are unfused, and, for many muscles, the naturally elicited active tension will increase continuously with length up to the maximum limit in the body.

Most of the description is this section has referred to experiments on whole muscle contractions, but many of the observations have been confirmed in motor units. The tetanic optimum lengths are not uniform (Olson and Swett, 1966) but are distributed around the muscle optimum, the slowest units having the longest optima (Bagust *et al.*, 1973). In some fast-twitch muscles, the twitch responses of many motor units reach maximal tension at lengths well beyond the optimum length for whole muscle twitches (Lewis *et al.*, 1972); indeed, no optimum can be found for some, without inflicting damage on the fibers by overstretch. In a fast-twitch muscle, fast motor units show the greatest displacement of twitch optimum (Fig. 22). The cause of the discrepancy is not known; it is not due to neuromuscular failure, as the EMG is unaffected over this range of lengths (Fig. 22). Even if it were caused by a distortion produced by the fibers of the unit having to contract alongside inactive fibers, it does describe the response of the motor unit under natural conditions, and emphasizes the danger of inferring too much about motor-unit responses from whole muscle behavior.

PHYSIOLOGICAL CONTROL. The central nervous system controls the tension produced by a motor unit by regulating the rate of discharge of its axon (Adrian and Bronk, 1929). During a steadily increasing tension of a voluntary contraction in man, motor units are recruited at their own threshold (corresponding to a par-

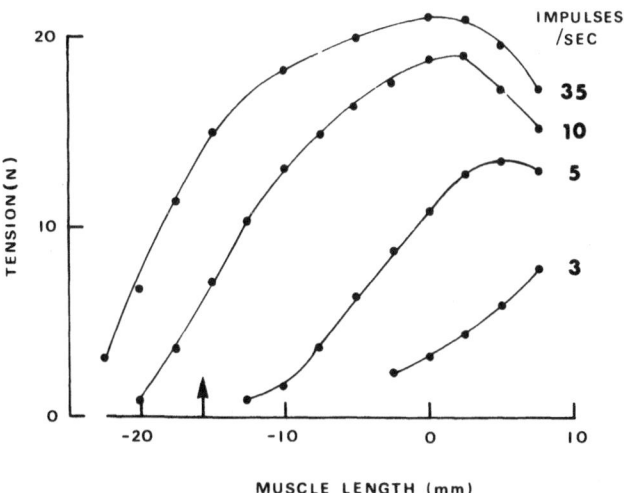

Fig. 21. Effect of muscle length on tetanic tension. Each curve is the response at a particular stimulus frequency (shown on the right) and applied in a distributed way (see Fig. 8) to obtain smooth records. From Rack and Westbury (1969).

ticular muscle tension). At its threshold, a unit will fire at a rate of about 8 impulses/sec in one of the hand muscles (Milner-Brown, Stein, and Yemm, 1973). Increase in force is produced by recruitment of more motor units and by an increase in firing frequency up to a recorded maximum of about 120 impulses/sec and probably higher. Experimental stimulation of the whole muscle indicated that the steep rising part of the stimulus frequency-response curve occurred between 10 and 20 impulses/sec; this is the mean response, and individual motor units would be expected to increase tension over a smaller range of frequencies (see below, Table III). Other studies have demonstrated a consistent order of recruitment of motor units, even when they discharge transiently in very rapid or ballistic movements (Desmedt and Godaux, 1977). In contrast, some have found that motor units come in and out at particular phases of a given muscle contraction (Hennerz, 1974), or that recruitment order may be reversed by suitable stimulation (Garnett and Stephens, 1978).

The general picture from animal experiments and recording simple human contractions is that motor units operate over that range of firing frequencies that will take them from a just-fused tetanus to sufficient fusion to give close to maximal tension. In other terms, the dynamic tension response of the muscle (increase in tension per unit increase in firing frequency) is large over the natural range of use.

The experiments reported so far do not allow one to judge if this large

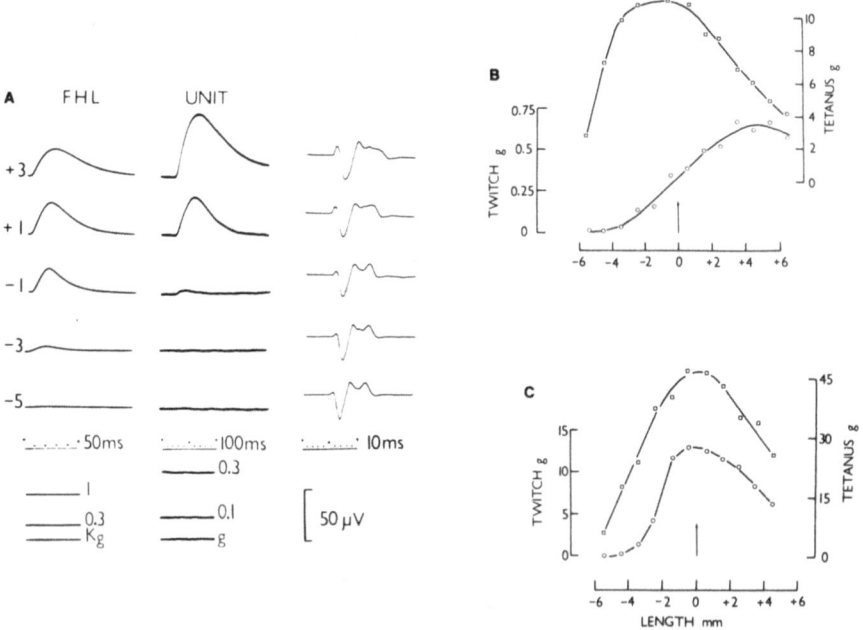

Fig. 22. (A) Myograms showing the effect of muscle length on a fast muscle twitch *(left)*, on its fastest motor units *(center)*, and on the motor unit EMG *(right)*. Muscle length indicated in mm relative to optimum length. In (B) and (C) twitch and tetanic tensions of two units are plotted against relative muscle length. (C) Slow unit behaves as the whole muscle. (B) Fast unit has a long optimum length for its twitch. From Lewis *et al.* (1972).

dynamic tension response occurs precisely, over a wide range of situations, in different animals and muscles (e.g., the postural slow muscles). The results do not necessarily imply that the nervous system can monitor and control the tension of individual motor units, although such control is not impossible, since each Golgi tendon organ is sensitive to only a small number of motor units (Houk and Henneman, 1967) and the number of their axons approaches half the number of motor α-axons. However, an alternative mechanism exists, as over long periods the contraction speed of a muscle is controlled by its innervation (Buller, Eccles, and Eccles, 1960b; Fig. 23) by means of the frequency of activation. An assumption could be made that there is a continuous relationship between firing frequency and the contraction speed of the muscle for all physiological rates. This has not been shown, the existing experimental evidence relates only to extreme rates of stimulation, either at 10/sec (Salmons and Vrbová, 1969) or at 100/sec (Lømo et al., 1974). It must also be remembered that motor units discharge over a range of frequencies so only the mean can be specified for any one unit, and some units may be brought into use very rarely. If the assumption were correct, when a characteristic rate of firing was established for a motor unit, its muscle fibers would adjust (over about four weeks) to produce an appropriate speed of contraction. This mechanism would automatically adjust the contractile properties of the muscle fibers to the firing rate imposed on them, and could produce maximum efficiency or maximum effectiveness.

ASYNCHRONY. Although the typical contraction of a motor unit is an unfused tetanus, the overall muscle response is fairly smooth; any tremors that may occur even in normal subjects are apparently due to characteristics of the nervous control system. The smoothness occurs because the discharges of motor units are rarely synchronized in normal subjects (Milner-Brown et al., 1973a). So in a contraction in which more than one unit is active, the ripples in individual tension responses

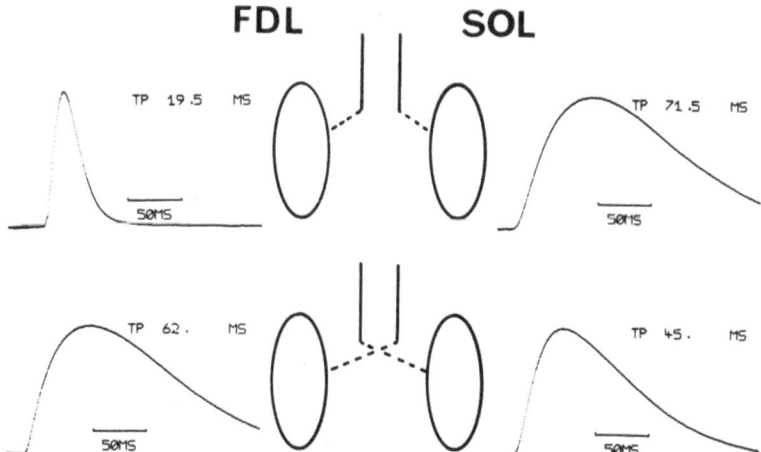

Fig. 23. The effects of cross-innervation. Diagrams represent self-reinnervation *(top)* and cross-reinnervation *(below)* of a fast-and slow-twitch muscle. Twitch myograms of the reinnervated muscle of the cat are shown on either side (with time to peak indicated at TP). Modified from Buller (1972).

will be averaged out by other units; beyond five units, smoothing will be almost complete (Fig. 8).

TETANIC TENSION. It was shown earlier that the number of motor axons innervating a muscle tends to be greater in the more highly developed vertebrates. A large number would allow of greater complexity of organization of motor units, and this has been found. In the frog (Knott, 1971; Luff and Proske, 1977), the largest motor unit develops a tension which is about ten times that of the smallest. In the rat, the difference is a hundredfold (Close, 1967), and in the cat more than a thousandfold (Bagust, Lewis, and Westerman, 1973a). Soleus, the one slow postural muscle that has been studied in the cat, is quite different from the fast muscles in that its motor units are relatively uniform in tension. This comparison between fast and slow muscles is made in Fig. 24 (top row), in which the fast muscle is FDL, which was chosen because it developed a tension very close to that of soleus. These two muscles are also convenient for comparison because they are innervated by approximately the same number of α-motor axons. Soleus has neither the very large nor the very small motor units that are found in FDL. A second difference between the two muscles is in the shapes of the distributions of tension. In soleus the shape is not distinguishable from a normal distribution, whereas that of FDL is grossly skewed. The skew is due to a predominance of small motor units, and a large proportion of these are of the slow type, with long contraction times, resistance to fatigue, and a characteristic histochemical profile. Examples of a large fast motor unit and a small slow one both from a cat fast-twitch muscle are illustrated in Fig. 25. The relationship between tetanic tension and twitch time to peak, illustrated in Fig. 26, shows that there is a continuous distribution of these two parameters. The slow motor units of FDL are shown as clear columns in Fig. 27, and

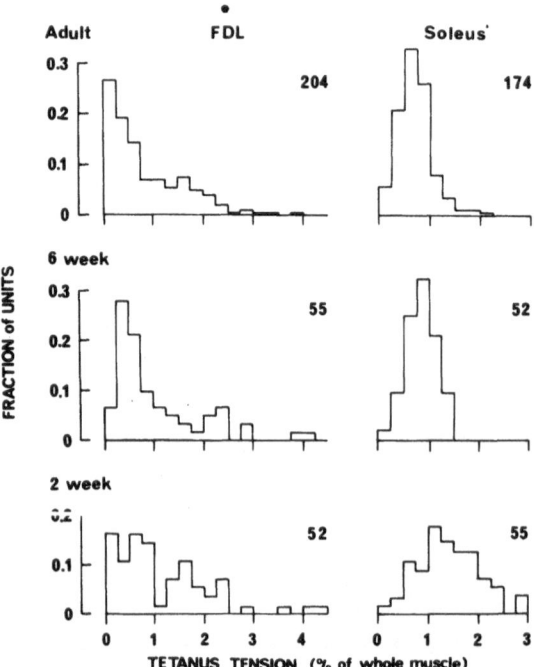

Fig. 24. Distributions of motor-unit tensions of fast- and slow-twitch muscles in adult cats and the changes during development. The number of motor units investigated is indicated to the right of each histogram. Motor-unit tetanic tension has been expressed as a percentage of the tetanic tension of the muscle from which it was isolated. Same series of motor units as in Fig. 16.

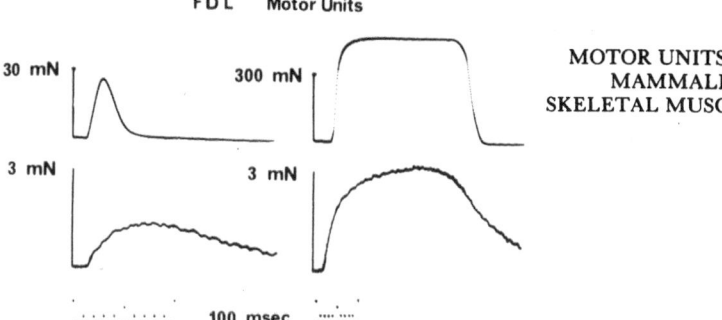

FDL Motor Units

Fig. 25. Twitch and tetanus myograms of a large fast motor unit and a small slow motor unit of a cat fast-twitch muscle. Vertical bars are individual tension calibrations. From Bagust *et al.* (1973).

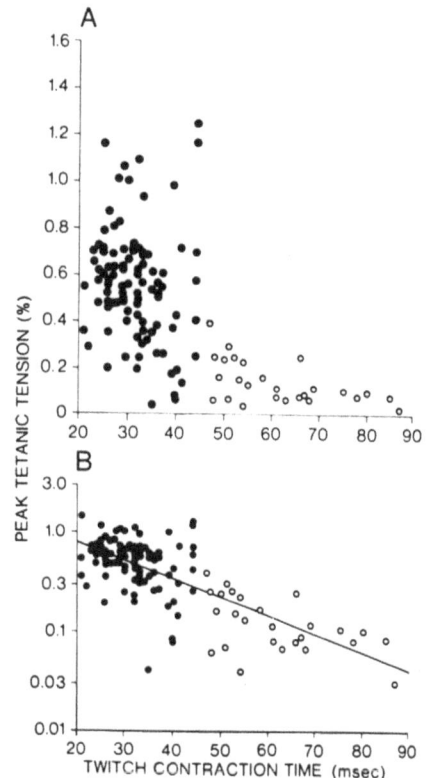

Fig. 26. Relation between tetanic tension and twitch time to peak of motor units from cat gastrocnemius muscle. Tension is expressed as a percentage of that of the parent muscle, and plotted both on linear (A) and logarithmic (B) scales. Open circles represent slow-twitch and filled circles fast-twitch motor units From Stephens and Stuart (1975).

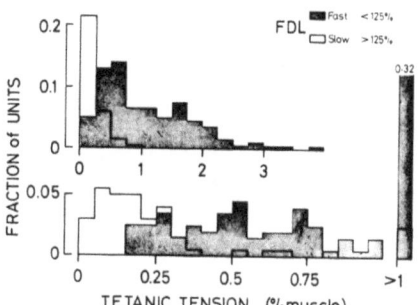

Fig. 27. Distribution of tensions of motor units in cat fast-twitch muscle; lower histogram expands the range of smaller tensions. In both histograms, two populations are superimposed, the division being on the basis of twitch time to peak. From Finol, Lewis, and Webb (unpublished observations).

although they are smaller on average, overlap with fast units is seen. The fast motor units can be subdivided: there is a group of smaller ones (lower histogram of Fig. 27), but the separation on tension alone is not absolute. Burke and his colleagues (1973) have distinguished the two groups of fast units by their resistance to fatigue, and described them by the terms FF (easily fatigued, also larger on average) and FR (resistant to fatigue). His organization into three groups in cat medial gastrocnemius on the basis of three criteria is illustrated in Fig. 28.

Some generalizations of the preceding paragraph must be qualified. Extensor digitorum communis (EDC) of the baboon forelimb has been studied by Eccles, Phillips, and Wu Chien-Ping (1968), and it has been found to have a limited range of motor unit tensions. The twitch tension of the largest unit was less than fifty times greater than the smallest. The muscle contained many motor units, and, although none of them was very large, the distribution was skewed toward the largest values as in other fast-twitch muscles. (A word of caution: the tension of the majority of units in baboon EDC was measured only in twitches, and if twitch rather than tetanic tension were used for cat FDL muscle, the range of sizes of motor units would be underestimated by 30%.) It is tempting to speculate that the rather uniform sets of small motor units in baboon EDC have evolved to serve fine manipulative movements. This speculation is supported by the observation of a relatively limited range in human interosseous muscle (Milner-Brown *et al.,* 1973a), although the argument is confused by this human muscle having a slightly larger range of twitch tensions (more than a hundredfold) than the baboon.

The largest motor unit in cat medial gastrocnemius developed 1.3 N tension (Burke *et al.,* 1973), and the largest in FDL 0.82 N (Bagust *et al.,* 1973). This difference is small when it is considered that the former muscle produces a tetanic tension that is about seven times that of the latter. Similarly, the small lumbrical muscle has motor units developing as much as 0.3 N. This is an indication that

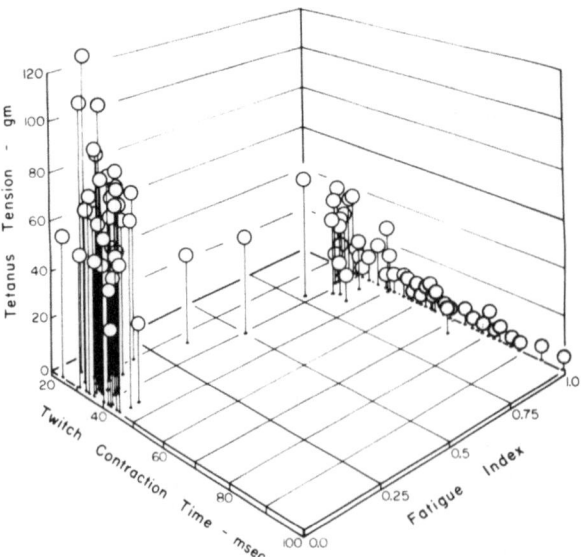

Fig. 28. Three-dimensional representation of relations between motor-unit tetanic tension, twitch time to peak, and extent of fatigue following repeated tetani. From Burke *et al.* (1973).

there is a general limit to the size of a motor unit, and to the number of fibers (or branches) that an axon can maintain.

The index of motor unit size has been taken in the preceding discussion as its tetanic tension, but it is also of interest to know the number of fibers innervated by individual axons. Systematic counts have been made in cat medial gastrocnemius after depletion of glycogen in fibers by repeated tetanization of the one axon (Burke and Tsairis, 1973). Exact counts are difficult because of the complex organization of the muscle fibers. It was concluded that FF units contained between 550 and 750 fibers and FR units 400 to 550 fibers. Three slow motor units contained between 14 and 164 fibers, but these estimates may be low because of the difficulty of completely depleting slow fibers of glycogen. Indirect evidence was considered that suggested that the number might be as high as 500. Burke preferred to accept the indirect estimate for slow units and concluded that innervation ratio was relatively uniform. The low tension of the slow motor units in fast-twitch muscle would be due to the small diameter of the fibers and the inferred low tension per unit area. Other authors had suggested that innervation ratios might vary considerably in a muscle. Eccles and Sherrington (1930) noted that large axons branched more frequently than small ones and concluded that they would innervate more fibers. Proske and Waite (1974) speculated from the fact that tension was linearly related to axonal area for FR and slow units. Knott and her colleagues (1971) found that slow motor units had fibers within a smaller area of the muscle and unless the density of fibers were higher for slow motor units, this would argue for a smaller innervation ratio for them. Further work needs to be done in this field with a glycogen depletion method more reliable for identifying slow-twitch fibers.

Relations between Muscle Fibers and Axon of a Motor Unit

The motor nerves supplying slow-twitch muscle have slower conduction velocities than those of fast-twitch muscle (Figs. 29 and 30). This was known from the early experiments of Eccles and Sherrington (1930), but it was not until 1963 that

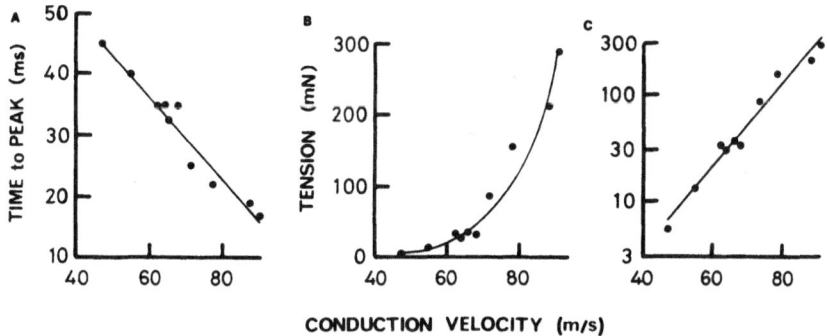

Fig. 29. Relations between axonal conduction velocity and motor unit twitch time to peak (A) or tetanic tension in a small muscle of the cat's foot. Tension plotted on linear (B) and logarithmic (C) scales. Data from Appelberg and Emonet-Dénand (1967).

Bessou, Emonet-Dénand, and Laporte showed that the relationship held at the motor-unit level. These first experiments were done in a deep lumbrical muscle of the cat, which contains only about six motor units. Later work (Appelberg and Emonet-Dénand, 1967, Emonet-Dénand *et al.*, 1971) was done on a superficial lumbrical muscle, which has two or three times as many units. An example is illustrated in Fig. 29. In Fig. 29C tension is logarithmically related to conduction velocity in motor axons, a relationship that fits the results in FDL muscle (see Fig. 31). These correlations have been found by many groups working with other muscles of the cat hind limb and might be inferred from human measurements.

All groups have accepted the general correlation between conduction velocity and tension in fast-twitch muscles, but several authors have suggested that if motor units of fast twitch gastrocnemius muscle are divided into fast and slow or FF, FR, and S groups, the correlation does not exist within groups (Stephens and Stuart, 1975) or only within the S group (Proske and Waite, 1976). More complex relations have been suggested by Olson and Swett (1966). In soleus, some authors (Bagust, 1974; McPhedran *et al.*, 1965) but not all (Moscher, Gerlach, and Stuart,

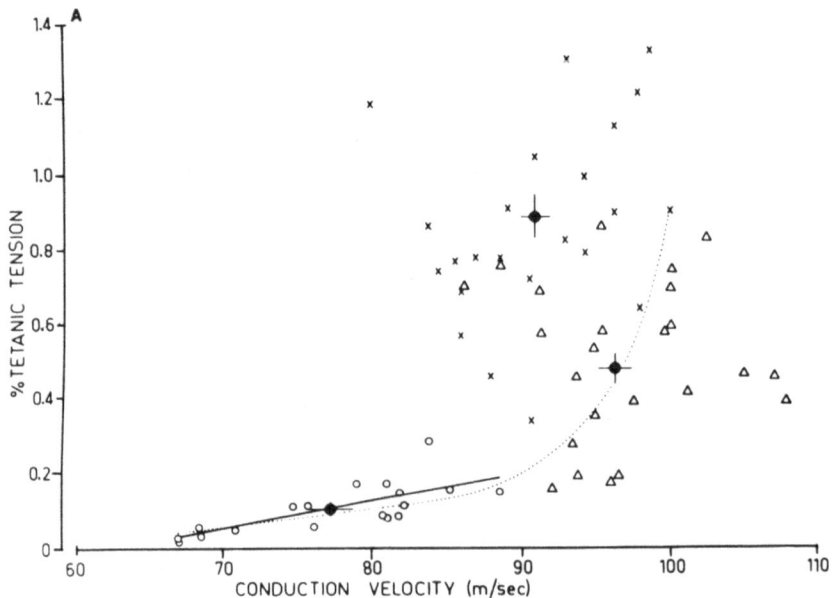

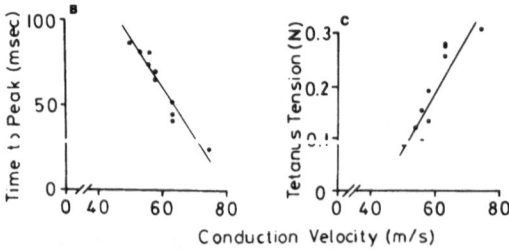

Fig. 30. Relations between axonal conduction velocity and motor-unit tetanic tension or time to peak in large limb muscles. (A) Cat gastrocnemius: (O) = slow fatigue-resistant; (△) = fast fatigue-resistant; and (X) fast easily-fatigued motor units. Filled circles indicate group means with standard error bars. The continuous line indicates a significant regression, and the dotted line is the curve of best fit to the data of Emonet-Dénand *et al.*, (1971) scaled to the tensions of gastrocnemius motor units. From Proske and Waite (1976). (B) Motor units from rabbit slow twitch muscle. From Bagust (1971).

1972) have seen the correlation between tension and conduction velocity. One difficulty in demonstrating the relationship is that the slope of the regression varies a little from one animal to another, so that pooled data lead to more scatter than is seen in a large sample from one muscle. The relationship between tension and conduction velocity in soleus can be adequately expressed as a linear rather than a log-linear one, possibly because of the very limited range of motor-unit tensions. There is a lot of scatter about the relationships in the large muscles of the cat (Fig. 30A). However, in rabbit soleus (Fig. 30B) the results are much closer to the consistency seen in the small lumbrical muscles.

MOTONEURON PROPERTIES

Correlates of axonal diameter may also be sought centrally, and it has already been suggested that motoneurons supplying slow-twitch muscles are more readily excited and discharge more slowly but more continuously than those innervating fast muscle. The relationships within the motoneuron pool of one muscle have not been examined, but inferences can be drawn from two types of experiments.

In the first group of experiments, motoneuron properties have been related to axonal conduction velocity. Input resistance decreases with increasing axonal velocity (Burke, 1967), and this effect is associated with (and in part is a cause of) a decrease in the size of the EPSP elicited by stretch of the muscle (Burke, 1968). These and other experiments indicate that slow conducting axons (and small, slow motor units) will be recruited most easily in a reflex or voluntary response (Henneman, Sonijen, and Carpenter, 1965). This was confirmed directly in experiments

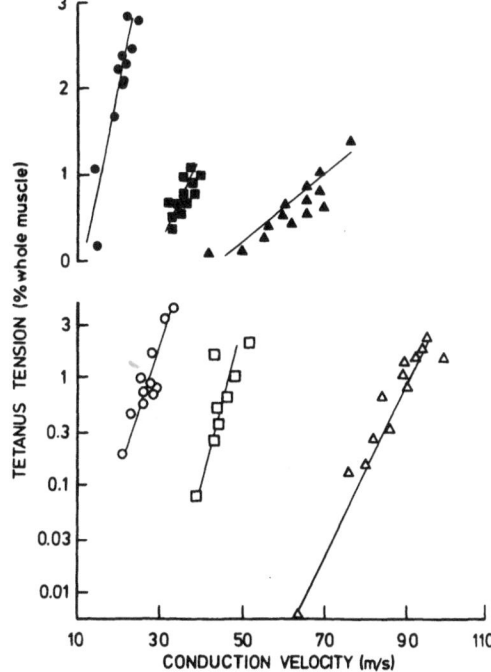

Fig. 31. Relationships between motor-unit tetanic tension and axonal conduction velocity during postnatal development. *Above:* slow twitch (linear ordinate), and *below:* fast twitch muscles (logarithmic ordinate). In each case examples are given at two weeks (circles) and at six weeks (squares) of age and in adults (triangles). Data from individual animals different than those of Fig. 33. From Bagust (1974); Bagust *et al* (1973, 1974); and Finol, Lewis, and Webb (unpublished observations).

of the second group in man (Milner-Brown *et al.,* 1973 a,b,c), in which it was shown that a weak contraction mainly brought in motor units with a small slow twitch; the faster, larger responses came in at stronger forces. No measurements of axonal conduction velocities were possible in these human experiments.

It may be concluded that the slow motor units of fast muscles resemble motor units of slow muscle in being used more continuously than fast motor units.

The second question which must be asked is about the frequency of discharge of motoneurons, but there is less evidence and some of it is conflicting. In the period around 1930, Sherrington and his colleagues produced some evidence that moto-neurons innervating slow muscles discharged at lower rates than those supplying fast muscles. Some twenty years later, this was confirmed by Eccles and by Granit, who showed a longer after-hyperpolarization following an action potential in small (slow-twitch muscle) motoneurons, which would limit their rate of discharge. A systematic study of EMG potentials in the rat (Fischbach and Robbins, 1969) confirmed different average rates of discharge in fast and slow muscle.

All these experiments concern differences between muscles, but there is little evidence about differences within muscles. Certainly Fischbach and Robbins (1969) found a sufficient range in the discharge rates within either muscle to cause overlap between extreme values in fast and slow, but did not include any observation of twitch characteristics. Inferences can be made from the experiments of Milner-Brown *et al.* (1973a,b,c). Twitch characteristics were related to threshold of recruitment, but neither the frequency of discharge of a motor unit at threshold nor the rate at which this frequency increased with total muscle force showed significant dependence on motor-unit threshold. This presumably means that the twitch time of a motor unit was not related to the range of frequencies at which it operated. Indeed, at moderate forces the low threshold units (slow twitch) would be firing at higher rates than the higher threshold units (fast twitch).

A recent brief report by Garnett and Stephens (1978) suggests that the recruitment order of motor units is not rigidly fixed but depends on factors other than spinal motoneuronal organization. They determined the recruitment threshold of motor units in a human hand muscle at rest and during electrical stimulation of a finger, a stimulus that produced a sensation similar to that of a firm grip. Such repetitive stimulation reversed the recruitment order; those units that originally had come in at low tensions were recruited only when the muscle was contracting strongly when tested during electrical stimulation. The large, high-threshold units had their thresholds reduced by the skin stimulation. Kanda, Burke, and Walmsley, (1977) also found that cutaneous stimulation in the cat caused large motor units to be excited in a vibration reflex of gastrocnemius muscle.

A wider range of direct experiments needs to be done. At present, it cannot be stated with any certainty that twitch speeds of motor units within one muscle are related to their frequency of discharge in the same way that fast and slow muscle can be said to be iinked causally with their characteristic frequency of discharge.

It is accepted that the slow frequency of activation of postural muscles induces the formation of the slow type of contractile proteins and possibly the high frequency imposed on phasic muscles is responsible for the fast type of protein. As an extension of this, it might be speculated that within all motor units with similar frequencies of activation, all fibers will contain the same proteins. But there may be a second modulating factor (such as the total number of impulses) that modifies the extent (or duration) of activation in the twitch and therefore brings about changes in twitch duration linked to differences in twitch:tetanus ratio. Many would not accept such speculation, although there is evidence from Pette and Schnez (1977), who found identical electrophoretic patterns of myosins from types A and C fibers.

PLASTICITY OF MOTOR UNITS

POSTNATAL DEVELOPMENTAL CHANGES

Motor units in cats have been studied at two stages of postnatal development (Bagust *et al.*, 1974). Figures 16 and 24 illustrate that characteristic differences between fast and slow muscle are established at two weeks after birth, when the eyes have just opened and the kitten is making only clumsy, crawling movements. Two major changes with age can be detected. First, there is a decrease in the mean motor-unit relative tension. Relative tension is calculated as the absolute tension divided by the parent muscle tension; this allows for growth in mass of muscle fibers, as there is no change in fiber numbers from birth onward. The decrease in relative tension with age is due to two factors. One factor is an increase in the number of motor axons reaching and innervating the muscle; innervation in this species is not complete until about two months after birth (Nystrom, 1968). The other factor is the polyneuronal innervation of muscle fibers at birth; a muscle fiber will be innervated by more than one axon, and will therefore contribute to more than one motor unit. The sum of motor-unit tensions will then be greater than the tension of the muscle (Fig. 2), and the mean motor-unit size relatively greater than in the adult. The second change with age is in the range of tensions as the very large motor units disappear, both by loss of fibers to new axons and by abolition of polyneuronal innervation by six weeks. The very small motor units also disappear, presumably because newly arrived axons can capture more muscle fibers. A linear or log-linear relationship is maintained between axon conduction velocity and tension, although the typical conduction velocities observed vary with age (Fig. 31).

The second change that can be seen during development is in the range of motor-unit times to peak (Fig. 16). This is related to changes in the whole muscle contractions (e.g. Buller *et al.*, 1960a; Fig. 32). At birth, slow and fast muscles have similar, slow-twitch contractions (although they are not identical in that the fast muscle has a high twitch:tetanus ratio and the slow muscle a low ratio). During postnatal development, the slow muscle transiently becomes a little faster, and then slow again. The fast muscle becomes progressively faster. The modifications of isometric twitches are accompanied by and also partly caused by changes in isotonic shortening velocities (Close, 1964) and in the biochemistry of the contracile proteins (Huszar, 1972). In the motor-unit pattern of the fast muscle, the twitch time changes are seen mainly in the group of fast motor units. The range of times to peak extends during development of both muscles; an increase in the complexity of organization is parallel with the increasing variety of movements performed by the muscles. Relationships to axonal conduction velocity are present at all stages of development, although no one formula can describe all ages (Fig. 33).

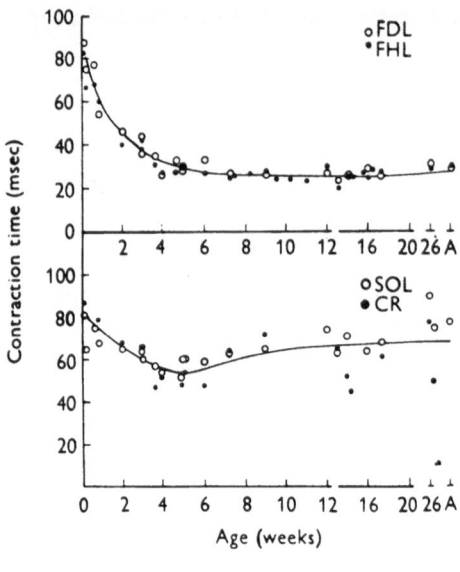

Fig. 32. Change in twitch time to peak during postnatal development in the cat. *Above:* two fast-twitch muscles, and *below:* two slow-twitch muscles. From Buller, Eccles and Eccles (1960a).

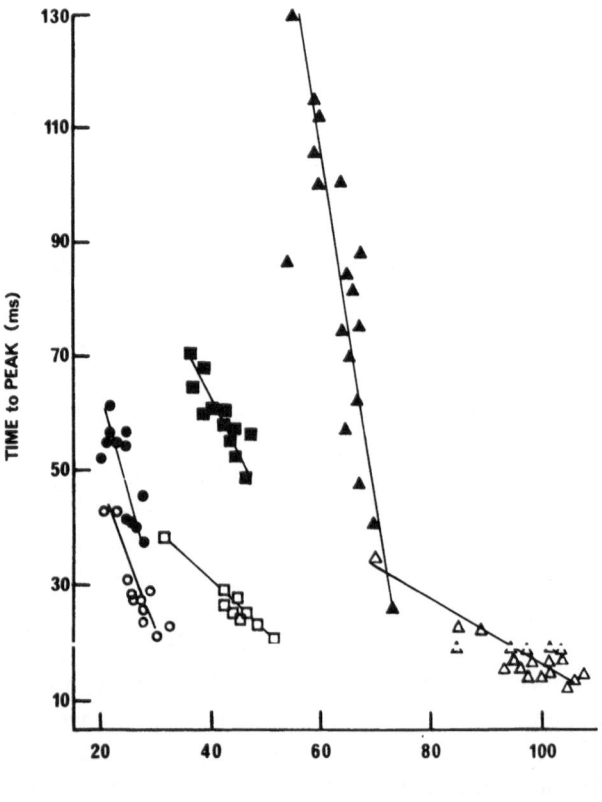

Fig. 33. Relationships between motor-unit twitch time to peak and axonal conduction velocity during postnatal development. Circles from two-week-old kittens, squares from six-week-old kittens, and triangles from adult cats. Open symbols designate fast-twitch and solid symbols slow-twitch muscles. From same series as Fig. 16.

In young adult rat Close (1967) described two clear groups of motor units, one group having slow and the other group having intermediate twitch contraction times. Kugelberg (1976) showed that the proportions of intermediate units changed from one-third to one-tenth between the fifth and thirty-fourth week of age. There was a corresponding change in the proportions of histochemical Type B and Type C fibers, but this change appeared to be via transitional fibers. Perhaps in the transitional fibers there is a mixture of the two myosin isoenzymes, which are assumed to be present in a pure state in the typical Type B and Type C fibers.

Unusual changes in developing pig muscle have already been discussed (Davies, 1972).

REINNERVATION IN THE ADULT

Postnatal development may be compared with the experimental condition in which a nerve is cut and allowed to regrow into a muscle. Guth (1962) showed that polyneuronal innervation, which does not occur in normal adult muscle, could occur transiently during reinnervation. Motor units have been studied after the phase of polyneuronal innervation. Bessou, Laporte, and Pagés (1966) showed that in cat lumbrical muscles the relationships between motor-unit tension or time to peak and axonal conduction velocity were reestablished, although with more scatter than in normal animals. Bagust and Lewis (1974) examined larger fast and slow muscles at an earlier stage of reinnervation than used for the lumbrical. Although they agreed that many of the interrelations in motor-unit properties were reestablished, there was an exception in that the fastest motor units were no longer the largest. It would appear that the motoneuron can remold the muscle fibers with which it makes connections during adult reinnervation. However, some authors believe that reinnervation is nonrandom and selective (Hoh, 1975).

Histochemical examination of motor units (Kugelberg et al., 1970) and human EMG recording indicate that the distribution of fibers in reinnervated muscles is different from that in the normal. The muscle fibers of a motor unit are often grouped together in bundles of up to several hundred fibers, rather than being scattered through wide areas of the muscle. The terms "type-grouping" and "checker-board" or "mosaic" patterns have been used to describe the histochemical pattern in reinnervated and normal muscle, respectively.

More recently, motor units have been examined in cross-reinnervated muscles (Bagust, Lewis, and Westerman, 1981; Egerton, Goslow, Rasmusen, and Spector, 1980).

INTERACTIONS OF MOTOR UNITS

SUMMATION OF CONTRACTIONS

Sherrington and his school had assumed that motor-unit responses would sum linearly. Their deductions about summation, facilitation, and occlusion of reflex responses all contain this implicit assumption. Relatively little work has been done to test the assumption. It has been shown by a number of authors that mean motor-unit tetanic tension is close to that predicted from total muscle tension and the number of motor axons. Also, groups consisting of many motor units sum together linearly. Knott and her colleagues (1971) found that this was not true of motor-unit twitches, which summed to only two-thirds the total muscle twitch in one fast muscle. However, physiological contractions are neither fully fused tetanic contractions nor twitches, although it might be guessed that unfused tetanic contractions are closer to twitches. It is not clear if the nonlinear summation of motor-unit twitches is due to mechanical interference with tension production by the mass or friction of inactive fibers, or to a more fundamental mechanism, such as less complete activation. Either mechanism would be of interest in understanding total muscle response.

NONUNIFORMITY OF MOTOR UNITS

CONSEQUENCES FOR TOTAL MUSCLE RESPONSE. Apart from nonlinear summation, the behavior of a muscle composed of a uniform set of muscle fibers would be different from that of the nonhomogeneous muscles found in all vertebrates so far investigated, even if the mean fiber properties were matched.

Some simple consequences of nonuniformity could be discussed by considering two theoretical muscles. In one, all the fibers have a twitch time to peak of 20 msec and a twitch:tetanus ratio of 0.2. Twitches would begin to fuse at about 20/sec and 95% complete tension would occur at 80/sec. All fibers achieve maximum tetanic tension at length L mm. The nonuniform muscle is assumed to consist of three equal-tension motor units with times to peak of 10, 20, and 30 msec; their other properties are set out in Table III. The final row of Table III shows the response of the whole muscle calculated by linearly summing the twitch and tetanic

TABLE III. NONUNIFORMITY IN MUSCLE[a]

Muscle	Twitch contraction time (msec)	Twitch tension (N)	Tetanic tension (N)	Optimum length (mm)	Stimulation rate for tension increase between twitch and tetanus (/sec) 10% increase	95% increase
A	20	0.2	1	L	20	80
I	10	0.15	1/3	L − 1	40	100
II	20	0.2	1/3	L	20	80
III	30	0.25	1/3	L + 1	15	50
B	20.5	0.15	1	L − 1 to L + 1	18	85

[a]Comparison responses of two theoretical muscles: A is composed of identical fibers and B is composed of three groups of fibers, the characteristics of which are shown as I–III. It is assumed that optimum length is related to twitch contraction time (Lewis *et al.*, 1972). The final row is the linear sum of the unit responses of B.

tensions of the motor units. The time to peak of the twitch of the nonuniform muscle will be close to that of the uniform one, but the twitch: tetanus ratio will be smaller since the peak twitch tensions of the three components do not occur at the same time. Tetanic tension will be approximately constant over a range of lengths from the optimum of Unit I to that of Unit III; that is, the muscle length–tension diagram will be flattened, with a broad optimum. Fusion of twitch will begin at the stimulation frequency characteristic of the slowest unit, but will not be complete until that of the fastest, so that the total response-frequency curve will be less steep.

Hill (1970) has made calculations of the force–velocity properties of a nonuniform muscle; the model was more complex than above, with a Gaussian distribution of numbers of fibers around the mean. His conclusion was that the velocities of shortening of a uniform and a nonuniform muscle would be similar over a wide range of loads. There would be a substantial difference, however, at loads less than 5% of maximum tetanic tension. With these small loads, the nonuniform muscle would be able to shorten at velocities greater than possible for the uniform muscle since the small proportion of very fast fibers could exert its full effect at low loads.

PHYSIOLOGICAL SIGNIFICANCE IN CONTROL OF TENSION. Henneman and Olson (1965) have emphasized one consequence of nonuniformity that is of great importance in the control of muscle contraction. These authors have pointed out that if small motor units were recruited early in a contraction and large ones late, then each motor unit would add approximately the same fraction of tension to that existing and the fineness of control would be proportionately the same over the whole range. This can be illustrated quantitatively by plotting the muscle tension at each motor unit increment against recruitment order. If each increment were exactly a constant fraction of the existing tension, then the graph would be linearized by use of a logarithmic scale for tension.

Henneman's evidence for postulating this mechanism was that small motoneurons had low thresholds and made connection with low tension motor units (see above). The extent to which constant fractional increments *could* be achieved may be tested by adding together motor-unit tensions and plotting the successive sums on a logarithmic scale (Fig. 34). The precise shape of the resulting cumulative curves depends on the criterion on which the orderly selection of motor units is based. If the criterion is tension (starting with the smallest unit), the result is shown in Fig. 34A. If the motor units are selected in order of increasing axon conduction velocity, different curves are obtained (Fig. 34B), which depend not only on the relationship between conduction velocity and tension but also on the numerical distribution of motor-unit conduction velocities. In both types of curves extensive linear segments indicate that Henneman's mechanism is possible over certain ranges of muscle tension and is possible without additional central organization.

Since little is known of the precise recruitment mechanism, detailed discussion is impossible. It is clear, however, that the plots of experimental results in Fig. 34 are considerable more log-linear than those that would have been obtained if motor units were uniform in tension (solid line curves in Fig. 34).

Stein and his colleagues (Milner-Brown *et al.*, 1973c) have shown that in voluntarily controlled contraction of one human muscle there is an orderly recruitment of motor units on a basis of twitch tension; the twitch tension of a motor unit was directly proportional to the muscle tension at which it was recruited. Of course, there was a normal amount of variability superimposed on this order. This finding has been confirmed in human jaw muscles (Goldberg and Derfler, 1972; Yemm, 1977) and in monkey jaw muscle (Clark, Luschei, and Hoffman, 1978). It may be concluded, therefore, that fineness of control is maintained constant over much of the range of at least one type of physiological contractions. If there is confirmation

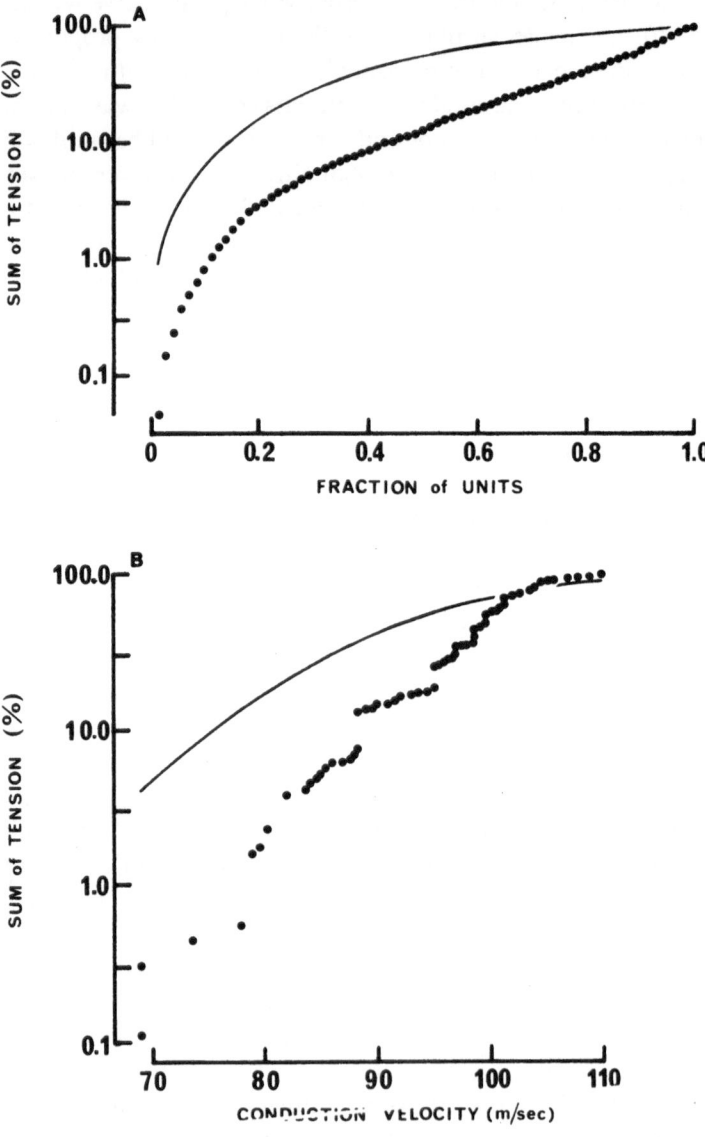

Fig. 34. Cumulative tension plots of motor units of cat fast-twitch muscle. In (A) the motor units have been ranked in order of increasing tension, and in (B) in order of increasing axonal conduction velocity. Each point represents a motor unit and the ordinate is the total tension developed by that unit plus all those to its left. Tensions are expressed as a percentage of parent muscle tension and plotted on a logarithmic scale. The continuous lines show the curves that would have been obtained if all motor units developed the same tension. From Bagust *et al.* (1973).

of the preliminary suggestions of Garnett and Stephens (1978) that large motor units are recruited early under certain conditions (see above), there will have to be reassessment of Henneman's hypothesis.

A proviso must be added from another paper by Milner-Brown *et al.* (1973c), who showed that recruitment of motor units accounts for only a part of the increase in tension at different levels; an increase in frequency and consequent increase in motor-unit tension is responsible for the remainder. Numerically, recruitment was most important only at low forces. Near maximum force, almost all motor units had been recruited, and increase in tension was due to increase in frequency, having been recruited last).

This is one aspect in which the nonuniform nature of muscle can be, and apparently is, used to produce the most efficient and precisely controlled response. It may be concluded that both in organization as well as in its overall properties muscle is arranged to serve and anticipate the functions of the nervous system.

Acknowledgments

I would like to thank those authors and publishers who have allowed me to reproduce their material in the figures. I am grateful to Drs. Ridge and Armstrong for reading the manuscript and for their valuable comments. Finally, my thanks to Ms. Hilary Lee and Ms. Jane Outtrim who typed the many drafts and to Mrs. Barbara Colfer for the photography.

REFERENCES

Adrian, E. D., and Bronk, D. W. The discharge of impulses in motor nerve fibres. Part II. The frequency of discharge in reflex and voluntary contractions. *Journal of Physiology (London)*, 1929, *67*, 119–151.

Al-Amood, W. S., and Pope, R. A comparison of the structural features of muscle fibres from a fast- and a slow-twitch muscle of the pelvic limb of the cat. *Journal of Anatomy (London)*, 1972, *113*, 49–60.

Andersen, P., and Sears, T. A. The mechanical properties and innervation of fast and slow motor units in the intercostal muscles of the cat. *Journal of Physiology (London)*, 1964, *173*, 114–129.

Appelberg, B., and Emonet-Dénand, F. Motor units of the first superficial lumbrical muscle of the cat. *Journal of Neurophysiology*, 1967, *30*, 154–160.

Aubert, X. *Le couplage énergétique de la contraction musculaire.* Thèse d'Agregation, Université Catholique de Louvain, 1956.

Bagust, J. *Motor unit studies in cat and rabbit solei.* Ph.D. Thesis, University of Bristol, Bristol, 1971.

Bagust, J. Relationships between motor conduction velocities and motor unit contraction characteristics in a slow twitch muscle of the cat. *Journal of Physiology (London)*, 1974, *238*, 269–278.

Bagust, J., and Lewis, D. M. Isometric contractions of motor units in self reinnervated fast and slow twitch muscles of the cat. *Journal of Physiology (London)*, 1974, *237*, 91–102.

Bagust, J., Lewis, D. M., and Westerman, R. A. Polyneural innervation of kitten skeletal muscle. *Journal of Physiology (London)*, 1973, 229, 241–255.

Bagust, J., Knott, S., Lewis, D. M., Luck, J. C., and Westerman, R. A. Isometric contractions of motor units in a fast twitch muscle of the cat. *Journal of Physiology (London)*, 1973, *231*, 87–104.

Bagust, J., Lewis, D. M., and Westerman, R. A. The properties of motor units in a fast- and a slow-twitch muscle during post-natal development in the kitten. *Journal of Physiology (London)*, 1974, *237*, 75–90.

Bagust, J., Lewis, D. M., and Luck, J. C. Post-tetanic effects of motor units of fast- and slow-twitch muscle of the cat. *Journal of Physiology (London)*, 1974, *237*, 115–121.

Bagust, J., Lewis, D. M., and Westerman, R. A. Motor units in cross-reinnervated fast and slow twitch muscle of the rat. *Journal of Physiology (London)*, 1981, *313*, 223–235.

Baldissera, F., and Campadelli, P. How motoneurons control development of muscle tension. *Nature (London)*, 1977, *268*, 146–147.

Bárány, M. ATPase activity of myosin correlated with speed of muscle shortening. *Journal of General Physiology*, 1967, *50*, 197–216.

Bárány, M., and Close, R. I. The transformation of myosin in cross-innervated rat muscles. *Journal of Physiology (London)*, 1971, *213*, 455–474.

Bessou, P., Emonet-Dénand, F., and Laporte, Y. Relation entre la vitesse de conduction des fibres nerveuses motrices et le temps de contraction de leurs unités motrices. *Comptes Rendus Hebdomadaires des Séances de l'Académie des Sciences (Paris), Serie D, Sciences Naturelles*, 1963, *256*, 5625–5627.

Bessou, P., Laporte, Y., and Pagés, B. Etude de la relation entre le temps de contraction d'unités motrices et la vitesse de conduction de leurs fibres motrices dans les muscles réinnervés. *Comptes Rendus Hebdomadaires des Séances de l'Académie des Sciences (Paris), Seris D, Sciences Naturelles*, 1966, *263*, 1486–1489.

Björkman, A., and Wohlfart, C. Faseranalyse der Nn. oculomotorius, trochlearis und abducens des Menschen und des N. abducens verchiedener tiere. *Zeitschrift fuer Mikroskopisch-Anatomische Forschung*, 1936, *39*, 631–647.

Boyd, I. A., and Davey, M. R. *The Composition of Peripheral Nerve.* Edinburgh: E. & S. Livingston, 1968.

Brooke, M. H., and Kaiser, K. K. Muscle fibre types: How many and what kind? *Archives of Neurology*, 1970, *23*, 369–379.

Brown, G. L., and von Euler, U. S. The after effects of a tetanus on mammalian muscle. *Journal of Physiology (London)*, 1938, *93*, 39–60.

Brown, M. C., and Matthews, P. B. C. An investigation into the possible existence of polyneural innervation of individual skeletal muscle fibres in certain hind-limb muscles of the cat. *Journal of Physiology (London)*, 1960, *151*, 436–457.

Browne, J. S. The contractile properties of slow muscle fibres in sheep extraocular muscles. *Journal of Physiology (London)*, 1976, *254*, 535–550.

Buchthal, F., and Schmalbruch, H. Contraction times of fibre types in intact human muscle. *Acta Physiologica Scandinavica*, 1970, *79*, 435–452.

Buchthal, F., Guld, Ch., and Rosenfalck, P. Multielectrode studies of the territory of a motor unit. *Acta Physiologica Scandinavica*, 1957, *39*, 83–103.

Buller, A. J. The neural control of some characteristics of skeletal muscle. In C. B. B. Downman (Ed.) *Modern Trends in Physiology.* London: Butterworths, 1972.

Buller, A. J. *The Contractile Behaviour of Mammalian Skeletal Muscle.* Oxford Biology Readers No. 36. London: Oxford University Press, 1975.

Buller, A. J., and Buller, N. B. *The Contractile Behaviour of Mammalian Skeletal Muscle.* Carolina Press (in press)

Buller, A. J., and Lewis, D. M. Factors affecting the differentiation of mammalian fast and slow muscle fibres. In E. Gutmann and P. Hnik (Eds.), *The Effect of Use and Disuse on Neuromuscular Functions.* Amsterdam: Elsevier, 1963.

Buller, A. J., and Lewis, D. M. The rate of tension development in isometric tetanic contractions of mammalian fast and slow skeletal muscle. *Journal of Physiology (London)*, 1965a, *176*, 337–342.

Buller, A. J., and Lewis, D. M. Further observations on mammalian cross-innervated skeletal muscle. *Journal of Physiology (London)*, 1965b, *178*, 343–358.

Buller, A. J., Eccles, J. C., and Eccles, R. M. Differentiation of fast and slow muscles in the cat hind limb. *Journal of Physiology (London)*, 1960a, *150*, 399–416.

Buller, A. J., Eccles, J. C., and Eccles, R. M. Interactions between motoneurones and muscles in respect of the characteric speeds of their responses. *Journal of Physiology (London)*, 1960b, *150*, 417–439.

Buller, A. J., Mommaerts, W. F. H. M., and Seraydarian, K. Neural control of myofibrillar ATPase activity in rat skeletal muscle. *Nature, New Biology (London)*, 1971, *233*, 31–32.

Buller, A. J., Lewis, D. M., and Ridge, R. M. A. P. Some electrical characteristics of fast-twitch and slow-twitch skeletal muscle fibres in the cat. *Journal of Physiology (London)*, 1965, *180*, 29–30P.

Burke, R. E. Motor unit types of cat triceps surae muscle. *Journal of Physiology (London)*, 1967, *193*, 141–161.

Burke, R. E. Firing patterns of gastrocnemius motor units in the decerebrate cat. *Journal of Physiology (London)*, 1968, *196*, 631–654.

Burke, R. E. The correlation of physiological properties with histochemical characteristics in single muscle units. *Annals of the New York Academy of Sciences*, 1974, *228*, 145–159.

Burke, R. E., and Tsairis, P. Anatomy and innervation ratios in motor units of cat gastrocnemius. *Journal of Physiology (London)*, 1973, *234*, 749–765.

Burke, R. E., Levine, D. N., Zajac, F. E., Tsairis, P., and Engel, W. K. Mammalian motor units: Physiological-histochemical correlation in three types in cat gastrocnemius. *Science*, 1971, *174*, 709–712.

Burke, R. E., Levine, D. N., Tsairis, P., and Zajac, F. E. Physiological types and histochemical profiles in motor units of the cat gastrocnemius. *Journal of Physiology (London)*, 1973, *234*, 723–748.

Burke, R. E., Levine, D. N., Saloman, M., and Tsairis, P. Motor units in cat soleus muscle: Physiological, histochemical and morphological characteristics. *Journal of Physiology (London)*, 1974, *238*, 503–514.

Burke, R. E., Rudomin, P., and Zajac, F. E. The effect of activation history on tension production by individual muscle units. *Brain Research*, 1976, *109*, 515–529.

Clark, D. A. Muscle Counts of Motor Units: A study in innervation ratios. *American Journal of Physiology*, 1931, *96*, 296–304.

Clark, R. W., Luschei, E. S., and Hoffman, D. S. Recruitment order, contractile characteristics and firing patterns of motor units in the temporalis muscle of monkeys. *Experimental Neurology*, 1978, *61*, 31–52.

Cliff, C. S., and Ridge, R. M. A. P. Innervation of extrafusal and intrafusal fibres in snake muscle. *Journal of Physiology (London)*, 1973, *233*, 1–18.

Close, R. Dynamic properties of fast and slow skeletal muscles of the rat during development. *Journal of Physiology (London)*, 1964, *173*, 74–95.

Close, R. Properties of motor units in fast and slow skeletal muscles of the rat. *Journal of Physiology (London)*, 1967, *193*, 45–55.

Close, R. Dynamic properties of fast and slow skeletal muscle of the rat after nerve cross-union. *Journal of Physiology (London)*, 1969, *204*, 331–346.

Close, R. Neural influences on physiological properties of fast and slow limb muscles. In R. J. Podolsky (Ed.), *Contractility of Muscle Cells and Related Processes*. Englewood Cliffs, N.J.: Prentice-Hall, 1971.

Close, R. I. Dynamic properties of mammalian skeletal muscle. *Physiological Reviews*, 1972, *52*, 129–197.

Close, R., and Hoh, J. The after-effects of repetitive stimulation on the isometric twitch contraction of rat fast skeletal muscle. *Journal of Physiology (London)*, 1968a, *197*, 461–477.

Close, R., and Hoh, J. R. Influence of temperature on isometric contractions of rat skeletal muscles. *Nature (London)*, 1968b, *217*, 1179–1180.

62

DONALD M. LEWIS

Close, R., and Luff, A. R. Dynamic properties of inferior rectus muscle of the rat. *Journal of Physiology (London)*, 1974, *236*, 259–270.

Cooper, S., and Eccles, J. C. The isometric responses of mammalian muscles. *Journal of Physiology (London)*, 1930, *69*, 377–385.

Crawford, G. N. C. Experimentally induced hypertrophy of growing voluntary muscle. *Proceedings of the Royal Society of London, Series B*, 1961, *154*, 130–138.

Davies, A. S. Postnatal changes in the histochemical properties of porcine skeletal muscle. *Journal of Anatomy (London)*, 1972, *111*, 487–489.

Davies, A. S., and Gunn, H. M. Histochemical fibre types in the mammalian diaphragm. *Journal of Anatomy (London)*, 1972, *112*, 41–60.

Denny-Brown, D. On the nature of postural reflexes. *Proceedings of the Royal Society of London, Series B*, 1929, *104*, 253–301.

Denslow, J. S. Double discharges in human motor units. *Journal of Neurophysiology*, 1948, *11*, 209–278.

Denslow, J. S., and Gutensohn, O. R. Distribution of muscle fibers in a single motor unit. *Federation Proceedings*, 1950, *9*, 31.

Denslow, J. S., and Gutensohn, O. R. Neuromuscular organization of single motor units. *Federation Proceedings*, 1951, *10*, 34.

Desmedt, J. E., and Godaux, E. Ballistic contractions in man: Characteristic recruitment pattern of single motor units of the tibialis anterior muscle. *Journal of Physiology (London)*, 1977, *264*, 673–693.

Deuanandan, M. S., Eccles, R. M., and Westerman, R. A. Single motor units of mammalian muscle. *Journal of Physiology (London)*, 1965, *178*, 359–367.

Dubowitz, V., and Pearse, A. G. E. A comparative histochemical study of oxidative enzyme and phosphorylase activity in skeletal muscle. *Histochemie*, 1960, *2*, 105–117.

Dunn, E. H. The number and size of the nerve fibers innervating the skin and muscles of the thigh in the frog *(Rana virescens brachycephala, Cope) Journal of Comparative Neurology*, 1900, *10*, 218–242.

Eccles, J. C., and Sherrington, C. S. Numbers and contraction values of individual motor units examined in some muscles of the limb. *Proceedings of the Royal Society of London, Series B*, 1930, *106*, 326–557.

Eccles, R. M., Phillips, C. G., and Wu Chien-Ping. Motor innervation, motor unit organization and afferent innervation of m. extensor digitorum communis of the baboon's forearm. *Journal of Physiology (London)*, 1968, *198*, 179–192.

Edgerton, V. R., and Simpson, D. R. The intermediate muscle fiber of rats and guinea pigs. *Journal of Histochemistry and Cytochemistry.* 1969, *17*, 828–838.

Edgerton, V. R., Goslow, G. E. Jr., Rasmussen, S. A., and Spector, S. A. Is resistance of a muscle to fatigue controlled by its motoneurones? *Nature (London)*, 1980, *285*, 589–590.

Edjtehadi, G. D., and Lewis, D. M. Histochemical reactions of fibres in a fast-twitch muscle of the cat. *Journal of Physiology (London)*, 1979, *287*, 439–454.

Edström, L., and Kugelberg, E. Histochemical composition, distribution of fibres and fatiguability of single motor units. Anterior tibial muscle of the rat. *Journal of Neurology, Neurosurgery and Psychiatry*, 1968, *31*, 424–433.

Emonet-Dénand, F., Laporte, Y., and Proske, V. Contraction of muscle fibers in two adjacent muscles innervated by branches of the same motor axon. *Journal of Neurophysiology*, 1971, *34*, 132–138.

Engberg, I., and Lundberg, A. An electromyographic analysis of muscular activity in the hind limb of the cat during unrestrained locomotion. *Acta Physiologica Scandinavica*, 1969, *75*, 614–630.

Engel, W. K. The essentiality of histo- and cyto-chemical studies of skeletal muscle in the investigation of neuromuscular disease. *Neurology*, 1962, *12*, 778–784.

Engle, A. G., and Stonnington, H. H. Morphological effects of denervation of muscle. A quantitative ultrastructural study. *Annals of the New York Academy of Sciences*, 1974, *228*, 68–88.

Erulkar, S. D., Shelanski, M. L., Whitsel, B. L., and Ogle, P. Studies of muscle fibers of the tensor tympani of the cat. *Anatomical Record*, 1969, *149*, 279–298.

Feinstein, B., Lindegard, B., Nyman, E., and Wohlfart, G. Morphologic studies of motor units in normal human muscles. *Acta Anatomica*, 1955, *23*, 127–142.

Fischbach, G. D., and Robbins, N. Changes in contractile properties of disused soleus muscles. *Journal of Physiology (London)*, 1969, *201*, 305–320.

Ford, L. E., Huxley, A. F., and Simmons, R. M. Tension responses to sudden length change in stimulated frog muscle fibres near slack length. *Journal of Physiology (London)*, 1977, *269*, 441–515.

Frank, E., Jansen, K. K. S., Lømo, T., and Westgaard, R. H. The interaction between foreign and original motor nerves innervating the soleus muscle of rats. *Journal of Physiology (London)*, 1975, *247*, 725–743.

Garnett, R., and Stephens, J. A. Changes in the recruitment on threshold of motor units in human first dorsal interosseous muscle produced by skin stimulation. *Journal of Physiology (London)*, 1978, *282*, 13P.

Garnett, R. A. F., O'Donovan, M. J., Stephens, J. A., and Taylor, A. Motor unit organisation of human medial gastrocnemius. *Journal of Physiology (London)*, 1979, *287*, 33–44.

Gasser, H. S., and Hill, A. V. The dynamics of muscular contraction. *Proceedings of the Royal Society of London, Series B*, 1924, *96*, 398–437.

Gauthier, G. F. On the relationship of ultrastructural and cytochemical features to color in mammalian skeletal muscle. *Zeitschrift fuer Zellforschung und Mikroskopische Anatomie*, 1969, *95*, 462–482.

Gauthier, G. F., Lowey, S., and Hobbs, A. W. Fast and slow myosin in developing muscle fibres. *Nature (London)*, 1978, *274*, 25–29.

George, J. C., and Susheela, A. K. A histophysiological study of the rat diaphragm. *Biological Bulletin*, 1961, *121*, 471–480.

Goldberg, L. J., and Derfler, B. Relationship among recruitment order, spike amplitude and twitch tension of single motor units in human masseter muscle. *Journal of Neurophysiology*, 1977, *40*, 879–890.

Goldspink, G., Tabary, C., Tabary, J. C., Tardieu, C., and Tardieu, G. Effect of denervation on the adaptation of sarcomere number and muscle extensibility to the functional length of the muscle. *Journal of Physiology (London)*, 1974, *236*, 733–742.

Goodwin, G. M. Hoffman, D., and Luschei, E. S. The strength of the reflex response to sinusoidal stretch of monkey jaw closing muscles during voluntary contraction. *Journal of Physiology (London)*, 1978, *279*, 81–111.

Gordon, A. M., Huxley, A. F., and Julian, F. J. The variation in isometric tension with sarcomere length in vertebrate muscle fibres. *Journal of Physiology (London)*, 1966, *184*, 170–192.

Gordon, G., and Holbourn, A. H. S. The mechanical activity of single motor units in reflex contractions of skeletal muscle. *Journal of Physiology (London)*, 1949, *110*, 26–35.

Gordon, G., and Phillips, C. G. Slow and rapid components in a flexor muscle. *Quarterly Journal of Experimental Physiology*, 1953, *38*, 35–45.

Grillner, S. The role of muscle stiffness in meeting the changing postural and locomotor requirements for force development by the ankle extensors. *Acta Physiologica Scandinavica*, 1972, *86*, 92–108.

Guth, L. Neuromuscular function after regeneration of interrupted nerve fibers into partially denervated muscle. *Experimental Neurology*, 1962, *6*, 129–141.

Guth, L., Samaha, F. J., and Albers, R. W. The neural regulation of some phenotypic differences between the fiber types of mammalian skeletal muscle. *Experimental Neurology*, 1970, *26*, 126–135.

Gutmann, E., and Hanziliková, V. Effect of accessory nerve supply to muscle achieved by implantation into muscle during regeneration of its nerve. *Physiologia Bohemoslovenica*, 1967, *16*, 244–250.

Hartree, W., and Hill, A. V. The nature of the isometric twitch. *Journal of Physiology (London)*, 1921, *55*, 389–411.

Henneman, E., McPhedran, A. M., and Wuerker, R. B. Tensions of single motor units in cat muscles. *Federation Proceedings*, 1963, *22*, 279.

64

DONALD M. LEWIS

Henneman, E., and Olson, C. Relations between structure and function in the design of skeletal muscles. *Journal of Neurophysiology*, 1965, *28*, 581–598.

Hess, A. Vertebrate slow muscle fibers. *Physiological Reviews*, 1970, *50*, 40–62.

Hess, A., and Pilar, G. Slow fibres in the extraocular muscles of the cat. *Journal of Physiology (London)*, 1963, *169*, 780–798.

Hill, A. V. The heat of shortening and the dynamic constants of muscle. *Proceedings of the Royal Society of London, Series B*, 1938, *126*, 136–195.

Hill, A. V. The design of muscles. *British Medical Bulletin*, 1956, *12*, 165–166.

Hill, A. V. *First and Last Experiments in Muscle Mechanics*. Cambridge: Cambridge: University Press, 1970.

Hoff, H. E., and Grant, R. S. The supernormal period in the recovery cycle of motoneurons. *Journal of Neurophysiology*, 1944, *7*, 305–322.

Hoh, J. F. Y. Selective and non-selective reinnervation of fast-twitch and slow-twitch rat skeletal muscle. *Journal of Physiology (London)*, 1975, *251*, 791–801.

Houk, J., and Henneman, E. Responses of Golgi tendon organs to active contractions of the soleus muscle of the cat. *Journal of Neurophysiology*, 1967, *30*, 466–481.

Howells, K. R., and Jordan, T. C. The myofibril content of histochemically characterized rat muscle fibre types. *Journal of Physiology (London)*, 1978, *284*, 35P.

Hudlická, O., Brown, M., Cotter, M., Smith, M., and Vrbová, G. The effect of long-term stimulation of fast muscles on their blood flow, metabolism and ability to withstand fatigue. *Pfluegers Archiv fuer die Gesamte Physiologie*, 1977, *369*, 141–149.

Hunt, C. C., and Kuffler, S. W. Motor innervation of individual muscle fibres and motor function. *Journal of Physiology (London)*, 1954, *126*, 293–303.

Huszar, G. Developmental changes of the primary structure and histidine methylation in rabbit skeletal muscle myosin. *Nature, New Biology (London)*, 1972, *240*, 260–264.

Huxley, A. F., and Simmons, R. M. Mechanical properties of the cross-bridges of frog striated muscle. *Journal of Physiology (London)*, 1971, *218*, 59–60.

Joyce, G. C., Rack, P. M., and Westbury, D. R. The mechanical properties of cat soleus muscle during controlled lengthening and shortening movements. *Journal of Physiology (London)*, 1969, *204*, 461–474.

Kanda, K., Burke, R. E., and Walmsley, B. Differential control of fast and slow twitch motor units in the decerebrate cat. *Experimental Brain Research*, 1977, *29*, 57–74.

Karpati, G., and Engel, W. K. Transformation of the histochemical profile of skeletal muscle by "foreign" innervation. *Nature (London)*, 1967, *215*, 1509–1510.

Katz, B. The relation between force and speed in muscular contraction. *Journal of Physiology (London)*, 1939, *96*, 45–64.

Kean, C., Lewis, D. M., and McGarrick, J. Dynamic properties of denervated fast and slow twitch muscle of the cat. *Journal of Physiology (London)*, 1974, *237*, 103–113.

Knott, S. *A Study of Stretch Receptors and Motor Units in Frog Muscle*. Ph. D. Thesis, University of Bristol, Bristol, 1971.

Knott, S., Lewis, D. M. and Luck, J. C. Motor unit areas in a cat limb muscle. *Experimental Neurology*, 1971, *30*, 475–483.

Krnjevic, K., and Miledi, R. Motor units in the rat diaphragm. *Journal of Physiology (London)*, 1958, *140*, 427–439.

Kuffler, S. W., and Vaughan Williams, E. M. Properties of the "slow" skeletal muscle fibres of the frog. *Journal of Physiology (London)*, 1953, *121*, 318–340.

Kugelberg, E. Adaptive transformation of rat soleus motor units during growth. *Journal of Neurological Sciences*, 1976, *27*, 269–289.

Kugelberg, E., and Edstrom, L. Differential histochemical effects of muscle contractions on phosphorylase and glycogen in various types of fibres. Relation to fatigue. *Journal of Neurology, Neurosurgery and Psychiatry*, 1968, *31*, 415–423.

Kugelberg, E., Edstrom, L., and Abbruzzese, M. Mapping of motor units in experimentally reinnervated rat muscle. Interpretation of histochemical and atrophic fibre patterns in neurogenic lesions. *Journal of Neurology, Neurosurgery and Psychiatry*, 1970, *33*, 319–329.

Kühne, W. Ueber den Farbstoff der Muskeln. *Archiv fuer Pathologische Anatomie und Physiologie*, 1865, *33*, 79–94.

Lewis, D. M. The effect of denervation on the mechanical and electrical properties of fast and slow mammalian twitch muscle. *Journal of Physiology (London)*, 1972, *222*, 51–75.

Lewis, D. M. Motor units in fast- and slow-twitch muscles, and the effects of reinnervation. In W. G. Bradley, D. Gardner-Medwin, and J. N. Walton (Eds.), *Recent Advances in Myology*, New York: American Elsevier Publishing, 1975.

Lewis, D. M., and Webb, S. N. Adrenaline on isotonic contractions of mammalian skeletal muscle. *British Journal of Pharmacology*, 1976, *58*, 467P.

Lewis, D. M., Luck, J. C., and Knott, S. A comparison of isometric contractions of the whole muscle with those of motor units in a fast-twitch muscle in the cat. *Experimental Neurology*, 1972, *37*, 68–85.

Liddell, E. G. T., and Sherrington, C. S. Recruitment and some other features of reflex inhibition. *Proceedings of the Royal Society of London, Series B*, 1925, *97*, 488–518.

Lømo, T., Westgaard, R. H., and Dahl, H. A. Contractile properties of muscle: Control by pattern of muscle activity in the rat. *Proceedings of the Royal Society of London, Series B*, 1974, *187*, 99–103.

Luff, A., and Proske, U. Properties of motor units of the frog sartorius muscle. *Journal of Physiology (London)*, 1977, *258*, 673–686.

Margreth, A., Angelini, C., Valfre, G., and Salvaiti, G. Developmental patterns of LDH isoenzymes in fast and slow muscles of the rat. *Archives of Biochemistry and Biophysics*, 1970, *141*, 374–377.

Margreth, A., Salviati, G., and Garraro, U. Neural control of the activity of calcium transport system in sarcoplasmic reticulum of rat skeletal muscle. *Nature (London)*, 1973, *241*, 285–286.

Matthews, P. B. C. Evidence that the secondary as well as the primary endings of the muscle spindles may be responsible for the tonic stretch reflex. *Journal of Physiology (London)*, 1969, *204*, 365–395.

McPhedran, A. M., Weurker, R. B., and Henneman, E. Properties of motor units in a homogeneous red muscle (soleus) of the cat. *Journal of Neurophysiology*, 1965, *28*, 71–84.

Milner-Brown, H. S., Stein, R. B., and Yemm, R. The contractile properties of human motor units during voluntary isometric contractions. *Journal of Physiology (London)*, 1973a, *228*, 285–306.

Milner-Brown, H. S., Stein R. B., and Yemm, R. The orderly recruitment of human motor units during voluntary isometric contractions. *Journal of Physiology (London)*, 1973b, *230*, 359–370.

Milner-Brown, H. S., Stein, R. B., and Yemm, R. Changes in firing rate of human motor units during linearly changing voluntary contractions. *Journal of Physiology (London)*, 1973c *230*, 371–390.

Mines, G. R. On the summation of contractions. *Journal of Physiology (London)*, 1913, *46*, 1–27.

Mussini, I., Margreth, A., and Salviati, G. On the criteria for characterization of calcium oxalate in sarcoplasmic reticulum fragments. *Journal of Ultrastructure Research*, 1972, *38*, 459–465.

Morgan, D. L., Proske, U., and Warren, D. Measurements of muscle stiffness and the mechanism of elastic storage of energy in hopping kangaroos. *Journal of Physiology (London)*, 1978, *282*, 253–262.

Moscher, C. G., Gerlach, R. L., and Stuart, D. G. Soleus and anterior tibial motor units of the cat. *Brain Research*, 1972, *44*, 1–11.

Nichols, T. R., and Houk, J. C. Improvement in linearity and regulation of stiffness that results from actions of stretch reflex. *Journal of Neurophysiology*, 1976, *39*, 119–142.

Nystrom, B. Fibre diameters increase in nerves to "slow-red" and "fast-white" cat muscles during postnatal development. *Acta Neurologica Scandinavica*, 1968, *44*, 265–294.

Ogata, T. A histochemical study of the red and white muscle fibers. Part I. Activity of the succinoxydase system in muscle fibers. *Acta Medica Okayama*, 1958, *12*, 216–227.

Olson, C. B., and Swett, C. P. A functional and histochemical characterization of motor units in a heterogeneous muscle (flexor digitorum longus) of the cat. *Journal of Comparative Neurology,* 1966, *128,* 475–498.

Olson, C. B., and Swett, C. P. The effect of prior activity on properties of different types of motor units. *Journal of Neurology,* 1971, *34,* 1–16.

Parry, D. G., Mainwood, G. W., and Chan, T. The relationship between surface potentials and the number of active motor units. *Journal of the Neurological Sciences,* 1977, *33,* 283–296.

Paukul, E. Die Zuchungsformen von Kaniuchenmuskeln veschiedener Faber und Structur. *Archiv fuer Anatomie und Physiologie,* 1904, 100–120.

Peter, J. B., Barnard, R. J., Edgerton, V. R., Gillespie, C. A., and Stempel, K. E. Metabolic profiles of three fiber types of skeletal muscle in guinea pigs and rabbits. *Biochemistry,* 1972, *11,* 2627–2633.

Pette, D., and Schnez, U. Myosin light chain patterns of individual fast and slow-twitch fibres of rabbit muscles. *Histochemistry,* 1977, *54,* 97–107.

Pette, D., Smith, M. E., Standte, H. W., and Vrbová, G. Effects of long-term electrical stimulation on some contractile and metabolic characteristics of fast rabbit muscle. *Pfluegers Archiv fuer die Gestamte Physiologie,* 1973, *338,* 257–272.

Porter, E. L. Evidence that the postural tonus of decerebrate rigidity increases in amount by the successive innervation of single motor neurones. *American Journal of Physiology,* 1929, *91,* 345–361.

Porter, E. L., and Hart, V. W. Reflex contraction of an all-or-none character in the spinal cat. *American Journal of Physiology,* 1923, *66,* 391–403.

Prewitt, M. A., and Salafsky, B. Enzymic and histochemical changes in fast and slow muscles after cross innervation. *American Journal of Physiology,* 1970, *218,* 69–74.

Proske, U., and Waite, P. M. E. Properties of types of motor units in the medial gastrocnemius muscle of the cat. *Brain Research,* 1974, *67,* 89–101.

Proske, U., and Waite, P. M. E. The relation between tension and axonal conduction velocity for motor units in the medial gastrocnemius muscle of the cat. *Experimental Brain Research,* 1976, *26,* 325–328.

Prochazka, A., Westerman, R. A., and Ziccone, S. P. Ia afferent activity during a variety of voluntary movements in the cat. *Journal of Physiology (London),* 1977, *268,* 423–448.

Rack, P. M., and Westbury, D. R. The effects of length and stimulus rate on tension in the isometric cat soleus muscle. *Journal of Physiology (London),* 1969, *204,* 443–460.

Rack, P. M., and Westbury, D. R. The short range stiffness of active mammalian muscle and its effect on mechanical properties. *Journal of Physiology (London),* 1974, *240,* 331–350.

Ranatunga, K. W. Influence of temperature on the characteristics of summation of isometric mechanical responses of mammalian skeletal muscle. *Experimental Neurology,* 1977, *54,* 513–532.

Ranvier, L. Proprietés et structures différentes des muscles rouges et des muscles blanc chez les lapin et chez les raies. *Comptes Rendus Hebdomadaires des Séances de l'Académie des Sciences (Paris),* 1873, *77,* 1030–1043.

Redfern, P. A. Neuromuscular transmission in new-born rats. *Journal of Physiology (London),* 1970, *209,* 701–709.

Robinson, D. A. Oculomotor unit behaviour in the monkey. *Journal of Neruophysiology,* 1970, *33,* 393–404.

Romanul, F. C. A. Enzymes in muscle. I. Histochemical studies of enzymes in individual muscle fibers. *Archives of Neurology,* 1964, *11,* 355–369.

Romanul, F. C. A., and van der Meulen, J. P. Reversal of the enzyme profiles of muscle fibers in fast and slow muscles by cross innervation. *Nature (London),* 1966, *212,* 1369–1370.

Rossi, G., and Cortesina, G. Multimotor end-plate muscle fibres in the human vocalis muscle. *Nature,* 1965, *206,* 629–630.

Rutherford, W. A new theory of hearing. *Journal of Anatomy and Physiology,* 1887, *21,* 166–168.

Salmons, S., and Vrbová, G. The influence of activity on some contractile characteristics of mammalian fast and slow muscles. *Journal of Physiology (London),* 1969, *201,* 535–549.

Samaha, F. J., Guth, L., and Albers, R. W. The neural regulation of gene expression in the muscle cell. *Experimental Neurology*, 1970, *27*, 276–282.

Schmalbruch, H., and Kamieniecka, Z. Histochemical fiber typing and staining intensity in cat and rat muscles. *Journal of Histochemistry and Cytochemistry*, 1975, *23*, 395–401.

Sherrington, C. S. Remarks on some aspects of reflex inhibition. *Proceedings of the Royal Society of London, Series B*, 1925, *97*, 519–545.

Sica, R. E. P., and McComas, A. J. Fast and slow twitch units in a human muscle. *Journal of Neurology, Neurosurgery and Psychiatry*, 1971a, *34*, 113–120.

Sica, R. E. P., and McComas, A. J. An electrophysiological investigation of limb-girdle and facioapulohumeral dystrophy. *Journal of Neurology, Neurosurgery and Psychiatry*, 1971b, *34*, 469–474.

Spurway, N. C. Objective typing of mouse ankle-extensor muscle fibres based on histochemical photometry. *Journal of Physiology (London)*, 1978, *277*, 47–48.

Sréter, F. A., Seidel, J. C., and Gergely, J. Studies on myosin from red and white skeletal muscles of the rabbit. I. Adenosine triphosphatase activity. *Journal of Biological Chemistry*, 1966, *241*, 5772–5776.

Stalberg, E., and Thiele, B. Motor unit fibre density in the extensor digitorum communis muscle. *Journal of Neurology, Neurosurgery, and Psychiatry*, 1975, *38*, 874–880.

Steg, G. The function of muscle spindles in spasticity and rigidity. *Acta Neurologica Scandinavica*, 1962, *38:* Supplementum 3, 53–59.

Stein, J. M., and Padykula, H. A. Histochemical classification of individual skeletal muscle fibers of the rat. *American Journal of Anatomy*, 1962, *110*, 103–124.

Stein, R. B., French, A. S., Mannard, A., and Yemm, R. New methods for analyzing motor unit function in man and animals. *Brain Research*, 1972, *40*, 187–192.

Stephens, J. A., and Stuart, D. G. The motor units of cat medial gastrocnemius: Speed-size relations and their significance for the recruitment order of motor units. *Brain Research*, 1975, *91*, 177–195.

Tanji, J., and Kato, M. Recruitment of motor units in voluntary contraction of a finger muscle in man. *Experimental Neurology*, 1973, *40*, 759–770.

Thiele, B., and Stalberg, E. The biomodal jitter: A single fibre electromyographic finding. *Journal of Neurology, Neurosurgery and Psychiatry*, 1974, *37*, 403–411.

Twarog, B. M., and Muneoka, Y. Calcium and the control of contraction and relaxation in a molluscan catch muscle. *Cold Spring Harbor Symposia of Quantitative Biology*, 1972, *37*, 489–503.

Walmsley, B., Hodgson, J. A., and Burke, R. E. The forces produced by medial gastrocnemius and soleus muscles during locomotion in freely moving cats. *Journal of Neurophysiology*, 1978, *41*, 1203–1216.

Weeds, A. Myosin: polymorphism and promiscuity. *Nature (London)*, 1978, *274*, 417–418.

Weeds, A. G., Trentham, D. R., Kean, C. J. C., and Buller, A. J. Myosin from cross-reinnervated cat muscles. *Nature (London)*, 1974, *247*, 135–139.

Weeds, A. G., Hall, R., and Spurway, N. C. Characterization of myosin light chains from histochemically identified fibres of rabbit psoas muscle. *FEBS Letters*, 1975, *49*, 320–324.

Westerman, R. A., Lewis, D. M., Bagust, J., Edjtehadi, G., and Pallot, D. Communication between nerves and muscles: Postnatal development in kitten hind limb fast and slow twitch muscle. In H. P. Zippel (Ed.), *Memory and Transfer of Information*, New York: Plenum, 1974.

Weuker, R. B., McPhedran, A. M., and Henneman, E. Properties of motor units in heterogeneous pale muscle (m. gastrocnemius) of the cat. *Journal of Neurophysiology*, 1965, *28*, 85–99.

Wilkie, D. R. *Muscle*. (Institute of Biology, Studies in Biology, No. 11). London: Arnold, 1976.

Yellin, H., and Guth, L. The histochemical classification of muscle fibers. *Experimental Neurology*, 1970, *26*, 424–432.

Yemm, R. The orderly recruitment of motor units of the masseter and temporal muscles during voluntary isometric contraction in man. *Journal of Physiology (London)*, 1977, *265*, 163–174.

2

Motor Units and Their Activity during Movement

H. Peter Clamann

Introduction

The motor unit is the smallest functional subdivision of the neuromuscular system and, therefore, the smallest part of that system which can be controlled with any degree of independence. A *motor unit* is defined as a motoneuron, its axon, and all the muscle fibers which that axon innervates (Burke and Edgerton, 1975), although the original definition included only the axon and adjunct muscle fibers (Creed, Denny-Brown, Eccles, Liddell, and Sherrington, 1932; Liddell and Sherrington, 1925). Accepting the terminology of Burke (e.g., Burke, 1967), we will refer to the muscle fibers supplied by a single motoneuron as a *muscle unit*. A group of motor units, acting together, produces muscle force. The vast variety of movements of which we are capable—from tiny, precisely controlled motor acts to powerful, ballistic flailing of limbs—are all produced by varying numbers of motor units cooperating in different ways. This versatile repertoire of movements is brought about by two mechanisms: varying the number of motor units that are active or varying the discharge rate, and hence the force output, of individual motor units.

To understand how motor units are utilized in different movements, it is first necessary to examine the properties of individual motor units. As we shall see, motor units vary widely in their speed of contraction, maximum force output, and resistance to fatigue. The nervous system, therfore, has a wide range of choice in selecting appropriate motor units to produce different movements.

H. Peter Clamann   Department of Physiology, Medical College of Virginia, Richmond, Virginia 23298.

H. PETER CLAMANN

Having seen the properties of individual motor units, we will then turn to the problem of how the motor units are assembled to produce a movement. As noted above, force can be varied in only two ways: by the addition of more motor units to increase force, a process called *recruitment,* or by varying the firing frequency, and hence the force output, of motor units already active. These two processes interact in all movements, and the relative importance of each is still controversial. More is known about each process separately: for example, the way motor units are recruited to perform different movements has been much studied. Considerable data are also available on the relationship between the firing frequency of an individual motor unit and the total force output of the muscle it serves.

Finally, we will take up the problem of motor-unit control by higher centers. If a variety of motor units can be assembled to produce a movement, can the selection process by modified by higher centers? To what degree is the activity of individual motor units under voluntary control? Although these questions cannot be answered categorically, considerable evidence exists to allow us to describe the recruitment process.

MOTOR-UNIT ANATOMY

LOCATION OF A MOTOR UNIT

By virtue of its definition as a motoneuron, its axon, and all the muscle fibers it innervates, a motor unit is both a neural and a muscular element. It also spans a large area. A typical gastrocnemius motor unit in man has its cell body in the lumbar enlargement of the spinal cord, located in the small of the back (approximately between T 11 and L 2 vertebrae). In the cat, this motoneuron may lie in the seventh lumbar of first sacral segment of the spinal cord, approximately adjacent to the fourth and fifth lumbar vertebrae (Romanes, 1951; Sherrington, 1892). It sends an axon via the sciatic nerve, in the back and leg, to the calf muscle. Along its course, the axon branches two to four times, usually within a few centimeters of the muscle (Eccles and Sherrington, 1930). As the axon enters the muscle, it branches many more times, finally supplying from 400 to 700 or more muscle fibers (Burke and Edgerton, 1975; Wuerker, McPhedran, and Henneman, 1965), depending on the type of motor unit.

The number of muscle fibers supplied by one motor axon is called its *innervation ratio.* More generally, innervation ratio is the quotient of the number of fibers in a muscle divided by the number of alpha motor axons supplying it. Much importance was once attached to the term, since it was believed to define the size of an average or typical motor unit. As we shall see, there is no such thing. Motor units may be classified into types with great differences in size (and therefore innervation ratio), mechanical properties, and modes of use. Since motor units are so diverse, and since different motor units of different types mediate particular kinds of movements, the notion of a "typical" motor unit has lost much of its usefulness.

All the motoneurons supplying a single muscle constitute a motoneuron pool. In the spinal cord, this pool forms an array of scattered motoneurons, less than a millimeter in cross section and from one to two spinal segmets long (Burke, Strick, Kanda, Kim, and Walmsley, 1977; Clamann and Kukulka, 1977; Romanes, 1951). The region of a spinal motoneuron pool is roughly cigar-shaped, with the greatest abundance of cells, and the densest packing, in the center of the pool. At the rostral and caudal poles, the motoneurons of a pool are few and widely separated. The exact position of a motoneuron pool can vary by as much as a segment from animal to animal (Burke *et al.*, 1977; Clamann and Kukulka, 1977; Romanes, 1951; Sherrington, 1892), and the segmental length of a motoneuron pool may also vary. Nevertheless, it is possible to assign reliably a position to any motoneuron pool, a task that has been done in greatest detail for the hind-limb motoneuron pools because of continuing experimental interest in hind-limb muscles (Marinesco, 1904; Romanes, 1951; Sherrington, 1892). This experimental interest is largely due to anatomical convenience: the hind-limb muscles are easy to dissect and isolate, and the long spinal roots of the lumbosacral region make motor units of these muscles easy to approach with stimulating and recording electrodes. Because of this convenience, there exists an almost overwhelming literature on hind-limb muscles and motor units; the literature on other muscles is considerably smaller.

The motoneuron pools of different skeletal muscles may be closely adjacent or intermingled. The motoneurons of a muscle cannot, therefore, be identified by position alone. Generally, the pools of close synergists are most closely adjacent or greatly overlapping in extent. The closer the degree of synergy, the greater is the degree of overlap. Antagonist muscles will also have motoneuron pools closely adjacent, but extensor motoneurons pools lie more ventral and lateral in the ventral horn than do the motoneurons pools of flexor muscles. The motoneurons pools of antagonist muscles are thus less likely to be mingled.

The motoneurons of a pool do not form a contiguous group, as has been said. The space between motoneurons of a single pool in the spinal cord may be taken up by the motoneurons of close synergists, or by a variety of other cells, such as gamma-motoneurons, and interneurons of diverse types. A great deal of signal processing, by way of Renshaw cells, Ia interneurons, and others can thus take place in the immediate vicinity of the motoneurons.

In the brain stem, the motoneurons of a single pool appear to be much more tightly packed and the shape of the pools more varied (Batini, Buisseret-Delmas, and Corvisier, 1976, for description of pools of masticatory muscles). Although interneurons are dispersed among the motoneurons, there may be fewer of them; "interneurons" have only recently been identified in the abducens nucleus (Goldberg, Hull, and Buchwald, 1974). These "interneurons" send their axons to target cells outside the abducens nucleus and thus are not interneurons in the strict sense; interneurons that influence motoneurons directly have not been identified in the abducens nucleus. Certain neural connections that have been much studied in spinal motoneurons, such as the monosynaptic spindle pathway, and recurrent inhibition via Renshaw cells, may be altogether absent in some cranial motoneuron pools (e.g., Matthews, 1972).

The Muscle Unit

Muscle fibers may be divided generally into three types according to their mechanical properties, responses to histological stains, or metabolic properties. Whether such a simple classification is valid is still being debated; a fairly detailed discussion is included in Chapter 1, in this volume. The division of muscle fibers and of motor units into three types is convenient, much accepted in the literature, and aids enormously in understanding how motor units are utilized in different movements. With an awareness of some of the limitations of the classification, we will proceed to discuss three types of muscle units, adopting the terminology of Burke (Burke, Levine, Zajac, Tsairis, and Engel, 1971; Burke, Levine, Tsairis, and Zajac, 1973).

A typical muscle containing three types of muscle fibers, and three types of motor units, is the medial gastrocnemius of the cat. The first thorough description of these motor units, and the recognition that three types could be defined, came from Henneman and co-workers in a series of papers in 1965 (Henneman and Olson, 1965; McPhedran, Wuerker, and Henneman, 1965; Wuerker, *et al.,* 1965). Generally, the three types of motor units found were: (1) motor units of great strength and producing a rapid twitch, but susceptible to fatigue; (2) motor units producing a rather weak contraction and characterized by much slower twitches, these units being fatigue-resistant; and (3) motor units producing rapid twitches but being fatigue-resistant. Histochemically, it is also possible to identify three distinct muscle fiber types by their reactions to certain stains.

A muscle unit possesses several hundred muscle fibers; if these fibers were selected at random from the three types available, one would expect all motor units to possess mechanical properties that are an average of the three available muscle fiber types. Instead, the mechanical properties of motor units may also be grouped into three types. Henneman, therefore, reasoned that the many muscle fibers (400–700 in MG) belonging to one motor unit must all be of the same type (Wuerker *et al.,* 1965), a hypothesis that was directly verified three years later (Edström and Kugelberg, 1968) and that has since been reconfirmed (e.g., Burke *et al.,* 1973; Burke, Levine, Salcman, and Tsairis, 1974). Histochemical study added another fact that applies to all motor units: the fibers of a motor unit are scattered throughout a considerable fraction of the cross section of a muscle. Thus, a muscle biopsy or an electrode sampling the electrical activity of a restricted region of muscle, will "see" at most a few fibers of a muscle unit whose territory encompasses one-fourth to one-third or more of a muscle's cross section. Only in muscles that have been reinnervated does one find clumps of muscle fibers of identical histochemical properties and belonging to the same motor unit (Jennekens, Tomlinson, and Walton, 1971; Kugelberg, Edström, and Abbruzzese, 1970). This is borne out by the pattern of staining of most mixed muscles: fibers of one histochemical type are randomly scattered, rather than being arranged in clumps. An odd exception is in the muscles of the pig, where clumps of uniformly red fibers lie among regions of white fibers. These clumps are believed not to be whole muscle units, but rather to represent fibers from a number of units, so that in the pig, muscle units are as scattered as in other animals and in man (Beerman and Cassens, 1977).

MUSCLE-UNIT TYPES AND THEIR PROPERTIES

It is now generally agreed that the muscle units of a mixed muscle fall into three types, and these three types have been named in various and sometimes contradictory ways. Some workers define only two types in man. It will be most convenient to describe the three types of motor units in the cat gastrocnemius muscle and then to examine the limitations of this scheme of classification.

A useful system of naming muscle units is that of Burke, which names them according to their most obvious mechanical properties; slow (type S), fast and fatigue-resistant (type FR), and fast and fatigable (FF) (Burke et al., 1971; 1973). The sizes and metabolic properties of these muscle units form a logical complement to their functional capabilities.

Type FF comprises the largest, fastest, and most powerful of the three muscle-unit types. Its individual muscle fibers have the largest cross-sectional area (5290 μ^2: Burke and Edgerton, 1975; 3100 μ^2: Henneman and Olson, 1965). The muscle fibers contain a large store of glycogen, which is converted to contractile energy by anaerobic pathways. They therefore contain few mitochondria, little myoglobin, and are not directly adjacent to any capillaries, unless they happen to share a vascular supply with muscle fibers of other types.

Type FF units produce rapid and powerful twitch contractions, as would be expected both from their large innervation ratios and the large sizes of their individual muscle fibers. Table I lists some important properties of these and of the other two types of muscle units. Because the twitch produced by an FF unit is so brief, the tetanic fusion frequency is high, usually in the range 50–100/sec (Wuerker et al., 1965). These muscle units are also the most fatigable, sometimes being

TABLE I. SOME PROPERTIES OF THREE MOTOR-UNIT TYPES IN CAT MEDIAL GASTROCNEMIUS MUSCLE[a]

Motor-unit property	FF	FR	S
	Motor-unit type		
Tetanic tension (gm)	30–130	4.5–55	1.2–12.6
Twitch contraction time (msec)	20–47	30–55	58–110
Tetanic fusion frequency (pps)	50–100	—	10–20
Fatigue resistance	poor	good	excellent
Capillary supply	sparse	very rich	rich
Mitochondrial content in muscle fiber	sparse, chiefly near surface	densely packed	dense, esp. near surface
Innervation ratio	750	440	450
Conduction velocity of axon	85–114	84–113	75–99
Motoneuron surface area (estimated) $\mu^2 \times 1000$	130–196	128–193	108–162

[a]Data collected or calculated from Barrett and Crill, 1974a; Burke and Edgerton, 1975; Henneman and Olson, 1965; and Wuerker, McPhedran, and Henneman, 1965.

unable to sustain a tetanic contraction for more than a few seconds (Burke and Edgerton, 1975; Wuerker *et al.*, 1965). This pronounced susceptibility to fatigue is directly related to the metabolic mechanisms of their muscle fibers: a high concentration of contractile material, few energy stores or mitochondria, and the presence of anaerobic metabolic pathways. Type FF units appear to be designed to produce rapid, powerful contractions that need not be long sustained.

At the opposite end of the spectrum of mechanical properties lies the Type S unit. It produces the slowest and weakest twitch of the three motor unit types. It also has a low innervation ratio, about half that of an FF unit. As one would expect, it produces a fused tetanic contraction at a low stimulus frequency, sometimes as low as 5–10/sec. Type S muscle units are extremely resistant to fatigue, being able to produce a tetanic contraction almost indefinitely. These units appear best suited for the slowly changing, weak contractions of long duration involved in posture and the support of limbs against gravity.

Type S muscle fibers are small to intermediate in cross-sectional area, averaging 2500–4300 μ^2 in soleus, and somewhat smaller in the medial gastrocnemius. Wuerker *et al.* (1965) caution against taking the actual fiber size too seriously, pointing out that the important and consistent finding is that FF muscle fibers are the largest, S fibers are intermediate in size, and FR the smallest. Later workers have generally supported these findings, although Burke (Burke and Edgerton, 1975) shows a marked difference between the sizes of S fibers in soleus and in medial gastrocnemius.

The fiber of a Type S muscle unit is rich in mitochondria, oxidative enzymes, and myoglobin. These fibers are also surrounded by many capillaries (see Fig. 2 in Wuerker *et al.*, 1965). Hence, they are ideally suited to the steady, sustained contractions they produce.

The Type FR muscle unit is in many respects intermediate between the other two. It produces rapid twitches and has a high tetanic fusion frequency, being somewhat slower than the Type FF unit. These twitches or tetani are rather weak, intermediate between those of the Type FF and Type S (see Table I). In fatigue resistance Type FR is second only to the Type S, being able to sustain a tetanic contraction for many seconds, and even for minutes.

The fiber of a Type FR unit is the smallest of the three types, and the darkest staining for mitochondrial ATPase. It is richly supplied with capillaries and contains much myoglobin; this muscle unit is well equipped to produce a sustained contraction.

A Comparison of Muscle Units in Different Muscles

It has been noted that there are differences between the Type S units of the medial gastrocnemius and soleus muscle of the cat. This sort of difference can cast doubt on the tripartite classification of motor units. If a table were drawn up listing the properties of motor-unit types in several muscles of one animal (Table II) or in the same muscle in many different species, it would not be possible to divide them into three types. For example, it has been suggested that in human muscles the Type FR muscle unit has an intermediate supply of mitochondria, whereas the

Type S has the richest supply (Peter, 1971). Still, the majority of mammalian muscles are mixed, and within one muscle of one species, a tripartite classification is readily made. Since the use and control of motor units in a muscle is directly related to their properties, the classification remains useful.

A remarkable illustration of the above principle is found in extraocular muscles of the cat. These muscles are mixed (Peachey, 1971), so motor-unit types can be identified as described above. However, the twitch strengths of these units are measured in milligrams instead of grams, and their twitch contraction times are an order of magnitude faster than those of skeletal muscles. A "slow" red muscle unit in cat extraocular muscle will have a twitch contraction time of 14 msec and a corresponding tetanic fusion frequency of 75/sec. It is impossible to compare it to a medial gastrocnemius muscle unit, yet it fits well into a tripartite muscle-unit classification for extraocular muscles, where all muscle units are much weaker, smaller, and faster than in limb muscles. Table II gives some properties of muscle units of several different muscles in the cat.

Most skeletal muscles are mixed, containing varying ratios of all three muscle-unit types. Several exceptions should be noted. The best-known is the soleus muscle of the cat, rabbit, and guinea pig, which is composed exclusively of Type S units. (In man and the rat, soleus is a mixed muscle, containing a predominance of Type S units.) The vastus intermedius or crureus of the rabbit is also a pure muscle, containing only Type S units. There are probably others, but these are the best known.

Interestingly, there are almost no known muscles in any animal that are totally pale, containing only Type FF units. The only examples of which I am aware are the retractor bulbi and the stapedius of the cat. (Lennerstrand, 1974; Teig, 1972). The stapedius inserts on a middle-ear bone, the stapes, and reduces the sensitivity of the auditory system when it contracts by blocking vibration of the

TABLE II. SOME MECHANICAL PROPERTIES OF MOTOR UNITS IN SEVERAL CAT MUSCLES[a]

Muscle	Tetanic tension (gm)	Twitch tension (gm)	Twitch contraction time (msec)	Tetanic fusion frequency (pps)
Medial gastrocnemius (FF unit)	30–130	—	20–47	50–100
Medial gastrocnemius (S unit)	1.2–12.6	—	58–110	10–20
Soleus	3.2–40.4	—	58–193	10–20
Flexor digitorum longus	.3–83	—	12.5–47	—
First superficial lumbrical	1–30	.080–2.31	15–45	—
Tensor tympani (fast unit)	(.1–2.24)[b]	.028–.640	20–40	75–100
Tensor tympani (slow unit)	(.075)[b]	.016–.034	55–90	(15–20)[b]
Stapedius	(.04–.59)[b]	.011–.168	10–40	100–150
Extraocular muscle	.05–.40	.007–.040	3–14	100–300
Retractor bulbi	.22–.775	.015–.085	7–12.5	140–200

[a]Data collected from the following sources: Appleberg and Emonet-Dénand, 1967; Bagust *et al.*, 1965; Burke and Edgerton, 1975; Goldberg, Lennerstrand and Hull, 1976; Lennerstrand, 1974; McPhedran, Wuerker, and Henneman, 1965; Teig, 1972; Wuerker, McPhedran, and Henneman, 1965.
[b]Numbers in parentheses are estimated or taken from single experiments.

stapes. Its motor units are much weaker, and slightly faster, than those of skeletal muscles. The retractor bulbi retracts the eyeball into the orbit. Motor units of this muscle are similar to those of the lateral rectus, and the motoneurons of the retractor bulbi are located in the abducens nucleus, along with those of the lateral rectus (Goldberg, Lennerstrand, and Hull, 1976; Lennerstrand, 1974).

MOTOR-UNIT FUNCTION: RECRUITMENT AND FIRING PATTERNS

MOTONEURONS AND THE SIZE PRINCIPLE

The motoneurons supplying different types of muscle units themselves differ in size, and this single fact has a considerable number of consequences in motor-unit structure and function. Motoneurons with the largest cell bodies possess axons of the largest diameter (Barrett and Crill, 1974a), which, in turn, conduct action potentials at the highest velocity and produce the largest number of axonal branches. Therefore, the largest motoneurons supply the largest and fastest muscle units. Conversely, the smallest motoneurons, whose axons conduct most slowly and produce the fewest axonal branches, innervate the smaller, slow Type S muscle units. The morphological pattern is thus complete.

It has long been known that excitatory postsynaptic potentials recorded in small motoneurons are usually bigger than those recorded in large ones (e.g., Granit, 1972). This can be directly attributed to the sizes of the motoneurons (Barrett and Crill, 1974b; Kernell, 1966). To see how size alone can determine the relative excitability of motoneurons, let us assume that a small and a large motoneuron receive identical synaptic input from a reflex source such as a muscle spindle primary afferent fiber. This means that each cell receives the same number of end plates from the afferent fiber, and that the surface area covered by these end plates is the same for each motoneuron. For simplicity we may consider that the afferent fiber makes a single synaptic connection (although this, in general, is not the case; see Mendell and Henneman, 1971) and that the synaptic boutons supplying the large and the small motoneuron are identical in size.

The same amount of neurotransmitter is released onto each motoneuron when the afferent fiber discharges, and this produces the same conductance change at the synaptic site in each cell. The current flowing into each cell is the same, but the change in the cell's transmembrane potential produced by the synaptic current depends on the total electrical resistance of the cell membrane, which varies with cell size (Henneman *et al*, 1975; Kernell, 1966). By Ohm's law, the voltage change produced by synaptic activity should be the greatest in the cell having the highest input resistance, that is, the smallest cell. It is this cell that produces the largest EPSP, and is therefore most likely to be discharged.

Red muscles, composed predominantly of small, slow, motor units, are more responsive to reflex inputs than are their paler synergists (Denny-Brown, 1928–1929). The same applies for the slower and faster parts of a single muscle (Gordon and Phillips, 1953). In a series of papers beginning in 1965, Henneman and his co-workers demonstrated that motor units are recruited in order of their sizes in response to a variety of excitatory and mixed excitatory and inhibitory reflex inputs. This observation complements the structural properties and mechanical capabilities of motor units. The slowest and most fatigue-resistant motor units are used most often, for postural support and finely graded, slow contractions. The fastest and most fatigable motor units are used rarely, and only for rapid, powerful, and brief contractions.

The relation of structure and function in motor units, stated as the Size Principle (Henneman, Somjen, and Carpenter, 1965), asserts that motor units are recruited in order of their sizes, from small to large, by increasing excitatory drive to a motoneuron pool; units fall silent in the reverse order of their sizes in response to decreasing excitation, or to increasing inhibition, of a motoneuron pool (Clamann, Gillies, and Henneman, 1974).

The functional result of such a fixed recruitment order is to provide great simplicity in the control of muscles. Although a wide spectrum of motor-unit sizes and capabilities exists, it is not necessary for the central nervous system to maintain an inventory of these capabilities. In a contraction of increasing strength, the motor unit that will produce the least change (the smallest remaining inactive unit) is the next to be recruited into activity. In this way, a central nervous system command need not be delivered selectively to motor units of different capabilities. Rather, any input to a motoneuron pool can simply be distributed equally throughout the pool, and the sizes of the affected motoneurons will determine recruitment order, in the absence of any additional neuronal circuitry. The motor nucleus is able to translate the intensity of a synaptic input into the appropriate number and types of motor units to produce a response. A reflex correction may modulate the input, and add or subtract the correct number and types of motor units, again automatically (Fig. 1).

There is a vast quantity of evidence that force is modulated by the recruitment of motor units in order of their sizes in many types of movement. Recruitment order has been most precisely measured in a variety of spinal reflexes in extensor

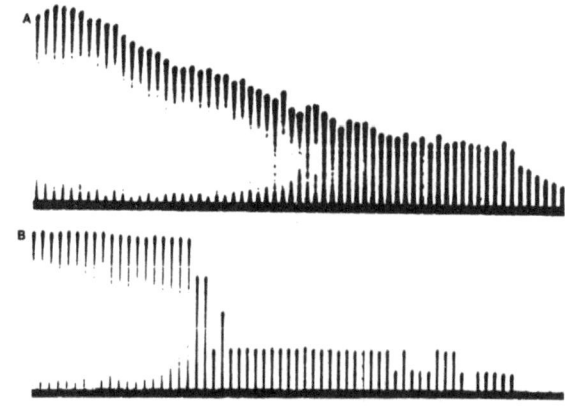

Fig. 1. Response of five plantaris motoneurons shown during steady decline in monosynaptic reflexes of the plantaris motoneuron pool. (A) Monosynaptic reflexes of plantaris motoneurons recorded in half of L7 ventral root, produced at a rate of one every 2 sec. Height of each pulse represents the magnitude of the reflex. (B) Combined response of five plantaris motoneurons in a ventral root filament, recorded synchronously with the reflexes above. Units with high reflex thresholds generate larger action potentials, and produce larger step changes in this trace when they drop out, than do low-threshold units. An example of two units falling silent together is seen in the left third of the record. From Clamann and Henneman (1976).

muscles of the cat (Clamann *et al.,* 1974; Clamann and Henneman, 1976; Henne-man *et al.,* 1965). Reflexes studied included polysynaptic stretch reflexes, the monosynaptic reflex, electrical driving of the muscle nerve, and inhibition from antagonist muscles, recurrent (Renshaw cell) pathways (Henneman *et al.,* 1965), cutaneous nerves, and the medullary reticular formation (Clamann *et al.,* 1974). There is also good evidence that motor units are recruited in order of size in voluntary contractions in man (Desmedt and Godaux, 1977; Freund, Büdingen, and Dietz, 1975; Hannerz, 1974; Milner-Brown, Stein, and Yemm, 1973a; Tanji and Kato, 1973; see also Fig. 2). There is no longer any dispute that the recruitment order is stable in steady or slowly varying voluntary contractions in a number of finger, forearm, and leg muscles. Whether motor units are recruited in the same pattern in very rapid movements is still disputed (see below).

The orderly recruitment of motor units has been directly or indirectly observed in a variety of other neuromuscular systems. Several workers (Fuchs and Luschei, 1970; Robinson, 1970) showed that motor units in monkey extraocular muscles become active, or fall silent, as a precise function of eye position. The force produced by extraocular muscle to hold the eye in a fixed position is a function of eye position, so that in this system, recruitment order is shown to be stable. Yemm (1977) demonstrated the orderly recruitment of motor units in the masseter and temporalis muscles in voluntary contractions in man, and this work has been confirmed and extended by Goldberg and Derfler (1977). Davis (1971) showed the orderly recruitment of motoneurons by size in the lobster swimmeret motor system. Numerous other examples can be given, all making the same point: in a large variety of reflex movements in experimental animals, and in most voluntary movements in man, motor units are recruited in order of increasing size, and silenced in the reverse of that order.

For the Size Principle to operate as postulated, it is necessary that each input to the motoneuron pool be uniformly distributed to all the motoneurons of the pool. This is not to say that all inputs must be equally potent. The requirement is merely that any one input, regardless of its potency, supply the same strength of synaptic input to all the motoneurons of the pool. This is a stringent requirement, which, to date, has been shown to be true only for the monosynaptic spindle afferent pathway by which the spindles of a muscle signal stretch to that same (homonymous) muscle (Mendell and Henneman, 1971; Mendell and Scott, 1975). It has been shown that all spindle Ia afferent fibers from a muscle make connections with all motoneurons of that same muscle. The connections that spindle afferent fibers of a muscle make with the motoneurons of a close synergistic muscle (a heteronymous

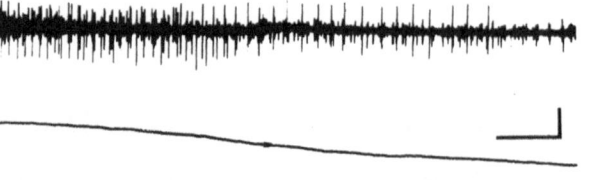

Fig. 2. Motor units slow their discharge rate and then cease firing *(upper trace)* during steady decline in force output of a muscle *(middle trace). Lower trace* represents zero force. Note that units producing large-amplitude spikes cease firing first. The low force produced at the right of the figure is sustained by units producing low-amplitude spikes; these have been shown to be small motor units. Calibration: vertical, 5 kg; horizontal, 0.5 sec.

muscle) are not as complete. For example, only about 60% of medial gastrocneuius motoneurons receive a monosynaptic input from any one lateral gastrocnemius spindle afferent fiber (Mendall and Scott, 1975). Therefore, stimulation of this heteronymous input might be expected to disturb the recruitment order of medial gastrocnemius motoneurons, and this has been demonstrated (Clamann, 1978; Clamann and Kukulka, 1976).

Several reflex inputs to a motoneuron pool have been described that select a subpopulation of that pool. These include cutaneous nerve inputs to the cat medial gastrocnemius motoneuron pool, the descending rubrospinal pathway (Burke, Jankowska, and ten Bruggencate, 1970a), the recurrent loop via Renshaw cells (Eccles, Eccles, Iggo, and Ito, 1961), and descending brain stem inputs (Sasaki and Tanaka, 1964). The effects of stimulating these pathways on recruitment order have not been studied in detail, with one exception. Clamann *et al.* (1974) found that Renshaw inhibition did not alter the recruitment order established by monosynaptic reflexes. This interesting issue is thus far from settled.

EXCEPTIONS TO THE SIZE PRINCIPLE

If the predictions of the Size Principle are borne out, severe restrictions are placed on the anatomy and function of inputs to a motor nucleus. It is no surprise, therefore, that these predictions are being put to experimental test. Several laboratories are testing whether a nonuniformly distributed input to a motoneuron pool can be clearly demonstrated, and whether such an input will alter the size-dependent recruitment order (see above). Because this work is preliminary, it will not be discussed here.

It has been argued that ballistic, powerful movement requires only the participation of the large, fast motor units of a muscle; activation of the smallest, slowest units also, as the Size Principle requires, would add too little force too late to be of any use (Granit and Burke, 1973) and would therefore be inefficient. Evidence for this view is indirect. It will be recalled that red and pale synergistic muscles are "recruited" in the order red-to-pale in many reflexes; this is often taken as a model for the more detailed recruitment of motor units. It may readily be shown that, in a rapid movement, this order of the recruitment of muscles can be reversed (Clamann, 1978; Smith, Edgerton, Betts, and Collatos, 1977), an observation that has its origin in the work of Denny-Brown (Creed *et al.*, 1932). Whether or not this reversal can be demonstrated in a single motoneuron pool is still debated.

This argument is answered (Henneman, Clamann, Gillies, and Skinner, 1974) by observing that the additional neural circuitry needed to select and control subpopulations of motor units is far more demanding of energy and information, and therefore it is much less economical to control select subpopulations of motor units than to employ the same recruitment order for all movements. Since the high contraction speed and conduction velocity of the large motor units would determine the speed of onset of the contraction in any case, no time is lost in recruiting the slow units as well, as the Size Principle demands.

Nevertheless, spurred by the evidence of some circuits that seem not to be uniformly distributed throughout a motoneuron pool and by the suggestion that rapid movements should be produced by fast motor units, recruitment reversals have been sought in rapid movements, particularly in human muscles. Reversals of recruitment order have been found in several studies (Ashworth, Grimby, and Kugelberg, 1967; Grimby and Hannerz, 1968; 1970; 1974; 1976; Hannerz, 1974; Wyman, Waldron, and Wachtel, 1974). (The issue of voluntary control of motor units and reversals of recruitment is discussed on pp. 86–88.)

There are several problems in the interpretation of these results. First, studies on the Size Principle have not shown recruitment order to be completely fixed. There are temporal fluctuations in the excitability of motoneurons, and this "noise" produces limited reversals of recruitment order. These fluctuations are most apparent in motoneurons of similar sizes and similar recruitment thresholds. Henneman *et al.* (1974) found regular reversals in recruitment order in motor units of nearly identical thresholds, but none of those thresholds differed by more than 5%. Threshold was measured as the size of the reflex, expressed as a percentage of maximum, at which the unit was just recruited or inactivated. A second problem appears in the definition of threshold during a rapid, powerful movement. Many motor units are activated almost simultaneously in such a movement, and the onset time need not reflect the recruitment order. Since a muscle is a heavily damped system, the force measured in a rapid contraction is only a fraction of the actual force generated by the muscle. Force threshold of a motor unit cannot be accurately measured. When these two factors are taken into account, force threshold has been estimated and a precise recruitment order during rapid movements has been consistently seen (Büdingen and Freund, 1976; Desmedt and Godaux, 1977; Freund *et al.*, 1975).

The final problem with a demonstration of recruitment reversals is the physiological significance. Occasional reversals of recruitment order in a few motor units cannot be expected to have much impact on the control of a muscle consonant with the Size Principle. Numerous changes in recruitment order of cat medial gastrocnemius muscle have been found by exciting the pool via lateral gastrocnemius and soleus spindle afferent fibers (Clamann and Kukulka, 1976), yet the recruitment order of MG motoneurons is still rather stable in response to mixed monosynaptic inputs. It may be that the distribution of inputs via any one pathway is not uniform, having random omissions. The excitation of a motoneuron pool by the summed excitation of several of these inputs produces, on the average, a uniform input to all the motoneurons, as the Size Principle requires. This is the normal physiological state. In this view, strong stimulation of any one reflex pathway will upset the recruitment order, but this event is unlikely to occur in normal behavior. This view is supported by the observation that the recruitment order is altered in CNS lesions (Ashworth *et al.*, 1967; Grimby and Hannerz, 1970) and when certain pathways are interrupted (Grimby and Hannerz, 1976).

MOTOR-UNIT FIRING FREQUENCY

The firing patterns of motor units have been much studied; nevertheless, numerous questions remain to be answered. For example, neither the maximum nor the minimum firing frequency of motoneurons has been unequivocally determined. What is clear is that motoneurons can discharge at rates much higher than those seen in normal use and, in fact, can generally maintain firing rates well below the tetanic fusion frequency of their muscle units (e.g., Milner-Brown *et al.,* 1973b). The discharge of a motoneuron can vary widely and rapidly; before discharge rate is measured, it is necessary to agree on some definitions.

The shortest interval during which discharge frequency can be measured in a spike train is the time interval between two successive spikes. This is called the *interspike interval,* and its reciprocal is the *instantaneous frequency.* The *mean* discharge frequency is found by counting the number of spikes occurring in a given time interval and dividing that number by the time interval. The firing rate of a motoneuron can vary during the time over which an average frequency is determined, so that the average discharge frequency and the instantaneous frequency can be very different. Similarly, values of average frequency can fluctuate widely, depending on the length of the time interval used to determine that average. Clamann (1969) described criteria for the number of spikes that need to be counted to arrive at an average firing rate with a given degree of confidence. Much confusion in the literature regarding frequency limits of motor units results from taking an inadequate number of samples, or from failure to distinguish average from instantaneous discharge frequency.

The range of discharge frequencies that a motoneuron can produce may be studied by injecting currents of varying strengths into the cell and recording the cell response. The result is a graph relating current strength to motoneuron discharge frequency. If, as has been shown (Schwindt and Calvin, 1973), the cell responds to current injection as it does to the current flow produced by synaptic input, the current-frequency graph gives the transfer function of the motoneuron, that is, the "rule" that the motoneuron uses in converting intensity of synaptic input to discharge rate.

For many motoneurons studied, the relationship between injected current and discharge frequency is piecewise linear. Below a threshold current, the cell does not respond. If the current exceeds approximately 10 nA, the cell begins to discharge steadily and increases its discharge frequency linearly with increasing injected current until the injected current is about 30–40 nA. The sensitivity of the cell, defined as the change in firing frequency per nanoampere of current is 1.7 (average of nine cells, Kernell, 1965). This range of currents defines the primary range of cell sensitivity. As the current strength is increased beyond 40 nA, the cell abruptly enters its secondary range, again showing a linear current-frequency relationship but with a greater sensitivity: 4.6 impulses sec^{-1} nA^{-1} (Kernell, 1965). The maximum steady discharge rate of a motoneuron in the secondary range may approach 200 impulses sec^{-1} at injected current strengths of 60 nA or greater. For nine cells studied extensively, the average maximum discharge rate was 125 impulses sec^{-1} (Kernell 1965). About half the cells studied did not exhibit a secondary firing range; these cells did not differ from the remainder of the population in primary frequency-current slope, time-course of after-hyperpolarization, spike amplitude, or threshold for continuous firing.

If the strength of injected current is increased beyond the secondary range, the sensitivity of the cell decreases again, and the firing pattern changes, exhibiting closely spaced pairs of spikes with longer intervals in between.

The experiments described above were performed by injecting steps of current, and the results described the *steady* discharge frequency of the motoneuron in response to injected current. The initial response of the cell to a current step is a series of 2–4 action potential spikes of high instantaneous frequency declining by about one third to reach the steady discharge just described. This decline is called the initial phase of adaptation. During maintained discharge, discharge frequency continues to decline slightly in a late phase of adaptation.

The instantaneous frequency of the first two spikes may be related to strength of injected current, and the result is again a piecewise linear relation showing a primary and a secondary range. However, the sensitivity (slope) of the relation is greater than that for steady firing. The slope of the primary range in an unadapted cell is 6.5 impulses sec^{-1} nA^{-1}, and that of the secondary range is 24.9 impulses sec^{-1} nA^{-1} (Kernell, 1965). The first few spikes of a motoneuron subjected to a current step may show a firing rate of up to 250 impulses sec^{-1}. We see, then, that the initial phase of adaptation produces both a slowing of discharges in the cell and a decrease in its sensitivity to changes of injected current.

The reason why motoneurons exhibit a piecewise linear response to steady injected currents, and to tonic synaptic inputs (Jack, Noble, and Tsien, 1975) are not well understood. Kernell, and others, have attempted to model the process using known electrical properties of motoneurons and the Hodgkin-Huxley equations; an excellent description is given in Chapter 11 of the book by Jack *et al.* (1975).

The average discharge rate of motoneurons in response to reflex inputs or in voluntary contractions is much lower, lying between 7 and perhaps 50/sec for sustained spike trains (Burke, Rudomin, and Zajac, 1976; Clamann, 1969; Denny-Brown, 1928–1929; Smith, 1934). Marsden, Meadows, and Merton (1971) took advantage of an anatomical anomaly in two subjects to demonstrate discharge rates in excess of 100/sec in a single motor unit. Experiments showed that in these subjects, a few fibers of the median nerve passed to the ulnar nerve and innervated muscle units of the adductor pollicis muscle. Normally adductor pollicis is innervated solely by ulnar nerve fibers. An ulnar nerve block at the elbow silenced all but the few anomalously innervated units of adductor pollicis, which discharged briefly at rates up to 150/ sec during an attempted maximal contraction of the muscle. This demonstration, although interesting, cannot be considered typical.

The average discharge rate increases with increased synaptic drive. Modulation of the firing rate of motor units is therefore an important mechanism in altering the force output of these units and hence of the whole muscle (Fig. 3). It is

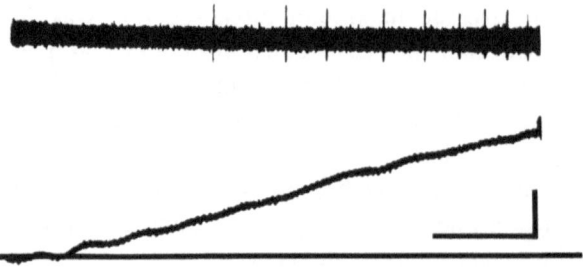

Fig 3. A single motor unit *(upper trace)* is recruited during steadily increasing force output of human brachial biceps *(middle trace)*. Note the increasing discharge rate of the unit. Calibration: vertical, 2 kg; horizontal, 0.5 sec.

interesting to note that in normal use, motoneurons discharge at rates well below the tetanic fusion frequency of their muscle units. A smooth contraction of a muscle is attained by numerous motor units discharging in an asynchronous manner. In this way, the rippling force outputs of individual motor units is averaged out, resulting in a smooth over-all contraction.

The discharge rate of motor units is related to the mechanical properties of their muscle units. Small motor units discharge slowly; large motor units discharge more rapidly. Small motoneurons produce a prominent after-hyperpolarization following each spike, a feature that is less prominent or absent in the larger cells (Granit, 1972). This is believed to be largely responsible for limiting discharge rate.

Extraocular motoneurons, as usual, are an exception. They are capable of discharging in bursts in excess of 600/sec (Fuchs and Luschei, 1970) and steadily at rates up to 300/sec (Fuchs and Luschei, 1970; Robinson, 1970). This fits well with the high tetanic fusion frequencies of muscle units in extraocular muscles.

Of great physiological and experimental interest is the firing pattern of motor units and its relation to force output. A motoneuron, when recruited, produces two or three spikes at a high instantaneous frequency (perhaps 200/sec), and then settles down to a much lower sustained rate of 7–50/sec (Burke *et al.*, 1976; Denny-Brown, 1928–1929; see Fig. 4). This is the source of a great deal of confusion in the literature regarding values of maximum discharge rate. It also appears that discharge rate of motor units differs for different muscles. Clamann (1970) did not see rates above 25/sec in the large brachial biceps, whereas discharge rates of 20–50/sec are easily observed in finger muscles (Clamann, 1970; Milner-Brown *et al.*, 1973b; Tanji and Kato, 1973b).

The discharge rate of a motor unit determines its force output: the more rapid the discharge rate, the greater the force, up to the tetanic fusion frequency. However, the rate of rise of force also increases with increasing discharge rate. It is likely that the rapid burst with which a motor unit begins to discharge allows it to reach a higher force level, and to reach it faster, than would be the case if the unit began firing at its average frequency (Burke *et al.*, 1976).

The relation between discharge frequency of a motor unit and total muscle

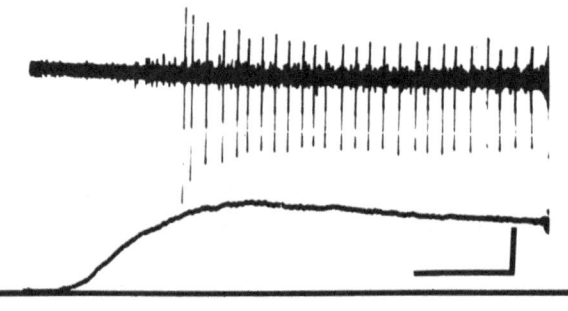

Fig. 4. Recruitment of a single motor unit *(upper trace)* during steadily increasing force *(middle trace)*. *Lower trace* is zero force. Note that the instantaneous frequency of the first two spikes is much higher than that of any later spike pairs. The signal producing the spike train has been passed through a high-pass filter so that the changes in amplitude reflect small changes in spike rise-time in the original record. The spike train is not readily visible in the unfiltered record. Calibration: vertical, 2 kg force; horizontal, 0.5 sec.

force is disputed. The general pattern appears to be this: a unit, when recruited, discharges at a low rate, say 7–12/sec. Initially, its firing rate increases rapidly with increasing force; later, the change is much less dramatic (Clamann, 1970; Bigland and Lippold, 1954; Tanji and Kato, 1973b). This is the pattern observed when steady isometric tension is related to discharge frequency.

In general, motor units discharging steadily also discharge regularly. The distribution of interspike intervals is approximately Gaussian and the duration of successive interspike intervals is statistically independent (Clamann, 1969; Freund *et al.*, 1975; Rosenfalck and Madsen, 1954). The more rapidly the motor unit discharges, the more regular is the discharge (Clamann, 1969; Tanji and Kato, 1973b). It has been suggested that the regularity of discharge can be used to classify motor-unit types into tonic and phasic (Tanji and Kato, 1973b; Tokizane and Shimazu, 1964). At high force levels, units fire less regularly (Tanji and Kato, 1973b) or in bursts (Norris and Gasteiger, 1955). Burst-like firing would account for the observation of Tanji and Kato that high-threshold units actually fire more slowly than low-threshold units, and less regularly. The instantaneous frequency within the burst would be high, and the longer intervals between bursts would reduce the average discharge rate.

The minimum discharge rate of motor units lies above 5–7/sec, whether the motoneurons are driven electrically or synaptically (Clamann, 1969; Tanji and Kato 1973b). However, it is possible for motoneurons to fire more slowly; lower rates could reflect a unit that is being turned on and off intermittently rather than firing *continuously* (i.e., with approximately constant instantaneous frequency). As discharge rate becomes less regular, the relation between rate and regularity of firing breaks down. The regularity (or lack of it) of motor-unit discharge at low rates, and even minimum rate, has been attributed to local mechanisms (DeLuca and Forrest, 1973), whereas it is probably owing to higher centers.

GRADING OF FORCE: THE IMPORTANCE OF RECRUITMENT AND FREQUENCY MODULATION

The discharge of a single motor unit may be recorded with electrodes inserted into an active muscle. If the muscle is producing slowly varying isometric contractions, the unit always becomes active when the muscle reaches a particular force level. This is called the *force threshold* of the unit. As the force output of the whole muscle increases, the unit is seen to discharge progressively more rapidly; this means that the motor unit is producing progressively more force as well. During this time, additional motor units are recruited. In this way, recruitment and increased discharge rate interact in the grading of tension. However, the relative importance of each is still disputed; it probably depends on the muscle studied and on the movement pattern produced. For example, a sudden step change in force is produced by the synchronous recruitment of several units; the force is maintained as some of these units fall silent again, while the firing rate of those still active is higher than it was before the step was produced (Büdingen and Freund, 1976; Freund *et al.*, 1975). Similarly, the force threshold of a motor unit depends on the rate of change of force (Desmedt and Godaux, 1977) since in rapid contractions, the heavily damped muscle must produce some force just to move. This force is, of course, not recorded by a force transducer, which measures only the net force output of the muscle. The limb produces a net force, but motor units respond to gross force (force required to move the muscle plus net force required to affect the external world).

Some early authors, such as Adrian and Bronk (1928), gave great emphasis to the importance of frequency change or rate coding in altering force output, a position reemphasized recently by Stein and co-workers (Milner-Brown *et al.*, 1973b). These authors argue that recruitment plays a minor role, and only at low force output. Other authors place a much heavier emphasis on recruitment (Bigland and Lippold, 1954; Clamann, 1970; Denny-Brown 1928–1929; Tanji and Kato, 1973b). It is generally agreed that motor units show their greatest change in firing frequency in a force range just above the recruitment threshold, implying a strong interaction between recruitment and rate coding at low force levels. It is suggested that recruitment ceases at a force level just below 100%, and that within some force range near maximum, rate coding is the sole method of force grading (Bigland and Lippold, 1954; Clamann, 1970; Milner-Brown *et al.*, 1973b; Robinson, 1970). The size of this range is in dispute. Robinson finds it to be about 25% (i.e., from 75–100%) in steady contractions of the extraocular muscles, a figure in agreement with estimates of Bigland and Lippold (1954) and Clamann (1970) on skeletal muscles. Stein and co-workers (Milner-Brown *et al.*, 1973b), in an ingenious model of force grading in the first dorsal interosseus muscle, report a much different result: that rate coding accounts for over 66% of the total force output of the muscle. This is not equivalent to saying that the last motor unit is recruited at a tension level 34% of maximum; to say so would require the clearly false assumption that recruitment acts alone at low force levels. Thus, the figures are not directly comparable; it is likely that fewer and fewer units, each producing a greater force increment, are recruited at progressively higher force levels. Although Milner-Brown *et al.* emphasize that they were unable to study motor-unit behavior throughout the force range, they nevertheless state that recruitment is the primary method of force grading at low force levels only, whereas rate coding is the predominant mechanism throughout, and is likely to be the sole method at intermediate and high force levels.

It is necessary to bear in mind that all these experiments are performed under the artificial constraint of a steady isometric force. In fact, few movements are made this way. Most steady isometric movements we make are weak and are produced to maintain posture; powerful movements are commonly brief and seldom isometric. In the latter instance, it is likely that a large number of motor units are recruited simultaneously, discharge a brief burst, and then fall silent again, making the measurement of either recruitment order or firing frequency modulation almost meaningless. Freund and co-workers have addressed this problem (Büdingen and Freund, 1976; Freund *et al,* 1975) in a careful study of motor-unit recruitment in phasic movements. Thus, our present knowledge of recruitment and rate coding in motoneurons can only be considered a first step to understanding the events that produce the variety of movements actually made in our normal existence.

To summarize, motor units show their greatest change in firing rate at low force levels, and this has led to one view that rate coding is the predominant method of grading force at low force levels, whereas recruitment is more important at high force levels. On the other hand, Stein and his co-workers argue that there are many more small than large motor units, all of which are recruited at low force levels, and all of which can vary their force output widely by rate coding. Therefore, rate coding is claimed to be the chief way of grading force, particularly at high force levels. The true answer must lie between these extremes, and awaits more detailed study of single motor units at high force levels, which is technically quite difficult. Meanwhile, a tantalizing question remains unanswered. At what force level is the last motor unit of a muscle recruited, and is this related to the type of muscle under study?

Voluntary Control of Human Motor Units

Beginning with the work of Harrison and Mortensen (1962), a number of reports have appeared showing that the discharge of individual motor units can be voluntarily controlled (Basmajian, 1963; Basmajian, Baeza, and Fabrigar, 1965; Gray, 1970; Wagman, Pierce, and Burger, 1965). Motor units could be made to discharge single action potentials, pairs of spikes, and even rhythms imitating drum cadences. How can this be reconciled to the regular, rhythmic discharges described earlier? In addition, it was reported that motor units could be recruited and silenced at will, and that several units could be separately and independently recruited and controlled (Basmajian *et al.,* 1965; Wagman *et al.,* 1965). How does this finding relate to the orderly recruitment of motor units, observed by so many authors?

To explain some of the apparent contradictions between the behavior of motor units as described previously and the experiments on voluntary motor unit control, it is first necessary to understand how the latter studies were carried out. Action potentials were recorded from muscle units with selective electrodes inserted in the appropriate muscle (muscles of the hand, arm, or leg), amplified, and displayed on a cathode ray oscilloscope or played through an audio monitor. The subject used information from the display as feedback to aid in controlling the motor units, which were recruited at force levels ranging from imperceptible to very low. This is an important point. It means that all motor units studied were small, with similar, if not identical, force thresholds.

In the range of forces studied, subjects were able to make delicate, precisely controlled movements. By varying the force output of a muscle, a motor unit could be recruited or silenced, made to discharge slowly, rapidly, or in bursts. During this time, other motor units recorded with adjacent electrodes produced similar patterns, showing that muscle force was indeed modulated (Smith, Basmajian, and Vanderstoep, 1974). To facilitate the delicate modulation of motor-unit firing patterns, the limb studied was not restrained in most of these studies, allowing movements as well as isometric contractions to be made. In such a setting, it is easy to see how a subject might produce varying patterns of muscle force to achieve the desired patterns of motor-unit discharges.

A much more dramatic claim, made by several investigators, is that the recruitment order of motor units can be controlled voluntarily (Basmajian *et al.*, 1965; Smith *et al.*, 1974; Wagman *et al.*, 1965). Although the control of the firing patterns of individual motor units with audio feedback presents no great conceptual problems, the voluntary recruitment of motor units does. In its broadest sense, it requires the direct neural control of individual motoneurons by some descending pathway (Basmajian *et al.*, 1965; Smith *et al.*, 1974). To my knowledge, no such pathway has yet been found, and, in fact, the wide distributions of many inputs to motoneuron pools that have been studied speak against such a possibility. Yet, the evidence is there: a few individuals are able to recruit any one of two to three and in exceptional cases as many as six motor units. How might this be done?

It is known that two motor units of similar thresholds, and therefore of similar sizes, may be recruited in either the "correct" or reverse order at random (Henneman *et al.*, 1974). The same thing may occur in the recruitment of units of similar sizes in man; it will be recalled that the experiments on motor-unit control were all done on a population of units of almost identical sizes. Therefore, simple reversals of recruitment order are not unexpected, and, in fact, may readily be seen (Ashworth *et al.*, 1967; Grimby and Hannerz, 1968, 1970; 1974; 1976), especially in the presence of altered proprioceptive input.

It was noted earlier that spindle afferent input from close synergists (lateral gastrocnemius and soleus) is not distributed equally to all the motoneurons of the medial gastrocnemius (Mendell and Henneman, 1971; Mendell and Scott, 1975). This nonuniform input can alter the recruitment order of medial gastrocnemius motoneurons (Clamann, 1978; Clamann and Kukulka, 1976). Several authors have noted the dependence of lability in the recruitment order on proprioceptive influences (Grimby and Hannerz, 1968; Wagman *et al.*, 1965), and particularly on adjustments in the lengths of other muscles. It is quite possible that the pathways of these proprioceptive influences are not distributed uniformly throughout the motoneuron pool of the muscle under study, though no specific pattern of distribution need be postulated. Since the motor units studied are almost identical in threshold to begin with, it would require only a slight difference in their input to alter their recruitment order. During the long training period required to learn to recruit specific units at will (Wagman *et al.*, 1965), a subject might learn to position his limb so as to produce the correct combination of proprioceptive inputs to excite a particular motor unit.

If this view is correct, it should be difficult or impossible to reverse the recruitment order of motor units that differ significantly in force threshold or size. A recent study by Henneman, Shahani, and Young (1976) has shown this to be the case. Several hundred attempts were made by nine subjects to silence the second of two units which had been recruited by steadily increasing the force output of the extensor indicis proprius muscle. Six of the subjects were unsuccessful. In the other three, occassional reversals occurred at random, but these reversals could not be voluntarily controlled.

The voluntary control of single motor-unit firing patterns is a demonstration that man can acquire control over physiological processes by biofeedback. It does not appear to support the hypothesis that individual spinal motoneurons may be controlled by selective inputs to them from higher centers.

The precise control over motor units illustrated in these studies does provide the hope that individuals may be taught to modulate accurately the electromyographic output of a muscle. This electromyographic signal may then be used to control a prosthetic device, or to provide feedback in training a person with a motor dysfunction to use a disabled muscle. Promising work in these directions is currently in progress in several laboratories.

CONCLUSION

We have seen that any one muscle contains motor units that differ widely in their contractile ability. In this way, nature has wisely supplied muscles with a wide range of elements, allowing them to perform a wide variety of tasks, from weak, precisely controlled contractions to movements of great vigor. Studies on the usage of individual motor units, and on their discharge patterns, reinforce the impression that the properties of these motor units are well suited to their diverse tasks, and are controlled in a logical manner by the neural systems within the brain and spinal cord.

Many questions remain to be answered on the topic of muscle organization and control; a few have been raised in the foregoing discussion. In particular, the usage of motor units in rapid or powerful contractions has not been fully explained. Nor have the differences in the control of diverse muscles, from the small finger muscles to large postural muscles of the hip, been scrutinized. One would think that muscles designed for posture would not be controlled in the same way as muscles used for manipulation.

Finally, it is becoming clear that neuromuscular diseases produce subtle changes in the structure and control of muscles. By increasing our detailed knowledge of the normal state, we can better understand the changes that occur in diseases.

REFERENCES

Adrian, E.D., and Bronk, D.W. Impulses in motor nerve fibres. I. Impulses in single fibres of the phrenic nerve. *Journal of Physiology (London)*, 1928, *66*, 81–101.

Appelberg, B., and Emonet-Dénand, F. Motor units of the first superficial lumbrical muscle of the cat. *Journal of Neurophysiology*, 1967, *30*, 154–160.

Ashworth, P., Grimby, L., and Kugelberg, E. Comparison of voluntary and reflex activation of motor units. Functional organization of motor neurons. *Journal of Neurology, Neurosurgery, and Psychiatry*, 1967, *30*, 91–89.

Bagust, J., Knott, S., Lewis, D.M., Luck, J.C., and Westerman, R.A. Isometric contractions of motor units in a fast twitch muscle of the cat. *Journal of Physiology (London)*, 1973, *231*, 87–104.

Barrett, J.N., and Crill, W.E. Specific membrane properties of cat motoneurones. *Journal of Physiology (London)*, 1974a, *239*, 301–324.

Barrett, J.N., and Crill, W.E. Influence of dendritic location and membrane properties on the effectiveness of snyapses on cat motoneurones. *Journal of Physiology (London)*, 1974b, *239*, 325–345.

Basmajian, J.V. Control and training of individual motor units. *Science,* 1963, *141,* 440–441.

Basmajian, J.V., Baeza, M., and Fabrigar, C. Conscious control and training of individual spinal motor neurons in normal human subjects. *Journal of New Drugs,* 1965, *5,* 78–85.

Batini, C., Buisseret-Delmas, C., and Corvisier, J. Horseradish peroxidase localization of masticatory muscle motoneurons in cat. *Journal of Physiology (Paris)* 1976, *72,* 301–310.

Beerman, D.H., and Cassens, R.G. Terminal innervation ratios and fiber type grouping in normal porcine skeletal muscle. *Anatomy and Embryology,* 1977, *150,* 123–127.

Bigland, B., and Lippold, O.C.J. Motor unit activity in voluntary contraction of human muscle. *Journal of Physiology (London),* 1954, *125,* 322–335.

Büdingen, H.J., and Freund, H.J. The relationship between the rate of rise of isometric tension and motor unit recruitment in a human forearm muscle. *Pflugers Archiv,* 1976, *362,* 61–67.

Burke, R.E. Motor unit types of cat triceps surae muscle. *Journal of Physiology (London),* 1967, *193,* 141–160.

Burke, R.E., and Edgerton, V.R. Motor unit properties and selective involvement in movement. In J.W. Wilmore and J.F. Keogh (Eds.), *Exercise and Sports Sciences Reviews* (Vol. 3). New York: Academic Press, 1975.

Burke, R.E., Jankowska, E., and ten Bruggencate, G. A comparison of peripheral and rubrospinal synaptic input to slow and fast twitch motor units of triceps surae. *Journal of Physiology (London)* 1970, *207,* 709–732.

Burke, R.E., Levine, D.N., Zajac, F.E., Tsairis, P., and Engel, W.K. Mammalian motor units: Physiological-histochemical correlation in three types in cat gastrocnemius. *Science,* 1971, *174,* 709–712.

Burke, R.E., Levine, D.N., Tsairis, P., and Zajac, F.E. Physiological types and histochemical profiles in motor units of the cat gastrocnemius. *Journal of Physiology (London),* 1973, *234,* 723–748.

Burke, R.E., Levine, D.N., Salcman, M., and Tsairis, P. Motor units in cat soleus muscle: Physiological, histochemical, and morphological characteristics. *Journal of Physiology (London),* 1974, *238,* 503–514.

Burke, R.E., Rudomin, P., and Zajac, F.E., III. The effect of activation history on tension production by individual muscle units. *Brain Research,* 1976, *109,* 515–529.

Burke, R.E., Strick, P.L., Kanda, D., Kim, C.C., and Walmsley, B. Anatomy of medical gastrocnemius and soleus motor nuclei in cat spinal cord. *Journal of Neurophysiology,* 1977, *40,* 667–680.

Clamann, H.P. Statistical analysis of motor unit firing patterns in a human skeletal muscle. *Biophysical Journal* 1969, *9,* 1233–1251.

Clamann, H.P. Activity of single motor units during isometric tension. *Neurology (Minneapolis),* 1970, *20,* 254–260.

Clamann, H.P. The influence of different inputs on the recruitment order of muscles and their motor units. Submitted to J.E. Desmedt (Ed.), *Proceedings of the International Symposium on Human Reflexes and Motor Disorders* (Vol. 9). Basel: Karger, 1978.

Clamann, H.P., and Henneman, E. Electrical measurement of axon diameter and its use in relating motoneuron size to critical firing level. *Journal of Neurophysiology,* 1976, *39,* 844–851.

Clamann, H.P., and Kukulka ,C.G. Evidence for selective inputs to a motoneuron pool. *Society for Neuroscience Abstracts,* 1976, *2,* 539.

Clamann, H.P., and Kukulka, C.G. The relation between size of motoneurons and their position in the cat spinal cord. *Journal of Morphology,* 1977, *153,* 461–466.

Clamann, H.P., Gillies, J.D., and Henneman, E. Effects of inhibitory inputs on critical firing level and rank order of motoneurons. *Journal of Neurophysiology,* 1974, *37,* 1350–1360.

Creed, R.S., Denny-Brown, D., Eccles, J.C., Liddell, E.G.T., and Sherrington, C.S. *Reflex Activity of the Spinal Cord.* London: Oxford University Press, 1932.

Davis, W.J. Functional significance of motoneuron size and soma position in swimmeret system of the lobster. *Journal of Neurophysiology,* 1971, *34,* 274–288.

DeLuca, C.J., and Forrest, W.J. Probability distribution function of the inter-pulse intervals of single motor unit action potentials during isometric contraction. In J.E. Desmedt (Ed.), *New Developments in Electromyography and Clinical Neuro-physiology* (Vol. 1). Basel: Karger, 1973.

Denny-Brown, D. On the nature of postural reflexes. *Proceedings of the Royal Society, Series B*, 1928–1929, *104*, 252–301.

Desmedt, J.E., and Godaux, E. Ballistic contractions in man: Characteristic recruitment pattern of single motor units of the tibialis anterior muscle. *Journal of Physiology (London)*, 1977, *264*, 673–693.

Eccles, J.C., and Sherrington, C.S. Numbers and contraction-values of individual motor units examined in some muscles of the limb. *Proceedings of the Royal Society, Series B*, 1930, *106*, 326–357.

Eccles, J.C., Eccles, R.M., Iggo. A., and Ito, M. Distribution of recurrent inhibition among motoneurons. *Journal of Physiology (London)*, 1961, *159*, 479–499.

Edström, L., and Kugelberg, E. Histochemical composition, distribution of fibers, and fatiguability of single motor units. Anterior tibial muscle of the rat. *Journal of Neurology, Neurosurgery, and Psychiatry*, 1968, *31*, 424–433.

Freund, H.J., Büdingen, H.J., and Dietz, V. Activity of single motor units from human forearm muscles during voluntary isometric contractions. *Journal of Neurophysiology*, 1975, *38*, 933–946.

Fuchs, A.F., and Luschei, E.S. Firing patterns of abducens neurons of alert monkeys in relationship to horizontal eye movements. *Journal of Neurophysiology*, 1970, *33*, 382–392.

Goldberg, L.J., and Derfler, B. Relationship among recruitment order, spike amplitude, and twitch tension of single motor units in human masseter muscle. *Journal of Neurophysiology*, 1977, *40*, 879–890.

Goldberg, S.J., Hull, C.D., and Buchwald, N.A. Afferent projections in the abducens nerve: An intracellular study. *Brain Research*, 1974, *68*, 205–214.

Goldberg, S.J., Lennerstrand, G., and Hull, C.D. Motor unit responses in the lateral rectus muscle of the cat: Intracellular current injection of abducens nucleus neurons. *Acta Physiologica Scandinavia*, 1976, *96*, 58–63.

Gordon, G., and Phillips, C.G. Slow and rapid components in a flexor muscle. *Quarterly Journal of Experimental Physiology*, 1953, *38*, 35–45.

Granit, R. *Mechanisma Regulating the Discharge of Motoneurons*. Sherrington Lectures IX, Springfield, Ill.: Charles C Thomas, 1972.

Granit, R., and Burke, R.E. The control of movement and posture. *Brain Research*, 1973, *53*, 1–28.

Gray, E.R. Conscious control of motor units in a tonic muscle. *American Journal of Physical Medicine*, 1971, *50*, 34–40.

Grimby, L., and Hannerz, J. Recruitment order of motor units in voluntary contraction: Changes induced by proprioceptive afferent activity. *Journal of Neurology, Neurosurgery, and Pyschiatry*, 1968, *31*, 565–573.

Grimby, L., and Hannerz, J. Differences in recruitment order of motor units in phasic and tonic flexion reflex in "spinal man." *Journal of Neurology, Neurosurgery, and Psychiatry*, 1970, *33*, 562–570.

Grimby, L., and Hannerz, J. Differences in recruitment order and discharge pattern of motor units in the early and late flexion reflex components in man. *Acta Physiologica Scandinavica*, 1974, *90*, 555–564.

Grimby, L., and Hannerz, J. Disturbances in voluntary recruitment order of low and high frequency motor units on blockades of proprioceptive afferent activity. *Acta Physiologica Scandinavia*, 1976, *96*, 207–216.

Hannerz, J. Discharge properties of motor units in relation to recruitment order in voluntary contraction. *Acta Physiologica Scandinavia*, 1974, *91*, 374–384.

Harrison, V.F., and Mortensen, O.A. Identification and voluntary control of single motor unit activity in the tibialis anterior muscle. *Anatomical Record*, 1962, *144*, 109–116.

Henneman, E., and Olson, C.B. Relation between structure and function in the design of skeletal muscles. *Journal of Neurophysiology*, 1965, *28*. 581-598.

Henneman, E., Somjen, G.G., and Carpenter, D.O. Functional significance of cell size in spinal motoneurons. *Journal of Neurophysiology*, 1965, *28*, 560-580.

Henneman, E., Clamann, H.P., Gillies, J.D., and Skinner, R.D. Rank order of motoneurons within a pool: Law of combination. *Journal of Neurophysiology*, 1974, *37*, 1338-1349.

Henneman, E., Shahani, B.T., and Young, R.R. Extent of voluntary control of human motor units. (Abstract). In: J.E. Desmedt (Ed.), *International Symposium on Human Reflexes and Motor Disorders*. Brussels. 1976.

Jack, J.J.B., Noble, D., and Tsien, R.W. *Electric Current Flow in Excitable Cells*. Oxford: Clarendon Press, 1975.

Jennekens, F.G.I., Tomlinson, B.E., and Walton, J.N. Histochemical aspects of five limb muscles in old age. An autopsy study. *Journal of Neurological Science*, 1971, *14*, 259-276.

Kernell, D. High-frequency repetitive firing of cat lumbosacral motoneurones stimulated by long-lasting injected currents. *Acta Physiologica Scandinavica*, 1965, *65*, 74-86.

Kernell, D. Input resistance, electrical excitability, and size of ventral horn cell in cat spinal cord. *Science*, 1966, *152*, 1637-1640.

Kugelberg, E., Edström, L., and Abruzzese, M. Mapping of motor units in experimentally reinnervated rat muscle. *Journal of Neurology, Neurosurgery, and Psychiatry*, 1970, *33*, 319-329.

Lennerstrand, G. Mechanical studies on the retractor bulbi muscle and its motor units in the cat. *Journal of Physiology (London)*, 1974, *236*, 43-55.

Liddell, E.G.T., and Sherrington, C.S. Recruitment and some other features of reflex inhibition. *Proceedings of the Royal Society, Series B*, 1925, *97*, 488-518.

Marinesco, G. Recherches sur les localisations motrices spinales. *La Semaine Médicale Paris*, 1904, *24*, 225-231.

Marsden, C.D., Meadows, J.C., and Merton, P. A. Isolated single motor units in human muscle and their rate of discharge during maximal voluntary effort. *Journal of Physiology (London)*, 1971, *217*, 12P-13P.

Matthews, P.B.C. *Mammalian Muscle Receptors and their Central Actions*. Baltimore: Williams and Wilkins, 1972.

McPhedran, A.M., Wuerker, R.B., and Henneman, E. Properties of motor units in a homogeneous red muscle (soleus) of the cat. *Journal of Neurophysiology*, 1965, *28*, 71-84.

Mendell, L.M., and Henneman, E. Terminals of single Ia fibers: Location, density, and distribution within a pool of 300 homonymous motoneurons. *Journal of Neurophysiology*, 1971, *34*, 171-187.

Mendell, L.M., and Scott, J.G. The effect of peripheral nerve cross-union on connections of single Ia fibers to motoneurons. *Experimental Brain Research*, 1975, *22*, 221-234.

Milner-Brown, H.S., Stein, R.B., and Yemm, R. The orderly recruitment of human motor units during voluntary isometric contractions. *Journal of Physiology (London)*, 1973a, *230*, 359-370.

Milner-Brown, H.S., Stein, R.B., and Yemm, R. Changes in firing rate of human motor units during linearly changing voluntary contractions. *Journal of Physiology (London)*, 1973b, *230*, 371-390.

Norris, F.H., and Gasteiger, E.L. Action potentials of single motor units in normal muscle. *Electroencephalography and Clinical Neurophysiology*, 1955, *7*, 115-126.

Peachey, L. The structure of the extraocular muscle fibers of mammals. In. P. Bach-y-Rita, C.C. Collins, and J.E. Hyde (Eds.), *The Control of Eye Movements*. New York: Academic Press, 1971.

Peter, J.B. Histochemical, biochemical, and physiological studies of skeletal muscle and its adaptation to exercise. In R.J. Podolski (Ed.), *Contractility of Muscle Cells and Related Processes*. Englewood Cliffs, N.J.: Prentice-Hall, 1971.

Robinson, D.R. Oculomotor unit behavior in the monkey. *Journal of Neurophysiology*, 1970, *33*, 393-404

Romanes, G.J. The motor cell columns of the lumbosacral spinal cord of the cat. *Journal of Comparative Neurology*, 1951, *94*, 313–364.

Rosenfalck, P., and Madsen, A. Preferred intervals of muscle action potentials in voluntary contraction. *Acta Physiologica Scandinavica*, 1954, *31*, Supplementum 114, 47–48.

Sasaki, K., and Tanaka, T. Phasic and tonic innervation of spinal alpha motoneurons from upper brain centers. *Japanese Journal of Physiology*, 1964, *14*, 56–66.

Schwindt, P.C., and Calvin, W.H. Equivalence of synaptic and injected current in determining the membrane potential trajectory during motoneuron rhythmic firing. *Brain Research*, 1973, *59*, 389–394.

Sherrington, C.S. Notes on the arrangement of some motor fibres in the lumbosacral plexus. *Journal of Physiology (London)*, 1892, *13*, 621–771.

Smith, H.Mc., Basmajian, J.V., and Vanderstoep, S.F. Inhibition of neighboring motoneurons in conscious control of single spinal motoneurons. *Science.* 1974, *183*, 975–976.

Smith, J.L., Edgerton, V.R., Betts, B., and Collatos, T.C. EMG of slow and fast ankle extensors of cat during posture, locomotion, and jumping. *Journal of Neurophysiology*, 1977, *40*, 503–513.

Smith, O.C. Action potentials from single motor units in voluntary contraction. *American Journal of Physiology*, 1934, *108*, 629–638.

Tanji, J., and Kato, M. Recruitment of motor units in voluntary contraction of a finger muscle in man. *Experimental Neurology*, 1973a, *40*, 759–770.

Tanji, J., and Kato, M. Firing rate of individual motor units in voluntary contraction of abductor digiti minimi muscle in man. *Experimental Neurology*, 1973b, *40*, 771–783.

Teig, E. Tension and contraction time of motor units of the middle ear muscle of the cat. *Acta Physiologica Scandinavica*, 1972, *84*, 11–21.

Tokizane, T., and Shimazu, H. *Functional Differentiation of Human Skeletal Muscle: Corticalization and Spinalization of Movement*, Tokyo: University of Tokyo Press, 1964.

Wagman, I.H., Pierce, D.S., and Burger, R.E. Proprioceptive influence in volitional control of individual motor units. *Nature*, 1965, *207*, 957–958.

Wuerker, R.B., McPhedran, A.M., and Henneman, E. Properties of motor units in a heterogenous pale muscle (m. gastrocnemius) of the cat. *Journal of Neurophysiology*, 1965, *28*, 85–99.

Wyman, R.J., Waldron, I., Wachtel, G.M. Lack of fixed order of recruitment in cat motoneuron pools. *Experimental Brain Research*, 1974, *20*, 101–114.

Yemm, R. The orderly recruitment of motor units of the masseter and temporal muscle during voluntary isometric contraction in man. *Journal of Physiology (London)*, 1977, *265*, 163–174.

3

Proprioceptors and the Regulation of Movement

Peter B. C. Matthews

Introduction—Wider Perspectives

By the beginning of the present century, histological work, particularly that of Sherrington and of Ruffini, had established that skeletal muscle is as fully supplied with afferent nerve fibers as it is with the motor nerve fibers which induce overt contraction and thus all movement. It was widely accepted that this afferent input played a crucial role in the regulation of movement both by its reflex actions and by feeding information to that "head ganglion of the proprioceptive system," the cerebellum, but precise information and understanding eluded all concerned; indeed, the debate was still in progress as to whether the familiar tendon jerk was really a reflex or just the direct response of muscle to a mechanical stimulus. Half a century later, the importance of the muscle proprioceptors for the processes of neural arithmetic was further highlighted by the demonstration that nearly a third of the motor fibers to a muscle were devoted exclusively to the intrafusal muscle fibers of the muscle spindles and thus concerned with the proprioceptive regulation of movement rather than with directly producing movement itself (Kuffler, Hunt and Quilliam, 1951; Leksell, 1945). Today, we have gathered a great deal more information about the particular workings of the proprioceptive system, but in many essential matters any deep understanding still eludes us, and as the present article will outline, debate continues on many topics that our successors may judge to be as elementary as that of the origin of the tendon jerk.

The difficulty of understanding the overall function of the proprioceptors may

Peter B. C. Matthews   University Laboratory of Physiology, Parks Road, Oxford OX1 3PT, England.

arise partly because it seems likely that the information that they provide on the state of affairs within the muscle may be put to different uses by the central nervous system (CNS) at one and the same time. For different muscles, or for the same muscle acting in different situations, the relative predominance of these different functions might vary. Also, information normally provided by the proprioceptors might on occasion be supplanted by signals provided by another sensory source or even replaced by an internal analysis within the CNS of the outgoing motor signals. Thus the search for a unique function for one or all of the proprioceptors is as likely to confuse the issue as to simplify it, and for the moment a catholic approach seems preferable. Various uses to which proprioceptive information might be put are readily suggested, virtually *ab initio*.

• First, and most classically, they can be supposed to contribute to the conscious awareness of the position of the limbs in space and whether and how the limbs are moving; they probably also contribute to the elaboration of relative Ia complex perceptions such as how "stiff" a spring feels. This view passed under a cloud for many years, but is once again attracting adherents on the basis of fresh experimental evidence (Goodwin, McCloskey, and Matthews, 1972; McCloskey, 1978; Matthews, 1977; Roland and Ladegaard-Pedersen, 1977).

• Second, muscle afferent signals may be used by feedback pathways, whether spinal or supraspinal, to provide a continuous control of movement and the immediate compensation for unexpected motor disturbances so as to correct movement from moment to moment in the course of its performance. The idea of "load compensation" provides a particular example of such closed loop control. For the extraocular muscles, however, any such a function appears to be in abeyance (Keller and Robinson, 1971).

• Third, developing from this approach, there is the suggestion that movement arises from the action of a "follow up length servo" in which some or all of the motor command is delivered as a biassing signal to the muscle spindles by their gamma motor fibers. It is then left to proprioceptive reflex action to ensure that the appropriate movement occurs and by negative feedback restores spindle activity to its initial level. The original pure form of this idea is now dead, but its successors march on.

• Fourth, still thinking at the level of rapid reflex compensation, the muscle receptors may be seen as linear transducers inserted into a nonlinear feedback loop so as to linearize its overall performance (Crago, Houk, and Hasan, 1976).

• Fifth, thinking on a slightly wider time scale, the proprioceptors may provide the higher centers with information on the state of affairs at the periphery so as to enable them to elaborate the detailed instructions for a movement in relation to its starting point; they could equally provide continuous up-dating to the control centers as the movement proceeds.

• Sixth, the proprioceptors might play a part in calibrating movements, especially rapid ones, that are generated and controlled by intrinsic CNS processes without any type of on-going external feedback but which need to be adjusted if they should fail to achieve their objective. The example is provided by the vestibuloocular reflex, which operates open loop without any immediate feedback from the eyes as to whether it is of appropriate magnitude to hold them on target, but when it fails to do so it is recalibrated over a period of hours to days (Melvill Jones, 1977). Proprioceptive feedback might well be utilized for any recalibration of somatic motor discharges that is required to compensate for muscular fatigue or weakness or even for the enhanced strength that occurs with training.

• Seventh, on the even longer term, proprioceptive activity might be suggested to have a special role to play in learning motor acts, which once acquired could then be reproduced ad infinitum without such feedback, though continuing feedback would permit further modification if the environment or the properties of the motor apparatus should change.

Such a list can doubtless be expanded. For obvious reasons most work has been concerned with the short-term aspects of regulation, and this article will follow the field in this respect simply because it is here that there is most to say, as it was for the tendon jerk in an earlier day. Closely related aspects of the matter will be found in Chapters 8 and 10 in this volume. It is hoped that in spite of a somewhat cursory approach to the mass of data that has now accumulated, the present account of the proprioceptors and their functional role will help engender a feeling for the shifting state of current understanding and so replace the simple dogmatism that is too commonly found in elementary texts. Fuller reference to the original literature may be found in a number of other accounts (Granit, 1970; Matthews, 1972; Stein, 1974).

THE PROPRIOCEPTORS THEMSELVES

Mammalian muscle contains two major types of specialized sense endings, namely, the muscle spindles and the Golgi tendon organs. Together these endings receive all the larger afferent fibers to muscle (above 5 μm diameter). In addition, there are numerous free nerve terminals that are supplied by smaller afferent fibers, both medullated and nonmyelinated. However, few of these are excited by stretch or by contraction of a muscle, and when they are, the response is often weak so that they are normally considered to subserve muscle pain and possibly circulatory and/or respiratory adjustments rather than the regulation of muscle contraction. Thus they need not occupy present attention.

TENDON ORGANS

Figure 1 shows the classical picture of the Golgi tendon organ. It consists of nerve branches derived from a large medullated axon, called the Ib afferent, which spread themselves over tendinous fascicles close to the musculotendinous junction, and which may be presumed to be deformed whenever the tendon is stressed. Recent histological work has emphasized that the tendinous fascicles underlying a single Golgi ending are connected to a relatively small number of muscle fibers (10–20), and thus the ending is unlikely to be influenced by what is happening in the rest of the muscle. Electrophysiological work, consisting of recording from single Ib afferent fibers while stimulating single motor fibers, confirms that this is indeed so and suggests, moreover, that the threshold of the afferent fiber is sufficiently low for it to be excited to discharge impulses on the contraction of a single one of the appropriately placed muscle fibers (Houk and Henneman, 1967). In spite of this limited sample of overall muscle tension provided by a single Ib fiber, it seems likely that the level of firing in the whole array of Ib fibers supplying a given muscle (some 50 for the intensively studied cat soleus) provides a reasonable measure of the overall tension in the muscle (cf. Binder, Kroin, Moore, and Stuart, 1977). The sample provided by a given tendon organ, moreover, provides contributions from muscle fibers of different contractile properties when a muscle contains a mixture of "red" and "white" fibers (Reinking, Stephens, and Stuart, 1975). Since the tendon organ lies "in series" with the muscle fibers, it is excited equally by active contraction of the muscle and by passive stretch. But the tension a muscle can develop on contraction is very much greater than the passive tension produced by stretching it to the physiological extreme so that most of the time the latter is physiologically insignificant; moreover, some passive tension may bypass the tendon organs by being borne by the fascial sheath while the muscle fibers pull nearly directly upon them. Thus for most practical purposes the tendon organs may be thought of as "contraction receptors" and ones with a low enough threshold to be involved in the continuous signaling of muscle contraction and not just brought into play when the tension becomes dangerously high, as sometimes suggested in the past.

The frequency of firing of a tendon organ increases monotonically with the tension in the muscle, but there is still debate as to how far the relation can be viewed as a linear one in the range of interest (Jami and Petit, 1976). The tendon organ, like the spindle secondary ending, shows a comparatively low sensitivity to dynamic stimuli over and above that to the same tension applied in the steady state, thus it signals tension *per se* rather more than the rate of change of tension (Matthews, 1933; Stauffer and Stephens, 1977). The tendon organ, of course, has no private motor supply so its pattern of discharge is relatively simple and can presumably be used as it stands by the CNS to indicate the loading of a contracting muscle, without the need for a knowledge of the quantity of motor discharge. It should be noted, moreover, that because of the force–velocity properties of muscle

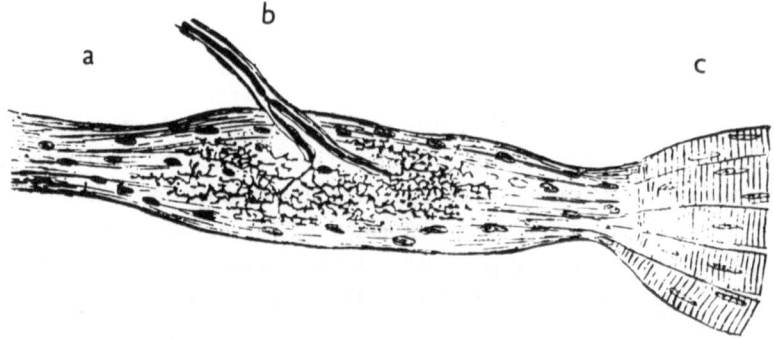

Fig. 1. The tendon organs as drawn by Cajal in 1909: (a) tendon; (b) large afferent fiber (group Ib); (c) muscle fibers. Note the small number of muscle fibers that directly influence the tendon organ.

this is a different thing from measuring the extent of muscular activation since a given motor discharge produces more tension under isometric conditions than it does when the muscle is shortening (Joyce, Rack, and Westbury, 1969).

MUSCLE SPINDLES

The muscle spindle is vastly more complicated than the Golgi tendon organ so that, although it has been much more intensively studied, it is less well understood. To begin with it contains two distinct types of afferent terminal, the primary and the secondary ending. Next, these receptor terminals lie on specialized contractile intrafusal muscle fibers rather than on passive tendon. Third, the intrafusal muscle fibers receive a complex motor supply, partly from the gamma motor fibers, which provide a private fusimotor supply, and partly from the beta fibers, which also supply the main muscle. Figure 2 shows an outline of the muscle spindle central region as recognized by Ruffini in 1898; the tails of the intrafusal muscle fibers may run on for a further 1–2 mm on either side. Figure 3 shows a simplified schema of the central region, which incorporates the essential new finding of the

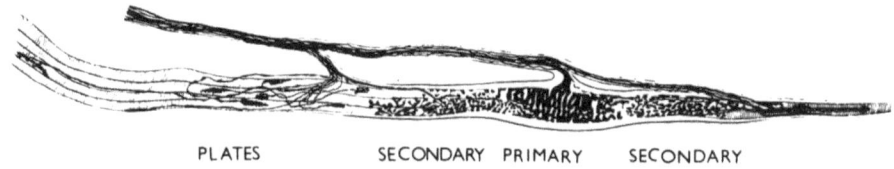

PLATES   SECONDARY PRIMARY   SECONDARY

Fig. 2. The muscle spindle as seen by Ruffini in 1898 with three morphologically separate types of nerve terminal (primary and secondary afferent endings, and motor plates).

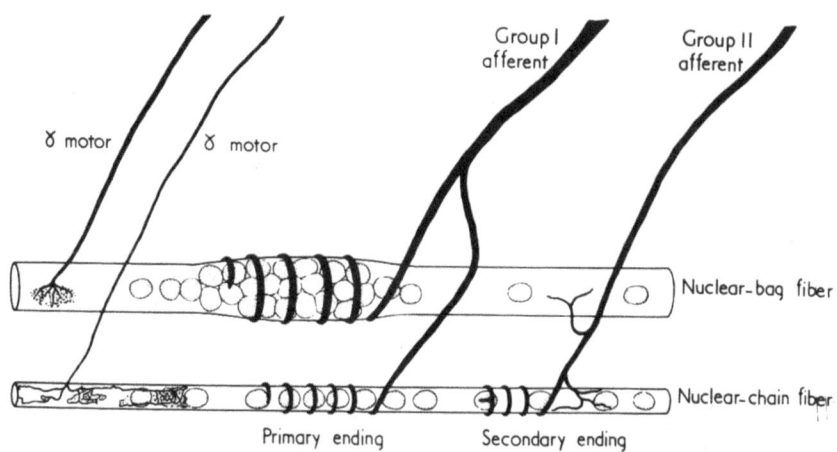

Fig. 3. Simplified diagram of the central region (about 1mm) of the muscle spindle as it was generally recognized from about 1962–1975. Two types of intrafusal muscle fiber are distinguished, each with their distinctive motor supply (plates to bag fibers and trail endings to chain fibers), and with different relationships to the two kinds of afferent ending. A given cat spindle contains 2–3 bag fibers and 4–6 chain fibers. From Matthews (1964).

early 1960s that the intrafusal muscle fibers are of two distinct types (nuclear bag and nuclear chain) with a motor supply that is to some degree independent.

SPINDLE AFFERENT ENDINGS. The primary afferent ending, of which there is only one per spindle, lies in the center of the equatorial region and spirals around every single one of the intrafusal muscle fibers occupying a length of about 0.3 mm in which the intrafusal muscle fibers are appreciably less well striated than at their poles. The ending arises from a large afferent fiber (of 12–20 μm diameter for the cat hind limb) which is given the name Ia, indicating that it is of group I diameter and to be distinguished from Ib fibers of similar size to the tendon organ.

The secondary endings lie to the side of the primary ending, arise from smaller group II afferents (4–12 μm diameter) and are a variable feature of the spindle. On average there is just over one secondary per muscle spindle but some spindles contain none and some may contain 4 or 5. Each secondary ending occupies a length of about 300 μm of the spindle. The secondary endings do not overlap each other or the primary ending and they fill up the juxta-equatorial space, starting from the primary ending, rather than occurring as scattered entities. They usually lie on only some of the several intrafusal fibers and the chain fibers are normally much more heavily innervated than are the bag fibers (recent evidence suggests that both bag$_1$ and bag$_2$ fibers may receive an innervation - Banks, Barker, and Stacey, 1977—see later). The secondary terminals are ultrastructurally similar to those of the primary ending but do not form as marked spirals around the intrafusal muscle fibers. The region of intrafusal fiber underlying the secondary ending is more markedly striated than that under the primary ending, but not always as well striated as the poles of the spindle. This suggests that there are regional variations in the strength of the intrafusal fibers along their length so that uniform longitudinal activation would lead to stretch of the sensorially innervated regions, as is certainly seen to occur when living spindles are observed microscopically on motor nerve stimulation (Boyd, 1976).

The overall length of a spindle may be up to 6 mm with its tails blending with the fascia around the extrafusal fibers, but no more than the central 1.5 mm receives an afferent innervation. This central region is enclosed in a fluid-filled capsule, possibly to provide mechanical and/or electrical insulation of the endings from the activity of the main extrafusal muscle fibers; the capsule is sometimes said to enclose a "lymph space," yet the fluid differs from lymph and is not contiguous with the main lymphatic system. Much of the motor innervation also lies inside the capsule, but especially some plate endings lie extracapsularly. In contrast to the tendon organ, the spindle lies "in parallel" with the main muscle fibers, so that it is unloaded by contraction of the extrafusal fibers and the afferent firing then diminishes unless the intrafusal muscle fibers are also contracting strongly. Since the beta supply to spindles (shared skeleto-fusimotor fibers) is comparatively weak in its action, sudden contraction of the main muscle silences its spindles, unless there is concomitantly a powerful gamma fusimotor discharge.

PATTERNS OF SPINDLE AFFERENT DISCHARGE. The existence of a morphological difference between the primary and secondary endings suggests that there should also be some functional difference. Yet under static conditions they behave broadly alike. They discharge tonically at comparable rates when the muscle length is held constant and show a similar sensitivity to increase of static length of the muscle (in the 5 cm long cat soleus they both have position sensitivities of about 4 impulses/sec firing per mm stretching). In some, but not all, muscles the threshold stretch required to elicit firing is greater for secondary endings than for primary endings. In the cat, especially when a degree of fusimotor activity is present, both types of ending would appear to be firing virtually throughout the physiological range of lengths. In conscious man, however, during voluntary relaxation of a muscle, both types of ending seem to be silent over an appreciable range of muscle lengths since fusimotor activity then seems to be in abeyance and both types of ending seem to have a relatively high threshold (cf. Burke, Hagbarth, Löfstedt, and Wallin, 1976; Vallbo, 1974a). One intriguing and unexplained difference between the two types of ending under static conditions is that the secondary ending fires its impulses in a more regular stream than does the primary ending, although the latter has long been held up as an example of a regularly firing afferent in comparison with many cutaneous afferents. When the spindle is de-efferented the coefficient of variation of the interspike interval distribution is 0.02 and 0.06 for the secondaries and primaries, respectively. This difference is accentuated during fusimotor activation of the spindle and means that the secondary ending would appear to be sending a more accurate signal, by being less "noisy," than does the primary ending.

A difference in the behavior of primary and secondary endings appears the moment the muscle is *moved* dynamically rather than just held at a constant length. The primary ending then shows itself very much more sensitive to any and every such dynamic stimulus than the secondary ending, as illustrated in Fig. 4. During the dynamic phase of stretch the frequency of firing of the primary ending is increased markedly above its subsequent static level, while that of the secondary ending is relatively much less affected by the dynamic stimulus although at the final length the secondary ending is firing at a slightly higher rate than the primary

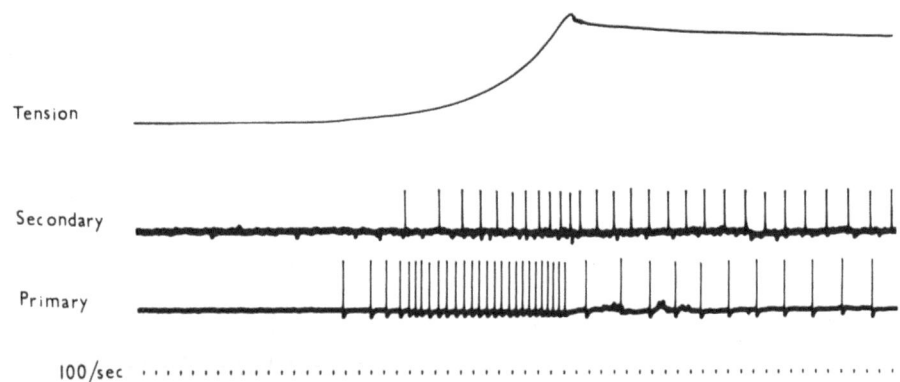

Fig. 4. The contrasting responses of spindle primary and secondary endings to a rapidly applied stretch (approximately 14 mm at 70 mm/sec to the 5 cm long cat soleus). The endings were identified by virtue of the conduction velocity of their afferent fiber, which was about twice as great for the primary ending (88 m/sec versus 44 m/sec). The ventral roots were cut to eliminate fusimotor activation. From Matthews (1972).

ending. Likewise, when the muscle is released from preexisting stretch, the primary ending slows much more markedly than the secondary ending and in the absence of fusimotor activity normally falls silent. The magnitude of both these stretch and release effects increases with the velocity of movement and thus may provide the CNS with information about the rate of movement as well as its absolute extent.

The greatest effect of a dynamic stimulus is, however, produced by its mere occurrence so that even with quite slow movements the response of the primary ending differs appreciably from that seen under genuine static conditions. The superadded response to increasing velocity then increases far more slowly than linearly when assessed as impulses/sec firing per unit increase in velocity. Such nonlinear signaling would not appear to present any obstacle to the interpretative skill of the CNS, and might even be suggested to be advantageous in permitting a wider range of signals to be transmitted. However, it has very recently been argued that in mathematical terms the response of both primary and secondary endings may be better presented as the product, rather than as the sum, of the length and velocity signals and that the difference in dynamic sensitivity between the two kinds of ending should be attributed to a difference in their overall sensitivity rather than to a difference in their "velocity" sensitivity (Rymer, Houk, and Crago, 1977). Although an interesting case can be made out for this being so in the special case of the response of the ending to a prolonged ramp stretch applied in the decerebrate cat, it seems unlikely to prove applicable to the behavior of the endings under all other conditions. At any rate, it remains fundamental that for any brief dynamic stimulus the primary ending gives a large response (whether positive or negative) in relation to the firing occurring in the absence of movement, whereas the secondary ending shows a much smaller effect. Such signaling of the same event by two sense endings of different relative sensitivity to the various components of the stimulus seems an important feature of the muscle spindle.

Another difference between the endings is that the primary ending behaves in a much more markedly nonlinear fashion to small stretches, applied either as repeated alternating steps or as sinusoidal movements (Hasan and Houk, 1975; Matthews and Stein, 1969). Thus the primary ending may be a hundredfold more sensitive to a small stretch than it is to a large one, expressing the response as impulses/sec firing per mm stretching. Because of this high sensitivity, a movement of 50 μm by a 5 cm long muscle may lead to appreciable modulation of firing of the primary ending. Such nonlinearity of behavior would seem to increase the versatility of the primary ending by allowing it to signal the occurrence of small stretches without its being prevented by saturation of its limited firing range from also signaling the magnitude of large stretches. When the muscle is held at a much greater length for a few seconds, the high sensitivity is reset to the new length. This behavior seems related to the way in which the ending gives a marked response to a dynamic stimulus independent of the precise velocity of movement; but the two phenomena are not precisely equivalent since the enhanced firing produced by a small step stretch can persist at the new length for several seconds.

THE INTRAFUSAL MUSCLE FIBERS. Over the last few years the evidence has been hardening that these should be subdivided into three distinct types, and not just two as commonly believed throughout the 1960s. The earlier subdivision of intrafusal fibers into nuclear bag and nuclear chain fibers retains its validity, but there are now recognized to be two different types of nuclear bag fibers which are most usually called the bag$_1$ and the bag$_2$ fibers. The bag and chain fibers are distinguished principally by the arrangement of nuclei in the central or equatorial region of the spindle. In the chain fibers, the nuclei lie in continuous single file in a "chain," whereas in the rather fatter bag fibers the nuclei lie two or three abreast in a solid column although not enclosed in any "bag" other than the intrafusal fiber itself. Elsewhere along their length the nuclei are sparsely placed in either kind of fiber, and the functional meaning of the equatorial concentration of nuclei beneath the sensory terminals remains obscure. The bag fibers tend to be appreciably longer than the chain fibers, which often end just outside the capsule. A given spindle contains only two or three bag fibers, but four to six chain fibers. Ultrastructurally, the chain fibers show a well-marked M line in the middle of the sarcomere whereas that of the bag fibers is poorly developed or absent. This is interesting because the presence of an M line in various muscles correlates with the speed of contraction and microscopic observation shows that the chain fibers do indeed contract more rapidly than the bag fibers.

The two types of bag fiber have only recently been separated and their recognition as functional entities hangs as much upon the physiological work of Boyd and his colleagues upon living isolated spindles (Boyd, 1976; Boyd, Gladden, McWilliam, and Ward, 1977) as it does upon their purely structural differences seen histologically (Banks, Harker, and Stacey, 1977). The morphological separation was first made histochemically by Ovalle and Smith (1972) who showed that the ATPase differed in the two kinds of fibers. That in the bag$_1$ fiber was stable on preincubation with acid but unstable on preincubation with alkali, whereas that of the bag$_2$ fiber was stable with both; the ATPase in the chain fibers differed from that in either of the bag fibers by being stable with alkali but labile with acid.

More recently, the M line has been recognized to be better developed in the bag$_2$ fiber, though the position is complicated by the occurrence of regional variations in the development of the line along the length of one and the same intrafusal fiber, with the prominence of the line increasing toward the poles. In accordance with the overall difference between them in this respect, microscopic observation shows that the bag$_1$ fiber does contract more slowly than the bag$_2$ fiber which, in its turn, contracts somewhat more slowly than the chain fibers. The contraction of the bag$_2$ fiber on motor nerve stimulation is markedly more powerful than that of the bag$_1$ fiber, at any rate under isometric conditions, as shown by the production of a much greater extension of the central region underlying the primary terminals (2–8% as opposed to 10–25%). The bag$_1$ fibers have the interesting property not shared by the other two types of fiber of showing a slow "creep" when the fiber is stretched to a new length, and this is particularly marked when the fiber is contracting (Boyd, 1976; Boyd *et al.*, 1977). The "creep" consists of a progressive "viscoelastic" yielding of the stretched poles of the fiber allowing a shortening of the central sensorially innervated region. It seems likely that it is responsible for a slow component of adaptation of the primary ending firing to a maintained stretch and which is particularly prominent during stimulation of "dynamic" fusimotor fibers (see below). A functionally intriguing morphological difference between the two kinds of bag fiber is that the poles of the bag$_2$ fiber are much more richly supplied with elastic fibers than are those of the bag$_1$ fiber. This has the important practical use of permitting ready identification of bag$_1$ and bag$_2$ fibers at the termination of an electrophysiological experiment, and so has helped in ascribing the various functional properties described above to the two types of bag fiber characterized morphologically. For the present, the chain fibers continue to be viewed as a structural and functional unity. This view is favored by the observation that the several chain fibers within a spindle seem to be attached together and move as a bundle on stimulation of single fusimotor fibers, whereas this bundle moves independently of the bag fibers and the two kinds of bag fiber slide freely past each other.

MOTOR TERMINALS. Three types of motor terminal have been described within the spindle on the basis of their general appearance. There are two types of discrete "plate" endings, with underlying junctional folding and broadly similar to those on extrafusal fibers, and "trail" endings consisting of fine terminals that meander for several hundred μm along the surface of the muscle fiber, probably with repeated points of synaptic contact. The p1 plates are distinguished from the p2 plates by being very much closer in appearance to normal motor end plates; p2 plates differ chiefly in being about twice as large and have a less well-marked subneural apparatus than ordinary plates (Barker, 1974). At the extreme, each type of ending is clearly distinguishable, but it has been questioned how far the two classes of ending may shade into each other in their appearance, and how securely a given ending may be identified on the basis of observing a limited number of histological sections. Related to this, it is hardly surprising that a certain amount of uncertainty and disagreement is to be found in the literature on the distribution of the various terminals. But it is clear that no one type of ending has the prerogative of controlling any one type of intrafusal muscle fiber. It has been held that a given stem motor axon terminates in the same type of ending at all its branches, but even this is open to question. However, the limited evidence available does support the view that when a given intrafusal muscle fiber has more than one type of motor termination upon it, then they all elicit the same type of contraction from it; it remains possible that differences in longitudinal distribution of the endings or in the strength of their synaptic coupling with the underlying fiber leads to some sort of difference in their action. The urgent current question is how far any one motor fiber branches to supply the different types of intrafusal fiber and thus whether these are obligatorily activated in parallel or are independently innervated, and hence potentially controllable on their own. As will be discussed in relation to fusimotor action, current evidence suggests that the bag$_2$ and chain fibers receive, by and large, a common motor supply but that the bag$_1$ fibers may be activated independently.

FUSIMOTOR STIMULATION. In the typical mammal, motor fibers below about 7 μm in diameter (gamma motor fibers) are specifically fusimotor and their stimulation evokes no significant tension in the muscle as a whole. Instead, the intrafusal muscle fibers contract and thereby both excite the sensory terminals by stretching the underlying poorly striated region of intrafusal fiber, and also influence their sensitivity to stimuli applied at its ends to the muscle spindle as a whole. By virtue of the effects on the sensitivity of the primary afferent to a ramp and hold stretch, the fusimotor fibers may be classified into the functionally distinct "static" axons (γ_S) and "dynamic" axons (γ_D), as illustrated in Fig. 5. Stimulation of a γ_S axon has a powerful excitatory action on the primary ending and tends to decrease its dynamic responsiveness to the ramp stretch so that there is then little change in the frequency of Ia firing at the beginning and end of the dynamic phase of stretching. In contrast, stimulation of a γ_D axon has little direct excitatory action but greatly enhances the response of the ending to the dynamic component of the stimulus as shown by the high frequency of firing during the stretching with subsequent fall on attaining the final length. Conventionally, these effects are assessed by measuring the arbitrary "dynamic index," which is defined as the fall in firing frequency of the ending in the first 0.5 sec on completing the ramp stretch. Stimulation of γ_D axons increases the dynamic index, whereas γ_S fusimotor stimulation decreases it or at least leaves it unchanged in spite of a considerable excitatory action. On release of stretch, corresponding to active muscle shortening, γ_D stimulation is largely ineffective at maintaining afferent firing so the primary ending

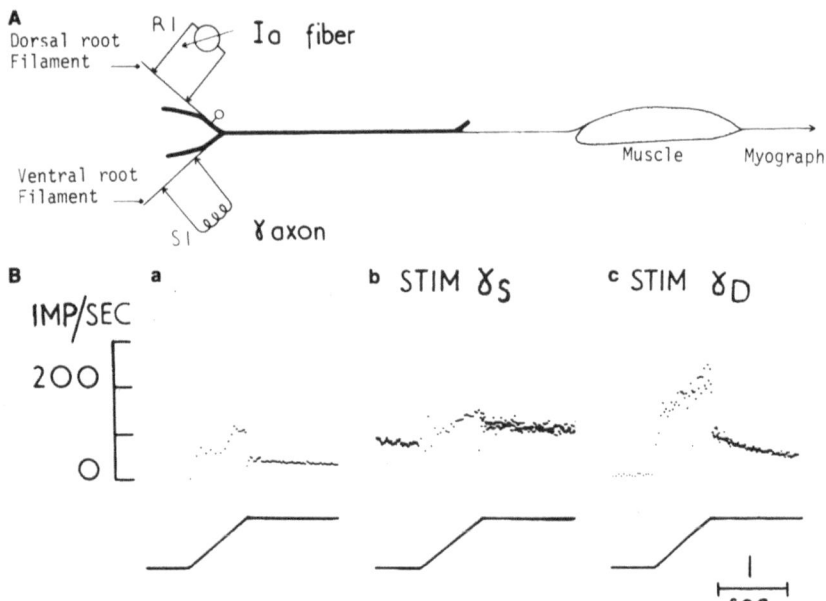

Fig. 5. The characterization of static and dynamic fusimotor action. (A) Preparation to record the discharge of a functionally single afferent from a spindle primary ending, isolated in a dorsal root filament, while stimulating a functionally single gamma motor fiber supplying the same spindle, isolated in a ventral root filament. (B) Afferent responses to stretch alone (a); during repetitive stimulation of a single static fusimotor axon (b); and during stimulation of a single dynamic fusimotor axon (c). The records show the "instantaneous frequency of firing", determined spike by spike; each dot represents an action potential and its height above zero gives the reciprocal of the time since the immediately preceding spike (i.e., frequency, see scale). Stretch, 6 mm at 30 mm/sec with the time scale linearly expanded during the dynamic phase of stretching. From Matthews (1972).

typically falls silent as it does in the absence of stimulation. Stimulation of γ_S fusimotor axons, however, maintains the firing of the primary ending during shortening at even quite high velocities.

γ_D stimulation has little or no action on the secondary ending even though the ending can be proved to lie in the spindle whose γ_D fiber is being stimulated. γ_S stimulation excites the secondary ending, though to a slightly lesser extent than it does the primary, and maintains its dynamic sensitivity at the low value characteristic of the passive ending. Thus during γ_S stimulation the primary and secondary endings behave much more similarly than they do in its absence or during γ_D stimulation. γ_S axons are several times as numerous as γ_D axons, though because a given axon may branch to supply several spindles, there is some uncertainty about the precise ratio. A crucial observation is that when a particular axon is found to have a given effect on one primary ending, then it is also found to have the same type of effect, be it static or dynamic, on any other primary ending that it can be shown to influence. Remembering that there is only one primary ending per spindle, this shows that the type of the action is specific for a given fusimotor fiber, and does not arise from any chance arrangement that it happens to enter into with one particular muscle spindle.

INTERRELATION BETWEEN MOTOR STRUCTURE AND FUNCTION. The relation between the functional dichotomy of the fusimotor axons and the various types of intrafusal muscle fiber has been a matter of long continued debate. For the decade 1962–1972, during which the intrafusal muscle fibers were subdivided just into the chain and undifferentiated bag fibers, it seemed most likely that the γ_D axons achieved their specificity by supplying the bag fibers and that the γ_S axons supplied the chain fibers. This hypothesis contained two separate elements: first, that the functional specificity arose from a particular kind of intrafusal muscle fiber being innervated by a fusimotor axon and, second, that the correspondence fell out one particular way round, as in Fig. 6A. The continued accumulation of evidence supports the view that γ_D axons supply only the bag fibers, but that γ_S axons commonly supply bag fibers as well as, or even to the exclusion of, the chain fibers. The chief lines of evidence have been (1) the histological observation of the simplified spindle in which all but one of the fusimotor axons have been caused to degenerate by prior ventral root section (Barker, Emonet-Dénand, Laporte, Proske, and Stacey, 1973), (2) the direct observation of the isolated living spindle on stimulating single functionally identified fusimotor fibers (Bessou and Pagès, 1975; Boyd et al., 1977), and (3) the application of the glycogen depletion method in which repetitive stimulation of a single motor fiber leads to a depletion of the glycogen in those muscle fibers that it innervates, as can be displayed by subsequent

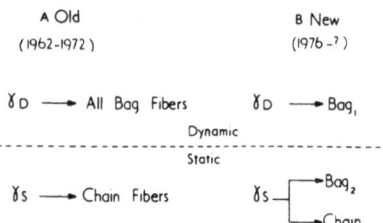

Fig. 6. Developing views on the relation between the *functional* classification into static and dynamic axons and the *morphological* classification of the intrafusal muscle fibers.

histological study (Barker, Emonet-Dénand, Harker, Jami, and Laporte, 1976; Brown and Butler, 1973).

Thus the question of the relation between static and dynamic action and bag and chain innervation was temporarily thrown back into the melting pot. Almost simultaneously the subdivision of bag fibers into two separate types offered a solution which has now gained widespread acceptance, namely, that it is the bag_1 fibers that are responsible for dynamic action and that the bag_2 fibers contribute to static action along with the chain fibers, as illustrated in Fig. 6B. The initial evidence for this view was that on direct observation of the isolated spindle, stimulation of a γ_S and of a γ_D axon was always found to excite separate bag fibers and not the same one (Bessou and Pagès, 1975; Boyd et al., 1977). Boyd and his colleagues originally gave these the names of Static Nuclear Bag (SNB) and Dynamic Nuclear Bag (DNB) by virtue of the source of their motor supply, but it has now proved possible to equate DNB with the histologically characterized bag_1 fiber and the SNB with the bag_2 fiber. The SNB was often supplied in parallel with the chain fibers, but the DNB was always activated on its own. Glycogen depletion experiments (Barker et al., 1976) confirm that the bag_1 fiber is selectively activated by dynamic axons so there is full agreement on this count. But γ_S axons may deplete bag_1 fibers as well as bag_2 and chain fibers. All would agree that the *specificity* of static action arises from the activation of bag_2 and/or chain fibers, but there is a degree of conflict as to whether or not γ_S axons also activate bag_1 fibers to a significant degree; either the glycogen method overestimates the degree of activation, or visual observation is insufficiently sensitive to detect a weak bag_1 contraction when it occurs at the same time as the stronger bag_2 or chain contractions.

The recent examination of some 160 examples of the action of a single fusimotor fiber on a primary ending suggested that a certain number of cases (up to ⅓ on the widest interpretation) showed signs of a combination of static and dynamic effects, rather than being "pure" examples of static and dynamic action (Emonet-Dénand, Laporte, Matthews, and Petit, 1977). In accordance with the glycogen evidence, such "contamination" was very much commoner for static than for dynamic action and manifested itself as enhancement of the response to large amplitude ramps, with preservation of the normal "static" features of response. Experiments in which a pair of fusimotor fibers was stimulated together to mimic such "contamination" show that each type of intrafusal fiber adds its own contribution to the overall response, so there is no question of breaching the idea that static and dynamic actions arise from activating different types of intrafusal fiber. But experiment has yet to show whether the specialization in intrafusal action arises directly from the differences in contractility of the various intrafusal muscle fibers, or whether from a specialization in the transducer or other properties of the primary terminations lying on the different intrafusal fibers. However, the possible admixture of dynamic with static effects does not invalidate the idea of a dual fusimotor system, since, as before, a given fusimotor axon has proved to be predominantly static or predominantly dynamic on its action on all the primary endings influenced, and there was no particular tendency for a given axon regularly to have a "mixed" effect.

A continuing question is why static action, which is currently viewed as a functional entity, should be mediated by two separate types of intrafusal muscle fiber. As illustrated in Fig. 7, in a given spindle some γ_s axons may supply chain fibers on their own, others the bag$_2$ fiber on its own, and others may supply both. But, on the limited evidence available, a given γ_s may distribute itself quite differently in different spindles. Thus there is no question of the bag$_2$ fiber and the chain fibers forming completely independent effector systems, though it remains possible that statistically they may prove to be activable to different relative extent under different conditions.

BETA ACTION. Reptiles and amphibians lack a specialized fusimotor system, and the motor supply to their muscle spindles is entirely provided by branches from the ordinary motor fibers to the extrafusal muscle fibers. It is now recognized that in the cat also such mixed skeleto-fusimotor or beta innervation may occur, but as a supplement to the specialized gamma fibers rather than as a replacement for them. According to Barker (1974), beta fibers end in the p1 plates (histologically similar to extrafusal plates) and conversely all p1 plates may be suggested to be derived from beta fibers.

The first experimental evidence for the existence of beta fibers was obtained from various rather small muscles, such as the dorsal interossei of the foot, whose limited motor supply facilitated the tedious search required to isolate the fibers electrophysiologically and prove their fusimotor action (Bessou, Emonet-Dénand, and Laporte, 1965; McWilliam, 1975); the concomitant extrafusal contraction may mask a weak fusimotor action and in any case complicates the demonstration that any excitatory effect is due to an intrafusal contraction rather than external mechanical stimulation of the spindle. Subsequently, large muscles such as the tibialis anterior and the soleus have been reduced to a simpler state for experimentation by severing all but a single one of their supplying nerve branches; beta innervation has then been demonstrable for them also (Emonet-Dénand, Jami, and Laporte, 1975). In one series studied electrophysiologically it was estimated that about 70% of spindles in the peroneus brevis muscle receive some beta innervation as well as some four or more gamma fibers (Emonet-Dénand and Laporte, 1976).

Beta fibers isolated in electrophysiological experiments have usually proved

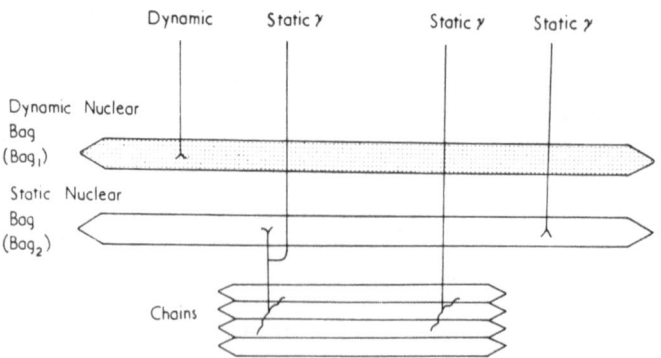

Fig. 7. The motor innervation of the spindle as currently described by Boyd, Gladden, McWilliam, and Ward (1977). The motor terminals are drawn to show their location, but without implication as to their type (i.e., whether plate or trail). Redrawn from their Fig. 8.

to have a typical dynamic action with static actions in abeyance, and the axons have had relatively low conduction velocities. Quite recently, however, glycogen depletion evidence suggests that there may be an appreciable number of "fast" betas with a static action since stimulating a number of fast motor fibers, without gamma fibers, may cause depletion of chain fibers in some spindles thus suggesting a static action; in this case the extrafusal contraction almost always seems to be of such strength as to prevent the electrophysiological recognition of beta action by the tests used hitherto (Harker, Jami, Laporte, and Petit, 1977). However, in an exhaustive search there was no sign of fast "α" motor fibers with a specifically fusimotor action (Ellaway, Emonet-Dénand, Joffroy, and Laporte, 1972). It may be concluded that there is a regular beta supply to many spindles and that it may be either static or dynamic in its action, but that it supplements rather than supplants the gamma motor supply to the spindle.

REFLEX CONNECTIONS OF MUSCLE AFFERENTS—FINDINGS WITH ELECTRICAL STIMULATION

GENERAL PATTERN OF CONNECTIVITY

Sherrington's work at the turn of the century with physiological stimuli in the decerebrate and chronic spinal preparations established, as now seems obvious, that the activity of sense endings in muscle can elicit multifarious reflex responses from muscle. But any physiological stimulus almost inevitably excites simultaneously both types of spindle afferents and also the tendon organs. Disentangling their respective reflex contributions to any normal motor act continues to present severe problems and was simply impossible in Sherrington's time. Lloyd opened the subject in the 1940s by using the then newly discovered monosynaptic reflex to test the excitability of motoneurons following graded electrical stimulation of a variety of muscle nerves in the acute spinal animal. Weak stimuli to a muscle nerve excite the largest afferent fibers, and these were found to elicit monosynaptic excitation of their own motoneurons and their synergists, and inhibition of the motoneurons of antagonistic muscles. This arrangement was found for both the flexor and for the extensor muscles at the knee and provided a particular example of Sherrington's principle of "reciprocal innervation," with the automatic regulation of opposing muscles so that on their reflexly elicited activation they should not interfere with each other.

Over the next few years various arguments led to the view that the afferents responsible for the monosynaptic reflex were the Ia fibers from the spindle primary endings, as has been amply demonstrated by subsequent experiment. Considerably stronger stimuli, of a strength suitable to excite the group II afferents from the spindle secondary endings and any other axons of a similar diameter, were found to produce an apparently nonspecific flexor action, like that produced by nociceptors, with excitation of the motoneurons to flexor muscles and inhibition of those to extensor muscles. This occurred irrespective of whether the nerve stimulated rose from a flexor or extensor muscle; but it remains an open question as to whether it is due to the spindle group II fibers themselves or to "contamination" of group II by the afferents from nociceptors. Still stronger stimuli, sufficient to excite the small group III fibers from free endings, produced a yet more powerful flexor action. All these effects from group II and III afferents were of such a central latency that they were demonstrably due to the excitation of polysynaptic pathways rather than to the monosynaptic route. Some years later, all such afferents with a relatively high threshold to electrical stimulation, including those from skin and joint, which produced the flexor reflex, were grouped together under the convenient term *flexor reflex afferents* (FRA), though without prejudice as to whether or not some smaller fibers might also prove to have a more specific reflex action at least under some conditions (Holmqvist and Lundberg, 1961).

The group I effects were further fractionated by Laporte and Lloyd (1952) who showed by careful grading of the strength of weak stimuli that autogenetic inhibitory actions could be produced as well as the monosynaptic excitation, and were accompanied by excitation of antagonists. These effects were disynaptic or trisynaptic and were attributed to the Ib fibers from the Golgi tendon organs. Lloyd had earlier equated the monosynaptic reflex with Liddell and Sherrington's (1924) stretch or myotatic reflex so Laporte and Lloyd called the apparently mirror image effect of the Ib fibers "the inverse myotatic reflex" (but see the section "Ib Action and Interneurons," pp. 111–112).

All these patterns of connectivity were more firmly established in the 1950s, particularly by Eccles and his collaborators. The afferents were still excited by graded stimulation of the muscle nerves but the indicator of motoneuronal activation was shifted to the intracellular recording of postsynaptic potentials from the motoneuron, with depolarization representing excitation (the EPSP) and hyperpolarization observed under appropriate conditions representing inhibition (the IPSP). With the more precise timing and analysis provided thereby, they were able to establish that the "direct" inhibition of antagonists elicited by Ia action was mediated disynaptically via a single interneuron, rather than monosynaptically as Lloyd had originally believed. Also, as a wider range of muscles came to be studied, it was appreciated that although the basic pattern of Ia excitation of synergists and inhibition of antagonists was nearly universal, minor deviations could be found at joints that were anatomically more complicated than the knee of the cat, such as the hip of the cat and the wrist of the primate. Detailed reference to all this standard work can be found in many sources (Granit, 1970; Matthews, 1972). Interestingly, the physiological analysis of all this synaptic connectivity ran far ahead of the anatomical analysis, by microscopic methods, of what is essentially an anatomical matter.

CURRENT STANDING OF THE MONOSYNAPTIC REFLEX

From all this work in the 1950s, the Ia evoked monosynaptic reflex emerged as an apparently unique phenomenon. First, the incoming afferent volley arrived, so it appeared, at the motoneuron without possibility of interference since there were no interneurons to be blocked by other inputs. Second, the monosynaptic reflex seemed to be the prerogative of the Ia fibers, with all other inputs required to run the gauntlet of interneuronal interference. On both counts, views have changed. The experimental demonstration of "presynaptic inhibition" showed that a monosynaptic input could indeed be altered in its efficacy by other inputs that did not need to play directly upon the motoneuron. It may be emphasized, however, that on present showing presynaptic inhibition differs from postsynaptic inhibition in its time scale, being far the slower, with its level rising and falling over tens and hundreds of msec as opposed to the few msec of postsynaptic action. Thus it seems unlikely to play much part in the moment-to-moment regulation of muscle contraction. It could, however, well play the part of a gain control for which it seems particularly suited by virtue of the fact that it acts in a multiplicative manner, reducing the efficacy of a given input channel however much traffic it should happen to be carrying; in contrast, postsynaptic inhibition acts in an additive manner, and simply annuls a certain quantity of excitation so that the inhibition can be counterbalanced simply by augmenting the excitatory signal. A description of the complicated pattern of which muscle afferents are presynaptically influenced by which other muscle afferents can be found elsewhere (Matthews, 1972); it has yet to have physiological meaning read into it in terms of motor control.

Next, the restriction of the monosynaptic pathway to Ia afferents has been challenged, and recent evidence suggests that the spindle group II afferents can also excite the motoneurons of their own muscle without an intervening interneuron. As illustrated in Fig. 8, this has been demonstrated by the "spike triggered averaging" of the intracellularly recorded potential of motoneurons, using as the trigger the natural discharge of a single group II afferent that has been left in continuity. Any synaptic potential related to that afferent is then raised above the noise level, whereas those from all other afferents are rejected by virtue of their bearing no

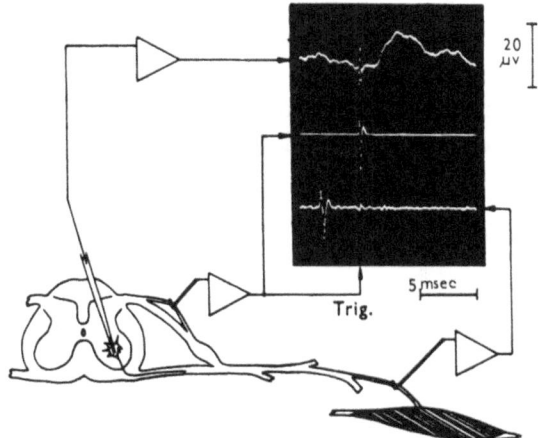

Fig. 8. Evidence obtained by "spike triggered averaging" for the occurence of monosynaptic excitatory action of the motoneurons of their own muscle by group II afferents from spindle secondary endings. Experiment on the triceps surae of the anaesthetized cat. Each trace is the average of 4,096 sweeps that were individually triggered by a spike in a single spindle group II axon that had been left in continuity, while being lifted up in a fine filament from the main mass of the dorsal root. *Middle trace:* the triggering spike. *Bottom trace:* the same spike recorded from the muscle nerve, thus permitting determination of its conduction velocity, which proved to be 56 m/sec (such pretrigger averaging is performed by temporarily storing the input signal to be averaged). *Top trace:* intracellular recording from a triceps motoneuron showing a potential in the depolarizing direction (i.e., excitatory) and of such a latency and rise time that it may be presumed to have been transmitted monosynaptically. The voltage calibration refers to the average value of the EPSP for a single sweep. From Kirkwood and Sears (1975).

regular phase relation to the triggering pulse. This has regularly shown EPSPs in extensor motoneurons from extensor group II afferents which are of appropriate latency and rise time to be monosynaptically elicited. They are up to about half the size of those elicited by single Ia fibers, though there is much scatter in both, and a given group II fiber probably only influences about half as many motoneurons as does a Ia fiber (Iles, 1976; Stauffer, Watt, Taylor, Reinking, and Stuart, 1976). The quantitative importance of this group II pathway has yet to be fully assessed, and that the group II fibers may have an autogenetic excitatory effect seems more firmly established than the belief that the connection is principally a monosynaptic one. Reassessment of the effects seen by the more classical methods of nerve stimulation coupled with monosynaptic testing or unaveraged intracellular recording suggests that there is no major conflict between the various findings (cf. Lundberg, Malgren, and Schomburg, 1975), though a certain puzzlement remains as to how the spindle group II effect can be so easily overlooked when using the grosser methods; this is perhaps a commentary on a relatively weak action and more work is awaited. At the same time as the uniqueness of the Ia monosynaptic pathway has been challenged, the evidence has been accumulating that the Ia fibers may well mediate some of their autogenetic excitation by polysynaptic pathways as well as monosynaptically.

EMERGENCE OF THE IMPORTANCE OF INTERNEURONS

Sherrington rightly emphasized the importance of the motoneuron as the "final common path" on which neural activity had eventually to impinge to produce overt action. Until recently, it was almost impossible to study spinal interneurons directly, so the motoneuron continued to occupy the forefront of experimental attention and interneurons were sometimes seen as little more than devices for producing a change of transmitter when inhibition was required rather than excitation, or as a way of introducing delay into a reflex pathway. Now, however, the spinal interneurons have emerged as integrative elements in their own right, on which varied inputs converge both from local sources and from higher centers. Present work is largely concerned with working out the wiring diagrams of this

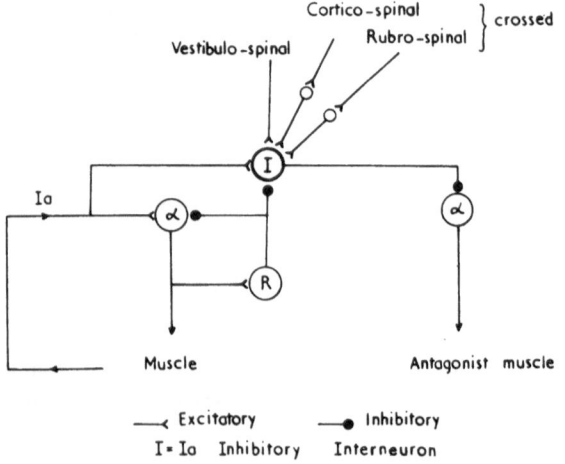

Fig. 9. Partial wiring diagram for the Ia inhibitory interneuron showing that it cannot help but have an "integrative" function by the summing of various inputs, and is far more than a simple "chemical commutator" in the conveyance of Ia inhibition to antagonistic muscles. α = a motoneuron; R = Renshaw cell.

neural circuitry and is disclosing a remarkable level of complexity. The tools available include the intracellular injection of dye into interneurons and, more especially, careful deduction from the pattern of convergence of activity onto the motoneuron since the intracellular recording from the interneurons themselves is a matter of some difficulty. For example, if two separate inputs are each ineffective at producing an EPSP in a motoneuron when given on their own, but succeed in doing so when given in combination, then the facilitation must have occurred at a common interneuron rather than on the motoneuronal soma. Likewise, the inhibition of a polysynaptically elicited EPSP unaccompanied by detectable direct action on the inhibited motoneuron argues for the inhibition having been exerted upon an intermediary interneuron, provided that presynaptic inhibition has been excluded.

Two interneurons that have recently occupied attention have been the Renshaw cell and the Ia inhibitory interneuron, both of which lend themselves to unequivocal identification by virtue of their best-known connections; the Renshaw cell is excited by the recurrent collaterals of the motor fibers and is unique in this respect, and the Ia interneurons are excited by the Ia fibers in a particular pattern. Both also lie in known restricted regions of the cross section of the spinal cord. On their discovery, the former was seen simply as a way station for conveying recurrent inhibition back from motoneurons to motoneurons to damp down any tendency to convulsive activity, and the latter were taken as the "chemical commutator" required for the Ia input to produce a very short latency inhibition of antagonists. Figure 9 illustrates a number of the connections that are now known to be made with these two interneurons and which show that they must indeed have an "integrative" function of some complexity. Most strikingly, the Renshaw cell is seen to play upon the Ia interneurons as well as the motoneurons (Hultborn, Jankowska, and Lindström, 1971a). Among other things, this pathway seems to provide a route by which the *same* Ia activity reaches a given interneuron in two opposite senses, since the Renshaw cells excited by the discharge of a given motoneuron pool then inhibit the very Ia interneurons that are excited by the Ia fibers that converge monosynaptically onto that motoneuron pool (contrary to early evidence, different populations of Renshaw cells are now known to be excited by flexor and extensor motoneurons, Hultborn, Jankowska, and Lindström, 1971b). Recent indirect evidence suggests that the gain via the two routes may be the same, though opposite in sign; there seems to be no change in the excitability of these interneurons with progressive stretch when the Ia fibers are successfully making the motoneurons fire, although their excitability increases with stretch when the Ia input is too weak to elicit the stretch reflex and thus is providing excitation without Renshaw inhibition (Hultborn and Lundberg, 1972). However, the Ia interneuron can be seen to receive a large number of inputs from higher levels of the nervous system, and this must be concerned with far more than just regulating the stretch reflex.

Ib Action and Interneurons

The Ib fibers influence the motoneurons solely via interneurons, and thus their spinal action can be expected to be even more subject to manipulation by the higher centers than is that of the Ia fibers. This is borne out by experiment. The inhibitory interneurons on the pathway from extensor Ib afferents to extensor motoneurons are played upon by the rubrospinal tract and the pyramidal tract, both probably monosynaptically, as well as by joint and cutaneous afferents, though here apparently not monosynaptically (Lundberg, 1975). Likewise, the excitatory interneurons on the pathways to antagonist muscles can be facilitated by supraspinal action. Moreover, by combining the stimulation of the red nucleus with that of the peripheral afferents, Hongo, Jankowska, and Lundberg, (1969) have shown that a number of short-latency spinal Ib pathways can be opened up that are not readily demonstrable in the acute spinal state, especially when the animal is also anesthetized. Ib reflex effects are then far more widespread and varied than in the simple spinal state. Thus the idea of an "inverse myotatic reflex" seems quite inadequate to cover all that happens. For example, during rubrospinal facilitation, Ib effects are commonly found between muscles acting at different joints, and sometimes an *inhibition* of extensor motoneurons by flexor Ib fibers is found, and of flexor motoneurons by extensor Ib fibers, rather than the normal excitation. Thus, as Hongo *et al.* said in 1969, "Clearly the Ib pathways are much more complex in their organization than has hitherto been assumed" (p. 384) and the concept of the "inverse myotatic reflex" provides at best a partial rather than a comprehensive generalization of the organization of Ib spinal reflex action.

REFLEX SWITCHING

The role for interneurons that has received the most attention is that of acting as "switches" enabling a reflex to be turned on or off at higher behest. This idea came into early prominence with the discovery that many of the simple inhibitory reflexes elicited by the Flexor Reflex Afferents in the spinal or lightly anesthetized state are virtually absent in the decerebrate state, though they can be restored by making the same preparation spinal (Eccles and Lundberg, 1959). Such switching is now attracting particular interest since it has recently been shown to occur phasically in the course of the walking cycle, so that a reflex that is present at one phase of the cycle may be missing or transmuted at another phase of the cycle (Duysens and Pearson, 1976; Forssberg, Grillner, and Rossignal, 1975). In this case, all the mechanisms would seem to be present in the cord itself, since the effect is also found in the chronic spinal state. Figure 10 shows an example of the switching of the effect of a given afferent input from zero in the decerebrate state, to inhibitory on making a midpontine lesion, to excitatory in the spinal state. In this case there is no evidence to show whether or not the opposing reflex effects were actually due to the same group of afferent fibers, since several kinds must have been excited with these strong stimuli; but with the example of the convergence of opposite effects from the Ia fibers onto the Ia interneuron (Fig. 9), there is every reason to follow the suggestion (Eccles and Lundberg, 1959) that in many cases a given afferent input may have two opposing reflex pathways available to it to influence a given motoneuronal pool, so that appropriate descending activity can reverse the sign of a reflex as well as simply blocking it! At the least, however, in spite of the shortness of their pathways, the standard spinal reflex actions of the muscle afferents would appear to be a permissive rather than an obligatory phenomenon, and to be operative only when the CNS wishes them to be so. The simple classical wiring diagrams deduced on the basis of monosynaptic testing in a single type of preparation, such as the acute spinal or barbiturate-anesthetized cat, can be seen to have provided a very limited appreciation of the reflex potentiality of the cord. There is no need to think that any of the standard findings are "wrong," merely that there may be other equally interesting connections that have failed to display themselves through being insufficiently facilitated.

The switching of spinal reflexes by supraspinal action can be expected to be only one facet of the convergence of higher and lower signals onto the same interneuron. In some cases, it may indeed perhaps prove to be a measure of economy in the conversion of an excitatory signal into an inhibitory one. But equally, it seems possible that the spinal inputs are switching the effects of supraspinal volleys so that a given descending motor command can have different effects depending upon the state of the peripheral muscular machinery, as signaled by the muscle afferents and presumably also the joint and cutaneous inputs. At any rate, such convergence of spinal and supraspinal inputs onto interneurons means that descending motor commands are subjected to peripheral influence well before they can begin to command the motoneurons to fire. *A priori*, it is as reasonable to talk

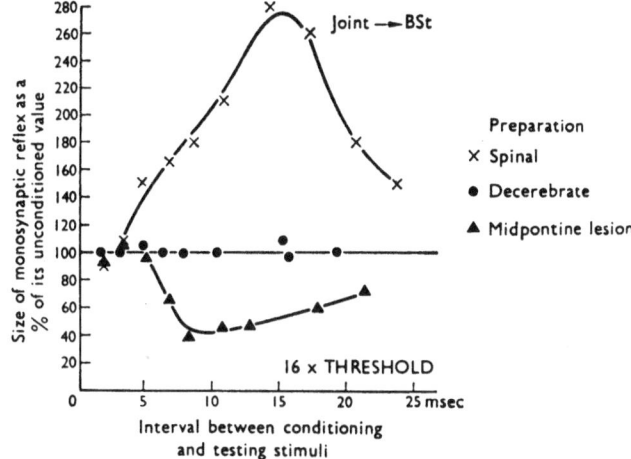

Fig. 10. The "switching" of spinal reflexes by changing the supraspinal bias (control) of varied spinal interneurons by making lesions in the higher parts of the CNS. The effects are shown by studying the action in the cat of a strong conditioning stimulus to the nerve to the knee joint (16 × the threshold of the most excitable fibers) on the monosynaptic reflex elicited by stimulating the nerve to the biceps-semitendinosus muscles (BSt) and altering the time between the stimuli. In the spinal state, the usual polysynaptic flexor reflex is observed with a latency of some 3 msec and rising to a peak at about 15 msec interval, as shown by the increase in the size of the monosynaptic reflex elicited from this flexor motoneuron pool by a standard testing shock to its nerve. In the decerebrate state, earlier in the experiment, all reflex activity is absent showing that the relevant spinal interneurons have been blocked by tonic discharges descending from supraspinal centers. On making a midpontine lesion in the decerebrate state, the same stimulus produces an inhibitory action (indicated by the reduction of the testing reflex) showing that the spinal interneurons have been switched to a very different functional configuration from that which they take up in the spinal state. From Holmqvist and Lundberg (1961).

of the function of the descending signals being the facilitation of the spinal reflexes as it is to talk, in the more usual sense, of the function of the spinal reflexes being the appropriate facilitation of the descending signals. What is required is a much better understanding of the spinal cord wiring, both in itself and also in terms of functional units conceived in broader terms of reference than simply the anatomical unit of the neuron. The appropriate way of doing this, however, remains an enigma. Electrical stimulation of nerves has proved to be an admirable tool for deducing direct neural connectivity, synapse by synapse, but it can be expected to provide only limited understanding of broader neural function. Success on this wider front probably requires the use of physiologically more natural inputs.

The Stretch Reflex and Its Occurrence during Voluntary Contraction

The degree of tone in a muscle, that is its resistance to passive stretching, has long been of clinical interest since it is altered in a variety of diseases. Sherrington put its study on a firm basis when he showed that the heightened tone of the decerebrate cat was maintained reflexly by afferent impulses arising segmentally in the rigid limb since the rigidity was abolished by dorsal root section. Later, the essential afferent source was traced to the muscle itself and the reflex by which stretching a muscle led to its contraction was named the stretch or myotatic reflex. Liddell and Sherrington (1924) made an extensive myographic study of the reflex and emphasized its tonic nature, with the reflex contraction persisting as long as the stretch is maintained, as illustrated in Fig. 11A. The familiar tendon jerk, elicited by a brief stretch applied to a muscle, they saw as but a "fractional manifestation" of the more comprehensive reflex since the afferent input was limited to but a few afferent volleys arriving in near synchrony at the spinal cord and thus leaving little time for neural integration.

Subsequent work, partly with electrical stimulation and partly with brief stretches, has established that the tendon jerk can be equated with the Ia-elicited monosynaptic reflex. Thus the briskness of the jerk depends on the responsiveness of the spindle, the excitability of the α motoneurons and the prevailing level of presynaptic inhibition on the Ia fibers, but it does not depend on the state of the variety of interneurons that may be brought into play when the stimulus is tonically maintained. For example, there is no opportunity for Renshaw inhibition to intervene, nor for the Ib or group II fibers to modulate the response, for their effect on the motoneurons will arrive too late. Yet during a prolonged stretch of a tonically contracting muscle, Ia, Ib, and II afferents will all be firing tonically and thus, by their central actions, they will influence the reflex outcome of maintained stretch, except for the special case where *all* the relevant spinal interneurons are totally inhibited by supraspinal action. Thus several mechanisms are available to explain the common failure of the stretch reflex to maintain itself tonically, as in Fig. 11B. Indeed, indirect evidence suggests that, in contrast with the findings with electrical stimulation, the spindle group II fibers contribute substantial excitation, not inhibition, to the stretch reflex, though the matter remains unresolved (Jack and Roberts, 1978; Kanda and Rymer, 1977; Matthews, 1969).

During tonic action it becomes largely irrelevant how many synapses happen to lie on a particular reflex pathway; what matters is the potency of the final action. In this respect, it may be noted that establishing the "latency" or even the existence of any delayed contributions of the stretch reflex, whether excitatory or inhibitory, is far from clear cut, especially if a short-latency action is also elicited by the same group of afferents. For example, debate continues as to the role of the polysynaptic pathways in mediating the "tonic vibration reflex" which is set up, for as long as the stimulus continues, by high-frequency vibration of an amplitude small enough for its excitatory action to be largely restricted to the Ia fibers (cf. Matthews, 1975). Viewed yet more generally, with the known projection of muscle afferents to the higher levels of the nervous system, any tonic stretch reflex in the whole animal may be suspected to have supraspinal as well as spinal contributions, as specifically suggested by Phillips (1969) on the basis of the establishment of a Ia input to cortical sensory area 3a. In the decerebrate cat, however, a cortical contribution is eliminated, and a cerebellar contribution seems unlikely since the reflex is basically similar after cerebellectomy, though usually with a lower threshold. It is, however, certain that the spinal cord itself contains all the basic machinery required to give a tonic stretch reflex provided that it is suitably facilitated; how far the requisite "facilitation" is that of the motoneurons and how far that of interneurons (with the switching off of inhibitory interneurons) remains an open question. Such "facili-

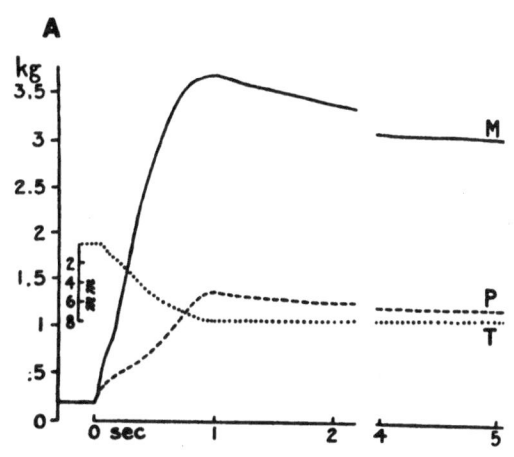

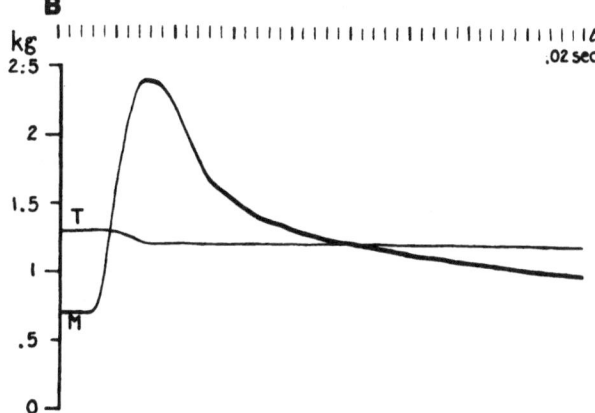

Fig. 11. Classical myographic records of the stretch reflex of the quadriceps muscle in the decerebrate cat showing different degrees of development of the tonic and phasic components of the reflex in different preparations. M = muscle tension; T = muscle extension ("less than 1 mm in B"), P = passive tension produced by the same extension after muscle denervated (not shown in B, but may be presumed to be small). From Liddell and Sherrington (1924).

tation" can be provided in the absence of the higher centers by giving the acute spinal preparation, which is normally totally flaccid and without a tonic stretch reflex, one or other of the drugs (DOPA or clonidine) that mimic or support the action of the catecholaminergic fibers found in the spinal cord (Grillner, 1969). The acute spinal preparation possesses a clear tendon jerk, so the failure to have a maintained stretch reflex or a tonic vibration reflex is another indication of the restriction of vision introduced by equating the "stretch reflex" with the tendon jerk.

THE PURELY MUSCULAR CONTRIBUTION

Reflex action, however, is not the sole source of the resistance of muscle to stretching; a contribution is provided by its own innate properties. First, there are the classical length–tension properties of muscle whereby the muscle comes to deliver its maximum contractile tension at a length close to the maximum length that it occupies *in situ*. Thus on starting from any shorter length, stretching a muscle that is contracting in response to a constant motor discharge automatically leads to an increase in its contractile force, without the intervention of nervous regulation. The precise form of the length–tension relation is complex and depends on the frequency of stimulation used to activate the muscle (see Chapter 1, in this volume). However, a linear approximation can often be fitted to the curve in regions well away from the plateau that occurs near physiological maximum; the lower the frequency of activation the more pronounced is this increase of muscle strength with muscle length, and the less pronounced is the plateau (Joyce *et al.,* 1969; Matthews, 1959). Second, muscle shows a short-range rigidity, so that it resists small pertubations applied dynamically much more vigorously than would be expected from the tension–length curve determined under static conditions (Joyce *et al.,* 1969; Nichols and Houk, 1976). Thus slow stretching leads to an initial rapid increase in tension followed by a yielding, as in Fig. 12, and slow release leads to a rapid initial fall in tension followed by a slower decline. This nonlinearity of muscle action is particularly well seen on low frequencies of activation, as often occur in life, and probably arises from the slowness with which the cross bridges between the actin and myosin filaments can rearrange themselves

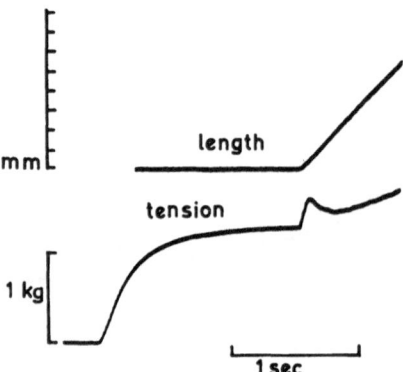

Fig. 12. The effect of forced lengthening of a contracting muscle in producing an initial overshoot of tension, showing that the muscle has an initial region of high stiffness that will help it resist deformation quite apart from the stretch reflex and before the reflex can "come to its help." The tension record shows the contraction of the soleus muscle of the cat elicited by stimulation of severed ventral root filaments. From Rack (1970).

with change of muscle length; while they remain attached the muscle is much stiffer than when they are moving their point of attachment to permit appreciable change in the length of the muscle. This short-range stiffness is applicable to about 1% of the length of the muscle, and thus should play a part in resisting the initial part of any applied mechanical disturbance. It will be particularly effective when the disturbance is applied by a system which is itself compliant since the relatively high rigidity of muscle means that the "displacement" will be absorbed in the external system as much as by the muscle, which can thus maintain its preexisting length more effectively than if it did not show this initial rigidity. Such an ability of muscle on its own to resist deformation may be seen as the first line of defense in its maintenance of a contant length, before reflex action has had time to come to its help.

The ensuing stretch reflex is normally and reasonably seen as a further device for maintaining the constancy of muscle length. However, Houk and his colleagues (Crago, Houk, and Hasan, 1976; Houk, 1978; Nichols and Houk, 1976) have recently argued forcefully that this is the wrong way of looking at things and that the function of the reflex is to linearize the behavior of muscle and maintain the initial high value of stiffness for ranges under which muscle itself is unable to do so. The advantage of such linearization, they suggest, is that any higher-level motor controller would have at its disposal a motor (i.e., the reflexly controlled muscle) with simple length–tension properties so that less information would have to be dispatched from above downward to achieve any particular mechanical end. In support of their view they emphasize the long-known remarkable linear relation often found between length and tension in the stretch reflex of the decerebrate cat, whether elicited dynamically or statically, which, as recognized before, is most simply explained by the reflex centers striking a balance between length feedback from the spindles and tension feedback from the tendon organs. Because of the complexity of the properties of muscle itself, linearity would not readily rise simply by augmenting the motor discharge in direct proportion to the degree of stretch (cf. Matthews, 1959). They also emphasize that for large stretches in both cat and man the "stretch reflex" must be very asymmetrical with regard to stretch and release to compensate for the quantitative asymmetry shown by the contractile properties of muscle. Thus, for Houk, the "function" of the stretch reflex is not to resist extension *per se* but to regulate reflexly the stiffness of muscle (which is a restatement of the experimental findings), and iconoclastically, it is suggested that the "main function" of autogenetic reflexes is to "compensate for variation in the properties of skeletal muscle rather than to oppose changes in load." To many, however, deprived of the philosophical trappings, these important ideas do not seem to exclude most that has been believed before.

THE STRETCH REFLEX DURING VOLUNTARY CONTRACTION

The stretch reflex has long been seen as a component of posture, particularly because in the decerebrate cat the overactive stretch reflex occurs in the postural antigravity muscles and leads to a caricature of standing. In addition, it may be suggested to be present during most normal muscular acts, so that unexpected change in the length of the muscle leads to reflex alteration of the strength of its contraction. With the muscle relaxed, there is little sign of this in the normal subject. When the muscle is active, however, the classical phenomenon of the unloading reflex, in which sudden release of a muscle leads to a brief cessation in its electromyographic activity, as in Fig. 13, argues that beforehand it was receiving a continuous stream of excitation reflexly initiated from within itself. In other words, when a muscle is contracting it does have a tonic stretch reflex.

In the last few years there has been an intensive study of the effect of unexpected muscular stretching, by limb displacement, in both man and the conscious monkey with a number of important findings. Such work was initiated by Hammond in the 1950s (1956, 1960), but a decade elapsed before the matter was taken up in more detail by others. Hammond asked his subjects to develop a steady force at the elbow and then unexpectedly pulled their forearm away by engaging a clutch on an electric motor so as to stretch the already contracting muscle and to see if it had a stretch reflex. This simple question did not, however, lead to a simple "answer." First, although the monosynaptic electromyographic wave of the tendon jerk made its expected appearance with a latency of about 20 msec, it failed to evoke any appreciable rise in tension, thereby suggesting that in mechanical terms it was insignificant. Second, following this initial wave, there was a period of electrical silence, although the muscle stretching continued, and it was not until 45 msec from the beginning that significant electromyographic activity developed and the muscle tension rose to resist the pull reflexly. Subsequently, on making a basically similar observation on stretching the triceps surae by sudden foot dorsiflexion, Melvill Jones and Watt (1971) gave the delayed EMG response the useful name of the *functional stretch reflex* (FSR). Third, the subject's main reflex response was influenced by the prior instructions given to him as to how he should meet the disturbance. If he was told to "resist" the pull, the main EMG response was large and prolonged; but if he was told to "let go," the main response was cut off almost as soon as it had begun. The two responses deviated at 50–60 msec from the begin-

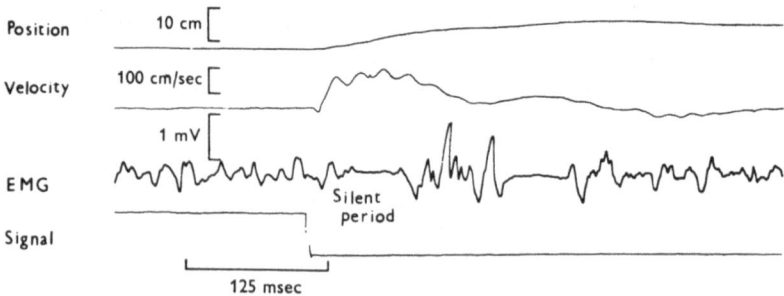

Fig. 13. The silent period that occurs on the unloading of a voluntary contracting muscle. At the beginning of the trace, the subject was using the flexor muscles of his forearm to hold up a weight. At the point shown in the bottom trace the weight was suddenly released so that the subject's forearm began to flex, as shown in the upper two traces. The EMG record, taken with surface leads, shows that about 40 msec after the beginning of the release the previous continuous electrical activity is replaced by a short-lasting "silent period" that is so early that it must have been mediated reflexly rather than voluntarily. The later responses, however, do not lend themselves to ready analysis. From Alston, Angel, Fink, and Hoffman (1967).

ning, whereas the latency for a truly voluntary response initiated by a tap of the arm was over 90 msec, thus making it clear that the "let go" and "resist" responses differed from a straightforward voluntary reaction. On all this Hammond suggested that much of the main "stretch reflex" could have been mediated by a supraspinal pathway rather than simply by the cord, and that the relevant reflex centers, be they spinal or supraspinal in their location, could be preset in their excitability by voluntary intent. The field has continued to be preoccupied with these points, and it is premature to attempt a final balance.

A major contribution to the continuing investigation of these "servo" effects has been provided by Marsden, Merton, and Morton (1972, 1976a,b, 1977) who have turned their attention to what happens when the course of a voluntary tracking *movement* is disturbed rather than simply studying the effect of perturbation on a steady isometric contraction. For muscles of the upper limb acting at the shoulder (supraspinatus, infraspinatus, pectoralis major) or at the elbow (biceps), the situation remains much as expected from Hammond's work. Release of the steady force that the muscle is working against during its tracking leads to a decrease of electromyographic activity, and forcibly stretching the muscle elicits increased EMG activity, as also does just halting the progress of the movement without actually elongating the muscle. This last effect especially can be viewed as demonstrating a degree of servo-assistance of the movement. In these human "proximal" muscles, the initial response to stretching normally occurs with a short latency and one that is approximately the same as that of the tendon jerk. At about twice this time, there is a further rather large EMG response that is thus a candidate for having been relayed supraspinally by a "long loop reflex." The responses to halt and to release could sometimes be observed at the short latency, but often were clearly detectable only at the longer latency. In two "distal" muscles, however, namely those controlling flexion of the thumb or of the big toe (flexor pollicis longus and flexor hallucis longus, respectively), the monosynaptic component was always absent for modest rates of stretch, as well as for "halt" and "release," whereas the delayed component persisted for all three maneuvers, thus apparently permitting the study of the "long loop reflexes" in isolation from spinal reflex effects (see Fig. 14).

The delayed servo response is shown to depend on intra-muscular receptors, presumably the muscle spindle afferents, because it may persist in both the thumb and the big toe after locally anesthetizing the moving digit so that the source of the feedback as to what is happening is restricted to the stretched muscle. The toe response is virtually unaffected by the anesthesia, but the thumb response is very greatly diminished and in the early experiments of this kind was totally abolished. Since these experiments have been restricted to very few subjects, the present detection of a thumb response on anesthesia is attributed to central "habituation" of the subjects on repeated anesthesia or else to a form of training produced by constant repetition of the task. The responses of the muscles acting at the shoulder are not influenced by making the arm anesthetic so, all in all, it would appear that the thumb is a special case. The locus of the lowering of "servo gain" produced by thumb anesthesia remains obscure; it seems unlikely to be the muscle spindle (by deprivation of fusimotor outflow) or the motoneuron (by lowered excitability) since the thumb jerk elicited by a very rapid stretch may persist unchanged when the delayed servo responses have been grossly diminished by local anesthesia. For the human thumb, the central "gain" of the reflex appears to increase *pari passu* with the level of preexisting motor activity, since the proportional change in the EMG for a given mechanical maneuver remains approximately constant when the background level of motor firing is increased. This is so both for a fatigued muscle when the firing is increased in order to maintain a standard tension, and for a normal muscle when the increased firing is producing a greater tension; *inter alia,* the

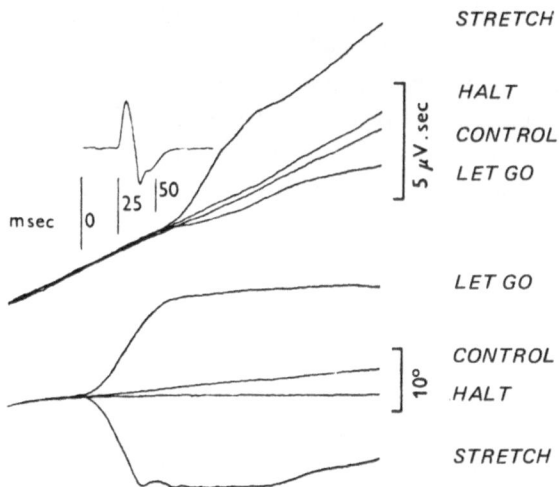

Fig. 14. "Servo" responses seen electromyographically in the long flexor of the human thumb, and contrasted in their latency with a finger jerk of the same limb ("tulips" in the jargon of the authors). *(Top left)* Finger jerk recorded electromyographically with surface leads showing a latency of 25 msec, which is the value appropriate for monosynaptic transmission. *(Middle)* Integrated electromyographic responses evoked by various mechanical maneuvers; their point of departure from each other at 50 msec from the beginning of the stimuli indicates that their central latency is too great to be attributed simply to Ia monosynaptic action. *(Bottom)* Mechanical stimuli used to elicit the various servo responses. The records show the angle at the interphalangeal joint of the thumb while the subject was making a voluntary flexion planned to be at a constant velocity of movement of about 20°/sec (the whole trace lasts 250 msec). In the control run, he did this against a load that remained constant throughout and was reasonably successful in achieving his objective. In the "stretch" run, the movement was thrown back on itself by suddenly increasing the load so as to stretch the contracting muscle, and in the "release" run the movement was brought to run away with itself by suddenly decreasing the load into which the muscle was working and so providing a "let go." In the "halt" run the load was made to simulate a very stiff spring so that the subject's thumb was brought to a virtual standstill. All these maneuvers were performed by varying the current in an electric motor, which provided the resistance against which the subject was working. The electromyographic traces *(middle)* were obtained by first rectifying and then integrating the responses obtained with surface electrodes (in their raw state these responses will have been like those in Fig. 13). Thus, continued steady activity, as in the control record, leads to a straight line. Increase of EMG activity is indicated by an increase of slope, as in the stretch record, and decrease of activity as a diminution of slope, as in the release record (complete silence cannot have occurred here since it would be indicated by a short horizontal segment). Vertical shift of a record from the control, without change of slope, merely indicates that the EMG activity in the two cases had differed at some earlier stage. Each trace represents the computer average of 32 separate trials of the testing procedure randomly interspersed among the other conditions. From Marsden *et al.* (1976b).

former observation shows that the requisite stimulus for the servo cannot be the tension changes produced by the maneuvers since these will be unchanged in the fatigued state.

The occurrence of a degree of servo action, occurring too early to be a normal voluntary response, seems amply established by all the electromyography. The mechanical effectiveness of the servo, however, awaits full delimitation and on present showing may prove to be relatively weak. In two quite separate situations the increment of tension produced by reflex activity has been compared with that produced simply as a consequence of the mechanical properties of muscle, and the two values found to be of the same order. This was done by measuring first the overall tension increment due to the reflex plus the muscle properties, and then repeating the measurement when the reflex had been abolished; the purely reflex contribution was then determined by subtraction. Marsden *et al.* (1977) did this for the human thumb during their "halting" maneuver, as in Fig. 14, and abolished the reflex by local anoxia. Goodwin, Hoffman, and Luschei (1978) did it for the jaw-closing muscles of the conscious monkey, which was trained to exert a constant mean force while the jaw was being moved sinusoidally at various frequencies, and they eliminated the reflex by destroying the mesencephalic nucleus of the Vth nerve. Analogously, for the human shoulder, Allum (1975) measured the tension evoked by unexpected stretching and found only a modest effect attributable, by virtue of the time of its occurrence, to the electromyographically observed "servo" response and which was comparable to that attributable to the innate visco-elasticity of the already contracting muscles, recognizable by being developed from the very onset of stretch.

Whether the main delayed servo response is transmitted supraspinally is occupying considerable attention. As required for this, its latency is found to vary with the central transmission distance from the point of entry of the segmental afferent volley to the motor cortex (jaw less than thumb less than big toe); also the latency for proximal arm muscles is only slightly less than distal arm muscles thus making it unlikely that the delay occurs peripherally or within the cord. In addition, lesions of the dorsal columns or of the sensorimotor cortex or of the internal capsule, all of which could reasonably be suggested to interfere with the long loop, presumed via the motor cortex, could block the delayed servo response, moreover often without abolishing the tendon jerk (Marsden, Merton, Morton and Adam, 1977a,b). The existence of suitable fast conduction paths from the muscle afferents to the cortex had earlier been amply established in the primate and had of itself already led to the suggestion that a stretch reflex servo response might be mediated by the motor cortex (Phillips, 1969). In other recent work, the human electromyographic responses to perturbation of a steady position have been divided into M1, M2, and M3 waves and provisionally and respectively attributed to the spinal reflex, a long-loop reflex via the cortex, and the beginning of a "voluntary" response (Tatton and Lee, 1975). In a similar situation in the trained conscious monkey, Evarts and Tanji (1976) have observed firing of neurons projecting in the pyramidal tract (PT cells) that occurred sufficiently long after the beginning of the perturbation, yet sufficiently before the delayed muscular servo response to make it probable that they were watching the cortical reflex in action. Moreover, Tatton,

Forner, Gerstein, Chambers, and Lin (1975) found that a cortical lesion eliminated the M2 response but not the M1 response in the trained monkey. Thus, although the matter is not proved, there is very reasonable evidence supporting the idea that a "long-loop reflex" via the cortex plays a part in mediating the delayed human servo response.

MATTERS SHOWING NEED FOR CAUTION IN THE INTERPRETATION OF DELAYED RESPONSES

The acceptance of the idea that a response depends in certain crucial respects on transmission around a "long loop reflex" in no way excludes the possibility that spinal reflexes also contribute essential elements to the overall response. Indeed, it would be remarkable if they did not do so, since the afferent barrage originating in the stretched muscle is continuing to arrive at the spinal cord at the time when the "long loop" volleys eventually get back to the spinal motoneurons. Thinking along these lines, Vilis and Cooke (1976) studied the effect in the conscious monkey of reversing the stretch almost as soon as it was begun so as to cut off the spinal input before the cortical volley returned. The M2 electromyographic response, attributed to cortical activation, was then greatly diminished in size although on latency considerations the "long loop" components of excitation should have been unaffected by truncation of the afferent input, thus arguing that there was a significant spinal contribution to the response. But the long-loop pathway can be powerful enough to produce some firing on its own, since occasionally a double EMG wave (representing presumed spinal and supraspinal components) has been seen in response to a single brief tendon tap (Iles, 1977; Marsden *et al.,* 1977). In addition, appropriate double responses may be seen in a muscle following a single electrical shock applied to its nerve during voluntary contraction (V1 and V2; Iles 1977; Upton, McComas, and Sica, 1971).

It requires emphasis, however, that the spinal cord acting on its own is quite capable of delivering a response with an appreciable latency, whether in isolation or more usually following an initial response. Thus, Ghez and Shinoda (1978) have observed M1, M2, and M3 responses from the forelimb of the conscious trained cat that appear comparable to those in man and monkey, yet in the cat three EMG waves of appropriate latency may persist after rendering the animal decerebrate or even spinal, thus ruling out cortical transmission as an essential for the delayed responses. This, of course, in no way precludes the suggestion that when present the cortex may contribute. Their work also throws emphasis on the elementary fact that the occurrence of several somewhat separate electromyographic waves in a response does not necessitate the postulating of a number of separate excitatory pathways each with the appropriately staggered delay. On the commencement of a steady afferent stream impinging on a motoneuron pool, the initial synchronized firing of a number of the most excitable motoneurons may lead to a Renshaw inhibition of the remainder, so ensuring a short delay before any further motoneurons fire, which is then followed by a further delay thus giving a series of waves. Moreover, any neurons that fire at the very beginning will, by virtue of their afterpotentials, be debarred from firing again for an appreciable time and may then come to fire again together in approximate synchrony. But, in addition to this, two cases have been established in which the spinal cord can either of itself or with its immediate supraspinal structures initiate firing in a given motoneuron pool with a long latency in response to a brief stimulus. First, after the intravenous injection of DOPA into the spinal cat stimulation of FRA may lead to motor firing after a delay of some tens of msec (Jankowska, Jukes, Lund, and Lundberg, 1967) instead of the normal msec or two. This may, perhaps, relate to the activation of coordinated pools of interneurons involved in walking. Second, in the decerebrate preparation both before and after cerebellectomy, a brief period of vibration of quadriceps (or Ia stimulation of its nerve) may initiate tonic firing in soleus with a latency of some 70 msec (Hultborn and Wigström, 1977). Thus there is no necessity to postulate a supraspinal and especially a cortical pathway simply on the basis of a prolonged latency of an EMG wave, particularly when it is preceded by earlier waves; but, of course, a long-loop reflex does display itself in this way, and so the matter remains open for detailed investigation in each particular case.

Another matter on which discussion continues is the extent to which the various components of the delayed motor response are under "voluntary" control, in the sense that their magnitude and appearance vary with the subject's intention whether or not to resist a perturbation when it occurs. In the monkey, Evarts and Tanji (1976) describe an initial PT neuron response that is directly related to the direction of the stimulus (muscle stretch or release) and largely unaffected by how the animal is required to respond; this may thus be considered simply as "reflex." Blending into this initial response is a later response that varies with the monkey's "intent," which may be considered more or less at choice either as a modifiable reflex or as a "voluntary response," albeit one that occurs with a shorter latency in response to the proprioceptive stimulus than it would to a visual or auditory stimulus. In the analysis of human electromyographic responses, the field is still at the stage of debating which of the various components can, in fact, be modified by prior instruction. The situation is complicated by the finding that responses of relatively short latency, below the supposed minimum voluntary reaction time, can occur in muscles quite other than the one being stretched, including contralaterally. Moreover, the effect depends on how the "target muscle" is being used (Marsden, Merton, and Morton, 1978; see also Chapter 10 in this volume). There has been an unwarranted tendency to interpret any delayed electromyographical wave, simply by virtue of its appearance, as the response of a "servo" with properties not unlike the tonic spinal stretch reflex and with its output graded in direct proportion to the stimulus. But, as Crago et al. (1976) have emphasized, some of the responses may be better seen as a triggered item of a behavioral response that has been preprogrammed and is then run off with relatively little reference to the size of the triggering proprioceptive stimulus. The latency of such triggering by a proprioceptive input that is immediately routed to the motor areas could well be appreciably less than the classical reaction time for other types of stimuli; the latter has tended to provide the touchstone as to whether or not a motor response could be considered as "voluntary" rather than "reflex," but this approach now appears unduly simple. Be this as it may, the signals from muscle afferents should be suspected of contributing far more to ongoing motor control than a simple spinal "resistance reflex," which in an evolutionary sense was probably their original purpose as such responses occur so widely in the animal kingdom. Moreover, that "electrophysiologists' mammal"—the cat—may well differ considerably from man in the degree of encephalization it has adopted for routine tasks.

ROLE OF THE FUSIMOTOR FIBERS

The value to the body of its possessing a dual system of fusimotor control, namely, the γ_S and γ_D axons, continues to defy precise understanding as indeed do the reasons for the existence of a specialized fusimotor system at all. On the discovery of the gamma system, Leksell (1945) wisely contented himself with the statement that "it is concluded that the gamma fibres serve as regulators of sensory activity originating in the muscle" (p. 78). Three main lines of thought may be distinguished in the subsequent attempts to read functional meaning into the working of the fusimotor system; they continue to coexist in relative harmony and indeed blend into each other, though at times some have talked as if they are totally irreconcilable, as they are in their extreme forms. In outline, in order of their historical appearance, these are as follows. First, fusimotor activity may serve to maintain the constancy of calibration of the muscle spindle, viewed as a measuring instrument, in the face of various disturbances and particularly of extrafusal contraction. Second, the fusimotor fibers may provide the input to some sort of servo, based on the length regulating properties of the stretch reflex, though in so far as they do so they are now recognized to work in collaboration with the α system to provide "servo-assistance" rather than acting on their own. Third, they may provide what can be called "parameter control" of the spindle and adjust its responsiveness to external mechanical stimuli to suit the prevailing motor task. These will now be discussed *seriatim*.

CALIBRATION MAINTENANCE

On the basis of their pioneering work stimulating single fusimotor fibers, Kuffler *et al.* (1951) suggested that fusimotor "activity would enable the same muscle stretch to cause a similar increment of sensory discharge at different initial tensions, thus providing a peripheral adjustment for maintaining the constancy of the reflex arc in the face of different conditions" (p. 52) and likewise that fusimotor action could be "interpreted as a compensatory mechanism to adjust for mechanical change and to ensure a more constant response to given amounts of stretch" (Kuffler and Hunt, 1952, p. 42). In essence, though this was never explicitly formulated, the fusimotor system was suggested to maintain the constancy of calibration of the muscle spindle as a measuring instrument, particularly during muscle contraction when the spindles are otherwise silenced by the internal muscular shortening that occurs even under conditions that are externally isometric. This view has gained a new lease of life with the recognition that the high sensitivity that the spindle primary ending shows to small stimuli is better maintained at different muscle lengths when the fusimotor fibers are active than when the spindle is behaving passively (Goodwin, Holliger, and Matthews, 1975). Thus fusimotor activity may also be presumed to help in the resetting of the spindle that occurs so as to maintain the initial region of high sensitivity when the length of the muscle is altered.

Servo Input

Such a restricted view of fusimotor function was almost immediately overtaken by Merton's (1953) persuasive and far-reaching suggestion that the gamma fibers were a normal route for the initiation of movement and, by biassing the spindle, frequently formed the sole input to a follow-up servomechanism in which the activity was then channelled to the α motoneurons by the Ia fibers acting monosynaptically, as illustrated in Fig. 15. The closed loop self-regulating properties of the spinal stretch reflex were seen as ensuring that a given fusimotor discharge was reflexly transmuted into a given shortening of the main muscle, namely, that required to restore the Ia firing almost back to its previous level. Provided that the gain of the servo (sensitivity of the stretch reflex) was sufficiently high, this would occur automatically, irrespective of any load on the muscle or its weakening by fatigue. This was suggested to contrast with the effect of a command signal that impinged directly on the α motoneurons, which was thought to produce a movement lacking these advantages because it was unmodulated by spindle feedback. But this would be so only in the special case in which the background fusimotor activity was insufficient to prevent silencing of the spindle, and thus servo control of a movement can be envisaged whether or not the prime command is delivered to the α or to the γ fibers; an input restricted to the α fibers just has to be made

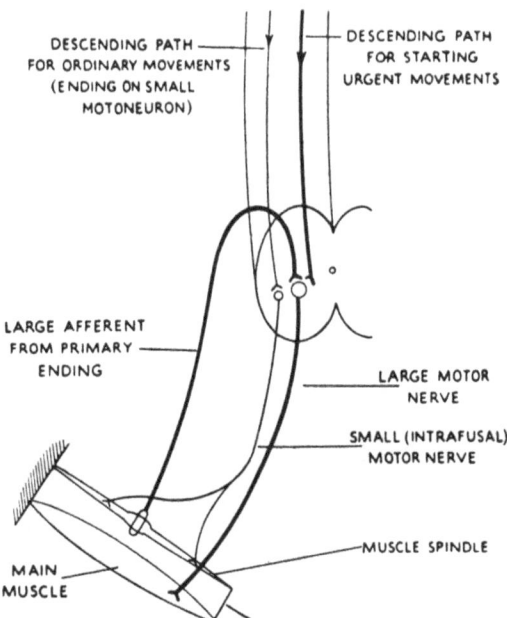

Fig. 15. Merton's original (1953) diagram of the two possible motor pathways by which the higher centers may be suggested to command a movement. The round-about route via the small gamma motoneurons and the large spindle afferents was suggested to form a "follow up length servomechanism," with a given movement occurring whatever the loading on the muscle or its state of fatigue.

DESCENDING PATH
FOR ORDINARY MOVEMENTS
(ENDING ON SMALL
MOTONEURON)

DESCENDING PATH
FOR STARTING
URGENT MOVEMENTS

LARGE AFFERENT
FROM PRIMARY
ENDING

LARGE MOTOR
NERVE

SMALL (INTRAFUSAL)
MOTOR NERVE

MUSCLE SPINDLE

MAIN
MUSCLE

sufficiently powerful to counteract a partial withdrawal of the Ia reflex activation on spindle slowing.

The evidence adduced in favor of the idea of servo control was twofold. First, at the inception of the hypothesis, detailed study of the silent period occurring in the human EMG during an evoked muscle twitch suggested that a maintained voluntary contraction was under continuous reflex control. Although this has been amply validated by subsequent experiemnt (see earlier), it can now be seen to be largely irrelevant to the question as to how far the command for movement is transmitted via the gamma fibers. Second, as illustrated in Fig. 16, experiments on decerebrate cats showed that Ia firing frequently preceded a reflexly elicited contraction and remained high during the time the muscle was contracting, when the passive spindle would have slowed or been silent. Moreover, interruption of the reflex arc by section of the dorsal roots could abolish the overt reflex but not the augmented Ia firing produced by the enhanced fusimotor activity. Various other experimental situations were described subsequently in which spindle firing was the opposite of that expected for the passive spindle or one with a constant degree of fusimotor activation, most notably for the respiratory muscles (Euler, 1966). Such findings, however, though necessary for the servo hypothesis, are not sufficient to establish it. The increase in spindle firing must, *inter alia,* be shown to occur at just the right time and be of just the right magnitude to elicit reflexly the observed contraction, and this has never been attempted in any quantitative manner. On present evidence in both man and cat, the gain of the stretch reflex is far too low for the limited increase in Ia firing to be held responsible for the concomitant contraction (Matthews, 1972; Vallbo, 1974b). The matter is complicated by the certainty that fusimotor activity excites secondary endings as well as primary endings; yet although it is improper to ignore their role, there is very little positive

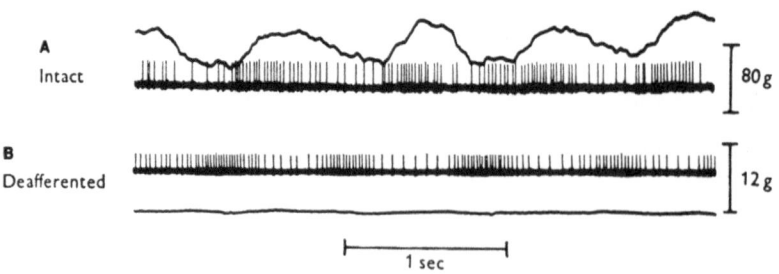

Fig. 16. Classical evidence for the occurrence of excitation via the γ route The contraction of the soleus muscle and the discharge of one of its Ia afferents were recorded in the decerebrate cat while the animal's head was turned rhythmically from side to side to activate the muscle reflexly. The ventral roots were intact throughout, but the dorsal roots were cut between (A) and (B). With the reflex arc intact, the Ia firing led the contraction and accelerated during it, instead of slowing as it would have in the passive spindle; this shows that the fusimotor neurons had been coactivated with the α motoneurons and sufficiently strongly to offset the unloading action of the overt contraction on the spindle (recording isometric). In (B), after deafferentation, the overt contraction was absent but the spindle response continued very much as before. The Ia discharge was recorded from a thin dorsal root filament, but for (A) the remainder of the appropriate dorsal roots were left intact. From Eldred, Granit, and Merton (1953).

that can be said in this respect. Moreover, the idea that any maintained stretch reflex compounds the activity traversing different routes within the CNS, including via the cortex, removes the very precision from the idea of servo control that helped to make the hypothesis so attractive, when regulated behavior seemed to flow from the very simplest of neural connections.

In spite of all this and because of its great explanatory power, the early version of the servo hypothesis gained much wider credence than was ever justified by the direct evidence in its favor. But by achieving such prominence, it stimulated quantitative thought about the working of the reflex arc and thus the performance of further experiments. Also, in its modified forms, it continues to provide a synthetic approach in a field that is so rich in disparate data. The recent spindle recordings in the conscious animal, which invalidate the simple follow-up servo hypothesis, will be dealt with shortly. Here it suffices to note that movement is well known to continue after local deafferentation by dorsal root section. In the particular case of chewing, the spindle afferents may be eliminated without all the other regional afferents by making a lesion in the mesencephalic nucleus of the Vth nerve, which is where their cell bodies lie; the effect on the normal performance of movement is then found to be negligible (Goodwin and Luschei, 1974). However, given the learning abilities of the CNS, the retention of very reasonable motor function after recovery from a lesion provides only limited evidence on the normal working of the system that has been interferred with; the stimulus of necessity may have led to its function being taken over by some other system although this seems unlikely in Goodwin and Luschei's experiments since motor function was retained immediately postoperatively.

PARAMETER CONTROL

With the subdivision of the gamma system into separable γ_S and γ_D axons, it was immediately obvious that they must be subserving quite separate functions. Not only were their effects on the spindle so different, but there was also evidence that they could be activated independently by the CNS. Up to this time (1962), the chief action of fusimotor stimulation had seemed to be a simple excitation of the afferent terminals that was not so very different from that produced by an externally applied stretch. But such a view was inapplicable to dynamic fusimotor action in which there is relatively little direct excitation of the primary ending yet a powerful sensitization of its response to a large dynamic stimulus, as described earlier. Moreover, dynamic action was seen to be incapable of acting in any way as a command for shortening in a length servo, since it was incapable of preventing the Ia afferent from falling silent during releases of quite moderate velocity (Lennerstrand and Thoden, 1968). Thus dynamic action was felt to exist to control the dynamic responsiveness of the primary ending without appreciably biasing its firing or influencing the behavior of the ·secondary ending. This may be called "parameter control" in that the fusimotor action changes the parameters of response of the ending. As a corollary, it follows that static fusimotor action also should be expected to be important for its "parameter control" action in setting the dynamic responsiveness of the afferent endings to a low value, as well as for its action in biasing the firing of both primary and secondary endings by direct excitatory action.

The next question is to ask just which parameter or parameters are controlled by fusimotor action. Unfortunately, there is still no precise answer because of the difficulties of comprehensively describing the nonlinear behavior of the endings by a simple equation with a limited number of separable parameters. Indeed, the attempt to do so with mathematical precision may even tend to obscure physiological reality, which at the present stage of understanding may perhaps be better approached qualitatively. Thus, for many purposes, it suffices to say that dynamic action enhances or at least maintains the responsiveness of the primary ending to dynamic stimuli of almost every kind, whereas static action tends to lower the sensitivity. Such a loose statement provides quite reasonable help in thinking of the possible role of the two kinds of fusimotor fibers in regulating, for example, say, posture. But fusimotor action does not regulate the relative sensitivity of the afferent endings to the conceptually separate stimuli of the instantaneous values of the length and of the velocity occurring during the course of a movement, as in the very earliest stages of study it seemed possible that they might do (Jansen and Matthews, 1962). This idea fitted in with the then fashionable "follow-up servo hypothesis" since the γ_S fibers could provide the requisite biassing signal to command a movement and the γ_D fibers might, so it was thought, independently regulate velocity sensitivity in relation to the length sensitivity of the primary ending and thereby control the "damping" of the servo loop and its tendency to go into oscillation. Linear engineered servos can be controlled in just this way. But such a change in relative sensitivity to length and velocity would entail a fusimotor induced change in the phase of the response of the ending to sinusoidal stretching; but this has now been found to remain constant, as has been suspected for some time (Crowe and Matthews, 1964b).

Figure 17 illustrates the effect of fusimotor stimulation in altering the responsiveness of the primary ending to sinusoidal stretching of a wide range of frequency but of restricted size; the amplitude of movement was adjusted at each frequency to be small enough to fall within the linear range of response of the ending. At frequencies up to 30 Hz, which may be considered to be the limit of the range of interest for motor control, the phase is unchanged by either static or dynamic stimulation, and the absolute value of the sensitivity of the ending to the stretching is reduced by a constant amount irrespective of the frequency (Goodwin *et al.*, 1975). At higher frequencies of stretching a variety of changes do occur in both phase and gain, but these seem of interest for understanding the internal mechanisms of the spindle rather than for appreciating its role in the body. Similar results have been obtained with the secondary ending (Cussons, Hulliger, and Matthews, 1977). For larger stretches, falling within the range of much normally occurring movement, similar analysis has been complicated by the nonlinear behavior of the endings, but again there is evidence against any appreciable change of the phase of the response occurring on fusimotor stimulation (Hulliger, Matthews, and Noth, 1977). Now, however, dynamic action increases the sensitivity of the primary ending and static action decreases it as would be expected from the classic findings with large ramps. Thus a paradoxical feature of these experiments is that, in comparison with the passive behavior of the ending, dynamic action slightly diminishes the sensitivity of the primary ending in its linear range (Fig. 17) but increases it for larger stretches. This may relate to the precise method of measurement but is probably immaterial from the functional point of view when what seems important is that during dynamic action the sensitivity is always very much higher than it is during static action, irrespective of the amplitude of movement, while the phase of the responses is the same, indicating that the relative balance of length and velocity sensitivity is the same in the two cases. It should be emphasized, however, that the length sensitivity under consideration is not the absolute value of the position sensitivity seen when the muscle is held still, and on which fusimotor stimulation has a highly variable action; rather it is the value of the length sensitivity measured under dynamic conditions, which is that component of the response that increases linearly with the elongation and which, owing to the complexity of spindle behavior, is usually larger than the static length sensitivity. Thus there is no particular

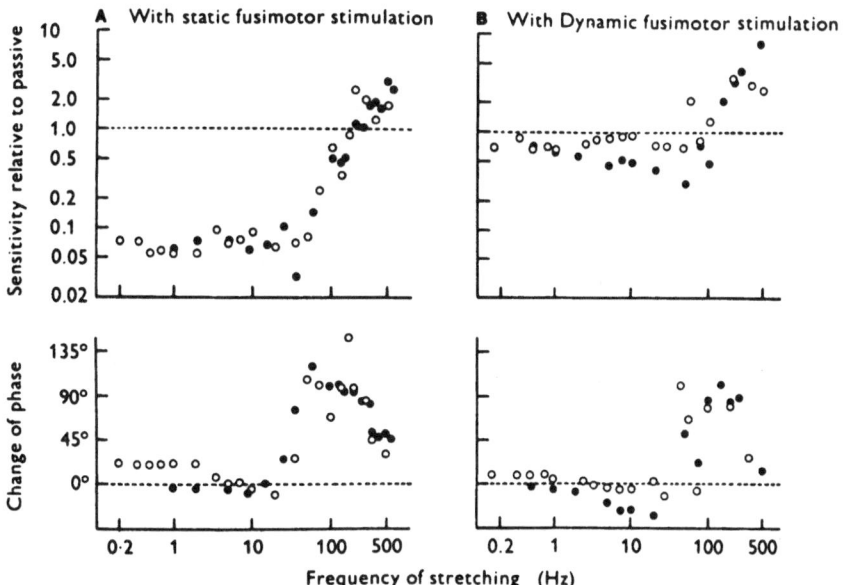

Fig. 17. The effect of fusimotor activation on the response of the spindle primary ending to small-amplitude sinusoidal stretching of a wide range of frequencies. At frequencies below 30 Hz, neither kind of fusimotor action influenced the phase of the response. The stretching modulated the spindle firing sinusoidally. The sensitivity of the spindle at each frequency was defined as the amplitude of this modulation divided by the amplitude of stretching; in the range studied, the endings behaved reasonably linearly. The phase of the response was determined relative to that of the sinusoidal stretching. The graphs show the *change* in these parameters induced by stimulating a single γ_S axon *(left)* or a single γ_D axon *(right)*: the results from two separate experiments have been included, in each of which the action of both a static and a dynamic axon were studied upon the same primary ending (equivalent symbols represent the same experiment). The upper graphs are a logarithmic plot of the ratio of the sensitivity of the activated spindle divided by the corresponding value for the passive spindle. The lower graphs are a linear plot of the difference between the phase in the two cases, with phase advance of the active response over the passive response being plotted upward. From Goodwin *et al.* (1975).

conflict between the present findings with sinusoidal stretching and experiments with ramp stretching in which the crude indicator of velocity sensitivity of the "dynamic index" is found to be increased on dynamic fusimotor stimulation, since such stimulation also markedly increases the position response measured dynamically (cf. Crowe and Matthews, 1964a; discussed in Matthews, 1978).

Thus, in summary as shown in Fig. 18, it seems clear that both static and dynamic action do have a significant "parameter control" action in regulating the overall sensitivity of the spindle to a wide range of dynamic stimuli. Static action reduces the sensitivity of the primary ending very appreciably and of the secondary ending rather less so, and in both cases also produces a significant biasing action of direct excitation. Dynamic action maintains the sensitivity of the primary ending at a high value unaccompanied by any appreciable biasing action, and is almost without effect on the secondary ending. The names "static" and "dynamic" still seem as reasonable for these two fusimotor systems as any others that come to mind, though care must be taken to use the term "dynamic sensitivity" in a broad context and to avoid equating it simply with "velocity sensitivity."

Biassing of Spindles during Motor Acts

Apart from its intellectual elegance, continued impetus was given to the servo hypothesis by the repeated observation that gamma firing often preceded alpha firing, and thus offered itself as the cause of the latter (cf. Fig. 16). It was, however, equally recognized that under many circumstances the alpha and the gamma fibers were coactivated (α–γ linkage) by reflex or central stimulation applied after the dorsal roots had been severed to eliminate the alpha firing having been elicited by activity traversing the spindle loop. The regular occurrence of coactivation is now coming to the fore, and indeed in man firm evidence has yet to be obtained for activation of γ fibers in isolation from α fibers. Thus there is now little or no possibility that gamma firing ever normally precedes alpha firing in sufficient degree to be the cause of the main motor discharge. A recent example is shown in Fig. 19, which documents the firing of three separate spindles lying in the triceps surae muscle of the cat (as in Fig. 16) during unrestricted walking in the fully

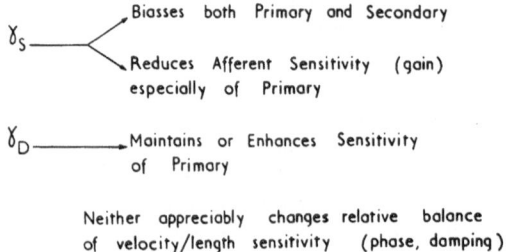

Fig. 18. Summarizing diagram of static and dynamic fusimotor actions that may be deemed to be of functional importance.

conscious state. Firing is always high during the initial part of the flexor phase (F) which is when the triceps muscle is being stretched by its antagonists and is itself relaxed. The expected stretch reflex at this phase would appear to have been annulled by inhibition. Spindle firing does occur in the extensor phases (E_1, E_2, E_3) when the muscle is contracting, but it varies markedly from example to example, and so its role in reflexly mediating the contraction is uncertain. In most of I, any servo drive by spindle biassing must have been negligible, but in III it remains a possibility. Similar results have been obtained for chewing in the monkey (Goodwin and Luschei, 1975). In isometric contractions in man, which provide a less searching test of servo ideas than isotonic movements, the biassing is comparatively moderate (Vallbo, 1974b). Even when a peripheral element of servo drive is present, as shown by augmented spindle firing, it remains unknown whether the central "gain" of the reflex centers is high enough for it to make an appreciable contribution to the movement. Thinking about this is, as already noted, complicated by the fact that such servo biassing must be applied by the static axons rather than the dynamic axons, and so equally affects the secondary endings, even though these may happen not to have been studied. The outcome of any enhanced firing produced by a fusimotor biassing command will thus depend crucially on the relative central actions of the two kinds of spindle afferent, and whether they are operating in synergy or opposition. This may well depend on the "set" of a variety of interneurons. Moreover, whatever the evidence for slow movements, the existence of appreciable servo drive has always seemed ruled out in the genesis of rapid "ballistic" movements.

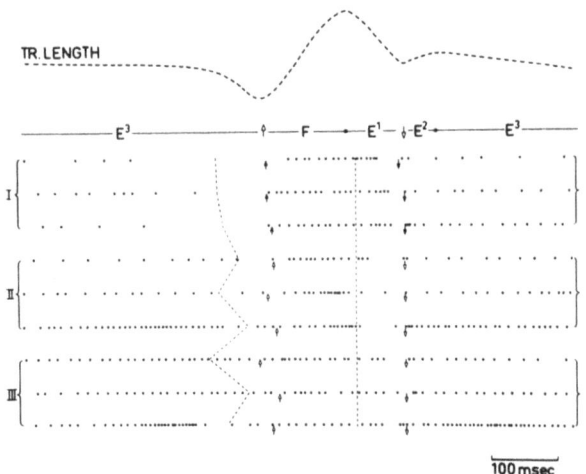

Fig. 19. Firing patterns of three ankle extensor spindle primary endings (I, II, III) during unrestricted walking of the conscious cat recorded with implanted electrodes and telemetry; they were not recorded simultaneously. Three separate examples are shown for each ending because of the variability. In contrast to Fig. 16, the maximum rate of spindle firing is often during muscle relaxation, with its consequent lengthening, rather than muscle contraction, though comparison of various responses shows clear signs of the coactivation of the α and the γ fibers. Each action potential is represented as a dot. The variation in the length of the muscle is shown schematically above, along with the division of the step cycle into three extensor phases (E_1, E_2, E_3) when the muscle was contracting, and a flexor phase when it was relaxed and electromyographically silent (the precise period of gastrocnemius EMG silence lies between the vertical dotted lines). The foot was in the air in the swing phase between the two arrows ($\uparrow$ lift off, $\downarrow$ touch down). From Prochazka, Westerman, and Ziccone (1976).

PETER B. C.
MATTHEWS

Thus, it remains an open question as to whether a partial servo drive ever commonly provides an appreciable contribution to a movement that is taking its expected course. In this sense there is very little evidence for the servo-assistance of movement. But in another sense, it would appear that servo-assistance is commonly present in that any change in the course of a movement leads to a change in spindle firing in such a direction that it will tend to oppose the change, so that the muscle servo will come to the assistance of the direct motor command and help in its due performance. As already noted, the existence of such assistance is now well established, although its neural routing and mechanical efficacy remain the subject of debate. From this point of view, what is perhaps principally asked of the gamma firing is not that it should provide a direct stimulus for the main contractile activity of the muscle, but that it should lead to the setting up of a simulacrum or model within the spindle to mirror the intended movement. At the same time, the higher centers would dispatch a direct α command that was appropriate for the expected external load. If the real load were unduly great, then the internal intrafusal model of the movement would, so to speak, get ahead of the real movement and so the spindle firing would increase and reflexly assist the movement. Conversely, if the real movement should proceed unexpectedly rapidly, then the normal level of spindle reflex excitation would be withdrawn. For such a mode of operation there would be no necessity for the spindle firing to increase during the course of the movement or even for it to be held at a particular value and the same for different spindles. It is merely required that the time course of fusimotor firing should reflect the intended movement sufficiently to ensure that the spindle does not fall too near to silent at any stage, else reflex servo control would inevitably fail.

A particular virtue of the idea that the spindle provides a model for the movement is that it provides a rationale for the independence of the alpha and the gamma systems. If the two are always to be coactivated in the same proportions, then the shared skeleto-fusimotor arrangement of the amphibian would appear to suffice for all biasing actions, although it could not provide for independent parameter control of sensitivity. But independence of the α and γ motor pathways would allow the message to the spindle to signal the desired trajectory of movement, whereas that to the extrafusal fibers would also have to take into account the external load and any fatigue they might be suffering (cf. Matthews, 1964). The idea that an important function of the spindle is to model planned movements retains its validity if the spindle messages simply inform the CNS how well a movement has been performed throughout its course and does not depend on whether or not this information is used for any immediate corrective action. The detection of an unexpected small signal that occurs in the presence of a large, but predictable, signal can be achieved expeditiously by a highly sensitive measuring instrument that is held within its working range by the application of a suitable backing signal to counteract the large signal whose time course is known. To attempt to achieve the same precision by having an instrument that would as effectively measure throughout the full range of the large signal would make very much greater demands on its performance and the accuracy of its calibration, as well as requiring much greater elimination of noise.

REFERENCES

Allum, J. H. J. Responses to load disturbances in human shoulder muscles: The hypothesis that one component is a pulse test information signal. *Experimental Brain Research,* 1975, *22,* 307–326.

Alston, W., Angel, R. W., Fink, F. S., and Hoffman, W. W. Motor activity following the silent period in human muscle. *Journal of Physiology (London),* 1967, *190,* 189–202.

Banks, R. W., Barker, D., and Stacey, M. J. Intrafusal branching and distribution of primary and secondary afferents. *Journal of Physiology (London),* 1977, *272,* 66–67.

Banks, R. W., Harker, D. W., and Stacey, M. J. A study of mammalian intrafusal muscle fibres using a combined histochemical and ultrastructural technique. *Journal of Anatomy,* 1977, *123,* 783–796.

Barker, D. The morphology of muscle receptors. In C. Hunt (Ed.), *Handbook of Sensory Physiology,* Vol. III/2, Muscle Receptors. Berlin: Springer, 1974.

Barker, D., Emonet-Dénand, F., Harker, D. W., Jami. L., and Laporte, Y. Distribution of fusimotor axons to intrafusal muscle fibres in cat tenuissimus spindles as determined by the glycogen depletion method. *Journal of Physiology (London),* 1976, *261,* 49–69.

Barker, D., Emonet-Dénand, F., Laporte, Y., Proske, U., and Stacey, M. J. Morphological identification and intrafusal distribution of the endings of static fusimotor axons in the cat. *Journal of Physiology (London),* 1973, *230,* 405–427.

Bessou, P., and Pagès, B. Cinematographic analysis of contractile events produced in intrafusal muscle fibres by stimulation of static and dynamic fusimotor axons. *Journal of Physiology (London),* 1975, *252,* 397–427.

Bessou, P., Emonet-Dénand, F. and Laporte, Y. Motor fibres innervating extrafusal and intrafusal muscle fibres in the cat. *Journal of Physiology (London),* 1965, *180,* 649–672.

Binder, M. D., Kroin, J. S., Moore, G. P., and Stuart, D. G. The response of golgi tendon organs to single motor unit contractions. *Journal of Physiology (London),* 1977, *271,* 337–349.

Boyd, I. A. The response of fast and slow nuclear bag fibres and nuclear chain fibres in isolated cat muscle spindles to fusimotor stimulation, and the effect of intrafusal contraction on the sensory endings. *Quarterly Journal of Experimental Physiology,* 1976, *61,* 203–254.

Boyd, I. A., Gladden, M. H., McWilliam, P. N., and Ward, J. Control of dynamic and static nuclear bag and nuclear chain fibres by β and γ axons in isolated cat muscle spindles. *Journal of Physiology (London),* 1977, *265,* 133–162.

Brown, M. C., and Butler, R. G. Studies on the termination of static and dynamic fusimotor fibres within muscle spindles of the tenuissimus muscle of the cat. *Journal of Physiology (London),* 1973, *233,* 553–573.

Burke, D., Hagbarth, K.-E., Löfstedt, L., and Wallin, B. G. The responses of human muscle spindle endings to vibration during isometric contraction. *Journal of Physiology (London),* 1976, *261,* 695–711.

Crago, P. E., Houk, J. C., and Hasan, Z. Regulatory actions of the human stretch reflex. *Journal of Neurophysiology,* 1976, *39,* 925–935.

Crowe, A., and Matthews, P. B. C. The effects of stimulation of static and dynamic fusimotor fibres on the response to stretching of the primary endings of muscle spindles. *Journal of Physiology (London),* 1964a, *174,* 109–131.

Crowe, A., and Matthews, P. B. C. Further studies of static and dynamic fusimotor fibres. *Journal of Physiology (London),* 1964b, *175,* 132–151.

Cussons, P. D., Hulliger, M., and Matthews, P. B. C. Effects of fusimotor stimulation on the response of the secondary ending of the muscle spindle to sinusoidal stretching. *Journal of Physiology (London),* 1977, *270,* 835–850.

Duysens, J., and Pearson, K. G. The role of cutaneous afferents from the distal hindlimb in the regulation of the step cycle of thalamic cats. *Experimental Brain Research*, 1976, *24*, 245–255.

Eccles, R. M., and Lundberg, A. Supraspinal control of interneurones mediating spinal reflexes. *Journal of Physiology (London)*, 1959, *147*, 565–584.

Eldred, E., Granit, R., and Merton, P. A. Supraspinal control of the muscle spindles and its significance. *Journal of Physiology (London)*, 1953, *122*, 498–523.

Ellaway, P., Emonet-Dénand, F., Joffroy, M., and Laporte, Y. Lack of exclusively fusimotor α axons in flexor and extensor leg muscles of the cat. *Journal of Neurophysiology*, 1972, *35*, 149–153.

Emonet-Dénand, F., and Laporte, Y. Proportion of muscle spindles supplied by skeleto-fusimotor axons (β-axons) in Peroneus Brevis muscle of the cat. *Journal of Neurophysiology*, 1976, *38*, 1390–1393.

Emonet-Dénand, F., Jami, L., and Laporte, Y. Skeleto-fusimotor axons in hind-limb muscles of the cat. *Journal of Physiology (London)*, 1975, *249*, 153–166.

Emonet-Dénand, F., Laporte, Y., Matthews, P. B. C., and Petit, J. On the sub-division of static and dynamic fusimotor actions on the primary ending of the cat muscle spindle. *Journal of Physiology (London)*, 1977, *268*, 827–861.

Euler, C. V. Proprioceptive control in respiration. In R. Granit (Ed.), *Muscular Afferents and Motor Control*. Stockholm: Almqvist & Wiksell, 1966.

Evarts, E. V., and Tanji, J. Reflex and intended responses in motor cortex pyramidal tract neurones of monkey. *Journal of Neurophysiology*, 1976, *39*, 1069–1080.

Forssberg, H., Grillner, S., and Rossignol, S. Phase dependent reversal during walking in chronic spinal cats. *Brain Research*, 1975, *85*, 103–107.

Ghez, C., and Shinoda, Y. Spinal mechanisms of the functional stretch reflex. *Experimental Brain Research*, 1978, *32*, 55–68.

Goodwin, G. M., and Luschei, E. S. Effects of destroying spindle afferents from jaw muscles on mastication in monkeys. *Journal of Neurophysiology*, 1974, *37*, 967–981.

Goodwin, G. M., and Luschei, E. S. Discharge of spindle afferents from jaw-closing muscles during chewing in alert monkeys. *Journal of Neurophysiology*, 1975, *38*, 560–571.

Goodwin, G. M., McCloskey, D. I., and Matthews, P. B. C. The contribution of muscle afferents to kinaesthesia shown by vibration induced illusions of movement and by the effects of paralysing joint afferents. *Brain*, 1972, *95*, 705–748.

Goodwin, G. M., Hulliger, M., and Matthews, P. B. C. The effects of fusimotor stimulation during small amplitude stretching on the frequency-response of the primary ending of the mammalian muscle spindle. *Journal of Physiology (London)*, 1975, *253*, 175–206.

Goodwin, G. M., Hoffman, D., and Luschei, E. S. The strength of the reflex response to sinusoidal stretch of monkey jaw closing muscles during voluntary contraction. *Journal of Physiology (London)*, 1978, *279*, 81–111.

Granit, R. *The Basis of Motor Control*. New York: Academic Press, 1970.

Grillner, S. The influence of DOPA on the static and dynamic activity to the triceps surae of the spinal cat. *Acta Physiologica Scandinavica*, 1969, *77*, 490–500.

Hammond, P. H. The influence of prior instruction to the subject on an apparently neuromuscular response. *Journal of Physiology (London)*, 1956, *132*, 17–18.

Hammond, P. H. An experimental study of servo action in human muscular control. *Proceedings of the III International Conference on Medical Electronics*, London: Institution of Electrical Engineers, 1960, pp. 190–199.

Harker, D. W., Jami, L., Laporte, Y., and Petit, J. Fast-conducting skeleto fusimotor axons supplying intrafusal chain fibers in the cat peroneus tertius muscle. *Journal of Neurophysiology*, 1977, *40*, 791–799.

Hasan, Z., and Houk, J. C. The transition in sensitivity of spindle receptors that occurs when the muscle is stretched more than a fraction of a millimeter. *Journal of Neurophysiology*, 1975, *38*, 673–689.

Holmqvist, B., and Lundberg, A. Differential supraspinal control of synaptic actions evoked by volleys in the flexion reflex afferents in alpha motoneurones. *Acta Physiologica Scandinavica*, 1961, *54*, Supplementum 186, 51pp.

Hongo, T., Jankowska, E., and Lundberg, A. The rubrospinal tract. II. Facilitation of interneuronal transmission in reflex pathways to motoneurones. *Experimental Brain Research*, 1969, *7*, 365-391.

Houk, J. C. Participation of reflex mechanisms and reaction-time processes in the compensatory adjustments to mechanical disturbances. In J. E. Desmedt. (Ed), *Cerebral Motor Control in Man*, Vol. 4. Basel: Karger, 1978.

Houk, J., and Henneman, E. Responses of Golgi tendon organs to active contractions of the soleus muscle of the cat. *Journal of Neurophysiology*, 1967, *30*, 466-481.

Hulliger, M., Matthews, P. B. C., and Noth, J. Static and dynamic fusimotor action on the response of Ia fibres to low frequency sinusoidal stretching of widely ranging amplitude. *Journal of Physiology (London)*, 1977, *267*, 811-838.

Hultborn, H., and Lundberg, A. Reciprocal inhibition during the stretch reflex. *Acta Physiologica Scandinavica*, 1972, *85*, 136-138.

Hultborn, H., and Wigström, H. Longlasting increase in motoneurone excitability caused by volleys in group Ia afferents. *Proceedings of the International Union of Physiological Sciences*, 1977, *13*, 994.

Hultborn, H., Jankowsa, E., and Lindström, S. Recurrent inhibition from motor axon collaterals of transmission in the Ia inhibitory pathway to motoneurones. *Journal of Physiology (London)*, 1971a, *215*, 591-612.

Hultborn, H., Jankowska, E., and Lindström, S. Relative contribution from different nerves to recurrent depression of Ia IPSPs in motoneurones. *Journal of Physiology (London)*, 1971b, *215*, 637-664.

Iles, J. F. Central terminations of muscle afferents on motoneurones in the cat spinal cord. *Journal of Physiology (London)*, 1976, *262*, 91-117.

Iles, J. F. Responses in human pretibial muscles to sudden stretch and to nerve stimulation. *Experimental Brain Research*, 1977, *30*, 451-470.

Jack, J. J. B., and Roberts, R. R. The role of muscle spindle afferents in stretch and vibration reflexes of the soleus muscle of the decerebrate cat. *Brain Research*, 1978, *146*, 366-372.

Jami, L., and Petit, J. Frequency of tendon organ discharge elicited by the contraction of motor units in cat leg muscles. *Journal of Physiology (London)*, 1976, *261*, 633-645.

Jankowska, E., Jukes, M. G. M., Lund, S., and Lundberg, A. The effect of DOPA on the spinal cord. 5. Reciprocal organisation of pathways transmitting excitatory action to alpha motoneurones of flexors and extensors. *Acta Physiologica Scandinavica*, 1967, *70*, 369-388.

Jansen, J. K. S., and Matthews, P. B. C. The central control of the dynamic response of muscle spindle receptors. *Journal of Physiology (London)*, 1962, *161*, 357-378.

Joyce, G. C., Rack, P. M. H., and Westbury, D. R. The mechanical properties of cat soleus muscle during controlled lengthening and shortening movements. *Journal of Physiology (London)*, 1969, *204*, 461-474.

Kanda, K., and Rymer, W. Z. An estimate of the secondary spindle receptor afferent contribution to the stretch reflex in extensor muscles of the decerebrate cat. *Journal of Physiology (London)*, 1977, *264*, 63-87.

Keller, E. L., and Robinson, D. A. Absence of a stretch reflex in extraocular muscles of the monkey. *Journal of Neurophysiology*, 1971, *34*, 908-919.

Kirkwood, P. A., and Sears, T. A. Monosynaptic excitation of motoneurones from muscle spindle secondary endings of intercostal and triceps surae muscles in the cat. *Journal of Physiology (London)*, 1975, *245*, 64-66.

Kuffler, S. W., and Hunt, C. C. The mammalian small-nerve fibres: A system for efferent nervous regulation of muscle discharge. *Research Publications. Association for Research in Nervous and Mental Diseases*, 1952, *30*, 24-37.

Kuffler, S. W., Hunt, C. C., and Quilliam, J. P. Function of medullated small-nerve fibers in mammalian ventral roots: Efferent muscle spindle innervation. *Journal of Neurophysiology*, 1951, *14*, 29-54.

Laporte, Y., and Lloyd, D. P. C. Nature and significance of the reflex connections established by large afferent fibers of muscular origin. *American Journal of Physiology*, 1952, *169*, 609–621.

Leksell, L. The action potential and excitatory effects of the small ventral root fibres to skeletal muscle. *Acta Physiologica Scandinavica*, 1945, *10*, Supplementum 31, 1–84.

Lennerstrand, G., and Thoden, U. Muscle spindle responses to concomitant variations in length and in fusimotor activation. *Acta Physiologica Scandinavica*, 1968, *74*, 153–165.

Liddell, E. G. T., and Sherrington, C. C. Reflexes in response to stretch. *Proceedings of the Royal Society, Series B*, 1924, *96*, 212–242.

Lundberg, A. Control of spinal mechanisms from the brain. In D. B. Tower (Ed.), *The Nervous System. Vol. I: The Basic Neurosciences*. New York: Raven Press, 1975.

Lundberg, A., Malmgren, K., and Schomburg, E. D. Characteristics of the excitatory pathway from group II muscle afferents to alpha motoneurones. *Brain Research*, 1975, *88*, 538–542.

Marsden, C. D., Merton, P. A., and Morton, H. B. Servo action in human voluntary movement. *Nature, (London)*, 1972, *238*, 140–143.

Marsden, C. D., Merton, P. A., and Morton, H. B. Servo action in the human thumb. *Journal of Physiology (London)*, 1976a, *257*, 1–44.

Marsden, C. D., Merton, P. A., and Morton, H. B. Stretch reflex and servo action in a variety of human muscles. *Journal of Physiology (London)*, 1976b, *259*, 531–560.

Marsden, C. D., Merton, P. A., and Morton, H. B. The sensory mechanism of servo action in human muscle. *Journal of Physiology (London)*, 1977, *265*, 521–535.

Marsden, C. D., Merton, P. A., Morton, H. B., and Adam, J. The effect of posterior column lesions on servo responses from the human long thumb flexor. *Brain*, 1977a, *100*, 185–200.

Marsden, C. D., Merton, P. A., Morton, H. B., and Adam, J. The effect of lesions of the sensorimotor cortex and the capsular pathways on servo responses from the human long thumb flexor. *Brain*, 1977b, *100*, 503–526.

Marsden, C. D., Merton, P. A., and Morton, H. B. Anticipatory responses in the human subject. *Journal of Physiology (London)*, 1978, *275*, 47–48P.

Matthews, B. H. C. Nerve endings in mammalian muscle. *Journal of Physiology (London)*, 1933, *78*, 1–53.

Matthews, P. B. C. The dependence of tension upon extension in the stretch reflex of the soleus muscle of the decerebrate cat. *Journal of Physiology (London)*, 1959, *147*, 521–546.

Matthews, P. B. C. Muscle spindles and their motor control. *Physiological Reviews*, 1964, *44*, 219–288.

Matthews, P. B. C. Evidence that the secondary as well as the primary endings of the muscle spindles may be responsible for the tonic stretch reflex of the decerebrate cat. *Journal of Physiology (London)*, 1969, *204*, 365–393.

Matthews, P. B. C. *Mammalian muscle receptors and their central actions*. London: Arnolds, 1972.

Matthews, P. B. C. The relative unimportance of the temporal pattern of the primary afferent input in determining the mean level of motor firing in the tonic vibration reflex. *Journal of Physiology (London)*, 1975, *251*, 333–361.

Matthews, P. B. C. Muscle afferents and kinaesthesia. *British Medical Bulletin*, 1977, *33*, 137–142.

Matthews, P. B. C. Developing views on the muscle spindle. In J. E. Desmedt (Ed.), *Spinal and Supraspinal Mechanisms of Voluntary Motor Control and Locomotion*. Basel: Karger, *Progress in Clinical Neurophysiology*, 1980, *8*, 12–27.

Matthews, P. B. C., and Stein, R. B. The sensitivity of muscle spindle afferents to small sinusoidal changes of length. *Journal of Physiology (London)*, 1969, *200*, 723–743.

McCloskey, D. I. Kinaesthetic sensibility. *Physiological Reviews*, 1978, *58*, 763–820.

McWilliam, P. N. The incidence and properties of β axons to muscle spindles in the cat hind limb. *Quarterly Journal of Experimental Physiology*, 1975, *60*, 25–36.

Melvill Jones, G. Plasticity in the adult vestibulo-ocular reflex arc. *Philosophical Transactions of the Royal Society (London)*, 1977, *278*, 319–334.

Melvill Jones, G., and Watt, D. G. D. Observations on the control of stepping and hopping in man. *Journal of Physiology (London)*, 1971, *219*, 709–727.

Merton, P. A. Speculations on the servo-control of movement. In G. E. W. Wolstenholme (Ed.), *The Spinal Cord.* London: Churchill, 1953.

Nichols, T. R., and Houk, J. C. The improvement in linearity and the regulation of stiffness that results from the actions of the stretch reflex, *Journal of Neurophysiology,* 1976, *39,* 119–142.

Ovalle, W. K., and Smith, R. S. Histochemical identification of three types of intrafusal muscle fibres in the cat and monkey based on myosin ATPase reaction. *Canadian Journal of Physiology and Pharmacy,* 1972, *50,* 195–202.

Phillips, C. G. Motor apparatus of the baboon's hand. *Proceedings of the Royal Society, Series B,* 1969, *173,* 141–174.

Prochazka, A., Westerman, R. A., and Ziccone, S. P. Discharges of single hindlimb afferents in freely moving cat. *Journal of Neurophysiology,* 1976, *39,* 1090–1104.

Rack, P. M. H. The significance of the mechanical properties of muscle in the reflex control of posture. In P. Andersen and J. K. S. Jansen (Eds.), *Excitatory Synaptic Mechanisms.* Oslo: Universitetsforlaget, 1970.

Reinking, R. M., Stephens, J. A., and Stuart, D. G. The tendon organs of cat medial gastrocnemius: Significance of motor unit type and size for the activation of Ib afferents. *Journal of Physiology (London),* 1975, *250,* 491–512.

Roland, P. E., and Ladegaard-Pedersen, H. A quantitative analysis of sensations of tension and of kinaesthesia in man. Evidence for a peripherally originating muscular sense and for a sense of effort. *Brain,* 1977, *100,* 671–692.

Rymer, W. Z., Houk, J. C., and Crago, P. E. The relation between dynamic response and velocity sensitivity for muscle spindle receptors. *Proceedings of the International Union of Physical Sciences,* 1977, *13,* 1922.

Stauffer, E. K., and Stephens, J. A. Responses of Golgi tendon organs to ramp-and-hold profiles of contractile force. *Journal of Neurophysiology,* 1977, *40,* 681–691.

Stauffer, E. K., Watt, D. G., Taylor, A., Reinking, R. M., and Stuart, D. G. Analysis of muscle receptor connections by spike-triggered averaging. 2. Spindle group II afferents. *Journal of Neurophysiology,* 1976, *39,* 1393–1402.

Stein, R. B. The peripheral control of movement. *Physiological Reviews,* 1974, *54,* 215–243.

Tatton, W. G., and Lee, R. G. Evidence for abnormal long loop reflexes in rigid Parkinsonian patients. *Brain Research,* 1975, *100,* 671–676.

Tatton, W. G., Forner, S. D., Gerstein, G. L., Chambers, W. W., and Liu, C. N. The effect of postcentral cortical lesions on motor responses to sudden upper limb displacements in monkeys. *Brain Research,* 1975, *96,* 108–113.

Upton, A. R. M., McComas, A. J., and Sica, R. E. P. Potentiation of "late" responses evoked in muscles during effort. *Journal of Neurology, Neurosurgery, and Psychiatry,* 1971, *34,* 699–711.

Vallbo, Å. B. Afferent discharge from human muscle spindles in non-contracting muscles. Steady state impulse frequency as a function of joint angle. *Acta Physiologica Scandinavica,* 1974a, *90,* 303–318.

Vallbo, Å. B. Human muscle spindle discharge during isometric voluntary contractions. Amplitude relations between spindle frequency and torque. *Acta Physiologica Scandinavica,* 1974b, *90,* 319–336.

Villis, T., and Cooke, J. D. Modulation of the functional stretch reflex by the segmental reflex pathway. *Experimental Brain Research,* 1976, *25,* 247–254.

Note Added in Proof

References for this chapter were completed in 1977. Current references to work on the muscle spindle may be found in a review lecture by the author to be published in the *Journal of Physiology (London),* 1981.

4

Control of Motoneuron Output by Pathways Descending from the Brain Stem

PETER C. SCHWINDT

IMPORTANCE OF THE BRAIN STEM EFFERENT SYSTEMS IN MOTOR CONTROL

This chapter will be concerned with the control of spinal motoneuron output by neuronal activity originating from the red nucleus, the vestibular nuclear complex, and the medial pontomedullary reticular formation. These brain stem nuclei give rise, respectively, to the rubrospinal tract and the various vestibulospinal and reticulospinal tracts which, together with the corticospinal tract, constitute the major descending fiber systems controlling motor output in mammals. Each of these tracts contains some fibers running the length of the spinal cord, and each influences motoneurons from cervical to lumbosacral levels. These brain stem nuclei together with their descending fibers will be referred to in this chapter as the *brain stem efferent systems*.

Historically, descending motor pathways have been divided into "pyramidal" and "extrapyramidal" systems. The brain stem efferent systems are part of the extrapyramidal system, which also includes the basal ganglia. Indeed, "extrapyramidal" is often understood to mean "basal ganglia" in clinical usage, and it seems best to avoid the term in discussing motor control by descending brain stem pathways. The latter are much more directly influenced by cerebral cortex and cere-

PETER C. SCHWINDT   Epilepsy Center, Veterans Administration Hospital, and Departments of Physiology and Biophysics and Medicine (Neurology), University of Washington School of Medicine, Seattle, Washington 98195. Supported by the Veterans Administration.

bellum than by the basal ganglia. A more fundamental reason for avoiding the pyramidal-extrapyramidal dichotomy is that it unjustifiably divides descending motor control into that mediated by the corticospinal component of the pyramidal tract and "everything else." This division seems to have arisen from the long-held assumption that the pyramidal tract is the principal mediator of voluntary movement, especially in higher mammals, whereas "everything else" plays a minor role or is at most concerned only with unconscious or reflex activity. The nineteenth-century findings that movement could be elicited by cortical stimulation and that the pyramidal tract, a large, highly visible structure, runs directly from cortex to spinal cord, seem to have initiated the assumption that the pyramidal tract must be the prime mediator of the corticomotor effects. The importance of the brain stem efferent pathways has only been elucidated more recently. The results of lesion studies over at least the past 20 years have indicated that the pyramidal tract may play a less than major role in control of voluntary movement. In the past decade, a great amount of evidence has accumulated, suggesting, on the one hand, that the corticospinal system plays more of a "fine tuning" role, lending fractionization, smoothness, and some power and speed to the basic motor reportoire, especially in the distal forelimbs and digits. On the other hand, the basic control of voluntary movement appears to be mediated by the brain stem efferent systems. Thus, understanding brain stem efferent function is essential to understanding motor control.

The most dramatic examples illustrating this point come from lesion studies, but it is interesting to consider first the results of some electrophysiological studies. Since this chapter will mainly be concerned with electrophysiology, it may be useful to compare the general type of information obtained from electrophysiological and lesion studies. One way an electrophysiological study may compare the amount of cortical motor control mediated by the pyramidal tract and the brain stem systems is to compare the effects produced by electrical stimulation of motor cortex with an intact and with a sectioned pyramidal tract. For example, Lewis and Brindley (1965) showed that movements produced by electrical stimulation of the precentral cortex in baboons with the bilateral pyramidal tracts sectioned were remarkably similar to those obtained with the pyramidal tracts intact, though the required electrical stimulation strength was somewhat higher and the number of discrete movements fewer and more fatiguable. The cortical "motor map" obtained was substantially the same as that with pyramids intact. Similar results were obtained by Jankowska and Tarnecki (1965) in the cat. In addition, Hongo and Jankowska (1967) demonstrated that stimulation of sensory-motor cortex in cats with transected pyramids produced effects in hind-limb motoneurons and some segmental reflex arcs that were very similar to those seen in cats with *only* the pyramids intact. Effects remaining after various lesions of the spinal cord suggested the cortical actions on motoneurons were transmitted in the ventral funiculus, mainly by reticulospinal fibers (see also Stewart, Preston, and Whitlock, 1968). These studies suggest not only that cortically initiated motor commands may be effectively transmitted to motoneurons by the brain stem pathways, but that these effects are potent and, interestingly, very similar in action to cortico-motoneuronal effects obtained with intact pyramids. Transmission of cortical motor effects by the brain stem systems is not so surprising since, as will be discussed below for each system, the cells of origin of all tracts except vestibulospinal receive direct cortical input, both pyramidal and nonpyramidal.

The electrophysiological studies cited above indicate that brain stem efferent systems are important in motor control but do not indicate in any detail the relative importance of the corticospinal system and the various components of the brain stem systems. Nor do such studies determine if the different systems subserve different types of musculature. This question has been most clearly answered by various lesion studies. Selective lesion of a tract or nucleus may, technical and interpretational difficulties permitting, indicate the role of the tract by observation of deficits produced by the lesion. Observation of such deficits may not reveal the complete function of the tract in the normal animal but can indicate functions for which an intact tract or nucleus is absolutely necessary. Consider now the results of some lesion studies. The potential for the control and initiation of movement by subcortical structures alone is indicated by the experiments of Travis and Woolsey (1956) who demonstrated that partially and totally decorticate monkeys can retain enough motor control for activities, such as righting, standing, and walking. Use of the hands was retained even after removal of all frontal cortex including the primary motor area. Clearly, the brain stem systems, which must transmit all descending motor commands in these animals, is capable of controlling this reportoire of movements. On the other hand, a series of lesion studies (reviewed in Chapter 8) have sought to determine the necessary function of the pyramidal tract by observing motor capabilities after selective pyramidal lesions in an otherwise intact animal. After an initial study of the cat (Kuypers, 1963), Lawrence and Kuypers (1968a) performed a careful qualitative study of the monkey. Subsequently, Beck and Chambers (1970) performed a more rigorous and quantitative study on the monkey. The major persistent deficit found after bilateral pyramidotomy in these studies was a failure to move individual fingers independently (which could, however, be usefully moved together). Some generalized slowness, fatigability, and clumsiness in limb movement was also observed by Lawrence and Kuypers (1968a), and all symptoms were lessened in cases of incomplete pyramidal section. Beck and Chambers (1970) could find no difference in absolute reaction time before and after pyramidotomy for a task requiring forelimb movement.

In contrast to the subtle effects on motor performance produced by pyramidotomy, lesions of the various brain stem pathways produced much more serious deficits (Lawrence and Kuypers, 1968b). Lesion of the dorsolateral spinal fasciculus, interrupting primarily rubrospinal fibers, produced persistent impairment of distal limb flexion movements (superimposed on the effects of preceding pyramidotomy) but little deficit in more proximal or extensor muscles. Axial and proximal limb movements were severely impaired, however, after lesion of the medial brain stem, interrupting primarily vestibulospinal and reticulospinal fibers. The work of Lawrence and Kuypers (1968a,b) not only points to the importance of the brain stem efferent systems in basic motor control, but also suggests that different components of the brain stem system have different domains of action; for example, the rubrospinal tract appears to be absolutely necessary for proper functions of distal flexor musculature but not proximal or extensor muscles that depend instead on intact vestibulospinal or reticulospinal tracts. The possibility of such differential control among components of the brain stem systems will be further examined in this chapter.

Before examining each brain stem system in detail, it is useful to consider a few more general comparisons between the corticospinal and brain stem systems that can be provided by electrophysiology. Because the corticospinal tract runs directly from cortex to spinal cord and directly to some motoneurons in primates, it is often assumed that cortically initiated activity traveling via fast pyramidal fibers (which, incidentally, are a very small fraction of pyramidal tract fibers; see Humphrey, Corrie, and Rietz, 1976; Lassek, 1954) must necessarily activate spinal motoneurons before impulses descending via the brain stem systems. An intact pyramidal tract does enable slightly more rapid movements of the distal forelimb or digits in some behavioral situations (Hepp-Reymond, Trouche, and Wiesendanger, 1974; Laursen and Wiesendanger, 1967) but apparently does not influence the absolute reaction time of the forelimb (Beck and Chambers, 1970). An unchanged absolute reaction time in the pyramidotomized animal is expected from the results of electrophysiology: postsynaptic potentials recorded in cat hind-limb motoneurons after stimulation of sensory-motor cortex can occur with shorter latencies if transmitted over nonpyramidal pathways than over the pyramidal tract (Endo, Araki, and Kawai, 1975; see also Hongo and Jankowska, 1967). Such a direct comparison of latencies from the cortex has not been made in the primate, but it is interesting that the latency of all rubrospinal monosynaptic EPSPs in monkey hind-limb motoneurons is shorter than those from the fastest pyramidal fibers (Shapovalov, Karamjan, Kurchavyi, and Repina, 1971). In fact, impulses originating in the nucleus interpositus of the cerebellum, relayed in the red nucleus, and descending in the rubrospinal tract, arrive at hind-limb motoneurons with about the same latency as impulses descending in the fastest corticospinal fibers (Shapovalov, Karamjan, Tamarova, and Kurchavyi, 1972). Since both rubrospinal and reticulospinal neurons, many of which are also fast conducting (Shapovalov, Grantzn, and Kurchavyi *et al.*, 1967), are monosynaptically activated from cortex (see below), there is no *a priori* assurance that cortically initiated activity traveling over the corticospinal tract arrives at the motoneurons first, or more than a few milliseconds before cortical activity transmitted via the brain stem systems. Therefore, it appears that the brain stem system might be as efficient as the corticospinal tract in initiating rapid movement.

The results of the lesion experiments outlined above, as well as other studies (see Chapter 8), have suggested that the corticospinal tract is mainly concerned with effecting fine, precise movements of the wrist and digits. It has also been found that motoneurons innervating the distal forelimb (wrist, hand, digits) in the primate receive the highest density of monosynaptic cortico-motoneuronal connections (Clough, Kernell, and Phillips, 1968), suggesting, perhaps, that the fine control exerted by the cortex over the distal musculature depends on the existence of the monosynaptic cortico-motoneuronal connections. If descending monosynaptic connections to motoneurons do affect precise muscular control, it should be recognized that the brain stem efferent systems may also be capable of precise control of the musculature they are primarily concerned with. Each descending tract in the primate has monosynaptic connections to *some* type of motoneuron. Furthermore, motoneurons having monosynaptic corticospinal connections also tend to have monosynaptic rubrospinal connections (Shapovalov *et al.*, 1971). In the cat, vestibulospinal and reticulospinal rather than corticospinal fibers are monosynaptic to some motoneurons. Whatever the actual role of descending monosynaptic connections to motoneurons may be, the bulk of the electrophysiological evidence currently available indicates that the most powerful effects from any descending tract are exerted over polysynaptic rather than monosynaptic pathways. It may be that the one or two additional synapses between the cortex and motoneurons in the shortest brain stem efferent pathways would make little difference in the final precision obtained in any sizeable movement.

In spite of the demonstrated importance of the brain stem efferent systems, it is still not possible to give a comprehensive picture of their normal function in motor control. A variety of approaches is required to determine neural function, but for the brain stem efferent systems, there are still little data available from two approaches that have been extremely useful in further elucidating corticospinal tract function, namely, the precise, quantitative measurement of deficit after localized lesions, and the analysis of neuronal firing in relation to movement in awake animals under different behavioral conditions. However, in the last 12 years, electrophysiological studies have produced a great deal of knowledge about synaptic connections of the brain stem systems. Although knowledge of synaptic circuitry by itself does not determine normal function, it does offer insight into the actual mechanisms available to the brain stem systems for motor control. These mechanisms are varied, but some general principles can be discerned. Although it would be desirable to have data available from different species, especially from primates for more direct comparison to humans, the great majority of experiments have been done in the cat. This chapter will thus be primarily concerned with the synaptic connections by which the brain stem systems influence α motoneuron output in cat. Comparison will be made with primates where data are available. Brain stem effects on fusimotor (γ) neurons will also be discussed, since these can affect α motoneuron output via the "gamma loop" (see below and Chapter 3). In recent years, a great deal of work has been done on the effect of the descending pathways on various spinal reflex arcs. It has been found that descending impulses impinging on interneurons of spinal reflex pathways can powerfully affect motoneuron output both by modifying spinal reflex transmission to motoneurons and by using the reflex arc interneurons to transmit the descending effects. The interaction of descending and spinal reflex effects will be discussed for each brain stem pathway.

The Rubrospinal Tract

Outline of Rubral Organization

The red nuclei are bilateral midbrain structures situated at the level of the colliculi. As discussed in the review of Massion (1967), the red nuclei are seen in reptiles, birds, and mammals. In marsupials and mammals, the rubral complex consists of a caudal magnocellular part, through which small cells are also scattered, and a rostral parvicellar part consisting entirely of small cells. The large- and small-celled parts are described as clearly divided, except in ungulates and carnivores. The magnocellular part is relatively large in species from marsupials to the monkey, but it is said to regress in apes and man, since few large cells are found in the rubral complex. This has been interpreted to mean that the rubrospinal tract may be of little importance or nonexistant in man, since many studies based both on older anatomical techniques as well as the more recent horseradish peroxidase method (Smith and Courville, 1976) all indicate that the rubrospinal tract arises only from the magnocellular part. If this inference is correct, it is unfortunate for the understanding of motor control in man, since the rubrospinal system is one of the better understood systems and appears to be very important in most mammals. However, both large and small cells in the magnocellular region give rise to the rubrospinal tract, which consists of fibers of various diameters (Kuypers, Fleming, and Farinholt, 1962; Pompeiano and Brodal, 1957), so that the absence of large cells does not necessarily imply the absence of a rubrospinal tract.

In contrast to the magnocellular part, the parvicellular red nucleus is small in marsupials but larger in monkeys and apes; in man, it apparently constitutes almost the entire red nucleus (see Massion, 1967, for references). The major projections of the parvicellular red nucleus are to the inferior olive, which sends climbing fibers to the cerebellar cortex (Brodal, 1969; Edwards, 1972; Massion, 1967). The increased importance of the parvicellular part thus parallels the development of the neocerebellum in phylogeny and seems to serve mainly as a component in a cerebellum-olive-cerebellum feedback loop. As will be discussed later in the chapter, there has been a recent suggestion that the parvicellular red nucleus may also be involved in descending control of dynamic fusimotor neurons and certain spinal reflexes.

The somatotopic organization of the red nucleus and the course and termination sites of the rubrospinal fibers are schematically indicated in Fig. 1. The rubrospinal fibers exist from the caudal magnocellular part of the red nucleus and cross the midline soon after leaving the nucleus. The tract sends some fibers to the superior sensory trigeminal, motor facial, and dorsal column nuclei, and projects more heavily to the lateral reticular nucleus, which itself projects to the cerebellar hemispheres (Brodal, 1969; Edwards, 1972; Massion, 1967). Most fibers project to the spinal cord, descending in the dorsolateral funiculus contralateral to the nucleus of origin and very near the path of the corticospinal fibers as indicated in Fig. 1C. In the cat and monkey, the fibers terminate in the "intermediate zone" of the spinal gray matter in Rexed's (1952, 1954) laminae V–VII (see Fig. 1C), largely overlapping the termination sites of the corticospinal tract component originating from motor cortex (Nyberg-Hansen and Brodal, 1963). As indicated in Fig. 1B, the cells of origin of the rubrospinal tract are arranged somatotopically: those in the dorsomedial red nucleus project to cervical levels; those in the ventrolateral part project to lumbar levels (Pompeiano and Brodal, 1957). There also appears to be considerable somatotopic organization of the afferents to the red nucleus (see below and Fig. 1A).

The major inputs to cells of origin of the rubrospinal tract in both the cat and the monkey come from the cerebellum and cerebral cortex. The cerebellar input derives almost entirely from the contralateral nucleus interpositus, which receives afferents from the intermediate region of the anterior lobe of the cerebellar cortex. The interposito-rubral fibers run in the brachium conjunctivum, and the great majority of these fibers terminate in the red nucleus (Brodal, 1969). Thus, the red nucleus appears to be the main target of nucleus interpositus projections. Cerebellar projections to the red nucleus are roughly somatotopically organized. The region of the nucleus interpositus innervated by that part of the cerebellar cortex that subserves hind-limb function sends its fibers to the "hind-limb" part of the red nucleus; the forelimb regions are similarly organized (Courville, 1966). In man, the main cerebellar input to the red nucleus is said to come from the contralateral

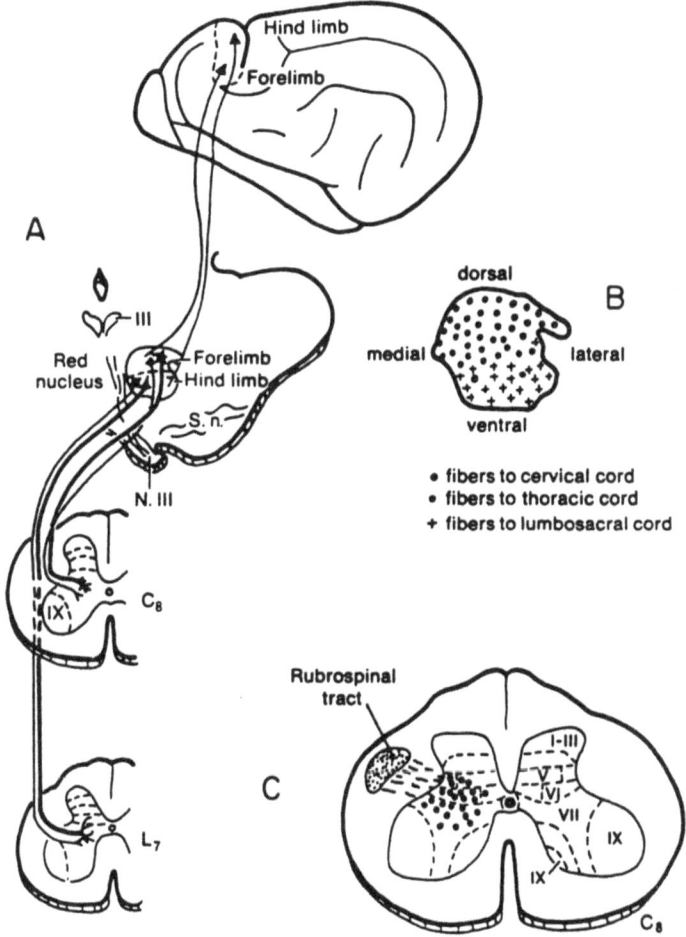

Fig. 1. The principal feature of the cortico-rubro-spinal pathway in the cat. (A) The corticorubral fibers originate from the anterior sigmoid gyrus ("motor cortex") and terminate in a somatotopical pattern in the red nucleus. The rubrospinal projection comes from small as well as large cells in the red nucleus. (B) The somatotopical pattern in the red nucleus as seen in a transverse section at middle levels of the nucleus. (C) The position of the rubrospinal tract in the spinal white matter and the sites of termination of rubrospinal fibers in relation to Rexed's laminae (indicated by dashed lines and Roman numerals). From Brodal (1969) with permission.

dentate nucleus, which projects almost exclusively to the parvicellular red nucleus (see Massion, 1967, for references).

The corticorubral input appears to come principally from ipsilateral "motor" cortex (Kuypers and Lawrence, 1967; Rinvik and Walberg, 1963). The projection consists of both pyramidal tract and nonpyramidal tract fibers, and again appears to be roughly somatotopically organized, with cortical hind-limb areas projecting mainly to the rubral hind-limb areas, and similarly for forelimb areas (see Fig. 1A).

As outlined above, anatomy indicates that the red nucleus is intimately related both to the cerebellum and the cerebral cortex; physiological data suggest that rubrospinal neurons are much more readily excited from the cerebellum than from cerebral cortex. The interposito-rubral pathway provides strong monosynaptic excitation of rubrospinal neurons in both cat and monkey. In cat, the red nucleus is more completely and easily activated by brachium conjunctivum stimulation than by stimulation within the red nucleus itself (Baldissera, Lundberg, and Udo, 1972b). In the monkey, the interposito-rubral connection is similarly found to be extremely potent and secure and able to follow repetitive stimulation at frequencies of 300–500/sec (Shapovalov et al., 1972).

Cortico-rubral connections of pyramidal tract collaterals have been extensively investigated in the cat (Tsukahara and Fuller, 1969; Tsukahara, Fuller, and Brooks, 1968; Tsukahara, Toyama, and Kosaka, 1967). Monosynaptic excitation is provided by slow pyramidal tract fibers and disynaptic inhibition by fast ones, but the monosynaptic corticorubral EPSP is weak and apparently located farther from the soma than the stronger interposito-rubral EPSP. In the monkey, Humphrey and Rietz (1976) have shown that collaterals from the pyramidal tract provide only a small fraction of the corticorubral projection from the wrist area of motor cortex; by far the greatest part of the corticorubral projection comes from small, slow, nonpyramidal tract cells. Electrical stimulation of these cells results in an initial facilitation and subsequent mild inhibition of rubrospinal cells (while only net facilitations are seen from cortico-rubro neurons with axons in the pyramidal tract). However, Humphrey and Rietz (1976) state that the corticorubral facilitation contrasts markedly with the powerful activation obtained from interposito-rubral fibers stimulated in the brachium conjunctivum. The general finding, then, is that the corticorubral pathway is direct and extensive. It may be capable of substituting for and aiding the corticospinal pathway in motor control, as suggested by the lesion studies of Lawrence and Kuypers (1968b). However, the usual role of the corticorubral projection appears to be facilitatory, perhaps firing rubral cells only if they are already depolarized by other inputs such as the powerful connection from the nucleus interpositus of the cerebellum.

RUBROSPINAL EFFECTS ON ALPHA MOTONEURONS

The most extensive study of rubrospinal effects has been performed by Hongo and his colleagues (Hongo, Jankowska, and Lundberg, 1969a,b, 1972a,b) in a series of studies on the control of cat hind-limb spinal mechanisms by the rubrospinal tract. The net effect of rubral stimulation on motoneurons innervating different hind-limb muscles was examined initially by observing the effect of a rubral conditioning stimulus on the amplitude of a test monosynaptic reflex recorded on the ventral roots and evoked from group Ia afferent fibers of the homonymous muscle nerve. This procedure uses the change in the conditioned monosynaptic reflex amplitude as a measure of the net change in excitability of the motoneuron population under study produced by volleys in the descending tract. It is a commonly used procedure in electrophysiological studies of motor control. Hongo *et al.* (1969a) found that the usual net effect of rubrospinal volleys was to facilitate flexor motoneurons, thus confirming the implications of older work (Pompeiano, 1957), but the character of the facilitation varied among different flexor motoneuron pools. Extensor motoneurons were usually inhibited (except for toe extensors, which were always facilitated), but facilitation or mixed effects were often seen in other extensor populations. When the ankle extensors, gastrocnemius, and soleus were examined separately, it was found that most gastrocnemius motoneurons were facilitated, whereas all soleus motoneurons were inhibited. Some functional implications of this particular observation will be discussed below. Thus, the rubrospinal tract gives net excitation to flexors, but does not consistently reciprocally inhibit extensors; extensor facilitation is also common, depending on the particular muscle.

These complex effects of rubrospinal stimulation are even more apparent on intracellular recording in motoneurons (Hongo *et al.*, 1969a). EPSPs were found to predominate generally in flexor motoneurons, especially those concerned with the knee and hip, but smaller IPSP components were also often seen, and the EPSP/IPSP mix could change with stimulus strength. In extensors, the IPSP component usually dominated, though EPSP components were also seen and sometimes dominated depending on the particular motoneuron and muscle (they always dominated in toe extensors). The time between arrival of the fastest descending tract wave and the onset of the fastest EPSPs or IPSPs in the motoneurons (the "segmental delay") was consistent with disynaptic and longer connections between fast rubrospinal fibers and the motoneurons. The interneurons mediating the di- and polysynaptic transmission apparently are in or near the recorded spinal segment since the segmental delay is referred to the arrival of the descending volley at the recorded segments. The use of segmental delay as a measure of number of synaptic links between terminals of descending fibers and motoneurons and as an indication of the location of those links is common practice in electrophysiological studies of descending systems and will be repeatedly referred to throughout the rest of the chapter. A recent study (Illert, Lundberg, and Tanaka, 1976b) suggests that short-latency rubrospinal effects in cat forelimb motoneurons are basically similar to those on hind-limb motoneurons. Disynaptic or longer EPSPs are common in flexors, whereas disynaptic or longer IPSP's are common in extensors. Mixed effects are also often seen.

Using a different approach, Ghez (1975) employed presumably localized stimulation through a microelectrode placed in the red nucleus of the awake cat while observing the effect of this stimulation at different points in the red nucleus on fore- and hind-limb muscle contraction and EMG activity. The somatopic organization of forelimb and hind-limb areas suggested by anatomy was confirmed. However, stimulation at different points in the nucleus could activate both flexors and extensors innervating both proximal and distal musculature. Flexors were activated a bit more often, especially knee flexors, and the effective stimulation zone for distal muscles was a bit narrower than for proximal muscles. In general, there was great overlapping of effective stimulation sites among synergists and antagonists in a given limb. Inhibitory effects of rubral stimulation were not reported.

Studies in the monkey have mainly been concerned with the distribution of the monosynaptic rubrospinal EPSP among hind-limb motoneurons; longer latency effects, though present, have not been so extensively studied. The monosynaptic rubrospinal EPSP often occurs in motoneurons of distal flexors and, apparently, in some extensor motoneurons subserving the ankle and digits. It is not observed in motoneurons subserving the hip and knee, which instead receive polysynaptic excitation and inhibition (Shapovalov, 1972; Shapovalov *et al.,* 1971, 1972). Motoneurons usually either receive the rubrospinal and corticospinal monosynaptic EPSP together or else lack both monosynaptic EPSPs (Shapovalov, 1972; Shapovalov *et al.,* 1971). Motoneurons receiving the monosynaptic EPSP were demonstrated, by injection of Procion yellow dye in the recorded cell, to lie dorsolaterally, nearest the area where rubro- and corticospinal fibers terminate; motoneurons not receiving the monosynaptic EPSP lie more ventromedially (Shapovalov, 1972; Shapovalov *et al.,* 1972). It seems that a strictly limited number of rubrospinal fibers contribute to the monosynaptic EPSP in each cell, since the EPSP appears almost full size near stimulation threshold with little further increase in size as stimulation intensity is raised. Apparently "unitary," all-or-none EPSPs are sometimes seen (Shapovalov *et al.,* 1971). These results in the monkey generally agree well with Lawrence and Kuypers' (1968b) interpretation that the corticospinal and rubrospinal tracts act similarly in motor control of distal flexors. It is unlikely that the monosynaptic effects provide the entire basis of rubrospinal control, however, since larger, polysynaptic potentials are also apparent upon rubral stimulation (Shapovalov *et al.,* 1971).

The interesting finding in the monkey that corticospinal and rubrospinal monosynaptic EPSPs occur together on certain motoneurons appears to have a counterpart in the cat. Both the rubro- and cortico-motoneuronal linkage is minimally disynaptic in cat forelimb motoneurons, and the single interneuron is shared by both the corticospinal and rubrospinal tracts (Illert and Tanaka, 1976; Illert, Lundberg, and Tanaka, 1976a,b; see also Bayev and Kostyuk, 1973).

It was mentioned above that in the cat the net action on the motoneurons of the soleus muscle is inhibitory whereas net excitation or mixed effects are found in motoneurons of the gastrocnemius. Similar reciprocal effects on gastrocnemius and soleus motoneurons are found in the monkey (but apparently not in cat) upon corticospinal tract stimulation (Preston, Schende, and Yemura, 1967). The soleus appears to be a slowly contracting muscle, whereas the gastrocnemius is a fast-contracting muscle that is probably only fully activated during phasic activity such as locomotion. It would thus appear that the net effects of the rubro- and corticospinal tracts on motoneurons subserving fast-contracting muscles may be different or even opposite to their effects on motoneurons subserving slowly contracting or tonically activated muscles. In fact, the gastrocnemius muscle itself contains slowly contracting, presumably tonically activated motor units as well as the more numerous fast-contracting type (Burke, 1967, 1968), and it has been found that gastrocnemius motoneurons innervating fast motor units are usually excited by the rubrospinal tract whereas those innervating slow motor units are usually inhibited (Burke, Jankowska, and ten Bruggencate, 1970). Differences in segmental reflex effects from cutaneous afferents are also seen when comparing fast- and slow-contracting motor units (Burke *et al.*, 1970; Kanda, Burke, and Walmsley, 1977). It thus appears that some descending and peripheral pathways can have quite different effects on motoneurons innervating the same muscle but innervating muscle fibers with different properties. As yet, it is unknown how general this principle is in motor control.

Rubrospinal Effects on Spinal Reflex Arcs

In a study on corticospinal effects on the cat hind limb, Lundberg and Voorhoeve (1962) showed that this system may have widespread and powerful control of transmission of spinal reflex activity to motoneurons. The hypothesis was advanced that the corticospinal tract not only modifies afferent activity transmitted to motoneurons over these reflex arcs but that the corticospinal tract may actually exert some or all of its effects on motoneurons (in the cat) by means of the interneurons of these spinal reflex arcs rather than by "private" interneurons subserving only corticospinal function. Knowledge of such mechanisms would obviously be of prime importance in understanding descending control of motoneuron output. The possibility that descending effects are mediated by reflex arc interneurons has been explored for almost all components of the brain stem efferent systems. Whether the corticospinal and brain stem efferent systems exert polysynaptic effects on motoneurons by segmental reflex interneurons or by "private" interneurons has been disputed (see, e.g., Vasilenko and Kostyuk, 1967/1968; Kostyuk and Pilyavsky, 1969), but there now appears to be ample evidence that many, and perhaps most, descending polysynaptic effects on motoneurons are mediated by interneurons shared with segmental reflex pathways.

Spinal Reflex Arcs and the Spatial Facilitation Technique. Since descending effects on spinal reflexes will be extensively discussed in this chapter, it may be useful first to describe briefly the reflex arcs studied and the technique of spatial facilitation used to study the interaction between descending pathways and spinal reflexes. The spinal reflex arcs that have been most often examined derive from group Ia afferent fibers from primary muscle spindles, group Ib afferent fibers from Golgi tendon organs, large cutaneous afferent fibers, and a variety of mostly small-diameter afferent fibers classified together as flexor reflex afferents (FRA).

Group IA fibers monosynaptically excite motoneurons of the homonymous muscle and its synergists acting at a joint and disynaptically inhibit motoneurons of the direct antagonist muscles acting at the same joint (Eccles, Fatt, and Landgren, 1956; Lloyd, 1946). The interneuron transmitting the disynaptic inhibition has been termed the Ia inhibitory interneuron. Group Ib fibers from an extensor muscle di- and trisynaptically inhibit motoneurons of the homonymous extensor muscle and its synergists acting at a joint, as well as many motoneurons innervating extensors acting at other joints. The inhibitory effect on the homonymous muscle and its synergists is often termed *autogenic inhibition.* Concomitant with the inhibition of extensor motoneurons, motoneurons innervating flexors at the same joint and at other joints are di- and trisynaptically facilitated (Eccles, Eccles, and Lundberg, 1957; Laporte and Lloyd, 1952). In the low-spinal cat, effects from stimulation of group Ib afferents from flexor muscles are rarely seen (Eccles *et al.,* 1957), but in the high-spinal cat, effects just opposite to those described above (i.e., inhibition of flexors and facilitation of extensors) are seen upon stimulation of group Ib fibers from flexor muscles (Laporte and Lloyd, 1952). Even more complex actions of group Ib afferents have been noticed. The principal effects of group Ib afferents in the low-spinal cat are schematically indicated in Fig. 2. As is also indicated in Fig. 2, it has recently been shown that some group Ib fibers from different muscles, including muscles acting at different joints and, in some cases, from extensors and flexors, share the same Ib interneurons (Lundberg, Malmgren, and Schomburg, 1975). For example, excitation of group Ib fibers from ankle extensors facilitates transmission of group Ib effects from knee extensors to various extensor and flexor motoneurons. As is further indicated in Fig. 2, stimulation of low-threshold cutaneous or joint receptor afferent fibers also facilitates transmission of group Ib effects to extensor and flexor motoneurons (Lundberg *et al.,* 1975; Lundberg, Malmgren, and Schomberg, 1977).

The flexor reflex afferents (FRA) consist of group II and III afferent fibers from muscle, cutaneous afferent fibers, and afferent fibers from joint receptors. These are all lumped into the same FRA category principally because they all give

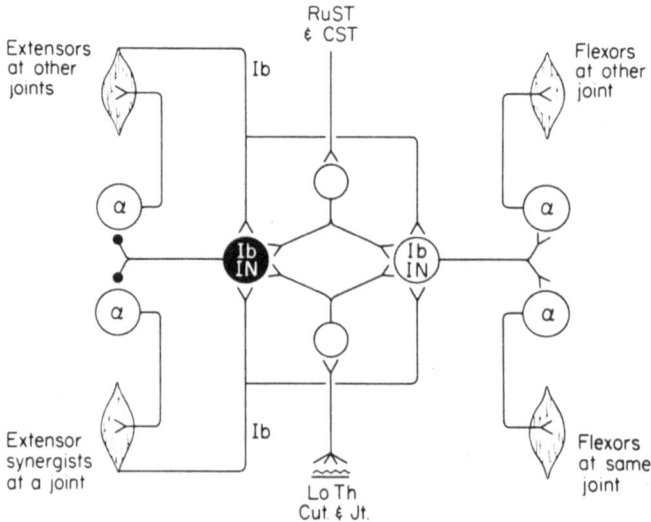

Fig. 2. The principal actions of group Ib afferent fibers observed in the low-spinal cat. The group Ib fibers monosynaptically excite inhibitory and excitatory interneurons (Ib IN), which mono- and disynaptically affect various α motoneurons (α). Note that Ib interneurons receive excitation from group Ib fibers originating from a variety of muscles. Low-threshold cutaneous and joint receptor fibers (Lo Th Cut. & Jt.) and fibers of the rubrospinal (RuST) and the corticospinal (CST) tracts also excite the Ib interneurons and thus facilitate transmission of impulses from the group Ib fibers to the motoneurons.

rise to synaptic actions in motoneurons, namely, excitation of flexors and inhibition of extensors (Eccles and Lundberg, 1959; Lloyd, 1943), that are consistent with the actions of the flexion reflex described by Sherrington (1910). Additional reasons for the common classification of these fibers have been presented (Holmqvist and Lundberg, 1962). The FRA response is mediated by polysynaptic circuits that are also highly divergent. Motoneuron activation over several segments may be obtained (Lloyd, 1943, 1946). Although the FRA is said to activate flexors and inhibit extensors, the actual response seen depends on the precise preparation used (Holmqvist and Lundberg, 1962). The FRA response is quite prominent in the spinal cat, but may be small or absent in the decerebrate cat. In an anesthetized cat with intact neuraxis, mixed facilitation and inhibition is common in both flexor and extensor motoneurons. Group Ib effects also vary with the precise preparation used, and this variation probably reflects the fact that these polysynaptic circuits are under control of descending tracts whose activity can vary among different experimental conditions (see below). Although they were originally included in the FRA category, a great deal of evidence has accumulated indicating that low threshold cutaneous and joint afferent fibers often cause different effects than the FRA on motoneurons under the same experimental conditions (Baldissera, ten Bruggencate, and Lundberg, 1971; Engberg, 1964; Grigg, Harrigan, and Fogarty, 1978; Hagbarth, 1952; Hongo *et al.,* 1969b; Hultborn, 1972; Illert *et al.,* 1976b). These afferents are thus thought to have a separate polysynaptic pathway to motoneurons in addition to, or instead of, the FRA pathway. Similarly, it has recently been shown that some group II fibers from secondary muscle spindles (as opposed to those with bare nerve endings in muscle) mono- and disynaptically excite the homonymous motoneurons (Kirkwood and Sears, 1974; Stauffer, Watt, Taylor, Reinking, and Stuart, 1976). Their synaptic actions thus resemble those of group Ia fibers, and they are presumably not part of the FRA system.

Because the reflex arcs outlined above, and thus the interneurons involved, are defined by their effects on the motoneurons, the activity of the motoneurons themselves must be used to infer the net effects of descending tracts on the interneurons composing the arc. This is most commonly done using intracellular recording from the motoneurons and the technique of spatial facilitation, as schematically indicated in Fig. 3. The postsynaptic potential produced over the spinal reflex pathway by stimulation of the primary afferents of interest are first observed (Fig.

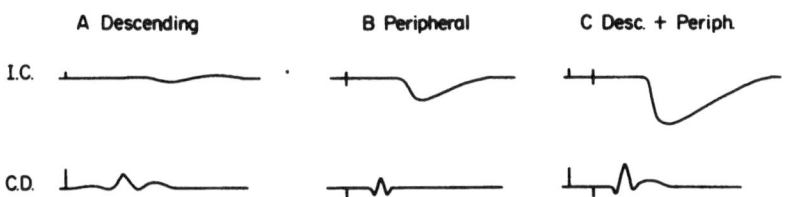

Fig. 3. The technique of spatial facilitation used to demonstrate the action of descending pathways on the interneurons of spinal reflex arcs. *(Upper traces)* Intracellularly (I.C.) recorded postsynaptic potentials in a motoneuron. *(Lower traces)* Monopolar extracellular recordings from the cord dorsum (C.D.) of the spinal segment containing the motoneuron. These are used to monitor the strength and time of arrival of the incoming nerve impulses evoked by electrical stimulation of the descending pathway (A) and the peripheral nerve (B). See text for further explanation.

3B). The strength of a conditioning descending volley is then adjusted so that it produces little or no postsynaptic potential in the motoneurons, though it may facilitate or inhibit interneurons of the reflex arc (Fig. 3A). The reflex is then conditioned by the descending volley (Fig. 3C). Any change now seen in the test reflex must be due to descending effects on interneurons of the reflex arc, since the descending volley was demonstrated to have no postsynaptic effect on the motoneuron itself. More dramatic descending facilitation of segmental reflexes can be demonstrated by adjusting the stimuli so that both the descending and afferent volleys produce no effect in the motoneuron when evoked separately, but reflex transmission is seen when the afferent volley is conditioned by the descending volley. The same kind of information can be obtained for the motoneuron population rather than from individual motoneurons by using the monosynaptic reflex recorded on the ventral roots produced by stimulation of the group Ia fibers from the homonymous muscle as a measure of net motoneuron excitability. This monosynaptic discharge may be conditioned by volleys over other reflex pathways and by descending volleys in a manner analogous to that used during intracellular recording (see, e.g., Hongo et al., 1969b; Lundberg and Voorhoeve, 1962). By varying the time interval between the conditioning and test volley and observing the time after the arrival of the peripheral volley that the descending volley produces its effect, it may be possible to obtain information on the length of the interneuron chains involved (Hongo et al., 1969b).

SIMILAR RUBRO- AND CORTICOSPINAL ACTIONS ON REFLEX ARCS. It is interesting to compare the effects of both the corticospinal and rubrospinal tracts on spinal reflex arcs, since their effects are so similar. Most work on the action of these descending pathways has been performed on lumbar reflexes and motoneurons controlling hind-limb muscles. The main results of these investigations are schematically indicated in Fig. 5. A recent series of investigations (Illert and Tanaka, 1976, 1978; Illert et al., 1976a,b; Illert, Lundberg, and Tanaka, 1977) has examined the shortest latency, corticospinal, and rubrospinal actions on cervical reflexes and motoneurons controlling forelimb muscles. These investigations have been able to determine the interaction between descending and spinal mechanisms in considerably more detail than is known at present for the lumbar spinal cord. The main results of these investigations on the cervical spinal cord are schematically indicated in Fig. 4. Part of the reason that the cervical interactions are known in such detail has to do with technical considerations such as the shorter conduction distance from the brain stem to cervical segments. This results in a descending volley less dispersed over time, a better measure of segmental delay of synaptic effects, and, thus, a more precise estimate of the length of interneuronal chains. There is also some indication that the pathways of the shortest latency effects on forelimb motoneurons may be somewhat less complex than the longer latency (perhaps more polysynaptic) effects on hind-limb motoneurons.

The shortest latency corticospinal EPSPs in forelimb motoneurons are disynaptic (Illert et al., 1976a), although these early EPSPs are very small compared to longer latency polysynaptic corticospinal EPSPs; the earliest corticospinal IPSPs are usually trisynaptic (Illert and Tanaka, 1978; Illert et al., 1976a). The shortest latency corticospinal EPSPs to hind-limb motoneurons are considered to be trisynaptic (Lloyd, 1941; Lundberg and Voorhoeve, 1962). As previously discussed (pp. 147–148), the earliest rubrospinal effects in forelimb or hind-limb motoneurons are di- and trisynaptic and appear to constitute the principal action of the rubrospinal tract on motoneurons (Hongo *et al.*, 1969a; Illert *et al.*, 1976b). There do not appear to be more potent long-latency rubrospinal effects like those of the corticospinal projection to forelimb motoneurons.

As indicated in Fig. 4, it has been determined that both the corticospinal and rubrospinal effects to forelimb motoneurons are transmitted by propriospinal neurons whose cell bodies lie in the C3-4 cervical segments and whose axons run in the ventrolateral funiculus to motoneurons in the cervical enlargement (Illert and Tanaka, 1976, 1978; Illert *et al.*, 1977). The technique of spatial facilitation has shown that some fibers of both the corticospinal and rubrospinal tract transmit their effects using the same population of propriospinal neurons (Illert *et al.*, 1976b; 1977). It appears that much of the characteristic action of the corticospinal tract on lumbar motoneurons is also transmitted by propriospinal neurons with cell bodies in more rostral segments and with axons descending in the ventrolateral funiculus (Stewart *et al.*, 1968). It is not known if these propriosponal neurons to lumbar segments are shared by rubrospinal fibers, however.

As shown in Fig. 4, the propriospinal neurons transmitting the short latency corticospinal and rubrospinal effects to forelimb motoneurons also transmit effects from these descending tracts to interneurons of various spinal reflex pathways. The propriospinal neuron is itself the last neuron in a pathway from low-threshold cutaneous afferents to spinal motoneurons. Fig. 5 shows that the descending actions on hindlimb spinal reflexes are generally similar to those on the forelimb reflexes, but the precise nature of the connections between descending fibers and identified spinal reflex interneurons is only partially known. One general feature of rubrospinal and corticospinal actions on forelimb or hind-limb reflex arcs is that transmission through these arcs is always facilitated and never inhibited by both

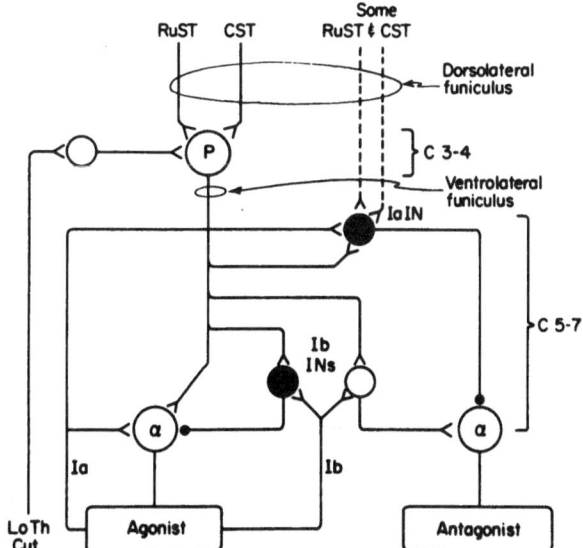

Fig. 4. The shortest rubrospinal (RuST) and corticospinal (CST) pathways to cat forelimb motoneurons and the interaction of these descending tracts monosynaptically excite propriospinal neurons (P) located in upper cervical segments. The propriospinal neurons send axons to the cervical enlargement and excite α motoneurons (α), group Ia inhibitory interneurons (Ia IN), and excitatory and inhibitory group Ib interneurons (Ib INs). The propriospinal neurons are themselves the last interneurons in a polysynaptic pathway from low-threshold cutaneous afferent fibers (Lo Th Cut.).

descending tracts (Hongo *et al.,* 1969b; Illert *et al.,* 1976b; Lundberg and Voorhoeve, 1962).

The general facilitory effect of rubro- and corticospinal conditioning is particularly noteworthy for the group Ia inhibitory pathway. Although both the corticospinal and rubrospinal tract usually excite flexor motoneurons and excite or inhibit different extensor motoneurons, each tract facilitates transmission of disynaptic inhibition through the Ia inhibitory interneuron, whether that inhibition be directed to flexor or extensor motoneurons. The time course of the facilitation (as well as intracellular recording from the Ia inhibitory interneurons, to be discussed later) suggests that some rubrospinal fibers and corticospinal fibers make monosynaptic connections with Ia inhibitory interneurons. Most of the facilitation is via a disynaptic or longer pathway, however (Hongo *et al.,* 1969b; Illert and Tanaka, 1978; Illert *et al.,* 1976b). Although the rubro- and corticospinal tracts greatly facilitate transmission of group Ia disynaptic inhibition, no previously unknown group Ia inhibitory pathways were revealed by rubro- or corticospinal conditioning. In particular, no reciprocal group Ia inhibition was seen between hip abductors and adductors, suggesting reciprocal Ia inhibition occurs strictly between flexors and extensors acting at the same joint (Hongo *et al.,* 1969b). It has been shown that trisynaptic inhibition of cat forelimb motoneurons by the corticospinal tract is actually mediated by the propiospinal neuron-Ia inhibitory interneuron pathway to the motoneurons (Illert and Tanaka, 1976, 1978). This is analogous to the transmission of disynaptic corticospinal inhibition of hind-limb motoneurons through the Ia inhibitory interneuron in the monkey (Jankowska, Padel, and Tanaka, 1976). Di- and trisynaptic rubrospinal inhibition of cat forelimb or hind-limb

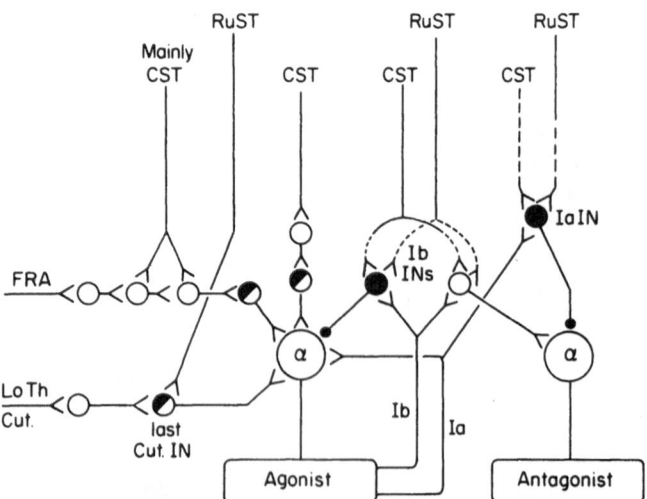

Fig. 5. The shortest latency rubrospinal and corticospinal pathways to cat hind-limb motoneurons and the interaction of these descending tracts with lumbar spinal reflex arcs. FRA refers to flexor reflex afferent fibers (see text). Other elements labeled as in Fig. 4. Dashed lines indicate that the actual connections are not monosynaptic as drawn in the diagram. Note that the descending actions are similar to those on the cervical segments diagrammed in Fig. 4. but the precise pathways in the lumbar segments have not been determined in such detail.

motoneurons may also be partially relayed through the Ia inhibitory interneuron (Hongo *et al.*, 1969b; Illert *et al.*, 1976b), but most inhibitory rubrospinal effects appear to be mediated by a different pathway to be discussed below.

As indicated in Figs. 2, 4, and 5, both the cortico- and rubrospinal tracts facilitate transmission through group Ib pathways to all kinds of motoneurons (Hongo *et al.*, 1969b; Illert *et al.*, 1976b; Lundberg and Voorhoeve, 1962). It is more difficult to identify effects from electrical stimulation as originating strictly from group Ib fibers in the forelimb as compared to the hind limb; thus, the most detailed results have come from studies on hind-limb motoneurons. The effects of rubrospinal conditioning on transmission through group Ib pathways has been particularly well studied (Hongo *et al.*, 1969b). In the experiments of Hongo *et al.* (1969b), all of the spinal cord except the dorsolateral funiculus containing rubrospinal (and corticospinal) fibers was sectioned at low thoracic levels, producing essentially a low-spinal preparation having the low-spinal group Ib reflex patterns (Fig. 2). (In some animals, cortical lesions were previously made to allow degeneration of corticospinal fibers in order to ensure they were not producing the observed effects by means of axon reflexes from corticorubral collaterals.) Rubrospinal conditioning not only facilitated transmission through the usual low-spinal group Ib pathways, but also resulted in inhibition from flexor group Ib fibers to flexor motoneurons and concomitant facilitation of many extensor motoneuron pools. This latter pattern is usually seen only in the high-spinal cat, as mentioned earlier. In addition, totally new patterns such as group Ib excitation of flexors from flexors and extensors from extensors could sometimes be demonstrated after rubrospinal conditioning. Corticospinal effects on hind-limb motoneurons appear to be generally similar but were not so extensively investigated (Lundberg and Voorhoeve, 1962).

These results indicate that the rubrospinal tract (and, probably, the corticospinal tract) has extensive control over the many group Ib reflex pathways and may be able functionally to turn on and off different combinations of group Ib reflex pathways under different behavioral conditions. Older hypotheses about group Ib function have focused on autogenic inhibition and have suggested this reflex serves mainly as "overload protection" of the homonymous muscle. This finding also seemed reasonable from the older finding that group Ib fibers were excited only at high, passive muscle tension. Recent findings indicate that group Ib afferents are activated at low tension during active contraction of muscle fibers (Houk and Henneman, 1967). This, together with the widespread synaptic effects of group Ib activation in addition to their autogenic effects, suggests that these reflex actions do much more than protect against muscle overload. As emphasized by Hongo *et al.* (1969b), the wide receptive field from which each motoneuron draws its group Ib effects, including muscles acting at different joints, suggests the many group Ib reflex pathways may be involved in more complex movements such as co-contraction of muscles of different actions at different or the same joint. The powerful control of Ib paths exerted by the rubrospinal and corticospinal tracts may serve differentially to facilitate the Ib pathways required to affect smoothly various complex movements while disfacilitating other Ib pathways that are inappropriate at the moment. Clearly, further work is needed to substantiate this hypothesis, but the available data suggest the appropriate neural circuits for such control exist.

As indicated in Fig. 5, the corticospinal tract significantly facilitates FRA excitation of flexors and inhibition of extensors (Lundberg and Voorhoeve, 1962), but the rubrospinal tract has a minor effect on FRA transmission (Hongo *et al.*, 1969b). Instead, the rubrospinal tract (and the corticospinal tract to forelimb motoneurons, Illert *et al.*, 1976b) effectively facilitates reflexes from low-threshold cutaneous and joint afferent fibers that are mediated by a pathway separate from the FRA (Baldiserra *et al.*, 1971; Hongo *et al.*, 1969b; Illert *et al.*, 1976b). Such differential effects of low- versus high-threshold cutaneous and joint afferents are, in fact, part of the evidence that these special, non-FRA paths are a general feature of spinal organization. Although non-FRA cutaneous effects have been demonstrated for cutaneous afferents from the foot (Engberg, 1964), they, as well as effects from low-threshold joint afferents, are not usually observed from stimulation elsewhere in the spinal cat but do appear after rubro- or corticospinal conditioning. Effects on motoneurons are both excitatory and inhibitory, but more commonly inhibitory to extensors.

In a study on hind-limb motoneurons, Baldiserra *et al.* (1971) have suggested that the low-threshold cutaneous pathways have a function in addition to transmitting information from low-threshold cutaneous afferent fibers. They have presented evidence indicating that both the disynaptic rubrospinal EPSPs and IPSPs are transmitted mainly through the last interneuron in these low-threshold cutaneous pathways. This is demonstrated by first adjusting the test rubral stimulus to be just subliminal for evoking a postsynaptic potential. Conditioning the subliminal rubral volley by a just-liminal, low-threshold cutaneous volley then results in the appearance of a disynaptic rubral postsynaptic potential of the same sign as the

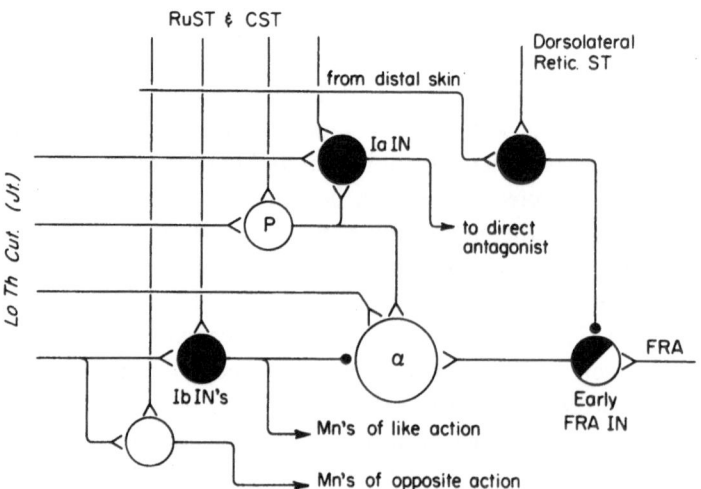

Fig. 6. Some actions of low-threshold fibers from skin and some joint receptors (Lo Th Cut. (Jt.)) on interneurons of other spinal reflex arcs. Interneurons are labeled as in Figs. 3, 4, and 5. Dorsolateral Retic. ST refers to reticulospinal fibers from the ventrocaudal medulla that descend in the dorsolateral funiculus of the spinal cord. Note that most of the actual pathways from the cutaneous fibers are *not* monosynaptic to the interneurons and motoneurons (as indicated in the diagram for the sake of clarity).

cutaneous one. Since there is only one interneuron in the rubrospinal path, this interneuron must be the last interneuron of the cutaneous path. Similar effects can sometimes be obtained by FRA preconditioning, but they are always obtained by low-threshold cutaneous conditioning. Thus, Baldiserra *et al.* (1971) suggest that the disynaptic rubrospinal IPSPs and EPSPs are mediated principally via the last interneuron of the cutaneous pathway rather than the FRA pathway. Similar results are obtained in cat forelimb motoneurons (Illert *et al.*, 1976b) from both the rubrospinal and corticospinal tract. Thus, as indicated in Figs. 4 and 5, rubrospinal and corticospinal fibers share and exert part of their effects through at least two types of interneurons: the group Ia inhibitory interneurons and the last interneurons of the low-threshold cutaneous afferent pathway.

It is apparent that the reflex pathways from low-threshold cutaneous fibers (and, often, from low-threshold fibers from joint receptors) play an important role in spinal cord neurophysiology because of their many interactions with descending motor pathways and other spinal reflex arcs. Interactions of low-threshold cutaneous pathways with rubrospinal, corticospinal, propriospinal, and group Ib pathways have been discussed so far, and other interactions will be discussed later in the chapter. The principal effects and interactions of low-threshold cutaneous pathways so far discovered are schematically illustrated in Fig. 6.

THE VESTIBULOSPINAL SYSTEMS

ORGANIZATION OF THE VESTIBULOSPINAL SYSTEMS

The vestibular complex occupies a relatively large region of the medulla just below the fourth ventricle at the level of the VIIIth nerve, from which it receives primary afferents. The complex is divided into four main nuclei: the superior vestibular nucleus, the lateral v.n. (commonly called Deiters' nucleus), the medial v.n., and the descending v.n. (also called the inferior v.n.), as well as several other minor subgroups (Brodal, 1969, 1972a; Brodal, Pompeiano, and Walberg, 1962). The great majority of anatomical and physiologycal work on vestibular organization has been done on the cat, but evidently the anatomical aspects of vestibular organization are identical in principle for all mammals, as exemplified by the observations that vestibular nuclear architectonics, nuclear subdivisions, and spinal projections in man are similar to those of the cat (Brodal, 1972a). There is, therefore, some basis for hoping that extensive knowledge of vestibular neurophysiology obtained from studies in the cat may also be applicable in principle to other species, including man.

The vestibular complex is involved in both oculomotor (see Chapter 6) and skeletomotor control. Some vestibular neurons act as powerful and secure relays of short latency effects from the labyrinthine receptors to the spinal cord and extraocular neurons. Other parts of the vestibular complex appear to function as integrative centers for more complex, though not yet fully defined, control of both the oculomotor and skeletomotor systems. One bit of anatomical evidence for a larger role of the vestibular system in motor control is the finding, emphasized by Brodal and co-workers (Brodal, 1969, 1972a; Brodal *et al.*, 1962; Walberg, 1972a) that relatively extensive areas in each of the four main vestibular nuclei do not receive primary afferent fibers from the labyrinth but do receive afferent fibers from "higher centers" such as the cerebellum and reticular formation. This suggestion will be seen below to be supported by the results of both anatomy and electrophysiology.

Figure 7 schematically indicates the subdivisions of the vestibular complex, the somatotopic organization of Deiters' nucleus, and the course and spinal ter-

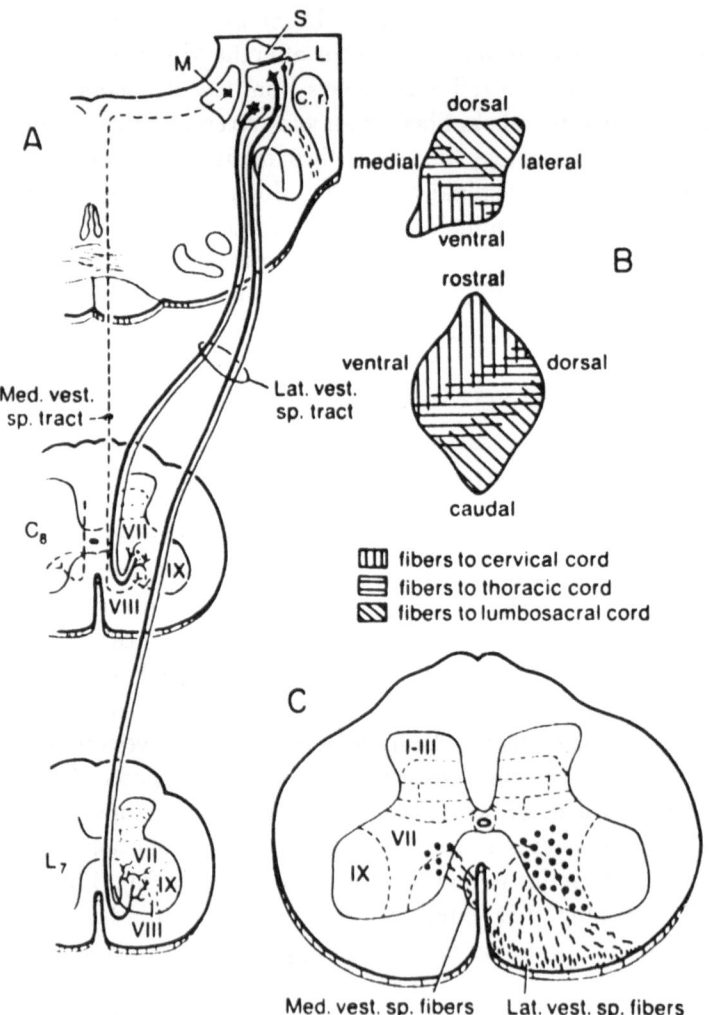

Fig. 7. The principal features of the vestibulospinal pathways in the cat. (A) Lateral vestibulospinal tract originates from large and small cells of the lateral vestibular nucleus (Deiters' nucleus). The medial vestibulospinal tract comes, in part, from the medial vestibular nucles and does not reach lumbar segments in the cat. (B) Somatotopical pattern in Deiters' nucleus seen in transverse section *(upper)* and in sagittal reconstruction *(lower)*. (C) Location in the spinal white matter of descending fibers of the lateral and medial vestibulospinal tracts and areas of their termination in the ventral horn. Dashed lines and Roman numerals refer to Rexed's laminae. From Brodal (1969) with permission.

mination of the vestibulospinal tracts. The nuclei that have been most often investigated in vestibulospinal function are Deiters' nucleus (the lateral vestibular
nucleus) and the medial vestibular nucleus. The lateral vestibulospinal tract
(LVST), often simply called the vestibulospinal tract, is the chief source of vestibulospinal fibers; it arises entirely from Deiters' nucleus (from both small and large
cells) and is composed of small and large fibers (Brodal, 1969; Brodal et al., 1962;
Nyberg-Hansen and Mascitti, 1964). These fibers leave the nucleus ventrocaudally
and descend ipsilaterally in the ventral-most part of the ventral fasciculus of the
spinal cord, where they terminate in the ventromedial spinal gray matter (see Fig.
7C) in Rexed's (1952, 1954) lamina VIII and the central and medial parts of
lamina VII; that is, rarely in lamina IX, the area of motoneuron somata (Nyberg-
Hansen and Mascitti, 1964). Both anatomical and physiological data indicate that
Deiters' nucleus is somatotopically organized (Fig. 7B) such that cells giving fibers
to the lumbosacral spinal cord are concentrated dorsocaudally (the main area of
giant, or Deiters's neurons); cells giving fibers to the cervical spinal cord are more
concentrated rostroventrally (Brodal, 1969; Brodal et al., 1962; Ito, Hongo, Yoshida, Okada, and Obata, 1964; Wilson, Kato, Peterson, and Wylie, 1967).

The medial vestibulospinal tract (MVST) is the second major vestibulospinal
pathway, though it contributes considerably fewer fibers than the LVST (Brodal,
1969; Nyberg-Hansen, 1964). Large and small fibers from the bilateral vestibulospinal complex form the MVST and descend together with reticulospinal fibers
in the medial longitudinal fasciculus (MLF) with few or no fibers descending
below thoracic levels (Akaike, 1973; Nyberg-Hansen, 1964). As shown in Fig. 7C,
the MVST runs in a circumscribed area in the most medial part of the ventral
white matter; the majority of fibers terminate ipsilaterally in the same general area
as those of the LVST (Nyberg-Hansen, 1964). There is conflicting data concerning
the precise origins of this tract. According to anatomical data from the cat, a large
contribution comes from the medial nucleus (Nyberg-Hansen, 1964), though contributions from elsewhere are not ruled out. According to electrophysiological data
from cat and rabbit (Akaike, 1973; Akaike and Westerman, 1973; Rapaport,
Susswein, Uchino, and Wilson, 1977), MVST fibers arise mainly from the ventral
part of Deiters's nucleus and from the rostral part of the descending nucleus with
a minor contribution from the medial nucleus. However, stimulation of the medial
nucleus in the cat is very effective in producing short latency effects in motoneurons
(Wilson and Yoshida, 1969c).

As mentioned above, primary afferent fibers from the labyrinth provide one
major input to the vestibular complex. The labyrinthine receptors are divided into
those responding to linear acceleration (including gravity) and those responding to
angular acceleration. The former are found in the utricle, sensitive to acceleration
in the horizontal plane, and the saccule, which is sensitive to vertical acceleration.
Each of the three semicircular canals (horizontal, anterior, posterior) on each side
is sensitive to angular acceleration in its own plane. The labyrinthine information
is transmitted via the vestibular branch of the VIIIth nerve, the fibers of which
terminate almost exclusively in the vestibular complex. Fibers from the semicircular canals terminate mainly in the superior nucleus (which receives only canal
input) and the rostral parts of the medial and descending nuclei. Utricular fibers

terminate in the rostroventral part of Deiters' nucleus (that part which projects mainly to cervical spinal segments, see Fig. 8A) and also in the rostral parts of the medial and descending nuclei (Gacek, 1969; Stein and Carpenter, 1967). Utricular fibers go mainly to the descending nucleus in the monkey (Stein and Carpenter, 1967) and mainly to the medial nucleus in the cat (Gacek, 1969).

Fibers from the cerebellum constitute the largest afferent contingent to the vestibular complex (Walberg, 1972c), and they are highly organized. The patterns of the cerebellar projections to Deiters' nucleus are schematically indicated in Fig. 8. There are two direct projections from the Purkinje cells of the cerebellar cortex (Fig. 8B) as well as projections from each fastigial nucleus (Fig. 8C). The mono-synaptic connections from Purkinje cells have been demonstrated to be inhibitory (Ito, 1972; Ito and Yoshida, 1964); those from the fastigial nucleus are thought to all be excitatory (Ito, 1972; Ito, Udo, and Mano, 1970). One set of direct Purkinje cell fibers comes from the "vestibulocerebellum" (that part receiving afferents from the vestibular complex, principally the flocculus, but also parts of the paraflocculus, nodulus, and uvula; see Walberg, 1972c). All four vestibular nuclei are supplied by these fibers. The floccular projection ends primarily in areas of the vestibular nuclei receiving primary vestibular afferent fibers, whereas afferents from the nodulus and uvula reach areas supplied by the contralateral fastigial nucleus. Another set of direct Purkinje cell fibers comes from the "spinal" part of cerebellar cortex (that part receiving afferents from the spinal cord, chiefly the vermis of the anterior lobe; see Fig. 8B). Remarkably, these fibers terminate only in two regions (Brodal, 1969; Walberg, 1972c), the dorsal part of the descending nucleus and the dorsal part of Deiters' nucleus (that part free of primary afferents and projecting mainly to lumbar spinal segments; see Fig. 8A,B). The fastigial nuclei also receive afferents from the "spinal" part of the cerebellar cortex as shown in Fig. 8C. As further indicated in Fig. 8C, the caudal part of each fastigial nucleus projects to particular regions of each vestibular nucleus on the contralateral side; the rostral part of each fastigial nucleus projects ipsilaterally, also to all four vestibular nuclei, but to different regions than supplied by the contralateral fastigial nucleus. In particular, the rostral part supplies most of the medial vestibular nucleus and the dorsal part of Deiters' nucleus (that part which also receives direct fibers from the "spinal" cerebellar cortex; see Fig. 8A,B) (Brodal, 1969; Walberg, 1972c). The connections from the cerebellum are, thus, extensive, and the detailed pattern of connections are very intriguing, but the precise functional consequences of these connections are largely unknown.

All four vestibular nuclei also receive a large projection from the reticular formation (Hoddevik, Brodal, and Walberg, 1975). This projection is rather diffuse and comes from both sides of the midline, somewhat more from the ipsilateral side. The projection is mainly from the central part of the medial pontomedullary reticular formation (nucleus reticularis gigantocellularis and n.r. pontis caudalis). The heaviest projections are to the medial vestibular nucleus and Deiters' nucleus. No direct projection is found from the mesencephalic reticular formation, except from the interstitial nucleus of Cajal (Hoddevik et al., 1975; Walberg, 1972b). The latter projection appears to be concerned with vertical and rotational eye movements (Markham, Precht, and Shimazu, 1966). As will be discussed later in the chapter, reticular cells in the area projecting most heavily to the vestibular complex receive direct connections from cerebral cortex, superior colliculus, and cerebellum, as well as indirect inputs from other sources. Thus, as emphasized by Hoddevik *et al.* (1975), these sources can act on the vestibular nuclei via the reticular projections; in principle, the vestibular complex is removed from direct cortical control by one reticular interneuron.

Akaike and co-workers (Akaike, 1973; Akaike and Westerman, 1973; Akaike,

A

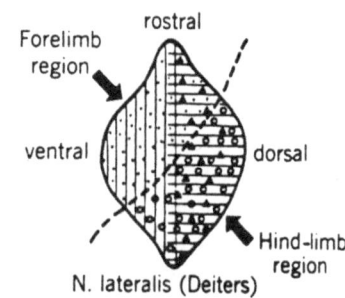

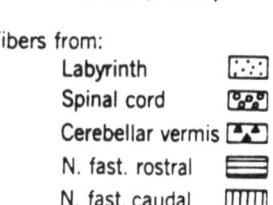

Fig. 8. The distribution of afferent fiber terminals in different regions of Deiters' nucleus in relation to its somatotopical organization in the cat. (A) Pattern of termination in Deiters' nucleus of afferent fibers from the indicated parts of the CNS and the labyrinth. Note that fibers from the labyrinth terminate mainly in the "forelimb region." (B) Region of cerebellar cortex containing Purkinje cells that project directly to Deiters' nucleus and the regions in Deiters' nucleus where their axons terminate. (C) The pattern of projections from the cerebellar cortex to the fastigial nucleus and from this nucleus to Deiters' nucleus. From Brodal (1969) with permission.

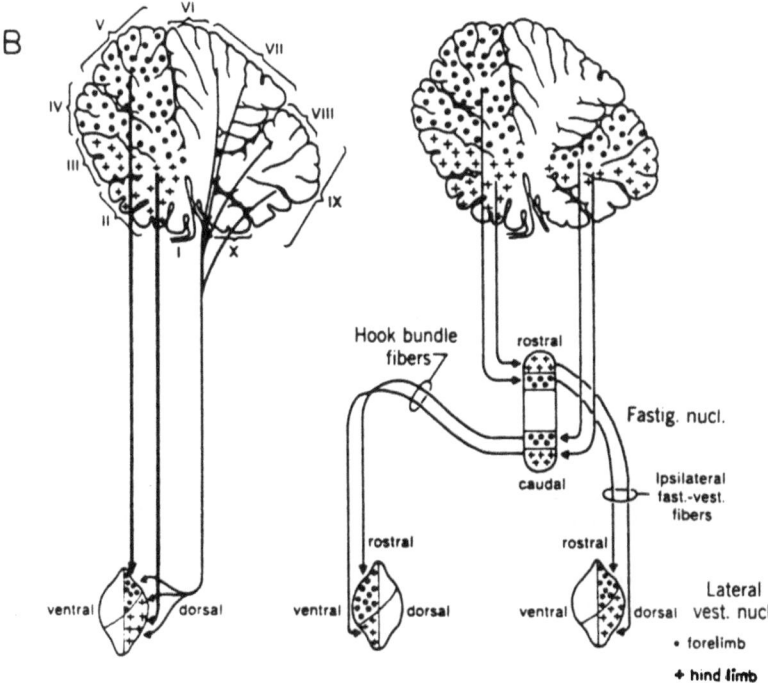

Fanardjian, Ito, Kamada, and Nakajima, 1973; Akaike, Fanardjian, Ito, and Ohno, 1973) have examined extensively the physiological properties of vestibular cells projecting to the spinal cord in both the cat and rabbit. In the rabbit, the distribution of conduction velocities for the sample of cells studied was found to be unimodal (mean: 97 m/sec; range: 30–140 m/sec), but they chose to classify the fibers as fast or slow using 75 m/sec as the dividing line. They used the criterion of lower antidromic stimulation threshold from electrodes placed medially or laterally on the ventral surface of the spinal cord at the C1 segment to decide if cells recorded in the vestibular nucleus belonged to the MVST or LVST, respectively. (The fibers constituting the MVST and LVST are fairly well separated at the C1 segment.) All fibers of the LVST were found to come from Deiters' nucleus. However, fibers of the MVST were also found to come from Deiters' nucleus, mainly from the ventral part, as well as from the descending nucleus in both the cat and the rabbit, rather than from the medial nucleus. These workers also noted if the recorded vestibular cells were monosynaptically activated from the VIIIth nerve (second-order cells) or not (non-second-order cells). The organization of the vestibulospinal projections found in the cat are diagrammatically indicated in Fig. 9A. In both the cat and rabbit, the LVST and MVST cell samples were approximately equally divided into second-order and non-second-order cells. Also, in both species, second-order LVST cells formed a distinct population having fast fibers, and non-second-order MVST cells formed another distinct population having slow fibers. Non-second-order LVST and second-order MVST samples were equally divided between fast and slow types in the rabbit, but many more slow, second-order MVST and fast, non-second-order LVST cells were seen in the cat. Where each fiber terminated in the spinal cord was also examined in the rabbit and cat. Most

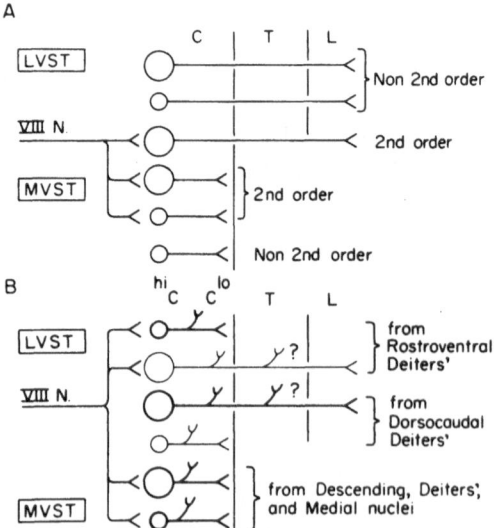

Fig. 9. The composition and spinal projections of the lateral (LVST) and medial (MVST) vestibulospinal tracts in the cat. (A) Organization of the LVST and MVST in the cat according to Akaike (1973). Each tract contains fast conducting (large circle) and slowly conducting (small circle) neurons. In each tract, some neurons are monosynaptically activated from the VIIIth nerve (second-order cells), whereas others are not (nonsecond-order cells). Most LVST neurons terminate in cervical (C) segments. (B) Schematic synthesis of the data of Akaike (1973), Abzug et al. (1974), and Rapaport et al. (1977) on the composition and projections of the LVST and MVST in the cat. Fast and slowly conducting neurons indicated as in (A). Most frequently encountered cell types indicated by darker lines. Branching axons indicate that most neurons terminating in lumbar segments (L) send collaterals to the ventral horn in cervical (C) and possibly in thoracic (T) segments. Similarly, most neurons terminating in low cervical (lo C) segments also send collaterals to the ventral horn in high cervical segments (hi C).

second-order MVST cells descended below the T1 segment in the rabbit, but only 11% of these cells did so in the cat, an observation in general agreement with anatomical data from the cat. Non-second-order MVST cells in both species terminated before reaching the L1 segment, but the great majority of both second-order and non-second-order LVST cells of both species terminated below T1, mainly in lumbar segments. It seems the main species difference between cat and rabbit consists of the level of termination of the second-order MVST fibers and the relative number of slow MVST and fast LVST cells. In both species, the medial vestibular nucleus was found to make a negligible contribution to the MVST, in disagreement with anatomical data (Nyberg-Hansen, 1964). The abundance of non-second-order cells, particularly in Deiters' nucleus, found in these studies again suggests the vestibular complex serves as more than a labyrinthine relay, though that role also is clearly important. The effects produced by the fast LVST population and the second-order MVST cells have been studied in detail and will be discussed below.

An interesting study on the course of LVST fibers has been made by Abzug, Maeda, Peterson, and Wilson (1974) in the cat. These investigators used microstimulation of the cervical gray matter to show that about half the LVST fibers reaching lumbar levels also send collaterals into the ventral gray matter at cervical levels. Most of the remaining cells projected only to cervical segments, and a small contingent projected only to lumbar levels. The cells projecting to lumbar segments but having cervical collaterals apparently constitute most of the "lumbar projecting" cells concentrated in the dorsocaudal part of Deiters' nucleus (which both anatomically and physiologically are found not to receive primary afferent fibers from the labyrinth). This population consists of both fast and slowly conducting fibers using the criterion of Akaike, Fanardjian, Ito, Kameda and Nakajima, (1973). Most fibers projecting only to cervical segments originated from the ventral part of Deiters' nucleus (that part receiving primary vestibular afferent fibers) and were mostly slowly conducting. Some cells projecting both to cervical and lumbar segments were also found in ventral Deiters' nucleus. These data have the intriguing implication that the same (presumably nonreflex) descending information is transmitted from the dorsal part of Deiters' nucleus to both cervical and lumbar segments, but the functional consequences of this are still unknown. It is noteworthy that such branching of descending axons to widely separated spinal segments has also been demonstrated to occur in the reticulospinal (Peterson, Maunz, Pitts, and Mackel, 1975) and corticospinal (Shinoda, Arnold, and Asanuma, 1976) tracts. Such branching has not yet been investigated in the rubrospinal tract.

Rapaport *et al.* (1977) have further investigated branching of vestibulospinal fibers projecting only to cervical segments in the cat and their distribution between the LVST and MVST. It was found that many fibers arising from either tract and terminating in the cervical enlargement (concerned with forelimb muscles) also send collaterals to the spinal gray matter in high cervical segments concerned with neck muscles. In agreement with the findings of Akaike, Fanardjian, Ito, Kamada, and Nakajima, (1973), the cell bodies of neurons sending fibers to the MVST were located mainly in the descending nucleus and Deiters' nucleus with a small contribution from the medial vestibular nucleus. Those from Deiters' nucleus mainly descended in the contralateral MVST. Fibers in the LVST arose almost entirely from Deiters' nucleus in agreement with anatomical studies. More than half of the cells contributing to the LVST, and essentially all those contributing to the MVST could be monosynaptically activated from the VIIIth nerve. A synthesis of the data of Akaike (1973), Abzug *et al.* (1974), and Rapaport *et al.* (1977) is presented in Fig. 9B.

VESTIBULOSPINAL EFFECTS ON ALPHA MOTONEURONS

Some comparison between the labyrinthine-relay and nonrelay function of the LVST and MVST paths to motoneurons has been obtained by comparing the quantity and latency of effects in motoneurons from central (e.g., Deiters' nucleus) stimulation with that evoked from the VIIIth nerve. As might be expected from the anatomical finding that primary vestibular afferents terminate in the area of Deiters' nucleus that projects mainly to cervical levels, synaptic connections to neck motoneurons are found to be monosynaptic from Deiters' nucleus and disynaptic from the VIIIth nerve, whereas connections even to forelimb motoneurons are found to be at least disynaptic from Deiters' nucleus and polysynaptic and very weak from VIIIth nerve stimulation. In this regard, the reflex function of the vestibular complex appears to be directed toward axial motoneurons, whereas it may play more of an integrative role for both proximal and distal limb musculature in the cat. A further functional dichotomy is suggested by the findings that the vestibulospinal tracts excite extensors and inhibit flexors, as earlier suggested by the finding that Deiters' nucleus is responsible for extensor tonus in decerebrate rigidity (Fulton, Liddell, and Rioch, 1930). The vestibulospinal tract appears to be much more consistent in this differential control than any other descending system. All work has focused on the shortest latency connections to spinal motoneurons, and most work has been concerned with effects from Deiters' nucleus. Unless otherwise noted, all results described below are from the cat.

EFFECTS ON AXIAL MOTONEURONS. In a study surveying the effects of stimulation of Deiters' nucleus on various spinal motoneurons, Wilson and Yoshida (1969b) found that stimulation of the ipsilateral Deiters' nucleus produced monosynaptic (followed by polysynaptic) EPSPs in most neck motoneurons, including populations of identified motoneurons of head elevators. In these cells, ipsilateral VIIIth nerve stimulation produced a disynaptic EPSP similar in shape to the monosynaptic EPSP from Deiters' nucleus (Wilson and Yoshida, 1969a,b). Thus, Deiters' nucleus appears to be a relay for this EPSP. However, other neck motoneurons receive a disynaptic IPSP from ipsilateral VIIIth nerve stimulation and a similarly shaped monosynaptic IPSP at low threshold from brain stem stimulating electrodes placed in the ipsilateral medial vestibular nucleus or the MLF (Wilson and Yoshida, 1969a,c). Thus, the IPSP appears to be produced by fibers that descend in the MVST component of the MLF.

A similar situation is found for motoneurons innervating vertebral column extensors in the T1–T10 spinal segments (Wilson, Yoshida, and Schor, 1970): an EPSP is obtained monosynaptically from the ipsilateral Deiters' nucleus and disynaptically from the ipsilateral VIIIth nerve. A disynaptic VIIIth nerve EPSP may additionally be transmitted by the MVST from the ipsilateral medial vestibular nucleus (Akaike, Fanardjian, Ito, and Ohno, 1973). Again, a disynaptic IPSP is seen, but in this case from the contralateral VIIIth nerve, and a similar monosynaptic IPSP is seen from MLF stimulation. The disynaptic VIIIth nerve IPSP is apparently relayed from the contralateral medial vestibular nucleus and, interestingly, is transmitted by a distinct population of slow fibers (69 m/s mean C.V.), possibly the slow fraction of the secondary MVST fibers studied by Akaike, Fanardjian, Ito, Kamada, and Nakajima (1973).

Because the postsynaptic potentials obtained from both the VIIIth nerve and central stimulation are similar in size and are of short latency, the vestibular nuclei appear to serve mainly as relays for excitatory and inhibitory effects from the labyrinthine receptors to these axial (neck and back) motoneurons. It is interesting for the theory of motor control that both excitatory and inhibitory vestibulospinal effects are monosynaptic to axial motoneurons and, as will be seen below, are partially reciprocal between different labyrinthine receptors on the same and opposite side. In contrast to the axial musculature, reciprocal inhibition of limb motoneurons from vestibulospinal and other descending pathways is mediated by excitatory fibers that activate inhibitory segmental interneurons. The independent, monosynaptic, excitatory, and inhibitory paths from the vestibular nuclei to axial motoneurons may ensure more precise control of the axial musculature by the labyrinthine reflexes. In addition, the direct reciprocal connections to axial motoneurons may be required for proper coordination of the activity of motor nuclei on each side of axial structures; whereas bilateral coordination of limbs is aided by crossed spinal reflexes, such as the crossed extension reflex (see below), such crossed reflexes are not seen in neck motoneurons (Anderson, 1977). The demonstration of the monosynaptic inhibition of neck and back motoneurons is also interesting because it provides the first demonstration of long descending inhibition. Previously, all effects on motoneurons or interneurons that were clearly monosynaptic from descending tracts were excitatory, and it was thought inhibition was mediated only by local interneurons.

Subsequent studies on neck motoneurons have sought to discover the effects from each type of labyrinthine receptor. These complex but stereotyped effects are summarized in Table I. Neck extensor (head elevator) motoneurons predominantly receive disynaptic EPSPs from each anterior semicircular canal (contralateral effects being stronger) and disynaptic IPSPs from each posterior canal. Lateral head flexor motoneurons instead receive predominantly disynaptic EPSPs from the contralateral horizontal canal and disynaptic IPSPs from the ipsilateral horizontal canal (Wilson and Maeda, 1974). The disynaptic EPSP from the ipsilateral anterior canal is mediated by the ipsilateral LVST; all other effects are mediated by the MVST, and crossing of contralateral fibers occurs in the brain stem. The saccule appears to exert di- and trisynaptic effects, generally exciting neck extensor motoneurons ipsilaterally and more strongly inhibiting them contralaterally; the utricle usually has just the opposite disynaptic effect (Wilson, Gacek, Maeda, and Uchino, 1977). The postsynaptic potentials from the utricle and saccule are very small. The exact pathways to motoneurons are unknown, but because some effects are disynaptic, and there are no disynaptic reticulospinal paths, the effects must be relayed in the vestibular nuclei. The short-latency effects from the semicircular canals appears to be stronger than those from the utricle and saccule (the saccular effects are weakest) even though the utricle and saccule are thought to play a major role in maintaining antigravity posture. The relation of these synaptic connections to the functional labyrinthine reflexes will be considered later in this section.

EFFECTS ON LIMB MOTONEURONS. During a survey of the effects of Deiters' nucleus stimulation on various spinal motoneurons, Wilson and Yoshida (1969b) failed to detect monosynaptic EPSPs in forelimb motoneurons. Stimulation of Deiters' nucleus did, however, produce polysynaptic EPSPs and IPSPs. EPSPs predominated in identified extensor motoneurons and mixed digit and wrist motoneurons; IPSPs predominated in flexor motoneurons, though the latter also sometimes received EPSPs. No effects at all from the VIIIth nerve were seen in these barbiturate-anesthetized animals, but in a subsequent study on decerebrate cats (Maeda, Maunz, and Wilson, 1975), effects from the VIIIth nerve and individual semicircular canal nerves could be seen. In this study, bilateral VIIIth nerve stimulation usually excited extensor motoneurons at least trisynaptically, whereas flexor motoneurons were inhibited through at least four synapses. Effects from the contralateral side appeared to be stronger, but the postsynaptic potentials from either side

TABLE I. PRINCIPAL SHORT LATENCY SYNAPTIC ACTIONS ON SPINAL MOTONEURONS FROM LABYRINTHINE RECEPTORS[a]

Neck motoneurons			
Receptor stimulated	Effect	Pathway	Type of motoneuron
Utricle, ipsilateral side	IPSP	Vestibulospinal	Motoneurons of head elevators
contralateral side	EPSP		
Saccule, ipsilateral side	EPSP	Vestibulospinal	
contralateral side	IPSP		
Anterior canal, either side	EPSP	LVST	
Posterior canal, either side	IPSP	MVST	
Horizontal canal, ipsilateral	IPSP	MVST	Motoneurons of lateral neck flexors
contralateral	EPSP	MVST	

Back motoneurons	
(Vertebral column extensors)	
Structure Stimulated	Effect
Ipsilateral VIIIth N. (all receptors)	di- and trisynaptic EPSP's via LVST
Contralateral VIIIth N. (all receptors)	di- and trisynaptic IPSP's via MVST

Limb motoneurons	
(Effects via LVST and local spinal interneurons)	
Structure stimulated	Effect
Either VIIIth N. (all receptors)	Polysynaptic EPSP's to extensor motoneurons
	Polysynaptic IPSP's to flexor motoneurons
Any canal nerve	Same pattern as above, smaller EPSP's and IPSP's

[a]From the results of Wilson and co-workers; see text for references.

were small and required repetitive stimulation to be observed. Stimulation of semi-circular canal nerves produced bilateral excitation of extensor and bilateral inhibition of flexor motoneurons, independent of the particular canal stimulated. The postsynaptic potentials from the semicircular canals were considerably weaker and more variable than obtained from VIIIth nerve stimulation, suggesting the saccule and utricle also make an important contribution to the VIIIth nerve effects. The results of various lesions of the medulla and spinal cord indicated that the contralateral effects were transmitted by the ipsilateral LVST with a relay to the contralateral side by interneurons that cross the midline in the spinal cord. Ipsilateral effects may also be partially transmitted by reticulospinal cells, but the MLF pathway is not involved. The contrast provided by these studies between the more powerful, short-latency effects obtained by Deiters' nucleus stimulation and the weak, long-latency effects from VIIIth nerve stimulation strongly suggest that Deiters' nucleus does not primarily serve as a labyrinthine reflex relay to forelimb motoneurons.

Similar direct comparisons of Deiters' nucleus and VIIIth nerve effects have not been made in the hindlimb, but Deiters' nucleus presumably serves other than labyrinthine relay functions there as well. The principal hind-limb effects from Deiters' stimulation are essentially similar to those observed in the forelimb, but more extensive investigations of the various motor nuclei have been made (Grillner, Hongo, and Lund, 1970; Wilson and Yoshida, 1969b). The principal effects are disynaptic EPSPs to extensor and IPSPs to flexor motoneurons, though IPSPs are also sometimes found in hip extensor motoneurons and EPSPs are often found in pretibial flexor motoneurons. In addition to these more powerful disynaptic effects, small monosynaptic EPSPs are seen in some hind-limb motor nuclei. These monosynaptic EPSPs occur principally in knee and ankle extensor, but never in flexor, motoneurons. The small monosynaptic EPSPs are not seen in all motoneurons nor in all cats (Grillner *et al.,* 1970; ten Bruggencate and Lundberg, 1974). However, they provide the only supraspinal monosynaptic input to extensor motoneurons in the cat (Lund and Pomperiano, 1968). In addition to the ipsilateral effects described above, similar effects from the contralateral Deiters' nucleus are also seen in hind-limb motoneurons, but these will be discussed in the section on segmental reflex control.

Only the monosynaptic EPSP to hind-limb motoneurons produced by stimulation of Deiters' nucleus has been studied in the monkey (Shapovalov, 1972). The monosynaptic EPSPs are said to occur principally in knee extensor motoneurons but also are seen in ankle extensor motoneurons. Cells receiving the monosynaptic vestibulospinal EPSP also often receive monosynaptic EPSPs from the reticular formation, but never from the corticospinal tract (in contrast to the convergence of cortico- and rubrospinal EPSPs in distal limb motoneurons). Injection of procion yellow dye in the recorded cells indicated that they were typically located ventromedially in the spinal gray matter, near the termination of vestibulospinal tract fibers. These findings in the monkey are generally compatible with the differential function of descending tracts deduced by Lawrence and Kuypers (1968b) from their lesion studies, in the sense that either rubrospinal and corticospinal or vestibulospinal and reticulospinal monosynaptic EPSPs are found in certain motoneurons. However, the fact that vestibulospinal and reticulospinal monosynaptic EPSPs occur in distal limb motoneurons (e.g., ankle extensors) do not support the idea that these descending tracts are important only for control of proximal musculature.

INTERACTION WITH SPINAL REFLEXES AND CONTRALATERAL VESTIBULOSPINAL EFFECTS

As mentioned above, motoneurons of the knee extensor, quadriceps, are one of the two main recipients of the small monosynaptic EPSP from Deiters' nucleus in the cat, and the antagonistic knee flexor motoneurons receive disynaptic inhibition. Spatial facilitation between the two disynaptic inhibitory effects (Grillner, Hongo, and Lund, 1966) has shown this inhibition is mediated by the interneuron in the disynaptic path from quadriceps group Ia afferent fibers to motoneurons of antagonistic flexors acting at the knee joint, as indicated in Fig. 10B. Excitation of these group Ia inhibitory interneurons by Deiters' nucleus appears to be mediated by the same population of large, fast fibers evoking the monosynaptic EPSP in quadriceps (Grillner *et al.*, 1966; Hultborn and Udo, 1972; Hultborn, Illert, and Santini, 1976c). The reverse effects, enhancement of the group Ia disynaptic IPSP from flexors to extensors by Deiters' nucleus is not seen. In fact, the knee flexors are the *only* group that have disynaptic inhibition from Deiters' mediated via group Ia inhibitory interneurons from antagonists, though all other flexors receive disy-

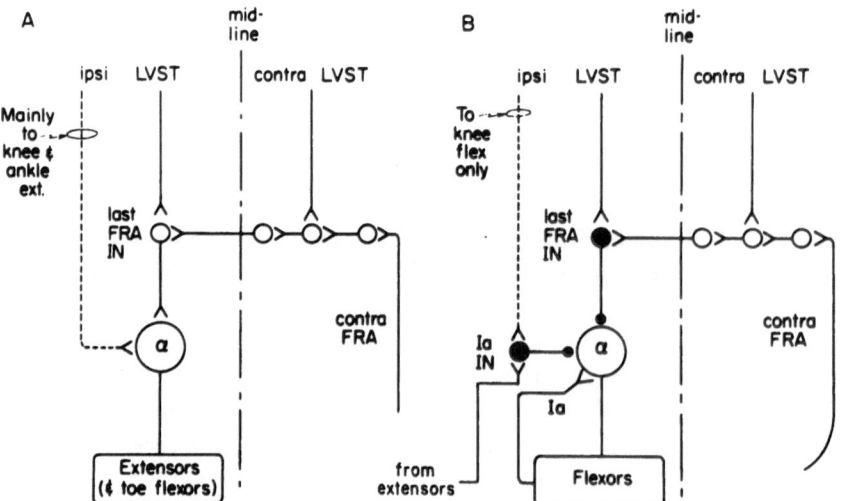

Fig. 10. The shortest pathways from the ipsilateral (ipsi LVST) and contralateral (contra LVST) lateral vestibulospinal tracts to hindlimb α motoneurons (α) in the cat. Pathways to extensor motoneurons are shown in (A) to flexor motoneurons in (B). The disynaptic effects from the ipsilateral LVST are exerted through the last interneurons of the chain from the contralateral flexor reflex afferent fibers (contra FRA). Effects from the contralateral LVST are also exerted through this interneuron chain, but it is not known precisely which interneurons of the chain receive contralateral LVST innervation. Note that only knee flexor motoneurons receive disynaptic inhibition mediated by the group Ia inhibitory interneuron (Ia IN).

naptic inhibition from Deiters' nucleus. Furthermore, the above example provides the *only* known polysynaptic interaction between the ipsilateral Deiters' nucleus and spinal reflex arcs originating from afferent fibers on the ipsilateral side (Grillner *et al.*, 1966; ten Bruggencate and Lundberg, 1974; ten Bruggencate, Burke, Lund, and Udo, 1969). As will be discussed below, and as diagrammatically indicated in Fig. 10, the ipsilateral Deiters' nucleus exerts its polysynaptic effects on ipsilateral motoneurons by using reflex arcs originating from afferent fibers on the *contralateral* side. This provides a contrast with the interactions of the cortico-, rubro-, and reticulospinal tracts with reflex arcs originating from ipsilateral afferent fibers.

The action of ipsilateral FRA fibers and their relation to the ipsilateral flexion reflex has been outlined in the section "Spinal Reflex Arcs and the Spatial Facilitation Technique." Related to the flexion reflex is the crossed extension reflex (Sherrington, 1910), which consists of activation of ipsilateral extensor motoneurons and inhibition of ipsilateral flexor motoneurons evoked by FRA fibers from the contralateral side of the animal. Both the flexion reflex and crossed-extension reflex were interpreted by Sherrington (1910) as being protective: the former withdraws the ipsilateral limb from noxious stimuli while the latter transfers support to the opposite limb. It will be recognized that the effects on ipsilateral motoneurons produced by stimulation of the ipsilateral (or contralateral, see below) Deiters' nucleus are identical to those of the crossed-extension reflex, perhaps suggesting a relationship between these pathways. In fact, as indicated in Fig. 10, the disynaptic ipsilateral effects are mediated by the last interneurons of the crossed-extension reflex pathway (ten Bruggencate and Lundberg, 1974; ten Bruggencate *et al.*, 1971). This was shown in a wide variety of hind-limb motoneurons by the spatial facilitation of disynaptic postsynaptic potentials from the ipsilateral Deiters' nucleus by stimulation of contralateral FRA fibers. Since the contralateral FRA pathway is highly polysynaptic, and facilitation occurred at long latency, special precautions were required to ensure the facilitation was due to interaction at segmental rather than brain stem levels. The facilitation of effects from the ipsilateral Deiters' nucleus by the contralateral FRA was produced only by high-threshold contralateral afferent fibers, not by group I or low-threshold cutaneous fibers. Effects produced in individual ipsilateral motoneurons by the contralateral FRA and by the ipsilateral Deiters' nucleus were identical, not only for the usual pattern of excitation and inhibition but also for the "exceptional" patterns seen in some hip extensors and most pretibial flexors.

It has also been found that stimulation of the *contralateral* Deiters' nucleus evokes effects in *ipsilateral* hind-limb motoneurons generally similar to those obtained in the same motoneurons from ipsilateral Deiters' nucleus stimulation (see Fig. 10), that is, excitation of extensor and inhibition of flexor motoneurons (Aoyama, Hongo, Kudo, and Tanaka 1971; Hongo, Kudo, and Tanaka, 1971, 1975). In these experiments, several precautions were taken to ensure that the effects from the contralateral Deiters' nucleus were solely from impulses descending in the contralateral LVST. Latency measurements indicated the EPSPs from the contralateral Deiters' nucleus were at least disynaptic, whereas the IPSPs were at least trisynaptic. Both are smaller (about half the amplitude) than the disynaptic postsynaptic potentials of like sign evoked from the ipsilateral Deiters' nucleus (Hongo *et al.*, 1971, 1975). Various lesion experiments indicated the chain of interneurons involved were located in and their axons crossed the midline at lumbar spinal segments (Hongo *et al.*, 1971, 1975). Furthermore, spatial facilitation of the effects on motoneurons from the contralateral Deiters' nucleus by the ipsilateral Deiters' nucleus and vice versa indicated that the effects from both the contralateral and ipsilateral Deiters' nucleus were transmitted to the motoneurons by a common final interneuron as indicated in Fig. 10 (Aoyama *et al.*, 1971). For example, stimulation of either the contra- or ipsilateral Deiters' nucleus could be adjusted so that the stimulation of each nucleus alone was just subliminal for a postsynaptic potential in a recorded motoneuron but subsequent conditioning of the ipsilateral by the contralateral Deiters' nucleus, or vice versa, resulted in a postsynaptic potential in the motoneuron. Since effects from the ipsilateral Deiters' nucleus are disynaptic, the common interneuron must be the last interneuron of the pathway from the contralateral Deiters' nucleus. As discussed earlier, this interneuron is *also* the last interneuron of the reflex pathway from the contralateral FRA fibers.

As discussed above, stimulation of the contralateral Deiters' nucleus generally produced excitation of ipsilateral extensor motoneurons and inhibiton of ipsilateral flexor motoneurons, similar to effects from the ipsilateral Deiters' nucleus. However, excitation of ipsilateral flexor motoneurons by the contralateral Deiters' nucleus could sometimes be observed. In fact, the dominant effect of the contralateral Deiters' nucleus on ipsilateral flexor motoneurons could vary under different types of anesthetics, during the course of an experiment, or even in the same motoneuron over time or as the stimulation parameters were varied (Hongo *et al.*, 1975). However, the effects could not be varied by stimulation at different points in the contralateral Deiters' nucleus. Alternative pathways from the contralateral FRA that produce effects opposite to those of the crossed extension reflex on ipsilateral motoneurons are known to exist (Holmqvist, 1961). The variable effects seen by Hongo *et al.* (1975) on some ipsilateral flexor motoneurons after stimulation of the contralateral Deiters' nucleus suggest that the contralateral LVST can transmit effects through interneurons of the alternative contralateral FRA pathway in addition to the crossed extension pathway. Which contralateral FRA pathway is effective at the moment is probably under descending control and can vary with the state of the preparation. The conditions determining which pathway may dominate, or under what conditions paths are switched, in the awake, behaving animal are unknown, but a general conclusion from these studies is that a given descending or segmental polysynaptic pathway may produce more than one effect; alternate pathways and effects may be available and may dominate in different behavioral and experimental conditions. Further examples of these alternative pathways will be given below and later in the chapter.

In summary, the results discussed above and diagrammed in Fig. 10 suggest that the effects on a given motoneuron from the ipsilateral or contralateral Deiters' nucleus are usually similar. The ipsilateral Deiters' nucleus exerts its disynaptic effects through the last interneuron of the pathways from the contralateral FRA; the results of Aoyama *et al.* (1971) suggest that the polysynaptic effects from the contralateral Deiters' nucleus and the disynaptic effects from the ipsilateral Deiters' nucleus are exerted through a common final interneuron. This final interneuron is that of the crossed-extension pathway from the contralateral FRA fibers. Vestibulospinal control thus depends on integration of descending information from the Deiters' nucleus on each side of midline with that from contralateral limb FRA afferent fibers. Increasing or decreasing activity in the contralateral FRA fibers may alternatively reinforce and disfacilitate the effects from each Deiters' nucleus under different behavioral conditions.

Effects of Natural Labyrinthine Stimulation: Comparison with Synaptology

As discussed earlier, one important function of the vestibular complex is to relay information from the labyrinthine receptors to the spinal cord. In theory, study of the labyrinthine reflexes provides an opportunity to compare known vestibulospinal synaptology with an important, normal, quantifiable mode of motor control. Such a comparison would be valuable for examining whether known synaptic circuits are necessary and sufficient to "explain" the reflex behavior and for understanding how these circuits are normally used. Unfortunately, such a comparison of known synaptology and reflex function is difficult, principally because most synaptology is avaiable for neck musculature, on which the least information is available about the normal labyrinthine reflexes, while the opposite situation holds for limb musculature. A comparison in limb musculature is additionally difficult because different workers have used somewhat different preparations and methods to demonstrate the reflexes and have often obtained somewhat different or even contradictory results on the nature of the reflexes. However, a brief review of available data will at least indicate where additional work would be useful.

Effects from the Semicircular Canals. The labyrinthine reflexes are usually considered to aid in maintaining postural equilibrium in the face of external perturbations; the reflexes from the semicircular canals, in particular, help to maintain head and eye fixation. It is first useful to consider the patterns of excitation of primary afferent fibers from the canals (see Fig. 11A). Angular acceleration to the ipsilateral side in the horizontal plane excites ipsilateral horizontal canal afferents. Afferent fibers from both the anterior and posterior canals on the ipsilateral side are excited by rotation toward the ipsilateral side in the frontal plane (called roll or tilt). Nose down rotation in the sagittal plane excites receptors in anterior canals on each side of the midline, while nose up rotation excites receptors in both posterior canals. The above patterns define the actions exciting canal afferent fibers for rotation in the three "natural," orthogonal planes about which the animal is commonly rotated in experimental work; effects from rotations in other planes can be deduced from these and from the diagram in Fig. 11A. It should also be realized that rotation in a direction exciting a given canal on one side will reduce activity in afferent fibers from the "antagonistic" canal on the contralateral side. Therefore, unless there are bilateral excitatory connections to the motoneurons involved, rotation should automatically excite certain muscles on one side of the body and relax their counterparts on the other side. In this context, it should be noted that vestibular neurons subserving canal function (but not oto-lith-activated vestibular neurons), in addition to relaying information from primary afferent information to the spinal cord and extraocular nuclei, also inhibit those contralateral vestibular neurons subserving the "antagonistic" contralateral canal (Kasahara and Uchino, 1974; Shimazu and Precht, 1966). However, this effect simply adds inhibition to contralateral vestibular neurons that are already disfa-cilitated from the antagonistic canal afferent fibers and would not be expected to modify the net actions of the shortest vestibulospinal pathways. It should also be mentioned that, because of the physical characteristics of the canals and the asso-ciated receptor organs, a mathematical integration of the angular acceleration sig-nal with respect to time is performed at the canal. The afferent fiber information actually transmitted to the brain stem is actually more related to head angular velocity than to angular acceleration (Fernandez and Goldberg, 1971; Melville-Jones and Milsum, 1970; Shinoda and Yoshida, 1973).

Reflexes to neck muscles will be considered first. Given the patterns of acti-vation of canal afferents listed above and the principal short-latency synaptic con-nections (Wilson and Maeda, 1974; see also Table I) from canal afferents to neck motoneurons elevating the head (bilateral excitation from anterior canals; bilateral inhibition from posterior canals) and to neck lateral flexor motoneurons (excitation

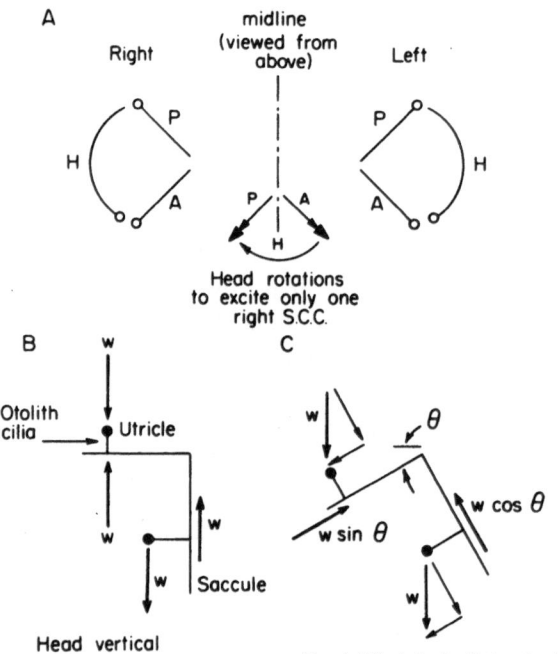

Fig. 11. The adequate natural stimuli that excite afferent fibers from the labyrinthine receptors. (A) General orientation of the horizontal (H) semicircular canals (S.C.C.), and the poste-rior (P) and anterior (A) vertical semicircular canals and their associated ampullae (circles). Arrows below the diagram indi-cate the directions of head rota-tion to excite only one right canal. If one's right thumb is pointed in the direction of the double arrows, rotation in the direction of the fingers of the right hand about the axis of the thumb will indicate the direc-tions of rotation to excite the A or P canals. (B) A force W resulting from the acceleration due to gravity is exerted on the otoliths and transmitted by the stereo- and kinocilia to the ver-tical macula of the saccule, and the macula of the utricle, which lies mostly near the horizontal. The adequate stimulus for the sensory hair cells in the maculae is the tangential component of the applied force. (C) With the head tilted $\theta°$ to the ipsilateral side, the tangential component of the force on the utricular macula (W sin θ) increases, whereas the tangential component on the saccular macula (W cos θ) decreases. See text for further explanation.

from contralateral horizontal canal; inhibition from ipsilateral horizontal canal), the synaptic actions will produce neck muscle contraction tending to maintain a fixed head position under a perturbing angular rotation in any direction. The only dubious situation is rotation in the frontal plane, since a variety of synaptic effects converge on motoneurons of head elevators. However, the strongest synaptic effect is usually excitation from the contralateral anterior canal, and this effect may be assumed to dominate. It is then predicted that ipsilateral neck extensors are activated during frontal plane rotation to the contralateral side. The various neck movements predicted from synaptology are, in fact, seen during sustained electrical stimulation of individual canal nerves (Suzuki and Cohen, 1964).

Several recent studies of the effects of natural labyrinthine stimulation on neck motoneurons (and forelimb extensor motoneurons, see below) have investigated the dynamic response of motor units to sinusoidally varying angular accelerations of different frequencies and amplitudes. Since the firing frequency of some individual motor units, or of multimotor unit rectified and smoothed EMG responses, were often found to be linearly related to the input angular acceleration, the powerful mathematical methods of linear systems analysis could be used to characterize quantitatively the motor output due to natural stimulation of the various labyrinthine receptors. The methods of linear systems analysis have also been successfully applied to the vestibulo-ocular reflexes (see Chapter 6). Before discussing the results of these studies, it should be mentioned that, unlike the vestibulo-ocular reflexes, the vestibulospinal reflexes function as part of a closed feedback loop in the freely moving animal. That is, contraction of neck or limb muscles due to neural activity originating from the labyrinthine receptors normally modifies head position or velocity and, thus, modifies the subsequent labyrinthine stimulation, motoneuron output, and muscle contraction. In the experiments to be discussed in the remainder of this section, the muscles studied were fixed and were contracting approximately isometrically so that the feedback loop could not function. The data obtained thus are not a complete description of normal labyrinthine-induced motoneuron output, but they do provide insight into several aspects of the conversion of labyrinthine receptor input into motoneuron output.

In studying labyrinthine reflexes, it is also important to ensure, as have the investigators in the studies to be discussed, that the effects seen upon head rotation arise solely from labyrinthine input and are not contaminated by effects from neck reflexes. These powerful reflexes arise from receptors in the C1 to C3 vertebral joints (McCouch, Deering, and Ling, 1951), which are excited by rotation of these joints (Magnus, 1926). Their effects on motoneurons are often just opposite to the labyrinthine effects (Erhardt and Wagner, 1970; Magnus, 1926).

Berthoz and Anderson (1971b) first used variable frequency sinusoidal angular acceleration to study the response of neck muscle motor units in the decerebrate cat. Most motor units fired phasically at angular accelerations above a certain threshold value. Because of this and other nonlinearities, linear systems analysis could only be applied to a few motor units. However, it was observed that ipsilateral neck motor units exhibited maximal firing during rotation toward the contralateral side in the frontal plance (cheek rotated up on recording side) and during rotation of the animal toward the contralateral side in the horizontal plane (nose rotated away from recording side). These results are as predicted from synaptology (see Table I and Fig. 11A).

Two recent studies (Ezure and Sasaki, 1978; Ezure, Sasaki, Uchino, and Wilson, 1978) have examined more extensively the response of neck motor units in the decerebrate cat to angular acceleration in the horizontal plane. In these studies, the response of many motor units and of the multimotor unit, rectified, smoothed EMG response was amenable to linear systems analysis. The maximal ipsilateral motor unit firing was found to occur for horizontal rotation toward the contralateral side, in agreement with the data of Berthoz and Anderson (1971b) and the predictions of synaptology. The peak motor-unit firing or EMG activity lagged the peak angular velocity to the contralateral side by nearly 90°; that is, the ipsilateral neck muscles were most active when the head was maximally rotated toward the contralateral side. This is the muscle response required to aid head fixation and counter the applied rotational perturbation, but the large phase shift between peak velocity and peak EMG activity strongly suggests this EMG response is not mediated by the short latency pathways from the labyrinth demonstrated by electrical stimulation. As mentioned earlier, the peak firing of primary afferent fibers from the canals and vestibular neurons in the brain stem are approximately in phase with angular velocity. The 90° phase lag of motor-unit firing with respect to velocity implies that somewhere between the afferent fibers or vestibular neurons and the motoneurons, additional neural processing occurs that is mathematically equivalent to an integration with respect to time of the velocity signal supplied by the canal afferents. (A similar finding has been made for the vestibulo-ocular reflexes, see Chapter 6). The monosynaptic pathway from the vestibular neurons to neck motoneurons cannot account for a 90° phase shift. Therefore, the additional neural processing must occur elsewhere, possibly in the cerebellum or reticular formation. It was, in fact, directly demonstrated (Ezure *et al.,* 1978) that the short latency pathways are not responsible for the naturally evoked reflex since it persisted unchanged after section of the MVST, fibers of which are known to transmit the short latency effects from the horizontal canals to neck motoneuron (Wilson and Maeda, 1974). These results do not imply that the role of the shortest vestibulospinal pathways is insignificant under all conditions. The results discussed above (and in other studies to be discussed below) were obtained at low frequencies of rotation (≤ 1 Hz), but, theoretically, the polysynaptic pathways mediating the low frequency effects would be expected to become less effective at high frequencies, and the short latency pathways would be expected to become relatively more effective. The results discussed above do, however, provide a caution against *a priori* assumptions that (1) motor effects qualitatively consistent with pathways discovered by electrophysiology must be mediated by those pathways, and (2) that the shortest latency pathways (which are the ones usually investigated by electrophysiology) are necessarily the most important ones under all circumstances.

It is equally difficult to find a correlation between the results of synaptology and the effects of natural semicircular canal stimulation on limb motoneurons. Electrical stimulation of any canal nerve was found to evoke EPSPs in forelimb extensor and IPSPs in forelimb flexor motoneurons on both sides of the spinal cord through at least three or four synapses (Maeda *et al.,* 1975; see also Table I). In light of these results, it is interesting that the type of horizontal-canal-activated vestibular neuron found most often to project to the spinal cord (even occasionally to lumbar segments) is the so-called Type III vestibular neuron, which is activated by horizontal rotation in either direction (Precht, Grippo, and Wagner, 1967). In contrast to the study of polysynaptic potentials in motoneurons referred to above, it has been found that sustained electrical stimulation of any canal nerve in the awake cat or monkey produces ipsilateral forelimb extension and contralateral forelimb flexion (Suzuki and Cohen, 1964). Although these latter results contrast with those from the synaptic studies, they are generally consistent with the results of natural labyrinthine stimulation (see below).

Gernandt and Thulin (1953) first showed that angular acceleration of the decerebrate cat in the horizontal plane toward the contralateral side (nose rotated away from recording side) results in facilitation of hind-limb extensor and inhibition of hind-limb flexor motoneurons; the opposite effects are seen for horizontal angular acceleration to the ipsilateral sides. These results were confirmed by Ehrhardt and Wagner (1970). Changes of the monosynaptic reflex evoked from group Ia afferent fibers of the various muscle nerves and recorded on the ventral roots were used to demonstrate facilitation and inhibition of the motoneurons by the horizontal canal input, but Ehrhardt and Wagner (1970) also found that the firing rate of individual identified hind-limb flexor and extensor motoneurons was only infrequently modulated by angular acceleration in the horizontal plane. A similar observation has been made by Berthoz and Anderson (1971b) for forelimb extensor motor units.

A recent series of experiments (Anderson, Berthoz, Soechting, and Terzuolo, 1977; Anderson, Soechting, and Terzuolo, 1977a; Anderson, Soechting, and Terzuolo, 1977b; Soechting, Anderson, and Berthoz, 1977) have examined quantitatively the dynamic response of forelimb extensor motor units to natural stimulation of the semicircular canals and otolith organs in the decerebrate cat. The labyrinthine input was sinusoidally varying angular or linear acceleration at different frequencies (≤ 1 Hz), and it was found that linear systems analysis could be applied if the output were taken to be the rectified, averaged, multimotor-unit EMG response of a forelimb extensor muscle (the triceps brachii). In order to obtain reliable EMG responses, it was necessary, however, to enhance the forelimb extensor rigidity by low-spinal cord section (the Shiff-Sherrington phenomena; Ruch and Watts, 1934). Ipsilateral motor units of the forelimb extensor were maximally active during rotation toward the contralateral side for angular acceleration in the horizontal plane, in agreement with the results of Ehrhardt and Wagner (1970) and Berthoz and Anderson (1971b). However, as found in the studies of neck motor units, the maximal forelimb extensor activation lagged the peak angular velocity by nearly 90°, suggesting that extensive neural processing mathematically equivalent to an integration with respect to time of the velocity signal supplied by primary canal afferents is performed somewhere in the CNS. As discussed for the neck reflexes, this phase shift cannot be accounted for by short latency synaptic pathways from the vestibular complex. Gernandt and Thulin (1953) concluded that the effects on hind-limb motoneurons obtained by angular accelerations in the horizontal plane were mainly transmitted by reticulospinal tracts since these effects persisted after section of the LVST in the spinal cord.

The results of angular acceleration in the frontal plane are more complex. Berthoz and Anderson (1971a) found that ipsilateral forelimb extensor motor units were activated by rotation toward the ipsilateral side (cheek down on the recording side), consistent with the data of Soechting *et al.* (1977), and maximal motor unit firing lagged angular velocity by almost 30°. However, Soechting *et al.* (1977) pointed out that, unlike angular rotation in the horizontal plane about an axis midway between the labyrinths, rotation in the frontal plane about the longitudinal axis of the cat would dynamically stimulate the otolith organs (the utricle and saccule) in addition to the vertical canals, since the former are off the axis of rotation in this case. Because the dynamic reflexes from the otolith organs had previously been determined (Anderson *et al.*, 1977a), it was possible to evaluate quantitatively the relative contributions of the vertical canals and the otolith organs to the forelimb extensor response. It was inferred that at low rotational frequency (ca. 0.1–0.5 Hz) the forelimb extensor motor units were mainly activated by impulses originating from the utricles, whereas the vertical canal input became important only at higher frequencies of rotation. The maximal forelimb extensor EMG acitivity occurred during ipsilateral cheek-up rotation at low frequencies and ipsilateral cheek-down rotation at higher frequencies, reflecting the dominant influence of the utricle and vertical canals, respectively, in each frequency range.

EFFECTS FROM THE OTOLITH ORGANS. The otolith organs, the utricle and saccule, appear to aid the animal to maintain an appropriate position in space. The physiology of these receptor organs is somewhat more complicated than that of the semicircular canals. Perhaps this fact accounts, in part, for the often contradictory descriptions of their actions on motoneurons. As indicated diagrammatically in Fig. 11B, a force W acting on the otoliths due to an imposed linear acceleration, including that due to gravity, may be resolved into components normal and tangential to the maculae of these organs. It is the tangential component of force, transmitted to the sensory hair cells by their kinocilia and stereocilia, that provides the adequate stimulus for the sensory hair cells (see, e.g., Fernandez and Goldberg, 1972; von Holst, 1950). Since most of the utricle is situated near the horizontal plane and the saccule in the vertical plane, it is expected that the utricle would be most sensitive to linear accelerations in the horizontal plane and the saccule to vertical accelerations. There is experimental evidence supporting these expectations in both the cat (Anderson *et al.*, 1977a) and monkey (Fernandez and Goldberg, 1972).

As indicated in Fig. 11B, a static rotation of the head will decrease the shear force resulting from gravitational acceleration on the saccular macula and increase the shear force on the utricular macula. Unfortunately, a simple analysis of shear forces cannot be used to predict the response of all afferent fibers from the otolith organs. There is a definite pattern of polarization of the cilia of the sensory hair cells, and there is a relation between this polarization and the nature of the response observed in the afferent nerve fibers from each receptor organ (Lowenstein and Wersäll, 1959). All the hair cells of a given semicircular canal are polarized in the same direction, but there is a systematic variation in the direction of polarization of the hair cells over the surface of both the utricle and saccule. As a very rough description, it may be said that one group of hair cells is oriented one way over a large part of the macula of each otolith organ, and another group is oriented the opposite way over the remainder of the macula. This is reflected in the response obtained from afferent fibers. In the monkey, some afferent fibers from both the utricle and sacculus are found to increase firing for static head rotation in a certain direction, whereas other afferents from each organ decrease firing for the same head rotation (Fernandez and Goldberg, 1972). In this study, the spontaneous firing rates of the afferents from either organ were modulated approximately as a trigonometric function of the angle of head rotation, as predicted from the simple analysis in Fig. 11B (see also Adrian, 1943). Notice that the slope near 0° of the sine function, predicted from the analysis in Fig. 11B to control the utricular firing rate, is steeper than that of the cosine function predicted to describe the saccular afferent firing. As predicted from this simple analysis, utricular afferents were much more sensitive to small static rotations from the vertical than were saccular afferents (Fernandez and Goldberg, 1972). Impulses from the utricle would therefore be expected to dominate motoneuron output for such rotations, and, as discussed earlier in connection with head rotation in the frontal plane, there is experimental support for this expectation (Soechting *et al.*, 1977).

Fernandez and Goldberg (1972) found that about 70% of the afferent fibers from the utricle increased firing in response to static, ipsilateral (cheek down) tilt in the frontal plane. The preponderance of such afferents appears to be reflected in the finding that the great majority of secondary vestibular neurons responding to static tilt increase firing upon ipsilateral tilt (Adrian, 1943; Fujita, Rosenberg, and Segundo, 1968; Peterson, 1970). Most vestibular neurons are sensitive to static rotation about only one axis, but a less common subset of vestibular neurons increases or decreases firing to tilt in either direction (Fujita *et al.*, 1968; Peterson, 1967, 1970).

Little information is available on the response of neck muscles to natural otilith organ input. Berthoz and Anderson (1971b) briefly reported that motor-unit activity increased in contralateral neck muscles upon ipsilateral tilt. This observation is consistent with the EPSPs in contralateral and IPSPs in ipsilateral neck motoneurons observed upon electrical stimulation of the utricular nerve (Wilson *et al.*, 1977), but it is not certain that the effects reported by Berthoz and Anderson (1971b) were mediated by the pathway demonstrated by electrophysiology.

Little information is available about the synaptic connections from the otolith organs to limb motoneurons except the inference that they contribute a significant fraction of EPSPs in forelimb extensor and IPSPs in forelimb flexor motoneurons observed upon electrical stimulation of either VIIIth nerve (Maeda *et al.*, 1975). The synaptic effects from VIIIth nerve stimulation appear to be consistent with the conclusions of Magnus (1926) that static tilt in any direction caused activation of all limb extensors. More recent work on the forelimbs has suggested, however, that static ipsilateral tilt activates ipsilateral extensor muscles and decreases activity in contralateral extensor muscles (Roberts, 1970; Rosenberg and Lindsay, 1973). Berthoz and Anderson (1971a) also reported such reciprocal forelimb extensor activation upon ipsilateral tilt, but this motor unit activity decreased to resting levels over 10–20 sec. This latter observation raises the possibility that the reciprocal motor-unit activity was mediated by the vertical semicircular canals, the afferents of which may decrease firing with time constants on the order of 8–10 seconds after a brief angular acceleration (Shimazu and Precht, 1965).

More in accord with the suggestion of Magnus (1926) is the finding that, on the one hand, some Deiters' neurons are sensitive to static tilt in either direction (Peterson, 1967, 1970), while on the other hand, the otolith organs tend to excite hind-limb extensor and inhibit hind-limb flexor motoneurons for static tilt in any direction (Ehrhardt and Wagner, 1970; Poppele, 1967). Static tilt to the ipsilateral (Ehrhardt and Wagner, 1970) or the contralateral side (Poppele, 1976) is variously reported to produce the strongest or most consistent effects. In these studies, it was also found that many motoneurons were not activated at all, and considerable variability was seen in the response of those that were activated. The principal results of these latter studies are, however, more in line with the bilateral effects observed on electrical stimulation of the VIIIth nerve (Maeda *et al.,* 1975) and inferred by Magnus (1926) from observation of limb movement.

Although the otolith organs are usually considered to be static gravity-sensitive receptors, they also have considerable dynamic sensitivity (Anderson *et al.,* 1977a; Goldberg and Fernandez, 1974; Schor, 1974). Parenthetically, the same is true of the neck-joint receptors alluded to earlier (Erhardt and Wagner, 1970), although the reflexes evoked are usually called tonic neck reflexes (Magnus, 1926). Early experiments on the frog (Ashcroft and Hallpike, 1934) indicated that the saccule was mainly sensitive to vibration, whereas the utricle was thought mainly to subserve static position sense. But, it is now clear that afferents from both the saccule and utricle in mammals can respond to static head rotation (Fernandez and Goldberg, 1972) and dynamic, low-frequency, linear acceleration (Anderson *et al.,* 1977a). Using linear systems analysis and the rectified, averaged, multimotor-unit EMG response as output, Anderson *et al.* (1977a) showed that the saccule can modulate forelimb extensor activity during linear acceleration in the vertical plane as effectively as the utricle modulates forelimb activity during linear acceleration in the horizontal plane. EMG activity increased during downward linear accelerations (exciting the saccule) and during linear acceleration both to the ipsilateral side and in the nose-forward direction in the horizontal plane (exciting the utricle). These results suggest that the otolith organs produce a net excitatory or inhibitory effect on the forelimb extensor motoneurons, even though it is known that some otolith afferents increase firing and some decrease firing during a given static adequate stimulus (Fernandez and Goldberg, 1972). The observation of forelimb extensor activation during linear acceleration to the ipsilateral side implies that forelimb extensors are also activated from the utricle during ipsilateral cheek-up static tilt in the frontal plane. This inference received experimental support (Soechting *et al.,* 1977) but it is at odds with results discussed earlier that indicated that most vestibular neurons as well as forelimb extensor muscles are activated by ipsilateral cheek-down static tilt. These results are also different than those discussed in connection with the activity of hind-limb extensor motoneurons under static tilt, where some extensor motor units were found to be activated for static tilt in any direction.

Overall, the studies of the various labyrinthine reflexes outlined above suggest that these reflexes cannot be accounted for by the simple, short-latency vestibulo-spinal synaptic connections, but more work is clearly required in that direction as well as additional quantitative work on the reflexes themselves. Further analysis of relatively stereotyped reflex behavior can provide one means of understanding vestibulospinal function. On the other hand, the difficulty found by all investigators in obtaining strong, consistent responses in limb motoneurons, unless special measurements were taken, points up one of the main conclusions from electrophysiology: the vestibulospinal system does not primarily serve as a relay of labyrinthine effects to limb musculature.

THE RETICULOSPINAL SYSTEMS

ORGANIZATION OF THE MEDIAL PONTOMEDULLARY RETICULAR FORMATION

It is worth considering the organization of the medial pontomedullary reticular formation in some detail because it is likely that the functional meaning of the descending effects obtained from this large brain stem area will ultimately be understood only in relation to the functional organization of their source. The following outline thus attempts to sketch what is known about the source of the reticulospinal fibers. A reticular formation is found in the medulla, pons, and midbrain, but anatomy indicates that the reticulospinal fibers derive only from the pons and medulla (Torvik and Brodal, 1957). Reticulospinal fibers derive principally from cells in the medial two-thirds of each area. It is in these two regions, illustrated in Fig. 12A, that most large reticular cells are found, but both small and large cells give rise to reticulospinal firbers (Torvik and Brodal, 1957). Although the reticular formation is characterized by diffuse aggregations of cells separated by fibers traveling in all directions, architectonic differences have been found between areas, and the reticular formation has been divided into various nuclei (Brodal, 1957; see also Fig. 13). Most of the reticulospinal fibers from the medulla come from the nucleus reticularis gigantocellularis, whereas the pontine fibers come from n.r. pontis caudalis and the caudal part of n.r. pontis oralis (Torvik and Brodal, 1957). There are also areas of maximal concentration of cells giving rise to ascending fibers that overlap but that are centered differently from the areas of concentration of reticulospinal cells (see Fig. 12A). However, many reticular cell axons divide into ascending and descending branches (Scheibel and Scheibel, 1958).

The reticulospinal tracts are composed of both large and small fibers, many of which descend to lumbosacral levels (Nyberg-Hansen, 1965). The position of the main reticulospinal fiber tracts in the spinal white matter and their termination in the spinal gray matter are illustrated in Fig. 12B. Fibers from the pons descend almost exclusively ipsilaterally in the ventral funiculus of the spinal cord in the same general region as the LVST. Fibers from the medulla are both crossed and uncrossed; both projections run in the ventral part of the lateral funiculus. The pontine fibers terminate more ventrally than the medullary fibers. The former end mainly in Rexed's (1952, 1954) lamina VIII and adjacent parts of lamina VII; the

PETER C. SCHWINDT

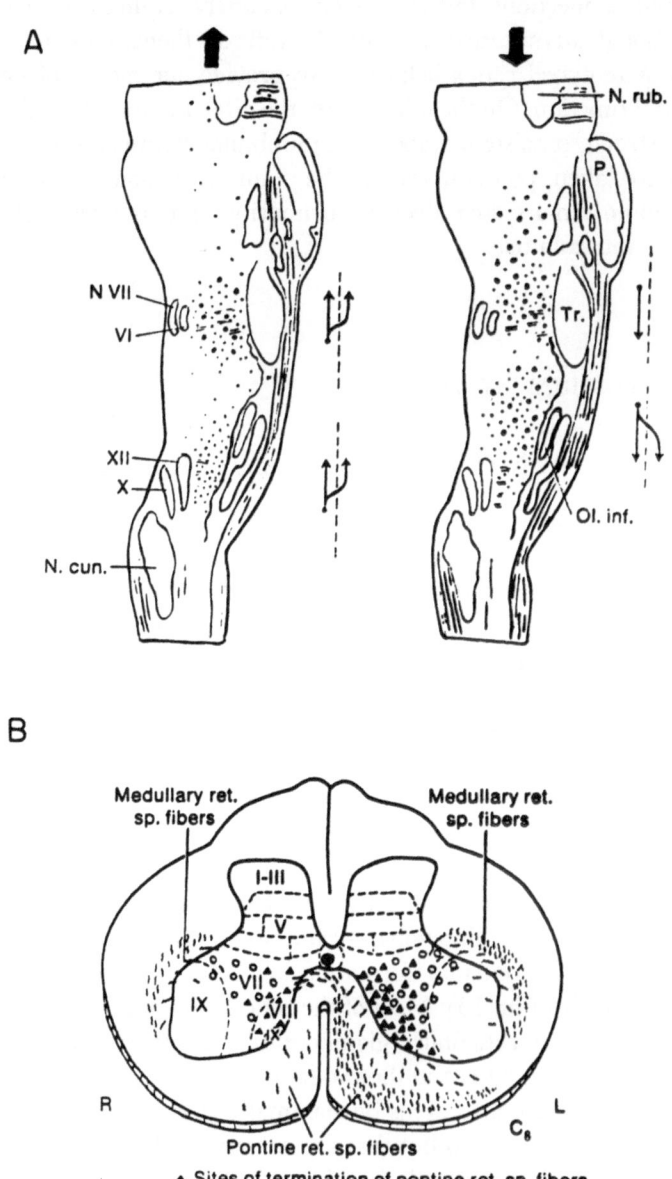

Fig. 12. Some anatomical features of the reticulospinal tracts in the cat. (A) Parasagittal section of the cat's brain stem indicating the distribution of reticular neurons (large and small dots) sending long axons to the spinal cord (*right*) and long ascending axons beyond the midbrain *(left)*. Arrows at right of each drawing indicate that the pontine reticulospinal fibers descend homolaterally, whereas all other projections are both crossed and uncrossed. (B) the location of the pontine and medullary reticulospinal fibers in the spinal white matter and the sites of termination of these fibers in the spinal gray matter. Dashed lines and Roman numerals refer to Rexed's laminae. From Brodal (1969) with permission.

latter terminate mainly in lamina VII with a few terminations in lamina VIII and in the region of motoneuron somata, lamina IX (Nyberg-Hansen, 1965). Another reticulospinal tract located in the dorsal part of the lateral funiculus has been identified physiologically (Engberg, Lundberg and Ryall, 1968a), and reticulospinal fibers in this region have been observed anatomically (Petras, 1967).

Reticulospinal cells receive input from many regions of the CNS. A large afferent projection from the spinal cord, both direct and indirect, is indicated anatomically (Brodal, 1969) and observed physiologically (Eccles, Nicoll, Rantucci, Táboříková, and Willey, 1976; Magni and Willis, 1964b; Peterson, Anderson, and Filion, 1974). Long-latency excitation, inhibition, or mixed effects are found from low-threshold cutaneous fibers, high-threshold cutaneous, and muscle afferent fibers, but not from group I muscle afferent fibers (Magni and Willis, 1964b; Peterson et al., 1974).

The projection from the cerebral cortex is also extensive. In the cat, fibers arise chiefly from sensorimotor cortex. These descend in the internal capsule and cerebral peduncle with corticospinal fibers, but leave the corticospinal tract in the brain stem to terminate bilaterally in the reticular formation. Interestingly. the largest fraction of corticoreticular fibers is found to terminate in the regions from which most reticulospinal fibers arise (Rossi and Brodal, 1956). The corticoreticular projection has also been studied electrophysiologically. Convergence from many areas of the ipsilateral and contralateral cerebral cortex is found in a given reticular neuron, with short-latency excitation the principal effect (Magni and Willis, 1964a). When the projections from sensorimotor cortex were studied in more detail (Peterson et al., 1974), bilateral short-latency excitation (followed by long-lasting inhibition) was found; about half the cells received monosynaptic EPSPs when the cerebral peduncles were stimulated. There was some differential distribution of effects: ipsilateral monosynaptic excitation was most prominent in the pontine reticular formation, and the latency of polysynaptic excitation was significantly shorter from ipsilateral compared to contralateral stimulation. Contralateral monosynaptic excitation became prominent in the central and rostral

Fig. 13. Parasagittal section of the cat's brain stem showing the location of most reticulospinal neurons found in electrophysiological studies (stippled area) and the two zones (dorsorostral and ventrocaudal) found by Peterson et al. (1975) to contain reticulospinal neurons with different patterns of projection and conduction velocities. The approximate longitudinal extent of each of the medial pontomedullary reticular nuclei described in anatomical studies is indicated by the brackets and labels below the brain stem section. XII = hypoglossal nucleus; P.H. = parahypoglossal nucleus; G VII = genu of the facial nerve; VI = Abducens nucleus; I.O. = inferior olive; T.B. = trapezoid body; NRTP = n.r. tegmenti pontis. After Peterson et al. (1975).

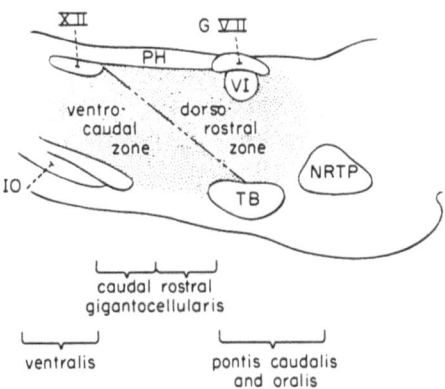

medulla (n.r. gigantocellularis) and was less common elsewhere. Apparently, short-latency crossed connections become more prevalent below the rostral pontine reticular formation; effects of about the same latency from each side are seen in n.r. gigantocellularis.

Connections from the superior colliculus are seen anatomically (Brodal, 1969) and have been studied physiologically (Peterson *et al.*, 1974). Most reticular cells receive short-latency EPSPs (followed by long duration inhibition) following stimulation at various contralateral and ipsilateral points in the tectum. Contralateral stimulation more often produces firing (rather than subthreshold facilitation) of pontine and rostrodorsal medullary cells, whereas ipsilateral stimulation is more effective for cells in the ventrocaudal medulla. The tectoreticular projection is consistent with a postulated reticular relay of the tectospinal projection to neck motoneurons (Anderson, Yoshida, and Wilson, 1971).

The medial pontomedullary reticular formation is also a major projection area for excitatory fibers from the ipsi- and contralateral fastigial nucleus of the cerebellum according to both anatomy (Brodal, 1969) and physiology (Eccles, Nicoll, Schwarz, Táboříková, and Willey, 1975). It has also recently been shown that reticulospinal cells may be monosynpatically activated from the dentate nucleus (Bantli and Bloedel, 1975a), previously thought to play a minor cerebelloreticular role, and that postsynaptic potentials may be evoked in hind-limb motoneurons via this route (Bantli and Bloedel, 1975b).

Reticulospinal cells also receive input from the vestibular complex on each side of the brain stem (Brodal, 1972b). Physiological studies have shown the vestibular input to come from both VIIIth nerve relay and nonrelay vestibular neurons (Peterson and Abzug, 1975; Peterson, Filion, Felpel, and Abzug, 1975). About half the vestibular neurons projecting to the medial pontomedullary reticular formation were acitvated only di- or polysynaptically from the VIIIth nerve and were thus not simple labyrinthine relay cells. About one-third of the reticulospinal cells were fired by repetitive VIIIth nerve stimulation, some at low threshold. Disynaptic or longer EPSPs and IPSPs from stimulation of either VIIIth nerve were found in about three-fourths of the reticular cells studied, but reticulospinal cells received only short latency EPSPs. These reticulospinal neurons were scattered throughout n.r. gigantocellularis and were rare elsewhere. Vestibular cells projecting to the medial pontine reticular formation were found in all vestibular nuclei but were less common in dorsal Deiters' nucleus and the central part of the superior vestibular nuclei.

The area where most reticulospinal cells are found in the electrophysiological studies (Eccles *et al.*, 1976; Ito *et al.*, 1970; Peterson *et al.*, 1974; Peterson, Maunz, Pitts, and Mackel, 1975) generally agrees with the anatomical studies, except that a uniform distribution of reticulospinal neurons is found throughout the central pontomedullary region rather than the two areas of concentration seen in the anatomical studies (this may be a microelectrode cell-sampling problem, however). Reticulospinal cells have been found mainly in the area extending from the rostral pole of the inferior olive to the n.r. tegmenti pontis, that is, mainly in the n.r. gigantocellularis of the medulla and n.r. pontis caudalis of the pons. In the study of Peterson *et al.* (1974), most reticulospinal cells were found between 1.5 to 4.5mm from the surface, whereas nonreticulospinal cells outnumbered reticulospinal cells more dorsally or ventrally; relatively few reticulospinal cells were found in n.r. ventralis. A useful dividing plane for classification of cell type and projections was found by Peterson, Maunz, Pitts, and Mackel (1975) to extend obliquely upward from about the middle of the trapezoid body (the rostroventral limit) to about the caudal end of the perihypoglossal nucleus (the dorsocaudal limit) and to run approximately through the center of n.r. gigantocellularis (see Fig. 13). The division about this oblique plane into a dorsorostral and a ventrocaudal zone also fits the data of Ito, Udo, Mano, and Kuwai (1970) better than the classification zones used by those workers. In the study of Peterson, Maunz, Pitts, and Mackel (1975), reticulospinal cells projecting ipsilaterally in the ventro*medial* fasciculus of the spinal cord were dominant in the dorsorostral zone; those projecting ipsilaterally in the ventro*lateral* fasciculus dominated in the ventrocaudal zone. These projections appear to correspond to the ipsilateral pontine and medullary projections, respectively, described anatomically (Nyberg-Hansen, 1965), but it seems that the "pontine" projection also includes fibers from dorsorostral n.r. gigantocellularis of the medulla. Cells with crossed fibers are seen only in the ventrocaudal zone, consistent with anatomy, but they constituted only 3 to 5% of the total recorded population (Ito *et al.*, 1970; Peterson, Maunz, Pitts and Mackel 1975). Cells projecting to cervical, thoracic, and lumbar spinal segments are scattered throughout the area where reticulospinal cells are located (Ito *et al.*, 1970; Peterson *et al.*. 1974; Petersen, Maunz, Pitts, and Mackel, 1975), but those cells projecting only as far as neck spinal segments are mainly situated in the caudal part of the dorsorostral zone (Peterson, Maunz, Pitts, and Mackel, 1975b).

Both ipsilaterally and contralaterally projecting reticulospinal cells in the ventrocaudal zone had significantly slower conduction velocities than found in the dorsorostral zone (mean CV: 70m/s vs. 101 m/s) (Eccles *et al.*, 1976; Peterson *et al.*, 1974; Peterson, Maunz, Pitts, and Mackel, 1975b). It is interesting that Ito *et al.* (1970) found that cells in this region often receive monosynaptic IPSPs when activated antidromically. From this observation, these investigators inferred that the IPSPs were probably from collaterals of antidromically activated neighboring neurons and, thus, that most cells in the region must have inhibitory function. Stimulation in the caudal part of the ventrocaudal region does indeed produce inhibitory effects, apparently mediated by slow fibers, on motoneurons and spinal reflex arcs, as will be discussed later in this section.

Peterson, Maunz, Pitts, and Mackel (1975) used local microstimulation of the spinal gray matter to show that 71% of the cells projecting to lumbar segments in the ventromedial fasciculus (the "pontine" projection) and 63% of these cells projecting in the ventrolateral fasciculus (the ipsilateral "medullary" projection) also give off collaterals to the cerivcal ventral horn. It will be recalled that a similar result was found for the vestibulospinal system (Abzug *et al.*, 1974), but the proportion of dually projecting reticulospinal neurons appears to be larger. Using microstimulation to follow the path of cervical collaterals, it was found that they ramify extensively on the ipsilateral side; crossing to the contralateral gray matter

184

PETER C. SCHWINDT

was also observed. Thus, the same descending information reaches spinal neurons controlling both forelimb and hind-limb musculature and possibly contralateral musculature as well.

The outline of medial pontomedullary reticular organization given above has shown that the area is not simply a diffuse, undifferentiated mass, but only the broad outlines of the detailed organization have so far been discovered. Reticulospinal fibers have varied effects on different types of motoneurons and reflex arcs, as will be discussed below, and it appears that further work on the organization of the brain stem source of these descending effects will be required to help unravel their functional meaning and normal use.

RETICULOSPINAL EFFECTS ON ALPHA MOTONEURONS

The role of the reticular formation in motor control received little attention until the publication of two papers by Magoun and Rhines in 1946. These investigators observed, on the one hand, that sustained stimulation at many points over a large area of the central brain stem, extending from medulla to thalamus (see Fig. 14), could facilitate limb movements (mainly flexion) induced by simultaneous stimulation of the cerebral cortex, and that such stimulation also greatly facilitated the tendon-jerk reflex from the extensor, quadriceps (Rhines and Magoun, 1946). It was ascertained that the facilitatory interactions did not occur at the cerebral level. On the other hand, sustained stimulation at other sites in a more limited area of the central, caudal medulla (see Fig. 14) inhibited limb flexion induced by the segmental flexion reflex and also inhibited the quadriceps knee jerk (Magoun and

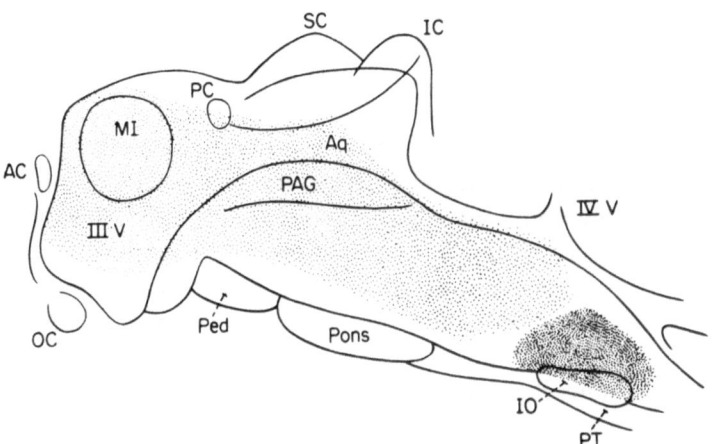

Fig. 14. Sagittal section of the cat's brain stem showing the regions where prolonged electrical stimulation gives rise to the characteristic effects of the "facilitatory reticular formation" of Rhines and Magoun (1946) (light stippling) and the "inhibitory reticular formation" of Magoun and Rhines (1946) (dark stippling). See text for further explanation. AC and PC = anterior and posterior commissures, respectively; OC = optic chiasm; III V and IV V = third and fourth ventricles, respectively; MI = massa inter-media, PAG = periquiductal gray; Aq = aqueduct; S.C. and I.C. = superior and inferior colliculi, respectively; Ped = cerebral peduncles; I.O. = inferior olive; PT = pyramidal tract. After Rhines and Magoun (1946) and Magoun and Rhines (1946).

Rhines, 1946). Stimulation in this caudal area could also "melt" extensor rigidity in all four limbs of decerebrate cats and inhibit limb flexion produced by cerebral cortex stimulation. The rostral brain stem region was termed the "facilitatory reticular formation" and assigned the function of generalized facilitation of all types of motoneurons; the caudal "inhibitory reticular formation" was assigned the role of generalized inhibition of all types of motoneurons. Subsequent studies (Bach, 1950; Sprague, Schreiner, Lindsley, and Magoun, 1948) showed that stimulation (at unspecified points) in the "facilitatory and inhibitory reticular formation" indeed produced the expected generalized effects on stretch reflexes and electrically evoked, crossed-extension, and other segmental reflexes. The results of these and other experiments have been summarized in a by-now-classical schematic diagram together with a hypothesis on the role of the facilitatory and inhibitory areas in the control of normal and pathological motor output (Lindsley, Schreiner, and Magoun, 1949; Schreiner, Lindsley, and Magoun, 1949).

From a present-day viewpoint, it is difficult to accept the results of Magoun and co-workers at face value for several methodological and technical reason. Many structures in the "facilitatory reticular formation" described by Rhines and Magoun (1946) are, in fact, not part of the reticular formation and have subsequently been found to have primary functions far removed from exerting generalized facilitation on motoneurons. More recent work, to be discussed below, indicates that specific inhibitory and excitatory effects may be obtained from more-or-less discrete regions of the rostral pontomedullary and mesencephalic reticular formation. However, the idea of a limited region in the caudal medullary reticular formation that produces generalized inhibition of motoneurons and reflex arcs has been supported by subsequent work.

The hypothesis of Magoun and co-workers concerning the existence of large brain stem areas producing only generalized facilitation and generalized inhibition was soon disputed by other investigators. Gernandt and Thulin (1955) used more controlled stimulation parameters and observed the effects of stimulating the medullary reticular formation on the monosynaptic reflex of hind-limb flexors and extensors. Although they did observe facilitation or inhibition of both flexors and extensors from stimulation at many points, in accordance with the results of Magoun, stimulation at many other points produced reciprocal effects. In particular, stimulation in medial regions, especially at the level of the vestibular complex, produced inhibition of extensors and facilitation of flexors in addition to nonreciprocal inhibition, whereas more lateral stimulation (including stimulation of Deiters' nucleus) often produced the opposite effect. The result of stimulation, even that producing nonreciprocal effects, was commonly rather complex in that the effects produced by reticular stimulation could grow, or diminish, or wax and wane, or even change sign over the ten second stimulation period used by these investigators. Longlasting "rebound" effects of opposite sign were always observed. In another series of experiments examining limb movement and EMG acitivity in decerebrate, barbituate-anesthetized, and awake cats with implanted electrodes, Sprague and Chambers (1954) failed to find the effects reported by Magoun and co-workers. Generalized facilitation could not be evoked, although flaccidity of one or more limbs was sometimes observed at stimulation intensities well above threshold for evoking limb movement from some sites. The most consistent effects found, especially near threshold stimulation, was ipsilateral flexion and contralateral extension during stimulation near the midline, whereas more lateral stimulation produced the opposite effect. These findings are thus consistent with the observations of Gernandt and Thulin (1955) on changes in the monosynaptic reflex of ipsilateral flexors and extensors upon reticular stimulation. In the light of present-day knowledge, it is tempting to speculate that medial stimulation mainly excited reticulospinal fibers in the medial longitudinal fasciculus (MLF), which can produce the observed ipsilateral effects, whereas lateral stimulation may have excited mainly LVST fibers from the Deiters' nuclei, which are known to give the opposite, reciprocal effects. It is now known that central stimulation more readily excites fibers than cell bodies (see, e.g., Baldissera, Lundberg, and Udo, 1972b; Gustafsson and Jankowska, 1976; Jankowska, Padel, and Tanaka, 1975a).

Subsequent investigations employed intracellular recording from motoneurons during reticular stimulation. Llinás and Terzuolo (1964, 1965; Terzuolo, Llinás, and Green, 1965) found that sustained EPSPs and IPSPs could be produced in both flexor and extensor motoneurons during repetitive stimulation at unspecified points in the "facilitatory" and "inhibitory" reticular formation, respecitvely. One implication of their data was that the IPSPs observed in flexor (but not extensor) motoneurons appeared to be produced in dendrites remote from the soma. This finding would suggest that there is a special significance to reticulospinal inhibition of flexor motoneurons since data obtained in spinal and brain stem motoneurons up to the present day suggest the great majority of inhibitory synaptic terminals of descending or segmental pathways are located on and near the soma, whereas excitatory terminals may be distributed over the dendritic tree. Jankowska, Lund, Lundberg, and Pompeiano (1968) made a thorough study of the inhibitory region of the caudal medulla. Stimulus trains applied to a region in the ventrocaudal medulla (see Fig. 15) were indeed found to produce IPSPs in a variety of hind-limb flexor and extensor motoneurons. Signs of inhibitory connections remote from the soma observed in flexor motoneurons by Llinás and Terzuolo (1965) were not found. IPSPs were produced in both flexor and extensor motoneurons from the effective area at similar stimulation thresholds. The most effective area was situated from 3mm to 5mm above the obex. Throughout this length, effects could be obtained from the region indicated in Fig. 15A. Measurements of segmental delay and the requirement of stimulus trains to evoke the IPSPs suggested the latter were transmitted polysynaptically. (Occasional disynaptic IPSPs were later attributed to stimulus current spread to other structures by ten Bruggencate *et al.*, 1969.) However, the findings of Peterson, Maunz, Pitts, and Mackel (1975) that mainly slow reticulospinal fibers originate from the effective region, combined with the suggestion of Ito *et al.* (1970) that most of these fibers may be inhibitory, raises the possibility that some of the IPSPs observed by Jankowska *et al.* (1968) may have been monosynaptic from slow fibers. The interneurons involved may have been in the brain stem rather than in the spinal cord. Segmental delay is referred to the tract wave of the fastest descending fibers and by itself gives no information on the latency of effects from slow fibers. As discussed in the vestibulospinal section, Wilson *et al.* (1970) found monosynaptic inhibition in back-muscle motoneurons from slow MVST fibers in the MLF; the effects would have appeared at least disynaptic if referred to the fastest tract wave. It is interesting that stimulation of a region of the brain stem overlapping that studied by Jankowska *et al.* (1968) inhibits certain spinal reflexes (Engberg, Lundberg, and Ryall, 1968b), but the fibers producing these effects apparently descend in the ipsilateral dorsolateral funiculus of the spinal cord, whereas the effects seen by Jankowska *et al.* (1968) were ascertained to descend in the ipsilateral ventrolateral quadrant of the spinal cord, almost certainly in the ipsilateral medullary reticulospinal tract.

Effects on motoneurons from stimulation of the medial longitudinal fasciculus (MLF) which contains fibers from the rostral pontomedullary reticular formation, have also been studied extensively by intracellular recording. The descending MLF contains fibers from many areas of the brain stem, including fibers of the MVST from the vestibular complex (see p. 159) and fibers from the midbrain tegmentum and the optic tectum as well as pontomedullary reticulospinal fibers (Petras, 1967). Lund and Pompeiano (1968) showed that the only fibers giving rise to monosynaptic EPSPs in hind-limb flexor motoneurons descended in the ipsilateral ventrolateral funiculus and were of supraspinal origin. Grillner and Lund (1968) subsequently recorded the monosynaptic EPSP in knee flexor motoneurons while stimulating the MLF in the brain stem at a series of rostrocaudal sites. One such effective site is illustrated in Fig. 15B. The monosynaptic EPSP weakened and finally disappeared at stimulation sites more rostral than the rostral n.r. gigantocellularis and caudal n.r. pontis caudalis (evidently, the caudal part of the dorsorostral zone of Peterson, Maunz, Pitts, and Mackel, 1975). It was inferred that fibers giving the monosynaptic EPSP in flexor motoneurons arose from cell bodies in the rostral medulla and caudal pons and descended to the spinal cord in the MLF. Only ipsilateral MLF stimulation proved effective at low stimulation

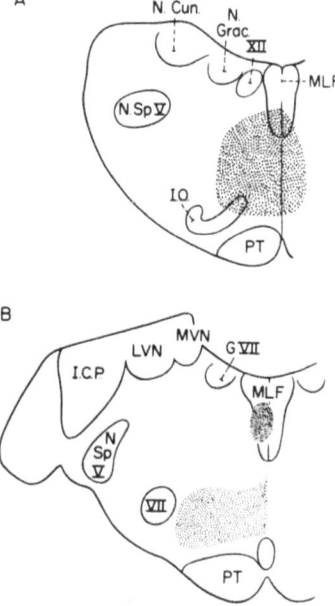

Fig. 15. Transverse sections of the cat's brain stem showing areas producing various reticulospinal effects on electrical stimulation. (A) Transverse section approximately 3 mm above the obex showing the area (stippling) found by Jankowska *et al.* (1968) to produce IPSPs in flexor and extensor motoneurons. (B) Transverse section approximately 6 mm above the obex showing a site stimulated by Grillner and Lund (1968) to produce effects in motoneurons from reticulospinal fibers descending in the MLF (dark stippling in MLF) and the more ventral area (light stippling) found by Engberg *et al.* (1968a) to produce inhibitory effects on certain reflex arcs. See text for further explanation. I.C.P. = inferior cerebellar peduncle; LVN = lateral vestibular nucleus of Deiters'; MVN = medial vestibular nucleus; G VII = genu of the facial nerve; MLF = medial longitudinal fasiculus; P.T. = pyramidal tract; VII = nucleus of the facial nerve; N. Sp. V = nucleus of the spinal tract of nerve V; I.O. = inferior olive; N. Cun. = cuneate nucleus; N. Grac = nucleus gracilis; XII = hypoglossal nucleus.

strength, and furthermore, disynaptic IPSPs were seen in knee and ankle extensor motoneurons at the same threshold stimulation values giving the monosynaptic EPSP in flexor motoneurons. Thus, the same population of fast fibers apparently gives rise to both effects. The results of a series of spinal cord lesions indicated that the monosynaptic effects were transmitted by fibers lying in the lateral part of the ventral funiculus and the ventral part of the lateral funiculus. Since the pontine and ipsilateral medullary reticulospinal tracts overlap in this area, the effects are probably transmitted by components of both tracts.

Grillner, Hongo, and Lund (1971) reconfirmed that these reticulospinal fibers in the MLF provide the only monosynaptic input to hind-limb flexor motoneurons and found them to comprise a rather uniform group of fast fibers (mean CV: 111 m/s; range: 90–130 m/s). These workers also examined the distribution of effects among various hind-limb motoneurons; the effects proved more complicated than first suspected. The monosynaptic MLF EPSP appears to have a somewhat wider distribution than the monosynaptic EPSP from Deiters' nucleus to extensor moto- neurons. The monosynaptic EPSP from reticulospinal fibers in the MLF is found mainly in knee, ankle, and toe flexor motoneurons but is also seen in some hip extensor and toe extensor motoneurons. Reciprocal, disynaptic IPSPs are seen mainly in knee and ankle extensor motoneurons but also in some knee flexor and hip extensor motoneurons. Furthermore, a larger disynaptic EPSP often follows the monosynaptic EPSP. This is seen mainly in toe extensor and other pretibial flexor motoneurons and occasionally in hip extensor motoneurons. Therefore, the monosynaptic and disynaptic effects can be of different sign in motoneurons inner- vating hip and toe extensors and sometimes in knee flexor motoneurons. It was discussed earlier how Deiters' nucleus usually excites extensors and inhibits flex- ors, but exceptions are often found in hip and toe motoneurons for these effects, too. Thus, the MLF reticulospinal and LVST effects are reciprocal in terms of monosynaptic and disynaptic connections only at the knee and ankle; effects of opposite sign from either tract can be observed in motoneurons innervating the hip and toe musculature. These nonuniform effects may mean that the MLF reticu- lospinal fibers do not comprise a homogeneous population but arise from reticular cells having different functions; the same may be inferred for the LVST arising from Deiters' nucleus to the extent that exceptions to the general innervation pat- tern are also found in hip and toe motoneurons. The interneuron in the MLF reticulospinal pathway producing disynaptic inhibition has been shown to be sit- uated in lumbar segments, but so far it has not been determined if it is a component in any spinal reflex pathway. The disynaptic inhibition from the MLF reticulos- pinal fibers is known, however, *not* to be mediated by the group Ia inhibitory interneurons (Hultborn *et al.*, 1976c; Hultborn and Udo, 1972).

In accordance with results in hind-limb motoneurons, MLF stimulation pro- duces monosynaptic EPSPs in many neck-muscle (Wilson and Yoshida, 1969b) and most back-muscle motoneurons (Wilson *et al.*, 1970). In both cases, these EPSPs were evoked by a population of fast-conducting fibers, and it was ascer- tained that the effects were not due to stimulus current spread to axon collaterals from Deiters' nucleus. MLF stimulation also produced monosynaptic EPSPs in some flexor, extensor, and mixed digit and wrist forelimb motoneurons, whereas

EPSPs from Deiters' nucleus were at least disynaptic, and no effect was obtained from VIIIth nerve stimulation (Wilson and Yoshida, 1969b). It may be assumed the effects seen in all these motoneurons were from reticulospinal fibers descending in the MLF.

It has recently been reported (Peterson, Pitts, Mackel, and Fukushima, 1976) that neck-muscle motoneurons may also receive monosynaptic EPSPs and IPSPs from reticular pathways not involving the MLF. The monosynaptic IPSPs are evoked only from the ipsilateral medial n.r. ventralis, and the overwhelming majority are to neck extensor motoneurons. The monosynaptic EPSPs are also mainly evoked ipsilaterally over the entire medial region from n.r. ventralis to n.r. pontis caudalis. The monosynaptic EPSPs are found in almost all neck extensor and about 80% of neck flexor motoneurons, but the EPSPs to extensor motoneurons are larger.

In spite of the exceptions to the general innervation pattern in hip and toe motoneurons, it appears the reticulospinal fibers descending in the MLF to hind-limb motoneurons are basically concerned with excitation of flexor and reciprocal inhibition of extensor motoneurons. This is reasonable considering the heavy innervation that reticulospinal neurons receive from cerebral cortex, which, at least in terms of corticospinal effects, also seems most concerned with excitation of flexor and inhibition (sometimes also excitation) of extensor motoneurons. Indeed, the effects from cortical stimulation on motoneurons in the cat and monkey with a sectioned pyramidal tract are remarkably similar to effects produced with the pyramidal tract intact (Hongo and Jankowska, 1967; Jankowska and Tarnecki, 1965; Lewis and Brindley, 1965); in the cat, the alternate pathway from the cortex to the spinal cord was deduced to be reticulospinal (Hongo and Jankowska, 1967). Thus, the reticulospinal MLF pathway provides effects on hindlimb musculature that are partially reciprocal with those of Deiters' nucleus and may represent an alternate pathway for corticospinal excitation of flexor motoneurons (in addition to that provided by the corticospinal and rubrospinal tracts). This conclusion contrasts with the inference of Lawrence and Kuypers (1968b) that the vestibulospinal and reticulospinal tracts act similarly. The studies outlined above also have shown that neither the LVST from Deiters' nucleus nor the reticulospinal fibers descending in the MLF show a marked difference in the strength of their effects on motoneurons of distal or proximal hind-limb muscles. These observations provide another contrast with the inference of Lawrence and Kuypers (1968b) that these tracts are most important in control of proximal musculature. However, the results of anatomical studies in the cat (Petras, 1967) suggest the reticulospinal MLF pathway provides only a small fraction of the total reticulospinal projection, and the action of most of the reticulospinal fibers remains to be discovered.

RETICULOSPINAL EFFECTS ON SPINAL REFLEX ARCS

As mentioned above, only negative results have so far been obtained from investigations of the MLF-reticulospinal action on spinal reflex arcs, but a great deal of work has been done on the control of polysynaptic group Ib and, especially, FRA reflexes by reticulospinal fibers descending in the dorsolateral funiculus of the spinal cord. This work was initially motivated by the finding that the effects of FRA stimulation (mainly excitation of flexors and inhibition of extensors) are weak or absent in the decerebrate cat, but strong FRA effects are seen in the spinal cat. The same holds true for polysynaptic group Ib effects, but not for group Ia disynaptic inhibition, which is apparent in both preparations (Holmqvist and Lundberg, 1959, 1962). These observations suggested that the FRA pathway is under supraspinal control by tonic descending influences on inhibitory interneurons, and it was shown by Holmqvist and Lundberg (1959) that these tonic inhibitory effects are transmitted in the dorsolateral quadrant of the spinal cord. Transection of this quadrant was sufficient to release all the FRA effects, whereas transection of the ventrolateral quadrants containing the LVST fibers from Deiters' nucleus responsible for maintaining extensor rigidity (Fulton *et al.,* 1930) had no effect on the FRA response. Similarly, the tonic inhibition of FRA effects remained after bilateral destruction of the vestibular nuclei, but the FRA effects were released after a medially placed section at the level of the obex. Holmqvist and Lundberg (1962) observed the effects on FRA release produced by a series of medially placed sections through the brain stem at different rostrocaudal levels (see Fig. 16). They used both electrical and natural stimulation to evoke the FRA response, and they evaluated the degree of FRA release produced by the various lesions by observation of FRA effects on the evoked monosynaptic reflex in many hind-limb muscle nerves as well as by means of intracellular recording in identified flexor and extensor motoneurons. Group Ib effects were similarly examined. It was found that a medially placed 3mm-wide section through the pons released only FRA inhibitory effects: FRA stimulation evoked inhibition not only to extensor but also to flexor motoneurons. Release of net excitation to flexor motoneurons occurred only after a similar medial lesion was made at the level of the obex or below. The medial obex lesion was as effective in releasing FRA excitation as transection of the whole brain stem or spinal cord. Similar results were obtained for release of group Ib effects. It thus appears that there are at least two FRA pathways in the spinal cord, one that provides net FRA inhibtion to both flexor and extensor motoneurons and another that additionally provides net FRA excitation to flexor motoneurons. Differential control of the alternate pathways, including the possibility of turning each on or off, is mediated by areas of the lower brain stem, undoubtedly the medial pontomedullary reticular formation, considering the lesion sites. Impulses controlling FRA release descend in the dorsolateral fasciculus of the spinal cord.

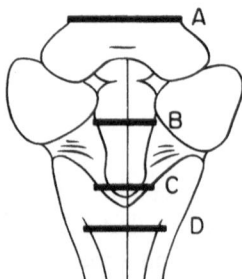

Fig. 16. View of the cat's brain stem from above with cerebellum removed, showing locations of lesions described by Holmqvist and Lundberg (1962) that release various FRA effects in the decerebrate cat. (A) Section of decerebration. (B) Lower medial pontine section releasing net FRA inhibition to all motoneurons. (C, D) Medial medullary (C) and spinal cord (D) sections equally effective in releasing net FRA excitation of flexor and inhibition of extensor motoneurons. See text for further explanation. After Holmqvist and Lundberg (1962).

One site in the lower brain stem that inhibits transmission of FRA and Ib effects to motoneurons was studied in some detail by Engberg *et al.* (1968a,b). The effective area is in the ventral one-third of the caudal, medial, medullary reticular formation; it extends from 3 to 8 mm above the obex and about 3 mm on each side of midline (see Fig. 15B). The area producing the effects at the lowest stimulation strength is centered about 6 mm above the obex and 5 mm deep. This region is thus more delimited but in the same region as Magoun's inhibitory area; it also overlaps, but extends more rostrally than, the area that Jankowska *et al.* (1968) found to inhibit both flexor and extensor motoneurons. However, Engberg *et al.* (1968a) found that stimulation strength could be adjusted to produce no postsynaptic potentials or observable conductance changes in motoneurons (nor dorsal root potentials in primary afferent fibers), as determined by monosynaptic reflex testing and intracellular recording, while, at the same time, transmission of FRA and group Ib postsynaptic potentials to both flexor and extensor motoneurons was greatly reduced or even completely eliminated. No effect was seen, however, on the monosynaptic group Ia EPSP or disynaptic IPSPs from direct antagonists or recurrent (Renshaw) inhibition. These results indicate that the reticular inhibition is exerted entirely on interneurons of the FRA and group Ib pathways.

The inhibition of FRA and group Ib paths was obtained by stimulation of the effective area on either side of the brain stem even though only the ipsilateral dorsolateral funiculus was left intact in these animals. The results of a series of partial lesions of the remaining dorsolateral funiculus suggested that the descending fibers producing the inhibitory effects were distributed throughout the funiculus. Demonstration of the inhibitory effects required repetitive stimulation; the effects were slow to build, and distinct tract waves could not be seen. Because of these factors, the precise latency of the effects could not be determined, nor could it be determined if the effects were relayed in rostral propriospinal neurons, or descended directly to the recorded segment. Assuming the latter case, it could only be determined that the conduction velocity of the fibers was greater or equal to 20 m/s.

The suppression of FRA and group Ib transmission produced by stimulation of this area of the ventrocaudal medulla is similar to that occurring "spontaneously" in the decerebrate cat. The release of FRA and Ib effects after a medial obex lesion seen by Holmqvist and Lundberg (1962) may be explained as due to interruption of tonically firing fibers from the regions stimulated by Engberg *et al.* (1968a,b). However, the relation between the observations of Engberg *et al.* (1968a,b) and the release of inhibitory FRA transmission produced by a more rostral, medial, low-pontine section (Holmqvist and Lundberg, 1962) is not readily apparent.

Subsequent work has traced the control of this dorsolateral reticulospinal tract to higher centers and has indicated that it is only one component in a complex control system. Baldissera, Lundberg, and Udo (1972a) noted that stimulation of an area just dorsal to the red nucleus produced a descending volley in the contralateral dorsolateral quadrant of the spinal cord. It was ascertained that the tract wave was not due to excitation of rubrospinal fibers since it was still present after a lesion of the rubrspinal tract in the medulla. The latency of the tract wave, together with the fact that multiple stimuli were required to evoke it, suggested that there was at least one synapse in the path between the midbrain and spinal cord. In addition, stimulation giving this tract wave inhibited FRA transmission to motoneurons in the same manner as the dorsolateral reticulospinal path of Engberg *et al.* (1968a,b), and it was suggested that the second-order neurons in this pathway from the midbrain were, in fact, those of the dorsolateral reticulospinal tract of Engberg *et al.* (1968a,b).

This idea was explored further by Jeneskog and Johansson (1977) in the latest of a series of experiments carefully examining the effect of stimulating a region just caudal and dorsal to the red nucleus on control of γ motoneurons (which will be discussed more fully in that section of this chapter). The regions stimulated by Baldissera *et al.* (1972a) and Jeneskog and Johansson (1977) appear to be identical. It has been suggested (Appelberg and Jeneskog, 1972; Jeneskog, 1974a) that stimulation in this area excites fibers from the rostral (mainly parvicellular) red nucleus, since these fibers are known to run in the effective area, and the effects obtained are quite different than those obtained from stimulation of the magnocellular portion of the red nucleus, as determined in carefully controlled experiments. It has been ascertained that the effects produced depend on fibers other than rubro- or corticospinal, which also descend in the contralateral dorsolateral funiculus of the spinal cord (Appelberg and Jeneskog, 1969). In addition, stimulation just rostral to the red nucleus introduces one synaptic delay into the tract waves or central effects produced, and this delay is not seen upon stimulation just dorsocaudally to the red nucleus. This observation suggests that the effects are further controlled from yet higher centers.

Jeneskog and Johansson (1977) confirmed the findings of Baldissera *et al.* (1972a) concerning the course of the descending impulses in the spinal cord and their effectiveness in inhibiting transmission of FRA effects to motoneurons in the absence of dorsal root potentials and postsynaptic effects in α motoneurons (the latter observation has been confirmed several times; see Appelberg and Molander, 1967; Appelberg, Jeneskog, and Johansson, 1975). In addition, they showed that FRA transmission to motoneurons is also effectively inhibited by stimulation of low-threshold cutaneous afferent fibers (see Fig. 6). This inhibition is produced by several hind-limb cutaneous nerves, but those innervating the most distal parts of the limb are most effective in suppressing the FRA effects. Furthermore, it was shown by the spatial facilitation technique that the descending pathway from the midbrain and the low-threshold cutaneous pathway share common interneurons that are inhibitory to the FRA interneuron chain. Precautions were taken to ensure that the spatial facilitation occurred at segmental rather than at bulbar levels. As diagrammatically illustrated in Fig. 17, the complete pathway for inhibtion of FRA (and group Ib) transmission to motoneurons consists of an excitatory path from "higher centers" to the rostral (possible parvicellular) red nucleus, or other cells in the area. From there, fibers descend and probably cross the midline to synapse with cells in the reticular formation of the ventrocaudal medulla. The reticular fibers descend in the dorsolateral funiculus and excite segmental inhibitory interneurons that are shared by low-threshold cutaneous afferents from the distal limbs and that inhibit interneurons of the FRA and group Ib pathways. The actual synaptic network of this system is even more complex and will be discussed in more detail in the section on descending control of γ motoneurons.

Just where in the FRA pathway the inhibition acts is not known for certain. Engberg *et al.* (1968b) found that certain FRA-activated segmental interneurons in the dorsal horn and intermediate gray matter could be effectively suppressed by stimulation of the dorsolateral reticulospinal pathway. However, the reticular stimulation produced no postsynaptic potentials in the great majority of these and other interneurons, but a small group of interneurons was found that received IPSPs from the dorsolateral reticulospinal tract and monosynaptic excitation from cutaneous (and, in one case, group I) afferent fibers. From these data, it has been inferred that the dorsolateral reticulospinal inhibition (and, presumably, the low-threshold cutaneous afferent inhibition) of FRA and group Ib transmission to motoneurons is produced by IPSPs in the first interneurons of the FRA pathway.

If the above inference is correct, it has important implications for motor control. It will be recalled that the corticospinal and rubrospinal pathways facilitate transmission through the FRA and group Ib pathways and possibly exert some of their polysynaptic effects using interneurons of these pathways. (Their disynaptic effects seem rather to be transmitted mainly by the last interneurons of low-threshold cutaneous pathways and group Ia inhibitory interneurons.) In addition, the

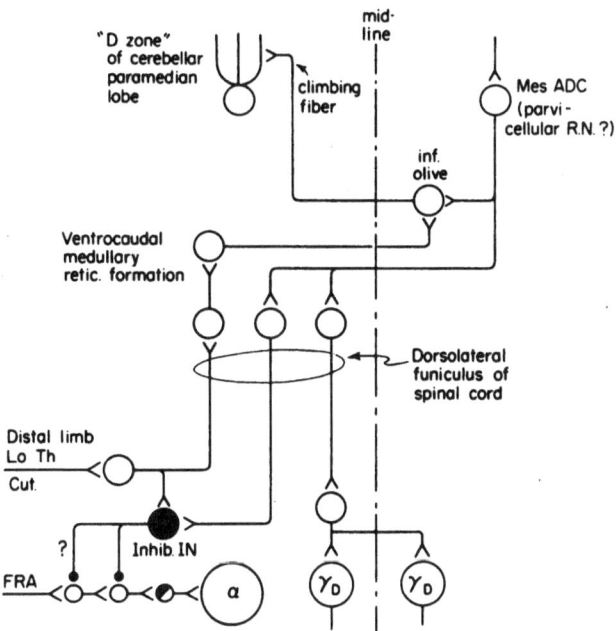

Fig. 17. A pathway from the midbrain producing effects on the spinal cord and cerebellum, based on the studies of Appelberg, Jeneskog, and co-workers (see text for references and further explanation). Electrical stimulation of a region of the midbrain near the red nucleus (Mes ADC) ultimately inhibits transmission of impulses from the flexor reflex afferent fibers (FRA) to α motoneurons (α), whereas dynamic γ motoneurons (γ_D) are concomitantly activated. Stimulation of low-threshold cutaneous fibers (Lo Th Cut.) from the distal limb also inhibits FRA transmission via an interneuron (Inhib. IN) shared with fibers of the descending pathway. The inhibition is presumed to occur early in the FRA interneuron chain. Stimulation of the same midbrain region also causes climbing fiber discharge of Purkinje cells in a certain region of cerebellar cortex. These effects are mediated by inferior olive (inf. olive) neurons, which also ultimately receive excitation from distal limb low-threshold cutaneous fibers.

ipsilateral, disynaptic LVST effects are exerted by the last interneurons of contralateral FRA pathways. Because the descending effects from these tracts are exerted through interneurons of spinal reflex arcs, it would be expected that the motoneurons would usually recieve the integrated peripheral and descending information. However, if the dorsolateral reticulospinal system and the distal low-threshold cutaneous afferent fibers do inhibit the first, or at least early, interneurons in the FRA (or group Ib) pathways, activation of the cutaneous afferent fibers and/or the reticulospinal system may render the FRA and group Ib pathways "private" for descending input alone (see Fig. 18). This could certainly occur for LVST effects, and possibly for polysynaptic cortico- and rubrospinal effects, if these are mediated by interneurons other than the first ones of the polysynaptic reflex pathways. Many interesting interactions among the various pathways are possible, but such interaction depends on cutaneous and reticular inhibition being exerted early in the polysynaptic chain and other descending effects being exerted late; this clearly needs to be examined more throughly.

A role in motor control has also been suggested for the noradrenalin and serotonin-containing reticulospinal neurons found in the medial pontomedullary reticular formation. Some or all of these are apparently the sole source of monoaminergic fibers to the spinal cord: no monoamine-containing cell bodies have been found in the spinal cord, while monoamine-induced fluorescence in fibers disappears caudal to the level of a chronic spinal transection (see Andén, Jukes, Lundberg, and Vyklicky, for references and further discussion). That these monoaminergic reticulospinal pathways can affect motoneuron output has been inferred from effects seen after intravenous injection of L-DOPA in acute low-spinal (L4 section) cats. L-DOPA (levo 3,4-dihydroxyphenylalanine) is a metabolic precursor of various putative monoaminergic transmitters. These are unable to pass the blood–barrier of the CNS, whereas L-DOPA can do so. The effects of L-DOPA injection are presumed to be mediated by increased synthesis and release of transmitter at the monoaminergic terminals in the presence of high concentrations of the L-DOPA precursor. Such release is known to occur in more well-defined monoaminergic systems, and this hypothesis for the action of L-DOPA in the spinal cord was supported by a series of pharmacological tests on the low-spinal cat (Andén, Jukes, and Lundberg, 1966).

Intravenous injection of L-DOPA was found, on the one hand, to produce

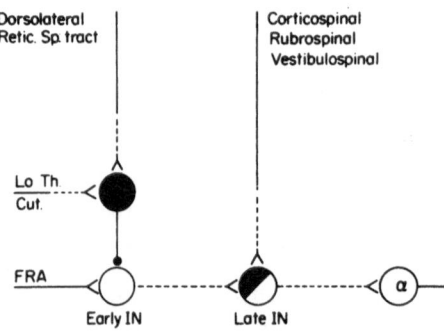

Fig. 18. How activation of the dorsolateral reticulospinal tract of Engberg *et al.* (1968a) and/or impulses in some low-threshold cutaneous fibers (Lo Th Cut.) may inhibit transmission through the flexor reflex afferent (FRA) interneuron chain (Early IN, Late IN). so that only the inputs from the various descending tracts are transmitted to α motoneurons (α) over the FRA pathway. Dashed lines indicate that the connections are not actually monosynaptic, as have been drawn for simplicity. After Jeneskog and Johansson (1977).

effects on spinal reflexes similar to decerebration: FRA and group Ib transmission were depressed, recurrent (Renshaw) inhibition was somewhat augmented, and no effects were observed on group Ia disynaptic inhibition. These effects were ascertained not to be due to tonic increase of presynaptic inhibtion or to be associated with changes in blood pressure due to L-DOPA injection. It was thus concluded that these effects were produced by the action of the monoaminergic pathway on segmental interneurons. On the other hand, after L-DOPA injection, stimulation (only) of the FRA produced long-latency (greater than 100 msec), long-lasting (at least 200 msec) motoneuron excitation; that is, the usual "early" FRA effects were depressed and were replaced by the long-latency, long-lasting, intense activation of flexor motoneurons by stimulation of the ipsilateral FRA and of extensor motoneurons by stimulation of the contralateral FRA. These effects were shown by continuous intracellular recording to occur in the same motoneurons before and after L-DOPA injection. The long-lasting effects were reciprocally organized: excitation of flexor motoneurons by the ipsilateral FRA resulted in long-lasting inhibition of the excitation of extensor motoneurons by the contralateral FRA and *vice versa* (Jankowska, Jukes, Lund, and Lundberg, 1967a). However, intracellular recording demonstrated that no prolonged IPSPs could be observed in flexor motoneurons corresponding to the crossed reciprocal inhibition, nor was there a prolonged IPSP in extensor motoneurons during the period when ipsilateral flexor motoneurons were excited. Both of these observations provide a contrast to IPSPs seen in motoneurons during the usual "early" FRA inhibition. The reciprocal inhibition of motoneuron output produced by these long-latency effects thus appeared to occur between FRA interneurons rather than at the motoneurons. In fact, a distinct population of interneurons were found in the lateral part of Rexed's (1952, 1954) lamina VII, exhibiting the characteristics required to classify them as mediating the long-lasting FRA effects seen in motoneurons after L-DOPA administration: one group of interneurons in the region was activated from the ipsilateral FRA and inhibited from the contralateral FRA, and another group received the opposite effects (Jankowska, Jukes, Lund, and Lundberg, 1967b). These interneurons exhibited the long-lasting effects only upon FRA stimulation and only after L-DOPA administration, and, in fact, L-DOPA sensitive interneurons were found only in this region.

The conclusions from these experiments are several. First, it appears that the descending monoaminaergic system can tonically inhibit the usual FRA and group Ib polysynaptic pathways and disinhibit or facilitate an alternate, long-latency FRA pathway as indicated in Fig. 19A. The long-latency pathway is organized so that its activation on one side mutually inhibits the long-latency pathway to antagonists on the contralateral size, and this crossed reciprocal inhibition occurs at the interneuron level, as indicated in Fig. 19B. The reciprocal crossed inhibition exhibited by the long-latency pathway is not usually seen during simultaneous stimulation of the "normal" ipsilateral and contra-lateral FRA (Graham-Brown, 1911), but it was recognized that such reciprocal crossed inhibition is exactly what is required for performance of stepping movements. The precise relation of the descending monoaminergic pathway to locomotion or normal motor function is not known, but it has been shown that injections of L-DOPA and other drugs potentiating monoaminergic transmitters does indeed aid stepping in low-spinal animals (see Grillner, 1973).

EFFECTS OF THE BRAIN STEM EFFERENT SYSTEMS ON γ MOTONEURONS

CONCEPTS OF γ MONTONEURON FUNCTION

Properties of fusimotor, or γ motoneurons, and their role in motor control are discussed in detail in the Chapter 3, but the following outline is presented to provide a context in which to view brain stem control of γ motoneurons. The γ motoneurons are small ventral horn cells that lie among the much larger α motoneurons in each motor nucleus. Their axons run with those of the α motoneurons in the motor nerve to the innervated muscle, but whereas the α motoneurons innervate the extrafusal fibers responsible for muscle contraction, the γ fibers innervate only the intrafusal fibers of the muscle spindles. The contraction of the spindles provides negligible muscle tension but does cause increased firing of the spindle afferent fibers (i.e., group Ia and group II fibers) at any muscle length over that occurring in the absence of fusimotor activity. In particular, γ motoneuron activity can cause the spindle afferent fibers to continue firing during muscle contraction when they would otherwise cease firing because of mechanical unloading of the spindles due to extrafusal fiber shortening.

After the above functions of the γ motoneurons were initially demonstrated, principally by the experiments of Leksell (1945) and Kuffler, Hunt, and Quilliam (1951), it appeared that the significance of the γ motoneurons was their ability to

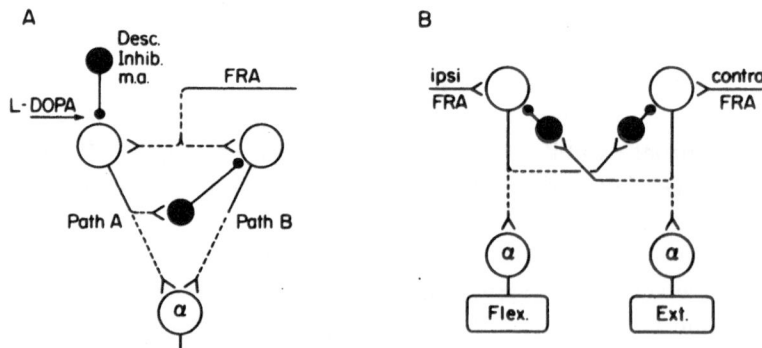

Fig. 19 The effects of L-DOPA on spinal flexor reflex afferent (FRA) pathways. Intravenous L-DOPA may produce its spinal effects as indicated in (A). L-DOPA causes increased release of transmitter from terminals of a descending inhibitory monoaminergic pathway (Desc. Inhib. m.a.). This suppresses the excitability of the FRA interneuron chain, usually activated in the low-spinal cat (Path A). Concomitantly, an alternative long-latency FRA pathway (Path B) is disinhibited, and it now transmits the FRA effects to α motoneurons (α). (B) In the alternative long-latency FRA pathway released by L-DOPA injection, inhibition of ipsilateral FRA effects by contralateral FRA stimulation and vice versa occurs at the interneuron level. No IPSPs are observed in the motoneurons. Dashed lines indicate that the actual connections are not monosynaptic, as drawn for simplicity. After Jankowska *et al.* (1976a).

preserve the reflex function of the spindle afferent fibers during muscle contraction (Kuffler and Hunt, 1952). How the CNS might use this property for motor control was suggested by the "follow-up length servo" hypothesis of Merton (1953). Although this hypothesis is now considered only in modified form, it still provides a useful general framework for discussing the significance of descending control of γ motoneurons. According to Merton's (1953) hypothesis, "urgent" movements might be produced by the action of descending impulses on α motoneurons, but "ordinary" movements could be produced by descending impulses acting solely on γ motoneurons (assuming that α and γ motoneurons can be activated independently by the CNS). It was proposed that activity in γ motoneurons evoked by descending impulses would cause group Ia spindle afferent fibers to increase their firing if the muscle were longer than a certain "command" length specified by the intensity of the descending barrage. The increased monosynaptic activation of α motoneurons by the Ia afferent input would then cause the muscle to contract until the "command" length was reached, at which point the Ia fiber input would have decreased to the "rest" firing rate existing before their activation by the γ motoneurons. Any further deviation from the "command" length would then be automatically corrected by an increase or decrease of Ia fiber activity according to the usual muscle length-firing rate relation of the Ia afferent fibers.

The advantage of this scheme for indirect control of movement is that it would greatly simplify the motor commands required of "higher centers" to specify limb position. Instead of having to compute and vary descending commands to obtain a certain muscle length in the face of uncertain and varying muscle loads, the "higher centers" would only need to transmit a barrage of impulses to the γ motoneurons that invariantly correspond to a certain muscle length. The γ motoneuron activity would then serve to bias the group Ia feedback loop in the manner just described, and the specified muscle length would be automatically obtained, since the spindle afferents are length, not load, sensitive.

This attractive scheme for control of limb position via the "γ route," or "γ loop" (Granit, 1955), was subsequently tested by Eldred, Granit, and Merton (1953) who recorded the response of spindle afferent fibers during contraction of the soleus muscle in the decerebrate cat under electrical stimulation and reflex activation of supraspinal structures. They found that supraspinal stimulation could indeed bias the group Ia muscle length-firing rate relation in a manner required by Merton's (1953) hypothesis and that contraction of the soleus evoked by the tonic neck reflex actually seemed to be mediated by the γ loop. Two basic findings were taken as validating the servo-loop hypothesis during the reflex contraction. First, spindle firing was found to accelerate prior to and during a reflex contraction obtained with dorsal roots intact, indicating a concomitant increase in γ activity that was more than sufficient to offset the intrafusal unloading. Thus, γ-mediated, group Ia activation of motoneurons *could* have caused the contraction. Second, cutting the dorsal roots (thus interrupting the γ loop) abolished the contractions although the Ia firing increased as before. Although these and other similar results were considered at the time to validate the servo hypothesis, it has subsequently been pointed out (Matthews, 1972) that they do not provide sufficient evidence to prove the hypothesis. For example, cessation of α motoneuron firing after dorsal root section could occur simply because an amount of facilitation necessary to bring the motoneurons to firing level had been removed; their lack of firing does not necessarily mean that they received excitatory input *only* over the γ route but. rather, that they did at least receive *some* facilitation by this route. Although it has been directly demonstrated that α motoneurons *can* be made to fire when activated solely over the γ loop (by stimulating the distal stump of a muscle nerve in a paralyzed animal—Granit, Kellerth, and Szomski, 1966), there is still no evidence that they actually are activated *solely* by this route under electrical or natural stimulation. On the contrary, there is now much evidence that γ motoneuron activation is not critical for descending control of reflex and other types of muscle contraction (see below).

The servo hypothesis has also been criticized on several other grounds (Matthews, 1972). For example, in order for the γ loop to provide the servo function proposed by Merton (1953), its feedback gain must be quite high, but there are several *a priori* reasons for expecting a low feedback gain in such a loop. Several lines of available evidence suggest that the actual gain is too low for the loop to function as a servo control; rather, it can only provide a "servo assist" to the main "α route" (Granit, 1955) control of movement. That the γ loop does at least provide a servo-assist function is suggested by the experiments of Eldred, Granit, and Merton (1953). Numerous other studies have also shown that electrical stimulation of various supraspinal structures often causes γ motoneurons to fire at a lower stimulation strength than for α motoneurons, whereas the afferents from spindles innervated by the γ motoneurons usually commence firing before the muscle contraction and accelerate firing during the contraction. The only exception to these findings is provided by the motoneurons innervating the wrist and digits of the baboon's hand (Koeze, 1968; Koeze, Phillips, and Sheridan, 1968), the α population of which is under strong monosynaptic control from motor cortex. The ubiquitous finding of concomitant α and γ motoneuron firing upon central stimulation was termed "α-γ linkage" (Granit, 1955; the term is also sometimes applied to the γ loop circuitry), but the more precise term, "α-γ coactivation," is more often used now.

In spite of the uniform results produced by electrical stimulation, it provides a rather artificial demonstration of α-γ coactivation. However, there now exist several studies indicating that coactivation occurs during more natural movements, such as normal respiration (Eklund, von Euler, and Rutkowski, 1964; Sears, 1964), labyrinthine reflexes (Ezure *et al.*, 1978; Poppele, 1967), stepping in the "mesencephalic" and low-spinal, deafferented cat (Severin, 1970; Sjöström and Zanger, 1975) as well as during voluntary contraction of human limb, wrist, and digit muscles (Hagbarth and Vallbo, 1968, 1969; Vallbo, 1970, 1971). In these studies, γ motoneuron behavior was usually examined indirectly by observing the acceleration of spindle afferent firing during muscle contraction. It should be noted, however, that most of the muscle contractions studied occurred under approximately isometric conditions so that little or no muscle shortening occurred. An investigation of spindle afferent behavior during natural chewing movements in the awake monkey has shown that spindle afferent firing ceases during most of the jaw muscle contraction and shortening associated with chewing (Goodwin and Luschei,

1975). Although γ motoneurons may be coactivated, they are not able to activate the spindle afferent fibers and thus provide afferent input to the motoneurons during the muscle-shortening phase of chewing. In the animal experiments, it could be verified that muscle contraction remained unimpaired after interruption of the γ route by dorsal root section, though the spindle afferent fibers firing during contraction remained unchanged. In the human studies, it was determined that the spindle acceleration occurred at or after the onset of muscle contraction and thus too late to be the cause of the α motoneuron activation (Vallbo, 1971). It is also interesting that the α-γ coactivation is apparent during locomotion in the low-spinal, deafferented cat (Sjöström and Zanger, 1975) after L-DOPA injections, since this suggests that a segmental interneuronal apparatus is capable of organizing the coactivation of α and γ flexor and extensor motoneurons at the appropriate times in the step cycle.

Early experimentation and theory focused on descending rather than segmental control of γ motoneurons, in contrast to early work on α motoneurons. This is because it was soon established that γ motoneurons are rather poorly activated from the periphery, compared to α motoneurons. The γ motoneurons are activated only by nocioceptive natural stimuli, or electrically only from the contralateral or ipsilateral FRA (Eldred and Hagbarth, 1954; Hunt, 1951); they receive little or no input from group I muscle afferents or muscle stretch. (They certainly receive no monosynaptic group I excitation; whether they receive significant polysynaptic group I inhibition is still debated.) In contrast, many studies have found that γ motoneurons are consistently and easily activated from various supraspinal structures. Although this view is perhaps biased by the relative lack of data available on segmental pathways to γ motoneurons, the general impression is that γ motoneurons primarily serve as relays of descending impulses to muscle spindles and are only secondarily involved in segmental reflexes.

It is apparent from preceding sections of this chapter that knowledge of the detailed connections to α motoneurons has been obtained largely by the technique of intracellular recording. Intracellular recording is very difficult in γ motoneurons because of their small size, and only a handful of such recordings have been made. Instead, their activity is recorded in dissected ventral root filaments or is inferred indirectly by recording the acitivity of spindle afferent fibers. These techniques do not allow the exploration of detailed synaptology, but they are suitable for examining more general questions about descending control of γ motoneurons and their role in motor control. Most of these questions concern the nature of α-γ coactivation in relation to the modified, "servo-assist" hypothesis of γ function. One question central to the servo-assist hypothesis is whether the CNS can discretely activate γ motoneurons to certain muscles as it can α motoneurons, or whether the γ motoneurons are diffusely activated simply as a means of increasing the "central excitatory state" by means of the γ loop. In fact, there is evidence for both discrete and diffuse control of γ motoneurons.

Concerning the discrete, reciprocal activation of γ motoneurons by descending impulses, a further question related to the servo-assist hypothesis is whether a given descending tract has like action on α and γ motoneurons. This is clearly required by the servo hypothesis and must be confirmed for each descending tract to establish the generality of the servo-assist function. A further question, implicit in the original "follow-up length servo" hypothesis, is whether the CNS is able independently to control α and γ motoneurons to a given muscle, and further, since there are both "static" and "dynamic" γ motoneurons (see below), whether each type of γ motoneuron can be controlled separately. A related question is whether any descending pathways may control only γ motoneurons or only one type of γ motoneuron. Finally, it would be desirable to assess quantitatively the contribution of the servo-assist function of the γ loop to muscle contraction. The progress made to date in answering these questions will be apparent upon considering the effect of each brain stem afferent system on the two types of fusimotor neuron.

As indicated above, the effects of each descending system must be evaluated on three functionally different types of motoneurons, α motoneurons, and static and dynamic γ motoneurons. The two types of γ motoneurons, first described by Matthews (1962), differ principally in that activation of dynamic γ motoneurons enhances and static motoneurons suppresses the natural dynamic sensitivity of the primary (group Ia) spindle afferent fibers (i.e., the spindle's response to the velocity of stretching). Both types of γ motoneurons enhance spindle firing under static stretch, though the static fibers sometimes provide greater enhancement than the dynamic fibers. In addition, static fusimotor activation also enhances firing of secondary (group II) spindle afferents, whereas dynamic fusimotor activation has no effect (thus providing one indirect test of whether a descending pathway activates static γ motoneurons). It should be noted that, by definition, the distinctive actions of the dynamic fibers are only apparent during muscle stretch and not during static muscle position, suggesting that the dynamic fibers would be particularly useful in resisting displacement from a fixed muscle length. However, as will be seen in Chapter 3, the effects of static and, especially, dynamic γ motoneurons on spindle firing are rather more complex and do not clearly specify a particular role for dynamic as opposed to static γ motoneurons on the basis of present knowledge. It is, nevertheless, useful to discover if the brain stem efferent systems exert differential or discrete control over the two types of fibers; such knowledge may ultimately aid in determining the precise role of each type in motor control.

Brain Stem Control of Static and Dynamic γ Motoneurons

The effects of the rubrospinal tract on γ motoneurons were first studied in detail by Appelberg (1962) and Appelberg and Kosary (1963). The activity observed in identified spindle afferent fibers upon rubral stimulation suggested that the red nucleus excited γ motoneurons innervating flexor muscles and generally inhibited those to extensor muscles, similar to rubrospinal actions on α motoneurons. These reciprocal effects were ascertained to occur in static γ motoneurons. Subsequent investigation, however, indicated that stimulation at various points in the area of the red nucleus was very effective in activating both static and dynamic fusimotor neurons (see below). Recently criteria have been developed to ensure fairly discrete stimulation of the rubrospinal tract (Baldissera *et al.*, 1972b), and, using these criteria for rubral stimulation, the rubrospinal control of γ motoneurons has been recently reexamined (Appelberg *et al.*, 1975). In this study, extracellular recordings were made from identified γ motoneurons in the spinal cord.

Dynamic γ motoneurons were identified by their activation from another midbrain site previously shown by many studies to activate only dynamic fusimotor neurons (see below); those not responding to such stimulation were classified as static fusimotor neurons. The results of this study confirmed those of the earlier studies: static γ motoneurons to a variety of hind-limb flexors were invariably excited from the red nucleus, whereas those to extensors were both excited and inhibited, depending on the particular muscle. However, dynamic fusimotor neurons were also found to be influenced by the red nucleus. The reciprocal excitation of flexor and inhibition of extensor dynamic fusimotor neurons appeared to be more complete since only 1 out of 13 identified dynamic γ motoneurons to extensors was found to be excited from the red nucleus. Overall, the percentage of extensor fusimotor neurons excited and inhibited were similar to those obtained by Hongo *et al.* (1969a) on extensor α motoneurons, though the fusimotor population studied was much smaller. The distribution of effects among different extensor muscles also was similar. Evidence was also obtained for a disynaptic coupling between the fastest rubrospinal fibers and some γ motoneurons. Thus, the red nucleus has an effect on static and dynamic neurons similar to that exerted on α motoneurons, and this effect may be exerted in part by the same population of fast fibers giving rise to disynaptic postsynaptic potentials in α motoneurons. Whether the red nucleus can independently control the three types of motoneurons is difficult to determine by electrical stimulation and has not yet been investigated by other means. It may be noted that the corticospinal tract, which is similar to the rubrospinal tract in many of its spinal actions, has also been shown to influence α and γ motoneurons (see Chapter 8).

Control of fusimotor neurons by the LVST originating from Deiters' nucleus has also been studied in some detail. It appears that Deiters' nucleus generally excites identified hindlimb γ motoneurons innervating extensors and inhibits those to flexors (Kato and Tanji, 1971), similar to its effects on α motoneurons. Although most of this excitation appears to be mediated by polysynaptic pathways, a significant fraction of γ motoneurons to knee and ankle extensors receives monosynaptic EPSPs from the same population of fast LVST fibers giving rise to monosynaptic EPSPs in α motoneurons (Grillner, Hongo, and Lund, 1969). The EPSPs in the γ motoneurons also are considerably larger than obtained in the α motoneurons, but this observation is based on a limited sample of intracellular recordings. Indirect evidence suggest that static γ motoneurons are excited from Deiters' nucleus: Carli, Diete-Spiff, and Pompeiano (1967) found that stimulation of Deiters' nucleus increased the firing of secondary endings of muscle spindles (which is caused only by activation of static fusimotor neurons), whereas Bergmans and Grillner (1968b) found that stimulation in the region of the LVST in a low-spinal preparation excited only γ motoneurons that did not fire spontaneously. Since it is believed that only dynamic fusimotor neurons spontaneously discharge in the spinal preparation (Alnaes, Jansen, and Rudjord, 1965), these latter observations suggest that *only* static fibers are excited by Deiters' nucleus. This evidence is rather tenuous, however, and a more direct test is required to exclude the possibility that Deiters' nucleus also controls dynamic γ motoneurons.

Fusimotor firing is also readily affected by VIIIth nerve stimulation, as first suggested by Andersson and Gernandt (1956) and confirmed by Diete-Spiff, Carli, and Pompeiano (1967). In both studies it was found that VIIIth nerve-evoked fusimotor firing occurs earlier than α motoneuron firing and can occur in the absence of such firing. Curiously, Diete-Spiff *et al.* (1967) found that fusimotor neurons were more effectively activated from the VIIIth nerve than from central stimulation of Deiters' nucleus or the reticular formation, but this may simply reflect the high physical density of effective fibers in the peripheral nerve.

The observation that fusimotor neurons may be effectively activated from electrical stimulation of the VIIIth nerve suggests they are also activated during natural labyrinthine reflex activity. Furthermore, the effect of interrupting the γ loop during these reflexes provides a means of quantitatively assessing the importance of the servo-assist function of the γ loop. Carli *et al.* (1967) examined the effect of dorsal root section on tension produced in ankle extensors upon stimulation of Deiters' nucleus. As mentioned above, such stimulation activated fusimotor neurons more readily and sooner than α motoneurons, but dorsal root section only somewhat reduced the rate of tension development and not the final tension obtained. A similar observation was reported by Poppele (1967) who found that tilt about the longitudinal axis of a decerebrate cat activated hind-limb extensor γ motoneurons earlier and more readily than α motoneurons. Both types of motoneurons were generally affected the same way: their firing rate increased for tilt in either direction and, during the phasic response, was related to the velocity of the roll. Again, dorsal root section had no effect on the final muscle tension developed during the reflex, though the rate of tension development was somewhat depressed.

These experiments clearly indicate that muscle contraction produced from stimulation of Deiters' nucleus and from certain labyrinthine reflexes does not depend on an intact γ loop. The γ loop did seem to affect the speed of contraction in these experiments, and it might, therefore, be expected that the servo-assist function may be more important in phasic than in tonic motor control. However, several recent experiments have quantitatively evaluated the dynamic control of motor units in neck muscles (Ezure *et al.*, 1978) and in forelimb extensor muscles in the decerebrate cat (Anderson, Berthoz, Soechting, and Terzuolo, 1977; Anderson *et al.*, 1977a,b; Soechting *et al.*, 1977) provided by various types of natural labyrinthine stimulation. In the experiments on neck-muscle motoneurons (Ezure *et al.*, 1978), it was ascertained that spindle afferent fibers are indeed modulated during natural semicircular canal stimulation in a manner suggesting α-γ coactivation. Nevertheless, interruption of the γ loop produced no detectable effect on the various natural dynamic reflex responses obtained in neck or forelimb extensor motor units. These results are rather disturbing since it is clear from the experiments cited above and earlier in this section that fusimotor neurons readily respond to natural or electrical VIIIth nerve stimulation in a manner suggesting a servo-assist function, but the quantitative investigations cited above indicate that the role of the γ loop is completely negligible. Thus, these investigations, like that of Goodwin and Luschei (1975), fail to support the idea that excitation of α motoneurons via the γ loop is an important factor in certain types of normal motor activity.

Granit and Kaada (1953) first showed that fusimotor neurons could be excited and inhibited by electrical stimulation of the reticular formation. They also noted that γ motoneurons were most effectively activated from the midbrain tegmentum, and subsequent investigation has indeed shown this area to be strongly involved in fusimotor control. Discrete, reciprocal effects from the red nucleus transmitted by the rubrospinal tract have been described earlier. An area ventral to the red nucleus has also been found which seems to activate mainly static γ motoneurons (Appelberg and Jeneskog, 1972), but it has not been extensively studied. Instead, attention has been focused on an area just dorsal to the red nucleus (see below) that, in contrast to the descending actions described so far, seems to rather diffusely activate only dynamic fusimotor neurons. Granit and Holmgren (1955) observed that, in addition to discrete fusimotor effects produced by stimulation at some midbrain site, a more prolonged diffuse activation could be produced by stimulation at other sites. That diffuse, nonreciprocal, fusimotor activation could be obtained by brain stem stimulation was confirmed by the experiments of Shimazu, Hongo, and Kubota (1962), who showed that such nonreciprocal activation remained after reciprocal effects were eliminated by various lesions. The fact that widespread fusimotor activation was obtained upon twisting the pinna of a decerbrate cat (Granit, Job, and Kaada, 1952) also suggest that diffuse fusimotor activation may be obtained under natural stimulation.

The area dorsal to the red nucleus producing such diffuse γ motoneuron activation has been extensively studied in a long series of experiments by Appelberg, Jeneskog, and co-workers (Appelberg, 1967; Appelberg and Jeneskog, 1969, 1972; Appelberg and Molander, 1967; Jeneskog, 1974a,b; Jeneskog and Johansson, 1977). This area has been suggested (Appelberg and Jeneskog, 1972) to be responsible for the diffuse γ activation observed by Shimazu *et al.* (1962) and has been termed (Appelberg, 1967) the "mesencephalic area for dynamic fusimotor control" (Mes ADC) because it has been repeatedly confirmed that stimulation in this area activates only dynamic γ motoneurons and does so effectively at stimulation strengths producing no postsynaptic potentials in α motoneurons. This area is identical to that producing generalized inhibition of transmission of FRA and group Ib reflexes via the dorsolateral reticulospinal tract (see p. 192). As will be recalled, and as indicated diagrammatically in Fig. 17, it was suggested that these effects are produced by stimulation of fibers originating from the rostral (parvicellular) red nucleus that cross in the brain stem and synapse at least once in the ventrocaudal medulla (in the inhibitory area of Engberg *et al.*, 1968a). The second-order fibers then descend in the dorsolateral funiculus (and have been shown to be separate from corticospinal and rubrospinal fibers in this funiculus) to the spinal cord where they concomitantly activate dynamic fusimotor neurons and inhibit FRA transmission to motoneurons. The latter action is mediated by inhibitory interneurons shared with low-threshold cutaneous afferents from the distal part of the limbs. In addition to these actions, stimulation of the Mes ADC at intensities similar to those producing the spinal effects also activate inferior olive neurons, which produce climbing fiber responses in Purkinje cells of the contralateral cerebellar cortex (see Fig. 17). These same inferior olive cells are activated by spinocerebellar pathways that ascend in the dorsolateral funiculus and that are most effectively activated by low-threshold cutaneous afferents from distal skin. It should be emphasized that the nature of the descending pathways and cerebellar feedback loops has not been proven to be exactly as described above. For example, it has not been proven that the same cell population in the Mes ADC region produces all three descending effects; it is possible that these are due to three separate cell populations occupying the same area. On the other hand, the described pathway is at least consistent with a considerable amount of data, and it might be expected that even separate cell populations occupying the same physical area would be activated together by "higher centers."

In spite of the rather detailed knowledge of these remarkable descending actions in connection with multiple-loop feedback to the cerebellum, the function of this pathway in motor control must still be determined. The multiple cerebellar feedback loops are consistent with a proposal by Oscarsson (1973) that such loops provide the cerebellum with a comparison between descending commands and actual effects produced at the spinal level. Why this should involve simultaneous activation of dynamic fusimotor neurons and climbing fiber responses in cerebellar Purkinje cells is unknown. Discovery of the function of the Mes ADC pathway may thus provide much needed insight into the precise function of dynamic fusimotor neurons as well as the meaning of the cerebellar climbing fiber response.

Another example of differential control of static and dynamic fusimotor neurons, though again acting somewhat nonreciprocally on flexor and extensor fusimotor neurons, is provided by the injection of L-DOPA in the low-spinal cat (Bergmans and Grillner, 1968a; Grillner, 1969). As mentioned previously, injection of L-DOPA is thought to mimic the effects of increased activity in the descending monoaminergic pathways. In the low-spinal cat, spontaneous fusimotor firing is largely restricted to dynamic γ motoneurons (Alnaes *et al.*, 1965), but after the injection of L-DOPA, the static fibers to both flexor and extensor muscles begin to fire spontaneously. Only those dynamic fusimotor neurons that supply flexor muscles become silent.

The reticulospinal system also appears to be capable of providing discrete control of fusimotor neurons, as exemplified by excitation of flexor γ motoneurons by reticulospinal fibers descending in the MLF. The studies of Grillner *et al.* (1969) also demonstrated that the population of fast reticulospinal fibers descending in the MLF provide monosynaptic EPSPs to some flexor γ motoneurons, as they do to α motoneurons in the same motor nuclei. Again, dynamic fusimotor neurons are believed to be unaffected on the basis of indirect evidence (Bergmans and Grillner, 1968b). The question of whether there is also reciprocal inhibition to extensor fusimotor neurons, similar to that existing in α motoneurons, has not been pursued.

It is useful to reconsider the general questions about α-γ coactivation in relation to the proposed servo-assist function of fusimotor neurons posed at the beginning of this section. There is no question that α and γ motoneurons are coactivated in a variety of motor acts, and it has been seen that the rubrospinal, vestibulospinal, and corticospinal systems generally provide the same discrete, reciprocal control of γ motoneurons as with α motoneurons. It has been shown that the MLF-reticulospinal neurons exert similar excitatory effects on α and γ motoneurons, but reciprocity has not been investigated. On the other hand, there is still no convincing evidence that these descending systems can, in general, independently control α and γ motoneurons. The fact that γ motoneurons may be made to fire in the absence of α activity by all these tracts may only indicate that the EPSPs are larger in the

γ motoneurons or that these are usually more depolarized than α motoneurons, which simultaneously receive subthreshold EPSPs but do not fire. Convincing evidence for control of dynamic motoneurons in the complete absence of α effects has been obtained from the Mes ADC system, and differential control of static from dynamic fusimotor neurons has been obtained only from the descending monoaminergic systems. Both of these systems appear to act rather diffusely compared to the other descending tracts, however. The semidifferential control of α and static, but not dynamic, γ motoneurons said to occur in the vestibulospinal and MLF reticulospinal systems rests on such indirect evidence that it cannot be seriously evaluated. There is also the disturbing findings that the labyrinthine reflexes, known to produce strong α-γ coactivation, are hardly affected by interruption of the γ loop and that any existing γ activation is not able to maintain high rates of spindle activity during the contraction of jaw muscles during chewing. In spite of the attractiveness of the servo-assist hypothesis, it must be concluded that the information presently available is far from sufficient to show that the hypothesis adequately describes the principal γ motoneuron function. It is possible that the principal function of the fusimotor neurons in many circumstances is quite different than envisioned in the servo-assist hypothesis. For example, it is possible that the facilitation of α motoneurons provided by the γ loop is useful but not particularly critical for many motor acts, while the γ-mediated firing of spindle afferent fibers may be critical in transmitting information on relative muscle length to the cerebellum via the dorsal spinocerebellar tract or in controlling alternating flexion/extension movements via the Ia inhibitory interneuron discussed in the next sections. These possibilities are suggested by the fact that impulses from group Ia afferent fibers appear to excite the neurons involved in these pathways much more readily than they excite α motoneurons.

Brain Stem and Segmental Effects on the Group Ia Inhibitory Interneuron

It has been emphasized in the sections on descending control of α motoneurons that the brain stem efferent systems greatly influence transmission of spinal reflex effects to motoneurons and probably exert most of their characteristic actions on motoneurons (in the cat) by means of the interneurons of these segmental reflex pathways. The next logical step in examining the synaptic connections of descending systems would then appear to involve recording from the interneurons themselves in order to observe the descending effects they receive and to examine their interactions with other reflex pathways. Such descending and segmental interactions will determine what information the interneurons pass on to the motoneurons. This step is difficult because the interneurons in each reflex pathway are defined only in terms of their effects on motoneurons but usually cannot be identified as belonging to a certain reflex pathway when recorded individually. One exception to this difficulty is provided by the inhibitory interneuron interposed in the path from group Ia afferent fibers to antagonistic motoneurons (the Ia inhibitory interneuron). It may be identified as such primarily by the fact that it receives monosynaptic excitation from group Ia afferents and monosynaptic inhibition from Renshaw cells, just as do the α motoneurons (Hultborn, Jankowska, and Lindström, 1971a,b). Earlier studies examined the convergence of descending and segmental effects on this interneuron by the indirect method of recording from hindlimb motoneurons and observing spatial facilitation of the disynaptic IPSP from antagonists by various sources (Hultborn, 1972; Hultborn and Udo, 1972). More recently, however, studies of the effects converging on the Ia inhibitory interneurons have been made by direct intra- and extracellular recordings (Hultborn, Illert, and Santini, 1976a,b,c). The results obtained offer considerable insight into the descending and segmental control of this interneuron.

The most remarkable finding is that stimulation of most segmental reflex pathways evokes similar effects in both the α motoneurons and in the Ia inhibitory interneurons receiving monosynaptic excitation from the same group Ia afferent fibers. Furthermore, the effects from the shortest latency pathways are identical in the motoneurons and the Ia inhibitory interneurons, as shown diagrammatically in Fig. 20. It has already been mentioned that the Ia inhibitory interneurons receive monosynaptic group Ia excitation and recurrent (Renshaw) inhibition, but they are also found to receive disynaptic group Ia inhibition from direct antagonists (Hultborn *et al.*, 1976a) just as do the motoneurons. Moreover, this disynaptic group Ia inhibition is mediated by the Ia inhibitory interneurons that are monosynaptically linked to the motoneurons of the antagonistic muscle (as demonstrated by the suppression of the disynaptic inhibition upon activation of Renshaw circuits to the antagonists). As shown in Fig. 21 for an extensor motoneuron pool, the associated Ia inhibitory interneurons also receive effects from ipsilateral and contralateral high-threshold muscle, joint, and cutaneous nerves over FRA pathways, as well as effects from low-threshold ipsilateral cutaneous afferents over a separate pathway (group Ib effects were not reported). However, the effects evoked in the Ia inhibitory interneurons over these polysynaptic segmental reflex pathways are somewhat different from those seen in the corresponding motoneurons. The FRA and low threshold cutaneous pathways generally excite flexor and inhibit extensor motoneurons in the low-spinal cat, but the Ia inhibitory interneurons were found to receive mixed excitation and inhibition whether they were flexor-linked or extensor-linked. A more detailed comparison revealed that the extensor-linked Ia inhibitory interneurons receive somewhat less excitation than the flexor-linked cells from FRA or cutaneous stimulation.

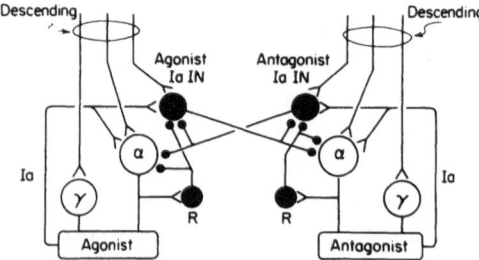

Fig. 20. Diagram indicating that the connections of the shortest spinal reflex pathways in the cat are identical for group Ia inhibitory interneurons (Ia IN) and α motoneurons (α). R indicates the Renshaw interneurons providing recurrent inhibition. The diagram also indicates that the Ia inhibitory interneurons, α motoneurons and γ motoneurons (γ), all receive effects from descending pathways. These descending effects are similar in α and γ motoneurons and sometimes in Ia inhibitory interneurons linked to the same muscle. After Hultborn *et al.* (1976c).

As outlined on p. 154, both the corticospinal and rubrospinal tracts facilitate transmission of group Ia disynaptic inhibition from flexor to extensor motoneurons and vice versa. Correspondingly, both flexor- and extensor-linked Ia inhibitory interneurons are predominantly excited disynaptically by each of these descending tracts, as indicated in Fig. 21, but mixed excitation and inhibition were also common. The disynaptic rubro-interneuronal IPSP was not mediated to any significant extent by the antagonistic Ia inhibitory interneurons. Rather, both the disynaptic excitation and inhibition seemed to be relayed mainly through last interneurons of ipsilateral low-threshold cutaneous pathways, as also found for rubrospinal effects to motoneurons (Baldissera *et al.*, 1971). FRA pathways seemed to play a minor role in transmission of rubrospinal effects to the interneurons, as is also true for rubrospinal effects on motoneurons (Hongo *et al.*, 1969b).

As also indicated in Fig. 21, generally similar effects were obtained from the corticospinal tract, but the exact latencies and synaptic linkages could not be determined. Inhibitory effects were, in part, transmitted via the antagonistic Ia inhibitory interneuron. FRA and cutaneous effects were facilitated by corticospinal stimulation, but it was not determined if these pathways transmit the corticospinal effects seen in the Ia inhibitory interneurons.

As discussed previously on p. 168, stimulation of Deiters' nucleus facilitates transmission of disynaptic group Ia inhibition only from knee extensors to knee flexor motoneurons (ten Bruggencate *et al.*, 1969; ten Bruggencate and Lundberg, 1974). Correspondingly, monosynaptic and disynaptic vestibulospinal EPSPs were seen in knee (and some hip) extensor-linked Ia inhibitory interneurons, whereas no effects were seen in flexor-coupled interneurons from Deiters' stimulation. Reticulospinal fibers descending in the MLF have actions reciprocal to those from Deiters' nucleus at the knee joint (Grillner *et al.*, 1971). However, no effect was observed in Ia inhibitory interneurons upon MLF stimulation, in confirmation of previous observations based on the indirect, spatial facilitation technique.

The experiments summarized above indicate that the Ia inhibitory interneurons receive effects over the same pathways as do the motoneurons with which they are monosynaptically linked by group Ia afferent fibers. They often receive descending effects mediated by the same type of segmental interneurons found to mediate the effects to the motoneurons. Such results imply that the information transmitted by the Ia inhibitory interneurons to motoneurons will depend on the

Fig. 21. The principal polysynaptic actions of descending pathways and polysynaptic spinal reflex arcs on extensor-linked Ia inhibitory interneurons. The net actions of these low-threshold cutaneous (Lo Th Cut.) flexor reflex afferent (FRA), rubrospinal (RuST), and corticospinal (CST) fibers are similar on both extensor-linked and flexor-linked Ia inhibitory interneurons (Ia IN). Note that the effects produced by each of these pathways on

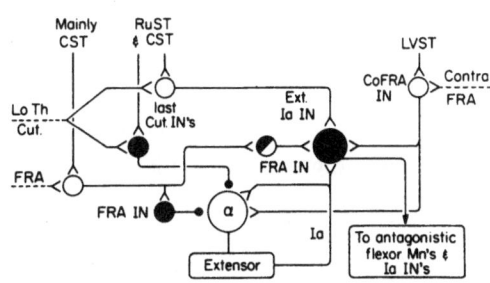

the associated α motoneurons (α) may be quite different. Also note that the same type of segmental interneurons (last Cut. IN, Co. FRA IN) transmits polysynaptic descending effects to both the Ia inhibitory interneurons and the α motoneurons.

summed activity of various descending pathways (principally the corticospinal and rubrospinal tracts) and the various polysynaptic segmental pathways. It would be expected that such extensive interactions also occur in other segmental reflex pathways.

These results also indicate the extensive control exerted on this particular interneuron by the CNS. It seems clear that its primary function is inhibition of antagonists during activation of agonists. According to older ideas, this reciprocal inhibition was sufficiently effected through activation of the inhibitory interneuron by group Ia afferent fibers from the agonist, but the current data suggest that this reciprocal inhibition is sufficiently important in normal function to require substantial reinforcement. This is effected at the segmental level by disynaptic inhibition of not only the antagonistic motoneuron but also of the antagonistic Ia inhibitory interneuron (see Fig. 20). Thus, the agonist motoneurons are both directly excited and indirectly disinhibited by the action of the homonymous group Ia afferents, while the antagonist is inhibited. Further reinforcement is derived from descending tracts since they may not only excite the agonist motoneurons but simultaneously provide disinhibition to them and inhibition to the antagonists by their action on the Ia inhibitory interneurons. As also indicated in Fig. 20, further reinforcement may be derived from the γ loop since descending tracts usually coactivate α and γ motoneurons as described in the preceding section. Increased fusimotor activation would further amplify the group Ia effects described above (such effects have been termed "α-γ linked inhibition" by Hongo *et al.*, 1969b). The powerful reciprocal inhibition effected by these various pathways to the Ia inhibitory interneuron would seem to be most important in alternate flexion/extension movements such as occur during locomotion. As will be discussed in the next section, there is now direct evidence that the Ia inhibitory interneurons are indeed activated at appropriate times in the step cycle.

Role of the Brain Stem Systems in "Mesencephalic Locomotion"

Various aspects of locomotion and the types of experimental preparation in which it is studied will be described in detail in Chapter 11. Some results of studies in the "mesencephalic cat" will be summarized here since they provide one of the few systematic studies of the action of the brain stem efferent systems during a complex motor act. The mesencephalic cat, introduced by Shik, Severin, and Orlovsky (1966), is basically a precollicular decerebrate cat. It can be made to step at different rates when a circumscribed region below the inferior colliculus is stimulated continuously while its feet are on a treadmill and most of its weight is supported from above (see Orlovsky and Shik, 1976; Shik and Orlovsky, 1976.) The rate of stepping from a "walk" to a "gallop" can be controlled by the speed of the treadmill and the strength of midbrain stimulation. If the treadmill is allowed to move freely, increased stimulation strength results in acceleration of the treadmill and increased frequency of stepping with limb movements that mimic those of the intact cat. The limb movements and patterns of muscle activation during walking are somewhat complex, but for the purposes of this summary, the step cycle may be divided into the stance phase, during which extensors are actively supporting the weight of the limb and developing the forward thrust, and the swing phase, during which flexors are activated to swing the limb up and forward for the next step. There is overlap of flexor and extensor activity at the end of one phase and beginning of another, and some muscles acting at two joints are activated during both phases.

Having an acute preparation in which complex motor acts can be obtained and varied, it is possible to inquire into the role of the various descending systems in controlling this activity. This has been done for the rubrospinal (Orlovsky, 1972c), lateral vestibulospinal (Orlovsky, 1972b), and MLF-reticulospinal systems (Orlovsky, 1970), as well as for various ascending pathways and cerebellar structures (see Shik and Orlovsky, 1976). Individual identified neurons reaching lumbar segments of the spinal cord from each of the brain stem efferent systems mentioned above have been recorded during stepping movements. Most neurons of each descending system rhythmically increase their activity during a particular phase of the step cycle and decrease activity, or cease firing, in other phases. The phase in which each type of neuron is active is as expected from the synaptic connections to flexor and extensor motoneurons described earlier. Thus, most (62%) rubrospinal neurons were found to increase their activity during the swing phase when flexors are activated, while the remainder were active during the stance or during both phases (21%), or showed no modulation (17%). Reticulospinal neurons descending in the MLF behave similarly. Most Deiters' nucleus neurons (52%) also behaved as would be predicted from synaptology, firing most rapidly at the beginning of the stance phase when extensors are most intensely activated; others (33%) displayed increased tonic firing during locomotion without further modulation. Interestingly, it was found that neurons in each descending tract would only increase their tonic firing (considerably) throughout both phases of locomotion if the hind limb controlled by the neurons was mechanically prevented from moving, although the other three limbs progressed as usual. The same increase of tonic neural activity without modulation could also be observed during walking after the cerebellum had been removed. This suggested that cerebellar modulation of the activity of the brain stem neurons was not a prerequisite for control of locomotion by the descending systems but was an aid to this control.

These results suggest that the minimal role of the brain stem efferent systems in locomotion is merely to provide a tonic descending barrage of impulses; the distribution of excitatory and inhibitory effects to the appropriate motoneurons at the appropriate times is taken care of by spinal mechanisms. Support for this idea comes from experiments in which the various brain stem nuclei are lesioned or stimulated during stepping movements (Orlovsky, 1972a). Continuous stimulation of the red nucleus or Deiters' nucleus results in increased EMG activity during the step phase normally controlled by each system but has no effect on step frequency or phase duration. A lesion of either nucleus decreases the EMG activity, again without effect on phase duration. The strength of the tonic descending input from each system thus appears to control the force of stepping, but perhaps more significantly for the theory of locomotion, the increased tonic excitatory input resulting from, for example Deiters' nucleus stimulation, does not prolong firing of extensors beyond the period that is normal for the rate of walking; the extensors are "shut off" at the appropriate time by spinal mechanisms.

One spinal pathway presumably capable of such powerful reciprocal control is that from the Ia inhibitory interneuron to antagonists, the synaptology of which was discussed in the preceding section. It has been established that these interneurons are active during the step cycle in a manner consistent with a role in controlling the step cycle duration (Feldman and Orlovsky, 1975). The most extensive data are from quadriceps-linked Ia inhibitory interneurons. All of these tested were active during the stance phase (when the quadriceps muscle is activated). The activity of most of these interneurons was modulated: they sometimes increased their firing rate in the middle of the stance phase and were silent during the swing phase. Evidence was presented suggesting they are actively inhibited during the swing phase. The maximum firing rates obtained during the stance phase were greater than could be produced by passive muscle stretch (which activates group Ia afferent fibers). Since these neurons fired maximally during contraction of the muscle from which they receive group Ia innervation, it was suggested they must be intensely activated over the γ loop in order to fire so well in the face of spindle unloading. However, it is now known that these, and only these, Ia inhibitory interneurons receive excitation from Deiters' nucleus, the neurons of which are active during the stance phase, while at the same time they are disinhibited from the antagonistic knee flexor Ia inhibitory interneurons. These synaptic actions could result in the observed modulation of interneuron firing during quadriceps contraction independent of γ loop enhancement of homonymous group Ia input, and, indeed, modulation is seen when the γ loop is interrupted. However, some modulation is also seen when the cerebellum is removed. Since this eliminates modulation of activity in all descending pathways, the interneurons may also be receiving input from the "spinal step generator" through synaptic connections as yet unspecified. Altogether, these results indicate that the Ia inhibitory interneuron plays a role in regulating the step cycle, as was suggested previously from its synaptic connections.

That fusimotor activation may offset the unloading of muscle spindles during stepping movements has been shown by Severin (1970). Group Ia afferent fibers from the gastrocnemius muscle were found to fire more rapidly during the stance phase when the muscle is contracting (though also being stretched during an active lengthening reaction during part of this phase) than during the swing phase when it is passively stretched. An attempt was made to evaluate the contribution of the γ loop on this muscle by applying novocaine to the muscle nerve. This procedure is said to block selectively small nerve fibers more rapidly than large fibers (Matthews, 1972). Spindle firing during the stance phase was seen to be greatly decreased subsequent to novocaine application, and EMG activity also decreased, but step cycle timing was not affected, and muscle contraction still persisted at the appropriate time. It was concluded that the γ loop was facilitatory but not critical in maintaining stepping behavior.

These studies of locomotion in the mesencephalic cat have indicated that the minimal role of the brain stem afferent systems is rather simple and straightforward during this complex motor act. As might be expected from some of the detailed spinal cord synaptology discussed in this chapter, most of the complex neural activity during this type of locomotion appears to occur at the spinal level. The synaptic connections of the Ia inhibitory interneuron deduced from electrophysiology appear to be fully utilized, and it is likely that descending control of other spinal reflex arcs is also involved, though demonstration of this control awaits further experimentation.

PROPERTIES OF SYNAPTIC TRANSMISSION FROM BRAIN STEM EFFERENT FIBERS

The bulk of this chapter has been concerned with questions of classical neurophysiology: what are the patterns of synaptic connections and effects on motoneurons from the brain stem efferent systems? Another question that may be asked is whether the EPSP produced by impulses in descending fibers is in any way more potent than the EPSP produced by the same number of primary afferent fibers. There are several ways this could occur. The synaptic terminals of descending fibers may, on the average, lie closer to the motoneuron initial segment where spikes are initiated. In this way, the descending monosynaptic EPSP might be less subject to electrotonic decrement than, for example, group Ia EPSPs occurring more distally on the dendrites and would, thus, more effectively excite the motoneuron. Another possibility is that each descending fiber may, on the average, give off more synaptic terminals per fiber or, for some other reason, liberate more transmitter per impulse than do primary afferents. These two possibilities have been carefully tested in the monkey by comparing the "minimal" group Ia and corticospinal monosynaptic EPSPs evoked by a limited number of fibers (Porter and Hore, 1969). The shapes of the EPSPs from each source may be compared, and, based on certain theoretical assumptions, the relative locations on the dendritic tree of the synaptic terminals from each source may be deduced. The EPSP amplitudes obtained from a limited number of group Ia and corticospinal fibers themselves give a rough idea of the relative amount of transmitter per impulse liberated by the two sets of fibers.

Similar thorough studies are not available for the brain stem efferent systems, but group Ia EPSP shapes have been compared with those of monosynaptic vestibulospinal and MLF-reticulospinal EPSPs in the cat (Grillner et al., 1970, 1971), and shapes together with amplitudes of EPSPs from a limited number of fibers have been reported for monosynaptic rubrospinal EPSPs in monkey (Shapovalov et al., 1971). There is room for debate concerning the relative location and single-fiber ("unitary") EPSP size from each source, but the descending synaptic terminals do not appear to have any overwhelming advantage over the group Ia synaptic terminals in terms of location on the motoneuron surface or transmitter liberated per impulse. In the case of the cortico-motoneuronal EPSP, where the most thorough investigation has been made (Porter and Hore, 1969), the corticospinal synaptic terminals may be, on the average, slightly more removed from the initial segment than the group Ia terminals, and the "minimal" corticospinal EPSPs are no larger, on the average, than the group Ia "minimal" EPSPs.

The conclusion that descending fibers do not liberate significantly more transmitter per impulse than group Ia fibers applies only to EPSPs evoked by single, isolated stimuli. When repetitive stimuli are applied at close intervals, the situation is different. Again, this has been studied most carefully using the monosynaptic cortico-motoneuronal EPSP in primates (Landgren, Phillips, and Porter, 1962; Muir and Porter, 1973; Porter, 1970). In this pathway, it has been found that when two monosynaptic EPSPs are evoked at close intervals, the second is significantly larger than the first; just how much larger depends on the precise interval at which they are evoked. Since EPSPs evoked at close intervals exhibit temporal summation, the size of, for example, the third EPSP of a train, is determined by subtracting the response to two stimuli from the response to three stimuli. The ratio of the amplitude of the third EPSP, so determined, to the amplitude of the first EPSP of the train is defined as the facilitation produced by three stimuli at a certain interstimulus interval. Such facilitation would not be remarkable if it were obtained from a polysynaptic or disynaptic pathway since temporal summation of EPSPs in the interneurons would be expected to recruit more interneuron firing with each successive stimulus, and the size of successive EPSPs observed at the motoneuron would, of course, grow geometrically. However, the cortico-motoneuronal EPSP is monosynaptic, and it also appears that the same number of descending fibers is excited by each stimulus, as nearly as can be determined by observing the size and shape of the descencding tract wave following each stimulus. (Of course, recruitment of additional descending fibers with each successive stimulus would also cause an expected and uninteresting geometrical increase in size of successive monosynaptic EPSPs.) It so far appears that the growth of successive monosynaptic EPSPs is not due to recruitment of interneurons or additional descending fibers but is, rather, a property of the synaptic terminals themselves. For some unknown reason, the occurrence of an impulse in the terminal makes it more likely that successive impulses will release even more transmitter if these impulses occur within a certain time period. The facilitation produced by each corticospinal impulse may reach 200% and decay with a time constant of about 10 msec. The total amount of facilitation produced by a train of impulses appears to be a linear sum of the facilitation produced by each impulse in the train (Muir and Porter, 1973). It should be noted that a precisely similar effect has been observed at the frog neuromuscular junction (Mallart and Martin, 1967).

This effect has not been studied so carefully or thoroughly in other descending pathways, but monosynaptic EPSP facilitation of at least 123% has been reported from the rubrospinal tract in the monkey (Shapovalov et al., 1971) for repetitive stimuli at an interval of 1.5 msec. The behavior of monosynaptic vestibulospinal and reticulospinal EPSPs on repetitive stimulation have also been studied in the cat (Grillner et al., 1970, 1971). These EPSPs do not appear to exhibit any facilitation, but they do summate well, approximately algebraically, even during stimulus trains applied at 500–600Hz. In terms of their ability to summate, they resemble the primate cortico-motoneuronal EPSPs more closely than cat group Ia EPSPs, which behave differently under repetitive stimulation.

The behavior of the monosynaptic group Ia EPSP upon repetitive stimulation has been studied in the cat by observing the amplitude of a second EPSP evoked at various intervals after the first and also by observing the size of individual EPSPs and the plateau potential reached by the summated EPSPs during trains of repetitive stimuli applied at different rates (Curtis and Eccles, 1960). It is likely that somewhat different synaptic properties and underlying mechanisms are involved in each case. The second of two group Ia EPSPs reaches a maximum size of about 120% of the first when evoked about 10 msec later; similar results have been obtained for EPSPs from single group Ia fibers (Kuno, 1964). This value is comparable to the facilitation observed in the monkey rubromotoneuronal EPSP and is greater than that observed in cat vestibulospinal or reticulospinal monosynaptic EPSPs. However, when stimulus trains are applied, the group Ia EPSPs do not summate well at all; the amplitude of individual EPSPs of the train remain at about 85% of the amplitude of a single, isolated EPSP up to stimulation rates of 250 Hz, after which they decrease further. The summated EPSP plateau potential never becomes more than about twice as large as the size of an isolated EPSP no matter what the stimulus rate. The results obtained from the stimulus trains seem more important physiologically because group Ia fibers are known to fire tonically, often at high rates. Recordings in unanesthetized animals also indicate that most cells in the cortex or brain stem nuclei more-or-less continuously fire with waxing and waning activity. High-frequency bursts of spikes are associated with certain motor acts, and the effects of these on EPSP size may be mimicked by the short, high-frequency stimulus trains employed in the study of cortico-motoneuronal EPSPs.

On the basis of the data summarized above, it would thus appear that descending EPSPs are superior to group Ia EPSPs in their ability to depolarize and fire the motoneurons under the usual physiological circumstances. Unfortunately, this conclusion cannot be made so readily. For one thing, only group Ia EPSPs from the cat have been systematically studied in this respect, and it is possible that primate group Ia EPSPs may be revealed to be more analogous to decending EPSPs if studied more throughly under repetitive stimulation. Secondly, it is still questionable if excitation of group Ib afferent fibers, or Renshaw inhibition due to firing of motoneurons other than that recorded, were excluded while obtaining the stimulus train data in the cat. The IPSPs produced by such inhibitory contamination might be masked by the larger EPSPs but could summate in the same way the EPSPs do and make the individual EPSPs and plateau potentials appear much smaller than they would if only excitation were present. Also, potentials produced by asynchronous firing of spindle afferent fibers during muscle stretch summate well to a more-or-less smooth depolarization, which is hundreds of times as large as the EPSP from individual fibers and which is certainly capable of firing the motoneurons (see, e.g., Burke, 1968). Until these questions are resolved, the conclusion that descending EPSPs can be more potent than group Ia EPSPs during repetitive stimulation must remain tentaive.

It has been suggested that facilitation of descending EPSPs plays a negligible role because maximal facilitation occurs only at stimulus intervals corresponding to high firing rates (500 Hz for corticospinal fibers and probably for rubrospinal fibers as well), and the facilitatory effect decays fairly rapidly (with a 10 msec time constant for corticospinal fibers). But large pyramidal tract neurons, which probably give rise to the fast corticospinal fibers whose EPSPs have been studied, are rarely found to fire faster than 100 Hz (see, e.g., Evarts, 1968). Even so, the cortico-motoneuronal EPSPs produced at a 100Hz firing rate would be larger than "normal," though the facilitation at this frequency would be reduced to about 40% of its peak value. Rubrospinal (and vestibulospinal) cells, on the other hand, normally fire at much faster rates. Their "rest" spontaneous activity is often greater than 50Hz, and firing rates may approach 200 Hz or more over periods of time during moderate motor activity (Ghez, 1975; Orlovsky, 1972a,b). Although monosynaptic EPSPs from primate rubrospinal fibers apparently exhibit less facilitation than EPSPs from corticospinal fibers, the faster normal firing rate of these fibers may ensure that some benefit is obtained from their facilitation property as well. Furthermore, this benefit would be most apparent during high-frequency bursts of firing associated with "urgent" motor activity. It is, therefore, clearly advantageous for the animal to possess even a moderate capability for facilitated synaptic transmitter release in descending motor control.

Concluding Remarks

In order to limit the scope of the chapter, two effects of the brain stem efferent systems have not been covered. It has been found that stimulation of each of the descending systems discussed in this chapter, with the exception of the lateral vestibulospinal tract, produces long-lasting depolarization of afferent fibers in the dorsal roots (see, e.g., Barnes and Pompeiano, 1970; Carpenter, Engberg, and Lundberg, 1966; Hongo *et al.,* 1972a). Observation of prolonged inhibition of the group Ia monosynaptic reflexes has been demonstrated in some cases to be associated with the dorsal root potentials. Both the appearance of dorsal root potentials and the prolonged inhibition of the monosynaptic reflexes are presumed to reflect presynaptic inhibition of synaptic transmission from primary afferent fibers (Eccles, 1964). It has been determined that stimulation of each brain stem system produces dorsal root potentials in some types of afferent fibers but not in others. The type of primary afferent fiber affected depends on the particular brain stem system stimulated, but the precise functional meaning of this is unclear. Presynaptic inhibition so far appears to be a generalized, diffuse mechanism for depressing transmission of primary afferent information to all types of flexor and extensor motoneurons. All inhibitory effects described in the body of this chapter, however, were ascertained by the experimenters involved to be postsynaptic rather than presynaptic.

It is also important to realize that in addition to controlling motor output, all the descending systems discussed in this chapter, with the exception of the monoaminergic system, have been shown to modulate transmission of information from primary afferent fibers to higher centers in the brain stem. This modulation is affected not only by the generalized presynaptic inhibition mentioned above, but also by discrete postsynaptic effects exerted on various spino-cerebellar tracts (see Oscarsson, 1973, for a discussion; for some recent work, see, e.g., Baldissera and ten Bruggencate, 1976; Baldissera and Roberts, 1976). Since the cells of origin of the brain stem efferent systems are influenced from the cerebellum, they appear to serve as components in long, cerebellar feedback loops, in addition to their more traditional roles in motor control.

In the first section of this chapter, the results of various lesion experiments were discussed. Those of Lawrence and Kuypers (1968a,b) provided a general picture of the necessary function and different types of musculature controlled by each of the brain stem efferent systems in the monkey. It was also mentioned that Kuypers (1963) performed a similar, though smaller scale lesion study in the cat and came to similar conclusions about the role of the corticospinal tract and brain stem efferent systems in this animal: the vestibulospinal and reticulospinal systems were deduced to be necessary for and to act together in proper control of axial and proximal limb musculature; the rubrospinal and corticospinal tracts were found to be necessary for and to act together in proper control of distal limb muscles, particularly distal flexor muscles. In the meantime, a great deal of detailed synaptology has been presented. It may be interesting to examine whether the synaptology generally supports the scheme of Kuypers (1963) and Lawrence and Kuypers (1968b), and whether synaptology provides any further generalizations about the role of the brain stem efferent systems in motor control. These questions will be considered below.

The rubrospinal tract in both the cat and monkey excites flexor motoneurons, but also many extensor motoneurons (as does the corticospinal tract in both the cat and monkey, Illert *et al.,* 1977; Jankowska, Padel, and Tanaka, 1975b; Lundberg and Voorhoeve, 1962). Furthermore, these effects are by no means limited to or necessarily stronger on motoneurons innervating the distal limb. It should also be pointed out that effects on axial motoneurons from the corticospinal and rubrospinal tracts have not yet been studied in the cat or monkey, so it is not ruled out that they may also be present. The corticospinal and rubrospinal tracts have a similar effect on motoneurons and spinal reflex arcs, however, which is one general implication of Lawrence and Kuyper's data (1968b). Fibers from each tract may monosynaptically excite some of the same interneurons in the cat, just as they tend to excite the same motoneurons monosynaptically in the monkey. Overall, it may certainly be concluded that the corticospinal and rubrospinal tracts have many similar actions. Both tracts excite flexor motoneurons and excite and inhibit various extensor motoneurons, and both tracts also have a similar facilitatory action on the same spinal reflex arcs. In the cat and monkey, however, such effects are not limited to distal limb musculature.

The lateral vestibulospinal tract originating from Deiters' nucleus appears to be rather uniform in excitation of extensor and inhibition of flexor motoneurons, though some exceptions were noted even for this tract. The large reticulospinal fibers descending in the MLF generally have many synaptic actions just opposite to those of Deiters' nucleus on certain motoneuron pools, but there are a number of exceptions to this reciprocal pattern. It should also be remembered that these MLF reticulospinal fibers constitute only a small fraction of all reticulospinal fibers. Neither the fibers from Deiters' nucleus nore the MLF-reticulospinal fibers appear primarily directed at motoneurons of proximal as opposed to distal hindlimb musculature. They seem to exert their most distinctive effects at knee and ankle joints. It is true that vestibulospinal connections to motoneurons innervating neck and back muscles are more direct than those to limbs, but that is probably because different functions are subserved in each case. The direct, bilateral effects to motoneurons of axial musculature mainly to subserve transmission of labyrinthine reflexes, while the less direct pathways to limbs seem to be concerned with transmitting more generalized and integrated motor commands. Furthermore, several studies of the effect of natural labyrinthine stimulation on neck and forelimb motor units suggested that the shortest latency pathways, even to axial muscles, are not necessarily the most important ones.

The cells of origin of both the rubrospinal and reticulospinal tracts receive direct and indirect connections from cerebral cortex, and vestibular neurons may receive indirect cortical effects via reticular cells. Thus, each brain stem system provides, in principle, an alternate pathway for transmission of neural activity from the cerebral cortex to the spinal cord. Each system can serve not only as a simple relay for cortical neural activity, but each can also serve as a center for integration with information from the cerebellum, which projects heavily to each of the brain stem systems.

The above statements describe the general effects of each brain stem system on different types of spinal motoneurons as deduced from electrophysiology, but they do not offer much insight into how motor control is actually carried out. The detailed mechanisms involved are described, in part, in the bulk of this chapter. These synaptic connections provide the actual circuitry by which motor control may be effected, and some general principles emerge from these studies. Most of the descending commands in the cat are transmitted to motoneurons through interneurons of segmental reflex pathways. For example, Deiters' nucleus exerts its disynaptic effects on motoneurons principally by means of the last interneurons of the pathway from the contralateral FRA. The rubrospinal tract exerts its disynaptic effects principally by means of the last interneurons of a polysynaptic pathway from low-threshold cutaneous afferent fibers separate from the FRA pathway. The corticospinal tract also uses this pathway, as well as the group Ia inhibitory interneuron from antagonists, to exert its shortest latency effects.

In other cases, the descending impulses may not be relayed to the motoneurons but, instead, affect the excitability of the spinal reflex arcs and, thus, the effectiveness with which they transmit information from primary afferent fibers to the motoneurons. This seems to be generally the case for rubrospinal actions on the group Ib pathways and group Ia inhibitory pathways from antagonists. The corticospinal tract also seems merely to facilitate transmission through group Ib and FRA pathways, but it is possible that both the corticospinal and the rubrospinal tract may actually exert their effects on motoneurons by means of these more polysynaptic segmental pathways under some conditions. It was also suggested that certain reticulospinal fibers and low-threshold cutaneous afferent fibers may exert generalized inhibition on first or early interneurons in the polysynaptic segmental reflex pathways, rendering them "private" for use by the other descending pathways under certain conditions.

The demonstration of descending actions on segmental reflex pathways was also useful in pointing up the larger role of these reflex arcs in motor control.

Rubrospinal effects on the group Ib pathways, in particular, suggested that the group Ib actions may be much more important in motor control than previously thought. The group Ia inhibitory interneurons have also been shown to play a much more important role in alternating flexion-extension movements than might be suggested by their group Ia connections alone. Recordings from the group Ia inhibitory interneurons have indicated particularly well how segmental reflex arcs may themselves interact. Clearly, there is a tremendous amount of integration of peripheral and descending information in the spinal cord. Most of this integration occurs in neurons other than, but ultimately relaying to, the motoneurons.

Though a great deal of information has been obtained on descending effects acting through segmental reflex arcs, additional information would be useful. The question of whether generalized reticulospinal inhibition is exerted on the first interneurons of polysynaptic reflex arcs seems especially important for the theory of motor control and needs to be investigated more thoroughly. Information is lacking on whether the rubrospinal and corticospinal tracts may exert effects on reflex arcs from the contralateral side, as the lateral vestibulospinal tract does. Information is also lacking on whether the reticulospinal fibers descending in the MLF exert their disynaptic effects through any reflex arcs. More generally, the action of the bulk of the reticulospinal fibers is unkown and is the largest remaining question to be answered by electrophysiology. Equally important is a clear demonstration of the role of the fusimotor neurons, though this is not likely to come from electrophysiology alone. In addition, the difficult task of obtaining recordings from identified interneurons, such as done for the Ia inhibitory interneurons, is required to provide further insight as to how descending and segmental pathways interact at the interneuron level.

This latter investigation would be especially useful if it shed light on the factors determining which effects are transmitted to motoneurons from descending impulses acting through reflex pathways. It was pointed out several times in the body of the chapter that it is difficult to assert that an effect transmitted over a polysynaptic pathway under one experimental condition will hold for all conditions. This was exemplified particularly well by recordings obtained in flexor motoneurons while stimulating the contralateral Deiters' nucleus. These recordings indicated that the effects observed may change sign over time, even while stimulating at the same strength at the same point and recording from the same motoneuron. Intracellular recordings in Ia inhibitory interneurons indicate that there is even some doubt about always obtaining the expected effect over this disynaptic pathway. Although the effects from these interneurons will never change from inhibitory to excitatory, they may themselves be inhibited from other segmental pathways. It seems that polysynaptic descending effects observed in any given motoneuron may be very much dependent on segmental and other descending activity.

A problem related to the variable effects that may be exerted by descending systems on a given motor nucleus was discussed in the section on the rubrospinal tract. It will be recalled that rubrospinal effects on individual gastrocnemius motoneurons were predominantly either excitatory or inhibitory, depending on the properties of the muscle fibers innervated by each motoneuron. The inhibited motoneurons had motor units like those of the soleus muscle, the motoneurons of which receive rubrospinal inhibition. Similar differential effects were seen from certain polysynaptic segmental reflex arcs. How general this phenomena is remains to be seen, but it could account for many of the "anomalous" exceptions found in the general reciprocal innervation patterns of all descending tracts. The phenomena certainly add one more complication to the descending control of motor output.

The above considerations make the statements given earlier in this section about the general actions of the various brain stem tracts seem tenuous indeed, since these effects are obtained under certain experimental conditions and, from the above considerations, may be different under other conditions. Only the monosynaptic effects will remain invariant. Unfortunately, the results of electrophysiology show that complicated interactions are possible, but shed no light on what the main actions of the tracts will be under any given condition. Furthermore, even the further discovery of synaptic connections and interactions is unlikely to help this situation significantly, though such discoveries will indicate additional mechanisms for motor control and will place some constraints on how the mechanisms may be used and interact. The fundamental problem with the electrophysiological approach is that knowledge of circuitry can suggest possible functions, but knowledge of circuitry alone cannot clearly or uniquely specify how that circuitry is employed by the CNS under varied conditions.

Knowledge of circuitry is clearly useful, but other approaches must also be employed to determine nervous system function. For example, the experiments of Lawrence and Kuypers (1968a,b) provide useful knowledge about the general role of the corticospinal and brain stem pathways in the normal life of the awake, behaving animal. As discussed above, such knowledge could only partially be deduced from known synaptic connections. Such lesion experiments have been further exploited in the case of the pyramidal tract by quantitatively examining deficits after selective lesions (see Chapter 8), but such data are lacking for the brain stem efferent systems. Another approach to understanding motor function is provided by the recording of brain stem neuronal activity during walking in the mesencephalic cat, described on pp. 208–211. Chronic recording from neurons in the awake, behaving animal during specified behavioral tasks has also been useful for exploring corticospinal and oculomotor functions (see Chapters 6, 8, and 9). Except for the mesencephalic cat preparation, only a few of these experiments have so far been done on the brain stem efferent systems (see, e.g., Ghez, 1975; Onoda and Burton, 1975; Otero, 1975). Such an approach may be even more successful for the brain stem efferent systems than for the corticospinal tract, especially if done in the cat, since the correlations between brain stem neuron firing and motor output may be more easily related to the spinal cord circuitry known in such detail for this animal. It is now clear that the brain stem systems are those responsible for the basic motor repertoire of the animal; recording from brain stem neurons should thus provide more direct information on that basic repertoire than recording from corticospinal neurons. It would seem that, in addition to further knowledge of synaptic pathways to motoneurons, approaches such as these will be required to elucidate further the role of the brain stem efferent systems in motor control.

Abzug, C., Maeda, M., Peterson, B. W., and Wilson, V. J. Cervical branching of lumbar vestibulospinal axons. *Journal of Physiology (London)*, 1974, *243*, 499–522.

Adrian, E. D. Discharges from vestibular receptors in the cat. *Journal of Physiology (London)*, 1943, *101*, 389–407.

Akaike, T. Comparison of neuronal composition of the vestibulospinal system between cat and rabbit. *Experimental Brain Research*, 1973, *18*, 429–432.

Akaike, T., and Westerman, R. A. Spinal segmental levels innervated by different types of vestibulo-spinal tract neurones in rabbit. *Experimental Brain Research*, 1973, *17*, 443–446.

Akaike, T., Fanardjian, V. V., Ito, M., Kumada, M., and Nakajima, H. Electrophysiological analysis of the vestibulospinal reflex pathway of rabbit. I. Classification of tract cells. *Experimental Brain Research*, 1973, *17*, 477–496.

Akaike, T., Fanardjian, V. V., Ito, M., and Ohno, T. Electrophysiological analysis of the vestibulospinal reflex pathway of rabbit. II. Synaptic actions upon spinal neurones. *Experimental Brain Research*, 1973, *17*, 497–515.

Alnaes, E., Jansen, J. K. S., and Rudjord, T. Fusimotor activity in the spinal cat. *Acta Physiologica Scandinavica*, 1965, *63*, 197–212.

Andén, N.-E., Jukes, M. G. M., and Lundberg, A. The effect of DOPA on the spinal cord. 2. A pharmacological analysis. *Acta Physiologica Scandinavica*, 1966, *67*, 387–397.

Andén, N.-E., Jukes, M. G. M., Lundberg, A., and Vyklicky, L. The effect of DOPA on the spinal cord. 1. Influence on transmission from primary afferents. *Acta Physiologica Scandinavica*, 1966, *67*, 373–386.

Anderson, J. H., Berthoz, A., Soechting, J. F., and Terzuolo, C. A. Motor output to deafferented forelimb extensors in the decerebrate cat during natural vestibular stimulation. *Brain Research*, 1977, *122*, 150–173.

Anderson, J. H., Soechting, J. F., and Terzuolo, C. A. Dynamic relations between natural vestibular inputs and activity of forelimb extensor muscles in the decerebrate cat. I. Motor output during sinusoidal linear accelerations. *Brain Research*, 1977a, *120*, 1–15.

Anderson, J. H., Soechting, J. F., and Terzuolo, C. A. Dynamic relations between natural vestibular inputs and activity of forelimb extensor muscles in the decerebrate cat. II. Motor output during rotations in the horizontal plane. *Brain Research*, 1977b, *120*, 17–33.

Anderson, M. E. Segmental reflex inputs to motoneurons innervating dorsal neck musculature in the cat. *Experimental Brain Research*, 1977, *28*, 175–187.

Anderson, M. E., Yoshida, M., and Wilson, V. J. Influence of the superior colliculus on cat neck motoneurons. *Journal of Neurophysiology*, 1971, *34*, 898–907.

Andersson, S., and Gernandt, B. E. Ventral root discharge to vestibular and proprioceptive stimulation. *Journal of Neurophysiology*, 1956, *19*, 524–543.

Aoyama, M., Hongo, T., Kudo, N., and Tanaka, R. Convergent effects from bilateral vestibulospinal tracts on spinal interneurons. *Brain Research*, 1971, *35*, 250–253.

Appelberg, B. The effect of electrical stimulation in nucleus ruber on the response to stretch in primary and secondary muscle spindle afferents. *Acta Physiologica Scandinavica*, 1962, *56*, 140–151.

Appelberg, B. A rubro-olivary pathway. II. Simultaneous action on dynamic fusimotor neurones and the activity of the posterior lobe of the cerebellar cortex. *Experimental Brain Research*, 1967, *3*, 382–390.

Appelberg, B., and Jeneskog, T. A dorso-lateral spinal pathway mediating information from the mesencephalon to dynamic fusimotor neurones. *Acta Physiologica Scandinavica*, 1969, *77*, 159–171.

Appelberg, B., and Jeneskog, T. Mesencephalic fusimotor control. *Experimental Brain Research*, 1972, *15*, 97–112.

Appelberg, B., and Kosary, I. Z. Excitation of flexor fusimotor neurones by electrical stimulation in the red nucleus. *Acta Physiologica Scandinavica*, 1963, *59*, 445–453.

Appelberg, B., and Molander, C. A rubro-olivary pathway. I. Identification of a descending system for control of the dynamic sensitivity of muscle spindles. *Experimental Brain Research,* 1967, *3,* 372–381.

Appelberg, B., Jeneskog, T., and Johansson, H. Rubrospinal control of static and dynamic fusimotor neurones. *Acta Physiologica Scandinavica,* 1975, *95,* 431–440.

Ashcroft, D. W., and Hallpike, C. A. On the function of the saccule. *Journal of Laryngology,* 1934, *49,* 450–460.

Bach, L. M. N. Effect of bulbar facilitation and inhibition on peripheral reflex inhibition. *Journal of Neurophysiology,* 1950, *13,* 260–264.

Baldissera, F., and Roberts, W. J. Effects from the vestibulospinal tract on transmission from primary afferents to ventral spino-cerebellar tract neurones. *Acta Physiologica Scandinavica,* 1976, *96,* 217–232.

Baldissera, F., and ten Bruggencate, G. Rubrospinal effects on ventral spino-cerebellar tract neurones. *Acta Physiologica Scandinavica,* 1976, *96,* 233–249.

Baldissera, F., Lundberg, A., and Udo, M. Activity evoked from the mesencephalic tegmentum in descending pathways other than the rubrospinal tract. *Experimental Brain Research,* 1972a, *15,* 133–150.

Baldissera, F., Lundberg, A., and Udo, M. Stimulation of pre- and postsynaptic elements in the red nucleus. *Experimental Brain Research,* 1972b, *15,* 151–167.

Baldissera, F., ten Bruggencate, G., and Lundberg, A. Rubrospinal monosynaptic connection with last order interneurones of polysynaptic reflex paths. *Brain Research,* 1971, *27,* 390–392.

Bantli, H., and Bloedel, J. R. Monosynaptic activation of a direct reticulospinal pathway by the dentate nucleus. *Pflügers Archiv.* 1975a, *357,* 237–242.

Bantli, H., and Bloedel, J. R. The action of the dentate nucleus on the excitability of spinal motoneurons via pathways which do not involve the primary sensorimotor cortex. *Brain Research,* 1975b, *88,* 86–90.

Barnes, C. D., and Pompeiano, O. The contribution of the medial and lateral vestibular nuclei to presynaptic and postsynaptic effects produced in the lumbar cord by vestibular volley. *Pflügers Archiv,* 1970, *317,* 1–9.

Bayev, K. V., and Kostyuk, P. G. Convergence of cortico- and rubrospinal influences on interneurones of cat cervical spinal cord. *Brain Research,* 1973, *52,* 159–171.

Beck, C. H., and Chambers, W. W. Speed, accuracy and strength of forelimb movement after unilateral pyramidotomy in rhesus monkeys. *Journal of Comparative Physiology and Psychology,* 1970, *70* (monograph 2, Pt. 2), 1–22.

Bergmans, J., and Grillner, S. Changes in dynamic sensitivity of primary endings of muscle spindle afferents induced by DOPA. *Acta Physiologica Scandinavica,* 1968a, *74,* 629–636.

Bergmans, J., and Grillner, S. Monosynaptic control of static γ-motoneurones from the lower brainstem. *Experientia,* 1968b, *24,* 146–147.

Berthoz, A., and Anderson, J. H. Frequency analysis of vestibular influence on extensor motoneurons. I. Response to tilt in forelimb extensors. *Brain Research,* 1971a, *34,* 370–375.

Berthoz, A., and Anderson, J. H. Frequency analysis of vestibular influence on extensor motoneurons. II. Relationship between neck and forelimb extensors. *Brain Research,* 1971b, *34,* 376–380.

Brodal, A. *The Reticular Formation of the Brain Stem. Anatomical Aspects and Functional Correlations.* Edinburgh: Oliver and Boyd, 1957.

Brodal, A. *Neurological Anatomy in Relation to Clinical Medicine.* London: Oxford University Press, 1969.

Brodal, A. Some features of the anatomical organization of the vestibular nuclear complex in the cat. *Progress in Brain Research,* 1972a, *37,* 31–53.

Brodal, A. Anatomy of the vestibuloreticular connections and possible "ascending" vestibular pathways from the reticular formation. *Progress in Brain Research,* 1972b, *37,* 553–566.

Brodal, A., Pompeiano, O., and Walberg, F. *The Vestibular Nuclei and Their Connections. Anatomy and Functional Correlations.* The Henderson Trust Lectures. Edinburgh: Oliver and Boyd, 1962.

Bruggencate, G. ten, and Lundberg, A. Facilitatory interaction in transmission to motoneurons from vestibulospinal fibers and contralateral primary afferents. *Experimental Brain Research,* 1974, *19,* 248–270.

Bruggencate, G. ten, Burke, R., Lundberg, A., and Udo, M. Interaction between the vestibulospinal tract, contralateral flexor reflex afferents and Ia afferents. *Brain Research,* 1969, *14,* 529–532.

Burke, R. E. Motor unit types of cat triceps surae muscle. *Journal of Physiology (London),* 1967, *193,* 141–160.

Burke, R. E. Firing patterns of gastrocnemius motor units in the decerebrate cat. *Journal of Physiology (London),* 1968, *196,* 631–654.

Burke, R. E., Jankowska, E., and ten Bruggencate, G. A comparison of peripheral and rubrospinal synaptic input to slow and fast twitch motor units of triceps surae. *Journal of Physiology (London),* 1970, *207,* 709–732.

Carli, G., Diete-Spiff, K., and Pompeiano, O. Responses of the muscle spindles and of the extrafusal fibers in an extensor muscle to stimulation of the lateral vestibular nucleus in the cat. *Archivio Italiano di Biologia,* 1967, *105,* 209–242.

Carpenter, D., Engberg, I., and Lundberg, A. Primary afferent depolarization evoked from the brain stem and the cerebellum. *Archivio Italiano di Biologia,* 1966, *104,* 73–85.

Clough, J. T. M., Kernell, D., and Phillips, C. G. The distribution of monosynaptic excitation from the pyramidal tract and from primary spindle afferents to motoneurones of the babboon's hand and forearm. *Journal of Physiology (London),* 1968, *198,* 145–166.

Courville, J. Somatotopical organization of the projection from the nucleus interpositus anterior of the cerebellum to the red nucleus. An experimental study in the cat with silver impregnation methods. *Experimental Brain Research,* 1966, *2,* 191–215.

Curtis, D. R., and Eccles, J. C. Synaptic action during and after repetitive stimulation. *Journal of Physiology (London),* 1960, *150,* 374–398.

Diete-Spiff, K., Carli, F., and Pompeiano, O. Comparison of the effects of stimulation of the VIIIth cranial nerve, the vestibular nuclei or the reticular formation on the gastrocnemius muscle and its spindles. *Archivio Italiano di Biologia,* 1967 *105,* 243–272.

Eccles, J. C. Presynaptic inhibition in the spinal cord. *Progress in Brain Research,* 1964, *12,* 65–91.

Eccles, J. C., Eccles, R. M., and Lundberg, A. Synaptic actions on motoneurones caused by impulses in Golgi tendon organ afferents. *Journal of Physiology (London),* 1957, *138,* 227–252.

Eccles, J. C., Fatt, P., and Landgren, S. Central pathway for direct inhibitory action of impulses in largest afferent nerve fibers to muscle. *Journal of Neurophysiology,* 1956, *19,* 75–98.

Eccles, J. C., Nicoll, A., Rantucci, T., Táboříková, H., and Willey, T. J. Topographic studies on medial reticular nucleus. *Journal of Neurophysiology,* 1976, *39,* 109–118.

Eccles, J. C., Nicoll, R. A., Schwarz, D. W. F., Táboříková, H., and Willey, T. J. Reticulospinal neurons with and without monosynaptic inputs from cerebellar nuclei. *Journal of Neurophysiology,* 1975, *38,* 513–530.

Eccles, R. M., and Lundberg, A. Synaptic actions in motoneurones by afferents which may evoke the flexion reflex. *Archivio Italiano di Biologia,* 1959, *97,* 199–221.

Edwards, S. B. The ascending and descending projections of the red nucleus in the cat: an experimental study using an autoradiographic method. *Brain Research,* 1972, *48,* 45–63.

Ehrhardt, K. J., and Wagner, A. Labyrinthine and neck reflexes recorded from single spinal motoneurons in the cat. *Brain Research,* 1970, *19,* 87–104.

Eklund, G., von Euler, C., and Rutkowski, S. Spontaneous and reflex activities of intercostal gamma motoneurones. *Journal of Physiology (London),* 1964, *171,* 139–163.

Eldred, E., and Hagbarth, K.-E. Facilitation and inhibition of gamma efferents by stimulation of certain skin areas. *Journal of Neurophysiology,* 1954, *17,* 59–65.

Eldred, E., Granit, R., and Merton, P. A. Supraspinal control of the muscle spindles and its significance. *Journal of Physiology (London),* 1953, *122,* 498–523.

Endo, K., Araki, T., and Kawai, T. Contra and ipsilateral cortical and rubral effects on fast and slow motoneurons of the cat. *Brain Research,* 1975, *88,* 91–98.

Engberg, I. Reflexes to foot muscles in cat. *Acta Physiologica Scandinavica,* 1964, *62,* Supplementum 235.

Engberg. I., Lundberg, A., and Ryall, R. W. Reticulospinal inhibition of transmission in reflex pathways. *Journal of Physiology (London),* 1968a, *194,* 201–223.

Engberg, I., Lundberg, A., and Ryall, R. W. Reticulospinal inhibition of interneurones. *Journal of Physiology (London),* 1968b, *194,* 225–236.

Evarts, E. V. Relation of pyramidal tract activity to force exerted during voluntary movements. *Journal of Neurophysiology,* 1968, *31,* 14–27.

Ezure, K., and Sasaki, S. Frequency-response analysis of vestibular-induced neck reflex in cat. I. Characteristics of neural transmission from horizontal semicircular canals to neck motoneurons. *Journal of Neurophysiology,* 1978, *41,* 445–458.

Ezure, K., Sasaki, S., Uchino, Y., and Wilson, V. J. Frequency-response analysis of vestibular-induced neck reflex in cat. II. Functional significance of cervical afferents and polysynaptic descending pathways. *Journal of Neurophysiology,* 1978, *41,* 459–471.

Feldman, A. G., and Orlovsky, G. N. Activity of interneurons mediating reciprocal Ia inhibition during locomotion. *Brain Research,* 1975, *84,* 181–194.

Fernandez, C., and Goldberg, J. M. Physiology of peripheral neurons innervating semicircular canals of the squirrel monkey. II. Response to sinusoidal stimulation and dynamics of peripheral vestibular system. *Journal of Neurophysiology,* 1971, *34,* 661–675.

Fernandez, C., and Goldberg, J. M. Response to static tilts of peripheral neurons innervating otolith organs of the squirrel monkey. *Journal of Neurophysiology,* 1972, *35,* 978–997.

Fujita, Y., Rosenberg, J., and Segundo, J. P. Activity of cells in the lateral vestibular nucleus as a function of head position. *Journal of Physiology (London),* 1968, *196,* 1–18.

Gacik, R. R. The course and central termination of first order neurons supplying vestibular end organs in the cat. *Acta Otolaryngologica,* 1969, *254,* 1–66.

Gernandt, B. E., and Thulin, C.-A. Vestibular mechanisms of facilitation and inhibition of cord reflexes. *American Journal of Physiology,* 1953, *172,* 653–660.

Gernandt, B. E., and Thulin, C.-A. Reciprocal effects upon spinal motoneurons from stimulation of bulbar reticular formation. *Journal of Neurophysiology,* 1955, *18,* 113–129.

Ghez, C. Input-output relations of the red nucleus in the cat. *Brain Research,* 1975, *98,* 93–108.

Goldberg, J. M., and Fernandez, C. Response dynamics of peripheral otolith neurons in barbiturate anesthetized squirrel monkey. *Society for Neuroscience Abstracts,* 1974, 231.

Goodwin, G. M., and Luschei, E. S. Discharge of spindle afferents from jaw-closing muscles during chewing in alert monkeys. *Journal of Neurophysiology,* 1975, *38,* 560–571.

Graham-Brown, T. Studies in the physiology of the nervous system. IX. Reflex termination phenomena-rebound, rhythmic rebound and movements of progression. *Quarterly Journal of Experimental Physiology,* 1911, *4,* 331–397.

Granit, R. *Receptors and Sensory Perception.* New Haven: Yale University Press, 1955.

Granit, R., and Holmgren, B. Two pathways from brainstem to gamma ventral horn cells. *Acta Physiologica Scandinavica,* 1955, *35,* 93–108.

Granit, R., and Kaada, B. R., Influence of stimulation of central nervous structures on muscle spindles in cat. *Acta Physiologica Scandinavica,* 1953, *27,* 130–160.

Granit, R., Job, C., and Kaada, B. R. Activation of muscle spindles in pinna re-reflex. *Acta Physiologica Scandinavica,* 1952, *27,* 161–168.

Granit, R., Kellerth, J.-O., and Szumski, A. J. Intracellular recordings from extensor motoneurons activated across the gamma loop. *Journal of Neurophysiology,* 1966, *29,* 530–544.

Grigg, P., Harrigan, E. P., and Fogarty, K. E. Segmental reflexes mediated by joint afferent neurons in cat knee. *Journal of Neurophysiology,* 1978, *41,* 9–14.

Grillner, S. The influence of DOPA on the static and the dynamic fusimotor activity to the triceps surae of the spinal cat. *Acta Physiologica Scandinavica,* 1969, *77,* 490–509.

Grillner, S. Locomotion in the spinal cat. In R. B. Stein, K. G. Pearson, R. S. Smith, and J. B. Redford (Eds.), *Control of Posture and Locomotion.* New York: Plenum, 1973.

Grillner, S., and Lund, S. The origin of a descending pathway with monosynaptic action of flexor motoneurones. *Acta Physiologica Scandinavica,* 1968, *74,* 274–284.

Grillner, S., Hongo, T., and Lund, S. Interaction between the inhibitory pathways from the Deiters' nucleus and Ia afferents to flexor motoneurons. *Acta Physiologica Scandinavica,* 1966, *69,* Supplementum 177, 1–61.

Grillner, S., Hongo, T., and Lund, S. Descending monosynaptic and reflex control of γ-motoneurones. *Acta Physiologica Scandinavica,* 1969, *75,* 592–613.

Grillner, S., Hongo, T., and Lund, S. The vestibulospinal tract. Effects on alpha-motoneurones in the lumbosacral spinal cord in the cat. *Experimental Brain Research,* 1970, *10,* 94–120.

Grillner, S., Hongo, T., and Lund, S. Convergent effects on alpha motoneurones from the vestibulospinal tract and a pathway descending in the medial longitudinal fasciculus. *Experimental Brain Research,* 1971, *12,* 457–479.

Gustafsson, B., and Jankowska, E. Direct and indirect activation of nerve cells by electrical pulses applied extracellularly. *Journal of Physiology (London)* 1976, *258,* 33–61.

Hagbarth, K.-E. Excitatory and inhibitory skin areas for flexor and extensor motoneurones. *Acta Physiologica Scandinavica,* 1952, *26,* Supplementum 94, 1–58.

Hagbarth, K.-E., and Vallbo, Å. B. Discharge characteristics of human muscle afferents during muscle stretch and contraction. *Experimental Neurology,* 1968, *22,* 674–694.

Hagbarth, K-E., and Vallbo, Å. B. Single unit recordings from muscle nerves in human subjects. *Acta Physiologica Scandinavica,* 1969, *76,* 321–324.

Hepp-Raymond, M.-C., Trouche, E., and Wiesendanger, M. Effects of unilateral and bilateral pyramidotomy on a conditioned rapid precision grip in monkeys *(macaca fascicularis)*. *Experimental Brain Research,* 1974, *21,* 519–527.

Hoddevik, G. H., Brodal, A., and Walberg, F. The reticulovestibular projections in the cat. An experimental study with silver impregnation methods. *Brain Research,* 1975, *94,* 383–399.

Holmqvist, B. Crossed spinal reflex actions evoked by valleys in somatic afferents. *Acta Physiologica Scandinavica,* 1961, 52, Supplementum 181, 1–66.

Holmqvist, B., and Lundberg, A. On the organization of the supraspinal inhibitory control of interneurones of various spinal reflex arcs. *Archivio Italiano di Biologia,* 1959, *97,* 340–356.

Holmqvist, B., and Lundberg, A. Differential supraspinal control of synaptic actions evoked by volleys in the flexion reflex afferents in alpha motoneurones. *Acta Physiologica Scandinavica,* 1962, *54,* Supplementum 186, 1–51.

Holst. E. von. Die Tätigkeit des Statolithenapparates im Wirbeltierlabyrinth. *Naturwissenschaften,* 1950, *37,* 265–272.

Hongo, T., and Jankowska, E. Effects from the sensorimotor cortex on the spinal cord in cats with transsected pyramids. *Experimental Brain Research,* 1967, *3,* 117–134.

Hongo, T., Jankowska, E., and Lundberg, A. The rubrospinal tract. I. Effects on alpha-motoneurones innervating hindlimb muscles in cats. *Experimental Brain Research,* 1969a, *7,* 344–364.

Hongo, T., Jankowska, E., and Lundberg, A. The rubrospinal tract. II. Facilitation of interneuronal transmission in reflex paths to motoneurones. *Experimental Brain Research,* 1969b. *7,* 365–391.

Hongo, T., Jankowska, E., and Lundberg, A. The rubrospinal tract. III. Effects on primary afferent terminals. *Experimental Brain Research,* 1972a, *15,* 39–53.

Hongo, T., Jankowska, E., and Lundberg, A. The rubrospinal tract. IV. Effects on interneurones. *Experimental Brain Research,* 1972b, *15,* 54–78.

Hongo, T., Kudo, N., and Tanaka, R. Effects from the vestibulospinal tract on the contralateral hindlimb motoneurones in the cat. *Brain Research,* 1971, *31,* 220–223.

Hongo, T., Kudo, N., and Tanaka, R. The vestibulospinal tract: Crossed and uncrossed effects on hindlimb motoneurones in the cat. *Experimental Brain Research,* 1975, *24,* 37–55.

Houk, J., and Henneman, E. Responses of Golgi tendon organs to active contractions of the soleus muscle of the cat. *Journal of Neurophysiology,* 1967, *30,* 466–481.

Hultborn, H. Convergence on interneurones in the reciprocal Ia inhibitory pathway to motoneurones. *Acta Physiologica Scandinavica,* 1972, Supplementum 375, 1–42.

Hultborn, H., and Udo, M. Convergence in the reciprocal Ia inhibitory pathway of excitation from descending pathways and inhibition from motor axon collaterals. *Acta Physiologica Scandinavica*, 1972, *84*, 95–108.

Hultborn, H., Illert, M., and Santini, M. Convergence on interneurones mediating the reciprocal Ia inhibition of motoneurones. I. Disynaptic Ia inhibition of Ia inhibitory interneurons. *Acta Physiologica Scandinavica*, 1976a, *96*, 193–201.

Hultborn, H., Illert, M., and Santini, M. Convergence on interneurones mediating the reciprocal Ia inhibition of motoneurones. II. Effects from segmental flexor reflex pathways. *Acta Physiologica Scandinavica*, 1976b, *96*, 351–367.

Hultborn, H., Illert, M., and Santini, M. Convergence on interneurones mediating the reciprocal Ia inhibition of motoneurones. III. Effects from supraspinal pathways. *Acta Physiologica Scandinavica*, 1976c, *96*, 368–391.

Hultborn, H., Jankowska, E., and Lindström, S. Recurrent inhibition from motor axon collaterals of transmission in the Ia inhibitory pathway to motoneurones. *Journal of Physiology (London)*, 1971a, *215*, 591–612.

Hultborn, H., Jankowska, E., and Lindström, S. Recurrent inhibition of interneurones monosynaptically activated from group Ia afferents. *Journal of Physiology (London)*, 1971b, *215*, 613–636.

Humphrey, D. R., and Rietz, R. R. Cells of origin of corticorubral projections from the arm area of primate motor cortex and their synaptic actions in the red nucleus. *Brain Research*, 1976, *110*, 162–169.

Humphrey, D. R., Corrie, W. S., and Rietz, R. Sizes, intracortical locations and properties of major neuronal output populations in the arm area of primate motor cortex. *Society for Neuroscience Abstracts*, 1976, *2*, 522.

Hunt, C. C. The reflex activity of small nerve fibres. *Journal of Physiology (London)*, 1951, *115*, 456–469.

Illert, M., and Tanaka, R. Transmission of corticospinal IPSP's to cat forelimb motoneurones via high cervical propriospinal neurones and Ia inhibitory interneurones. *Brain Research*, 1976, *103*, 143–146.

Illert, M., and Tanaka, R. Integration in descending motor pathways controlling the forelimb in the cat. 4. Corticospinal inhibition of forelimb motoneurones mediated by short propriospinal neurones. *Experimental Brain Research*, 1978, *31*, 131–141.

Illert, M., Lundberg, A., and Tanaka, R. Integration in descending motor pathways controlling the forelimb in the cat. 1. Pyramidal effects on motoneurones. *Experimental Brain Research*, 1976a, *26*, 509–519.

Illert, M., Lundberg, A., and Tanaka, R. Integration in descending motor pathways controlling the forelimb in the cat. 2. Convergence on neurones mediating disynaptic cortical-motoneuronal excitation. *Experimental Brain Research*, 1976b, *26*, 521–540.

Illert, M., Lundberg, A., and Tanaka, R. Integration in descending motor pathways controlling the forelimb in the cat. 3. Convergence on propriospinal neurones transmitting disynaptic excitation from the corticospinal tract and other descending tracts. *Experimental Brain Research*, 1977, *29*, 323–344.

Ito, M. Cerebellar control of vestibular neurones: Physiology and pharmacology. *Progress in Brain Research*, 1972, *37*, 377–390.

Ito, M., and Yoshida, M. The cerebellar-evoked monosynaptic inhibition of Deiters' neurones. *Experientia*, 1964, *20*, 515–516.

Ito, M., Hongo, M., Yoshida, Y., Okada, Y., and Obata, K. Antidromic and transsynaptic activation of Deiters' neurones from the spinal cord. *Japanese Journal of Physiology*, 1964, *14*, 638–658.

Ito, M., Udo, M., and Mano, N. Long inhibitory and excitatory pathways converging onto cat reticular and Deiters' neurons and their relevance to reticulofugal axons. *Journal of Neurophysiology*, 1970, *38*, 210–226.

Ito, M., Udo, M., Mano, N., and Kawai, N. Synaptic action of the fastigiobulbar impulses upon neurones in the medullary reticular formation and vestibular nuclei. *Experimental Brain Research*, 1970, *11*, 29–47.

Jankowska, E., and Tarnecki, T. Extrapyramidal activation of muscles from sensorimotor cortex in cats. *Experientia*, 1965, *21*, 656–657.

Jankowska, E., Jukes, M. G. M., Lund, S., and Lundberg, A. The effect of DOPA on the spinal cord. 5. Reciprocal organization of pathways transmitting excitatory action to alpha motoneurons of flexors and extensors. *Acta Physiologica Scandinavica*, 1967a, *70*, 369–388.

Jankowska, E., Jukes, M. G. M., Lund, S., and Lundberg, A. The effect of DOPA on the spinal cord. 6. Half-centre organization of interneurones transmitting effects from the flexor reflex afferents. *Acta Physiologica Scandinavica*, 1967b, *70*, 389–402.

Jankowska, E., Lund, S., Lundberg, A., and Pompeiano, O. Inhibitory effects evoked through ventral reticulospinal pathways. *Archivio Italiano di Biologia*, 1968, *106*, 124–140.

Jankowska, E., Padel, Y., and Tanaka, R. The mode of activation of pyramidal tract cells by intracortical stimuli. *Journal of Physiology (London)*, 1975a, *249*, 617–636.

Jankowska, E., Padel, Y., and Tanaka, R. Projections of pyramidal tract cells to α-motoneurones innervating hind-limb muscles in the monkey. *Journal of Physiology (London)*, 1975b, *249*, 637–667.

Jankowska, E., Padel, Y., and Tanaka, R. Disynaptic inhibition of spinal motoneurones from the motor cortex in the monkey. *Journal of Physiology (London)*, 1976, *258*, 467–487.

Jeneskog, T. Parallel activation of dynamic fusimotor neurones and a climbing fiber system from the cat brain stem. I. Effects from the rubral region. *Acta Physiologica Scandinavica*, 1974a, *91*, 223–242.

Jeneskog, T. Parallel activation of dynamic fusimotor neurones and a climbing fiber system from the cat brain stem. II. Effects from the inferior olivary region. *Acta Physiologica Scandinavica*, 1974b, *92*, 66–83.

Jeneskog, T., and Johansson, H. The rubro-bulbospinal path. A descending system known to influence dynamic fusimotor neurones and its interaction with distal cutaneous afferents in the control of flexor reflex afferent pathways. *Experimental Brain Research*, 1977, *27*, 161–179.

Kanda, K., Burke, R. E., and Walmsley, B. Differential control of fast and slow twitch units in the decerebrate cat. *Experimental Brain Research*, 1977, *29*, 57–74.

Kasahara, M., and Uchino, Y. Bilateral semicircular canal inputs to neurons in cat vestibular nuclei. *Experimental Brain Research*, 1974, *20*, 285–296.

Kato, M., and Tanji, J. The effects of electrical stimulation of Deiters' nucleus upon hindlimb γ-motoneurones of the cat. *Brain Research*, 1971, *30*, 385–395.

Kirkwood, P. A., and Sears, T. A. Monosynaptic excitation of motoneurones from secondary endings of muscle spindles. *Nature*, 1974, *252*, 243–244.

Koeze, T. H. The independence of corticomotoneuronal and fusimotor pathways in the production of muscle contraction by motor cortex stimulation. *Journal of Physiology (London)*, 1968, *197*, 87–105.

Koeze, T. H., Phillips, C. G., and Sheridan, J. P. Thresholds of cortical activation of muscle spindles and α motoneurones of the baboon's hand. *Journal of Physiology (London)*, 1968, *195*, 419–449.

Kostyuk, P. G., and Pilyavsky, A. I. A possible direct interneuronal pathway from rubrospinal tract to motoneurones. *Brain Research*, 1969, *14*, 158–166.

Kuffler, S. W., and Hunt, C. C. The mammalian small nerve fibers. A system for efferent nervous regulation of muscle spindle discharge. *Research Publications. Association for Research in Nervous and Mental Diseases*, 1952, *30*, 24–37.

Kuffler, S. W., Hunt, C. C., and Quilliam, J. P. Function of medullated small-nerve fibers in mammalian ventral roots: Efferent muscle spindle innervation. *Journal of Neurophysiology*, 1951, *14*, 29–54.

Kuno, M. Mechanisms of facilitation and depression of the excitatory synaptic potential in spinal motoneurones. *Journal of Physiology (London)*, 1964, *175*, 100–112.

Kuypers, H. G. J. M. The organization of the "motor system." *International Journal of Neurology*, 1963, *4*, 78–90.

Kuypers, H. G. J. M., and Lawrence, D. G. Cortical projections to the red nucleus and the brainstem in the rhesus monkey. *Brain Research*, 1967, *4*, 151–188.

Kuypers, H. G. J. M., Fleming, W. R., and Farinholt, J. W. Subcorticospinal projections in the rhesus monkey. *Journal of Comparative Neurology*, 1962, *118*, 107–137.

Landgren, S., Phillips, C. G., and Porter, R. Minimal synaptic actions of pyramidal impulses in some alpha motoneurones of the baboon's hand and forearm. *Journal of Physiology (London)*, 1962, *161*, 91–141.

Laporte, Y., and Lloyd, D. P. C. Nature and significance of the reflex connections established by large afferent fibers of muscular origin. *American Journal of Physiology*, 1952, *169*, 609–621.

Lassek, A. M. *The Pyramidal Tract: Its Status in Medicine*. Springfield, Ill. Charles C Thomas, 1954.

Laursen, A. M., and Wiesendanger, M. The effect of pyramidal lesions on response latency in cats. *Brain Research*, 1967, *5*, 207–220.

Lawrence, D. G., and Kuypers, H. G. J. M. The functional organization of the motor system in the monkey. I. The effects of bilateral pyramidal lesions. *Brain*, 1968a, *91*, 1–14.

Lawrence, D. G., and Kuypers, H. G. J. M. The functional organization of the motor system in the monkey. II. The effects of lesions of the descending brain stem pathway. *Brain*, 1968b, *91*, 14–33.

Leksell, L. The action potential and excitatory effects of the small ventral root fibers to skeletal muscle. *Acta Physiologica Scandinavica*, 1945, *10*, Supplementum 31, 1–84.

Lewis, R., and Brindley, G. S. The extrapyramidal motor map. *Brain*, 1965, *88*, 397–406.

Lindsley, D. B., Schreiner, L. H., and Magoun, H. W. An electromyographic study of spasticity. *Journal of Neurophysiology*, 1949, *12*, 197–205.

Llinás, R., and Terzuolo, C. A. Mechanisms of supraspinal actions upon spinal cord activities. Reticular inhibitory mechanisms on alpha-extensor motoneurons. *Journal of Neurophysiology*, 1964, *27*, 570–591.

Llinás, R., and Terzuolo, C. A. Mechanisms of supraspinal actions upon spinal cord activities. Reticular inhibitory mechanisms upon flexor motoneurons. *Journal of Neurophysiology*, 1965, *28*, 413–422.

Lloyd, D. P. C. The spinal mechanism of the pyramidal system in cats. *Journal of Neurophysiology*, 1941, *4*, 525–546.

Lloyd, D. P. C. Neuron pattern controlling transmission of ipsilateral hindlimb reflexes in cat. *Journal of Neurophysiology*, 1943, *6*, 292–315.

Lloyd, D. P. C. Integrative pattern of excitation and inhibition in two-neuron arcs. *Journal of Neurophysiology*, 1946, *9*, 439–444.

Lowenstein, O., and Wersäll, J. A functional interpretation of the electron-microscopic structure of the sensory hair cells in the cristae of the elasmobranch *Raja clavata* in terms of directional activity. *Nature*, 1959, *184*, 1807–1808.

Lund, S., and Pompeiano, O. Monosynaptic excitation of alpha motoneurones from supraspinal structures in the cat. *Acta Physiologica Scandinavica*, 1968, *73*, 1–21.

Lundberg, A., and Voorhoeve, P. Effects from the pyramidal tract on spinal reflex arcs. *Acta Physiologica Scandinavica*, 1962, *56*, 201–219.

Lundberg, A., Malmgren. K., and Schomburg, E. D. Convergence from Ib, cutaneous and joint afferents in reflex pathways to motoneurones. *Brain Research*, 1975, *87*, 81–84.

Lundberg, A., Malmgren, K., and Schomburg, E. D. Cutaneous facilitation of transmission in reflex pathways from Ib afferents to motoneurons. *Journal of Physiology (London)*, 1977, *265*, 763–780.

McCouch, G. P., Deering, I. D., and Ling, T. H. Location of receptors for tonic neck reflexes. *Journal of Neurophysiology*, 1951, *14*, 191–195.

Maeda, M., Maunz, R. A., and Wilson, V. J. Labyrinthine influence on cat forelimb motoneurons. *Experimental Brain Research*, 1975, *22*, 69–86.

Magni, F., and Willis, W. D. Cortical control of brain stem reticular neurons. *Archivio Italiano di Biologia*, 1964a, *102*, 418–433.

Magni, F., and Willis, W. D. Subcortical and peripheral control of brain stem reticular neurons. *Archivio Italiano di Biologia*, 1964b, *102*, 434–448.

Magnus, R. Some results of studies in the physiology of posture. *Lancet*, 1926, *211*, 531–536, 585–588.

Magoun, H. W., and Rhines, R. An inhibitory mechanism in the bulbar reticular formation. *Journal of Neurophysiology*, 1946, *9*, 165–171.

Mallart, A., and Martin, A. R. An analysis of facilitation of transmitter release at the neuromuscular junction of the frog. *Journal of Physiology (London)*, 1967, *193*, 679–694.

Markham, C. H., Precht, W., and Shimazu, H. Effect of stimulation of interstitial nucleus of Cajal on vestibular unit activity in the cat. *Journal of Neurophysiology*, 1966, *29*, 493–507.

Massion, J. The mammalian red nucleus. *Physiological Reviews*, 1967, *47*, 383–436.

Matthews, P. B. C. The differentiation of two types of fusimotor fibre by their effects on the dynamic response of muscle spindle primary endings. *Quarterly Journal of Experimental Physiology and Psychology*, 1962, *47*, 324–333.

Matthews, P. B. C. *Mammalian Muscle Receptors and Their Central Actions*. Baltimore: Williams and Wilkins, 1972.

Melville-Jones, G., and Milsum, J. H. Characteristics of neural transmission from the semicircular canals to the vestibular nuclei of cats. *Journal of Physiology (London)*, 1970, *209*, 295–316.

Merton, P. A. Speculations on the servo-control of movement. In G. E. W. Wolstenholme (Ed.), *The Spinal Cord*. London: Churchill, 1953.

Muir, R. B., and Porter, R. The effect of a preceding stimulus on temporal facilitation at cortico-motoneuronal synapses. *Journal of Physiology (London)*, 1973, *228*, 749–763.

Nyberg-Hansen, R. Sites and mode of transportation of reticulospinal fibers in the cat. An experimental study with silver impregnation methods. *Journal of Comparative Neurology*, 1965, *124*, 71–100.

Nyberg-Hansen, R. Origin and termination of fibers from the vestibular nuclei descending in the medial longitudinal fasciculus. An experimental study with silver impregnation methods in the cat. *Journal of Comparative Neurology*, 1964, *122*, 355–367.

Nyberg-Hansen, R., and Brodal, A. Sites and mode of termination of rubrospinal fibers in the cat. An experimental study with silver impregnation methods. *Journal of Anatomy*, 1963, *98*, 235–253. •

Nyberg-Hansen, R., and Mascitti, T. A. Sites and termination of the vestibulospinal tract in the cat. An experimental study with silver impregnation methods. *Journal of Comparative Neurology*, 1964, *122*, 369–387.

Onoda, N., and Burton, J. E. Discharge of red nucleus neurons in relation to a voluntary elbow flexion. *Society for Neuroscience Abstracts*, 1975, *1*, 178.

Orlovsky, G. N. Work of the reticulo-spinal neurones during locomotion. *Biophysics*, 1970, *15*, 761–771.

Orlovsky, G. N. The effect of different descending systems of flexor and extensor activity during locomotion. *Brain Research*, 1972a, *40*, 359–371.

Orlovsky, G. N. Activity of vestibulospinal neurons during locomotion. *Brain Research*, 1972b, *46*, 85–98.

Orlovsky, G. N. Activity of rubrospinal neurons during locomotion. *Brain Research*, 1972c, *46*, 99–112.

Orlovsky, G. N., and Shik, M. L. Control of locomotion: A neurophysiological analysis of the cat locomotor system. In R. Porter (Ed.), *International Review of Physiology: Neurophysiology II* (Vol. 10). Baltimore: University Park Press, 1976.

Oscarsson. O. Functional organization of spinocerebellar paths. In A. Iggo (Ed.), *Handbook of Sensory Physiology. Vol. II. Somatosensory System*. Berlin: Springer, 1973.

Otero, J. B. Activity of red nucleus (RN) and motor cortex (MC) neurons during a continuous performance task in the monkey. *Federation Proceedings*, 1975, *34*, 446.

Peterson, B. W. Effect of tilting on the activity of neurons in the vestibular nuclei of the cat. *Brain Research*, 1967, *6*, 606–609.

Peterson, B. W. Distribution of neural responses to tilting within vestibular nuclei of the cat. *Journal of Neurophysiology*, 1970, *33*, 750–767.

Peterson, B. W., and Abzug, C. Properties of projections from vestibular nuclei to medial reticular formation in the cat. *Journal of Neurophysiology*, 1975, *38*, 1421–1435.

Peterson, B. W., Anderson, M. E., and Filion, M. Responses of ponto-medullary reticular neurons to cortical, tectal and cutaneous stimuli. *Experimental Brain Research*, 1974, *21*, 19–44.

Peterson, B. W., Filion, M., Felpel, L. P., and Abzug, C. Responses of medial reticular neurons to stimulation of the vestibular nerve. *Experimental Brain Research*, 1975, *221*, 335–350.

Peterson, B. W., Maunz, R. A., Pitts, N. G., and Mackel, R. G. Patterns of projection and branching of reticulospinal neurons. *Experimental Brain Research*, 1975, *23*, 333–351.

Peterson, B. W., Pitts, N. G., Mackel, R. G., and Fukushima, K. Monosynaptic excitation and inhibition of neck motoneurons by a reticulospinal pathway. *Society for Neuroscience Abstracts*, 1976, *2*, 528.

Petras, J. M. Cortical, tectal and tegmental fiber connections in the spinal cord of the cat. *Brain Research*, 1967, *6*, 275–324.

Phillips, C. G., and Porter, R. The pyramidal projection to motoneurones of some muscle groups of the babboon's forelimb. *Progress in Brain Research*, 1964, *12*, 222–242.

Pompeiano, O. Analisi degli effecti della stimulazione elletrica del nucleo rosso nel gatto decerebrato. *Atti dell'Accademia Nazionale dei Lincei. Memorie. Sez. III*, 1957, *22*, 100–103.

Pompeiano, O., and Brodal, A. Experimental demonstration of a somatotopical origin of rubrospinal fibers in the cat. *Journal of Comparative Neurology*, 1957, *108*, 225–252.

Poppele, R. E. Response of gamma and alpha motor systems to phasic and tonic vestibular inputs. *Brain Research*, 1967, *6*, 535–547.

Porter, R. Early facilitation at corticomotoneuronal synapses. *Journal of Physiology (London)*, 1970, *207*, 733–745.

Porter, R., and Hore, J. Time course of minimal corticomotoneuronal excitatory postsynaptic potentials in lumbar motoneurons of the monkey. *Journal of Neurophysiology*, 1969, *32*, 443–451.

Precht, W., Grippo, J., and Wagner, A. Contribution of different types of central vestibular neurons to the vestibulospinal system. *Brain Research*, 1967, *4*, 119–128.

Preston, J. B., Schende, M. C., and Uemura, K. The motor cortex-pyramidal system: Patterns of facilitation and inhibition on motoneurons innervating limb musculature of cat and baboon and their possible adaptive significance. In M. D. Yahr and D. P. Purpura (Eds.), *Neurophysiological Basis of Normal and Abnormal Motor Activities*. New York: Raven Press, 1967.

Rapaport, S., Susswein, A., Uchino, Y., and Wilson, V. J. Properties of vestibular neurones projecting to neck segments of the cat spinal cord. *Journal of Physiology (London)*, 1977, *268*, 493–510.

Rexed, B. The cytoarchitectonic organization of the spinal cord in the cat. *Journal of Comparative Neurology*, 1952, *96*, 415–495.

Rexed, B. A cytoarchitectonic atlas of the spinal cord in the cat. *Journal of Comparative Neurology*, 1954, *100*, 297–379.

Rhines, R., and Magoun, H. W. Brain stem facilitation of cortical motor responses. *Journal of Neurophysiology*, 1946, *9*, 219–229.

Rinvik, E., and Walberg, F. Demonstration of a somatotopically arranged corticorubral projection in the cat. An experimental study with silver methods. *Journal of Comparative Neurology*, 1963, *120*, 393–407.

Roberts, T. D. M. Changes in stretch reflexes in limb extension muscles during position reflexes from the labyrinth in the cat. *Journal of Physiology (London)*, 1970, *211*, 5p.

Rosenberg, J. R., and Lindsay, K. W. Asymmetric tonic labyrinthine reflexes. *Brain Research*, 1973, *63*, 347–350.

Rossi, G. F., and Brodal, A. Corticofugal fibres to the brainstem reticular formation. An experimental study in the cat. *Journal of Anatomy*, 1956, *90*, 42–62.

Ruch, T. C., and Watts, J. W. Reciprocal changes to reflex activity of the fore limbs induced by post-brachial "cold block" of the spinal cord. *Journal of Physiology (London)*, 1934, *110*, 362–375.

Scheibel, M. E., and Scheibel, A. B. Structural substrate for integrative patterns in the brainstem reticular core. In H. H. Jasper *et al.* (Eds.), *Reticular Formation of the Brain*. Boston: Little, Brown, 1958.

Schor, R. H. Responses of cat vestibular neurons to sinusoidal roll tilt. *Experimental Brain Research*, 1974, *20*, 347–362.

Schreiner, L. H., Lindsley, D. B., and Magoun, H. W. Role of brainstem facilitatory systems in maintenance of spasticity. *Journal of Neurophysiology*, 1949, *12*, 207–216.

Sears, T. A. Efferent discharges in alpha and fusimotor fibres of intercostal nerves of the cat. *Journal of Physiology (London)*, 1964, *174*, 295–315.

Severin, V. The role of the gamma motor system in the activation of the extensor alpha motor neurones during controlled locomotion. *Biophysics*, 1970, *15*, 1138–1145.

Shapovalov, A. I. Extrapyramidal monosynaptic and disynaptic control of mammalian alpha-motoneurons. *Brain Research*, 1972, *40*, 105–115.

Shapovalov, A. I., Grantyn, A. A., and Kurchavyi, G. G. Short latency reticulospinal synaptic projections on α-motoneurons. *Bulletin of Experimental Biology and Medicine*, 1967, *64*, 685–690.

Shapovalov, A. I., Karamjan, O. A., Kurchavyi, G. G., and Repina, Z. A. Synaptic actions evoked from the red nucleus in the spinal alpha-motoneurons in the rhesus monkey. *Brain Research*, 1971, *32*, 325–348.

Shapovalov, A. I., Karamjan, O. A., Tamarova, Z. A., and Kurchavyi. G. G. Cerebello-rubrospinal effects on hindlimb motoneurons in the monkey. *Brain Research*, 1972, *47*, 49–59.

Sherrington, C. S. Flexion-reflex of the limb, crossed extension reflex and reflex stepping and standing. *Journal of Physiology (London)*, 1910, *40*, 28–121.

Shik, M. L., and Orlovsky, G. N. Nuerophysiology of locomotor automatism. *Physiological Reviews*, 1976, *56*, 465–501.

Shik, M. L., Severin, V., and Orlovsky, G. N. Control of walking and running by means of electrical stimulation of the mid-brain. *Biophysics*, 1966, *11*, 756–765.

Shimazu, H., and Precht, W. Tonic and kinetic responses of cat's vestibular neurons to horizontal angular accelerations. *Journal of Neurophysiology*, 1965, *28*, 991–1013.

Shimazu, H., and Precht, W. Inhibition of central vestibular neurons from the contralateral labyrinth and its mediating pathway. *Journal of Neurophysiology*, 1966, *29*, 467–492.

Shimazu, H., Hongo, T., and Kubota, K. Two types of central influences on gamma motor system. *Journal of Neurophysiology*, 1962, *25*, 309–323.

Shinoda, Y., and Yoshida, K. Dynamic characteristics of responses to horizontal head angular acceleration in vestibulo-ocular pathway in the cat. *Journal of Neurophysiology*, 1973, *37*, 643–673.

Shinoda, Y., Arnold, A., and Asanuma, H. Spinal branching of corticospinal axons in the cat. *Experimental Brain Research*, 1976, *26*, 215–234.

Sjöström, A., and Zanger, P. α-γ linkage in the spinal generator for locomotion in the cat. *Acta Physiologica Scandinavica*, 1975, *94*, 130–132.

Smith. A. M., and Courville, J. The origin of the rubrospinal tract in primate as shown by the retrograde transport of horseradish peroxidase. *Society for Neuroscience Abstracts*, 1976, *2*, 551.

Soechting, J. F., Anderson, J. H., and Berthoz, A. Dynamic relation between natural vestibular inputs and activity of forelimb extensor muscles in the decerebrate cat. III. Motor output during rotations in the vertical plane. *Brain Research*, 1977, *120*, 35–47.

Sprague, J. M., and Chambers, W. W. Control of posture by reticular formation and cerebellum in the intact, anesthetized and unanesthetized and in the decerebrate cat. *American Journal of Physiology*, 1954, *176*, 52–64.

Sprague, J. M., Schreiner, L. H., Lindsley, D. B., and Magoun, H. W. Reticulospinal influences on stretch reflexes. *Journal of Neurophysiology*, 1948, *11*, 501–507.

Stauffer, E. K., Watt, D. G. D., Taylor, A., Reinking, R. M., and Stuart, D. G. Analysis of muscle receptor connections by spike-triggered averaging. 2. Spindle group II afferents. *Journal of Neurophysiology,* 1976, *39,* 1393-1402.

Stein, B. M., and Carpenter, M. B. Central projections of portions of the vestibular ganglia innervating specific parts of the labyrinth in the rhesus monkey. *American Journal of Anatomy,* 1967, *120,* 281-318.

Stewart, D. H., Preston, J. B., and Whitlock, D. G. Spinal pathways mediating motor cortex excitability changes in segmental motoneurons in pyramidal cats. *Journal of Neurophysiology,* 1968, *31,* 928-937.

Suzuki, J. I., and Cohen, B. Head, eye, body and limb movements from semicircular canal nerves. *Experimental Neurology,* 1964, *10,* 393-405.

Terzuolo, C. A., Llinás, R., and Green, K. T. Mechanisms of supraspinal actions upon spinal cord activities. Distribution of reticular and segmental inputs in cat's alpha motoneurons. *Archivio Italiano di Biologia,* 1965, *103,* 635-651.

Torvik, A., and Brodal, A. The origin of the reticulospinal fibers in the cat. An experimental study. *Anatomical Record,* 1957, *128,* 113-137.

Travis, A. M., and Woolsey, C. N. Motor performance of monkeys after bilateral partial and total cerebral decortications. *American Journal of Physical Medicine,* 1956, *35,* 273-310.

Tsukahara, N., and Fuller, D. R. G. Conductance changes during pyramidally induced postsynaptic potentials in red nucleus neurons. *Journal of Neurophysiology,* 1969, *32,* 34-42.

Tsukahara, N., Fuller, D. R. G., and Brooks, V. P. Collateral pyramidal influences on corticorubrospinal systems. *Journal of Neurophysiology,* 1968, *31,* 467-484.

Tsukahara, N., Toyama, K., and Kosaka, K. Electrical activity of red nucleus neurons investigated with intracellular microelectrodes. *Experimental Brain Research,* 1967, *4,* 18-33.

Vasilenko, D. A., and Kostyuk, P. G. Functional properties of interneurons activated monosynaptically by the pyramidal tract. *Neuroscience Translations,* 1967/68, *1,* 66-72.

Vallbo, Å. B. Slowly adapting muscle receptors in man. *Acta Physiologica Scandinavica,* 1970, *78,* 315-333.

Vallbo, Å. B. Muscle spindle response at the onset of isometric voluntary contractions in man. Time difference between fusimotor and skeletomotor effects. *Journal of Physiology (London),* 1971, *318,* 405-431.

Walberg, F. Light and electron microscopical data on the distribution of primary vestibular fibers. *Progress in Brain Research,* 1972a, *37,* 79-88.

Walberg, F. Descending and reticular relations to the vestibular nuclei: Anatomy. *Progress in Brain Research,* 1972b, *37,* 585-588.

Walberg, F. Cerebellovestibular relations: Anatomy. *Progress in Brain Research,* 1972c, *37,* 361-376.

Wilson, V. J., and Maeda, M. Connections between semicircular canals and neck motoneurons in the cat. *Journal of Neurophysiology.* 1974, *37,* 346-357.

Wilson, V. J., and Yoshida, M. Bilateral connections between labyrinths and neck motoneurons. *Brain Research,* 1969a, *13,* 603-607.

Wilson, V. J., and Yoshida, M. Comparison of effects of stimulation of Deiters' nucleus and medial longitudinal fasciculus on neck, forelimb and hindlimb motoneurons. *Journal of Neurophysiology,* 1969b, *32,* 743-758.

Wilson, V. J., and Yoshida, M. Monosynaptic inhibition of neck motoneurons by the medial vestibular nucleus. *Experimental Brain Research,* 1969c, *9,* 365-380.

Wilson, V. J., Gacek, R. R., Maeda, M. M., and Uchino, Y. Saccular and utricular input to cat neck motoneurons. *Journal of Neurophysiology,* 1977, *40,* 63-73.

Wilson, V. J., Kato, M., Peterson, B. W., and Wylie, R. M. A single-unit analysis of the organization of Deiters' nucleus. *Journal of Neurophysiology,* 1967, *30,* 603-619.

Wilson, V. J., Yoshida, M., and Schor, R. H. Supraspinal monosynaptic excitation and inhibition of thoracic back motoneurons. *Experimental Brain Research,* 1970, *11,* 282-295.

5

Cerebellar Control of Movement

Rodolfo R. Llinás and John I. Simpson

General Description of the Cerebellum

Morphologically, the term "cerebellum" may be unambiguously used to denote the neuronal mass composed by the cerebellar cortex, the cerebellar nuclei, and the intervening white matter (the fibers that lie between these two sites). This mass is clearly separated from the rest of the central nervous system by three large fiber bundles, the cerebellar peduncles (superior, middle, and inferior). However, until now no such unambiguous description could be given for its overall physiology because owing to its intermediate location between the sensory and motor realm, its function could only be considered within the context of that of the rest of the nervous system. One may say, in fact, that in absolute terms the cerebellum does not play a primary role in either sensory or motor function; its destruction does not produce alteration of sensation or paralysis.

On the other hand, it is clear that lesions of the cerebellum produce well-defined and often devastating changes in the ability of the rest of the nervous system to generate both the simple and the elegant motor sequences that normal animals utilize to attain motor goals. The ability to produce such well-defined movements is known as motor coordination, in the sense that movements are generated in the context of the total motor state of the individual at a given moment. Because this "setting into context" relates to many different levels of brain function, as evidenced by the elaborate connectivity of the cerebellum, the understanding of its function is not reducible to a set of simple neuronal loops but, rather, requires the devel-

Rodolfo R. Llinás and John I. Simpson   Department of Physiology and Biophysics, New York University Medical Center, New York, New York 10016. Research was supported by United States Public Health Service grant NS-13742 from the National Institute for Neurological and Communicative Disorders and Stroke.

opment of more general views. Thus, our understanding of the function of the cerebellum is intimately related to our understanding of motor coordination and to the properties of the cerebellar circuits that bring about such a global property. We shall refer to this problem at the end of the chapter.

This chapter is divided into two parts: the first part is concerned with the structure and function of the neuronal elements in the cerebellar cortex and related cerebellar and olivary nuclei, and the second part with the structure and function of cerebellar efferents and afferents. We shall also discuss the relations between these two levels and some of the general theories regarding overall cerebellar function.

Morphology and Physiology of Cerebellar Cortical Neurons

Without a doubt, the morphology of the cellular elements in the cerebellar cortex is one of the most richly studied and best known in the vertebrate phyla. Probably the outstanding feature of neuronal organization of this cortex is the stereotyped character of the cellular types and connectivity in its neuronal circuits. Equally intriguing is the similarity of the overall connectivity and the similarity of neuronal elements across different vertebrate species. These general morphological features were recognized by the early workers and emphasized by Ramón y Cajal, who initially described the organization of the cellular elements in the cerebellar cortex of avians with amazing accuracy and completeness (Ramón y Cajal, 1888). In subsequent years, he and his pupils proceeded to describe similarities across species and concluded that the cerebellar cortex is one of the least variable of the central structures of the vertebrate brain, at least with regard to the neuronal elements that constitute it (Estable, 1923; Ramón y Cajal, 1904).

More recent studies of the comparative aspects of the morphology and physiology of this cortex have indicated the possibility that the cerebellar cortex be viewed as having two stages of morphological and functional organization, one designated as the "basic cerebellar circuit" and the second related to the organization of the cortical interneurons (Llinás and Hillman, 1969).

The Basic Cerebellar Circuit. We assume the basic cerebellar cortical circuit to be present in all vertebrate species. This circuit is composed of the Purkinje cell, as the single output system of the cortex (Ramón y Cajal, 1888), and two inputs: a monosynaptic input to the Purkinje cell, the climbing fiber, and a disynaptic input, the mossy fiber–granule cell–Purkinje cell system (see Fig. 1, upper left diagram).

Spatially, Purkinje cell somata are organized as a continuous sheet of closely spaced neuronal elements. Their dendritic trees are planar, arise from the outer pole of the somata, are spatially organized such that they overlap like playing cards held in a "hand", and are stacked transversely to the main direction in the cerebellar folia. The Purkinje cell sheet divides the cerebellar cortex into two main layers: the level peripheral to the Purkinje cell somata known as the molecular layer and the layer central to the Purkinje cells (i.e., toward the white matter), the granular layer. Central to the granular layer is the white matter formed by the input and output nerve fiber systems of this cortex.

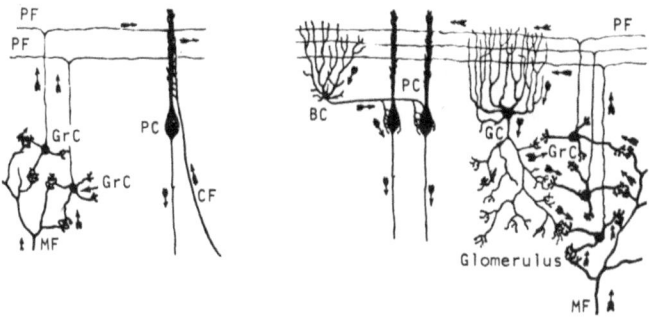

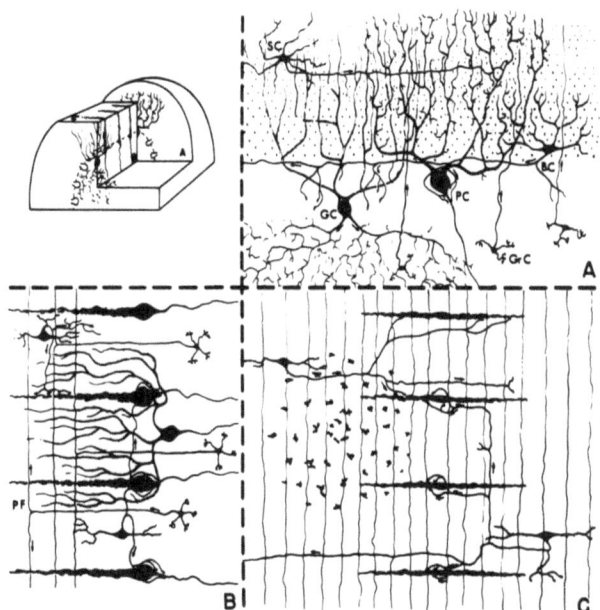

Fig. 1. *(Top left)* Diagram of the basic cerebellar circuit comprising the mossy and climbing fiber inputs to Purkinje cells. Activity in the mossy fiber (MF) is relayed through the granule cells (GrC) via the parallel fibers (PF) to the Purkinje cell (PC). The axons of Purkinje cells are the only output system from the cerebellar cortex. The second afferent is a climbing fiber (CF) which establishes a monosynaptic input to Purkinje cell dendrites. *(Top right)* Diagram of the two basic inhibitory systems of the cerebellar cortex. On the left, a basket cell axon contacts the somata of Purkinje cells. These axons contact the dendrites of these cells as well. The stellate cell (not shown) establishes direct contact with Purkinje cell dendrites. Both basket (BC) and stellate cells receive input via parallel fibers (PF) and constitute the inhibitory systems of the molecular layer. The Golgi cell (GC to the right) is seen to receive input from parallel fibers, mossy fibers (MF) and climbing fibers (not shown) and to relay inhibition to the dendrites of the granule cells (GrC) in the granular layer. *(Bottom)* Detail of the geometrical organization of the neuronal elements of the cerebellar cortex. The drawing demonstrates different sections through a cerebellar folium. (A) transverse plane; (B) saggital plane; (C) tangential plane. The cellular elements are displayed in drawings A, B, and C, as though the cerebellum were transparent. The orthogonal organization of the parallel fibers, with respect to the isoplanar characteristics of the dendrites of Purkinje cells and basket and stellate cells (SC), is self-explanatory. Axons of the basket and stellate cells run at right angles to the parallel fibers and the dendritic tree of Golgi cells is close to cylindrical rather than isoplanar. From Llinás (1974).

The two main types of afferents to Purkinje cells indicate that the cerebellar cortex has two basic functional organizations (Eccles, Llinás, and Sasaki, 1966a; Llinás and Hillman, 1969; Marr, 1969; Oscarsson, 1969): (1) the climbing fiber–Purkinje cell system, which, as we will see later, is organized into groups of specifically and synchronously activated Purkinje cells having quite different spatial locations (Armstrong, 1974), and (2) the mossy fiber–granule cell-parallel, fiber–Purkinje cell system, where Purkinje cells are activated, at given loci, in very specific geometric patterns owing to the particular spatial relationship between the parallel fibers and the Purkinje cell dendrites. In the latter case, rather than the one-to-one relationship seen between the Purkinje cell and its climbing fiber afferent, a many-to-many relationship is present. Moreover, the directionality of the parallel fibers has been shown to be all important in determining the peculiar orthogonal organization of these fibers with respect to the main direction of dendritic lateral spread in Purkinje, basket, and stellate cells. Thus, in the climbing fiber mode of activation, specific Purkinje cells may be fired, while mossy and parallel fiber input is bound, of necessity, to activate large numbers of Purkinje cells in rows.

CORTICAL INTERNEURONS. Superimposed on this basic cerebellar circuit, a second more variable neuronal system has evolved with characteristics that probably depend on phylogeny and adaptation (Fig. 1, upper right diagram). This system is represented by the inhibitory interneurons of the molecular and granular layers. Fundamentally, these neurons may be categorized into four classes (Ramón y Cajal, 1888, cf. Palay and Chan-Palay, 1974): those interneurons in the molecular layer that receive mostly from the parallel fibers and terminate on Purkinje cell dendrites and somata (forming an elaborate synapse around the somata and axon hillock), the so-called basket cells; second, the stellate cells, located more superficially in the molecular layer, whose axons terminate in contact with Purkinje cell dendrites; third, the uncommon cells of Lugaro (cf. Fox, 1959) of unknown function; and fourth, the interneurons with somata in the granular layer and occasionally in the molecular layer (cf. Fox, Siegesmund, and Dutta, 1964; Palay and Chan-Palay, 1974), which receive inputs from parallel, climbing, and mossy fibers directly. Their axons terminate in the granular layer producing inhibition on the granule cell dendritic element at the granular layer glomerulus.

In short, the cerebellar cortex is spatially organized in a set of three mutually orthogonal directions (Fig. 1, lower diagram): first, the dendrites of all molecular layer neurons (Purkinje, basket, stellate, Golgi, and Lugaro cells) that run in the vertical direction with respect to the cerebellar surface; second, the parallel fibers that run in the direction of the cortical folia, parallel to the surface of the cortex and to each other and at right angles to the planar molecular layer dendrites; and third, the axons of basket and stellate cells that run parallel to the surface of the cortex but at right angles to the direction of spread of the parallel fibers.

ANATOMICAL AND PHYSIOLOGICAL DESCRIPTION OF THE PURKINJE CELL

MORPHOLOGY. The Purkinje cell is the largest neuron in the brain. In mammals, their dendrites expand the thickness of the molecular layer (400 μm) and branch profusely as they extend peripherally into the molecular layer (Hillman, 1969a,b; Mugnaini, 1972; Palay and Chan-Palay, 1974; Sotelo, 1969). Computer measurements of the dendritic arbor indicate a total dendritic length of approximately 4 mm if all dendritic segments are added (Llinás and Hillman, 1975). This extensive dendritic tree is close to isoplanar; that is, flattened such that if viewed from the surface of the molecular layer, a Purkinje cell dendritic tree would appear (see Fig. 1C) as a rather thin structure very much like a fan having a length of approximately 400 μm. Thus, the fanlike dendrites are spatially organized so as to form rows of parallel planes (like stacks of plates) occupying minimum volume. Although the overlap of dendrites in the same plane is very small, neighboring dendrites not in the same plane overlap as much as 80%. The Purkinje cells receive their major synaptic input from the parallel fibers via dendritic spines on the spiny branchlets located in the periphery of the dendritic tree (Fox, Hillman, Siegesmund, and Dutta, 1967; Ramón y Cajal, 1888) (Fig. 2). These dendrites may be viewed as organized into three main levels: the smooth dendrites, which are generally a single first-order stem arising from the soma; a set of second-order branches (Hillman, 1969a), which may be as many as 50 to 100 in number; and a set of third-order branches, which exist only in the more peripheral dendritic ramifications and represent the main postsynaptic sites of the neuron. In cats they number approximately 80,000 (Palkovits, Magyar, and Szentágothai, 1971b).

The climbing fiber afferents synapse on specialized short-necked clusters of spines on the smooth dendritic branches (Larramendi and Victor, 1967). Terminals arising from the basket and stellate cell interneurons terminate directly on the smooth dendrites and on the soma of the Purkinje cells, no postsynaptic spine being involved at this junction.

ELECTROPHYSIOLOGY. Purkinje cell electrical activity may be recorded under *in vivo* or *in vitro* conditions. Since the most reliable recordings may now be obtained *in vitro,* we shall describe the electrical properties of the mammalian Purkinje cell as seen in cerebellar slices (Llinás and Sugimori, 1980a,b). Antidromic or direct activation of a Purkinje cell is characterized by a large spike having an IS–SD (initial segment/soma dendritic) break that is in many ways similar to that obtained *in vivo* for motoneurons (Eccles *et al.,* 1966a) and other central neurons (cf. Spencer, 1977). These somatic spikes, however, do not seem to invade the dendrites actively, as will be shown below. Direct stimulation of the Purkinje cell somata via the the recording microelectrode demonstrates that these cells fire in a manner slightly different from that seen in other neurons. Indeed, square current pulses lasting approximately 1 sec (Fig, 3A–C) produce, at just threshold depolarization, a repetitive activation of the cell. That is to say, with long pulses the neuron fires, but a single isolated spike cannot be generated by this type of stimulus. This burst of activity (similar to Class 2 repetitive firing of crab nerve, described by Hodgkin, 1948) is produced by a low-threshold, sodium-dependent conductance that does not inactivate within several seconds and serves to trigger the fast action potentials (Llinás and Sugimori, 1980a). With increased stimulation the onset of the repetitive firing moves earlier. At the end of the initial pulse of firing, a reduction in the amplitude of the spikes is followed by a rhythmic bursting (Fig. 3B). This bursting is clearly seen in Fig. 3C.

RODOLFO R.
LLINÁS AND
JOHN I. SIMPSON

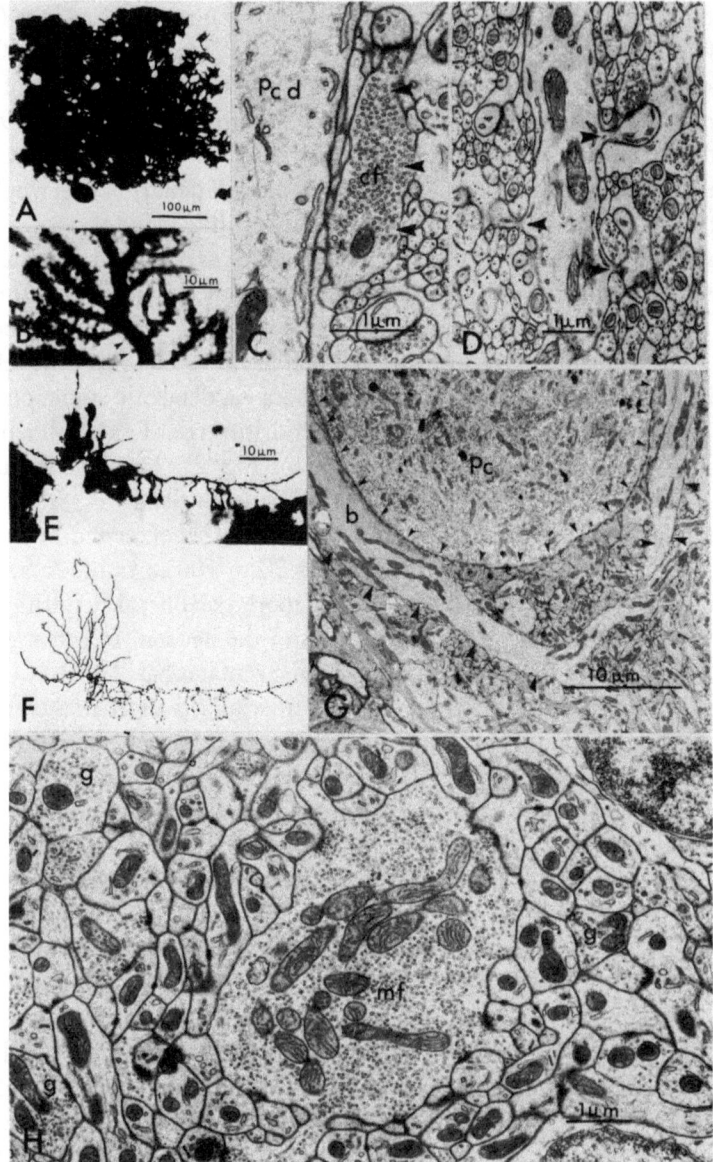

Fig. 2. Morphology of some neuronal elements in the cerebellar cortex. (A) Golgi stain of a mammalian Purkinje cell showing cell body and full dendritic tree. (B) Detail of smooth and spiny branchlets. Smooth branch gives three spiny branchlets to the left. The smooth branch has itself short stubby spines (two arrows), which are the site for the climbing fiber junction. (C) Electron micrograph showing Purkinje cell dendrite (Pcd) and a climbing fiber (cf) terminal contacting a spine. Arrows indicate preterminal enlargement with synaptic vesicles. (D) Spiny branchlet ultrastructure showing three spines marked with arrows and a parallel fiber synapsing on a spine. (E) Golgi stain of a basket cell. (F) Computer reconstruction of the same cell, showing the dendrites as a continuous line and the axons as a broken line. The main axon generates descending terminals that envelope the soma of the Purkinje cells as shown in G. (G) Electron micrograph of a Purkinje cell soma and the basket formation (b) which envelopes the lower part of the Purkinje cells and their initial segments. Large arrows indicate synaptic terminals. Small arrows indicate postsynaptic site of inhibitory synaptic contacts on the Purkinje cell. (H) Electron micrograph of mossy fiber

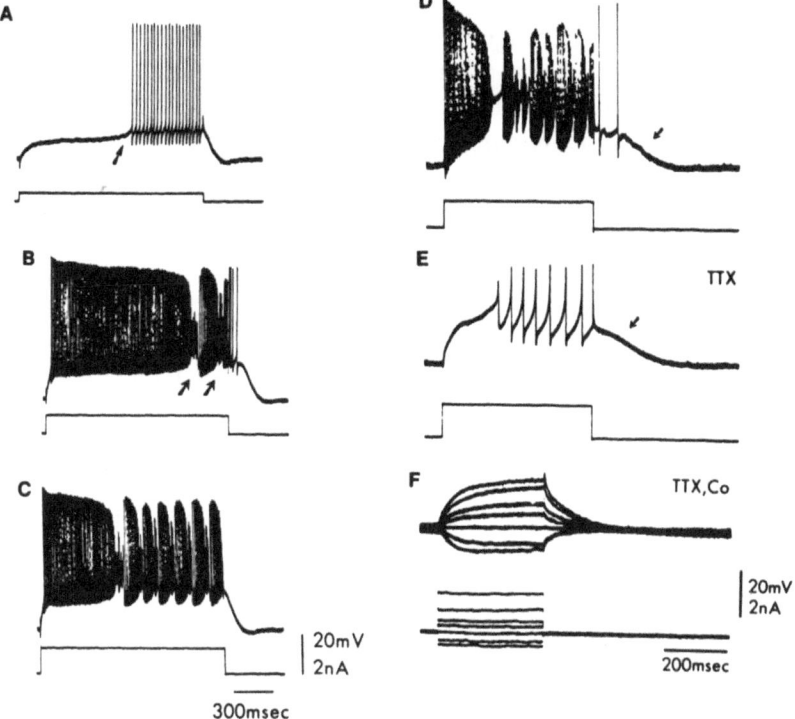

Fig. 3. Ionic mechanisms for Purkinje cell firing. (A–F) Recordings from mammalian Purkinje cell somata *in vitro*. (A–C) Repetitive firing obtained with prolonged current pulses. In (A) a "threshold" current stimulus produces a repetitive activation of the Purkinje cell after an initial local response (arrow). In (B) and (C) increases in current injection amplitude produce high-frequency firing and an oscillatory behavior marked with arrows in (B). (D and E) TTX sensitivity of Purkinje cell spikes. (D) Control response to square pulse depolarization. (E) A similar pulse after the addition of TTX to the bath. Note that the fast (ss) spikes are blocked while the slower oscillations and the after-depolarization (arrow) remain. (F) Addition of cobalt chloride to the TTX saline removes all electroresponsiveness. Modified from Llinás and Sugimori (1980a).

Pharmacological studies of this preparation indicate that these two types of responses on Purkinje cells (i.e., the fast action potentials and the bursting) have different ionic mechanisms (Llinás and Sugimori, 1980a). Typical repetitive firing and bursting are illustrated in Fig. 3D. Removal of extracellular sodium or the application of tetrodotoxin (a sodium conductance blocker) to the bath causes a complete abolition of the fast action potentials but leaves a late, slow-rising burst potential intact, as shown in Fig. 3E. Because these spikes are obtained in the absence of sodium, and because they are blocked by ions that block the slow calcium conductance (cobalt, cadmium, manganese, or the drug D600) and by the removal

---

rosette. The large presynaptic mossy fiber terminal (mf) contacts several granule cell dendrites that surround the preterminal. Other presynaptic structures in this neuropile are the Golgi cell axons (g) which also contact the granule cell dendrites. To the right, portions of two granule cell somata are seen. Courtesy of D. E. Hillman.

of extracellular calcium from the extracellular medium, we have concluded that this slow bursting of Purkinje cells is generated by a voltage-dependent calcium conductance followed by a calcium-dependent potassium conductance change. Indeed, replacement of calcium by barium tends to reduce the after-hyperpolarization and converts the bursting response into a prolonged single action potential (Llinás and Sugimori, 1980a) since barium does not activate the late potassium conductance. The blockage of this oscillatory behavior following calcium blockage is shown in Fig. 3F after the application of both tetrodotoxin (TTX) and cobalt to the extracellular medium. The results shown above imply that at the somatic level three main mechanisms for spike generation are observed: (1) a low-threshold, sodium-dependent spike similar to that seen in other cells, which is blocked by the absence of extracellular sodium or by the application of TTX, (2) a noninactivating sodium spike, and (3) a calcium-dependent action potential which has a low rising time and a rather rapid return to baseline.

Dendritic Recordings. The electrical activity in Purkinje cells may also be recorded at the dendritic level (cf. Llinás and Nicholson, 1971; Llinás and Hess, 1976). In mammals, the type of spontaneous action potentials that may be seen at different levels in a Purkinje cell soma and dendrites is illustrated for an *in vitro* experiment in Fig. 4B to E. The typical bursting is seen at the somatic level in Fig. 4E. Recordings obtained at different levels in the dendritic tree are shown in Figs. 4B, C, and D, and indicate clearly that the rather fast sodium action potentials

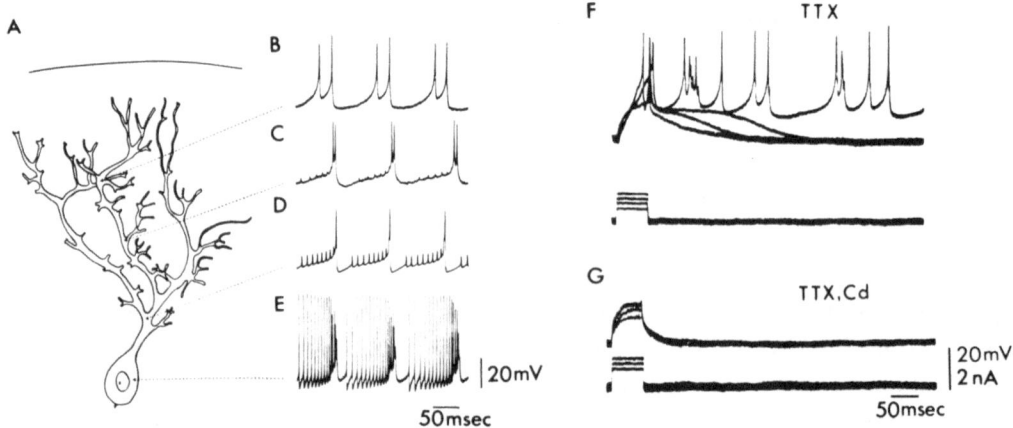

Fig. 4. Dendritic recording from mammalian Purkinje cells *in vitro*. (A) Composite picture showing the relationship between somatic and dendritic action potentials following d.c. depolarization through the recording electrode. A clear shift in amplitude of the fast action potentials and the dendritic calcium-dependent spikes is seen when comparing the more superficial recording in (B) with the somatic recording in (E). At increasing distances from the soma, the fast spikes are reduced in amplitude and are barely noticeable in the more peripheral recordings. However, the prolonged and slow-rising burst spikes are more prominent at dendritic level. (F) Calcium-dependent plateau and burst spikes. These plateau potentials seen intradendritically with short depolarizations are recorded in the presence of TTX. As the stimulus is increased, prolonged local responses are observed that ultimately result in full dendritic spike bursts. (G) Addition of cadmium chloride to the TTX solution produces a complete blockage of the plateau and the burst response recorded intradendritically. Modified from Llinás and Sugimori (1980b).

seen at the soma do not actively invade dendrites but rather are electrotonically conducted and can only be seen to about mid-dendritic level, their amplitude decrementing rather quickly with distance. The bursting calcium-dependent spike, on the other hand, can be seen to be large and rather prominent in upper dendrites, indicating a differential distribution for sodium and calcium conductances. Furthermore, direct stimulation of dendrites after application of TTX, as shown in Fig. 4F, indicates that local stimulation at that level produces two types of calcium-dependent electroresponsiveness. In the first case, a small stimulus can generate a plateau-like response, as seen in the first three stimuli in Fig. 4F. The stronger stimulus produces, in addition to the plateau, a burst of action potentials. Because this response can be blocked by cobalt, cadmium, or D600 (Fig. 4G) (Llinás and Sugimori, 1980b), it must be concluded that the dendrites of the Purkinje cell are capable of generating calcium-dependent spikes, which may be either of a prolonged plateau form or clear all-or-none action potentials.

The Purkinje cells thus demonstrate the following set of voltage-dependent ionic conductances. At the somatic level, there is a rapid sodium current that inactivates and a fast voltage-dependent potassium current that generates the afterhyperpolarization following the fast spike. In addition, somatic membrane displays a noninactivating voltage-dependent sodium conductance capable of generating a repetitive firing of the Purkinje cell following prolonged depolarization. At the dendritic level, on the other hand, excitability seems mainly to be a voltage-dependent calcium conductance. This conductance may be followed by potassium activation due to both a voltage dependence and a calcium-dependent conductance to potassium.

Climbing Fiber–Purkinje Cell Synapse. One of the most powerful synaptic junctions in the CNS is that between the climbing fiber afferent and the dendrite of a Purkinje cell. Although, for the most part, a one-to-one relationship exists between a climbing fiber and a given Purkinje cell (each Purkinje cell receives one climbing fiber), the cell of origin of this afferent (the inferior olivary neuron) is probably capable of producing more than one climbing fiber afferent and probably as many as ten. It was demonstrated electrophysiologically in 1966 (Eccles *et al.*, 1966a) that stimulation of the inferior olive produces a powerful activation of the Purkinje cell. This synaptic excitation is characterized by an all-or-none burst of spikes with little variability from one activation to the next (Fig. 5A). These spikes are produced at the dendrite by the voltage-dependent calcium conductances and at the somatic and axonic levels by the usual sodium-potassium spikes (Llinás and Sugimori, 1978, 1980b). The climbing fiber afferent twines around the dendritic tree of the Purkinje cell and establishes as many as 200 to 300 synaptic contacts (Hillman, 1969a) on specialized short-neck spines found on the smooth branches (Larramendi and Victor, 1967). Following a presynaptic action potential, these synapses are activated simultaneously and produce a very large unitary EPSP in the postsynaptic dendrite. The all-or-none character of the climbing fiber burst (Fig. 5A) actually represents the all-or-none character of the presynaptic spike in the climbing fiber. If the Purkinje cells are depolarized far enough to reduce the sodium and calcium spike components, the all-or-none character of the EPSPs may be seen. Under these conditions, the chemical nature of the synapse may be studied in detail and its distributed character clearly demonstrated. Depolarization of the soma or dendrite can produce a reduction and an actual reversal of the climbing fiber EPSP, as shown in Fig. 5B. A large increase in the EPSP is seen when the membrane potential is moved in the hyperpolarizing direction. This reversal is then necessary and sufficient evidence to indicate that a synaptic junction is chemical in nature. Furthermore, the fact that different parts of the EPSP (the peak and falling phase) reverse at different levels of depolarization (see biphasic reversal at 22.1 nA in Fig. 5B), indicates that the synapse is distributed (Llinás and Nicholson, 1976) (i.e., occurring at multiple sites having different distances from the site of recording). Since a current point source, a microelectrode, was utilized to change the membrane potential, the potential change along the dendrite is maximum at the site of impalement and decreases with distance. Because the synapses closest to the site of recording generate most of the rising phase of the recorded EPSP, this component is the first to reverse. Those synapses located at a distance generate the slower components (owing to the cable properties of the dentrites) and are less affected by the current injection. Recordings similar to those obtained *in vitro* can also be shown *in vivo*.

A second demonstration of the chemical nature of the synapse is afforded by removing the extracellular calcium from the bathing solution to block the climbing fiber EPSP and the associated dendritic spikes. Although there has been much debate regarding the role of this input in cerebellar function, it does seem clear that its activation is capable of producing a burst of spikes in many Purkinje cells simultaneously and thus generating a sizeable IPSP on cerebellar nuclear cells (c.f. ten Bruggencate, Teichman, and Weller, 1972).

Mossy Fiber-Parallel Fiber Synapse. This synaptic input can best be studied *in vivo* since in the'*in vitro* preparation parallel fibers are for the most part severed as the slices are made at right angles to the parallel fibers themselves. Parallel fibers generate a well-graded synaptic input as opposed to the all-or-none depolar-

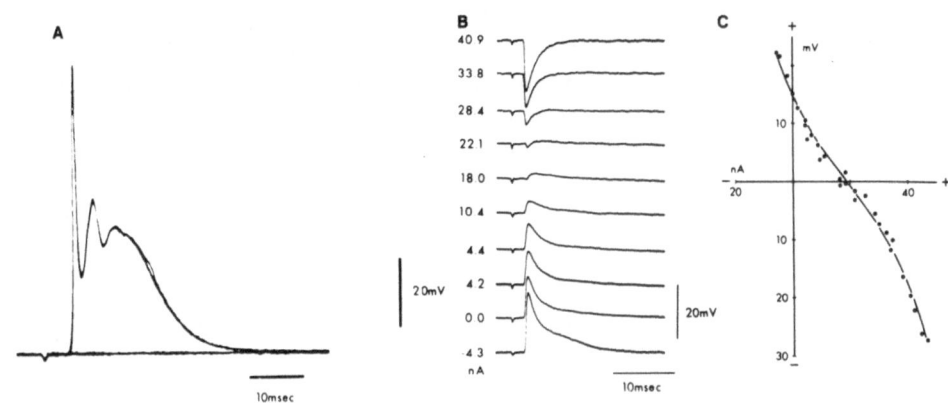

Fig. 5. Climbing fiber activation of mammalian Purkinje cell *in vitro*. (A) All-or-none Purkinje cell activation following white matter stimulation. (B) Reversal of the climbing fiber-evoked synaptic potential. The properties of the reversal are particularly clear at 18, 22, and 28 nA where the biphasic nature of the reversal is clearly observed. (C) Voltage–current relationship for the synaptic potential. Modified from Llinás and Sugimori (1980a).

ization seen for the climbing fiber. This graded depolarization generally evokes simple, sodium-dependent spikes and only when the input is massive is the complex spike seen. As stated above, the parallel fiber–Purkinje cell input, unlike the climbing fiber synapse, may be as large as 80,000 synapses (Palkovits *et al.*, 1971b) making the Purkinje cell the neuron with the greatest number of afferents. Because each parallel fiber establishes maximally two synapses with each Purkinje cell, the system may be said to be organized for both maximum divergence and maximum convergence. That is to say, a single Purkinje cell receives 80,000 inputs probably from as many as 80,000 parallel fibers (maximum convergence) while each parallel fiber touches as many Purkinje cells as possible (maximum divergence). However, it must be remembered that as many as 400,000 parallel fibers cross the dendritic tree of a given Purkinje cell so that less than one in four parallel fibers contacts a given Purkinje cell in its path.

Inputs from a climbing fiber and from the parallel fibers represent the two types of excitatory afferents terminating on a Purkinje cell. The Purkinje cell, on the other hand, receives three inhibitory systems: one subserved by the basket cell, the second by the stellate cell, and the third by the catecholamine system arising from locus coeruleus (cf. Bloom, Hoffer, and Siggins, 1971). Activation of the basket cells generates a graded inhibition at each side of the activated bundle of parallel fibers (Andersen, Eccles, and Voorhoeve, 1964; Eccles *et al.*, 1966b) (Fig. 6). This inhibition is produced via direct contacts on Purkinje cell dendrites and soma. Even

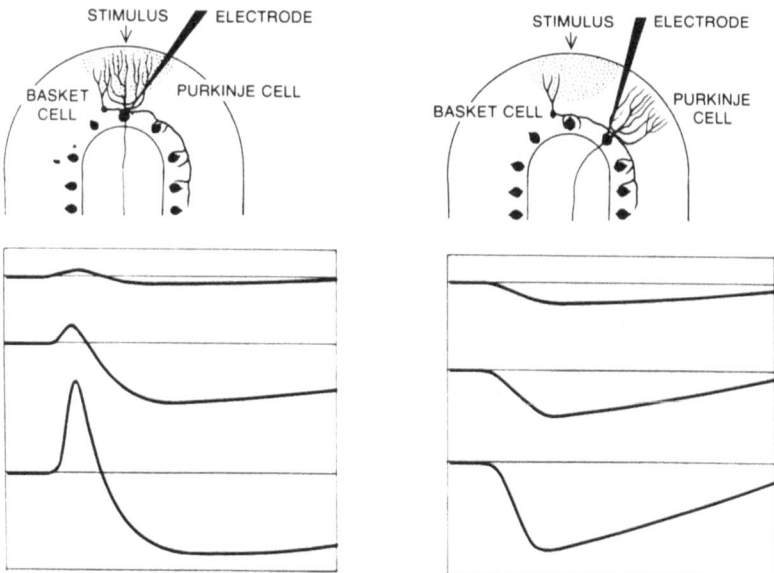

Fig. 6. Synaptic depolarization and hyperpolarization of Purkinje cells following parallel fiber stimulation. When the response of a cell directly under the stimulated area is recorded *(left)*, a brief EPSP is observed, followed by a long IPSP. The EPSP results from the direct depolarization of the Purkinje cell by parallel fibers, the inhibition from the action of basket cells and stellate cells. The magnitude of the response varies with the intensity of the stimulation. When the response of a laterally located Purkinje cell is monitored *(right)*, only the inhibition is observed since only basket-cell axons reach Purkinje cells. Modified from Llinás (1975).

though basket cell terminals cover not only the somatic but also the axon hillock of the Purkinje cells, only a few inhibitory synapses have been observed at the axon hillock level; however, a rather impressive morphological structure known as the *pinso terminale* may be found at this level (Ramón y Cajal, 1888). As many as 20 to 30 basket cells are believed to wrap their axon terminals around each Purkinje cell soma, forming a basketlike meshwork resembling that on a Chianti bottle (Hámori and Szentágothai, 1965). The basket cell IPSP is generated by a membrane conductance increase to chloride most probably by the release of gamma-aminobutyric acid. The second inhibitory system is that represented by the stellate cells that synapse mainly on Purkinje cell dendrites. Although the inhibitory transmitter of this second system is not finally determined, it may be taurine (Frederickson, Neuss, Morzorati, and McBride, 1978). The third inhibitory system is that of the locus coeruleus; its catecholamine-mediated inhibition generates a large hyperpolarization that seems to be related to the activation of an electrogenic sodium pump (Bloom *et al.,* 1971).

Granule Cell System. The most common neuronal elements in the granular layer are the granule and Golgi cells and the mossy fibers. The granule cells represent the most numerous group of cells in the central nervous system, probably numbering on the order of 10^{10} or 10^{11} in the adult human being (Braitenberg and Atwood, 1958). Their axons project to the molecular layer where they branch in a T-shape and run side by side to form the parallel fibers (which are parallel to each other in all planes). The embryological organization of the granule cell layer is such that the axons of the most superficial granule cells (which are generated late in embryogenesis) comprise the most superficial layer of parallel fibers whereas axons of deep granule cells are found at deeper levels of the molecular layer (Mugnaini, 1969). Granule cells receive excitatory input from the mossy fibers, which are the mode of termination of most afferents to the cerebellar cortex. The anatomy of the mossy fiber–granule cell synapse is shown in Fig. 2. As indicated in that diagram, the mossy fibers terminate in the form of large *en passant* expansions that contact many granule cell dendrites. This excitatory synapse, as originally demonstrated by Eccles *et al.* (1966a), is organized so that an action potential in a single fiber probably releases transmitter at all its rosettes. The rosette is defined as a presynaptic mossy fiber enlargement and the granule cell dendrites that it contacts. Each mossy fiber generates, on the average, 16 rosettes and, given that each rosette makes contact with 28 granule cells, activation of a single mossy fiber would probably generate activity in 448 different granule cells (Palkovits, Magyar, and Szentágothai, 1971a, 1972). However, physiological measurements from the granular and the Purkinje cell layers suggest that for a granule cell to fire, probably more than one synaptic input to one dendrite must be activated. Although the evidence here is very indirect, and other variables must be taken into account (for instance, the inhibitory effect of Golgi cells onto particular groups of granule cells), it is nevertheless a fact that a particular granule cell can respond with a variable number of action potentials. This suggests a definite degree of spatial and temporal summation, which can only occur if the dendritic input is large enough to produce a graded summation (Eccles *et al.,* 1966b). It does appear that the granular layer represents a summing center for a large number of action potentials arriving through the mossy fiber afferent system. Further, considering that granule cells have a characteristic background activity, the Purkinje cells themselves must receive a continuous synaptic barrage from the parallel fibers; thus explaining the well-known background activity of the Purkinje cells.

The third element in the granuler layer is the Golgi cell. As shown in Fig. 1, the Golgi cell is a large neuron with dendrites ramifying in the granular layer and commonly reaching the molecular layer; some Golgi cells, however, are confined to the granular layer (Fox *et al.*, 1964). As opposed to the Purkinje, basket, and stellate cells, the Golgi cell dendrites distribute in space to form a cylinder rather than the planar distribution seen in the other cells. Their axons cover a certain area of the granule cell layer, demarcating a second cylinder (Palkovits *et al.*, 1972). This axon is inhibitory onto the granule cells (Eccles *et al.*, 1966c) generating a graded, rather powerful, inhibition either by Golgi cell activation via the parallel fiber–Golgi cell dendritic system or by activation of the direct mossy fiber–Golgi cell input.

Integrative Properties of Purkinje Cells. From the above description of the electrophysiological properties of neurons in the cerebellar cortex, it is apparent that Purkinje cell activity will be regulated, basically, by modulation of the background activity produced by the mossy fiber–granule, cell-parallel fiber system. Thus, activation of specific inputs modulate in a continuous manner the firing frequency of the Purkinje cells. On the other hand, climbing fiber input generates burst activity in the Purkinje cell axon at an intraburst frequency of 300 to 800 impulses per second. This implies that, due to their distribution, the climbing fibers serve to produce a space-clamp type of depolization of the dendrites and thus, by diminishing internal longitudinal current flow, to allow the Purkinje cells to fire at a high frequency.

Overall Purkinje cell activity is also modulated by somatic and dendritic inhibition. Simple spikes, however, are generated by either electrotonic or by local dendritic calcium responses, which are themselves sculptured by dendritic inhibition via stellate and basket cell terminals. In addition to the postsynaptic events mentioned above, the mossy fiber–granule cell input is deeply influenced by the inhibition generated by the Golgi cells. Although in most experimental physiological descriptions it is emphasized that inhibition produces a deep silencing of Purkinje cell and granule cell activity, in the normally functioning cerebellum all these inhibitory systems probably modulate and restrict the spatial distribution of activity in the cortex rather than prevent the actual firing of cells.

CEREBELLAR NUCLEI

The cerebellar nuclei are central neuronal ensembles formed by a collection of neurons grouped into three main nuclei on each side of the midline: the medial or fastigial, the lateral or dentate, and an intermediate nucleus known as the interpositus with its two components, anterior and posterior. The description of the different inputs to these nuclei will be presented in detail below. Suffice it to say here that these nuclei receive peripheral input via collaterals of both mossy and climbing fibers and send axons to the rest of the nervous system via the peduncles and to the cerebellar cortex through ascending collaterals that terminate as mossy fibers. This topic will be reviewed below. In addition, the cerebellar and Deiters'

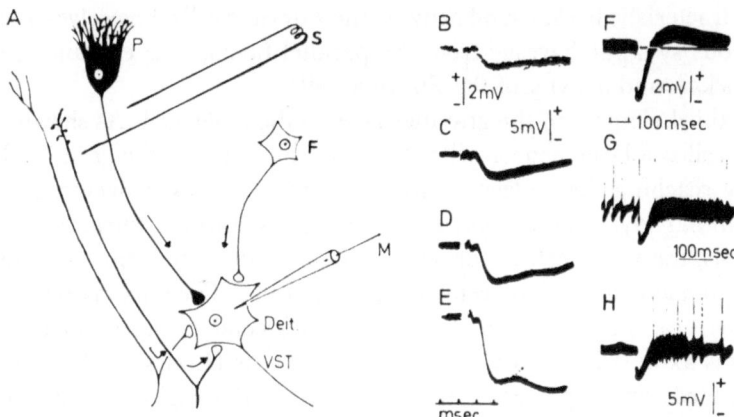

Fig. 7. Diagrammatic illustration of the monosynaptic connections between cerebellar cortex and Deiters' neurons. In (A) Deiters' nucleus (Deit), which generates the vestibulospinal tract (VST), receives excitatory inputs from both mossy and climbing fiber collaterals; in addition, it receives excitatory input from the fastigial nucleus (F). Purkinje cells produce direct inhibition on Deiters' neurons. (B–E) IPSP in Deiters' neuron, recorded with microelectrode (M), produced by stimulation of the ipsilateral anterior lobe of the cerebellar cortex (S). In (B–E) stimulation was increased from 1.9 to 30 V to the vermal cortex at lobule IV. Dotted line in (E) indicates the time course of the potential change if it were similar to that in (D). (F) IPSP shown at slower sweep speed following activation of lobule III. (G) Suppression and rebound facilitation of the spontaneous discharge induced by stimulation of lobule III. (H) An illustration of this effect in the absence of spontaneous firing. Modified from Ito and Yoshida.

nuclei are the main target for the Purkinje cell axons, which are inhibitory (Ito and Yoshida, 1966). Given their connectivity, these nuclei integrate the input to the cerebellar cortex (the climbing fibers and the mossy fibers) with the inhibitory output produced by the activity of the Purkinje cells (Fig. 7). The organization of this system seems to be such that the anterior vermis projects and gives inhibition onto the anterior fastigial nucleus while the posterior vermis projects posteriorly onto the fastigial nucleus. The intermediate cortex (between the vermis and the lateral cerebellar hemispheres) projects basically onto the interpositus nuclei while the hemispheres project for the most part onto the dentate nuclei. Details of the projection of Purkinje cells will be given later in the section on afferent and efferent systems.

INFERIOR OLIVE

Because the inferior olive is a central nucleus in the economy of the cerebellum, and because it is the place of origin of the climbing fiber system (Eccles *et al.*, 1966a; Szentágothai and Rajkovits, 1959), a few words will be devoted to the organization of this nucleus. The nucleus lies in the ventromedial part of the bulbar region flanked ventrally by the pyramids and laterally, through most of its course, by the perforating axons of the hypoglossal nerves. Axons from these neurons cross the midline and project via the contralateral inferior peduncle to the contralateral cerebellum. These axons give collaterals to all cerebellar nuclei and terminate, as stated above, as climbing fiber afferents to Purkinje cells. The neurons in the inferior olive are small and the dendritic tree of most of these cells has a spherical symmetry, with the cell's nucleus located in the center of the sphere formed by the radiating dendrites. This arrangement is only modified at the periphery of the olive, where the dendrites simulate hemispheres as they project toward the olive, leaving their soma near the periphery (Ramón y Cajal, 1904). Electrophysiologically, the inferior olivary cells have two interesting properties. First, their action potentials are quite complicated (Crill, 1970) and, as in the case of the Purkinje cell, have two basic components: an early sodium spike followed by prolonged dendritic calcium action potentials (Llinás and Yarom, 1981a). (Fig. 8). As in the case

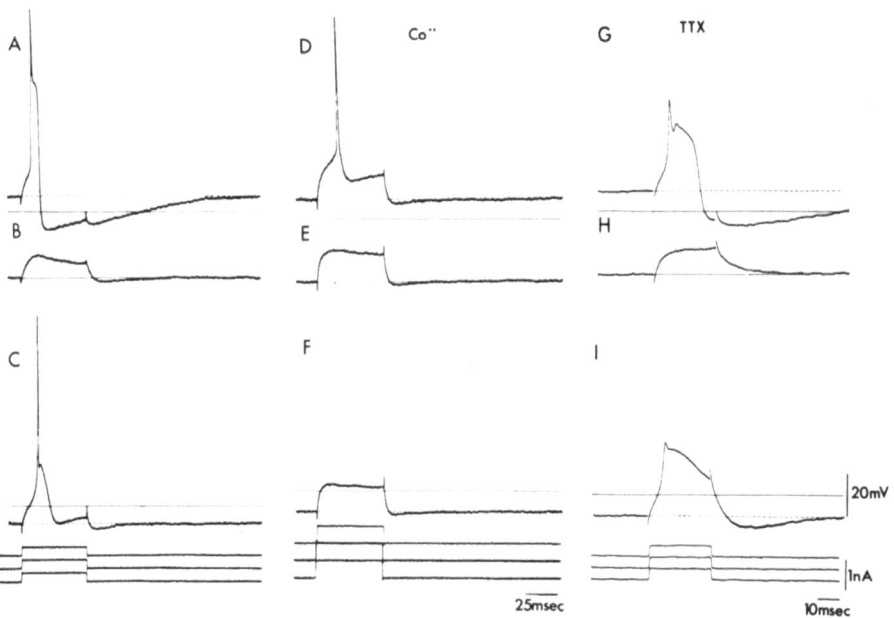

Fig. 8. Electrical excitability of mammalian inferior olivary cells tested *in vitro*. In (A, B, and C) normal electroresponsiveness; (D, E, and F) the electrophysiology following calcium blockage by the extracellular cobalt; (G, H, and I) electrophysiological properties after blockage of sodium conductance by extracellular TTX, indicating two different sets of calcium-dependent action potentials. In (B) a subthreshold current pulse is given at rest membrane potential. In (A) the subthreshold stimulus superimposed on a small dc depolarization (dotted line, with respect to solid line) generates a fast-action potential followed by an after-depolarization and a prolonged after-hyperpolarization. In (C) a hyperpolarization about 8 mV from rest (dotted line) also produces an increase in excitability as seen by the action potential generated by the otherwise subthreshold stimulus. Records (A), (B), and (C) have been separated for clarity. (D), (E), and (F) show a similar sequence as in (A–C) but following blockage of calcium conductance by extracellular cobalt. Notice that in (D) the action potential lacks the after-depolarization and the prolonged after-hyperpolarization seen in (A). In (F) the subthreshold stimulus riding on the hyperpolarization is now incapable of generating an action potential. In (G), (H), and (I), the sodium spike has been blocked by TTX. Again, a subthreshold stimulus (H) can generate an action potential by either a depolarizing (G) or hyperpolarizing (I) membrane potential change. Although the fast action potentials seen in (A), (C), and (D) are generated by sodium-dependent conductances, the after-depolarizations in (A) and (C) and in (G) and (I) are generated by calcium conductances. Those of (A) and (G) are generated by high-threshold, calcium-dependent action potentials from dendrites. Those in (C) and (I) are generated by inactivating calcium conductances at the somatic level. For further details, see text. Llinás and Yarom (1981a).

of the Purkinje cells, the calcium action potentials can be blocked by calcium conductance blockers, such as cadmium, cobalt, manganese, and D600 (Yarom and Llinás, 1979). The activity of these calcium spikes is such that it generates a very prolonged after-hyperpolarization, which gives the inferior olive its basic low-firing frequency. This after-hyperpolarization is produced by a calcium-dependent conductance change since blockage of calcium conductance produces a disappearance of the prolonged after-hyperpolarization and an increase in the firing frequency of the cell (Fig. 8D). *In vitro* studies have demonstrated that the application of TTX does not change the basic properties of these neurons as far as the activation of calcium spikes or the basic low frequency of firing is concerned. This suggests that,

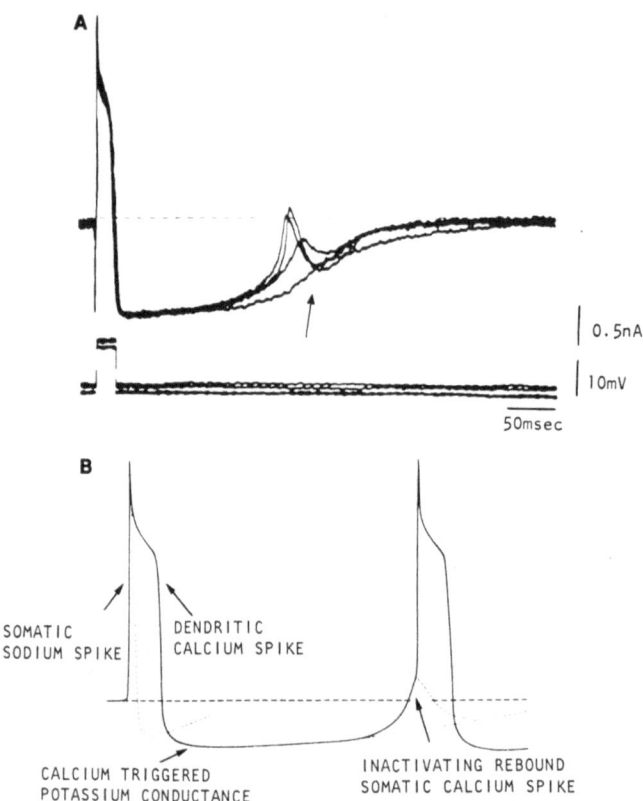

Fig. 9. Rhythmic firing of inferior olivary neurons. (A) Rebound calcium spike in the presence of TTX. Direct stimulation of an inferior olivary neuron produces a dendritic calcium spike, followed by an after-hyperpolarization and a rebound spike (arrow). Small changes in d.c. hyperpolarizations (note current record) facilitates the rebound spike, which becomes larger and moves to the left. (B) Diagram of the sequence of events that generates inferior olivary cell rebound leading to oscillation. Antidromic or direct stimulation generates a somatic sodium spike having a fast rise and about 1 msec duration (broken line). At the appropriate membrane potential, this action potential generates a dendritic calcium spike, which produces a plateau after-depolarization followed by a sizeable potassium conductance. The potassium conductance generates the prolonged after-hyperpolarization. This membrane hyperpolarization removes inactivation from the somatic calcium conductance, which can then produce a rebound depolarization and can start the sequence once again. Llinás and Yarom (1981b).

as in the case of the Purkinje cells, the low-frequency activation is a property of the calcium conductance system (Fig. 8G).

Furthermore, inferior olivary neurons demonstrate a voltage-dependent calcium conductance at the somatic level. This conductance, as opposed to the dendritic calcium conductance, inactivates at resting membrane potential level (Figs. 8A to C, and 8G to I). This novel conductance (Llinás and Yarom, 1980; Yarom and Llinás, 1979) allows the inferior olivary cells to behave as a single cell oscillator (Fig. 9). Thus, in normal circumstances, an action potential consists of a prolonged spike followed by a large hyperpolarization as a rebound response, often big enough to generate a second train of events (Fig. 9A). A schematic diagram of the different ionic components in the firing of inferior olivary cells is seen in Fig. 9B. The morphology of the inferior olive is such that synaptic input to these cells is basically restricted to the glomerulus. These glomeruli are formed by a central core of inferior olivary dendrites surrounded by a large number of presynaptic terminals (Sotelo, Llinás, and Baker 1974). A second interesting organizational property of this nucleus is the fact that their neurons are electrotonically coupled through gap junctions in the center of this glomerulus (King, Andrezik, Falls, and Martin, 1976; Sotelo et al., 1974). It is therefore not surprising that drugs that seem to influence the electrical activity of the olive, such as harmaline or locally applied drugs (de Montigny and Lamarre, 1973; Llinás and Volkind, 1973), can produce synchronous activity of the climbing fiber system, implying that the organization of this system is such that synchronicity may be of importance in the generation of coordination via the climbing fiber system. An interesting possibility here, of course, is that electrotonic coupling between these neurons may be modulated by chemical synapses (Sotelo et al., 1974). By the strategic localization of this input on the dendrites of the inferior olivary cells, the activity of a chemical synapse may serve to shunt (by reducing dendritic input resistance) the coupling of neurons, much as has been shown by Spira and Bennett (1972) in Navanax.

THE AFFERENT AND EFFERENT SYSTEMS

The majority of anatomical and physiological research studies on cerebellar afferents and efferents can be characterized as a search for "the organizational principles" of cerebellar localization. Three general categories of localization are commonly distinguished: topographic, somatotopic, and functional. Topographic localization refers to a discrete mapping of one neuronal population onto a second, whereas somatotopic localization refers to a discrete mapping of the body onto a neuronal population. Functional localization refers to the spatial segregation, within a neuronal population, of defined functional properties. The functional properties of the cerebellum as a whole have usually been defined in terms that are descriptive of the different motor abnormalities and deficits resulting from lesions of its various parts. These descriptive terms run the gamut of motor activity from posture and simple reflexes to skilled volitional movements. However, the finding of correlations of regional cerebellar damage with specific motor abnormalities speaks more to the question of the ability of the CNS to cope with cerebellar lesions than to the question of how the cerebellum functions. Furthermore, although anatomical and many physiological studies have established a rather impressive body of knowledge about the connectivity patterns of cerebellar afferents and efferents, most of this knowledge has not yet proven of great use in understanding the role of the cerebellum in the organization of motor function, even despite the very impressive and regularly organized cortical and corticonuclear structure. In comparison to the number of studies dealing with cerebellar connectivity, few studies have been aimed at defining the functional significance of afferent and efferent information and at analyzing the signal processing in the cerebellar cortex and nuclei. As a consequence, we still have but limited knowledge of how neuronal operations of the cerebellum actually effect the coordination of motor behavior.

In attempting a description of cerebellar connectivity, we rapidly recognized that a coverage of afferent and efferent connections of the cerebellar cortex and nuclei that would do justice to presently available knowledge would be voluminous. The recent advent of neuroanatomical methods based on incorporation and axonal transport of marker substances has resulted in a dramatic increase in the available information on cerebellar connectivity. Many previously held beliefs about the "wiring diagram" of cerebellar afferents and efferents have been revised, and some long-discarded anatomical observations have been retrieved and restored to good standing. Even though a consensus has not been achieved on many points, the overall tendency is for a closer and more realistic agreement between hodology and electroanatomy. Finally, we wish to warn that our treatment of the afferent and efferent connections of the cerebellum is admittedly biased, but we do hope, nevertheless, to convey the essence of the organizational pattern of cerebellar connectivity. The reader interested in the details of particular pathways and the differences existing among different species, as well as the disagreements among investigators, is, as usual, well advised to consult the original papers. Before considering the various afferent and efferent systems specifically, some general comments on cerebellar localization may help to put the later descriptions into perspective.

GENERAL ATTEMPTS TO DEFINE GLOBAL PLANS OF LOCALIZATION IN THE CEREBELLAR CORTEX

The cerebellar cortex has basically the same neuronal organization throughout, so attempts to understand the cerebellum by decomposition into parts (localization) must be based on other than a cytoarchitectonic approach. Subdivision of the cerebellum by various combinations of transversely and sagittally oriented components has been repeatedly proposed. These organizational schemes have been based on (1) phylogenetic and ontogenic criteria, (2) the spatial distribution of afferents and/or efferents, (3) myeloarchitectonics, and (4) somatotopy (e.g., Bolk, 1906; Brodal, 1940; Chambers and Sprague, 1955a,b; Ingvar, 1918, 1923, 1928; Jansen and Brodal, 1940, 1942; Larsell, 1934, 1936, 1937; Voogd, 1964, 1969; see also reviews by Dow, 1942, 1961; Jansen and Brodal, 1954; Larsell and Jansen, 1972; Voogd, 1967). Let it be said at the outset that a simple and universally applicable organizational scheme has not as yet emerged.

TRANSVERSE LOBULES. Gross inspection of the cerebellum of mammals directly suggests the possibility that the transversely oriented fissures could form the basis of an organizational principle. Many early morphologists were attracted by this possibility, which was pursued most completely by Larsell (1967, 1970; and Larsell and Jansen, 1972) in an extensive series of phylogenetic and ontogenetic studies encompassing many vertebrate species. Since the posterolateral fissure is phylogenetically and ontogenetically the first to appear, Larsell (1934, 1937) argued that there are two fundamental divisions of the cerebellum: (a) the flocculonodular lobe, posterior to the posterolateral fissure and (b) the remainder of the cerebellum, the corpus cerebelli (Fig. 10). In addition, Larsell and others have emphasized the importance of the primary fissure, which divides the corpus cerebelli into two lobes, the anterior lobe and the posterior lobe. As a result of Larsell's work (e.g., Larsell, 1948, 1952, 1953), it is currently accepted that the cerebellar cortex is divided transversely into ten principal lobules, each lobule in turn being subdivided into folia. Of these lobules, five are in the anterior lobe and four are in the posterior lobe (Fig. 10). The tenth lobule, consisting of the flocculus and nodulus, ia simultaneously a lobule and a lobe. At one time, it was believed that the cerebellar lobules represented a natural mode of subdivision which also defined the boundaries of projection areas of individual afferent systems. and, by extension, also defined functional subdivisions. This concept, however, has not been well supported by more contemporary investigations (e.g., Chambers and Sprague, 1955a,b; Voogd, 1964, 1967). At the present time, transverse division of the cere-

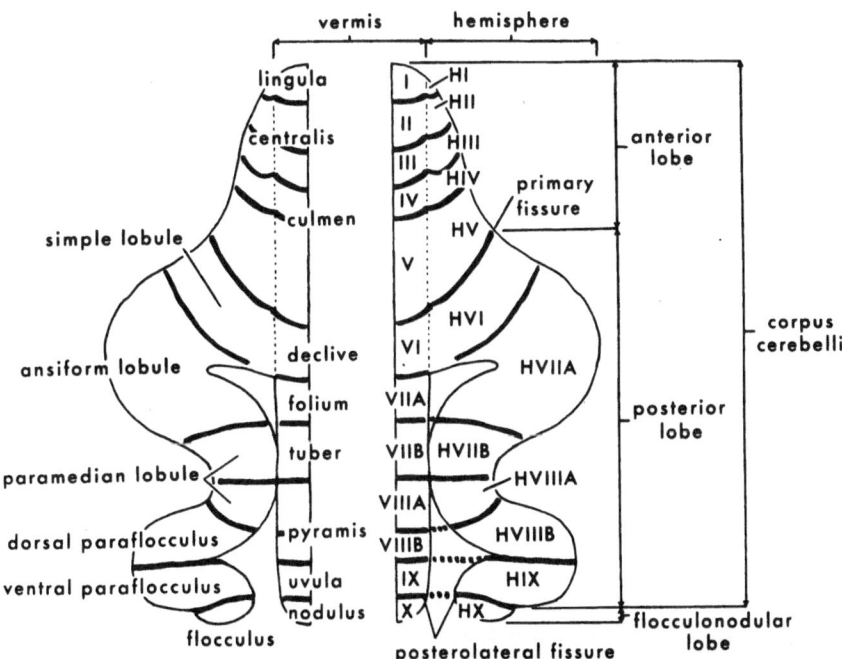

Fig. 10. Nomenclatures used in referring to the lobes and lobules of the mammalian cerebellum. *(Right)* Roman numeral terminology of Larsell; *(left)* terminology derived from gross morphological appearance. For the majority of the posterior lobe, the scheme of Larsell indicates a closer relation between the hemisphere and the vermis than does the system of common names. Division of the hemisphere into intermediate and lateral zones is not indicated. In referring to certain parts of the human cerebellum, other names are frequently used. Modified from Voogd (1964).

bellum into ten lobules finds its greatest usefulness as a broadly based, convenient naming scheme.

LONGITUDINAL ZONES. Just as gross inspection of the cerebellum raises the possibility of transverse division into lobules, so too does it suggest division of the cerebellum into two parasagittal or longitudinal zones on each side of the midline, especially in higher mammals. Within the posterior and flocculonodular lobes, two longitudinal zones can be so distinguished; a medial or vermal zone and a hemispherical zone lateral to the vermis (Fig. 10). This divisional scheme has been extended to include also the anterior lobe (Riley, 1929). Approaches other than that of gross morphology have shown that, in fact, more than two longitudinal zones can be discriminated. For example, studies of the histogenesis of the human cerebellum (Hayashi, 1924; Jakob 1928a,b; Langelaan, 1919) raised the possibility that three longitudinal zones are present on each side of the midline. In the tripartite scheme, the hemispherical zone of earlier investigators is subdivided into two parts. The part adjacent to the vermis is called the *pars intermedia* or intermediate zone, whereas the remaining lateral part of the hemisphere is called the *lateral zone*. Division into three longitudinal zones received support from the anatomical studies of Jansen and Brodal (1940, 1942; see also Hohman, 1929) who demonstrated that on the basis of the projection pattern of Purkinje cells to the deep cerebellar nuclei one could argue for the existence of three longitudinal zones: a medial zone projecting to the fastigial (and vestibular) nucleus, an intermediate zone projecting to the (undivided) interpositus nucleus, and a lateral zone projecting to the lateral (dentate) nucleus. In this scheme based on the corticonuclear projection pattern, the preciseness of the boundary between the intermediate and lateral zone can be only as precise as that between the interpositus and lateral nuclei, which in most mammals is not particularly well demarcated. The behavioral studies of Chambers and Sprague (1955a,b) provided physiological validation of division of the cerebellum into three longitudinal zones; a different constellation of motor deficits resulted from lesions placed in each of the three anatomically defined zones. As we shall see, however, division of the cerebellar cortex into longitudinal zones has not stopped at three.

The increase in the number of recognized longitudinal zones began with the myeloarchitectonic studies of Voogd (1964), which revealed that the lobular white matter is divided, in the mediolateral direction, into a series of compartments by repeating local differences in axon diameters. Corresponding compartments can, in general, be identified in successive lobules, and collectively the compartments establish a mediolaterally oriented series of rostrocaudally (longitudinally) directed zones. The more lateral zones are not strictly parasagittal but rather run perpendicular to the long axis of the lobules. Not all the zones run the entire rostrocaudal length of the cerebellum, but typically 6 to 7 zones are present on each side of the midline (Voogd, 1967, 1969). Each zone has been associated with a specific target site for the Purkinje cells located within that zone. Thus, each longitudinal zone defined on the basis of myeloarchitectonics is in effect a distinct corticonuclear projection zone. The corticonuclear zonal pattern appears to be equivalent also to the zonal pattern established by the projection of climbing fibers from the inferior olive, as originally pointed out by Brodal (1940; see also Jansen and Brodal, 1940) and later revealed in detail by Oscarsson and collaborators and by Voogd and collaborators. More will be said of this parallelism later.

EVOLUTIONARY AND AFFERENT DIVISIONS. Attempts to subdivide the cerebellum on evolutionary grounds are worth mentioning, if only for purely historical reasons. From its inception (Comolli, 1910; Edinger, 1910), the scheme of dividing the mammalian cerebellum into phylogenetically old (paleocerebellar) and new (neocerebellar) parts has had one foot in phylogeny and the other in the distribution pattern of afferents from different sources. Initially, the vermis plus the flocculus of the mammalian cerebellum was believed to have fiber connections identical to the entire cerebellum of birds, and therefore these parts of the mammalian cerebellum were said to comprise a paleocerebellum. The mammalian cerebellar hemisphere (intermediate and lateral zones) was considered to be a phylogenetically recent acquisition, connected principally with the cerebral cortex via the pontine nuclei. Thus, this part of the cerebellum was termed the neocerebellum. Subsequently, Ingvar (1918, 1923) proposed division of the cerebellum into three parts on the basis of the distribution pattern of vestibular, spinal and pontine afferents, considering that these three afferent sources were the most prominent. Although the presence of some overlap was recognized even then, the emphasis was on the creation of the concept of a vestibular and spinal cerebellum, on the one hand, and a corticoponto-cerebellum on the other. The pontocerebellum of Ingvar differed from the previously described neocerebellum on a number of points but most notably by inclusion of the middle part of the vermis (Lobule VII of Larsell) and by exclusion of the paraflocculus, which was allocated to the spinocerebellum. These changes are but an example of the many successive and ultimately fruitless revisions in the definitions of paleo- and neocerebellum necessitated by an expanding body of knowledge of the anatomical projection patterns (see for example, Brodal and Jansen, 1964).

To indicate the primacy of the relation between the cerebellum and the vestibular system and to establish a parallel between division on phylogenetic grounds and division on the basis of the afferent projection patterns, Larsell (1937) introduced the additional term, archicerebellum. This term designated those cerebellar regions receiving a direct vestibular nerve input. On the basis of anatomical evidence available at that time, distinction of these three divisions, archi-, paleo-, and neocerebellum and their approximate equivalents, vestibular, spino-, and pontocerebellum was reasonably justified. Gradual refinement of our knowledge of cerebellar afferent connectivity has shown, however, extensive overlap of spinal and pontine projections to many parts of the cerebellum formally considered to be the predominant preserve of one or the other of these afferent systems. Furthermore, the flocculonodular lobe, previously the "backbone" of the vestibular or archicerebellum, is now known to receive a wide variety of afferents (e.g., visual, neck, oculomotor, etc.) in addition to those from the vestibular system (Hess and Simpson, 1978; Maekawa and Simpson, 1973; Miles and Fuller, 1975; Wilson, Maeda, and Frank, 1975; Wilson, Maeda, Frank, and Shimazu, 1976). While various revisions in the definition of paleo- and neocerebellum have been attempted to try to keep this concept viable and in harmony with changing knowledge of afferent connectivity, ultimately this scheme has proven not to be very useful. The opinion voiced by Brodal and Jansen in 1954 that "it seems expedient to dispose of the terms paleo- and neocerebellum, as they are apt to confuse rather than clarify our conception of the cerebellum," is even clearer today than formerly and should be extended to include the term "archicerebellum." The roughly equivalent and equally dubious terms "vestibulo-," "spino-," and "pontocerebellum" also no longer indicate, in any deep sense, a basic organizational principle of the cerebellum. Regional differences do, of course, exist but they are no longer adequately or usefully indicated by unidimensional labels.

SOMATOTOPY. One of the most perplexing problems associated with understanding the operation of the cerebellum has been how to analyze a structure whose uniform organization, both from species to species and from one region to another, suggests a uniform function. One approach toward solving this problem was thought to be the concept of somatotopic localization, and the search for somatotopic localization in the cerebellum has been a central theme of many investigations since Bolk (1906) proposed that specific areas of the cerebellum are related to specific muscle groups of the body. This approach could be viewed as equivalent to a kind of functional localization, not in the sense used in the beginning of this section, but rather in the sense that each part of the cerebellum exerts its action on a discrete region of the body. In such a scheme, somatotopic localization should be valid for both afferents and efferents. The question of afferent somatotopy may be answered unambiguously in the negative (see below) but that of efferent somatotopy is more difficult to resolve. Although Purkinje cells and cerebellur nuclear cells do not show a remarkable somatotopy (see below), their ultimate influence, funneled through successive synaptic chanels, may ultimately find a specific somatotopic expression.

The specific afferent somatotopic map presented by Bolk has long been known to be incorrect, but other maps, dealing principally with the vermal and intermediate zones, have been offered in its stead (Adrian, 1943; Hampson, 1949; Hampson, Harrison, and Woolsey, 1952; Snider, 1952; Snider and Eldred, 1952; Snider and Stowell, 1944). These maps and others, which depict forelimb, hind limb, head areas, etc., are based largely on the distribution of activity evoked by afferent systems in anesthetized or surgically reduced preparations. The somatotopy seen in each of the maps individually is not preserved when they are superimposed (see Bloedel, 1973). Furthermore, in unanesthetized preparations, where a larger number of afferent systems are operational, the case for afferent somatotopy becomes nonexistent (e.g., Combs, 1954; Dow, 1939; Leicht and Schmidt, 1977). The point is simply that some afferent pathways are not somatotopically organized while others may be. Furthermore, those that may be somatotopically organized do not all yield the same mapping of the body onto the cerebellum.

Results of ablation and stimulation studies of the cerebellum have been presented as evidence for efferent somatotopic organization (e.g., Chambers and Sprague, 1955a,b; Pompeiano, 1967). Certain regions of the cerebellum, particularly within the intermediate zone, have been found to show a preponderance or maximal representation of influence on one part of the body or another. However, such studies are, by design, not really investigating the somatotopy of cerebellar efferents exclusively (Purkinje cells and cerebellar and vestibular nuclei cells), but are, in effect, investigating the organization of the entire motor system. Thus, the results of these studies reflect the organizational properties of the nervous systems, in general, and not the somatotopy of the cerebellar efferents, in particular. More-over, recent electrophysiological recordings from Purkinje cells and cerebellar nuclei cells indicate that somatotopy is a far less common finding than would be expected from previously proffered somatotopic organizational schemes. An exam-ple of the difficulties encountered in the search for efferent somatotopy is presented in a series of papers by Eccles and collaborators (Eccles, Faber, Murphy, Sabah, and Táboríková, 1971; Eccles, Nicoll, Schwarz, Táboríková, and Willey, 1975; Eccles, Provini, Strata, and Táboríková, 1968a,b; Eccles, Rantucci, Sabah, and Táboríková, 1974; Eccles, Sabah, and Táboríková, 1974a,b; Provini, Redman, and Strata, 1968), which treats the anterior lobe vermis, the fastigial nucleus, and the medial reticular formation. The initial observations from the vermis already sug-gested that the somatotopy was less than striking. Subsequent observations in the fastigial nucleus and medial reticular formation indicated an ever-decreasing emphasis on somatotopy, which was ultimately "sacrificed to integration." Com-parable investigation of the interpositus nucleus (Eccles, Rantucci, Rosén, Scheid, and Táboríková, 1974; Eccles, Rosén, Scheid, and Táboríková, 1974a,b) in which somatotopy would be expected to be more apparent as this nucleus is within the intermediate zone, did show greater somatotopical localization than had been observed in the fastigial nucleus. Even in the interpositus nucleus, however, well over half the neurons were without clear somatotopic relations. More recently, investigation of the dentate nucleus has produced many examples of cells receiving inputs converging from many parts of the body (Fig. 11; Bantli and Bloedel, 1977). Arguments against investing heavily in somatotopic localization as a means of understanding the function of the cerebellum have been presented previously (e.g.,

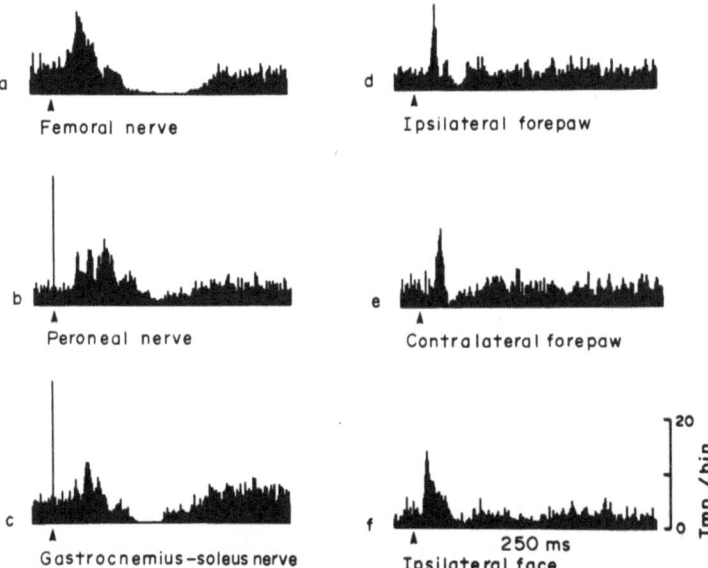

Fig. 11. Effects of electrically stimulating peripheral inputs from various parts of the body on the activity of a dentate (lateral) nucleus neuron in a monkey anesthetized with alpha-chloralose. All stimuli, even those from widely separated parts of the body, evoked an initial increase in the dis-charge rate of the neuron. The histograms were obtained from 64 consecutive responses to stimuli applied at the location indicated under each histogram. From Bantli and Bloedel (1977).

Bloedel, 1973; Dow and Moruzzi, 1958). More recent observations mentioned above provide additional evidence for arguing against somatotopy as a concept deeply underlying the organization of the functioning cerebellum. Indeed, somatotopy could be considered antithetical or of only limited importance to a neural structure subserving coordination of motor behavior.

The message that emerges from the above consideration of cerebellar localization is that one-dimensional labels such as "vestibulocerebellum" or "forelimb area" really are clearly wrong. What seems to be the case is that the afferents are basically distributed widely over the cerebellar cortex, but *such distribution is not uniform in density.* There is no doubt that particular areas have higher densities of given afferent terminations than others. Nevertheless, inhomogeneity of density, even with areas of high afferent density, does not mean exclusive territoriality; rather, it negates in principle the very concept of projection boundaries, which constituted the conceptual basis for the search for somatotopic mapping in the first place.

Mossy Fiber Afferents

In describing the cerebellar afferents, we have adopted, for the sake of simplicity, the traditional but arbitrary approach of grouping afferents according to their place of origin within the nervous system and considering, in turn, their cerebellar projection pattern. This plan emphasizes the divergence of cerebellar afferents; an alternative approach considering, in turn, separate parts of the cerebellum would emphasize the convergence of afferents. In either case, the great interconnectedness of the nervous system (its distributed and parallel organization) makes it imperative to realize that the notion of boundaries between the various afferent groups is a misconception. Coupling among the afferent groups occurs not only in the cerebellum but also in the nuclei that project to the cerebellum. Therefore, one should be very wary of simple one-dimensional labeling of afferent groups, since single terms such as "vestibular" or "spinal," which can have functional connotations, suggest a distinction that is more apparent than real. We shall now proceed to describe first the mossy fiber afferents and then the climbing fiber afferents. The aminergic projections to the cerebellum from the locus coeruleus and the raphe nuclei are not treated here; the relevant literature is reviewed by Bloom *et al.,* (1971), Shinnar, Maciewicz, and Shofer (1975), Eller and Chan-Palay (1976), and Freedman, Hoffer, Woodward, and Puro (1977).

Vestibular Afferents

The Vestibular Nerve. The Marchi degeneration study of Ingvar (1918) revealed that the distribution of vestibular nerve fibers to the cerebellum is not restricted to only one of the three cerebellar lobes (anterior, posterior, or flocculonodular) later distinguished by Larsell (1934, 1937). Ingvar found that primary vestibular afferents project not only to the ipsilateral flocculus and nodulus but also to the uvula, lingula, and the fastigial nucleus. Subsequent studies (Brodal and Høivik, 1964; Carpenter, Stein, and Peters, 1972; Dow, 1936, 1939) confirmed and added to the list of cerebellar regions receiving primary vestibular afferents, expanding it to include the ipsilateral ventral paraflocculus, the parvicellular part of the lateral cerebellar nucleus, and a portion of the dorsal paraflocculus. In addition, a few primary vestibular fibers were found to reach some of the corresponding contralateral cerebellar regions. Electrophysiological studies have confirmed the sometimes questioned projection of primary vestibular fibers to the fastigial nucleus (Furuya, Kawano, and Shimazu, 1975; Precht and Llinás, 1968) and to the contralateral flocculus (Shinoda and Yoshida, 1975). Most recently, both physiological (Precht, Volkind, and Blanks, 1977) and anatomical (Kotchabhakdi and Walberg, 1978a) evidence has been obtained for a primary vestibular afferent projection throughout the vermis, not only to the nodulus, uvula, and lingula. Moreover, the intensity of the projection to the uvula and nodulus is far greater than to other parts of the vermis or even to the flocculus (Korte and Mugnaini, 1979).

The Vestibular Nuclei. Prior to the advent of the horseradish peroxidase (HRP) technique for determining neuronal connectivity, the projections from the vestibular nuclear complex to the cerebellum were thought to be restricted to the flocculus, nodulus, uvula, lingula, and fastigial nucleus, and to arise from restricted parts of both the ipsilateral and contralateral medial and inferior vestibular nuclei and also the x group (Brodal and Torvik, 1957; Dow, 1936, 1939). [The x group, which is juxtaposed lateral to the inferior vestibular nucleus, receives no primary vestibular input but does receive an input from the spinal cord (Brodal and Pompeiano, 1957; Wilson *et al.,* 1976).] Recent anatomical (Frankfurter, Weber, and Harting, 1977; Gould and Graybiel, 1976; Kotchabhakdi and Walberg, 1978b; Shinnar, Maciewicz, and Shofer, 1975) and physiological (Precht *et al.,* 1977) studies have revealed a far more extensive set of connections from the vestibular nuclei to the cerebellum, on both the projecting and the receiving sides. The list of cerebellar areas receiving an input from the vestibular nuclei has been expanded to include the entire vermis and the interposed nuclei, whereas the list of vestibular nuclei areas projecting to the cerebellum has been expanded to include the superior and lateral vestibular nuclei and the y group. For the intricacies of these newly found projections, the original papers must be consulted because a consensus has not been reached on such matters as the laterality and the intensity of the projections.

The Perihypoglossal Nuclei. The perihypoglossal nuclei are considered here because they are spatially and functionally closely associated with the vestibular nuclei. Collectively, the three subdivisions of the perihypoglossal nuclei (the nucleus prepositus hypoglossi, the nucleus intercalatus of Staderini, and the nucleus of Roller) project bilaterally to all parts of the cerebellum except for the ansiform lobule, the paramedian lobule, and the lateral cerebellar nucleus (Alley, Baker, and Simpson, 1975; Brodal, 1952; Frankfurter *et al.,* 1977; Kotchabhakdi, Hoddevik, and Walberg, 1978; McCrea, Baker, and Delgado-Garcia, 1979; Ruggiero, Batton, Jayaraman, and Carpenter, 1977; Torvik and Brodal, 1954). In addition to receiving a projection from the fastigial nuclei (Walberg, 1961; cf. Batton, Jayaraman, Ruggiero, and Carpenter, 1977), the perihypoglossal nuclei, like the vestibular nuclei, also receive a direct Purkinje cell input from some cerebellar cortical areas, particularly the flocculonodular lobe, the uvula, and the paraflocculus (Angaut and Brodal, 1967; McCrea *et al.*, 1979). Although a topographical organization of the cerebellar projection from each of the three subdivisions of the perihypoglossal nuclei can be distinguished anatomically, the physiological significance of these patterns has not been determined. It is clear, however, that the prepositus subdivision is intimately involved with eye and head movements, for in addition to its cerebellar projection, it also projects to all of the extraocular motor nuclei (Baker and Berthoz, 1975; Baker, Berthoz, and Delgado-Garcia, 1977; Graybiel and Hartweig, 1974; Maciewicz, Eagen, Kaneko, and Highstein, 1977). Furthermore, physiological studies (Baker, Gresty, and Berthoz, 1976; Blanks, Volkind, Precht, and Baker, 1977; Gresty and Baker, 1976) have shown that prepositus neurons receive vestibular, visual, eye movement, and neck signals, indicating that they are connected with the control of gaze.

SPINAL AFFERENTS. Traditionally, four direct mossy fiber afferent systems are held to originate from neurons in the spinal cord, including the dorsal column nuclei (see reviews by Bloedel, 1973; Larsell and Jansen, 1972; Oscarsson, 1965, 1973). Two of these systems, the dorsal spinocerebellar tract (DSCT) and the ventral spinocerebellar tract (VSCT) are related to the lower trunk and hind limbs; the other two systems, the cuneocerebellar tract (CCT) and the rostral spinocerebellar tract (RSCT) are related to the upper trunk, forelimbs, and neck. Although other direct spinocerebellar pathways are present (e.g., Aoyama, Hongo, and Kudo, 1973; Cummings and Petras, 1977; Matsushita and Ikeda, 1975; Snyder, Faull, and Mehler, 1978; Wiksten, 1975), we shall concern ourselves here only with the four classical direct spinocerebellar tracts. It should be noted further that information from the spinal cord is also routed indirectly to the cerebellum via mossy fibers that arise from supraspinal structures, including the lateral reticular nucleus, the paramedian reticular nucleus, the vestibular nuclei, and some nuclei adjacent to them.

Dorsal Spinocerebellar Tract. The DSCT arises from cells of Clarke's column (nucleus dorsalis), which extends from spinal segments T_{1-2} to L_{3-4}. The axons of these cells ascend uncrossed in the spinal cord and reach the cerebellum via the inferior peduncle (restiform body). The DSCT terminates ipsilaterally in the rostral part of the anterior lobe (the intermediate zone and lateral part of the vermal zone), the caudal part of the paramedian lobule and parts of the pyramis and dorsal paraflocculus (Grant, 1962b; Oscarsson, 1965, 1973). Although some DSCT axons most likely send branches to both the anterior and the posterior lobes, the local termination pattern of DSCT axons is comparatively restricted.

Neurons of Clarke's column are tightly and selectively coupled to primary spinal afferents (principally from dorsal roots T_2 to S_1) and are capable of sending modality specific and spatially discriminative signals to the cerebellum. Modulation of DSCT neurons is strongly dependent on changes in the periphery, as succinctly demonstrated by Arshavsky, Berkinblit, Fukson, Gelfand, and Orlovsky (1972a) who showed that transection of the dorsal roots in the mesencephalic locomoting cat (see Chapter 11) abolished DSCT modulation even though the animal continued to locomote. In this respect the DSCT differs greatly from the VSCT which continued to modulate during locomotion when the hindlimbs were deafferented (Arshavsky *et al.*, 1972b).

Two components of the DSCT can be distinguished: a proprioceptive component and an exteroceptive component (Mann, 1973; Oscarsson, 1965, 1973). The proprioceptive component is comprised of neurons receiving monosynaptic excitation from either group Ia or group Ib muscle afferents, either from a single muscle or from a few synergists. Many of these neurons are also monosynaptically excited from group II muscle afferents. Disynaptic inhibition related to activity in group I muscle afferents from antagonist muscles may also be present. In short, the proprioceptive component of the DSCT conveys information to the cerebellum about the length, velocity, and tension in one or a few interrelated muscles. Some neurons of Clarke's column are monosynaptically activated by joint afferents (Kuno, Munoz-Martinez, and Randic, 1973; Lindstrom and Takata, 1972); these neurons may be more properly placed in the exteroceptive component because they also receive cutaneous inputs. The exteroceptive component of the DSCT is comprised of neurons postsynaptically excited by primary afferents from cutaneous mechanoreceptors. Some neurons are excited by receptors of only one type, whereas others are excited by convergence of receptors of different types. The receptive fields vary in size from 1 to 100 cm^2.

The Cuneocerebellar Tract. The CCT, regarded as the functional forelimb (upper body) equivalent of the DSCT, originates from the external (lateral) cuneate nucleus and the rostral part of the main cuneate nucleus (Cooke, Larson, Oscarsson, and Sjölund, 1971a,b; Grant, 1962a; Holmqvist, Oscarsson, and Rosén, 1963; Rosén and Sjölund, 1973a, b). The uncrossed CCT reaches the cerebellum via the inferior cerebellar peduncle and terminates ipsilaterally in the caudal part of the anterior lobe (intermediate zone and lateral part of the vermal zone), part of the lobule simplex (intermediate zone), the rostral part of the paramedian lobule, and parts of the pyramis. Like the DSCT, the CCT can be divided into a proprioceptive and an exteroceptive component. The proprioceptive component arises from the external cuneate nucleus and terminates predominantly at the bottom of the cerebellar folia. The exteroceptive component arises from the rostral part of the main cuneate nucleus and terminates mainly at the top of the folia (Ekerot and Larson, 1972; Rinvik and Walberg, 1975). The proprioceptive component of CCT is organized similarly to that of the DSCT except that the receptive fields tend to be smaller. The exteroceptive component of CCT differs from that of DSCT in that it is disynaptically rather than monosynaptically coupled to cutaneous afferents (Cooke *et al.*, 1971b).

The Ventral Spinocerebellar Tract. The VSCT arises from spinal border cells and from cells at the base of the dorsal horn in spinal segments L_3-L_6 (Burke, Lundberg, and Weight, 1971; Hubbard and Oscarsson, 1962; Oscarsson, 1957). Axons of these cells cross the spinal cord near their level of origin and, after attaining the cerebellum via the superior peduncle, they terminate bilaterally but with a predominance toward the side of origin. The lobular distribution pattern of the VSCT is basically the same as that of DSCT in that the regions innervated are the rostral anterior lobe (intermediate and vermal zones), posterior paramedian lobule and parts of the pyramis, and dorsal paraflocculus. The VSCT termination in the posterior lobe is less dense than that of the DSCT. The modest termination of VSCT in the contralateral anterior lobe is due at least in part to axons branching to both sides of the cerebellum. Locally, each branch arborizes to innervate a larger cortical area than is the case for the DSCT. This difference appears concordant with the more extensive receptive field sizes of VSCT cells.

The signal content of VSCT fibers is considerably more complex than that of DSCT or CCT because of (1) convergence from wide areas of the periphery of afferents serving many sensory submodalities, and (2) a greater influence of descending inputs from supraspinal levels (see reviews by Oscarsson, 1965, 1973). Even though nearly all VSCT cells receive a strong polysynaptic inhibitory input from the ipsilateral hind-limb flexor-reflex afferents (FRA), they differ greatly in regard to the specifics of their relation to Ia and Ib muscle afferents and to descending pathways (reticulospinal, rubrospinal, and vestibulospinal). [The term *flexor-reflex afferents* was introduced (R. Eccles and Lundberg, 1959; Holmqvist and Lundberg, 1961) to identify the collection of spinal afferents that evoke the flexion reflex in spinalized animals. Studies of the spinal cord have shown that these afferents have additional features in common. The FRA includes groups II and III muscle afferents, high-threshold joint afferents and low- and high-threshold cutaneous afferents]. The FRA input to VSCT cells does not necessarily originate from only the ipsilateral hind limb; the FRA receptive field can include various combinations drawn from all the limbs as well as the trunk. The comparatively weaker FRA effects from regions of the body other than the ipsilateral hind limb may be excitory and inhibitory in various combinations. As a consequence of the FRA inputs from very extensive areas of the body, VSCT neurons lack modality specificity and show only very limited spatial discrimination.

A number of subgroups of the VSCT system can be formed on the basis of distinctive relations with group I muscle afferents and descending pathways. For example, some VSCT neurons are monosynaptically activated by Ib afferents that converge from one synergistic set of muscles at each of several joints. Each group of muscles so established, which may include flexors at one joint but extensors at another, is presumed to co-contract in the performance of certain movements or postures. A different group of VSCT neurons receives monosynaptic excitation from Ia afferents from muscles acting at only one joint. Splitting of the VSCT population into subgroups can also be done with reference to the differing influences from descending pathways. The diversity of the connectivity patterns of afferents to VSCT neurons makes functional interpretation difficult.

Though the meaning of the signals carried by the VSCT is not specifically understood, several general proposals have been made (Lundberg, 1971; Oscarsson, 1965, 1973). Basically, the VSCT, like other ascending FRA pathways, is considered to inform the cerebellum about active processes occurring within spinal motor centers that are simultaneously, links in spinal reflex paths and in descending paths to motoneurons (Lundberg, 1959, 1964; Miller and Oscarsson, 1970; Oscarsson, 1967, 1968). This arrangement is in contrast to that of the DSCT and CCT, which appear to report directly peripheral measures of sensorimotor behavior. The view of the VSCT proposed by Oscarsson emphasizes the convergence of FRA pathways and descending pathways in spinal motor centers whose output to motoneurons is "monitored" by the VSCT. To the extent that VSCT neurons and motoneurons receive the same inputs, the VSCT signals could be considered as efference copy signals. As a result of searching for a common central theme among the diverse connectivity patterns of the VSCT system (Lundberg and Weight, 1971), Lundberg (1971) hypothesized that some VSCT neurons are monitoring the state (level of excitability) of certain inhibitory interneurons in spinal motor centers. Other VSCT neurons in Lundberg's scheme are postulated to compare effects from excitatory and inhibitory pathways converging onto motoneurons and, as such, could be considered to be creating a "limited" efference copy signal.

The Rostral Spinocerebellar Tract. The RSCT, which was originally discovered physiologically rather than anatomically, is regarded as the functional forelimb equivalent of the VSCT (Oscarsson and Uddenberg, 1964, 1965). The location of the cell bodies giving rise to the RSCT has not yet been determined precisely, but the RSCT reaches the cerebellum via both the ipsilateral inferior and superior peduncles. The RSCT terminates bilaterally mainly in the anterior lobe, although some axons send branches to the pyramis and ipsilateral paramedian lobule. Within the anterior lobe, the RSCT presents a more complex termination pattern than any of the other three spinocerebellar paths described above. Although the RSCT carries signals related to the ipsilateral upper body ("forelimb"), its termination is unlike that of CCT, for it terminates not only bilaterally (with ipsilateral predominance) in the forelimb area of the anterior lobe, but also bilaterally within the hind-limb area, overlapping with the termination areas of the DSCT and the VSCT. Like the other direct spinocerebellar tracts, axon branching contributes significantly to the termination pattern.

The RSCT is activated synaptically by group Ib muscle afferents from the ipsilateral forelimb (Oscarsson, 1965, 1973). Individual neurons receive convergent monosynaptic excitation from synergistic muscle groups at several joints, indicating a relation to the motor behavior of the whole limb rather than to discrete segments of the limb. The FRA system provides polysynaptic excitation of the RSCT from large receptive fields on the ipsilateral forelimb; the receptive fields of the FRA show no obvious relation to the group Ib receptive fields. RSCT neurons are under the influence of descending pathways, as is the case for VSCT neurons. Furthermore, the RSCT, like the VSCT, is probably more related to monitoring the state of spinal cord motor centers than to signaling peripheral events.

RETICULAR AFFERENTS. The projections from the reticular formation to the cerebellum are commonly considered to originate from three nuclei: the lateral reticular nucleus, the paramedian reticular nucleus, and the nucleus reticularis tegmenti pontis (Brodal, 1957; Jansen and Brodal, 1954). Connections from other reticular nuclei have also been identified (e.g., Avanzino, Hösli, and Wolstencroft, 1966; Batini, Buisseret-Delmas, Corvisier, Hardy, and Jassik-Gerschenfeld, 1978) but little more than that is known about them.

Lateral Reticular Nucleus. Of the three classical precerebellar reticular nuclei, the lateral reticular nucleus (LRN) has been the most extensively investigated. The LRN receives a strong projection from the spinal cord (Brodal, 1949; Burton, Bloedel, and Gregory, 1971; Corvaja, Grofová, Pompeiano, and Walberg, 1977; Künzle, 1973) with additional inputs from the contralateral magnocellular red nucleus (Corvaja *et al.*, 1977; Courville, 1966a; Walberg, 1958), the contralateral fastigial nucleus (Corvaja et al., 1977; Thomas, Kaufman, Sprague, and Chambers, 1956; Walberg and Pompeiano, 1960), the lateral vestibular nucleus (Ladpli & Brodal,

1968; but see Corvaja *et al.*, 1977), the primary sensorimotor cortex (mainly contralateral) (Brodal, Marsala, and Brodal, 1967; Bruckmoser, Hepp-Raymond, and Wiesendanger, 1970a; Kitai, DeFrance, Hatada, and Kennedy, 1974; Walberg, 1958; Zangger and Wiesendanger, 1973) and the trigeminal system (Clendenin, Ekerot, and Oscarsson, 1975). Each of these contingents terminates in a specific but not exclusive, part of the LRN and thus the probability of convergence from a number of regions of the brain is quite high.

Fibers from the LRN reach the cerebellum through the inferior peduncle and terminate predominantly ipsilaterally, largely within the same areas of the cerebellar cortex that receive direct connections from the spinal cord: the intermediate and vermal zones of the anterior lobe and lobule VI, the pyramis and the paramedian lobule (P. Brodal, 1975; Clendenin, Ekerot, Oscarsson, and Rosén, 1974a; Künzle, 1975; Matsushita and Ikeda, 1976; Voogd, 1964). Collaterals of LRN fibers also terminate in the deep cerebellar nuclei, especially the ipsilateral fastigial and interposed nuclei (Chan-Palay, Palay, Brown, and Van Itallie, 1977; Eccles, Sabah, and Táboříková, 1974b; Eller and Chan-Palay, 1976; Matsushita and Ikeda, 1976; McCrea, Bishop, and Kitai, 1977).

Several components of the LRN can be distinguished on anatomical grounds. The largest component, termed the pars principalis, can be divided into a parvocellular part located lateroventrally and a magnocellular part located dorsomedially. A separate, small pars subtrigeminalis can also be distinguished, both morphologically and by its absence of spinal inputs (Brodal, 1943; Künzle, 1973). The parvocellular LRN projects to the rostral part of the anterior lobe and the caudal part of the paramedian lobe while the magnocellular LRN projects to the caudal part of the anterior lobe and the rostral part of the paramedian lobule (P. Brodal, 1975). The magno- and parvocellular parts of the LRN are thus related, respectively, to the classical "forelimb" and "hind-limb" divisions of the cerebellum, but as we shall see below, the LRN cannot be said to be somatotopically organized. The subtrigeminal portion of the LRN projects mainly to the "forelimb" regions of the anterior lobe and paramedian lobule and to the flocculonodular lobe (Alley *et al.*, 1975; Brodal, 1943).

Of the various afferents to LRN, those from the spinal cord have been the most studied. About one-half the LRN neurons are influenced by the FRA system and commonly have receptive fields that often include all four limbs (Clendenin, Ekerot, Oscarsson, and Rosén, 1974b; Grant, Oscarsson, and Rosén, 1966; Oscarsson and Rosén, 1966; Rosén and Scheid, 1972a,b,c). The receptive fields have both excitatory and inhibitory portions, and the great diversity in receptive field composition has so far precluded the formation of a few simple response categories. Furthermore, the large size of the receptive fields negates the possibility of a somatotopic arrangement within the LRN (Crichlow and Kennedy, 1967; Grant *et al.*, 1966; Rosén and Scheid, 1973a). The spinoreticular pathway that underlies these LRN responses is termed the bilateral ventral flexor-reflex tract (bVFRT) (Grant *et al.*, 1966; Lundberg and Oscarsson, 1962). The bVFRT originates from contralaterally projecting neurons located throughout the length of the spinal cord and, as the name of the pathway implies, the neurons are modulated by activation in the FRA system and have bilateral receptive fields. For a given bVFRT neuron, the receptive field is by no means limited to that part of the body corresponding to the spinal cord segment in which the neuron is located. In fact, the receptive field often includes all four limbs; much of the convergence of FRA pathways from different areas of the body seen in LRN neurons occurs in the spinal cord (Rosén and Sheid, 1973b). Although LRN neurons and bVFRT neurons are not somatotopically organized, there does exist a topographic relation between bVFRT neurons and their projection to the lateral reticular nucleus (Künzle, 1973). The projection to the parvocellular part of LRN arises from the lumbar region of the spinal cord whereas the projection to the magnocellular part arises from the cervical region of the spinal cord. This important distinction between topographic mapping and somatotopic mapping is illustrated in Fig. 12. Neurons of the bVFRT are monosynaptically excited by descending fibers from the lateral vestibular nucleus

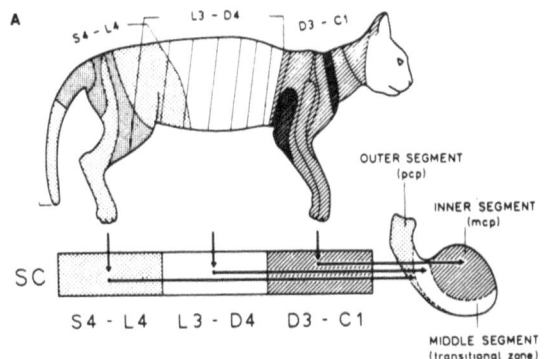

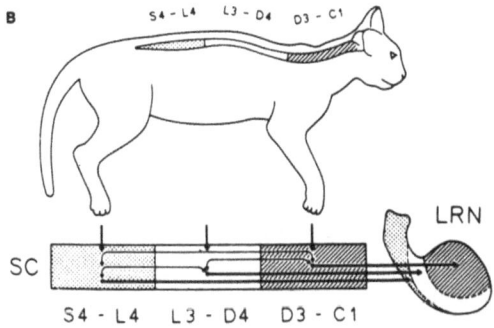

Fig. 12. Illustration of the distinction between somatotopic (A) and topographic (B) organization, using as an example the relations of the spinal and peripheral inputs to the lateral reticular nucleus (LRN). (A) Somatotopic organization, requiring segregation in the projection from the periphery to the LRN. The shading of the cat's body corresponds to the spinal cord (SC) and LRN segments. (The depiction of the dermatomes is based on Hekmatpanah, 1961; Kuhn, 1953; and Pubols, Welker, and Johnson, 1965). (B) Topographic organization of the spinal cord projection to the LRN but without somatotopic organization. The convergence of spinal afferents upon LRN neurons is as postulated by Rosén and Scheid (1973b). In view of the well-defined correspondence between spinal cord and LRN segments (shading as in A), it is assumed that the convergence takes place at spinal levels by the intermediary of propriospinal connections indicated by the two-way arrows. C = cerivcal; D = thoracic; L = lumbar; S = sacral; mcp = magnocellular portion; pcp = parvocellular portion. From Künzle (1973).

(Grillner, Hongo, and Lund, 1968) and this pathway likely contributes to the modulation of LRN neurons that occurs with tilting of the head (Coulter, Mergner, and Pompeiano, 1977; Pompeiano, 1975).

Spinal pathways other than the bVFRT also reach the LRN (Burton *et al.*, 1971; Clendenin Ekerot, and Oscarsson, 1974, 1975). For example, some LRN neurons that are not activated by descending vestibular pathways are activated from the FRA, but only from the ipsilateral forelimb. Such LRN neurons are found mainly within a part of the magnocellular division that projects to the ipsilateral intermediate zone of lobule V and the ipsilateral paramedian lobule (Clendenin *et al.*, 1974). In addition, some LRN neurons may be influenced by activity in group I muscle afferents by collateral projections of the DSCT and CCT (Burton *et al.*, 1971; Kitai, Kennedy, Morin, and Gardner, 1967; Künzle, 1973).

Physiological studies of the nonspinal pathways to LRN have essentially been limited to electrophysiological demonstrations of patterns of synaptic action and convergence (Bruckmoser, Hepp-Raymond, and Wiesendanger, 1970b; Kitai *et al.*, 1974). The cerebrocortical influence on the LRN is in part mediated by collaterals from both fast and slow pyramidal tract fibers (Bruckmoser *et al.*, 1970a; Zangger and Wiesendanger, 1973).

Paramedian Reticular Nucleus. At this time our understanding of the relation of the paramedian reticular nucleus (PRN) to the cerebellum is mostly limited to the anatomical sphere. A similar situation holds for the nucleus reticularis tegmenti pontis. The PRN is composed of three cell groups situated at the level of the inferior olive in the medial reticular formation of the medulla (Brodal, 1953; Jansen and Brodal, 1954). The cells project bilaterally with an ipsilateral predominance via the inferior peduncle to nearly all parts of the cerebellar cortex and to all the cerebellar nuclei. There are, of course, regional differences, and the work of Somana and Walberg (1978) should be consulted for details of the projection pattern. Afferents to the PRN arise from the spinal cord, the pontomesencephalic part of the brain stem, the fastigial nucleus, the superior colliculus, the trigeminal system, and the cerebral cortex bilaterally and chiefly from primary sensorimotor regions (Batton *et al.*, 1977; Brodal and Gogstad, 1957; Kawamura, Brodal and Hoddevik, 1974; Sousa-Pinto, 1970; Thomas *et al.*, 1956; Voogd, 1964; Walberg, Pompeiano, Westrum, and Hauglie-Hanssen, 1962).

Nucleus Reticularis Tegmenti Pontis. At present, the importance of the nucleus reticularis tegmenti pontis (NRTP) in cerebellar operations is testified to chiefly by anatomical studies. The NRTP, which is now considered a nucleus in its own right apart from the pontine nuclei (Brodal and Brodal, 1971), was shown by Brodal and Jansen (1946) to project bilaterally to much of the vermis and to project contralaterally to the simple, the paramedian, and the ansiform lobules. This projection is via the middle cerebellar peduncle. A more recent study (Hoddevik, 1978) has further shown that the greatest density of NRTP terminations occurs in the contralateral vermal lobules VI and VII and the flocculus; both of these cerebellar areas receive visual signals (Maekawa and Simpson, 1973; Snider and Stowell, 1944) and participate in the control of eye movements. Although the question of an NRTP projection to the cerebellar nuclei has not been fully answered, it appears that the greatest projection is to the lateral nucleus (Eller and Chan-Palay, 1976; Kitai, Kocsis, and Kiyohara, 1976; McCrea *et al.*, 1977).

The cerebral cortex, the cerebellar nuclei, and certain vestibular nuclei are the predominant afferent sources for the NRTP. Each of these afferent systems terminates in specific, but overlapping, regions of the nucleus, and the convergence of these three inputs on single NRTP cells has been observed at the synaptic level (Kitai *et al.*, 1976). In comparison to other mossy fiber precerebellar nuclei, a striking anatomical feature of the NRTP is the massive input from the cerebellar nuclei. This input arises chiefly contralaterally from the fastigial nucleus, the anterior interposed nucleus, and most of the lateral nucleus (Brodal and Szikla, 1972; Brodal, Lacerda, Destombes, and Angaut, 1972). Interestingly, the posterior interposed nucleus does not appear to project to the NRTP (Brodal, Lacerda, Destombes, and Angaut, 1972; McCrea, Bishop, and Kitai, 1978). The cerebrocortical projection to the NRTP is bilateral, but mainly ipsilateral, and arises from areas of the frontal and parietal lobes (Brodal and Brodal, 1971); the NRTP transmits a greater variety of cerebral signals to the cerebellum than either the LRN or the PRN. Additional afferents to NRTP arise from the oculomotor nuclei (Graybiel, 1977; Maciewicz and Spencer, 1977), the superior colliculus (Graham, 1977; Kawamura and Brodal, 1973; Kawamura *et al.*, 1974) the pretectum (Berman, 1977; Graybiel, 1974a; Mizuno, Mochizuki, Akimoto, and Matsushima, 1973); and the inferior olive (Courville and Faraco-Cantin, 1978). The NRTP differs from the other precerebellar reticular nuclei (LRN and PRN) in that it apparently does not receive a direct spinal input.

There are too little physiological data available on the reticular nuclei to make a convincing statement of their function in cerebellar operations. The complexity of the convergence in the reticular nuclei of inputs from diverse sources is so overwhelming that at best many investigators have proposed that the reticular nuclei are contributing a tonic background level to the cerebellar cortex and the cerebellar nuclei. Rather than subscribe to this view, we would prefer simply to say that the functions of the reticular nuclei in cerebellar operations are unknown.

PONTINE AFFERENTS. The pontine afferent system is the largest of those afferent systems that most directly route signals from the cerebral cortex to the cerebellum. A direct input from each of the four main cerebral lobes (frontal, parietal, occipital, and temporal) reaches the pontine nuclei but in differing degrees. The majority of the corticopontine fibers arise from the frontal and parietal cortices, including both primary and association sensorimotor areas. Subcortical structures also project directly to the pontine nuclei; most prominent among them are the superior and inferior colliculi and the cerebellar nuclei.

The contemporary view of the anatomical organization of the cerebral cortical input to the pontine nuclei is derived largely from the work of P. Brodal (1968a,b; 1971a,b; 1972a,b; 1978; see also Abdel-Kader, 1968; Mizuno, Mochizuki, Akimoto, Matsushima, and Sasaki, 1973). We shall not enumerate the details of the topographical relation between each of the typical subdivisions of the cerebral cortex (e.g., premotor area, sensorimotor areas, visual areas, etc.) and the pontine nuclei. Simply to describe the general pattern found in the cat will suffice. Basically, the termination pattern from each cerebrocortical area defines one or usually several (2–4) longitudinal columns in the ipsilateral pontine nuclei. Overlap of columns associated with different cortical areas is not uncommon, and, indeed, convergence from different areas has been observed at the single cell level (Ruegg and

Wiesendanger, 1975). The somatotopy observed within the sensorimotor cortical areas projecting to the pontine nuclei is to some extent preserved in the termination pattern, and it is these cortico-pontocerebellar projections, rather than the cortico-reticulocerebellar projections, which underlie the somatotopic mapping between the sensorimotor cortices and parts of the cerebellum seen under some experimental conditions (Adrian, 1943; Allen, Azzena, and Ohno, 1974; Allen, Gilbert, Marini, Schultz, and Yin, 1977; Allen, Gilbert, and Yin, 1978; Provini *et al.*, 1968; Sasaki, Oka, Kawaguchi, Jinnai, and Yasuda, 1977; Snider and Eldred, 1952).

The pontine input from the superior and inferior colliculi terminates in a single column in the dorsolateral part of the ipsilateral pontine nuclei (Altman and Carpenter, 1961; Kawamura, 1975; Kawamura and Brodal, 1973). This area is also the termination area of the ipsilateral auditory cortex (P. Brodal, 1972c). The cerebellar nuclear projection to the pons arises from the contralateral fastigial, anterior interposed and lateral nuclei and terminates according to the general plan of pontine afferents in the form of several fairly restricted longitudinal columns Brodal, Destombes *et al.*, 1972; Voogd, 1964; Brodal and Szikla, 1972). The posterior interposed nucleus is distinctive in that it appears not to project to the pontine nuclei (Brodal, Destombes *et al.*, 1972; McCrea *et al.*, 1977). The fastigial nucleus projects to that part of the pontine nuclei that also receives inputs from the colliculi while the interposed-lateral nuclei projection to the pontine nuclei overlaps especially with that from the sensorimotor cortex (Brodal, Destombes *et al.*, 1972).

All parts of the cerebellar cortex except the nodulus and ventral uvula have been found to receive afferents from the pontine nuclei (Brodal and Jansen 1946; Brodal and Hoddevik, 1978; P. Brodal and Walberg, 1977; Hoddevik, 1975, 1977; Hoddevik, Brodal, Kawamura, and Hashikawa, 1977). Of course, the density and specific origin of this projection varies from one cerebellar region to another. The pontine fibers reach the cerebellum via the brachium pontis (middle cerebellar peduncle) and terminate largely contralaterally in the intermediate and lateral zones. The pontine projection to most of the vermis is less massive than that to the rest of the cerebellar cortex. The vermis receives a more substantial ipsilateral pontine input than most of the other parts of the cerebellum, but some nonvermal regions (e.g., flocculus, paramedian lobule, paraflocculus) also receive a notable ipsilateral pontine input. Each cerebellar cortical region usually receives input from several groups of pontine cells arranged in longitudinal columns extending for varying distances along the rostrocaudal axis. This pattern is reminiscent of the arrangement of the termination of afferents to the pons, and columns defined on these two bases are, in some cases, roughly equivalent. A given pontine group may project to more than one cerebellar cortical area, but the question of whether individual cells within such a group do so has not been resolved. The matter of a pontine projection to the cerebellar nuclei is hardly resolved. Some investigations indicate an extensive projection to the lateral cerebellar nucleus Bishop, McCrea, and Kitai, 1976a; Eller and Chan-Palay, 1976) but others question the strength of this and other pontine projections to cerebellar nuclei (Courville and Coulombe, 1977; Hoddevik, 1975, 1977; Hoddevik *et al.*, 1977).

Anatomical studies have revealed connections through the pons whereby a number of specific regions of the cerebral cortex can access a specific region of the cerebellum and whereby a specific cerebral cortical region can access a number of cerebellar areas. In confirming these converging and diverging patterns, electrophysiological studies have also given an indication of the relative strengths of pontine coupling between various cerebral regions and various regions of the cerebellar cortex and nuclei (Allen *et al.*, 1977, 1978; Sasaki, Oka, Matsuda, Shimono, and Mizuno, 1975; Sasaki *et al.*, 1977). Although preferential relations between certain parts of the cerebral cortex and certain parts of the cerebellum are apparent, the degree of convergence from many cerebral areas is too pronounced to permit a firm interpretation, especially on a single dimension, of what the distribution differences mean in functional terms.

With the exception of some of the visual pathways through the pons, little is understood of the nature of pontine information transmission. Glickstein and collaborators (J. Baker, Gibson, Glickstein, and Stein, 1976; Gibson, Baher, Mower, and Glickstein, 1978; Mower, Gibson,and Glickstein, 1979) have determined the receptive field properties of two groups of visually responsive pontine cells: those that receive inputs from the visual cortex and those that receive inputs from the superior colliculus. For both groups of cells, only visual stimuli were effective in modulating their activity; neither somatosensory nor auditory stimuli were effective, although neighboring cells responding to such stimuli were found. The vast majority of pontine cells driven through connections from the visual cortex were direction-selective, usually over a wide range of speed, and they often responded best to movement of large, textured targets. Even though many of the pontine cells driven through connections from the superior colliculus were also direction-selective, they were most effectively modulated by a single target spot and had a comparatively narrow speed selectivity range (25–100°/sec). The focus of the projection of the dorsolateral pontine visual cells, which receive from the superior colliculus, is the cerebellar vermis, while those pontine visual cells receiving from the visual cortex project, for the most part, to the more lateral parts of the cerebellum (Mower *et al.*, 1979). Since the pontine visual cells receiving from the visual cortex responded best to moving full-field textured stimuli, they could provide visual feedback about self-induced movements. The properties of the visual pontine cells that project to the vermis, on the other hand, are more suggestive of a role in orienting toward or maintaining foveation of a moving object. Two other visual areas, the ventral lateral geniculate body (Edwards, Rosenquist, and Palmer 1974; Graybiel, 1974b; Swanson, Cowan, and Jones, 1974) and the pretectum (Graybiel, 1974a; Ito, 1977) also project to the pons, and investigation of these other visual projections should provide additional clues useful for understanding information transfer by the pontine nucleus to the cerebellum.

CLIMBING FIBER AFFERENTS AND THE INFERIOR OLIVE

Since the original anatomical demonstration (Szentágothai and Rajkovits, 1959) that the inferior olive is a major source of climbing fibers, additional evidence (e.g., Courville and Faraco-Cantin, 1978; Desclin, 1974a) has accumulated indicating that, in mammals, the inferior olive is the sole source of climbing fibers. Furthermore, inferior olive axons most likely project exclusively as climbing fibers to the cerebellar cortex. The inferior olivary projection to the cerebellum is, almost without exception, contralateral, reaching the cerebellum via the contralateral inferior cerebellar peduncle. In addition to providing the climbing fiber input to Purkinje cells, axons of inferior olivary neurons send collaterals to interneurons in the cerebellar cortex (Desclin, 1976; Hámori and Szentágothai, 1966; Palay and Chan-Palay, 1974; Scheibel and Scheibel, 1954). Collaterals also reach the deep cerebellar nuclei (e.g., Brodal, 1940; Ito, Yoshida, Obata, Kawai, and Udos, 1970; Matsushita and Ikeda, 1970) and the lateral vestibular nucleus (e.g., Allen, Sabah, and Toyama, 1972; Andersson and Oscarsson, 1978a,b; Desclin, 1974b; Ito, Obata, and Ochi, 1966). Because, in general, each Purkinje cell receives a climbing fiber input from only a single olivary neuron (Eccles *et al.*, 1966a), the 7–15 to 1 ratio of Purkinje cells to olivary neurons (Escobar, Sampedro, and Dons, 1968; Moatamed, 1966) is indicative of the degree of branching of inferior olivary axons (Armstrong, Harvey, and Schild, 1973a; Faber and Murphy, 1969). Indeed, branching of climbing fibers has been well demonstrated electrophysiologically (Armstrong *et al.*, 1969, 1973b,c,d, 1974; Faber and Murphy, 1969); some olivary neurons branch to innervate regions of both the anterior and posterior lobes.

One of the major endeavors of both anatomical and physiological studies of the inferior olive has been to map the various afferent systems onto the olive and to map, in turn, the olive onto the cerebellum. These maps are now known in some detail, but, in general, their functional implications have been formulated in only the vaguest of terms. Therefore, at this time, an exhaustive presentation of the mapping of afferents onto the cerebellum through the olive might be simply exhausting rather than clarifying. Even though a given afferent system may project to more than one olivary region and overlap with other projections, each afferent system shows an unusually precise termination pattern. As we shall see below, this organizational "crispness" is also characteristic of the olivocerebellar projections.

OLIVARY AFFERENTS. Although afferents terminating in the inferior olive arise from a number of regions of the brain, certain inputs are more pervasive than others. The more extensive and massive inputs originate from the spinal cord (Boesten and Voogd, 1975; Brodal, Walberg, and Blackstad, 1950; Mizuno, 1966) the dorsal column nuclei (Berkley and Hand, 1978; Berkley and Worden, 1978; Boesten and Voogd, 1975; Groenewegen, Boesten, and Voogd, 1975) the cerebellar nuclei (Dom, King, and Martin, 1973; Graybiel, Nanta, Lasek, and Nanta, 1973; Martin Henkel, and King, 1976; Tolbert, Massupost, Murphy, and Young, 1976) and a collection of nuclei (including the subparafascicular nucleus, the nucleus of Darkschewitsch and the interstitial nucleus of Cajal) at the junction of the diencephalon and mesencephalon (Brown, Chan-Palay, and Palay, 1977; Henkel, Linauts, and Martin, 1975; Linauts and Martin, 1978; Mabuchi and Kusama, 1970; Walberg, 1956; Walberg, 1974). Surprisingly, the direct input from the cerebral cortex is modest and arises mostly from the motor cortex (Bishop, McCrea, and Kitai, 1976b; Sousa-Pinto, 1969; Sousa-Pinto and Brodal, 1969). Other inputs terminate in comparatively small, restricted parts of the olive and arise from the vestibular nuclei (Saint-Cyr and Courville, 1979), the superior colliculus (Graham, 1977; Harting, 1977; Webert, Partlow, and Harting, 1978), the trigeminal system (Berkley and Hand, 1978; Cook and Wiesendanger, 1976; Stewart and King,

1963), the pretectum (Brown *et al.,* 1977; Martin, Dom, King, Robards, and Watson, 1975; Mizuno, Mochizuki, Akimoto, and Matsushima, 1973), the accessory optic system (Maekawa and Simpson, 1973; Maekawa and Takeda, 1977; Takeda and Maekawa, 1976), the red nucleus (Courville and Otabe, 1974; Edwards, 1972; Walberg, 1956) and the reticular formation (Edwards, 1975; Martin, Beattie, Hughes, Linauts, and Panneton, 1977; Walberg, 1974).

OLIVARY EFFERENTS. The existence of a precise topographic projection from the inferior olive to the cerebellum has long been recognized, but within the last decade our understanding of the basic organizational scheme of this projection has undergone significant changes. The classical study of Brodal (1940) emphasized a lobular organization of the olivo-cerebellar projection with different subdivisions of the inferior olive projecting to different cerebellar lobules. In addition, however, Brodal's results for the olivary projection to the anterior lobe also provided an indication of a longitudinal zonal organization which, in a greatly modified form, is now accepted as characterizing the olivo-cerebellar projection throughout the cerebellum. In the contemporary view, the inferior olive can be mapped onto the cerebellar cortex so as to form a number of longitudinal zones, each of which receives its climbing fiber projection from a single subdivision of the inferior olive. The zones are oriented perpendicularly to the long axis of the folia and, in general, run nearly the entire rostrocaudal length of the cerebellum. In most species of experimental animals the folia are oriented perpendicularly to the sagittal plane throughout much of the cerebellum, therefore, the climbing fiber zones are referred to as longitudinal or parasagittal zones. A summary diagram depicting, in schematic form, the zonal organization of the olivo-cerebellar projection is illustrated in Fig. 13. The zonal pattern of the olivo-cerebellar projection found with electrophysiological methods (Armstrong *et al.,* 1974; Ekerot and Larson, 1979a,b; Oscarsson, 1969, 1973; Oscarsson and Sjölund, 1977a,b) anterograde transport and degeneration methods (Groenewegen and Voogd, 1977; Groenewegen, Voogd, and Freedman, 1979; Kawamura and Hashikawa, 1979) and retrograde HRP transport methods (e.g., Brodal, Walberg, and Hoddevik, 1975; Brodal and Walberg, 1977a,b; Kotchabhakdi, Walberg, and Brodal, 1978b) are all in remarkable agreement. Within the zonal subdivisions, an indication of a finer topographic organization can be discerned in the sense that a certain segment of a zone may be innervated by a spatially distinctive group of cells within the olivary subdivision determining the entire projection zone. As a consequence of branching, however, some olivary cells project to widely separated segments within a zone.

Knowledge of the projection pattern of the inferior olive to the cerebellar cortex has ramifications beyond that of simply describing an afferent system. As first noted by Brodal in regard to the anterior lobe (1940; see also Jansen and Brodal, 1940, 1942), there is a marked similarity between the zonal pattern of the olivo-cerebellar projection and that of the Purkinje cells projecting to the cerebellar nuclei. At the time when attention was first called to this parallel, both the olivary projection to the anterior lobe and the corticonuclear projection for the entire cerebellum were considered to be divisible into three zones on each side of the midline. In this scheme, the vermal zone projects to the fastigial and vestibular nuclei, the intermediate zone projects to the (undivided) interposed nucleus, and the lateral zone projects to the lateral nucleus (Jansen and Brodal, 1949, 1942). Prior to the finding that the olivo-cerebellar projection defines more than three zones in the anterior lobe, Voogd (1964, 1969) had argued on myeloarchitectonic grounds that more than three longitudinal zones are present on each side of the cerebellum. Voogd's studies showed a mediolaterally directed repeating alteration of differences in axon size within the cerebellar white matter. With refinement of the anatomical methods, six to seven such repetitions or compartments have been distinguished on each side of the cerebellum. Voogd also obtained anatomical evidence indicating that more than three corticonuclear zones are present (see also van Rossum, 1969). A contemporary view of the corticonuclear projection pattern is summarized in Fig. 13 and will be discussed in the next section. The remarkable similarity of the zonal

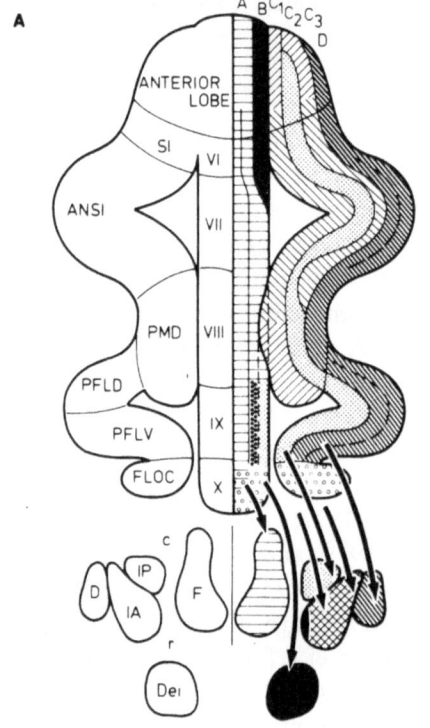

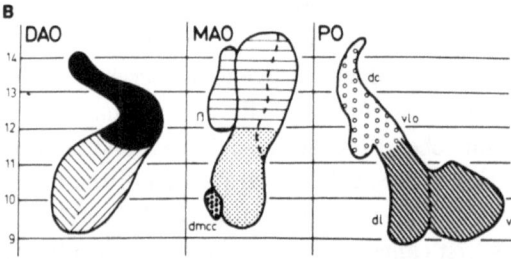

Fig. 13. Schematic diagram summarizing the projections of the inferior olive to the cerebellum of the cat. (A. *top*) Diagram of the longitudinal zonal organization of the inferior olivary projection to the cerebellar cortex. Each cortical zone receives from the correspondingly marked subdivision of the inferior olive. Some of the finer subdivisions of the olivo-cerebellar projection, such as that to the flocculus, are not illustrated. (A. *bottom*) Diagrammatic representation of a horizontal section through the deep cerebellar nuclei (fastigial, anterior and posterior, interposed and lateral) and the lateral vestibular (Deiters') nucleus showing the collateral projections of the olivo-cerebellar system. The corticonuclear relations are indicated with arrows (Purkinje cells of Zone A project to the fastigial nucleus; Purkinje cells of Zone B project to Deiters' nucleus, etc.). (B) Horizontal reconstruction of the inferior olive. Stereotactic levels are indicated to the left (level 14 is caudal; level 9 is rostral). ANSI-ansiform lobule; β-subnucleus beta; D-dentate (lateral cerebellar) nucleus; DAO-dorsal accessory olive; DC-dorsal cap of Kooy; Dei-Deiters' nucleus (lateral vestibular nucleus); dl-dorsal lamella of the principal olive; dmcc-dorsomedial cell column; IP-posterior interposed nucleus; MAO-medial accessory olive; PFLD-dorsal paraflocculus; PFLV-ventral paraflocculus; PMD-paramedian lobule; PO-principal olive; SI-simple lobule; vl-ventral lamella of the principal olive; vlo-ventrolateral outgrowth; A,B,C₁,C₂,C₃ and D, designations of the longitudinal cortical zones; c-caudal; r-rostral. (See also Fig. 10.) From Groenewegen *et al.* (1979).

pattern determined on the basis of myeloarchitectonics to that determined on the basis of olivo-cerebellar projections strongly suggests that these two bases are, in fact, delineating identical zones (Groenewegen and Voogd, 1977; Groenewegen *et al.*, 1979). Furthermore, the idea that the olivo-cerebellar projection zones and the corticonuclear projection zones are identical is receiving increasing experimental support, as discussed below.

The majority of inferior olivary neurons have collateral projections either to one of the cerebellar nuclei or to the lateral vestibular nucleus. The collateral pathways, in general, conform to the zonal organization scheme, with a given zonal subdivision of the inferior olive sending collaterals to that cerebellar nucleus receiving Purkinje cell afferents from the cortical zone to which the parent climbing fiber projects (Groenewegen and Voogd, 1977; Groenewegen *et al.*, 1979; however, see also Kitai, McCrea, Preston, and Bishop, 1977). Within the anterior lobe, Andersson and Oscarsson (1978a,b) have recently shown that this organizational plan holds even to the level of the microzones comprising the vermal zone B projecting to the lateral vestibular nucleus (Fig. 14). In view of the apparent equivalence of the olivo-cerebellar zones and the corticonuclear zones, it would not be unreasonable to expect that the projection from the cerebellar nuclei back to the inferior olive (Dom *et al.*, 1973; Graybiel *et al.*, 1973; Tolbert, Massopust *et al.*, 1976) would maintain this organizational plan. To some extent, such is the case; the projections from the interposed and lateral cerebellar nuclei to the inferior olive do, in general, regionally reciprocate with the olivary projections to them (Beitz, 1976; Eller and Chan-Palay, 1976; Martin *et al.*, 1976; Tolbert, Massopust *et al.*, 1976). However, the relation between the fastigial nucleus and the inferior olive is an exception to this reciprocal relation, at least in the cat and in the rat, for although the fastigial nucleus receives a projection from the inferior olive (Brodal, 1976; Courville, Augustine, and Martel, 1977; Hoddevik, Brodal, and Walberg, 1976) a fastigial nucleus projection to the olive has not been observed (Brown *et al.*, 1977; Graybiel *et al.*, 1973; Tolbert, Massopust *et al.*, 1976). The relation between the lateral vestibular nucleus and the inferior olive provides another example of a similar type of exception, since a projection from the lateral vestibular nucleus to the olive has not been found (Groenewegen and Voogd, 1977; Saint Cyr and Courville, 1979.

COMMENTS ON FUNCTION. The olivo-cerebellar system has several remarkable anatomical and physiological characteristics that make it something of a curiosity in the sensorimotor operations of the brain. On the anatomical side, the fact that each Purkinje cell is in extensive synaptic contact with but a single climbing fiber is most notable. On the physiological side, two conspicuous features are (1) the extremely powerful excitatory action of the climbing fiber on the Purkinje cell with the resulting burst discharge of the Purkinje cell, and (2) the extremely low firing frequency of olivary neurons. In both anesthetized and quiescent awake animals, olivary neurons discharge either a single spike or burst of spikes (2–5 spikes with an interspike interval of 2–3 msec) about once every second on the average (Armstrong and Harvey, 1966; Armstrong and Rawson, 1979b; Crill, 1970). In response to natural stimuli, olivary neurons typically have a maintained upper-firing limit of 4–5 per s. Even with application of chemical excitants, such as harmaline, only 8–10 discharges per sec are observed (de Montigny and Lamarre, 1973; Lamarre *et al.,* 1971; Llinás and Volkind, 1973). The range of olivary discharge is thus rather restricted, being limited to 0–10 per sec. The fact that the olivo-cerebellar system is in many ways unusual invites many diverse speculations as to its function(s). This invitation has been accepted by many (see review by Armstrong, 1974), but the puzzle of the olivo-cerebellar system remains to be solved experimentally.

Although it is evident that climbing fiber activity can be influenced by a variety of "natural" peripheral sensory stimuli (e.g., Eccles, Sabah, Schmidt, and Táboříková, 1972, Ishikawa, Kawaguchi, and Rowe, 1972; Leicht, Rowe, and

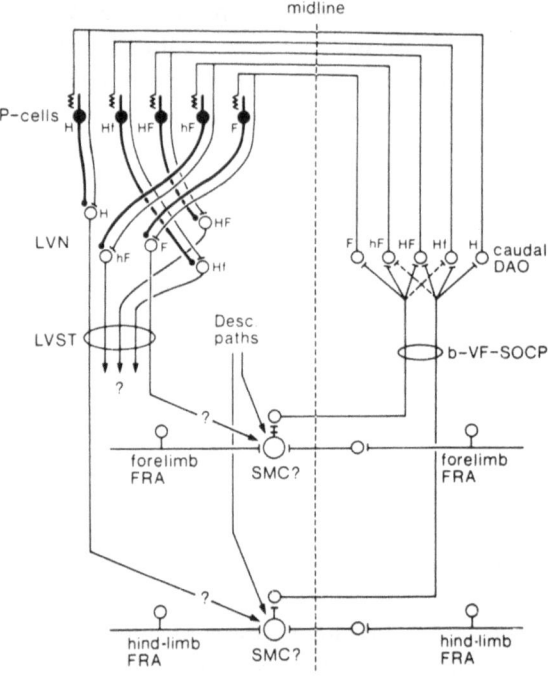

Fig. 14. Tentative diagram of the spino-olivo-cerebello-vestibulo-spinal path, including its microzonal organization. The spino-olivary tract (b-VF-SOCP), coursing in the ventral funiculus of the spinal cord and synapsing in the caudal part of the dorsal accessory olive (DAO), consists of forelimb and hind-limb components that are bilaterally activated from the flexor reflex afferents (FRA) of limb nerves. The activation is through interneurons, which are assumed to form spinal motor centers (SMC?) influenced by descending motor paths (Desc. paths). The minimal number of interneurons described by Oscarsson and Sjölund (1977a) is shown. Unknown mechanism responsible for long segmental delay in the fore-limb path is indicated by interneurons with horizontal bar. This delay compensates for the longer conduction distance in the hind-limb path, suggesting that the information forwarded might be concerned with the exact timing of movement in the four limbs (Oscarsson and Sjölund, 1977c). The forelimb and hind-limb components of the spino-olivary tract are assumed to terminate in partly overlapping regions in the caudal part of the DAO. The interrupted lines indicate long-latency connections. Five groups of olivary neurons are formed: F and H indicate short-latency forelimb and hind-limb activation; f and h, long-latency activation. The five groups of olivary neurons project to five corresponding groups of lateral vestibular nuclei (LVN) neurons. The projection is monosynaptic and excitatory through collaterals of climbing fibers, and disynaptic and inhibitory through climbing-fiber-activated Purkinje cells (P-cells) in the b-zone. The topographical order in DAO is retained in the b-zone, where five microzones are formed, but in LVN the neurons of the five groups occur intermingled. The LVN neurons receiving forelimb (F) and hind-limb (H) inputs project through the lateral vestibulospinal tract (LVST) to rostral and caudal levels of the cord; the termination of the LVN neurons receiving a mixed input is unknown. The LVST excites extensor motoneurons and inhibits flexor motoneurons of the limbs, largely through interneurons (not shown). Some observations suggest that these interneurons might belong to the spinal motor centers monitored by the b-VF-SOCP, as suggested in the figure. From Andersson and Oscarsson (1978b).

Schmidt, 1973a,b; Murphy, MacKay, and Johnson, 1973; Rubia and Kolb, 1978; Rushmer, Roberts, and Augter, 1976; Sedgwick and WIlliams, 1967; Simpson and Alley, 1974) their low discharge frequency has, in general, made it difficult to understand the "message" in the climbing fiber activity. The fact that climbing fiber signals can be less than cryptic was clearly established in an investigation of the responses of visually activated climbing fibers to the flocculonodular lobe (Simpson and Alley, 1974). The visual climbing fibers signal the direction and speed of movement of large parts of the visual world and are optimally responsive at low speeds, as illustrated in Fig. 15. Such signals report on the movement of the visual world on the retina and could be useful in controlling the stability of the

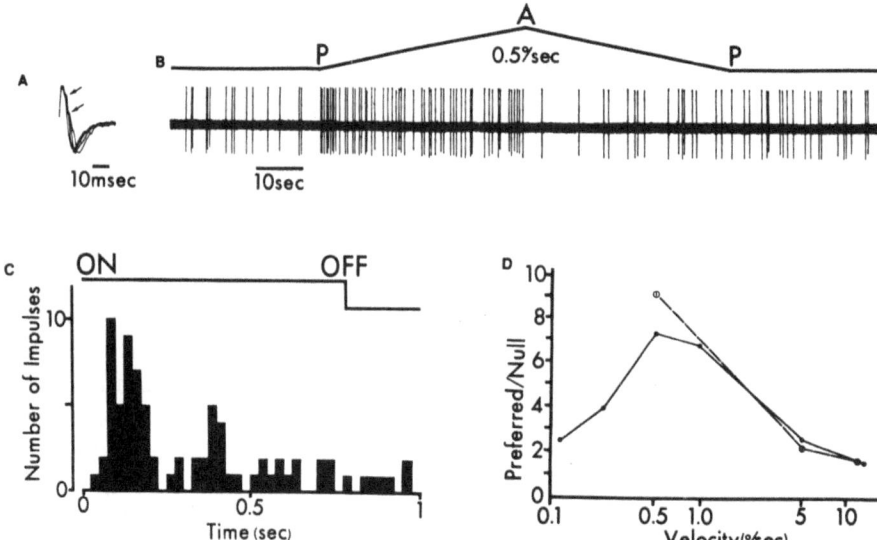

Figure 15. Trigger features of the predominant class of visual climbing fiber inputs to the flocculus, nodulus, and uvula of the rabbit. (A and B) Extracellular recordings of the climbing fiber responses of an individual Purkinje cell in the nodulus. In (A) the rising phase of the Purkinje cell response initiated the oscilloscope sweep. The multiple deflections (arrows) on the falling phase identify the responses as climbing fiber activation of the Purkinje cell. (B) Effect on climbing fiber activity of movement (0.5°/sec) of a large (100° × 75°), textured pattern rear projected onto a tangent screen centered on the pupillary axis. The most effective direction of movement for increasing the firing rate, the "preferred direction," was horizontal from temporal (posterior, P) to nasal (anterior, A) in the external world. The most effective direction of movement for decreasing the firing rate, the "null" direction, was horizontal from nasal to temporal. Movement of the pattern was preceded and followed by a control period during which the pattern was stationary. (C) Histogram showing the response of the climbing fiber unit in (A) and (B) to turning the illumination of the stationary pattern on and off. The unit showed only "on" activation. The ordinate represents total counts per 25 msec bin for 75 presentations of the stationary pattern. The interval between projection periods was 5 sec. (D) Two examples of the speed selectivity typical of the visual climbing fiber input to the flocculus and nodulus. The speed selectivity was determined for movement along the horizontal preferred-null axis. The index of selectivity is the ratio of firing frequency for movement in the preferred direction to firing frequency for movement in the null direction. Modified from Simpson and Alley (1974).

retinal image (Ito, 1972a). The capability of climbing fiber activity to signal the parameters of movement (position, velocity, and acceleration) has recently been demonstrated for passive movement of the limbs (Rubia and Kolb, 1978). With the realization that climbing fibers can signal the usual parameters describing movement, some of the mystique surrounding the olivo-cerebellar system has been removed, but the use made of these climbing fiber signals in cerebellar operations remains unknown.

Insight into olivo-cerebellar operations may be gained by examining the physiological correlates of the longitudinal zonal organization of the olivo-cerebellar system and by considering the identity of these zones with the corticonuclear projection zones. Early upon the initial electrophysiological demonstrations of a zonal pattern in the spino-cerebellar projections, it was suggested that each zone "would receive information related to a different motor control mechanism and its primary function would be to correct motor acts handled by this mechanism" (Oscarsson, 1969, p. 531; see also Miller and Oscarsson, 1970). Continuing work on the spino-olivary paths to the anterior lobe and paramedian lobule has shown that each of the four major spino-olivary cerebellar systems can be divided on the basis of receptive field location and response latency into components that conform to the zonal patterns determined anatomically (Ekerot and Larson, 1979a,b; Oscarsson and Sjölund, 1977a,b). Each zone typically receives from two of the four spino-olivary systems (Oscarsson, 1976) with the convergence likely occurring at the level of individual olivary neurons. Most components of the spino-olivary systems are activated by the FRA systems and have large receptive fields (Oscarsson, 1973); these components can be classified according to the degree to which descending pathways from supraspinal structures can influence transmission of the FRA signals. Some components are relatively free from descending control and are believed to forward information related mainly to peripheral events, whereas others, under descending control, are believed to monitor interneuronal activity in segmental motor centers (Andersson and Sjölund, 1978; Oscarsson and Sjölund, 1977c; Sjölund, 1978). This latter function is not unlike that proposed for certain spinal mossy fiber pathways (VSCT and RSCT). Although the zonal patterns of the spino-olivary cerebelllary system are known in some detail, even to the point of resolving microzones (Andersson and Oscarsson, 1978b; Ekerot and Larson, 1973, 1979b; see also Fig. 14), experimental identification of these spinal cord related zones with specific behavioral functions has not yet been achieved, undoubtedly because of the great complexity and diversity of motor behavior evolving through the spinal cord.

In an extensive theoretical treatment of anterior lobe operations in relation to locomotion, Boylls (1975a,b) has indicated that each longitudinal corticonuclear zone influences a specific functional group of muscles (flexors or extensors acting as agonistic multijoint groups) and has argued that climbing fiber inputs to these zones effect redistribution of activity among the zones so as to "bias cerebellar outflow to accord with muscle usage in different gaits." The "tonic" changes found in muscle activity with electrical stimulation in the inferior olive (Barmack, 1979; Boylls, 1978) are consistent with these theoretical predictions.

At the present time, perhaps the clearest description in behavioral terms of the climbing fiber zones and the significance of their correspondence with corticonuclear zones can be obtained by combining the results of several investigations relating to the cerebellar flocculus. Anatomical studies (Groenewegen and Voogd, 1977; see also Yamamoto, 1979a) have shown that different parts of the major olivary subdivision (dorsal cap–ventrolateral outgrowth) projecting to the flocculus (Alley *et al.*, 1975; Hoddevik and Brodal, 1977) terminate in separate zones in the flocculus. Each visually activated olivary neuron projecting to the flocculus responds best to movement of the visual world in one particular direction (Simpson and Hess, 1977; Simpson, Soodak, and Hess, 1979) and cells responding to a particular direction are grouped together within different parts of the dorsal cap–ventrolateral outgrowth division of the inferior olive (Simpson, 1979). Taken collectively, the preferred direction for the olivary neurons describe three directions in visual space. For the present purposes, some sacrifice in accuracy of description of these directions can be made, and these directions will be called horizontal, vertical, and rotational. Microstimulation of Purkinje cells in the flocculus has been shown to evoke eye movements in one of these three directions, depending on the location of the stimulation site (Dufossé, Ito, and Miyashita, 1977; Ito, 1979, Yamamoto, 1979b). The three directions established by the evoked eye movements have an internally consistent relation to the three directions established by the climbing fibers. That is, the evoked eye movements are in directions that would best activate one of the three groups of visual climbing fibers if the eye were viewing a stationary visual world. Thus, each of the three visual climbing fiber zones appears to be matched in a function space with one of three efferent zones. Although the intricacies of physical correspondence in the flocculus between the climbing fiber zones and corticovestibular efferent zones are not precisely known, these zones may be viewed as coordinate axes of sensorimotor space for compensatory eye movements. For the flocculus, the coordinate system represented by the climbing fiber zones can be readily referred to the external world geometry, but for most other cerebellar regions, interpretation of climbing fiber zones in terms of coordinate axes of sensorimotor space will likely require transformations from external to internal coordinate system using methods described in the final section of this chapter.

EFFERENT SYSTEMS

CORTICONUCLEAR AND CORTICOVESTIBULAR PROJECTIONS. Purkinje cells provide the only output from the cerebellar cortex. They project ipsilaterally and inhibit their target neurons (Ito and Yoshida, 1964; Ito, Yoshida, and Obata, 1964; Ito *et al.*, 1970) which, for the great majority of Purkinje cells, are in one of the cerebellar nuclei. The target neurons for the other Purkinje cells are located in the vestibular nuclei, the perihypoglossal nuclei and certain other neighboring nuclei. Since the classical studies of Jansen and Brodal (1940, 1942), the Purkinje cell projection pattern has provided a basis for considering that one organizational plan of the cerebellum is that of rostrocaudally oriented (longitudinal) zones, with Purkinje cells in each cortical zone projecting to a single specific cerebellar or vestibular nucleus. In the corticonuclear projection scheme formulated by Jansen and Brodal, three longitudinal zones on each side of the cerebellum were distinguished: a vermal or medial zone projecting to the medial (fastigial) nucleus, an intermediate zone (pars intermedia) projecting to the (undivided) interposed nucleus, and a lateral zone projecting to the lateral (dentate) nucleus. The corticovestibular projection arising from parts of the vermal zone (e.g., Corvaja and Pompeiano, 1979; Hohman, 1929; Walberg and Jansen, 1961) and the flocculus (e.g., Angaut and Brodal, 1967; Dow, 1936, 1938; Yamamoto, 1979b) was not set apart as a fourth separate cortical zone. A certain flexibility in the boundary between the intermediate and lateral zones was available since neither the nuclear nor the cortical boundary between these two zones was independently established. This tripartite organizational scheme for the Purkinje cell projection has had a strong influence on concepts of functional organization of the cerebellum (e.g., Allen and Tsukahara, 1974; Chambers and Sprague, 1955a,b; Evarts and Thach, 1969). It must be admitted, however, that division of the cerebellum into three zones on the basis of the corticonuclear projection is logically viable only as long as only three cerebellar nuclei are distinguished. With the recognition by most investigators of the distinction between an anterior interposed nucleus and a posterior interposed nucleus (e.g., Flood and Jansen, 1961; Voogd, 1964) came the need to modify the original scheme.

The contemporary view that more than three zones underlie the longitudinal organization of the cerebellum was initially championed by Voogd (1964, 1967, 1969). Anatomical evidence derived from studies of the myeloarchitectonics of the cerebellar white matter and of the Purkinje cell projection to cerebellar nuclei indicates that, in general, six major longitudinal zones are present, as illustrated in Fig. 13. Within the anterior lobe and lobule VI, the original vermal zone is now recognized as divisible into a medial zone projecting to the fastigial nucleus and a lateral zone projecting largely to the lateral (Deiters) vestibular nucleus. The projection to various of the vestibular nuclei from the posterior vermis, particularly from the uvula (lobule IX) and the nodulus (lobule X) is not indicated specifically in Fig. 13 but has been presented in a zonal configuration by van Rossum (1969; see also Haines, 1975, 1976). The original intermediate zone is now seen as divisible into three zones, a central zone projecting to the posterior interposed nucleus and bordered throughout most of its rostrocaudal length by zones projecting to the anterior interposed nucleus. At some levels, the original lateral zone is still represented as a single zone projecting to the lateral nucleus, but at other levels, two divisions of the lateral zone are distinguishable. A multizonal organization of the Purkinje cell projections from the flocculus has also been found recently (Yamamoto, 1978; Yamamoto and Shimoyama, 1977).

Within the anterior lobe, the various longitudinal zones are fairly precise and orderly; consequently, the particular efferent zonal organization is generally agreed upon (Haines and Rubertone 1977, 1979; van Rossum, 1969; Voogd, 1969). Within the posterior lobe, however, the geometrical form of the zonal divisions is more complicated and less precise, making agreement on the proposed zonal pattern less than unanimous (e.g., Bishop, McCrea, Lighthall, and Kitai, 1979; Brodal and Courville, 1973; Courville, Diakiw, and Brodal, 1973).

At present, the concept of a multizonal representation of the Purkinje cell projection is still in a state of flux, and a number of revisions in fine structure are to be expected. For example, it has been shown electrophysiologically that the vermal zone projecting to Deiters' nucleus can be divided into five microzones (see Fig. 14; see also Andersson and Oscarsson, 1978a,b), but a question to be addressed in considering further subdivision of the zones is how finely can the divisions be made before any organization and simplification derived from the grouping of individual Purkinje cells together is lost. Nevertheless, a longitudinal zonal organization pattern appears to be fundamental to cerebellar structure and function since such a pattern can be defined independently on the basis of the myeloarchitectonics of the white matter, on the basis of the cerebellar corticonuclear projection and on the basis of afferent fiber projections, particularly those of the olivo-cerebellar system. Even though we have focused here on the longitudinal zonal organization of the cerebellum, it should be remembered that the parallel fibers constitute a perpendicularly organized system capable of cutting across several zones. Determination of the functional meaning of this orthogonal relation is a major challenge to investigators of cerebellar function.

For each longitudinal zone, the Purkinje cells of individual lobules terminate within a specific region of the corresponding nucleus, but the degree to which that region is exclusive for a single lobule or coextensive with the termination regions of other lobules varies greatly depending on the zone, the lobular location within the zone, and the species (Armstrong and Schild, 1978a,b; Bishop *et al.*, 1979; Courville and Diakiw, 1976; Walberg and Jansen, 1964). The initial notion (Clarke and Horsely, 1905; Jansen and Brodal, 1942) that each lobule projects to that part of the cerebellar nuclei closest to it appears to hold, but only roughly, for the projection to the fastigial nucleus. This arrangement does not generally hold for the projection to the interposed nuclei because anterior lobe and paramedian lobule projections converge onto the same nuclear areas (Bishop *et al.*, 1979).

Electrophysiological recording from the cerebellar nuclei (e.g., Armstrong and Rawson, 1979a; Armstrong, Cogdell, and Harvey, 1975; Bantli and Bloedel, 1977; Eccles *et al.*, 1974a,b; Eccles, Rantucci *et al.*, 1974; Eccles, Rosén *et al.*, 1974a,b; Eccles, Rantucci, Rosén *et al.*, 1974) of responses to both electrical and natural stimulation have primarily addressed questions of somatotopy (see above) and of temporal patterning. These studies have revealed temporal patterns of discharge that, in general, can be accounted for in terms of convergence of inhibitory Purkinje cell signals with excitatory signals carried by collaterals of climbing fibers and certain mossy fibers. Although recordings from the cerebellar nuclei in behaving animals (e.g., Burton and Onoda, 1977; Robertson and Grimm, 1975; Thach, 1968, 1970) have revealed various correlations of neural activity with movement, the functional significance of the interactions between signals processed in the cerebellar cortex with those not processed in the cortex remains to be determined.

CEREBELLAR NUCLEAR EFFERENTS. The cerebellar cortical influences on the rest of the brain are, for the most part, mediated by cerebellar nuclear neurons, many of which project to more than one extracerebellar region (e.g., Ban and Ohno, 1977; McCrea *et al.*, 1978; Tolbert, Bantili, and Bloedel, 1978b). At present, the output of the cerebellar nuclei is believed to be excitatory (e.g., Angaut, Guilbaud, and Reymond, 1968; Ito *et al.*, 1970; Sasaki, Kawaguchi, Matsuda, and Mizuno, 1972; Uno, Yoshida, and Hirota, 1970). It should be mentioned that a finding of some inhibitory neurons would break no precedent because some Purkinje cells of the flocculus are known to synapse with some vestibular nuclei neurons that inhibit extraocular motoneurons (Baker, Precht, and Llinás, 1972; Fukuda, Highstein, and Ito, 1972; Highstein, 1973; Ito, Nisimaru, and Yamamoto, 1977). Before describing the extracerebellar projections of the cerebellar nuclei, it should be pointed out that some cerebellar nuclear neurons also project back to the cerebellar cortex, apparently ending as mossy fibers (Chan-Palay, 1977; Gould, 1979; Gould and Graybiel, 1976; Tolbert et al., 1976, 1977, 1978a). In the cat, the termination pattern of the nucleocortical projection is basically in accord with the zonal pattern of the corticonuclear projection since each cerebellar nucleus projects back to that cortical zone from which it receives a Purkinje cell input (Gould, 1979; Gould and Graybiel, 1976; Tolbert et al., 1976, 1978a). In primates, however, a reciprocal relation does not appear to hold, for the lateral nucleus projects to both vermal and lateral cortical zones (Chan-Palay, 1977; see also Haines, 1978; Tolbert et al., 1978a).

Axons from the cerebellar nuclei exit the cerebellum by way of both the juxtarestiform body and the superior cerebellar peduncle. The juxtarestiform body is a collection of certain cerebellar afferents and efferents coursing medial to the restiform body proper. Axons from the lateral, interposed and fastigial nuclei that leave the cerebellum within the *ipsilateral* superior peduncle are grouped together in a conspicuous fiber bundle called the brachium conjunctivum and are thus distinguished from other fibers traveling in the superior peduncle (Voogd, 1964). The initial routing from the cerebellum of fastigial nucleus fibers is considerably more complicated than that of interposed or lateral nuclear fibers. The fastigial nucleus gives rise to both ipsilateral and contralateral output paths in both the superior peduncle and the juxtarestiform body, whereas the lateral and interposed nuclei give rise to a single output path in the ipsilateral brachium conjunctivum.

Fastigial (medial) nucleus efferents leave the cerebellum by two main tracts: the direct fastigiobulbar tract and the uncinate tract. The uncinate tract originates from all parts of the fastigial nucleus, crosses the midline within the cerebellum, and then divides into an ascending and a descending component. The ascending component originates from the caudal half of the contralateral fastigial nucleus and exits via the superior cerebellar peduncle (Angaut and Bowsher, 1970; Batton *et al.*, 1977) but not within the brachium conjunctivum (Voogd, 1964). The ascending component of the uncinate tract is further distinguished from the brachium conjunctivum since it does not descussate with the brachium conjunctivum in the midbrain, although some crossing of the midline does occur at other places. The ascending uncinate tract terminates largely contralaterally to the fastigial nucleus of origin within the midbrain (central gray, superior colliculus, nucleus of the posterior commissure) and the diencephalon, mainly in the ventral tier nuclei of the thalamus (e.g., Angaut and Bowsher, 1970; Batton *et al.*, 1977; Carpenter, Britton, and Pines, 1958; Cohen, Chambers, and Sprague, 1958; Thomas *et al.*, 1956; Voogd, 1964). Axons comprising the descending component of the uncinate tract leave the cerebellum within the juxtarestiform body and distribute without recrossing to pontine reticular and medullary reticular nuclei, certain vestibular nuclei, perihypoglossal nuclei, the spinal cord (particularly cervical levels), the lateral reticular nucleus, the nucleus parasolitarius, the nucleus reticularis tegmenti pontis, the dorsolateral part of the pontine nucleus and the paramedian reticular nucleus (Batton *et al.*, 1977; Carpenter *et al.*, 1958; Cohen *et al.*, 1958; Thomas *et al.*, 1956; Voogd, 1964; Walberg, 1961; Walberg and Pompeiano, 1960; Walberg, Pompeiano, Brodal, and Jansen, 1962, Walberg, Pompeiano *et al.*, 1962; Wilson, Uchino, Maunz, Susswein, and Fukushima, 1978).

Axons from the fastigial nucleus leaving the cerebellum on the side of origin comprise the direct fastigiobulbar tract. This tract, like the uncinate tract, has both ascending and descending components. Fibers of the ascending component exit via the ipsilateral superior cerebellar peduncle within a restricted part of the brachium conjunctivum and cross in the midbrain with fibers from the interposed and dentate nuclei (Voogd, 1964). The ascending component of the direct fastigiobulbar tract apparently originates from the rostrolateral part of the fastigial nucleus (Thomas *et al.*, 1956; but see Angaut and Bowsher, 1970; Batton *et al.*, 1977; Cohen *et al.*, 1958; Voogd, 1964). After crossing in the midbrain, the ascending fibers of the direct fastigiobulbar tract eventually join the ascending fibers of the uncinate tract, but the particular termination sites of the ascending component of the direct fastigiobulbar tract have not been determined. The descending component of the fastigiobulbar tract originates from cells in the rostral portion of the fastigial nucleus, leaves the cerebellum within the ipsilateral juxtarestiform body, and projects to certain ipsilateral vestibular, perihypoglossal, pontine reticular, and medullary reticular nuclei (Batton *et al.*, 1977; Carpenter *et al.*, 1958; Thomas *et al.*, 1956; Voogd, 1964; Walberg, 1961; Walberg, Pompeiano, Brodal, and Jansen, 1962; Walberg, Pompeiano *et al.*, 1962).

Axons of cells in the lateral and interposed nuclei leave the cerebellum via the ipsilateral superior cerebellar peduncle within the brachium conjunctivum and cross to the contralateral side of the midbrain tegmentum. After decussating, the brachium conjunctivum divides into an ascending and a descending component. Many of the axons of the brachium conjunctivum bifurcate after crossing and send branches into both components (Ban and Ohno, 1977; McCrea *et al.*, 1978; Tolbert *et al.*, 1978b). The principal termination sites of the interposed nuclear axons in the ascending component of the brachium conjunctivum are the magnocellular (from the anterior interposed nucleus [NIA]) and the parvocellular (from the posterior interposed nucleus [NIP]) red nucleus and the ventral tier thalamic nuclei (e.g, Angaut, 1970; Cohen *et al.*, 1958; Courville, 1966b; Flumerfelt, Otabe, and Courville, 1973). The principal target nuclei for the lateral nucleus projection ascending in the brachium conjunctivum are the ventral tier thalamic nuclei and the parvocellular part of the red nucleus (e.g., Cohen *et al.*, 1958; Flumerfelt *et al.*, 1973). In addition to these major termination areas, ascending fibers from the interposed and lateral nuclei reach a number of other areas, including the interstitial nucleus of Cajal, the nucleus of Darkschewitch, the central gray, and the nucleus subparafascicularis. Descending components of the brachium conjunctivum reach primarily the nucleus reticularis tegmenti pontis (except from the NIP), the cervical spinal cord (from only the NIP), parts of the pontine nucleus (except from the NIP), the inferior olive and the ventromedial medullary reticular formation (Bantli and Bloedel, 1975, 1976; Beitz, 1976; Brodal, Destombes *et al.*, 1972; Brodal, Lacerda *et al.*, 1972; Brown *et al.*, 1977; Chan-Palay, 1977; Cohen *et al.,* 1958; Dom *et al.,* 1973; Graybiel *et al.,* 1973; Kitai *et al.,* 1976; Martin *et al.,* 1976; Matsushita and Hosoya, 1978; McCrea *et al.,* 1978; Tolbert, Massopust *et al.,* 1976; Tolbert *et al.,* 1978b; Voogd, 1964).

The cerebellar nuclear projections to the magnocellular part of the red nucleus, the reticular formation, and the vestibular nuclei allow for interaction of the cerebellum with the descending rubrospinal, reticulospinal and vestibulospinal motor systems. The cerebellar nuclei projections to the ventral thalamic nuclei allow for interaction of the cerebellum with a variety of cerebral cortical areas, including primary motor, premotor, and association cortical regions (Bava, Cicirata, Licciardello, Volsi, and Pantó, 1979; Rispal-Palel and Gangretto, 1977; Rispal-Padel and Latreille, 1974; Sasaki, Kawaguchi *et al.,* 1972; Sasaki, Matsuda, Kawaguchi, and Mizuno, 1972; Sasaki, Kawaguchi, Aka, Sakai, and Mizuno, 1976). In the next section, we will consider how these cerebellar pathways to the rest of the motor system effect motor coordination.

Cerebellar Coordination as a Transformation via a Metric Tensor

Understanding general cerebellar function has been a continuing endeavor in neurobiology for several reasons. First, the central role of the cerebellum in motor coordination makes it an essential link in understanding the general organization of movement. Second, and more important, because the cerebellar anatomy and physiology at the single-cell level is well understood, this system represents one of the prime challenges to the basic assumption that the function of any particular brain region may be reduced to, and defined by, the properties of the neuronal elements that constitute it. This view is indeed one of the basic tenets in neurobiology and constitutes implicitly, and often explicitly, the single most compelling reason for electrophysiological and morphological studies of single cells. Thus, the assumption that understanding of single cells and the networks they form will ultimately describe the overall properties of the brain will likely be first tested with regard to the cerebellum.

The question remains, nevertheless, of how information regarding the properties of single cells and their connectivities relate to the overall properties of the cerebellum in its role as a motor coordinator. Many different hypotheses have been proposed that relate electrical activity of single cells to cerebellar function. Basically, rather than formal theories or models, what is commonly available are loosely defined ideas of how such a system could work. For instance, it has been considered that the cerebellum may be, basically, a device to modulate, via Purkinje cell inhibition, the peripheral and central feedback loops which relate the activity of different parts of the brain to motor activity. Such types of hypotheses are really nothing but rudimentary restatements of the anatomical pathways involved in cerebellar connectivity rather than being formal hypotheses of how the cerebellum may function as the center for motor coordination. The second kind of hypothesis considers only certain aspects of motor control and often their bases are purely speculative. One such hypothesis regards the cerebellum as a learning machine having a perceptron-like function, where the firing of the Purkinje cell is assumed to be related to the mossy fiber input by a modifiable weighting factor such that the output of the cerebellum represents an elaboration of the input based on past experience (Ito, 1972a; Marr, 1969; Robinson, 1976). For this view, the climbing fiber system is seen as altering the weighting factor of the mossy fiber/parallel fiber system and thus the firing of Purkinje cells.

A third hypothesis suggests that the cerebellum is a distributed system having as a central function the setting of motor orders into the context of the rest of the nervous system (Llinás, 1974). This latter view has been recently formalized by considering the cerebellum as a tensorial network (Pellionisz and Llinás, 1979). Here it is assumed that the networks of the cerebellum cannot be described as a loop but, rather, the cerebellum must be described as a distributed neuronal network such that information is simultaneously sent to many Purkinje cell neurons via the mossy fiber input. Rather than describing the system as a set of serially connected neurons, the system is described as a "geometric object"; that is, as a system capable of generating particular vectorial transformations.

The parallel and distributed nature of brain organization has been recently approached by means of tensor network theory (Pellionisz and Llinás, 1979). This view implies that electrical activity in large sets of neurons (either frequency of firing or membrane potential), must be treated as a vectorial entity. This vector would be comprised of the frequency of firing or by the membrane potential, as a scalar, for each one of the neurons. Thus, if we assume a brain region or a pathway to be composed of n elements, the vector would be described in an n dimensional frequency space (a hyperspace). Although a vector in n dimensions is something like the general notion of "pattern of activity," an important difference is that vectors in n dimensions can be treated formally. Multidimensional space is familiar to us in examples, such as a television picture. Each dot on the screen has a light intensity and the television "picture" would be an n dimensional vector. A vectorial representation of such a picture in neuronal terms would be a set of frequencies in the different nerve fibers comprising a nerve bundle, such as the optic nerve. Once the above view is considered, the nervous system becomes endowed, by definition, with the ability to handle vectors in the frequency domain. This domain (really a space) has, as stated above, as many dimensions as there are neurons in the system. That the nervous system treats electrical activity as vectors has deeper implications, for it allows the brain to be treated or considered as a geometrical entity. This vectorial property (for instance, like that of a mirror that allows it to transform vectors and form images) makes it a geometric object. Continuing with this metaphor, we could assume that the internal "manipulation" of these vectors (a bit like the distorting mirrors at amusement parks) are really tensorial properties; that is, a vector → vector transformation that is reference–frame invariant. In this manner, one can regard the nervous system as a set of pathways carrying vectors and the neuropile like a mirror or lens tensorial system.

From the above statements, we may conclude that the generation of motor movements must arise in the CNS from a set of components of an intended movement vector, probably assembled by many structures of the brain at a particular time. These vectorial components are, however, specified in a n-dimensional space of firing frequency of neurons and must be translated into the activation of muscles whose forces implement the intended movement in three dimensions. Such transformation requires, first of all, that the intended movement vector must be placed in the context of the functional state of the body at a given moment. Such "putting into context" requires that this vector interact with other vectors having information regarding the *status quo* of the motor system such that the intended movement be adequate to achieve the desired goal. Thus, for instance, to reach for and grasp an object in space may require very different sets of muscles to be activated, depending on the relative position of the object and the subject executing the movement. If one is sitting, one's arm and hand may have to go up to reach the object whereas, if standing, the hand may have to be lowered. Even more interesting, the way of achieving the same goal may be different, depending on the initial posture, as in writing a letter on the blackboard when one's hand is immediately in front, and in writing a letter when one's arm is almost totally stretched out. The sets of muscles utilized under these two conditions are different; however, the letters are very much the same, indicating that the intended movement is the same whereas the mechanism of execution is rather different. In order to perform in such a manner, we must be able to have a constant intended movement vector, which generates the letter, capable of implementing it with different sets of muscles. A similar statement may be made regarding different sizes of handwriting. The shapes and styles of the letters on a small piece of paper are very similar to those made by the same individual on a large scale. There is, therefore, an intended movement vector which is independent of size, position, or coordinate system.

A second aspect of execution of movement is that the number of possible solutions for attaining a particular goal is very large, if not actually infinite. Assume that, as in Fig. 16A, a hand must be moved from point a to point b (by a vector $\overline{U}(x,y)$). This x,y displacement can be executed by changing, in this case, the angle

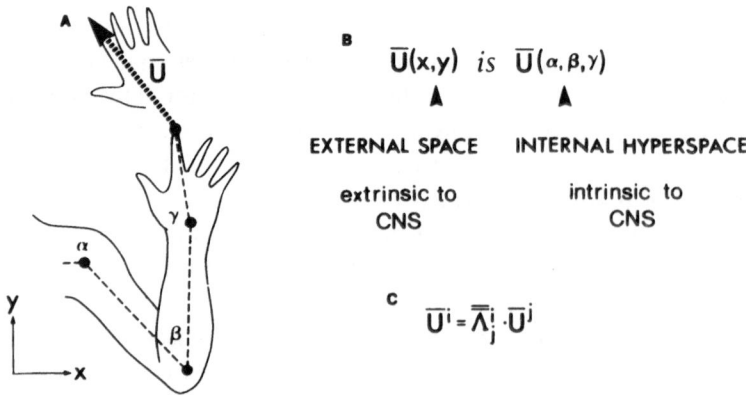

Fig. 16. Limb movement as a tensorial entity. (A) An upward displacement vector $\overline{U}$ is a physical entity that can be expressed in different reference frames: e.g., by the x, y coordinate system, or by the α, β, γ-ordered set of three quantities. (B) The two reference frames shown are of fundamentally different kinds. x, y applies to the CNS-independent external space; the α, β, γ to the space inherently connected to the CNS. The limb-displacement vector occurs in both spaces. (C) Different expressions of the one $\overline{U}$ vector are related by the limb-displacement tensor, $\overline{\overline{\Lambda}}{}_j^i$. Modified from Pellionisz and Llinás (1980).

of three joints of the limb (α, β, and γ). Note that the vector $\overline{U}(x,y)$ can be performed by an infinite set of combinations of angles of these three joints. This α, β, γ system is said to be "overcomplete" compared to the x,y. It is important to realize that both the $\overline{U}(x,y) = \overline{U}^i$ and the $\overline{U}(\alpha, \beta, \gamma) = \overline{U}^j$ vectors are expressions, in different systems of coordinates, of the same displacement that is invariant to reference frames. Such reference-frame invariant vectorial relations are *tensors;* e.g., as in the equation

$$\overline{U}^i = \Lambda^i_j \cdot \overline{U}^j$$

Following Kron (1939), those objects with the property of transforming vectors independently of coordinate systems, we call *geometric objects*. In these terms, the function of the cerebellum is understandable in a global sense. Previously, all one could say was that if the cerebellum is removed, no gross abnormality occurs (no paralysis, no loss of sensation); however, there is a rather strange inability to perform movements in a coordinated manner (i.e., smoothly, precisely, and rapidly). It was considered, therefore, that to understand the cerebellum one would ultimately require also that the rest of the nervous system be understood because the role of the cerebellum is intermediate between the sensory and the motor realms. Our present perception is that the damage to the cerebellum produces a motor abnormality that can be described in exact terms. The cerebellum is indeed the seat of motor coordination, and coordination can be better defined than as being the opposite of ataxia. Motor coordination is the transformation, in n-dimensional frequency hyperspace, of a vector of motor *intent* into a vector of motor *execution*. To be more precise, intended movement vectors are given by the so-called covariant vectorial components; however, its coordinated execution requires the contravariant components of the same vector (Pellionisz and Llinás, 1980). In order to utilize the covariant intended movement vector, its components must be transformed into their contravarient counterparts, which describe not the features of the intended movement but rather the steps necessary to implement the movement. Geometrically, this is well understood: in order to transform the covariant vector (which basically describes the features of the movement goal) into the contravariant vector (which achieves the goal), a metric tensor is required. The contravariant components are capable of generating the desired displacement vector by the limb. As shown in Fig. 17, the orthogonal projections of a vector onto oblique coordinate axes (the covariant vector components) may be transformed into parallelogram components (the contravariant components) via a metric tensor. The tensorial properties of the cerebellum have been treated in detail elsewhere (Pellionisz and Llinás, 1979, 1980). In short, motor coordination may be regarded as a geometrical transformation of what we want to do (the covariant intended vector $\overline{U}$ in Fig. 18A) into how to do it (the contravariant execution vector in Fig. 18C). Indeed, the activity of mossy and climbing fiber systems constantly bombarding the cerebellum serves two different purposes. The mossy fiber system serves the purpose of constantly describing the *status quo* and, in addition, carrying the *intended movement vector*. This intended movement vector is transformed, via the metric tensor that the cerebellar circuitry represents, into a contravariant vector in the context of the total motor state of the system. The climbing fiber system may serve the purpose of adjusting the metric tensor according to the *status quo* and intention. Thus, as

RODOLFO R.
LLINÁS AND
JOHN I. SIMPSON

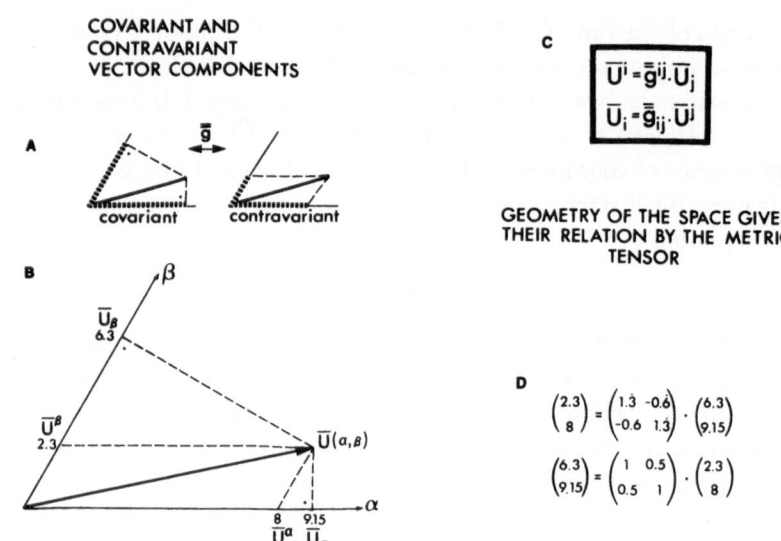

COVARIANT AND
CONTRAVARIANT
VECTOR COMPONENTS

C

$$\bar{U}^i = \bar{\bar{g}}^{ij} \cdot \bar{U}_j$$
$$\bar{U}_i = \bar{\bar{g}}_{ij} \cdot \bar{U}^j$$

GEOMETRY OF THE SPACE GIVES
THEIR RELATION BY THE METRIC
TENSOR

D

$$\begin{pmatrix} 2.3 \\ 8 \end{pmatrix} = \begin{pmatrix} 1.3 & -0.6 \\ -0.6 & 1.3 \end{pmatrix} \cdot \begin{pmatrix} 6.3 \\ 9.15 \end{pmatrix}$$

$$\begin{pmatrix} 6.3 \\ 9.15 \end{pmatrix} = \begin{pmatrix} 1 & 0.5 \\ 0.5 & 1 \end{pmatrix} \cdot \begin{pmatrix} 2.3 \\ 8 \end{pmatrix}$$

Fig. 17. Covariant and contravariant vectorial components and their relation established by the geometry of the space. In (A) the covariant components can be established independently from one another but do not physically compose the vector. (These components, also called resolved parts, are established by perpendicular projections.) Contravariant components physically add up to the resultant $\bar{U}$; however, they cannot be established independently from one another. In (B) the numerical relation of the two sets of components is determined by the metric tensor that describes the geometry of the space. (C) Relation of covariant and contravariant components, as determined by the metric tensor. (D) Numerical example of the relation of covariant and contravariant components shown in (A) and (B). Modified from Pellionisz and Llinás (1980).

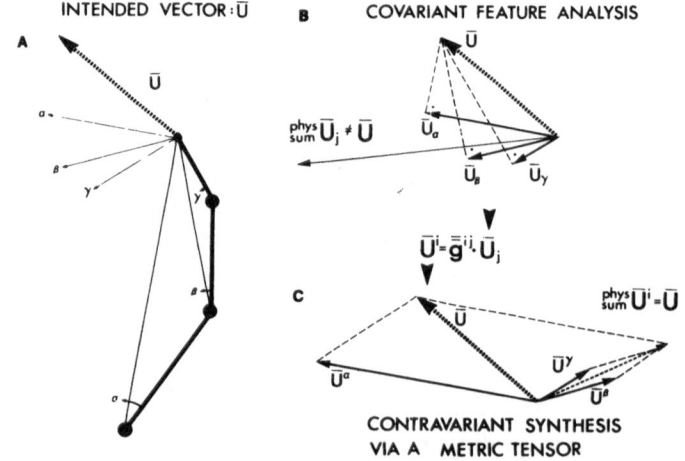

Fig. 18. Covariant analysis and contravariant synthesis via a metric tensor. (A) Given a two-dimensional intended vector $\bar{U}$ and three α, β, γ axes of an overcomplete reference frame, the decomposition could be performed by a two-step operation. (B) First, covariant components of $\bar{U}$ can be established, using the geometry of the two-space, to any number of directions independently. (The perpendicular projections, i.e., the inner products, provide the "features" of the desired $\bar{U}$ vector in any coordinate direction.) However, the physical sum of covariant components is not equal to the displacement. (C) Second, provided that the metric tensor is available (in contravariant expression) for the α, β, γ space, the corresponding set of contravariant components can be established. The physical sum of the contravariant components physically generates the displacement vector $\bar{U}$. Modified from Pellionisz and Llinás (1980).

illustrated in Fig. 18A, given a 3-segment limb and an intended movement vector $\overline{U}$, the implementation of such a movement, utilizing directly the covariant components of the intended movement vector, will generate an erroneous physical vector $\overline{U}_j$. That is, if the covariant components U_α, U_β and U_γ are allowed to sum directly as in Fig. 18B, the physical sum $\overline{U}_j$ does not generate the desired vector $\overline{U}$. However, as indicated in Fig. 18C, the intended movement vector $\overline{U}_j$ can be transformed by the metric tensor $\overline{\overline{\mathbf{g}}}^{ij}$ into a contravariant vector $\overline{U}^i$. If now the sum of the contravariant vector components $\overline{U}^\alpha$, $\overline{U}^\beta$ and $\overline{U}^\gamma$ is executed, the resultant $\overline{U}^i$ is the actual desired physical vector $\overline{U}$. Thus, motor coordination seems to call for a covariant to contravariant transformation from the intended movement vector to the execution vector, and the cerebellum is a neuronal system capable of implementing this transformation.

REFERENCES

Abdel-Kader, G.A. The organization of the cortico pontine system of the rabbit. *Journal of Anatomy*, 1968, *102*, 165–181.

Adrian, E.D. Afferent areas in the cerebellum connected with the limbs. *Brain*, 1943, *66*, 289–315.

Allen, G.I., and Tsukahara, N. Cerebrocerebellar communication systems. *Physiological Reviews*, 1974, *54*, 957–1006.

Allen, G.I., Sabah, N.H., and Toyama, K. Synaptic actions of peripheral nerve impulses upon Deiters' neurones via the climbing fibre afferents. *Journal of Physiology (London)*, 1972, *226*, 311–333.

Allen, G.I., Azzena, G.B., and Ohno, T. Somatotopically organized input from fore and hind limb areas of sensorimotor cortex to cerebellar Purkinje cells. *Experimental Brain Research*, 1974, *20*, 255–272.

Allen, G.I., Gilbert, P.F.C., Marini, R., Schultz, W., and Yin, T.C.T. Integration of cerebral and peripheral inputs by interpositus neurons in monkey. *Experimental Brain Research*, 1977, *27*, 81–99.

Allen, G.I., Gilbert, P.F.C., and Yin, T.C.T. Convergence of cerebral inputs onto dentate neurons in monkey. *Experimental Brain Research*, 1978, *32*, 151–170.

Alley, K., Baker, R., and Simpson, J.I. Afferents to the vestibulo- cerebellum and the origin of the visual climbing fibers in the rabbit. *Brain Research*, 1975, *98*, 582–589.

Altman, J., and Carpenter, M.B. Fiber projections of the superior colliculus in the cat. *Journal of Comparative Neurology*, 1961, *116*, 157–178.

Andersen, P., Eccles, J.C., and Voorhoeve, P.E. Postsynaptic inhibition of cerebellar Purkinje cells. *Journal of Neurophysiology*, 1964, *27*, 1138–1153.

Andersson, G., and Oscarsson, O. Projections to lateral vestibular nucleus from cerebellar climbing fiber zones. *Experimental Brain Research*, 1978a, *32*, 549–564.

Andersson, G., and Oscarsson, O. Climbing fiber microzones in cerebellar vermis and their projection to different groups of cells in the lateral vestibular nucleus. *Experimental Brain Research*, 1978b, *32*, 565–579.

Andersson, G., and Sjölund, B. The ventral spino-olivo cerebellar system in the cat. IV. Spinal transmission after administration of clonidine and L-dopa. *Experimental Brain Research*, 1978, *33*, 227–240.

Angaut, P. The ascending projections of the nucleus interpositus posterior of the cat cerebellum: An experimental anatomical study using silver impregnation methods. *Brain Research*, 1970, *24*, 377–394.

Angaut, P., and Bowsher, D. Ascending projections of medial cerebellar (fastigial) nucleus: An experimental study in the cat. *Brain Research*, 1970, *24*, 49–68.

Angaut, P., and Brodal, A. The projection of the "Vestibulocerebellum" onto the vestibular nuclei in the cat. *Archives Italiennes de Biologie*, 1967, *105*, 441–479.

Angaut, P., Guilbaud, G., and Reymond, M.-C. An electrophysiological study of the cerebellar projections to the nucleus ventralis lateralis of thalamus in the cat. I. Nuclei fastigii et interpositus. *Journal of Comparative Neurology,* 1968, *134,* 9-19.

Aoyama, M., Hongo, T., and Kudo, N. An uncrossed ascending tract originating from below Clarke's column and conveying group I impulses from the hindlimb muscles in the cat. *Brain Research,* 1973, *62,* 237-241.

Armstrong, D.M. Functional significance of the inferior olive. *Physiological Reviews,* 1974, *54,* 358-417.

Armstrong, D.M., and Harvey, R.J. Responses in the inferior olive to stimulation of the cerebellar and cerebral cortices in the cat. *Journal of Physiology (London),* 1966, *187,* 553-574.

Armstrong, D.M., and Rawson, J.A. Responses of neurons in nucleus interpositus of the cerebellum to cutaneous nerve volleys in the awake cat. *Journal of Physiology (London),* 1979a, *289,* 403-423.

Armstrong, D.M., and Rawson, J.A. Activity patterns of cerebellar cortical neurons and climbing fibre afferents in the awake cat. *Journal of Physiology (London),* 1979b, *289,* 425-448.

Armstrong, D.M., and Schild, R.F. An investigation of the cerebellar corticonuclear projections in the rat using an autoradiographic tracing method. I. Projections from the vermis. *Brain Research,* 1978a, *141,* 1-9.

Armstrong, D.M., and Schild, R.F. An investigation of the cerebellar corticonuclear projections in the rat using an autoradiographic tracing method. II. Projections from the hemisphere. *Brain Research,* 1978b, *141,* 235-249.

Armstrong, D.M., Harvey, R.J., and Schild, R.F. Branching of individual olivo-cerebellar axons to terminate in more than one subdivision of the feline cerebellar cortex. *Journal of Physiology (London),* 1969, *202,* 106P-108P.

Armstrong, D.M., Harvey, R.J., and Schild, R.F. Branching of inferior olivary axons to terminate in different folia, lobules or lobes of the cerebellum. *Brain Research,* 1973a, *54,* 365-371.

Armstrong, D.M., Harvey, R.J., and Schild, R.F. Spino-olivocerebellar pathways to the posterior lobe of the cat cerebellum. *Experimental Brain Research,* 1973b, *18,* 1-18.

Armstrong, D.M., Harvey, R.J., and Schild, R.F. Cerebello-cerebellar responses mediated via climbing fibres. *Experimental Brain Research,* 1973c, *18,* 19-39.

Armstrong, D.M., Harvey, R.J., and Schild, R.F. The spatial organization of climbing fibre branching in the cat cerebellum. *Experimental Brain Research,* 1973d, *18,* 40-58.

Armstrong, D.M., Harvey, R.J., and Schild, R.F. Topographical localization of the olivo-cerebellar projection: An electrophysiological study in the cat. *Journal of Comparative Neurology,* 1974, *154,* 287-302.

Armstrong, D.M., Cogdell, B., and Harvey, R.J. Effects of afferent volleys from the limbs on the discharge patterns of interpositus neurones in cats anesthetized with alpha chloralose. *Journal of Physiology (London),* 1975, *248,* 489-517.

Arshavsky, Y., Berkinblit, M.B., Fukson, O.I., Gelfand, I.M., and Orlovsky, G.N. Recordings of neurones of the dorsal spinocerebellar tract during evoked locomotion. *Brain Research,* 1972a, *43,* 272-275.

Arshavsky, Y., Berkinblit, M.B., Fukson, O.I., Gelfand, I.M., and Orlovsky, G.N. Origin of modulation in neurones of the ventral spinocerebellar tract during locomotion. *Brain Research,* 1972b, *43,* 276-279.

Avanzino, G.L., Hösli, L., and Wolstencroft, J.H. Identification of cerebellar projecting neurones in nucleus reticularis gigantocellularis. *Brain Research,* 1966, *3,* 201-203.

Baker, J., Gibson, A., Glickstein, M., and Stein, J. Visual cells in the pontine nuclei of the cat. *Journal of Physiology (London)* 1976, *255,* 415-433.

Baker, R., and Berthoz, A. Is the prepositus hypoglossi nucleus the source of another vestibulo-ocular pathway? *Brain Research,* 1975, *86,* 121-127.

Baker, R., Gresty, M., and Berthoz, A. Neuronal activity in the prepositus hypoglossi nucleus correlated with vertical and horizontal eye movement in the cat. *Brain Research,* 1976, *101,* 366-371.

Baker, R., Berthoz, A., and Delgado-Garcia, J. Monosynaptic excitation of trochlear motoneurons following electrical stimulation of the prepositus hypoglossi nucleus. *Brain Research,* 1977, *121,* 157-161.

Baker, R.G., Precht, W., and Llinás, R. Cerebellar modulatory action on the vestibulo-trochlear pathway in the cat. *Experimental Brain Research,* 1972, *15,* 364–385.

Ban, M., and Ohno, T. Projection of cerebellar nuclear neurones to the inferior olive by descending collaterals of ascending fibres. *Brain Research,* 1977, *133,* 156–161.

Bantli, H., and Bloedel, J.R. Monosynaptic activation of a direct reticulospinal pathway by the dentate nucleus. *Pflügers Archiv,* 1975, *357,* 237–242.

Bantli, H., and Bloedel, J.R. Characteristics of the output from the dentate nucleus to spinal neurons via pathways which do not involve the primary sensorimotor cortex. *Experimental Brain Research,* 1976, *25,* 199–220.

Bantli, H., and Bloedel, J.R. Spinal input to the lateral cerebellum mediated by infratentorial structures. *Neuroscience,* 1977, *2,* 555–568.

Barmack, N.H. Immediate and sustained influence of visual olivocerebellar activity on eye movement. In R.E. Talbot and D.R. Humphrey (Eds.), *Posture and Movement: Perspective for Integrating Sensory and Motor Research on the Mammalian Nervous System.* New York: Raven, 1979.

Batini, C., Buisseret-Delmas, C., Corvisier, J., Hardy, O., and Jassik-Gerschenfeld, D. Brain stem nuclei giving fibers to lobules VI and VII of the cerebellar vermis. *Brain Research,* 1978, *153,* 241–261.

Batton, R.R., III, Jayaraman, A., Ruggiero, D., and Carpenter, M.B. Fastigial efferent projections in the monkey: An autoradiographic study. *Journal of Comparative Neurology,* 1977, *174,* 281–306.

Bava, A., Cicirata, F., Licciardello, S., Volsi, G.L., and Pantó, M.R. Fastigial nuclei projections on the ventralis lateralis (VL) thalmic nucleus neurons. *Brain Research,* 1979, *168,* 169–175.

Beitz, A.J. The topographical organization of the olivo-dentate and dentate-olivary pathways in the cat. *Brain Research,* 1976, *115,* 311–317.

Berkley, K.J., and Hand, P.J. Projections to the inferior olive of the cat. II. Comparisons of input from the gracile, cuneate and the spinal trigeminal nuclei. *Journal of Comparative Neurology,* 1978, *180,* 253–264.

Berkley, K.J., and Worden, I.G. Projections to the inferior olive of the cat. I. Comparisons of input from the dorsal column nuclei, the lateral cervical nucleus, the spino-olivary pathways, the cerebral cortex and the cerebellum. *Journal of Comparative Neurology,* 1978, *180,* 237–252.

Berman, N. Connections of the pretectum in the cat. *Journal of Comparative Neurology,* 1977, *174,* 227–254.

Bishop, G.A., McCrea, R.A., and Kitai, S.T. Afferent projections to the nucleus interpositus anterior and the lateral nucleus of the cat cerebellum. *Anatomical Record,* 1976a, *184,* 360.

Bishop, G.A., McCrea, R.A., and Kitai, S.T. A horseradish peroxidase study of the cortico-olivary projection in the cat. *Brain Research,* 1976b, *116,* 306–311.

Bishop, G.A., McCrea, R.A., Lighthall, J.W., and Kitai, S.T. An HRP and autoradiographic study of the projection from the cerebellar cortex to the nucleus interpositus anterior and nucleus interpositus posterior of the cat. *Journal of Comparative Neurology,* 1979, *185,* 735–756.

Blanks, R.H.I., Volkind, R., Precht, W., and Baker, R. Responses of cat prepositus hypoglossi neurons to horizontal angular accelerations. *Neuroscience,* 1977, *2,* 391–403.

Bloedel, J.R. Cerebellar afferent systems: A review. *Progress in Neurobiology,* 1973, *2,* 1–68.

Bloom, F.E., Hoffer, B.J., and Siggins, G.R. Studies on norepinephrine-containing afferents to Purkinje cells of rat cerebellum. I. Localization of the fibers and their synapses. *Brain Research,* 1971, *25,* 501–521.

Boesten, A.J.P., and Voogd, J. Projections of the dorsal column nuclei and the spinal cord on the inferior olive in the cat. *Journal of Comparative Neurology,* 1975, *161,* 215–238.

Bolk, L. *Das Cerebellum der Säugetiere.* Jena: Gustav Fischer, 1906.

Boylls, C.C., Jr. A theory of cerebellar function with applications to locomotion. I. The physiological role of climbing fiber inputs in anterior lobe operation. *COINS Technical Report 75C–6, Computer and Information Science,* University of Massachusetts, Amherst, 1975a.

Boylls, C.C., Jr. A theory of cerebellar function with applications to locomotion. II. The relation of anterior lobe climbing fiber function to locomotor behavior in the cat. *COINS Technical*

Report 76–1, Computer and Information Science, University of Massachusetts, Amherst, 1975b.

Boylls, C.C., Jr. Prolonged alterations of muscle activity induced in locomoting premammillary cats by microstimulation of the inferior olive. *Brain Research,* 1978, *159,* 445–450.

Braitenberg, V., and Atwood, R.P. Morphological observations on the cerebellar cortex. *Journal of Comparative Neurology,* 1958, *109,* 1–34.

Brodal, A. Experimentelle Untersuchungen über die olivocerebellare Lokalisation. *Zeitschrift für die Gesamte Neurologie und Psychiatrie,* 1940, *169,* 1–153.

Brodal, A. The cerebellar connections of the nucleus reticularis lateralis (nucleus funiculi lateralis) in the rabbit and the cat. Experimental investigations. *Acta Psychiatrica Scandinavica,* 1943, *18,* 171–233.

Brodal, A. Spinal afferents to the lateral reticular nucleus in the medulla oblongata of the cat. An experimental study. *Journal of Comparative Neurology,* 1949, *91,* 259–295.

Brodal, A. Experimental demonstration of cerebellar connexions from the perihypoglossal nuclei (nucleus intercalatus, nucleus praepositus hypoglossi, and nucleus of Roller) in the cat. *Journal of Anatomy (London),* 1952, *86,* 110–120.

Brodal, A. Reticulo-cerebellar connections in the cat. An experimental study. *Journal of Comparative Neurology,* 1953, *98,* 113–153.

Brodal, A. *The Reticular Formation of the Brain Stem. Anatomical Aspects and Functional Correlations.* Springfield, Ill.: Charles C Thomas, 1957.

Brodal, A. The olivocerebellar projection in the cat as studied with the method of retrograde axonal transport of horseradish peroxidase. II. The projection to the uvula. *Journal of Comparative Neurology,* 1976, *166,* 417–426.

Brodal, A., and Brodal, P. The organization of the nucleus reticularis tegmenti pontis in the cat in the light of experimental anatomical studies of its cerebral cortical afferents. *Experimental Brain Research,* 1971, *13,* 90–110.

Brodal, A. and Courville, J. Cerebellar corticonuclear projection in the cat Crus II. An experimental study with silver methods. *Brain Research,* 1973, *50,* 1–23.

Brodal, A., and Gogstad, A.A. Afferent connexions of the paramedian reticular nucleus of the medulla oblongata in the cat. *Acta Anatomica,* 1957, *30,* 133–151.

Brodal, A., and Hoddevik, G.H. The pontocerebellar projection to the uvula in the cat. *Experimental Brain Research,* 1978, *32,* 105–116.

Brodal, A., and Høivik, B. Site and mode of termination of primary vestibulo-cerebellar fibres in the cat. An experimental study with silver impregnation methods. *Archives Italiennes de Biologie* 1964, *102,* 1–21.

Brodal, A., and Jansen, J. The pontocerebellar projection in the rabbit and cat. Experimental investigations. *Journal of Comparative Neurology,* 1946, *84,* 31–118.

Brodal, A., and Jansen, J. Structural organization of the cerebellum. In J. Jansen and A. Brodal (Eds.), *Aspects of Cerebellar Anatomy.* Oslo: John Grundt Tanum Forlag, 1954.

Brodal, A., and Pompeiano, O. The vestibular nuclei in the cat. *Journal Anatomy (London),* 1957, *91,* 438–454.

Brodal, A., and Szikla, G. The termination of the brachium conjunctivum descendens in the nucleus reticularis tegmenti pontis. An experimental anatomical study in the cat. *Brain Research,* 1972, *39,* 337–351.

Brodal, A., and Torvik, A. Über den Ursprung der sekundären vestibulocerebellaren Fasern bei der Katze. Eine experimentell-anatomische Studie. *Archiv für Psychiatrie und Nervenkrankheiten,* 1957, *195,* 550–567.

Brodal, A., and Walberg, F. The olivocerebellar projection in the cat studied with the method of retrograde axonal transport of horseradish peroxidase. IV. The projection to the anterior lobe. *Journal of Comparative Neurology,* 1977a, *172,* 85–108.

Brodal, A., and Walberg, F. The olivocerebellar projection in the cat studied with the method of retrograde axonal transport of horseradish peroxidase. VI. The projection onto longitudinal zones of the paramedian lobule. *Journal of Comparative Neurology,* 1977b, *176,* 281–294.

Brodal, A., Walberg, F., and Blackstad, T. Termination of spinal afferents to inferior olive in cat. *Journal of Neurophysiology,* 1950, *13,* 431–454.

Brodal, A., Destombes, J., Lacerda, A.M., and Angaut, P. A cerebellar projection onto the pontine nuclei. An experimental anatomical study in the cat. *Experimental Brain Research*, 1972, *16*, 115–139.

Brodal, A., Lacerda, A.M., Destombes, J., and Angaut, P. The pattern in the projection of the intracerebellar nuclei onto the nucleus reticularis tegmenti pontis in the cat. An experimental anatomical study. *Experimental Brain Research*, 1972, *16*, 140–160.

Brodal, A., Walberg, F., and Hoddevik, G.H. The olivocerebellar projection in the cat studied with the method of retrograde axonal transport of horseradish peroxidase. *Journal of Comparative Neurology*, 1975, *164*, 449–470.

Brodal, P. The corticopontine projection in the cat. I. Demonstration of a somatotopically organized projection from the primary sensorimotor cortex. *Experimental Brain Research*, 1968a, *5*, 210–234.

Brodal, P. The corticopontine projection in the cat. Demonstration of a somatotopically organized projection from the second somatosensory cortex. *Archives Italiennes de Biologie*, 1968b, *106*, 310–332.

Brodal, P. The corticopontine projection in the cat. I. The projection from the proreate gyrus. *Journal of Comparative Neurology*, 1971a, *142*, 127–140.

Brodal, P. The corticopontine projection in the cat. II. The projection from the orbital gyrus. *Journal of Comparative Neurology*, 1971b, *142*, 141–152.

Brodal, P. The corticopontine projection from the visual cortex in the cat. I. The total projection and the projection from area 17. *Brain Research*, 1972a, *39*, 297–317.

Brodal, P. The corticopontine projection from the visual cortex in the cat. II. The projection from areas 18 and 19. *Brain Research*, 1972b, *39*, 319–335.

Brodal, P. The corticopontine projection in the cat. The projection from the auditory cortex. *Archives Italiennes de Biologie*, 1972c, *110*: 119–144.

Brodal, P. Demonstration of the somatotopically organized projection onto the paramedian lobule and the anterior lobe from the lateral reticular nucleus. An experimental study with the horseradish peroxidase method. *Brain Research*, 1975, *95*, 221–239.

Brodal, P. The corticopontine projection in the rhesus monkey. Origin and principles of organization. *Brain*, 1978, *101*, 251–283.

Brodal, P., and Walberg, F. The pontine projection to the cerebellar anterior lobe. An experimental study in the cat with retrograde transport of horseradish peroxidase. *Experimental Brain Research*, 1977, *29*, 233–248.

Brodal, P., Maršala, J., and Brodal, A. The cerebral projection to the lateral reticular nucleus in the cat, with special reference to the sensorimotor cortical area. *Brain Research*, 1967, *6*, 252–274.

Brown, J.T., Chan-Palay, V., and Palay, S.L. A study of afferent input to the inferior olivary complex in the rat by retrograde axonal transport of horseradish peroxidase. *Journal of Comparative Neurology*, 1977, *176*, 1–22.

Bruckmoser, P., Hepp-Raymond, M.C., and Wiesendanger, M. Cortical influence on single neurons of the lateral reticular nucleus of the cat. *Experimental Neurology*, 1970a, *26*, 239–252.

Bruckmoser, P., Hepp-Raymond, M.C., and Wiesendanger, M. Effects of peripheral, rubral and fastigial stimulation on the lateral reticular nucleus of the cat. *Experimental Neurology*, 1970b, *27*, 388–398.

Bruggencate, G. ten, Teichmann, R., and Weller, E. Neuronal activity in the lateral vestibular nucleus of the cat. III. Inhibitory actions of cerebellar Purkinje cells evoked via mossy and climbing fiber afferents. *Pflügers Archiv*, 1972, *337*, 147–162.

Burke, R., Lundberg, A., and Weight, F. Spinal border cell origin of the ventral spinocerebellar tract. *Experimental Brain Research*, 1971, *12*, 283–294.

Burton, J.E., and Onoda, N. Interpositus neuron discharge in relation to a voluntary movement. *Brain Research*, 1977, *121*, 167–172.

Burton, J.E., Bloedel, J.R., and Gregory, R.S. Electrophysiological evidence for an input to lateral reticular nucleus from collaterals of dorsal spinocerebellar and cuneocerebellar fibers. *Journal of Neurophysiology*, 1971, *34*, 885–897.

Carpenter, M.B., Brittin, G.M., and Pines, J. Isolated lesions of the fastigial nuclei in the cat. *Journal of Comparative Neurology*, 1958, *109*, 65–90.

Carpenter, M.B., Stein, B.M., and Peter, P. Primary vestibulocerebellar fibers in the monkey: Distribution of fibers arising from distinctive cell groups of the vestibular ganglia. *American Journal of Anatomy*, 1972, *135*, 221–250.

Chambers, W.W., and Sprague, J.M. Functional localization in the cerebellum. I. Organization in longitudinal corticonuclear zones and their contribution to the control of posture, both extrapyramidal and pyramidal. *Journal of Comparative Neurology*, 1955a, *103*, 105–129.

Chambers, W.W., and Sprague, J.M. Functional localization in the cerebellum. II. Somatotopic organization in cortex and nuclei. *A.M.A. Archives of Neurology and Psychiatry*, 1955b, *74*, 653–680.

Chan-Palay, V. *Cerebellar Dentate Nucleus, Organization, Cytology and Transmitters*. Berlin: Verlag, 1977.

Chan-Palay, V., Palay, S.L., Brown, J.T., and Van Itallie, C. Sagittal organization of olivocerebellar and reticulocerebellar projections: Autoradiographic studies with ^{35}S-methionine. *Experimental Brain Research*, 1977, *30*, 561–576.

Clarke, R.H., and Horsley, V. On the intrinsic fibers of the cerebellum, its nuclei and its efferent tracts. *Brain*, 1905, *28*, 13–29.

Clendenin, M., Ekerot, C.F., Oscarsson, O., and Rosen, I. The lateral reticular nucleus in the cat. I. Mossy fiber distribution in cerebellar cortex. *Experimental Brain Research*, 1974a, *21*, 473–486.

Clendenin, M., Ekerot, C.F., Oscarsson, O., and Rosen, I. The lateral reticular nucleus in the cat. II. Organization of component activated from bilateral ventral flexor reflex tract. *Experimental Brain Research*, 1974b, *21*, 487–500.

Clendenin, M., Ekerot, C.F., and Oscarsson, O. The lateral reticular nucleus in the cat. III. Organization of component activated from ipsilateral forelimb tract. *Experimental Brain Research*, 1974, *21*, 501–513.

Clendenin, M., Ekerot, C.F., and Oscarsson, O. The lateral reticular nucleus in the cat. IV. Activation from dorsal funiculus and trigeminal afferents. *Experimental Brain Research*, 1975, *24*, 131–144.

Cohen, D., Chambers, W.W., and Sprague, J.M. Experimental study of the efferent projections from the cerebellar nuclei to the brainstem of the cat. *Journal of Comparative Neurology*, 1958, *109*, 233–259.

Combs, C.M. Electro-anatomical study of cerebellar localization-stimulation of various afferents. *Journal of Neurophysiology*, 1954, *17*, 122–143.

Comolli, A. Per una nuova divisione del cervelletto dei mammiferi. *Archive Italiano di Anatomia e di Embriologia*, 1910, *9*, 247–273.

Cook, J.R., and Wiesendanger, M. Input from trigeminal cutaneous afferents to neurones of the inferior olive in rats. *Experimental Brain Research*, 1976, *26*, 193–202.

Cooke, J.D., Larson, B., Oscarsson, O., and Sjölund, B. Origin and termination of cuneocerebellar tract. *Experimental Brain Research*, 1971a, *13*, 339–358.

Cooke, J.D., Larson, B., Oscarsson, O., and Sjölund, B. Organization of afferent connections to cuneocerebellar tract. *Experimental Brain Research*, 1971b, *13*, 359–377.

Corvaja, N., Grofová, I., Pompeiano, O., and Walberg, F. The lateral reticular nucleus in the cat. I. An experimental anatomical study of its spinal and supraspinal afferent connections. *Neuroscience*, 1977, *2*, 537–553.

Corvaja, N., and Pompeiano, O. Identification of cerebellar corticovestibular neurons retrogradely labeled with horseradish peroxidase. *Neuroscience*, 1979, *4*, 507–515.

Coulter, J.D., Mergner, T., and Pompeiano, O. Effect of tilting on the responses of lateral reticular nucleus neurons to somatic afferent stimulation. *Archives Italiennes de Biologie*, 1977, *115*, 294–331.

Courville, J. Rubrobulbar fibres to the facial nucleus and the lateral reticular nucleus (nucleus of the lateral funiculus). An experimental study in the cat with silver impregnation methods. *Brain Research*, 1966a, *1*, 317–337.

Courville, J. Somatotopical organization of the projection from the nucleus interpositus anterior of the cerebellum to the red nucleus. An experimental study in the cat with silver impregnation methods. *Experimental Brain Research*, 1966b, *2*, 191–215.

Courville, J., and Coulombe, G. Pontine projection to cerebellar cortex and nuclei. *Society for Neuroscience Abstracts*, 1977, *3*, 56.

Courville, J., and Diakiw, N. Cerebellar corticonuclear projection in the cat. The vermis of the anterior and posterior lobes. *Brain Research*, 1976, *110*, 1–20.

Courville, J., and Faraco-Cantin, F. On the origin of the climbing fibers of the cerebellum. An experimental study in the cat with an autoradiographic tracing method. *Neuroscience*, 1978, *3*, 797–809.

Courville, J., and Otabe, S. The rubro-olivary projection in the macaque: An experimental study with silver impregnation methods. *Journal of Comparative Neurology*, 1974, *158*, 479–494.

Courville, J., Diakiw, N., and Brodal, A. Cerebellar corticonuclear projection in the cat. The paramedian lobule. An experimental study with silver methods. *Brain Research*, 1973, *50*, 25–45.

Courville, J., Augustine, J.R., and Martel, P. Projections from the inferior olive to the cerebellar nuclei in the cat demonstrated by retrograde transport of horseradish peroxidase. *Brain Research*, 1977, *130*, 405–419.

Crichlow, E.C., and Kennedy, T.T. Functional characteristics of neurons in the lateral reticular nucleus with reference to localized cerebellar pontentials. *Experimental Neurology*, 1967, *18*, 141–153.

Crill, W.E. Unitary multiple-spiked responses in the cat inferior olive nucleus. *Journal of Neurophysiology*, 1970, *33*, 199–209.

Cummings, J.F., and Petras, J.M. The origin of spinocerebellar pathways. I. The nucleus cervicalis centralis of the cranial cervical spinal cord. *Journal of Comparative Neurology*, 1977, *173*, 655–692.

Desclin, J.C. Histological evidence supporting the inferior olive as the major source of cerebellar climbing fibers in the rat. *Brain Research*, 1974a, *77*, 365–384.

Desclin, J.C. Démonstration en microscopie optique de la dégénérescence terminale d' afférences d'origine olivarire inférieure dans le noyau vestibulaire latéral (Deiters') chez le rat. *Comptes Rendus de L'Académie des Sciences, Série D (Paris)*, 1974b, *278*, 2931–2934.

Desclin, J.C. Early terminal degeneration of cerebellar climbing fibers after destruction of the inferior olive in the rat. Synaptic relationships in the molecular layer. *Anatomical Embryology*, 1976, *149*, 87–112.

Dom, R., King, J.S., and Martin, G.F. Evidence for two direct cerebello-olivary connections. *Brain Research*, 1973, *57*, 498–501.

Dow, R.S. The fiber connections of the posterior parts of the cerebellum in the rat and cat. *Journal of Comparative Neurology*, 1936, *63*, 527–548.

Dow, R.S. Efferent connections of the flocculonodular lobe in *Macaca mulatta*. *Journal of Comparative Neurology*, 1938, *68*, 297–305.

Dow, R.S. Cerebellar action potentials in response to stimulation of various afferent connections. *Journal of Neurophysiology*, 1939, *2*, 543–555.

Dow, R.S. The evolution and anatomy of the cerebellum. *Biological Reviews*, 1942, *17*, 179–220.

Dow, R.S. Some aspects of cerebellar physiology. *Journal of Neurosurgery*, 1961, *18*, 512–530.

Dow, R.S., and Moruzzi, G. *The Physiology and Pathology of the Cerebellum*. Minneapolis: University of Minnesota Press, 1958.

Dufossé, M., Ito, M., and Miyashita, Y. Functional localization in the rabbit's flocculus determined in relationship with eye movements. *Neuroscience Letters*, 1977, *5*, 273–277.

Eccles, J.C., Llinás, R., and Sasaki, K. The excitatory synaptic action of climbing fibres on the Purkinje cells of the cerebellum. *Journal of Physiology (London)*, 1966a, *182*, 268–296.

Eccles, J.C., Llinás, R., and Sasaki, K. Parallel fibre stimulation and the responses induced thereby in the Purkinje cells of the cerebellum. *Experimental Brain Research*, 1966b, *1*, 17–39.

Eccles, J.C., Llinás, R., and Sasaki, K. The mossy fibre-granule cell relay of the cerebellum and its inhibitory control of Golgi cells. *Experimental Brain Research*, 1966c, *1*, 82–101.

Eccles, J.C., Provini, L., Strata, P., and Tábořiková, H. Analysis of electrical potentials evoked in the cerebellar anterior lobe by stimulation of hindlimb and forelimb nerves. *Experimental Brain Research*, 1968a, *6*, 171-194.

Eccles, J.C., Provini, L., Strata, P., and Tábořiková, H. Topographical investigations on the climbing fiber inputs from forelimb and hindlimb afferents to the cerebellar anterior lobe. *Experimental Brain Research*, 1968b, *6*, 195-215.

Eccles, J.C., Faber, D.S., Murphy, J.T., Sabah, N.H., and Tábořiková, H. Investigations on integration of mossy fiber inputs to Purkinje cells in the anterior lobe. *Experimental Brain Research*, 1971, *13*, 54-77.

Eccles, J.C., Sabah, N.H., Schmidt, R.F., and Tábořiková, H. Integration of Purkyne cells, mossy and climbing fiber inputs from cutaneous mechanoreceptors. *Experimental Brain Research*, 1972, *15*, 498-520.

Eccles, J.C., Rantucci, T., Rosén, I., Scheid, P., and Tábořiková, H. Somatotopic studies on cerebellar interpositus neurons. *Journal of Neurophysiology*, 1974, *37*, 1449-1459.

Eccles, J.C., Rantucci, T., Sabah, N.H., and Tábořiková, H. Somatotopic studies of the cerebellar fastigial cells. *Experimental Brain Research*, 1974, *19*, 100-118.

Eccles, J.C., Rosén, I., Scheid, P., and Tábořiková, H. Temporal patterns of responses of interpositus neurones to peripheral afferent stimulation. *Journal of Neurophysiology*, 1974a, *37*, 1424-1437.

Eccles, J.C., Rosén, I., Scheid, P., and Tábořiková, H. Patterns of convergence onto interpositus neurones from peripheral afferents. *Journal of Neurophysiology*, 1974b, *37*, 1438-1448.

Eccles, J.C., Sabah, N.H., and Tábořiková, H. Excitatory and inhibitory responses of neurons of the cerebellar fastigial nucleus. *Experimental Brain Research*, 1974a, *19*, 61-77.

Eccles, J.C., Sabah, N.H., and Tábořiková, H. The pathways responsible for excitation and inhibition of fastigial neurones. *Experimental Brain Research*, 1974b, *19*, 78-99.

Eccles, J.C., Nicoll, R.A., Schwarz, D.W., Tábořiková, H., and Willey, T.J. Reticulospinal neurons with and without monosynaptic inputs from cerebellar nuclei. *Journal of Neurophysiology*, 1975, *38*, 513-530.

Eccles, R.M., and Lundberg, A. Synaptic action in motoneurons by afferents which may evoke the flexion reflex. *Archives Italiennes de Biologie*, 1959, *97*, 199-221.

Edinger, L. Ueber die Einteilung des Cerebellums. *Anatomischer Anzeiger*, 1910, *35*, 319-338.

Edwards, S.B. The ascending and descending projections of the red nucleus in the cat: An experimental study using an autoradiographic tracing method. *Brain Research*, 1972, *48*, 45-64.

Edwards, S.B. Autoradiographic studies of the projections of the midbrain reticular formation. Descending projections of the nucleus cuneiformis. *Journal of Comparative Neurology*, 1975, *161*, 341-358.

Edwards, S.B., Rosenquist, A.C., and Palmer, L.A. An autoradiographic study of ventral lateral geniculate projections in the cat. *Brain Research*, 1974, *72*, 282-287.

Ekerot, C.F., and Larson, B. Differential termination of the exteroceptive and proprioceptive components of the cuneocerebellar tract. *Brain Research*, 1972, *36*, 420-424.

Ekerot, C.F., and Larson, B. Correlation between sagittal projection zones of climbing and mossy fibre paths in cat cerebellar anterior lobe. *Brain Research*, 1973, *64*, 446-450.

Ekerot, C.F., and Larson, B. The dorsal spino-olivocerebellar system in the cat. I. Functional organization and termination in the anterior lobe. *Experimental Brain Research*, 1979a, *36*, 201-217.

Ekerot, C.F., and Larson, B. The dorsal spino-olivocerebellar system in the cat. II. Somatotopical organization. *Experimental Brain Research*, 1979b, *36*, 219-232.

Eller, T., and Chan-Palay, V. Afferents to the cerebellar lateral nucleus: Evidence for retrograde transport of horseradish peroxidase after pressure injections from micropipettes. *Journal of Comparative Neurology*, 1976, *166*, 285-302.

Escobar, A., Sampedro, E.D., and Dow, R.S. Quantitative data on the inferior olivary nucleus in man, cat and vampire bat. *Journal of Comparative Neurology*, 1968, *132*, 397-404.

Estable, C. Notes sur la structure comparative de l'écorce cérébelleuse et dérivées physiologiques possibles. *Trabajos del Laboratorio de Investigaciónes Biologicas de la Universidad de Madrid*, 1923, *21*, 169-256.

Evarts, E.V., and Thach, W.T. Motor mechanisms of the CNS: Cerebro-cerebellar interrelations. *Annual Review of Physiology*, 1969, *31*, 451-498.

Faber, D.S., and Murphy, J.T. Axonal branching in the climbing fiber pathway to the cerebellum. *Brain Research*, 1969, *15*, 262-267.

Flood, S., and Jansen, J. On the cerebellar nuclei in the cat. *Acta Anatomy*, 1961, *46*, 52-72.

Flumerfelt, B.A., Otabe, S., and Courville, J. Distinct projections to the red nucleus from the dentate and interposed nuclei in the monkey. *Brain Research*, 1973, *50*, 408-414.

Fox, C.A. The intermediate cells of lugaro in the cerebellar cortex of the monkey. *Journal of Comparative Neurology*, 1959, *112*, 39-54.

Fox, C.A., Siegesmund, K.A., andDutta, C.R. The Purkinje cell dendritic branchlets and their relation with the parallel fibers: Light and electron microscopic observations. In M.M. Cohen and R.S. Snider (Eds.), *Morphological and Biochemical Correlates of Neural Activity*. New York: Harper and Row, 1964.

Fox, C.A., Hillman, D., Siegesmund, K.A., and Dutta, C.R. The primate cerebellar cortex: A Golgi and electron microscopical study. *Progress in Brain Research*, 1967, *25*, 174-225.

Frankfurter, A., Weber, J.T., and Harting, J.K. Brain stem projections to lobule VII of the posterior vermis in the squirrel monkey: As demonstrated by the retrograde axonal transport of tritiated horseradish peroxidase. *Brain Research*, 1977, *124*, 135-139.

Frederickson, R.C.A., Neuss, M., Morzorati, S.L., and McBride, W.J. A comparison of the inhibitory effects of taurine and GABA on identified Purkinje cells and other neurons in the cerebellar cortex of the rat. *Brain Research*, 1978, *145*, 117-126.

Freedman, R., Hoffer, B.J., Woodward, D.J., and Puro, D. Interaction of norepinephrine and climbing fibers. *Experimental Neurology*, 1977, *55*, 269-288.

Fukuda, J., Highstein, S.M., and Ito, M. Cerebellar control of the vestibulo-ocular reflex investigated in rabbit 3rd nucleus. *Experimental Brain Research*, 1972, *14*, 511-526.

Furuya, N., Kawano, K., and Shimazu, H. Functional organization of vestibulofastigial projection in the horizontal semicircular canal system in the cat. *Experimental Brain Research*, 1975, *24*, 75-87.

Gibson, A., Baker, J., Mower, G., and Glickstein, M. Corticopontine cells in area 18 of the cat. *Journal of Neurophysiology*, 1978, *41*, 484-495.

Gould, B.B. The organization of afferents to the cerebellar cortex in the cat: Projections from the deep cerebellar nuclei. *Journal of Comparative Neurology*, 1979, *184*, 27-42.

Gould, B.B., and Graybiel, A.M. Afferents to the cerebellar cortex in the cat: Evidence for an intrinsic pathway leading from the deep nuclei to the cortex. *Brain Research*, 1976, *110*, 601-611.

Graham, J. An autoradiographic study of the efferent connections of the superior colliculus in the cat. *Journal of Comparative Neurology*, 1977, *173*, 629-654.

Grant, G. Projection of the external cuneate nucleus onto the cerebellum in the cat: An experimental study using silver methods. *Experimental Neurology*, 1962a, *5*, 179-195.

Grant, G. Spinal course and somatotopically localized termination of the spinocerebellar tracts: An experimental study in the cat. *Acta Physiologica Scandinavica*, 1962b, *56*,(193), 45.

Grant, G., Oscarsson, O., and Rosén, I. Functional organization of the spino-reticulocerebellar path with identification of its spinal component. *Experimental Brain Research*, 1966, *1*, 306-319.

Graybiel, A.M. Some efferents of the pretectal region in the cat. *Anatomical Record*, 1974a, *178*, 365.

Graybiel, A.M. Visuo-cerebellar and cerebello-visual connections involving the ventral lateral geniculate nucleus. *Experimental Brain Research*, 1974b, *20*, 303-306.

Graybiel, A.M. Organization of oculomotor pathways in the cat and rhesus monkey. In R. Baker and A. Berthoz (Eds.), *Control of Gaze by Brain Stem Neurons*. New York: Elsevier/North Holland, 1977.

Graybiel, A.M., and Hartwieg, E.A. Some afferent connections of the oculomotor complex in the cat: An experimental study with tracer techniques. *Brain Research*, 1974, *81*, 543-551.

Graybiel, A.M., Nauta, H.J.W., Lasek, R.J., and Nauta, W.J.H. A cerebello-olivary pathway in the cat: An experimental study using autoradiographic tracing techniques. *Brain Research,* 1973, *58,* 205–211.

Gresty, M., and Baker, R. Neurons with visual receptive field, eye movement, neck displacement sensitivity within and around the nucleus prepositus hypoglossi in the alert cat. *Experimental Brain Research,* 1976, *24,* 429–433.

Grillner, S., Hongo, T., and Lund, S. The origin of descending fibers monosynaptically activating spinoreticular neurons. *Brain Research,* 1968, *10,* 259–262.

Groenewegen, H.J., and Voogd, J. The parasagittal zonation within the olivocerebellar projection. I. Climbing fiber distribution in the vermis of the cat cerebellum. *Journal of Comparative Neurology,* 1977, *174,* 417–488.

Groenewegen, H.J., Boesten, A.J.P., and Voogd, J. The dorsal column nuclear projections to the nucleus ventralis posterior lateralis thalami and the inferior olive in the cat: An autoradiographic study. *Journal of Comparative Neurology,* 1975, *162,* 505–518.

Groenewegen, H.J., Voogd, J., and Freedman, S.L. The parasagittal zonation within the olivocerebellar projection. II. Climbing fiber distribution in the intermediate and hemispheric parts of cat cerebellum. *Journal of Comparative Neurology,* 1979, *183,* 551–602.

Haines, D.E. Cerebellar cortical efferents of the posterior lobe vermis in a prosimian primate (Galago) and tree shrew (Tupaia). *Journal of Comparative Neurology,* 1975, *163,* 21–40.

Haines, D.E. Cerebellar corticonuclear and corticovestibular fibers of the anterior lobe vermis in a prosimian primate *(Galago senegalensis). Journal of Comparative Neurology,* 1976, *170,* 67–96.

Haines, D.E. Contralateral nucleocortical cells of the paraflocculus of tree shrew (Tupaia). *Neuroscience Letters,* 1978, *8,* 183–190.

Haines, D.E., and Rubertone, J.A. Cerebellar corticonuclear fibers: Evidence of zones in the primate anterior lobe. *Neuroscience Letters,* 1977, *6,* 231–236.

Haines, D.E., and Rubertone, J.A. Cerebellar corticonuclear fibers of the dorsal culminate lobule (anterior lobe, lobule V) in a prosimian primate, *Galago senegalensis. Journal of Comparative Neurology,* 1979, *186,* 321–342.

Hámori, J., and Szentágothai, J. The Purkinje cell baskets: Ultrastructure of an inhibitory synapse. *Acta Biologica Academiae Scientiarum Hungaricae,* 1965, *15,* 465–479.

Hámori, J., and Szentágothai, J. Identification under the electron microscope of climbing fibers and their synaptic contacts. *Experimental Brain Research,* 1966, *1,* 65–81.

Hampson, J.L. Relationships between cat cerebral and cerebellar cortices. *Journal of Neurophysiology,* 1949, *12,* 37–50.

Hampson, J.L., Harrison, C.R., and Woolsey, C.N. Cerebro-cerebellar projections and the somatotopic localization of motor function in the cerebellum. In P. Bard (Ed.), *Research Publications of the Association for Research in Nervous and Mental Disease.* Vol. 30: *Patterns of Organization in the Nervous System.* Baltimore: Williams & Wilkins, 1952.

Harting, J.K. Descending pathways from the superior colliculus: An autoradiographic analysis in the rhesus monkey *(Macaca mulatta). Journal of Comparative Neurology,* 1977, *173,* 583–612.

Hayashi, M. Einige wichtige Tatsachen aus der ontogenetischen Entwicklung des menschlichen Kleinhirns. *Deutsche Zeitschrift für Nervenheilkunde,* 1924, *81,* 74–82.

Hekmatpanah, J. Organization of tactile dermatomes C_1 through L_4 in cat. *Journal of Neurophysiology,* 1961, *24,* 131–140.

Henkel, C.K., Linauts, M., and Martin, G.F. The origin of the annulo-olivary tract with notes on other mesencephalo-olivary pathways. A study by the horseradish peroxidase method. *Brain Research,* 1975, *100,* 145–150.

Hess, R., and Simpson, J.I. Visual and somatosensory messages to the rabbit's cerebellar flocculus. *Neuroscience Letters, Supplement,* 1978, *1,* 146.

Highstein, S.M. Synaptic linkage in the vestibulo-ocular and cerebello-vestibular pathways to the VIth nucleus in the rabbit. *Experimental Brain Research,* 1973, *17,* 301–314.

Hillman, D.E. Morphological organization of the frog cerebellar cortex: A light and electron microscopic study. *Journal of Neurophysiology,* 1969a, *32,* 818–846.

Hillman, D.E. Neuronal organization of the cerebellar cortex in amphibia and reptilia. In R. Lli-nás (Ed.), *Neurobiology of Cerebellar Evolution and Development.* Chicago: American Medical Association, 1969b.

Hoddevik, G.H. The pontocerebellar projection onto the paramedian lobule in the cat: An experimental study with the use of horseradish peroxidase as a tracer. *Brain Research,* 1975, *95,* 291–307.

Hoddevik, G.H. The pontine projection to the flocculus studied by means of retrograde axonal transport of horseradish peroxidase in the rabbit. *Experimental Brain Research,* 1977, *30,* 511–526.

Hoddevik, G.H. The projection from nucleus reticularis tegmenti pontis onto the cerebellum in the cat. *Anatomy and Embryology,* 1978, *153,* 227–242.

Hoddevik, G.H., and Brodal, A. The olivocerebellar projection studied with the method of retrograde axonal transport of horseradish peroxidase. V. The projections to the flocculonodular lobe and the paraflocculus in the rabbit. *Journal of Comparative Neurology,* 1977, *176,* 269–280.

Hoddevik, G.H., Brodal, A., and Walberg, F. The olivocerebellar projection in the cat studied with the method of retrograde axonal transport of horseradish peroxidase. III. The projection to the vermal visual area. *Journal of Comparative Neurology,* 1976, *169,* 155–170.

Hoddevik, G.H., Brodal, A., Kawamura, K., and Hashikawa, T. The pontine projection in cerebellar vermal visual area studied by means of retrograde axonal transport of horseradish peroxidase. *Brain Research,* 1977, *123,* 209–227.

Hodgkin, A.L. The local electric changes associated with repetitive action in a nonmedulated axon. *Journal of Physiology (London),* 1948, *107,* 165–181.

Hohman, L.B. The efferent connections of the cerebellar cortex: Investigations based upon experimental extirpations in the cat. In F. Tilney, T.K. Davis, and H.A. Riley (Eds.), *Research Publications of the Association for Research in Nervous and Mental Disease* (Vol. VI). Baltimore: Williams & Wilkins Co., 1929.

Holmqvist, B., and Lundberg, A. Differential supraspinal control of synaptic actions evoked by volleys in the flexion reflex afferents in alpha motoneurons. *Acta Physiologica Scandinavica* 1961, *54* Supplementum 186, 1–51.

Holmqvist, B., Oscarsson, O., and Rosén, I. Functional organization of the cuneocerebellar tract in the cat. *Acta Physiologica Scandinavica,* 1963, *58,* 216–235.

Hubbard, J.I., and Oscarsson, O. Localization of the cell bodies of the ventral spinocerebellar tract in lumbar segments of the cat. *Journal of Comparative Neurology,* 1962, *118,* 199–204.

Ingvar, S. Zur Phylo- und Ontogenese des Kleinhirns. *Folia Neurobiology,* 1918, *11,* 205–495.

Ingvar, S. On cerebellar localization. *Brain,* 1923, *46,* 301–335.

Ingvar, S. Studies in neurology. I. The phylogenetic continuity of the central nervous system. II. On cerebellar function. *Bulletin of the Johns Hopkins Hospital,* 1928, *43,* 315–362.

Ishikawa, J., Kawaguchi, S., and Rowe, M.J. Actions of afferent impulses from muscle receptors on cerebellar Purkyne cells. II. Responses to muscle contraction: Effects mediated via the climbing fiber pathway. *Experimental Brain Research,* 1972, *16,* 104–114.

Ito, M. Neural design of the cerebellar motor control system. *Brain Research,* 1972a, *40,* 81–84.

Ito, M. Cerebellar control of the vestibular neurons: Physiology and pharmacology. *Progress in Brain Research,* 1972b, *37,* 377–390.

Ito, M. Adaptive modification of the vestibulo-ocular reflex in rabbits affected by visual inputs and its possible neuronal mechanisms. *Progress in Brain Research,* 1979, *50,* 757–761.

Ito, M., and Yoshida, M. The cerebellar-evoked monosynaptic inhibition of Deiters' neurons. *Experimentia,* 1964, *20,* 515.

Ito, M., and Yoshida, M. The origin of cerebellar-induced inhibition of Deiters' neurons. I. Monosynaptic initiation of the inhibitory postsynaptic potentials. *Experimental Brain Research,* 1966, *2,* 330–349.

Ito, M., Yoshida, M., and Obata, K. Monosynaptic inhibition of the intracerebellar nuclei induced from the cerebellar cortex. *Experimentia,* 1964, *20,* 575–576.

Ito, M., Obata, K., and Ochi, R. The origin of cerebellar-induced inhibition of Deiters' neurons.

II. Temporal correlation between the trans-synaptic activation of Purkinje cells and the inhibition of Deiters neurons. *Experimental Brain Research*, 1966, *2*, 350–364.

Ito, M., Yoshida, M., Obata, K., Kawai, N., and Udo, M. Inhibitory control of intracerebellar nuclei by the Purkinje cell axons. *Experimental Brain Research*, 1970, *10*, 64–80.

Ito, M., Nisimaru, N., and Yamamoto, M. Specific patterns of neuronal connexions involved in the control of the rabbit's vestibuloocular reflexes by the cerebellar flocculus. *Journal of Physiology (London)*, 1977, *265*, 833–854.

Itoh, K. Efferent projections of the pretectum in the cat. *Experimental Brain Research*, 1977, *30*, 89–105.

Jakob, A. Däs Kleinhirn. In von Möllendorff (Ed.), *Handbuch der mikroskopischen Anatomie des Menschen*, IV:1. Berlin: Springer, 1928a.

Jakob, A. Zur Problem der morphologischen und funktionellen Gliederung des Kleinhirns. *Deutsche Zeitschrift für Nervenheilkunde*, 1928b, *105*, 217–233.

Jansen, J., and Brodal, A. Experimental studies on the intrinsic fibers of the cerebellum. II. The cortico-nuclear projection. *Journal of Comparative Neurology*, 1940, *73*, 267–321.

Jansen, J., and Brodal, A. Experimental studies on the intrinsic fibers of the cerebellum. The cortico-nuclear projection in the rabbit and the monkey (Macacus rhesus). *Skrifter Utgill av det Norske Videnskaps-Academi i Oslo, I. Matematisk-Naturvidenskapelig Klasse*, 1942, *3*, 1–50.

Jansen, J., and Brodal, A. (Eds.). *Aspects of Cerebellar Anatomy*. Oslo: J.G. Tanum, 1954.

Kawamura, K. The pontine projection from the inferior colliculus in the cat. An experimental anatomical study. *Brain Research*, 1975, *95*, 309–322.

Kawamura, K., and Brodal, A. The tectopontine projection in the cat: An experimental anatomical study with comments on pathways for teleceptive impulses to the cerebellum. *Journal of Comparative Neurology*, 1973, *149*, 371–390.

Kawamura, K., and Hashikawa, T. Olivocerebellar projections in the cat studied by means of anterograde axonal transport of labeled amino acids as tracers. *Neuroscience*, 1979, *4*, 1615–1633.

Kawamura, K., Brodal, A., and Hoddevik, G. The projection of the superior colliculus onto the reticular formation of the brain stem. An experimental anatomical study in the cat. *Experimental Brain Research*, 1974, *19*, 1–19.

King, J.S., Andrezik, J.A., Falls, W.M., and Martin, G.F. Synaptic organization of cerebello-olivary circuit. *Experimental Brain Research*, 1976, *26*, 159–170.

Kitai, S.T., Kennedy, D.T., Morin, F., and Gardner, E. The lateral reticular nucleus of the medulla oblongata of the cat. *Experimental Neurology*, 1967, *17*, 65–73.

Kitai, S.T., DeFrance, J.F., Hatada, K., and Kennedy, D.T. Electrophysiological properties of lateral reticular nucleus cells: II. Synaptic activation. *Experimental Brain Research*, 1974, *21*, 419–432.

Kitai, S.T., Kocsis, J.D., and Kiyohara, T. Electrophysiological properties of nucleus reticularis tegmenti pontis cells: Antidromic and synaptic activation. *Experimental Brain Research*, 1976, *24*, 294–309.

Kitai, S.T., McCrea, R.A., Preston, R.J., and Bishop, G.A. Electrophysiological and horseradish peroxidase studies of precerebellar afferents to the nucleus interpositus anterior. I. Climbing fiber system. *Brain Research*, 1977, *122*, 197–214.

Korte, G.E., and Mugnaini, E. The cerebellar projection of the vestibular nerve in the cat. *Journal of Comparative Neurology*, 1979, *184*, 265–278.

Kotchabhakdi, N., and Walberg, F. Primary vestibular afferent projections to the cerebellum as demonstrated by retrograde axonal transport of horseradish peroxidase. *Brain Research*, 1978a, *142*, 142–146.

Kotchabhakdi, N., and Walberg, F. Cerebellar afferent projections from the vestibular nuclei in the cat: An experimental study with the method of retrograde axonal transport of horseradish peroxidase. *Experimental Brain Research*, 1978b, *31*, 591–604.

Kotchabhakdi, N., Hoddevik, G.H., and Walberg, F. Cerebellar afferent projections from the perihypoglossal nuclei: An experimental study with the method of retrograde axonal transport of horseradish peroxidase. *Experimental Brain Research*, 1978a, *31*, 13–29.

Kotchabhakdi, N., Walberg, F., and Brodal, A. The olivocerebellar projection in the cat studied with the method of retrograde axonal transport of horseradish peroxidase. VII. The projection to lobules simplex, Crus I and II. *Journal of Comparative Neurology,* 1978b, *182,* 293–314.

Kron, G. *Tensor Analysis of Networks.* New York: Wiley, 1939.

Kuhn, R.A. Organization of tactile dermatomes in cat and man. *Journal of Neurophysiology,* 1953, *16,* 169–182.

Kuno, M., Munoz-Martinez, E.J., and Randic, M. Sensory inputs to neurones in Clarke's column from muscle, cutaneous and joint receptors. *Journal of Physiology,* 1973, *228,* 327–342.

Künzle, H. The topographic organization of spinal afferents to the lateral reticular nucleus of the cat. *Journal of Comparative Neurology,* 1973, *149,* 103–116.

Künzle, H. Autoradiographic tracing of the cerebellar projections from the lateral reticular nucleus in the cat. *Experimental Brain Research,* 1975, *22,* 255–266.

Ladpli, R., and Brodal, A. Experimental studies of commissural and reticular formation projections from the vestibular nuclei in the cat. *Brain Research,* 1968, *8,* 65–96.

Lamarre, Y., deMontigny, C., Dumont, M., and Weiss, M. Harmaline-induced rhythm activity of cerebellar and lower brain stem neurons. *Brain Research,* 1971, *32,* 246–250.

Langelaan, J.W. On the development of the external form of the human cerebellum. *Brain,* 1919, *42,* 130–170.

Larramendi, L.M.H., and Victor, T. Synapses on spines of the Purkinje cell of the mouse. An electron microscopic study. *Brain Research,* 1967, *5,* 15–30.

Larsell, O. Morphogensis and evolution of the cerebellum. *A.M.A. Archives of Neurology and Psychiatry,* 1934, *31,* 373–395.

Larsell, O. The development and morphology of the cerebellum in the opossum. Part II. Later development and adult. *Journal of Comparative Neurology,* 1936, *63,* 251–291.

Larsell, O. The cerebellum: A review and interpretation. A.M.A. Archives of Neurology and Psychiatry, 1937, *38,* 580–607.

Larsell, O. The development and subdivisions of the cerebellum of birds. *Journal of Comparative Neurology,* 1948, *89,* 123–189.

Larsell, O. The morphogenesis and adult pattern of the lobules and fissures of the cerebellum of the white rat. *Journal of Comparative Neurology,* 1952, *97,* 281–356.

Larsell, O. Cerebellum of cat and monkey. *Journal of Comparative Neurology,* 1953, *99,* 135–200.

Larsell, O. *The Comparative Anatomy and Histology of the Cerebellum from Myxinoids through Birds.* J. Jansen (Ed.). Minneapolis: University of Minnesota Press, 1967.

Larsell, O. *The Comparative Anatomy of Histology of the Cerebellum from Monotremes through Primates,* J. Jansen (Ed.). Minneapolis: University of Minnesota Press, 1970.

Larsell, O., and Jansen, J. *The Comparative Anatomy and Histology of the Cerebellum: The Human Cerebellum, Cerebellar Connections, and Cerebellar Cortex.* Minneapolis: University of Minnesota Press, 1972.

Leicht, R., and Schmidt, R. F. Somatotopic studies on the vermal cortex of the cerebellar anterior lobe of unanesthetized cat. *Experimental Brain Research,* 1977, *27,* 479–490.

Leicht, R., Rowe, M.J., and Schmidt, R. F. Cutaneous convergence onto the climbing fiber input to cerebellar Purkinje cells. *Journal of Physiology* (London), 1973a, *228,* 601–618.

Leicht, R., Rowe, M.J., and Schmidt, R.F. Cortical and peripheral modification of cerebellar climbing fibre activity arising from cutaneous mechanoreceptors. *Journal of Physiology* (London), 1973b, *228,* 619–635.

Leicht, R., Rowe, M.J., and Schmidt, R.F. Mossy and climbing fiber inputs from cutaneous mechanoreceptors to cerebellar Purkyne cells in unanesthetized cats. *Experimental Brain Research,* 1977, *27,* 459–477.

Linauts, M., and Martin, G.F. An autoradiographic study of midbrain-diencephalic projections to the inferior olivary nucleus in the opossum *(Didelphis virginiana). Journal of Comparative Neurology,* 1978, *179,* 325–354.

Lindstrom, S., and Takata, M. Monosynaptic excitation of dorsal spinocerebellar tract neurones from low threshold joint afferents. *Acta Physiologica Scandinavica,* 1972, *84,* 430–432.

Llinás, R. Eighteenth Bowditch lecture: Motor aspects of cerebellar control. *Physiologist,* 1974, *17,* 19–46.

Llinás, R.R. The cortex of the cerebellum. *Scientific American,* 1975, *232,* 56–71.

296

RODOLFO R.
LLINÁS AND
JOHN I. SIMPSON

Llinás, R., and Hess, R. Tetrodotoxin-resistant dendritic spikes in avian Purkinje cells. *Proceedings of the National Academy of Science, (U.S.A.)*, 1976, *73*, 2520–2523.

Llinás, R., and Hillman, D.E. Physiological and morphological organization of the cerebellum circuits of various vertebrates. In R. Llinás (Ed.), *Neurobiology of Cerebellar Evolution and Development.* Chicago: American Medical Association, 1969.

Llinás, R., and Hillman, D.E. A multipurpose tridimensional reconstruction computer system for neuroanatomy. In M. Santini (Ed.), *Golgi Centennial Symposium.* New York: Raven Press, 1975.

Llinás, R., and Nicholson, C. Electrophysiological properties of dendrites and somata in alligator Purkinje cells. *Journal of Neurophysiology*, 1971, *34*, 534–551.

Llinás, R., and Nicholson, C. Reversal properties of climbing fiber potential in cat Purkinje cells: An example of a distributed synapse. *Journal of Neurophysiology*, 1976, *39*, 311–323.

Llinás, R., and Sugimori, M. Dendritic calcium spiking in mammalian Purkinje cells: *In vitro* study of its function and development. *Society for Neuroscience Abstracts*, 1978, *4*, 66.

Llinás, R., and Sugimori, M. Electrophysiological properties of *in vitro* Purkinje cell somata in mammalian cerebellar slices. *Journal of Physiology (London)*, 1980a, *305*, 171–195.

Llinás, R., and Sugimori, M. Electrophysiological properties of *in vitro* Purkinje cell dendrites in mammalian cerebellar slices. *Journal of Physiology (London)*, 1980b, *305*, 197–213.

Llinás, R., and Volkind, R.A. The olivo-cerebellar system: Functional properties as revealed by harmaline-induced tremor. *Experimental Brain Research*, 1973, *18*, 69–87.

Llinás, R., and Yarom, R. Electrophysiology of mammalian inferior olivary neurons *in vitro*. Different types of voltage-dependent ionic conductances. *Journal of Physiology (London)*, 1981a, *315*.

Llinás, R., and Yarom, R. Properties and distribution of ionic conductances generating electroresponsiveness of mammalian inferior olivary neurones *in vitro*. *Journal of Physiology (London)*, 1981b, *315*.

Lundberg, A. Integrative significance of patterns of connections made by muscle afferents in the spinal cord. *Symposia and Special Lectures, XXI International Physiological Congress*, pp. 100–105, 1959.

Lundberg, A. Ascending spinal hindlimb pathways in the cat. *Progress in Brain Research*, 1964, *12*, 135–163.

Lundberg, A. Function of the ventral spinocerebellar tract. A new hypothesis. *Experimental Brain Research*, 1971, *12*, 317–330.

Lundberg, A., and Oscarsson, O. Two ascending spinal pathways in the ventral part of the cord. *Acta Physiologica Scandinavica*, 1962, *54*, 270–286.

Lundberg, A., and Weight, F. Functional organization of connexions to the ventral spinocerebellar tract. *Experimental Brain Research*, 1971, *12*, 295–316.

Mabuchi, M., and Kusama, T. Mesodiencephalic projections to the inferior olive and the vestibular and perihypoglossal nuclei. *Brain Research*, 1970, *17*, 133–136.

Maciewicz, R.J., and Spencer, R.F. Oculomotor and abducens internuclear pathways in the cat. In R. Baker and A. Berthoz (Eds.), *Control of Gaze by Brain Stem Neurons*. New York: Elsevier/North Holland, 1977.

Maciewicz, R.J., Eagen, K., Kaneko, C.R.S., and Highstein, S.M. Vestibular and medullary afferents to the abducens nucleus in the cat. *Brain Research*, 1977, *123*, 229–240.

Maekawa, K., and Simpson, J.I. Climbing fiber responses evoked in vestibulocerebellum of rabbit from visual system. *Journal of Neurophysiology*, 1973, *36*, 649–666.

Maekawa, K., and Takeda, T. Afferent pathways from the visual system to the cerebellar flocculus of the rabbit. In R. Baker and A. Berthoz (Eds.), *Control of Gaze by Brain Stem Neurons, Developments in Neuroscience* (Vol. 1). New York: Elsevier/North Holland, 1977.

Mann, M.D. Clarke's column and the dorsal spinocerebellar tract. A review. *Brain, Behavior and Evolution*, 1973, *7*, 34–83.

Marr, D. A theory of cerebellar cortex. *Journal of Physiology, (London)*, 1969, *202*, 437–470.

Martin, G.F., Beattie, M.S., Hughes, H.C., Linauts, M., and Panneton, M. The organization of reticulo-olivocerebellar circuits in the North American opossum. *Brain Research*, 1977, *137*, 253–266.

Martin, G.F., Dom, R., King, J.S., Robards, M., and Watson, C.R.R. The inferior olivary nucleus of the opossum *(Didelphis marsupialis virginiana)*. Its organization and connections. *Journal of Comparative Neurology*, 1975, *160*, 507–534.

Martin, G.F., Henkel, C.K., and King, J.S. Cerebello-olivary fibers: Their origin, course and distribution in the North American opossum. *Experimental Brain Research*, 1976, *24*, 219–236.

Matsushita, M., and Hosoya, Y. The location of spinal projection neurons in the cerebellar nuclei (cerebellospinal tract neurons) of the cat. A study with the horseradish peroxidase technique. *Brain Research*, 1978, *142*, 237–248.

Matsushita, M., and Ikeda, M. Olivary projections to the cerebellar nuclei in the cat. *Experimental Brain Research*, 1970, *10*, 488–500.

Matsushita, M., and Ikeda, M. The central cervical nucleus as cell origin of a spinocerebellar tract arising from the cervical cord: A study in the cat using horseradish peroxidase. *Brain Research*, 1975, *100*, 412–417.

Matsushita, M., and Ikeda, M. Projection from the lateral reticular nucleus to the cerebellar cortex and nuclei in the cat. *Experimental Brain Research*, 1976, *24*, 403–421.

McCrea, R.A., Bishop, G.A., and Kitai, S.T. Electrophysiological and horseradish peroxidase studies of precerebellar afferents to the nucleus interpositus anterior. II. Mossy fiber system. *Brain Research*, 1977, *122*, 215–228.

McCrea, R.A., Bishop, G.A., and Kitai, S.T. Morphological and electrophysiological characteristics of projecting neurons in the nucleus interpositus of the cat cerebellum. *Journal of Comparative Neurology*, 1978, *181*, 397–420.

McCrea, R.A., Baker, R., and Delgado-Garcia, J. Afferent and efferent organization of the prepositus hypoglossi nucleus. *Progress in Brain Research*, 1979, *50*, 653–665.

Miles, F.A., and Fuller, J.H. Visual tracking and the primate flocculus. *Science*, 1975, *189*, 1000–1002.

Miller, S., and Oscarsson, O. Termination and functional organization of spino-olivocerebellar paths. In W.S. Fields and W.D. Willis, Jr. (Eds.), *The Cerebellum in Health and Disease*. St. Louis: Green, 1970.

Mizuno, N. An experimental study of the spino-olivary fibers in the rabbit and the cat. *Journal of Comparative Neurology*, 1966, *127*, 267–292.

Mizuno, N., Mochizuki, K., Akimoto, C., and Matsushima, R. Pretectal projection to the inferior olive in the rabbit. *Experimental Neurology*, 1973, *39*, 498–506.

Mizuno, N., Mochizuki, K., Akimoto, C., Matsushima, R., and Sasaki, K. Projections from the parietal cortex to the brain stem nuclei in the cat, with special reference to the parietal cerebrocerebellar system. *Journal of Comparative Neurology*, 1973, *147*, 511–522.

Moatamed, F. Cell frequencies in the human inferior olivary nuclear complex. *Journal of Comparative Neurology*, 1966, *128*, 109–116.

Montigny, C., de, and Lamarre, Y. Rhythmic activity induced by harmaline in the olivo-cerebello-bulbar system of the cat. *Brain Research*, 1973, *53*, 81–95.

Mower, G., Gibson, A., and Glickstein, M. Tectopontine pathway in the cat: Laminar distribution of cells of origin and visual properties of target cells in dorsolateral pontine nucleus. *Journal of Neurophysiology*, 1979, *42*, 1–15.

Mugnaini, E. Ultrastructural studies on the cerebellar histogenesis. II. Maturation of nerve cell populations and establishment of synaptic connections in the cerebellar cortex of the chick. In R. Llinás (Ed.), *Neurobiology of Cerebellar Evolution and Development*. Chicago: American Medical Association, 1969.

Mugnaini, E. The histology and cytology of the cerebellar cortex. In O. Larsell and J. Jansen (Eds.), *The Comparative Anatomy and Histology of the Cerebellum: The Human Cerebellum, Cerebellar Connections and Cerebellar Cortex*. Minneapolis: University of Minnesota Press, 1972.

Murphy, J.T., MacKay, W.A., and Johnson, F. Differences between cerebellar mossy and climbing fibre responses to natural stimulation of forelimb muscle proprioceptors. *Brain Research*, 1973, *55*, 263–290.

Oscarsson, O. Primary afferent collaterals and spinal relays of the dorsal and ventral spino-cerebellar tracts. *Acta Physiologica Scandinavica*, 1957, *40*, 222–231.

Oscarsson, O. Functional organization of the spino- and cuneo-cerebellar tracts. *Physiological Reviews*, 1965, *45*, 495–522.

Oscarsson, O. Functional significance of information channels from the spinal cord to the cerebellum. In M.D. Yahr and D.P. Purpura (Eds.), *Neurophysiological Basis of Normal and Abnormal Motor Activities*. New York: Raven Press, 1967.

Oscarsson, O. Termination and functional organization of the ventral spino-olivocerebellar path. *Journal of Physiology (London)*, 1968, *196*, 453–478.

Oscarsson, O. The sagittal organization of the cerebellar anterior lobe as revealed by the projection patterns of the climbing fiber system. In R. Llinás (Ed.), *Neurobiology of Cerebellar Evolution and Development*. Chicago: American Medical Association, 1969.

Oscarsson, O. Functional organization of spinocerebellar paths. In A. Iggo (Ed.), *Handbook of Sensory Physiology. Somatosensory System*. (Vol. II). New York: Springer-Verlag, 1973.

Oscarsson, O. Spatial distribution of climbing and mossy fibre inputs into the cerebellar cortex. *Experimental Brain Research*, 1976, *1*, 36–42.

Oscarsson, O., and Rosén, I. Response characteristics of reticulo-cerebellar neurones activated from spinal afferents. *Experimental Brain Research*, 1966, *1*, 320–328.

Oscarsson, O., and Sjölund, B. The ventral spino-olivocerebellar system in the cat. I. Identification of five paths and their termination in the cerebellar anterior lobe. *Experimental Brain Research*, 1977a, *28*, 469–486.

Oscarsson, O., and Sjölund, B. The ventral spino-olivocerebellar system in the cat. II. Termination zones in the cerebellar posterior lobe. *Experimental Brain Research*, 1977b, *28*, 487–503.

Oscarsson, O., and Sjölund, B. The ventral spino-olivocerebellar system in the cat. III. Functional characteristics of the five paths. *Experimental Brain Research*, 1977c, *28*, 505–520.

Oscarsson, O., and Uddenberg, N. Identification of a spino-cerebellar tract activated from forelimb afferents in the cat. *Acta Physiologica Scandinavica*, 1964, *62*, 125–136.

Oscarsson, O., and Uddenberg, N. Properties of afferent connections to the rostral spino-cerebellar tract in the cat. *Acta Physiologica Scandinavica*, 1965, *64*, 143–153.

Palay, S.L., and Chan-Palay, V. *Cerebellar Cortex. Cytology and Organization*. Berlin: Springer-Verlag, 1974.

Palkovits, M., Magyar, P., and Szentágothai, J. Quantitative histological analysis of the cerebellar cortex in the cat. II. Cell numbers and densities in the granular layer. *Brain Research*, 1971a, *32*, 15–30.

Palkovits, M., Magyar, P., and Szentágothai, J. Quantitative histological analysis of the cerebellar cortex in the cat. III. Structural organization of the molecular layer. *Brain Research*, 1971b, *34*, 1–18.

Palkovits, M., Magyar, P., and Szentágothai, J. Quantitative histological analysis of the cerebellar cortex in the cat. IV. Mossy fiber-Purkinje cell numerical transfer. *Brain Research*, 1972, *45*, 15–29.

Pellionisz, A., and Llinás, R. Brain modeling by tensor network theory and computer simulation. The cerebellum: Parallel processor for predictive coordination. *Neuroscience*, 1979, *4*, 323–348.

Pellionisz, A., and Llinás, R. Tensorial approach to the geometry of brain function: Cerebellar coordination via metric tensor. *Neuroscience*, 1980, *5*, 1125–1136.

Pompeiano, O. Functional organization of the cerebellar projections to the spinal cord. *Progress in Brain Research*, 1967, *25*, 282–321.

Pompeiano, O. Macular input to neurons of the spino-reticulocerebellar pathway. *Brain Research*, 1975, *95*, 351–368.

Precht, W., and Llinás, R. Direct vestibular afferents to cat cerebellar nuclei. *Proceedings of the International Union of Physiological Sciences*, 1968, *8*, 1063.

Precht, W. Volkind, R., and Blanks, R.H.I. Functional organization of the vestibular input to the anterior and posterior cerebellar vermis of cat. *Experimental Brain Research*, 1977, *27*, 143–160.

Provini, L., Redman, S., and Strata, P. Mossy and climbing fiber organization on the anterior lobe of the cerebellum activated by forelimb and hindlimb areas of the sensorimotor cortex. *Experimental Brain Research*, 1968, *6*, 216–233.

Pubols, B.H., Jr., Welker, W.I., and Johnson, J.I. Somatic sensory representation of forelimb in dorsal root fibers of racoon, coatimundi and cat. *Journal of Neurophysiology*, 1965, *28*, 312–341.

Ramón y Cajal, S. Estructura de los centros nerviosos de las aves. *Revista Trimestral de Histologia Normal y Patológica*, 1888, *1*, 305–315.

Ramón y Cajal, S. *La Textura del Sistema Nervioso del Hombre y los Vertebrados*. Madrid: Moya, 1904.

Riley, H.A. The arbor vitae and the folial pattern of the mammalian cerebellum. In F. Tilney, T.K. Davis, and H.A. Riley (Eds.), *Research Publication of the Association for Research in Nervous and Mental Disease. Vol. VI: The Cerebellum*. Baltimore: Williams & Wilkins, 1929.

Rinvik, E., and Walberg, F. Studies on the cerebellar projections from the main and external cuneate nuclei in the cat by means of retrograde axonal transport of horseradish peroxidase. *Brain Research*, 1975, *95*, 371–381.

Rispal-Padel, L., and Grangetto, A. The cerebello-thalamo-cortical pathway. Topographical investigation at the unitary level in the cat. *Experimental Brain Research*, 1977, *28*, 101–123.

Rispal-Padel, L., and Lartreille, J. The organization of projections from the cerebellar nuclei to the contralateral motor cortex in the cat. *Experimental Brain Research*, 1974, *19*, 36–60.

Robertson, L.T., and Grimm, R.J. Responses of primate dentate neurons to different trajectories of the limb. *Experimental Brain Research*, 1975, *23*, 447–462.

Robinson, D.A. Adaptive gain control of vestibulo-ocular reflex by the cerebellum. *Journal of Neurophysiology*, 1976, *39*, 954–969.

Rosén, I., and Scheid, P. Patterns of afferent input to the lateral reticular nucleus of cat. *Experimental Brain Research*, 1973a, *18*, 242–255.

Rosén, I., and Scheid, P. Responses to nerve stimulation in the bilateral ventral flexor reflex tract (bVFRT) of the cat. *Experimental Brain Research*, 1973b, *18*, 256–267.

Rosén, I., and Scheid, P. Responses in the spino-reticulocerebellar pathway to stimulation of cutaneous mechanoreceptors. *Experimental Brain Research*, 1973c, *18*, 268–278.

Rosén, I., and Sjölund, B. Organization of group I activated cells in the main and external cuneate nuclei of the cat: Identification of muscle receptors. *Experimental Brain Research*, 1973a, *16*, 221–237.

Rosén, I., and Sjölund, B. Organization of group I activated cells in the main and external cuneate nuclei of the cat: Convergence patterns demonstrated by natural stimulation. *Experimental Brain Research*, 1973b, *16*, 238–246.

Rossum, J., van. *Corticonuclear and Corticovestibular Projections of the Cerebellum*. Ph.D. thesis. Van Gorcum, Leiden, 1969.

Rubia, F.J., and Kolb, F.P. Responses of cerebellar units to a passive movement in the decerebrate cat. *Experimental Brain Research*, 1978, *31*, 387–401.

Ruegg, D.G., and Wiesendanger, M. Corticofugal effects from sensorimotor area I and somatosensory area II on neurons of the pontine nuclei in the cat. *Journal of Physiology, (London)*, 1975, *247*, 745–758.

Ruggiero, D., Batton, R.B., Jayaraman, A., and Carpenter, M.B. Brain stem afferents to the fastigial nucleus in the cat demonstrated by transport of horseradish peroxidase. *Journal of Comparative Neurology*, 1977, *172*, 189–210.

Rushmer, D.S., Roberts, W.J., and Augter, G.K. Climbing fiber responses of cerebellar Purkinje cells to passive movement of the cat forepaw. *Brain Research*, 1976, *106*, 1–20.

Saint-Cyr, J., and Courville, J. Projection from the vestibular nuclei to the inferior olive in the cat: An autoradiographic and horseradish peroxidase study. *Brain Research*, 1979, *165*, 189–200.

Sasaki, K., Kawaguchi, S., Matsuda, Y., and Mizuno, N. Electrophysiological studies on cerebello-cerebral projections in the cat. *Experimental Brain Research*, 1972, *16*, 75–88.

Sasaki, K., Matsuda, Y., Kawaguchi, S., and Mizuno, N. On the cerebello-thalamo-cerebral pathway for the parietal cortex. *Experimental Brain Research*, 1972, *16*, 89–103.

Sasaki, K., Kawaguchi, S., Oka, H., Sakai, M., and Mizuno, N. Electrophysiological studies on the cerebello-cerebral projections in monkeys. *Experimental Brain Research*, 1976, *24*, 495–507.

Sasaki, K., Oka, H., Matsuda, Y., Shimono, T., and Mizuno, N. Electrophysiological studies of the projections from the parietal association area to the cerebellar cortex. *Experimental Brain Research*, 1975, *23*, 91–102.

Sasaki, K., Oka, H., Kawaguchi, S., Jinnai, K., and Yasuda, T. Mossy fibre responses produced in cerebellar cortex by stimulation of the cerebral cortex in monkeys. *Experimental Brain Research*, 1977, *29*, 419–428.

Scheibel, M.E., and Scheibel, A.B. Observations on the intracortical relations of the climbing fibers of the cerebellum. *Journal of Comparative Neurology*, 1954, *101*, 733–763.

Sedgwick, E.M., and Williams, T.D. Responses of single units in the inferior olive to stimulation of the limb nerves, peripheral skin receptors, cerebellum, caudate nucleus and motor cortex. *Journal of Physiology (London)*, 1967, *189*, 261–279.

Shinnar, S., Maciewicz, R.J., and Shofer, R.J. A raphe projection to the cat cerebellar cortex. *Brain Research*, 1975, *97*, 139–143.

Shinoda, Y., and Yoshida, K. Neural pathways from the vestibular labyrinths to the flocculus in the cat. *Experimental Brain Research*, 1975, *22*, 97–111.

Simpson, J.I. Erroneous zones of the cerebellar flocculus. *Society for Neuroscience Abstracts*, 1979, *5*, 107.

Simpson, J.I., and Alley, K.E. Visual climbing fiber input to rabbit vestibulocerebellum: A source of direction-specific information. *Brain Research*, 1974, *82*, 302–308.

Simpson, J.I., and Hess, R. Complex and simple visual messages in the flocculus. In R. Baker and A. Berthoz (Eds.), *Control of Gaze by Brain Stem Neurons*. Amsterdam: Elsevier/North Holland, 1977.

Simpson, J.I., Soodak, R.E., and Hess, R. The accessory optic system and its relation to the vestibulo-cerebellum. *Progress in Brain Research*, 1979, *50*, 715–724.

Sjölund, B. The ventral spino-olivocerebellar system in the cat. V. Supraspinal control of spinal transmission. *Experimental Brain Research*, 1978, *33*, 509–522.

Snider, R.S. Interrelations of cerebellum and brain stem. In P. Bard (Ed.), *Research Publications of the Association for Research in Nervous and Mental Disease. Vol. XXX: Patterns of Organization in the Nervous System*. Baltimore: Williams & Wilkins, 1952.

Snider, R.S., and Eldred, E. Cerebro-cerebellar relationships in the monkey. *Journal of Neurophysiology*, 1952, *15*, 27–40.

Snider, R.S., and Stowell, A. Receiving areas of the tactile, auditory, and visual systems in the cerebellum. *Journal of Neurophysiology*, 1944, *7*, 331–358.

Snyder, R.L., Faull, R.L.M., and Mehler, W.R. A comparative study of the neurons of origin of the spinocerebellar afferents in the rat, cat, and squirrel monkey based on the retrograde transport of horseradish peroxidase. *Journal of Comparative Neurology*, 1978, *181*, 833–852.

Somana, R., and Walberg, F. Cerebellar afferents from the paramedian reticular nucleus studied with retrograde transport of horseradish peroxidase. *Anatomy and Embryology*, 1978, *154*, 353–368.

Sotelo, C. Ultrastructural aspects of the cerebellar cortex of the frog. In R. Llinás (Ed.), *Neurobiology of Cerebellar Evolution and Development*. Chicago: American Medical Association, 1969.

Sotelo, C., Llinás, R., and Baker, R. Structural study of the inferior olivary nucleus of the cat. Morphological correlates of electrotonic coupling. *Journal of Neurophysiology*, 1974, *37*, 541–559.

Sousa-Pinto, A. Experimental anatomical demonstration of a cortico-olivary projection from area 6 (supplementary motor area?) in the cat. *Brain Research* 1969, *16*, 73–83.

Sousa-Pinto, A. The cortical projection onto the paramedian reticular and perihypoglossal nuclei (nucleus praepositus hypoglossi, nucleus intercalatus and nucleus of Roller) of the medulla oblongata of the cat. An experimental-anatomical study. *Brain Research*, 1970, *18*, 77–91.

Sousa-Pinto, A., and Brodal, A. Demonstration of a somatotopical pattern in the cortico-olivary projection in the cat. An experimental-anatomical study. *Experimental Brain Research*, 1969, *8*, 364–386.

Spencer, W.A. The physiology of supraspinal neurons in mammals. In E.R. Kandel (Ed.), *The*

Nervous System, Vol. I, Part 2, Handbook of Physiology. Washington, D.C.: American Physiological Society, 1977.

Spira, M.E., and Bennett, M.V.L. Synaptic control of electrotonic coupling between neurons. *Brain Research,* 1972, *37,* 294-300.

Stewart, W.A., and King, R.B. Fiber projections from the nucleus caudalis of the spinal trigeminal nucleus. *Journal of Comparative Neurology,* 1963, *121,* 271-286.

Swanson, L.W., Cowan, W.M., and Jones, E.G. An autoradiographic study of the efferent connections of the ventral lateral geniculate nucleus in the albino rat and the cat. *Journal of Comparative Neurology,* 1974, *156,* 143-164.

Szentágothai, J., and Rajkovits, K. Uberden Ursprung der Kletterfasern des Kleinhirns. *Zeitschrift für Anatomie und Entwicklungsgeschichte,* 1959, *121,* 130-141.

Takeda, T., and Maekawa, K. The origin of the pretecto-olivary tract. A study using the horseradish peroxidase method. *Brain Research,* 1976, *117,* 319-325.

Thach, W.T. Discharge of Purkinje and cerebellar nuclear neurons during rapidly alternating arm movements in the monkey. *Journal of Neurophysiology,* 1968, *31,* 785-797.

Thach, W.T. Discharge of cerebellar neurons related to two maintained postures and two prompt movements. I. Nuclear cell output. *Journal of Neurophysiology,* 1970, *33,* 527-536.

Thomas, D., Kaufman, R., Sprague, J.M., and Chambers, W.W. Experimental studies of the vermal cerebellar projections in the brain stem of the cat (fastigiobulbar tract). *Journal of Anatomy,* 1956, *90,* 371-385.

Tolbert, D.L., Bantli, H., and Bloedel, J.R. Anatomical and physiological evidence for a cerebellar nucleocortical projection in the cat. *Neuroscience,* 1976, *1,* 205-217.

Tolbert, D.L., Massopust, L.C., Jr., Murphy, M.G., and Young, P.A. The anatomical organization of the cerebello-olivary pathway in the cat. *Journal of Comparative Neurology,* 1976, *170,* 525-544.

Tolbert, D.L., Bantli, H., and Bloedel, J.R. The intracerebellar nucleocortical projection in a primate. *Experimental Brain Research,* 1977, *30,* 425-434.

Tolbert, D.L., Bantli, H., and Bloedel, J.R. Organizational features of the cat and monkey cerebellar nucleocortical projection. *Journal of Comparative Neurology,* 1978a, *182,* 39-56.

Tolbert, D.L., Bantli, H., and Bloedel, J.R. Multiple branching of cerebellar efferent projections in cats. *Experimental Brain Research* 1978b, *31,* 305-316.

Torvik, A., and Brodal, A. The cerebellar projection of the perihypoglossal nuclei (nucleus intercalatus, nucleus praepositus and nucleus of Roller) in the cat. *Journal of Neuropathology and Experimental Neurology,* 1954, *13,* 515-527.

Uno, M., Yoshida, M., and Hirota, I. The mode of cerebello-thalamic relay transmission investigated with intracellular recording from cells of the ventrolateral nucleus of cat's thalamus. *Experimental Brain Research,* 1970, *10,* 121-139.

Voogd, J. *The Cerebellum of the Cat. Structure and Fibre Connections.* Ph.D. thesis. Van Gorcum, Leiden, 1964.

Voogd, J. Comparative aspects of the structure and fibre connexions of the mammalian cerebellum. *Progress in Brain Research,* 1967, *25,* 94-134.

Voogd, J. The importance of fiber connections in the comparative anatomy of the mammalian cerebellum. In R. Llinás (Ed.), *Neurobiology of Cerebellar Evolution and Development.* Chicago: American Medical Association, 1969.

Walberg, F. Descending connections to the inferior olive: An experimental study in the cat. *Journal of Comparative Neurology,* 1956, *104,* 77-173.

Walberg, F. Descending connections to the lateral reticular nucleus. An experimental study in the cat. *Journal of Comparative Neurology,* 1958, *109,* 363-389.

Walberg, F. Fastigiofugal fibers to the perihypoglossal nuclei in the cat. *Experimental Neurology,* 1961, *3,* 525-541.

Walberg, F. Descending connections from the mesencephalon to the inferior olive: An experimental study in the cat. *Experimental Brain Research,* 1974, 20, 145.

Walberg, F., and Jansen, J. Cerebellar corticovestibular fibers in the cat. *Experimental Neurology,* 1961, *3,* 32-52.

Walberg, F., and Jansen, J. Cerebellar corticonuclear projection studied experimentally with silver impregnation methods. *Journal für Hirnforschung*, 1964, *6*, 338–354.

Walberg, F., and Pompeiano, O. Fastigiofugal fibers to the lateral reticular nucleus: An experimental study in the cat. *Experimental Neurology*, 1960, *2*, 40–53.

Walberg, F., Pompeiano, O., Brodal, A., and Jansen, J. The fastigiovestibular projection in the cat. An experimental study with silver impregnation methods. *Journal of Comparative Neurology*, 1962, *118*, 49–75.

Walberg, F., Pompeiano, O., Westrum, L.E., and Hauglie-Hanssen, E. Fastigioreticular fibers in cat. An experimental study with silver methods. *Journal of Comparative Neurology*, 1962, *119*, 187–199.

Weber, J.T., Partlow, G.D., and Harting, J.K. The projection of the superior colliculus upon the inferior olivary complex: An autoradiographic and horseradish peroxidase study. *Brain Research*, 1978, *144*, 369–377.

Wiksten, B. The central cervical nucleus—A source of spinocerebellar fibres, demonstrated by retrograde transport of horseradish peroxidase. *Neuroscience Letters*, 1975, *1*, 81–84.

Wilson, V.J., Maeda, M., and Franck, J.I. Inputs from neck afferents to the cat flocculus. *Brain Research*, 1975, *89*, 133–138.

Wilson, V.J., Maeda, M., Franck, J.I., and Shimazu, H. Mossy fiber neck and second-order labyrinthine projections to cat flocculus. *Journal of Neurophysiology*, 1976, *39*, 301–310.

Wilson, V.J., Uchino, Y., Maunz, R.A., Susswein, A., and Fukushima, K. Properties and connections of cat fastigiospinal neurons. *Experimental Brain Research*, 1978, *32*, 1–17.

Yamamotô, M. Localization of rabbit's flocculus Purkinje cells projecting to the cerebellar lateral nucleus and the nucleus prepositus hypoglossi investigated by means of the horseradish peroxidase retrograde axonal transport. *Neuroscience Letters*, 1978, *7*, 197–202.

Yamamoto, M. Topographical representation in rabbit cerebellar flocculus for various afferent inputs from the brain stem investigated by means of retrograde axonal transport of horseradish peroxidase. *Neuroscience Letters*, 1979a, *12*, 29–34.

Yamamoto, M. Vestibulo-ocular reflex pathways of rabbits and their representation in the cerebellar flocculus. *Progress in Brain Research*, 1979b, *50*, 451–457.

Yamamoto, M., and Shimoyama, I. Differential localization of rabbit's flocculus Purkinje cells projecting to the medial and superior vestibular nuclei, investigated by means of the horseradish peroxidase retrograde axonal transport. *Neuroscience Letters*, 1977, *5*, 279–283.

Yarom, Y., and Llinás, R. Electrophysiological properties of mammalian inferior olive neurons in *in vitro* brain stem slices and *in vitro* whole brain stem. *Society for Neuroscience Abstracts*, 1979, *5*, 109.

Zangger, P., and Wiesendanger, M. Excitation of lateral reticular nucleus neurones by collaterals of the pyramidal tract. *Experimental Brain Research*, 1973, *17*, 144–151.

Additional Recommended Reading

Armstrong, D.M. The mammalian cerebellum and its contribution to movement control. In R. Porter (Ed.), *International Review of Physiology, Neurophysiology III* (Vol. 17). Baltimore: University Park Press, 1978.

Chan-Palay, V. *Cerebellar dentate nucleus. Organization, Cytology and Transmitters*. New York: Springer-Verlag, 1977.

Dow, R.S. Cerebellar syndromes. In P.J. Vinken and G.W. Bruyn (Eds.), *Handbook of Clinical Neurology* (Vol. 2). New York: John Wiley Interscience, 1969.

Eccles, J.C., Ito, M., and Szentágothai, J. *The Cerebellum as a Neuronal Machine*. New York: Springer-Verlag, 1967.

Fadiga, E., and Pupilli, G.C. Teleceptive components of the cerebellar function. *Physiological Reviews*, 1964, *44*, 432–486.

Holmes, G. The cerebellum of man. *Brain*, 1939, *62*, 1–30.

Llinás, R. (Ed.). *Neurobiology of Cerebellar Evolution and Development*. Chicago: American Medical Association, 1969.

6

Eye–Head Coordination

ALBERT F. FUCHS

BEHAVIORAL CONSIDERATIONS

When an object of interest appears in our visual world, we usually turn our eyes in its direction to have a look. In all mammals, each eye is controlled by three pairs of extraocular muscles that provide eye motility with three degrees of freedom within the head. In turn, the head is mounted on a very versatile neck that provides head movements about a naso-occipital axis (roll), an interaural axis (pitch), and a vertical axis (yaw). Finally, the head movements are superimposed upon the movements of a trunk, which also has three degrees of freedom. Therefore, a shift of gaze, especially to a distant eccentric target, can be considered as the sum of three components: (1) the eye movement that would be elicited by the visual stimulus alone; (2) the eye movement elicited by the head movement alone; and (3) the eye movement elicited by rotation of the body alone.

EYE MOVEMENT WITH THE HEAD STATIONARY

SACCADES. With the head held immobile, the appearance of an interesting target in the visual periphery elicits a rapid shift in the direction of gaze (for methodology, see Appendix) to bring the fovea (a small patch of retina with densely packed photoreceptors for high visual acuity) of each eye onto the target. These rapid movements, named *saccades* by Landolt in 1891, are the most rapid movements that the somatic musculature can produce. For example, a 30-deg simian saccade reaches maximum velocities of over 800 deg/sec and lasts only 60 msec.

ALBERT F. FUCHS   Department of Physiology and Biophysics and Regional Primate Research Center, University of Washington, Seattle, Washington 98195. Supported in part by NIH grants RR00166 and EY00745.

Not only is the saccade very rapid, but it is remarkably accurate and in most cases brings the eyes onto the target with little overshoot, undershoot, or oscillations. If a central target jumps to an eccentricity greater than about 20 deg, the saccade often falls short of the target and a second saccade is required after a latency of about 250 msec to bring the eye on target (Fig. 1A). In monkeys, saccade duration increases linearly with amplitude at a rate of 1 msec/deg; maximum saccadic velocity also increases linearly with amplitude to about 20 deg, after which a gradual velocity saturation occurs with an asymptote around 1000 deg/sec (Fuchs, 1967). Human saccades are qualitatively similar to monkey saccades except that they have quantitatively longer durations and reach lower peak velocities (asymptote around 600 deg/sec; see Fuchs, 1976 for review).

Several other characteristics of saccades are worth mentioning. First, successive saccades generally seem to have a refractory period or latency ranging from 100 to 250 msec. Second, except under special circumstances, saccades in flight cannot be altered by voluntary effort and are, therefore, said to be ballistic in nature. Third, vision seems to be suppressed during saccades, and one is unaware that they are occurring at up to two or three times/sec during some tracking tasks such as reading. Fourth, saccades do not require a visual target to be present since they can be executed (at slightly slower velocities) even in complete darkness. Evidence for these conclusions is reviewed elsewhere (Fuchs, 1976).

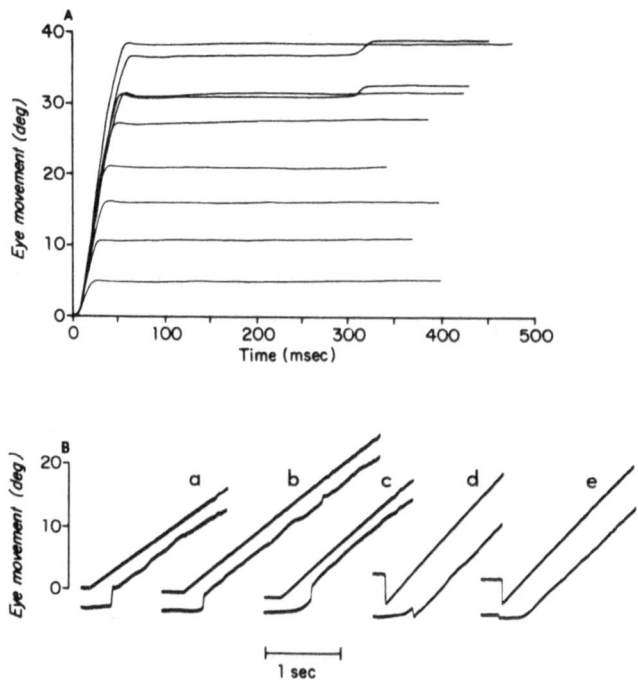

Fig. 1. (A) Family of typical monkey saccades to targets located at increasing eccentricities. (B) Smooth pursuit eye movements (bottom traces) to moving targets (top traces) which either start from rest (a, b, c) or are preceded by a target step in the opposite direction (d, e). Composite from Fuchs, 1967.

SMOOTH PURSUIT MOVEMENTS. If, rather than jumping, a central target moves to the side at a constant velocity, the eye, after an initial saccade to "catch" the target, executes a smooth pursuit eye movement that keeps the fovea very nearly on target (Fig. 1B). The eye can match target velocities of up to 50 deg/sec for monkeys (Fuchs, 1967), but only about 30 deg/sec for humans (Westheimer, 1954). With more predictable target trajectories, the simian maximum smooth pursuit velocity increases to 140 deg/sec whereas man's maximum smooth pursuit velocity reaches 90 deg/sec (Barmack, 1970). The large majority of responses to "ramp" target trajectories first have a short segment of smooth pursuit (Fig. 1B, b and c) before the "catch up" saccade, demonstrating that smooth pursuit movements have shorter latencies than saccades. A smooth pursuit movement may be elicited in isolation by first stepping the target to one side before moving it at constant velocity to the other side (Fig. 1B, d and e). The time required to accelerate the eye from rest to the correct target velocity (smooth pursuit duration) is relatively constant for most target velocities. If the initial target step is a bit too large (Fig. 1B, d) the smooth pursuit movement actually takes the eye away from the target (in fact, it apparently would be better if the eye had not moved at all) and a saccade backward is necessary to place the fovea on target. Such responses suggest that smooth pursuit is elicited by target velocity (Rashbass, 1961) or "slip" of the visual image over the retina, whereas saccades are elicited by displacements of the target image off the fovea. (However, more recent experiments in our laboratory suggest that eye acceleration rather than velocity is most reliably related to retinal slip.)

In addition to differences in maximum velocity, duration, and adequate stimulus, saccades and smooth pursuit movements differ in other important ways. First, unlike the reduction of visual sensitivity during saccades, vision is acute during smooth pursuit. Second, although saccades can be executed without a visual target, smooth pursuit, except under very special circumstances, requires a smoothly moving target; for example, attempts to track an imaginary swinging pendulum produce a succession (or staircases) of saccades. Third, if small amounts of anesthesia are given to human subjects, they first lose their ability to make smooth pursuit movements before losing their ability to make saccades (Rashbass, 1961).

All these data suggest that saccades and smooth pursuit movements are probably subserved by separate branches of the oculomotor system. This suggestion gains creditability since lesions of certain parts of the central nervous system (CNS) can selectively affect either saccades or smooth pursuit eye movements (Hoyt and Daroff, 1971).

EYE–HEAD–NECK COORDINATION DURING NATURAL HEAD ROTATION

If the head is also free to move, most attempts to track an eccentric target are accomplished by a combination of eye *and* head movements. The relative contributions of eye, head, and body movements to the shift of gaze under natural conditions have been investigated by Bizzi and his colleagues in trained monkeys (Bizzi, 1974; Bizzi, Kalil, and Tagliasco, 1971; Bizzi, Kalil, Morasso, and Tagliasco, 1972). If the head is held immobile, a typical monkey saccade to a 30-deg eccentric target resembles that in Fig. 2A. If the head is free to move, the same eccentric target first elicits an eye movement (E) that proceeds along a saccadic trajectory (Fig. 2B). After 20-40 msec, the head (H) begins to move in the same direction at a considerably slower speed (160 deg/sec peak head velocity for a 30-deg movement; Bizzi *et al.*, 1972) so that at the end of the 60-msec saccade the head has moved only about 5 deg. However, the usual 30-deg saccade has been foreshortened by about 5 deg (Morasso, Bizzi, and Dichgans, 1973) so that the sum of head and eye movement (the gaze, G) accurately brings the eye on target. Although the eye is now on target, the head continues to turn, and the eyes must rotate in the opposite direction to compensate for the head movement so that the gaze remains on target (Fig. 2B). The final head position (which is attained after 300 msec) is only slightly short of the target so that the eye must return almost to its initial straight-ahead position in the head.

The source of the compensatory eye movements has been examined by measuring the relative contributions of eye and head movements in normal and labyrinthine lesioned monkeys (Dichgans, Bizzi, Morasso, and Tagliasco, 1973). In normal monkeys, the fidelity of the eye movements is essentially perfect over the normal range of head movements since the gaze remains very constant during the entire head rotation. It is possible that the compensatory eye movements, and hence the accurate change of gaze, are preprogrammed by a set of neural commands that are unaffected by sensory feedback during the movement. However, this is not the case since, if head movement is prevented from occurring on random trials by activating a lightweight clutch applied to the animal's head, the compensatory eye movements do not occur (Fig. 2C; Bizzi *et al.*, 1971). Therefore, the compensatory eye movements are a consequence of visual, neck- and head-movement feedback signals that arise during the course of the head rotation. Since the saccade in the "braked" condition (Fig. 2C) has the same magnitude as normal saccades with the head fixed (Fig. 2A), the foreshortening of saccade amplitude during free head movements can also be attributed to local feedback (Morasso *et al.*, 1973).

For natural head movements, the greatest contribution to the compensatory eye movement is from the head-movement afferents rather than from visual or neck-movement feedback signals. When the visual component is eliminated by extinguishing the target light just before any movement begins, the compensatory

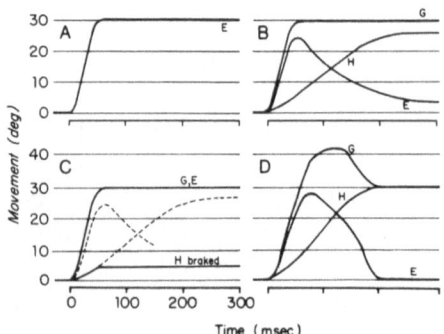

Fig. 2. Eye–head coordination in the monkey. (A) Normal 30-deg saccade with the head fixed. (B) Normal shift of gaze (*G*) to a 30-deg eccentric target is composed of a coordinated head (*H*) and eye (*E*) movement. (C) When normal head movement is prevented on random trials (*H* braked), a correct shift of gaze is accomplished by a saccade alone (dashed lines represent normal, unbraked trajectories). (D) When head movement receptors are surgically impaired, the gaze overshoots the target because the initial eye movement is too large and because the eye movement that compensates for the head movement is too small. Data abstracted from Bizzi and colleagues.

eye movements are qualitatively unchanged. The component arising from head movement alone, tested by passively rotating the whole animal in complete darkness with his head held fixed relative to his body, produces nearly perfect compensation (95% ± 5%). Eye movements due to head rotation alone are called the *vestibulo-ocular reflex* (VOR). Finally, the component arising from neck stimulation alone, tested by rotating the animal's body while keeping his head fixed in space, produces hardly any compensatory eye movements of measurable amplitude (Dichgans *et al.,* 1973).

Under the duress introduced by chronic lesions, other pathways and tracking strategies can assume more importance. After bilateral removal of the head-movement receptors in the vestibular apparatus (a labyrinthectomy), the gaze consistently overshoots the target (Fig. 2D). The overshoot occurs both because the saccade has not undergone its usual foreshortening (Fig. 2, cf. B and D) and because the compensatory eye movements are reduced in amplitude (Dichgans *et al.,* 1973). In the case illustrated in Fig. 2D, compensatory eye movements 10 days after labyrinthectomy correct only 40% of the head movement. After 2 months, however, compensation has improved to 90%. During this time, the compensatory eye movements resulting from passive rotations of the head increased markedly to provide 30% of the improvement. In contrast to findings in normal animals, a compensatory eye movement *is* elicited if the head movement of labyrinthectomized animals is "braked," indicating that some of the remaining compensation must now also be centrally programmed.

EYE–HEAD–NECK COORDINATION DURING FORCED HEAD ROTATION

Instead of studying eye movements in response to natural head rotations, most other investigators have artificially imposed head and body oscillations on subjects required to perform a variety of tracking tasks. When a human being undergoes sinusoidal whole body oscillations in the dark (many of these studies—e.g., Benson and Barnes, 1978; Meiry, 1971; and others—have been done in pilot-training simulators used for the space program), a very characteristic pattern of eye movements called nystagmus results (Fig. 3, inset E*). Clockwise head rotations about a vertical axis (yaw) cause counterclockwise compensatory eye movements that are interrupted at irregular intervals by rapid saccades in the opposite direction. The saccades bring the eye back toward its zero position after a compensatory movement drives it eccentrically; however, the precise stimulus for these resetting saccades is still unknown. If the eye-movement signal is differentiated, the saccades are removed, the remaining smooth movements connected with straight-line segments, and the smoothed velocity trace is integrated to produce position, the resulting trajectory is reasonably sinusoidal at the same frequency as the input so that the techniques of linear system theory can be applied (Fig. 3, inset). Therefore, a complete description of the compensatory eye-movement response (E) to head movements (H) can be obtained by plotting the gain (E/H) and the phase shift of eye movement relative to head rotation (Φ) as a function of the frequency of oscillation.

During whole body rotation with the subject in the dark, the fidelity of the compensatory eye movements (the VOR) is dependent on the frequency of head rotation. At frequencies between 0.1 and 1.0 Hz (a range that probably includes the majority of frequencies in natural head movements), the gain of the VOR is relatively constant, and the eye movement is almost exactly out of phase with the head movement (Fig. 3). The absolute magnitude of the gain varies considerably (from 0.42 to 0.75) in the three human studies presented, probably because of the instructions given to the subject. Barr, Shultheis, and Robinson (1976) measured a gain of 0.65 while subjects performed mental arithmetic, 0.95 when they fixated imaginary targets stationary in space, and 0.35 when they imagined targets rotating with them. Since the gain in every situation was less than 1, the VOR alone did not stabilize the visual world on the retina perfectly. Therefore, under normal viewing conditions in the light, if the VOR were the only compensatory mechanism available, the image of the target would still move across the retina in the direction of head rotation. The stabilization of gaze became even worse at frequencies less

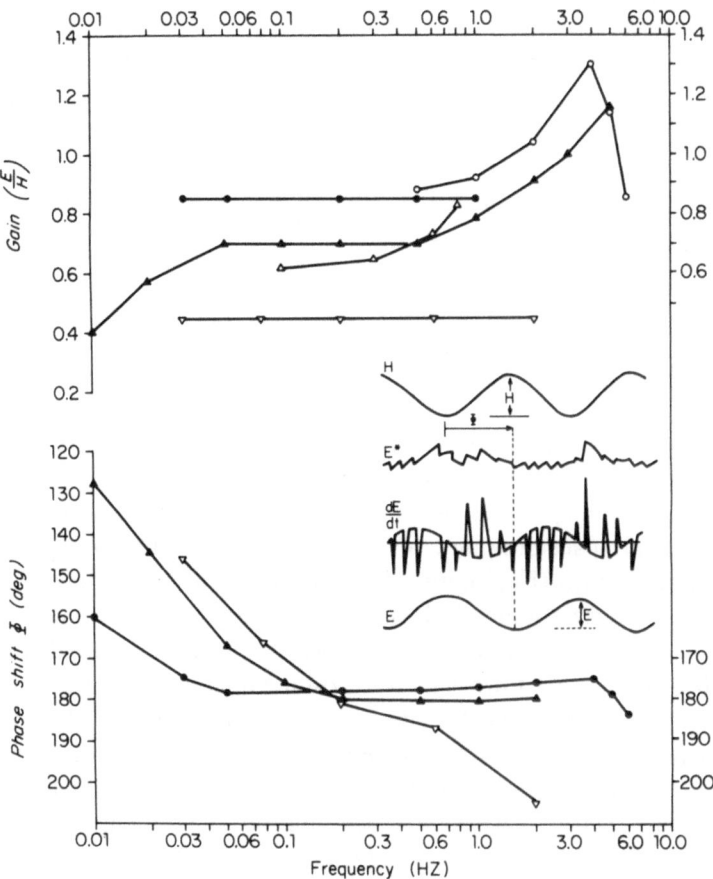

Fig. 3. Gain and phase shift of the VOR in monkey and man as a function of stimulus frequency. Human data obtained from Benson and Barnes, 1978 (▲), Meiry, 1971 (▽), and Barr *et al.*, 1976 (△). Monkey data from Skavenski and Robinson, 1973 (●) and Keller, 1978 (○). Inset shows typical nystagmic eye movements (E^*) resulting from sinusoidal whole body rotation (H) in the dark. The compensatory eye movement alone is obtained by differentiating E^*, removing the saccades and integrating the resulting signal to produce eye position (E).

than 0.1 Hz, where the gain decreased and phase advanced. In similar experiments on the monkey (Skavenski and Robinson, 1973), the gain ranged from 0.8 to 0.9 (a value lower than that reported by Dichgans *et al.* (1973) for natural, simian head movements) and the phase showed a slight, almost constant lead over the range 0.1 to 1.0 Hz (Fig. 3). At frequencies higher than 1 Hz the gains of both the monkey and the human VOR showed a marked increase, with the monkey exhibiting a resonance at about 4 Hz (Benson and Barnes, 1978; Keller, 1978; Fig. 3).

If the subject is rotated in a well-lighted room and asked to fixate a stationary target on the wall, the compensatory eye movements become virtually perfect. The improvement with vision is especially noticeable at low frequencies where the gain approaches 1 (Fig. 4, SIW) and the phase lag is essentially 180 deg (not shown). Although the VOR alone cannot completely correct for the head movement, espe-

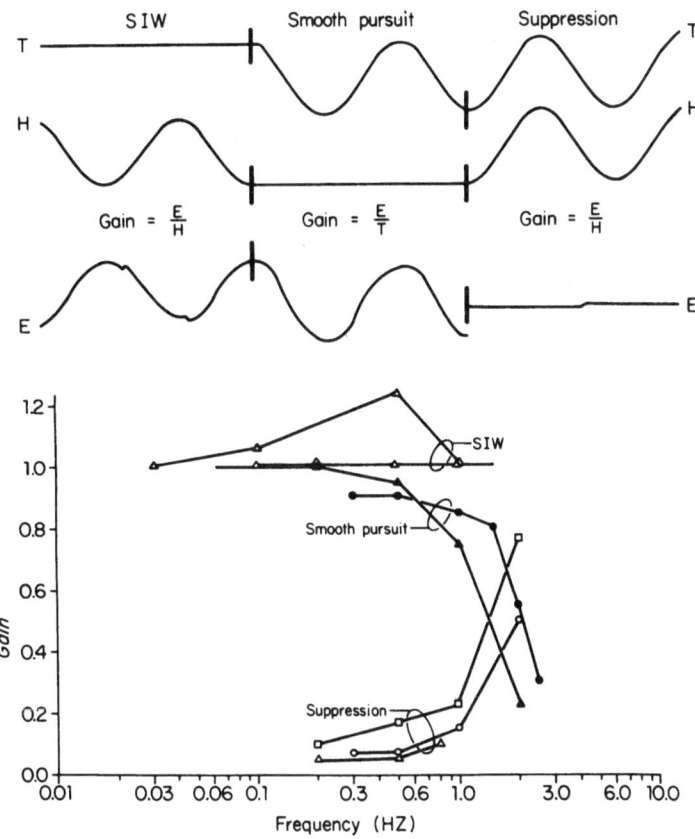

Fig. 4. Gain characteristics of the eye-movement response during different tracking conditions as a function of stimulus frequency. *T* and *H* represent target and head position in space (stimuli at 0.5 Hz, ± 10 deg); *E* represents eye position in the head. During SIW, the rotating subject fixates a target that is Stationary In (the) World. During smooth pursuit, the stationary subject tracks a target moving sinusoidally in space. During suppression, the rotating subject tracks a target that is rotating precisely with him. Human data obtained from Meiry, 1971 (∇), Barr *et al.*, 1976 ($\triangle$) and Benson and Barnes, 1978 ($\blacktriangle$), ($\square$). Monkey data from Lisberger, Evinger and Fuchs, unpublished ($\bullet$), ($\circ$).

cially at low frequencies, the target is maintained relatively stationary on the retina by the addition of continuous, smooth pursuit eye movements. The stimuli guiding the smooth pursuit movements are produced by the residual movements of the retinal image of the target. In fact, with the head held stationary, the smooth pursuit system alone is quite capable of tracking targets accurately at low frequencies. Figure 4 shows that smooth pursuit gain for both monkey and man is greater than 0.9 for frequencies up to 0.5 Hz. Above 1.0 to 1.5 Hz, the gain falls rapidly. However, at these frequencies, the VOR gain is increasing toward 1 (Fig. 3) and would be capable of providing the entire compensation. The difference in smooth pursuit and VOR tracking performance can be demonstrated qualitatively by comparing your own ability to track a moving finger while holding your head stationary with your ability to fixate a stationary finger while shaking your head. The improved tracking at higher frequencies in the latter situation is provided by the VOR (also see Benson and Barnes, 1978).

The experimental situation in which a rotating subject fixates a target stationary in visual space simulates the condition most usually encountered in our daily lives. Perhaps the second most common natural situation occurs when we are examining something carried in our hands and consequently moving with us (Barr *et al.*, 1976). This situation is created experimentally by requiring the subject to fixate a target rotating with him so that he has to suppress the VOR. In the suppression condition, the gain is less than 0.1 for both man and monkey at frequencies less than 0.3 Hz (Fig. 4); at frequencies above 1 Hz the gain increases steeply. A quantitative comparison shows that the increase in gain during suppression is closely matched by the decrease in gain of the smooth pursuit system (Fig. 4). Therefore, at low frequencies smooth pursuit is equal to the task of defeating the VOR, but at higher frequencies the VOR wins out and drives the gain toward 1.0. After a barbiturate is administered in low doses to eliminate smooth pursuit eye movements, human subjects lose the ability to suppress the VOR (Rashbass and Russell, 1961), supporting the notion that suppression is effected by the smooth pursuit system. By controlling the target with a signal proportional to head position, one can create a wide variety of visual-head interactions (Lisberger and Fuchs, 1978a). In almost every case requiring the cooperation of smooth pursuit and the VOR, the eye-movement response is accurately predicted by the linear addition of the smooth pursuit and VOR responses alone (Lisberger, Evinger, and Fuchs, unpublished observations).

Finally, the eye movements due to body rotation alone have also been investigated in man by rotating the trunk under a stationary head. As in the monkey, eye movements elicited by trunk rotations at the frequencies of normal head turning (0.3–1.0 Hz) are very small (gain ca. 0.08). At lower frequencies, the gain increases to between 0.2 and 0.32 (Barlow and Freedman, 1978; Meiry, 1971). Although there is general agreement on the gain characteristics of the neck-ocular reflex, the phase characteristics are in dispute. Meiry (1971) claims that eye movements are in phase with the trunk movements and are indeed compensatory since a clockwise trunk rotation with a fixed head would be equivalent to a counterclockwise head rotation on a stationary trunk, a situation that requires a clockwise compensatory eye movement. On the other hand, Takemori and Suzuki (1971) and Barlow and Freedman (1980) demonstrated that eye movements are in the opposite direction of body movements and would increase the slip of an object over the retina. Because of its low gain and dubious role in visual stabilization, the neck-ocular reflex will not receive much attention here.

In summary, at the head-movement frequencies achieved during normal shifts in gaze, both man and monkey rely most heavily on smooth pursuit eye movements elicited by visual stimuli and on the VOR elicited by head rotations. Movements of the trunk relative to the head, which would stimulate neck receptors to provide a neck-ocular reflex, seem to be relatively unimportant. Even at the lower frequencies where the possible contribution of the neck-ocular reflex might become significant, the smooth pursuit system alone is more than adequate to eliminate the retinal slip.

ANATOMY OF VESTIBULO-OCULAR PATHWAYS

In studying the neural substrates for the control of head and eye movements, and particularly their interaction, neurophysiologists have concentrated almost exclusively on pathways subserving eye movements resulting from head rotation. Relatively little is known about the neck-ocular reflex, and smooth pursuit movements have received very little neurophysiological attention. Consequently, the following sections will concentrate most heavily on the VOR (see also Cohen, 1974). Furthermore, attention will be focused on data from the cat and the monkey not only because they have been studied in most detail but also because their neurophysiology can probably be extrapolated to man.

Several recent developments have established the VOR as a very useful model to study the behavior of mammalian reflexes and their modification. First, as we have already seen in the behavioral section, the stimuli applied to the head and the resulting eye movements can be quantitatively controlled and accurately measured. Second, the mathematical techniques of linear-system theory can be applied to describe the neural processing that occurs between the vestibular end organs and the extraocular muscles. Third, single-unit activity can now be recorded from the brain stem and cerebellum of either the alert "encéphale isolé" cat or the alert monkey. The former preparation allows a more flexible placement of stimulus electrodes, the administration of transmitter blocking agents, etc., while still emitting relatively typical and brisk eye movements on vestibular stimulation. The alert monkey, on the other hand, can be trained to track moving targets (as in the experiments described in the behavioral section) so that the discharge patterns of neurons can be obtained during normal tracking. Finally, these neurophysiological techniques have been integrated with the recent development of modern anatomical tracers to reveal the neural pathways subserving the oculomotor system. In the last decade or so, all these influences have converged on the VOR to provide one of the most complete and detailed pictures of a complex movement system yet studied.

RECEPTORS AND PRIMARY AFFERENTS

The earliest demand for head–eye coordination probably evolved in simple aquatic forms that used undulations of the body for swimming. Since these body movements also moved the head, such an animal required a simple "hold" mechanism that prevented the slip of the visual image over the homogeneous mosaic of retinal photoreceptors. In these acorticate animals, the neural circuitry for compensatory eye movements was completely laid down in the brain stem and formed a bulbar reflex (the VOR), which was as powerful and important as the spinal reflexes. As specialized subareas of high visual acuity developed in mammals with frontal vision, the requirement for simple visual stability anywhere on the retina was replaced by the necessity to place the moving visual image precisely on the area of most acute vision (an area centralis in cat or fovea in primates). This demand led to the development of smooth pursuit eye movements which, as we have already seen, cooperate closely with the VOR when both target tracking and head rotation occur simultaneously and, as we shall see later, probably accomplish this cooperation by using some of the phylogenetically ancient pathways subserving the VOR.

Information from the vestibular apparatus can reach the oculomotor nuclei by a variety of pathways involving, in various combinations, the vestibular nuclei, the medullary, pontine, and mesencephalic reticular formations, the nuclei prepositus hypoglossi, and the cerebellum. The shortest possible pathway involves only three neurons: the primary 8th nerve fibers, the secondary vestibular neurons (relay interneurons) in the VIII cranial nucleus, and the oculomotor nuclei. Although this direct pathway has received almost all the attention of electrophysiologists and anatomists, we shall see that it, alone, cannot account for the compensatory eye movements of the VOR.

THE END ORGANS. Movement of the head is sensed by transducers that are basically mechanoreceptors. In their earliest form, they were simply hair cells whose apical cilia projected directly out into the animal's aqueous environment. As the animal swam, the cilia were bent according to the movement of the animal relative to its environment and the orientation of its body in space. In time, the pit that housed the hair cells was first closed off by cartilage, as in most fishes, and eventually by bone, as in man. The cavity that houses the hair cells is now filled with an internally generated fluid called the *endolymph* which fills the scala media of the cochlea as well. There are three different vestibular mechanoreceptors located bilaterally in each temporal bone: (1) the semicircular canals, which sense angular acceleration, (2) the utricle, which senses linear acceleration and (3) the saccule, which only recently has also been shown to sense linear acceleration.

The three semicircular canals lie in roughly orthogonal planes (Fig. 5). In man, the lateral canal lies in a plane tipped about 25 deg above the horizontal, the anterior (also called *superior*) canal is oriented anterolateral at about 41 deg with the sagittal plane and the posterior canal is oriented posterolateral at an angle of about 56 deg with the sagittal plane (Blanks, Curthoys, and Markham, 1975). The left anterior and right posterior canals are roughly in the same plane (within 23 deg), and the bilateral horizontal canals are also roughly coplanar (within 19 deg). In the cat, the orthogonality between ipsilateral canals is virtually perfect (Blanks, Curthoys, and Markham, 1972). The canals are not semicircular but are, in fact, complete torroids that communicate at the utricle (Fig. 5). The overall diameter of human canals ranges from 4 to 7 mm with a lumen diameter on the order of 0.5

mm. Before entering the utricle, the canal expands by twice its size into an ampulla that houses its sensory epithelium (the crista ampullaris). The ampullae of the horizontal and anterior canals lie close together on the anterolateral aspect of the utricle, whereas the posterior canal ampulla lies distant on the posterior aspect of the utricle. The lateral canal is physically independent of the other two, which communicate over 15% of their circumference (Fig. 5), allowing for the possibility of functional interaction. With the head in its normal, upright position, the sensory epithelia (the maculae) of the utricle and the saccule lie approximately in the horizontal and sagittal planes, respectively (Fig. 6C).

The sensory epithelium of all three end organs is very similar. Each has two types of hair cell with different shapes and innervation patterns (Fig. 6A). Type I cells (which first appear in birds and mammals) resemble a goblet and possess chalice-shaped afferent terminations from large- and medium-sized axons. The older Type II cells are more cylindrical in shape with plexus nerve endings from medium-sized axons. Each cell contains one larger, longer, true cilium (the kinocilium) and many (up to 70) shorter, smaller, so-called stereocilia. The hair cells possess conventional synaptic vesicles, suggesting that transmission to the 8th nerve

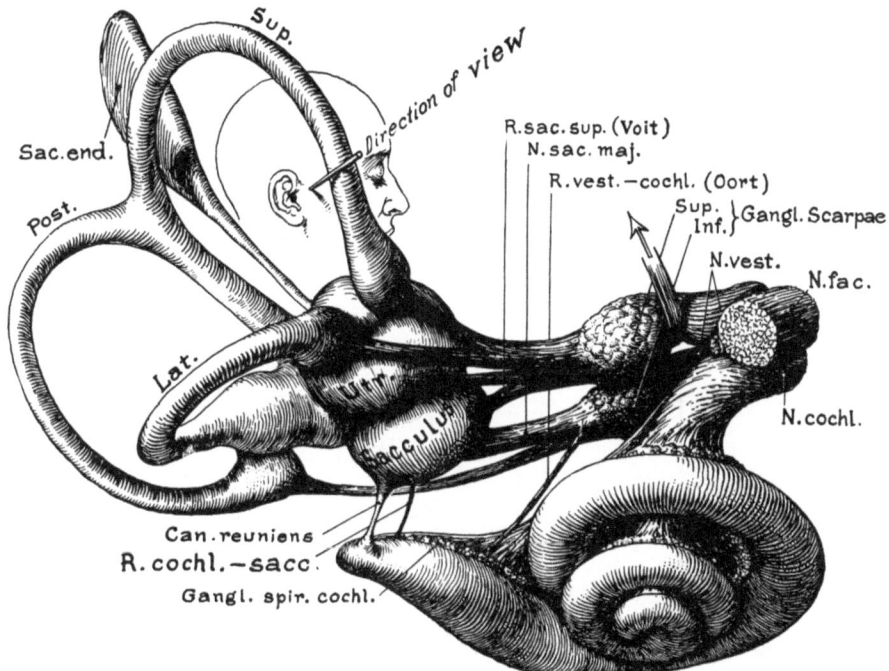

Fig. 5. Location and innervation of the vestibular end organs in the human temporal bone (after Hardy, 1934). The vestibular nerve (N. vest.) is composed of axons of bipolar cell somata lying in the superior (Sup.) and inferior (Inf.) vestibular (Scarpa's) ganglia. Distal processes of bipolar cells divide into major and minor [e.g., R. sac. sup. (Voit)] branches to innervate the three canals and the two otolith organs. Endolymph is produced in the sacculus endolymphaticus (Sac. end.) and reaches the cochlea through the canal reuniens via the otolith cavities. The facial (N. fac.) and cochlear divisions of the 8th nerve (N. cochl.) are also shown.

endings is mediated by chemical transmitters (Lowenstein, 1974; Wersäll and Bagger-Sjöbäck, 1974).

In the canals, the cilia insert into a gelatinous tongue that stretches across the lumen of the ampulla and reaches to its roof. Although early experiments in the pike suggested that head movements cause the cupula to swing like a trap door (Steinhausen, 1933), more recent studies on the frog suggest it billows like an elastic diaphragm (Hillman, 1972). In both the utricle and saccule, the cilia also insert into a gelatinous medium; however, these gelatinous substrates support imbedded CaCo₃ particles (called otoliths) with a lower specific gravity than the bathing

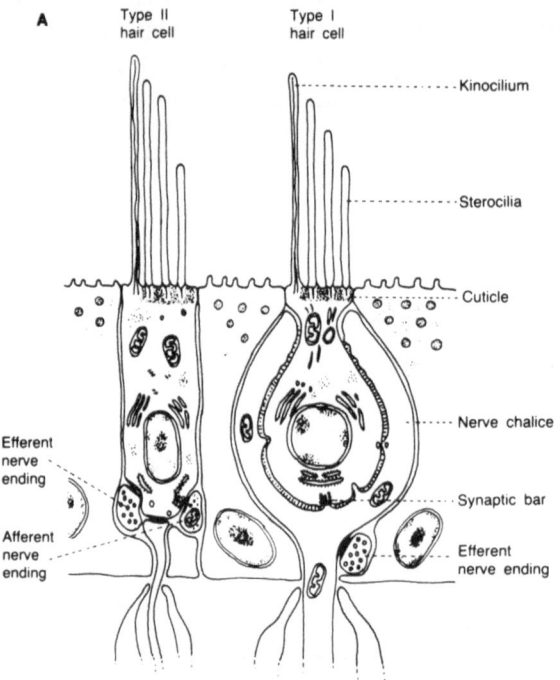

Fig. 6. Microanatomy of the vestibular end organs. (A) Structure and innervation of Type I and II hair cells (Wersäll and Bagger-Sjöbäck, 1974). Direction of activation (depolarization) of hair cells indicated by polarization vectors pointing toward the kinocilia in the three semicircular canals (B) and the utricle and saccule (C). In (B) and (C) A-anterior, P-posterior, S-superior, I-inferior, M-medial, and L-lateral directions with the head in its normal position. After Spoendlin, 1966.

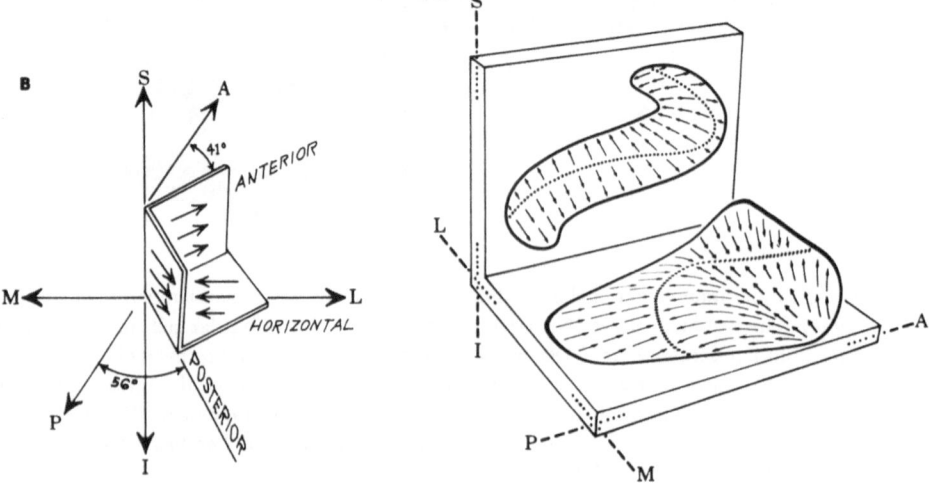

endolymph so that static forces in pitch and roll cause the particles (and hence the substrate) to be deflected. The bending of the cilia in either the canals or otolith organs causes an as yet unresolved change in the membrane properties of the hair cell (Precht, 1975).

Each hair cell is morphologically polarized since not only is the kinocilium set off to one end of the ciliary array, but the stereocilia exhibit ever-decreasing lengths with distance from the kinocilium (Fig. 6A). In the horizontal canal, all the kinocilia face the utricle. Since deflection of the kinocilia toward the utricle is associated with an increase in 8th nerve discharge, whereas deflection away from the utricle causes a decrease, the morphological polarization indicates the direction of the adequate stimulus. In the vertical canals, all the kinocilia face away from the utricle and bending the hairs away from the utricle causes an increase in 8th nerve discharge. Therefore, it is the orthogonal locations of the canals in the skull together with the regular orientation of the hair cells that allow the three degrees of rotational freedom to be sensed. On the other hand, in the utricle the hair cells are morphologically polarized in different directions, with the kinocilia facing the striola, a curved dividing ridge running through the middle of the spoon-shaped macula (Fig. 6C). In the saccule, the hair cells are polarized away from the striola (Lindeman, 1969). Therefore, in the otolith organs, directionality is imparted largely by hair-cell orientation.

The first satisfactory recordings from any vertebrate vestibular hair cell have only recently been obtained from the frog saccule (Hudspeth and Corey, 1977). Resting membrane potentials were usually 50 to 60 mv and mechanical deflections of the hair bundles elicited graded potentials between 5 and 10 mv. In all cells, bending the hair bundle toward the kinocilium caused a depolarization, which was associated with a decrease in membrane resistance.

THE 8TH NERVE. Afferents from the vestibular end organs in both the cat (Gacek, 1969) and the monkey (Stein and Carpenter, 1967) are bipolar cells located in either of two ganglia. The superior ganglion contains the perikarya of cells that supply the anterior canal, lateral canal, and utricle, whereas the inferior ganglion provides the afferent innervation of the saccule and posterior canal (Fig. 5). There are two major destinations for the central branches of vestibular ganglion cells. The 8th nerve enters the brain stem at the border of the pons and medulla, and many primary afferents terminate in the four major ipsilateral vestibular nuclei lying on the floor of the 4th ventricle. Some ascending branches of the vestibular root fibers traverse the superior vestibular nucleus to enter the cerebellum via the juxtarestiform body (Brodal and Hoivik, 1964; Carpenter, Stein, and Peter, 1972). No primary vestibular fibers cross the midline to the contralateral vestibular nuclei, and collaterals of 8th nerve fibers to the reticular formation are scarce (Brodal, 1974).

SECONDARY VESTIBULAR NEURONS

THE MAJOR VESTIBULAR NUCLEI. Projections of primary afferents to the vestibular nuclei are organized according to receptor. In the monkey, fibers from the three canals all terminate in largely overlapping areas of the superior vestibular nucleus (SVN) and the rostral medial vestibular nucleus (MVN), whereas the utricle and saccule project largely to the caudal MVN and the descending vestibular nucleus (DVN). Within these projection loci, individual canal nerves may have preferential distributions (e.g., fibers from the lateral and anterior canals project to the rostrolateral SVN, whereas the posterior canal projects to the medial SVN; Stein and Carpenter, 1967). On the whole, the distribution of primary afferents to the feline vestibular nuclei corresponds to those in the monkey, although the feline utricle and saccule project to the lateral vestibular nucleus (LVN) as well as MVN and DVN.

Electrophysiological studies generally support the anatomical distribution of primary afferents in the vestibular nuclei. In the cat, electrical stimulation of the labyrinth produces monosynaptic excitation of cells in the ipsilateral MVN, mainly in its rostral 60% (Wilson, Wylie, and Marco, 1968b). In the SVN of the squirrel monkey, 93% of the units respond to angular acceleration, mainly from stimulation of the anterior or posterior canals (Abend, 1977). Only 12% of the units (some of which also respond to angular acceleration) respond to static tilts. In the rostral MVN, 81% of all units are activated by angular acceleration. On the other hand, units whose discharge rates are deeply modulated by static tilts in the cat are most often encountered in the DVN and MVN and less often in the SVN (Peterson, 1967). In the cat, units activated by stimulation of the saccular nerve are confined to LVN and DVN, with no saccular projections to SVN (Wilson, Gacek, Uchino, and Susswein, 1978). However, some units sensitive to tilt are found in all four nuclei and, indeed, a small percentage of units in the monkey can be driven by utricular as well as canal inputs. Therefore, although there is not an absolute segregation, it is generally true that the rostral parts of the vestibular nuclei receive canal afferents while the caudal parts receive utricular and saccular afferents.

MINOR CELL GROUPS IN THE VESTIBULAR NUCLEI. Information about the smaller cell groups of the vestibular nuclei (e.g., the f, x, and y groups as well as the interstitial nucleus of the 8th nerve) is still rather sketchy (Brodal, 1974). However, the y group, which might mediate sacculo-ocular connections, will be dealt with later.

VESTIBULO-OCULOMOTOR CONNECTIONS

THE EXTRAOCULAR MUSCLES AND THEIR INNERVATION. Impulses from the vestibular nuclei can reach oculomotor neurons via one direct and several indirect routes. However, before these projections are considered, it will be necessary to describe the oculomotor nuclei themselves in some detail.

Each eye is moved by six extraocular muscles, which act as three synergistic pairs. The *horizontal rectus muscles* originate at the back of the orbit on the ligament of Zinn, a fibrous ring that surrounds the entering optic nerve. Both muscles attach approximately on the horizontal equator, with the medial rectus (MR) inserting onto the nasal and the lateral rectus (LR) onto the temporal pole of the globe. Their action causes primarily medial and lateral horizontal eye movement, respectively. Both *vertical rectus muscles* also originate on the annulus of Zinn, course forward and lateral in the orbit, and insert approximately on the vertical meridian at a 23-deg angle (in man) from the sagittal plane. The superior rectus (SR), which attaches at the top pole of the globe, causes primarily elevation of the eye, whereas the inferior rectus (IR), attached to the bottom pole, produces primarily depression. Finally, the *oblique muscles* insert approximately at right angles to the four recti, on the top (the superior oblique, SO) and bottom (the inferior oblique, IO) of the globe. Their isolated action causes primarily rotation (torsion) about the visual axis (the line through the center of the cornea and the fovea). In addition to producing intorsion of the eye, the SO produces depression, whereas the IO produces extortion and elevation. Since the muscles attach at angles to the visual axis, and since their attachments are not precisely on the equators and should be considered as distributed and not point insertions, the action of each muscle varies somewhat with eye position (see Robinson, 1975, for details).

The twelve extraocular muscles are innervated by motoneurons lying in three midline symmetric cranial nuclei. Each lateral rectus is supplied by the ipsilateral VIth cranial nucleus (the abducens), located below the floor of the 4th ventricle just rostral to the vestibular nuclei, and tucked under the genu of the facial (VIIth cranial) nerve. The superior oblique receives its innervation from the contralateral IVth cranial nucleus (the trochlear), which lies in the caudal midbrain under the inferior colliculus. All the remaining muscles are supplied by the IIIrd cranial nucleus (the oculomotor) within which motoneuron pools to the different muscles are anatomically segregated (Naito, Tanimura, Taga, and Hosoya, 1974; Warwick, 1964). Only the superior rectus motoneurons are crossed, with the decussation occurring within the nucleus. Other midbrain nuclei close to the oculomotor complex (e.g., interstitial nucleus of Cajal, INC; the nucleus of the posterior commissure, NPC; and possibly the nucleus of Darkschewitsch, ND) are thought to be preoculomotor and also receive secondary vestibular afferents.

DIRECT VESTIBULO-OCULOMOTOR PROJECTIONS

Anatomy. Following lesions restricted to individual vestibular nuclei, the distribution of degeneration in the various parts of the oculomotor, abducens, and trochlear nuclei is similar in monkey (McMasters, Weiss, and Carpenter, 1966) and in cat (Gacek, 1971; Tarlov, 1970) except for minor discrepancies. Apparently conflicting anatomical data are reconciled by noting that the MR and IR subdivisions in the oculomotor nucleus are reversed in the two species. Lesions in the SVN cause degeneration in the lateral wing of the ipsilateral ascending medial longitudinal fasciculus (MLF) and heavy terminal degeneration in the ipsilateral IVth nucleus. MLF fibers also terminate in the IIIrd nucleus, mostly the ipsilateral IR subdivision; in the cat, the ipsilateral IO subdivision is also supplied, and crossing fibers also innervate the contralateral nucleus. Lesions in the MVN and the ventral LVN cause degeneration in the ipsilateral and probably the contralateral VIth nucleus as well; this anatomy is difficult to interpret because fibers from the MVN and LVN cross at the level of the abducens to ascend in the contralateral dorsomedial portion of the MLF. The ascending contralateral fibers terminate profusely in the contralateral IVth nucleus and most subdivisions of the contralateral IIIrd nucleus. In both species, some fibers recross at the level of III to terminate in the ipsilateral IIIrd nucleus. In the cat, the MR subdivision, which is only sparsely provided with MLF inputs from the vestibular nuclei, appears to receive an ipsilateral input from the ventral LVN via the ascending tract of Deiters (Gacek, 1971). The connections from the SVN and MVN to the IIIrd nucleus have been confirmed by modern tracer techniques in both cat and monkey (Gacek, 1977; Graybiel, 1977b); more importantly, however, the injection of a retrograde label into the IIIrd nucleus has revealed sources for ascending projections arising outside the vestibular nuclei (see p. 322). Finally, INC and ND receive fibers from the SVN via the ipsilateral MLF and crossed fibers from the rostral MVN via the contralateral MLF. Both McMasters *et al.* (1966) and Tarlov (1970) point out that the regions of the vestibular nuclei that give fibers to the various oculomotor nuclei all originate within regions receiving primary afferents. Furthermore, neither the DVN nor the dorsal LVN provide oculomotor fibers. Therefore, the direct vestibulo-ocular pathways arise mostly from the rostral vestibular nuclei and therefore presumably provide largely canal inputs to oculomotor neurons. This organization is supported by intracellular studies showing that of the 41% of all MVN neurons antidromically activated by stimulation of the rostral MLF, virtually all were in the rostral two-thirds of the nucleus (Wilson *et al.,* 1968a). However, only just over one-third of these neurons were also monosynaptically activated from the labyrinth; on the other hand, 51% of cells sending fibers into the ascending MLF were polysynaptically activated. Although some of these later fibers may have cell bodies in the vestibular nuclei, others may originate in the pontine and medullary reticular formations (Remmel, Pola, and Skinner, 1978).

Electrophysiology. Electrophysiological studies on the cat and rabbit have provided more details about these connections and have revealed which end organs are involved. For example, electrical stimulation of the posterior canal (PC) nerve produces EPSPs with latencies in the disynaptic range in contralateral trochlear (SO) and IR motoneurons (Uchino, Hirai, and Watanabe, 1978). Disynaptic IPSPs are elicited in ipsilateral IO and SR motoneurons. Lesions placed in the SVN or the ipsilateral MLF eliminate the IPSPs while lesions in the MVN and the contralateral MLF eliminate the EPSPs. The neurons and pathways revealed by these studies are shown in Fig. 7A. Since stimulation of the PC simulates a backward pitch of the head, and remembering that the SO and SR motoneurons innervate the contralateral extraocular muscles, and that the SO, with its forward-placed "pulley" arrangement, depresses the eye, the synaptic patterns in Fig. 7A are appropriate to cause a depression of both eyes. Similar stimulation studies on anterior canal (AC) connections (Fig. 7B) reveal that both excitatory and inhibitory interneurons are located in the SVN (Precht, 1977; Wilson, 1972) with the excitatory pathway traveling in the contralateral brachium conjunctivum (BC) and the inhibitory pathway in the ipsilateral MLF. Here the synaptology is appropriate to produce a conjugate upward rotation of both eyes. Pharmacological studies suggest that GABA is the inhibitory transmitter of vestibular neurons projecting to trochlear motoneurons (Precht, Baker, and Okada, 1973; Roffler-Tarlov and Tarlov, 1975).

Finally, stimulation of the horizontal canal nerve produces a disynaptic EPSP in the contralateral VIth nucleus and a disynaptic IPSP in the ipsilateral VIth

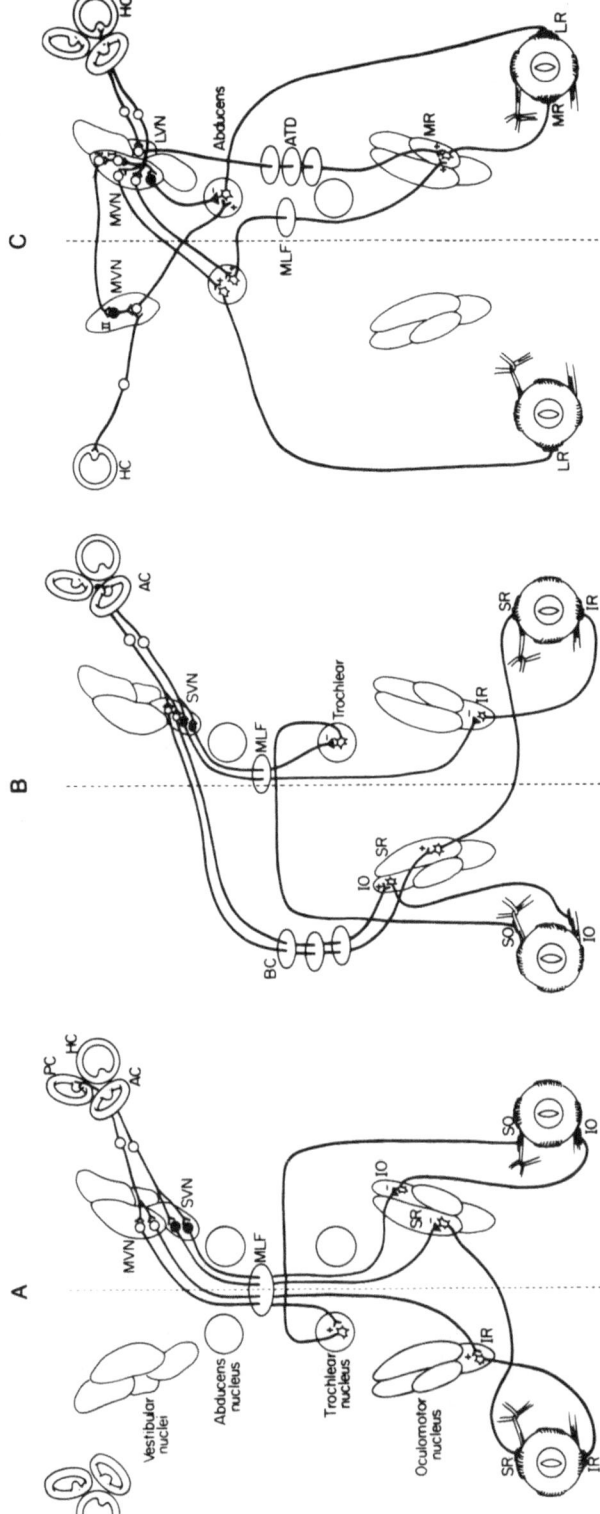

Fig. 7. The shortest pathways from the posterior (A), anterior (B), and horizontal (C) semicircular canals to extraocular motoneurons in the cat. Semicircular canals: HC-horizontal; PC-posterior; AC-anterior. Oculomotor nucleus subdivisions: IO-inferior oblique; SR-superior rectus; IR-inferior rectus; MR-medial rectus. Extraocular muscles: SO-superior oblique; IO-inferior oblique; SR-superior rectus; IR-inferior rectus. Fiber tracts: MLF-medial longitudinal fasciculus; BC-brachium conjunctivum; ATD-ascending tract of Deiters.

nucleus (Fig. 7C). The MVN is the relay nucleus for both effects since micro-stimulation there produces contralateral EPSPs and ipsilateral IPSPs at monosynaptic latencies (Baker, Mano, and Shimazu, 1969). Ipsilateral MR motoneurons receive large-amplitude, disynaptic EPSPs from 8th nerve stimulation; the axons of the secondary excitatory vestibular neurons, located in the rostral LVN, travel in the ascending tract of Deiters (ATD; Baker and Highstein, 1978). On the other hand, stimulation of the contralateral 8th nerve did not produce disynaptic IPSPs but in most neurons did produce small amplitude EPSPs with latencies around 1.4 ms. Inhibition to the MR subgroup is apparently polysynaptic and does not travel in the MLF (Baker and Highstein, 1978). The absence of a disynaptic, reciprocal excitation and inhibition from canals to motoneurons is unique for the MR subgroup. Another powerful excitatory pathway with one extra synapse has recently been revealed with the discovery of interneurons in the VIth nucleus (Baker and Highstein, 1975; Bienfang, 1978) that receive the same synaptic inputs as abducens motoneurons but whose axons cross the midline near the VIth nucleus to ascend in the MLF and innervate contralateral MR motoneurons (Highstein and Baker, 1978). By labeling motoneurons with one retrograde tracer deposited in the lateral rectus and interneurons with a tracer placed in the oculomotor nucleus, it was possible to show that ascending connections are not collaterals of motoneurons but a separate group of neurons comprising about one-third of all abducens neurons (Steiger and Büttner-Ennever, 1978). A similar pathway has been demonstrated anatomically in the monkey (Graybiel and Hartwieg, 1974; Büttner-Ennever, 1977). Furthermore, in both species, there are interneurons in the oculomotor nucleus that project back to the VIth nucleus (Büttner-Ennever, 1977; Maciewicz and Spencer, 1977).

In summary, each canal gives rise to two excitatory and two inhibitory VOR pathways. Except for the inhibitory innervation of the MR, all pathways contain three neurons. Most inhibitory neurons are in the SVN, with axons that ascend in the ipsilateral MLF; the exception is the inhibitory interneuron to the LR, which lies in the MVN. Excitatory interneurons lie in different vestibular nuclei according to the canal stimulated. Although the MLF carries eight of the twelve reflex arcs, three excitatory neuron connections are also mediated by the ascending tract of Deiters and the brachium conjunctivum. A similar specific organization of canal projections to particular extraocular muscles has also been revealed in the monkey by stimulating individual canal nerves and photographing the resulting eye movements (Suzuki, Cohen, and Bender, 1964). Since single-shock stimuli applied to the ampullary nerves evoke a twitch in extraocular muscles within 5 to 6 ms (Cohen and Suzuki, 1963), it is likely that the short-latency pathways demonstrated electrophysiologically in the cat and rabbit are also present in the monkey.

INDIRECT VESTIBULO-OCULOMOTOR PROJECTIONS. Indirect vestibulo-ocular pathways are probably limited only by one's imagination. However, several possibilities are worth discussing in some detail.

Pathways through the Reticular Formation. Lesions of all the vestibular nuclei in the cat produce significant degeneration in the ipsilateral (Gacek, 1971) and contralateral reticular formation (Ladpli and Brodal, 1968) although such studies are confounded by artifactual degeneration caused by damage to cerebellar afferents. However, it seems highly likely that SVN sends fibers to both the ipsilateral and contralateral nucleus reticularis pontis caudalis (NRPC) and tegmenti pontis. Lesions in the MVN also produce bilateral degeneration in the reticular formation, but since both SVN and LVN fibers pass through MVN, these projections might be suspect (Ladpli and Brodal, 1968). However, many neurons in the vestibular nuclei can be antidromically activated at long latencies by stimulating the contralateral NRPC and nucleus reticularis gigantocellularis (NRG); 44% of these neurons are activated either mono- or polysynaptically from the 8th nerve (Abzug and Peterson, 1973). Many axons of secondary vestibular neurons do terminate in the reticular formation, since stimulation in the vestibular nuclei produces monosynaptic EPSPs (from SVN) and IPSPs (from all vestibular nuclei) in many NRPC and NRG neurons (Peterson and Abzug, 1975).

Furthermore, there is some evidence that neurons from NRPC, nucleus reticularis pontis oralis (NRPO), and NRG project to abducens motoneurons. First, both retrograde and orthograde labeling techniques demonstrate that neurons in the feline NRG project to the VIth nucleus (Graybiel, 1977a; Maciewicz, Egan, Kaneko, and Highstein, 1977) and also to the IIIrd nucleus (Graybiel, 1977b). Second, orthograde label deposited into NRPC and NRPO of the cat (Graybiel, 1977a) or in the paramedian reticular formation of the monkey, which contains the same nuclei, produces labeling in the ipsilateral abducens nucleus (Büttner-Ennever and Henn, 1976). Third, neurons in NRPO and NRPC can be antidromically activated from the VIth nucleus with very low (5 μA) stimulus currents (Kaneko, Steinacker, Cohen, Maciewiez, and Highstein, 1975) and can be filled from horseradish peroxidase (HRP) deposits involving the abducens nucleus. Electrophysiological experiments have also suggested a connection from the pontine reticular formation to the oculomotor nuclei (Highstein, Cohen, and Matsunami, 1974), a pathway not thus far confirmed by modern labeling techniques (Graybiel and Hartwieg, 1974). Finally, neurons in NRPC, the paramedian medullary reticular formation, and the vicinity of the VIth nucleus project to MVN (Pompeiano, Mergner, and Corvaja, 1978) so that a reticular pathway that returns to vestibulo-ocular interneurons in the vestibular nuclei is possible (see later section on the brain stem integrator). Degeneration in the vestibular nuclei is also seen following lesions in NRPO (Hoddevik, Brodal, and Walberg, 1975).

Based on both the effects of lesions and the activity of single units related to eye movements, the pontine (and possibly the medullary) reticular formation seems most concerned with horizontal eye movements, whereas the mesencephalic reticular formation may be an analogous staging area for vertical eye movements. The pathways through the mesencephalic reticular formation and the accessory oculomotor nuclei to the IIIrd nucleus are very similar to those through the pons and medulla. Both the INC and the rostral interstitial nucleus of the MLF (iMLF) in the reticular formation (Büttner-Ennever and Büttner, 1978) receive afferents from the rostral vestibular nuclei in both the cat (Gacek, 1971; Tarlov, 1970) and the monkey (Büttner-Ennever and Büttner, 1978; McMasters et al., 1966). In turn, both the INC and iMLF project to the oculomotor nuclei (Carpenter, Stein, and Peter, 1970; Graybiel and Hartwieg, 1974; Steiger and Büttner-Ennever, 1979). The nucleus of Darkschewitsch and the nucleus of the posterior commissure

(NPC), which also receive vestibular nucleus afferents, do not project directly to the oculomotor nuclei of the monkey (Steiger and Büttner-Ennever, 1979), although the NPC may in the cat (Graybiel and Hartwieg, 1974).

In summary, there is good evidence for the existence of pathways from the rostral vestibular nuclei through both the pontine and medullary reticular formations to either the abducens (and possibly oculomotor) nuclei or back to the vestibular nuclei. Similar pathways from the rostral vestibular nuclei can also be traced through the mesencephalic reticular formation and accessory oculomotor nuclei to the IIIrd nucleus.

Pathways Utilizing the Perihypoglossal Nuclei. Long thought to be involved solely in tongue movements, the nucleus prepositus hypoglossus (NPH) of the cat has recently been shown to project to oculomotor neurons (Gacek, 1977; Graybiel and Hartwieg, 1974). Furthermore, NPH neurons receive disynaptic IPSPs from the ipsilateral 8th nerve and disynaptic EPSPs from the contralateral 8th nerve, apparently via the vestibular nuclei (Baker and Berthoz, 1975).

Since the projection from NPH is ipsilateral and provides "modest" excitatory inputs to trochlear and other oculomotor nuclei (Baker, Berthoz, and Delgado-Garcia, 1977), the sign of this pathway is similar to the three neuron routes through the vestibular nuclei and contains but one extra neuron. Connections from the vestibular nuclei to the NPH (McMasters *et al.*, 1966) and from NPH to the IIIrd nucleus (Graybiel, 1977b) have also been demonstrated in the monkey.

Pathways through the Cerebellum. In order of projection density, 8th nerve fibers terminate in the ipsilateral flocculus, the ipsilateral nodulus (vermis lobule X), and parts of the ipsilateral uvula (vermis lobule IX), and the ventral dentate nucleus (Brodal and Hoivik, 1964; Carpenter *et al.*, 1972). In the monkey, the vermis projection includes inputs from all vestibular receptors; however, the flocculus projection is organized according to receptor, with the rostral lobules (5–10) receiving anterior and lateral canal afferents, lobules 5 and 6 receiving mainly utricular afferents, and lobules 3 through 6 mainly saccular afferents; posterior canal afferents were not studied. The projections are only via mossy fibers in the cat (Brodal and Hoivik, 1964; Precht and Llinás, 1969; Shinoda and Yoshida, 1975) and probably also via mossy fibers in the monkey, although putative climbing fiber degeneration has been described (Carpenter *et al.*, 1972). In the cat, primary afferents that project to the flocculus apparently are collaterals of afferents that also terminate in the vestibular nucleus (Baker, Precht, and Llinás, 1972).

The flocculus, nodulus, and uvula (known together as the *vestibulocerebellum*) project in turn to the ipsilateral vestibular nuclei (Angaut and Brodal, 1967; Dow, 1938). In the cat, the efferents from the flocculus and nodulus/uvula to the vestibular nuclei are kept separate (Angaut and Brodal, 1967). The SVN is amply supplied from the entire vestibulocerebellum, with floccular efferents reaching its central and caudal parts, whereas the nodulus and uvular projections occupy the peripheral parts. The MVN receives fibers from the flocculus with the highest density rostrolaterally and from the nodulus chiefly caudomedially (the uvula provides no MVN fibers). In both these nuclei, the region of the floccular projection is coextensive with that from the 8th nerve, whereas the nodular projection is coextensive with that from the caudal contralateral fastigial nuclei. Considering the site of termination, it seems likely that the flocculus may affect ascending pathways to the oculomotor nuclei but the nodulus does not. Indeed, recent electrophysiological experiments in the cat have demonstrated that although the flocculus provides strong inhibition of secondary vestibular neurons participating in the VOR (Baker et al., 1972), special conditions have been required to demonstrate any nodulus inhibition of ascending pathways (Precht, Volkind, Maeda, and Giretti, 1976). In the primate *Galago*, a similar segregation of projections to the vestibular nuclei exists, with the flocculus supplying the central SVN and rostral-medial MVN, while the nodulus supplies the peripheral SVN and dorsocaudal MVN (Haines, 1977). Since, in *Macaca mulatta*, canal afferents terminate more rostrally in the vestibular nuclei than utricular afferents (Stein and Carpenter, 1967), it seems likely that the flocculus in primates also preferentially affects ascending vestibulo-ocular pathways.

Several other cerebellar pathways can influence neurons that carry vestibulo-ocular signals. First, vermis lobules X and IX also project to the underlying fastigial nucleus whose neurons project to the vestibular nuclei. Neurons caudal in the fastigial nucleus can be antidromically activated from the contralateral vestibular nuclei via the hook bundle of Russel (Furuya, Kawano, and Shimazu, 1975) in support of earlier anatomical studies (Walberg, Pompeiano, Brodal, and Jansen, 1962), whereas neurons in the rostral fastigial nucleus project to the ipsilateral vestibular nuclei (Walberg et al., 1962). Furthermore, electrical stimulation of the rostral two-thirds of the fastigial nucleus produces disynaptic IPSPs in ipsilateral trochlear motoneurons and disynaptic EPSPs in contralateral trochlear motoneurons (Hirai, Uchino, and Watanabe, 1977). Second, in the cat, fibers from all parts of the vestibulocerebellum terminate in the NPH (Angaut and Brodal, 1967). However, in *Galago*, there is little evidence for any projection from the vestibulocerebellum beyond the vestibular complex. Third, in the primate, fibers from the ventral dentate nucleus, which not only receives direct 8th nerve afferents but afferents from the ipsilateral flocculus and nodulus (Haines, 1977), travel in the brachium conjunctivum to terminate in the contralateral superior and inferior rectus subdivisions of the IIIrd nucleus (Carpenter and Strominger, 1964; Chan-Palay, 1977). In the cat, cells projecting to the IIIrd nucleus are located ventral to dentate in the infracerebellar nucleus (Gacek, 1977). Furthermore, the nucleus interpositus and not the dentate nucleus contains cell bodies projecting to the IIIrd nucleus. These vestibulo-ccrebello-ocular pathways could completely bypass the brain stem.

Finally, possible pathways involving the vestibulocerebellum and the reticular formation exist in both cat and monkey but, thus far, have not been extensively studied (Carpenter and Nova, 1960; Haines, 1975).

It should be pointed out that all the connections between the vestibulocerebellum and the brain stem are reciprocal (Kotchabhakdi and Walberg, 1978). In fact, the cerebellum provides a larger contingent of afferents to the vestibular nuclei than does the 8th nerve (Walberg, 1972). Therefore, neural routes through the vestibulocerebellum should be considered as interactive pathways that maintain a constant "dialog" with the vestibular nuclei rather than isolated feedforward pathways in parallel with the three-neuron arc.

ALBERT F. FUCHS

DISCHARGE PROPERTIES OF FIBERS IN THE VESTIBULAR NERVE

CANAL AFFERENTS. First-order canal neurons have a remarkably high resting discharge, which averages some 90 spikes/sec in the monkey (Goldberg and Fernandez, 1971) and 36 spikes/sec in the cat (Estes, Blanks, and Markham, 1975). The high resting rate has several functional consequences. First, it allows a bidirectional response. All horizontal canal neurons have their firing rate increased by head rotations causing utriculopetal flow of endolymph, and decreased by head rotations causing a utriculofugal flow. On the other hand, head rotations producing a utriculofugal flow of endolymph in vertical canals cause an increase in afferent activity, whereas utriculopetal flow causes a decrease. Therefore, the sign of the afferent discharge from a particular canal is always the same and can be predicted from the morphological hair-cell polarization. Second, a high resting rate theoretically allows a very low sensory threshold. In fact, no evidence of a threshold was detected in 8th nerve canal discharges from either monkey (Goldberg and Fernandez, 1971) or cat (Blanks *et al.,* 1975), in agreement with human psychophysical experiments that demonstrated that perceptual thresholds to angular acceleration may be as low as 0.1 deg/sec^2 (Clark and Stewart, 1968). Finally, a high resting rate causes a very strong neural barrage to be continually delivered to the vestibular nucleus and to the cerebellum. If a conservative estimate of half of the total 20,000 fibers in the simian 8th nerve (Gacek and Rasmussen, 1961) subserve the canals (of 12,000 total 8th nerve fibers in the cat, about 2,500 subserve each canal; Gacek, 1968), almost 10^6 spikes/sec reach the vestibular nucleus from each labyrinth with the head at rest.

The adequate stimulus for the canals is angular acceleration. If a squirrel monkey is accelerated from rest to a constant angular velocity, the discharge rate of a canal afferent increases rapidly during acceleration and then decreases after constant angular velocity has been attained (Fig. 8A). When the animal is again decelerated to zero velocity, the unit exhibits a decrease in discharge rate below resting level during deceleration and a return to resting discharge after zero velocity has been attained. Consequently, when undergoing a similar angular rotation, a human subject is aware of motion during the acceleration phases but soon loses the sensation of motion after constant velocity has been attained. A step change in angular acceleration (Fig. 8B) elicits an exponential rise in firing rate (*FR*; dotted curve), which can be described by the relation

$$FR = S_1(1 - e^{-t/T_1})$$

where T_1 averages 6 sec for the squirrel monkey (Goldberg and Fernandez, 1971) and 3.8 sec for the cat (Blanks, Estes, and Markham, 1975). For small magnitudes of acceleration, both accelerations and decelerations elicit roughly equal but opposite responses, whereas for larger accelerations, the increase in unit activity exceeds the decrease. For even larger accelerations, the unit is driven to zero firing rate during deceleration. Within the linear range, the average sensitivity (S_1) to angular acceleration is 2 spikes/sec/deg/sec^2 for both species. For brief accelerations (say, less than 5 sec), the exponential rise in firing rate can be approximated by a ramp trajectory (Fig. 8B, dashed line), which has the same form as the time integral of the step of applied acceleration. Therefore, for brief accelerations, changes in canal afferent activity accurately reflect changes in head velocity.

The above analysis suggests that a sinusoidal angular acceleration applied to the head should cause a periodic modulation in discharge rate that is in phase with head velocity (i.e., lags head acceleration by 90 deg). In fact, the phase lag of canal afferents is strongly dependent on the frequency of head oscillation. For example, for the rather low frequency shown in Fig. 8C, the unit activity lags acceleration

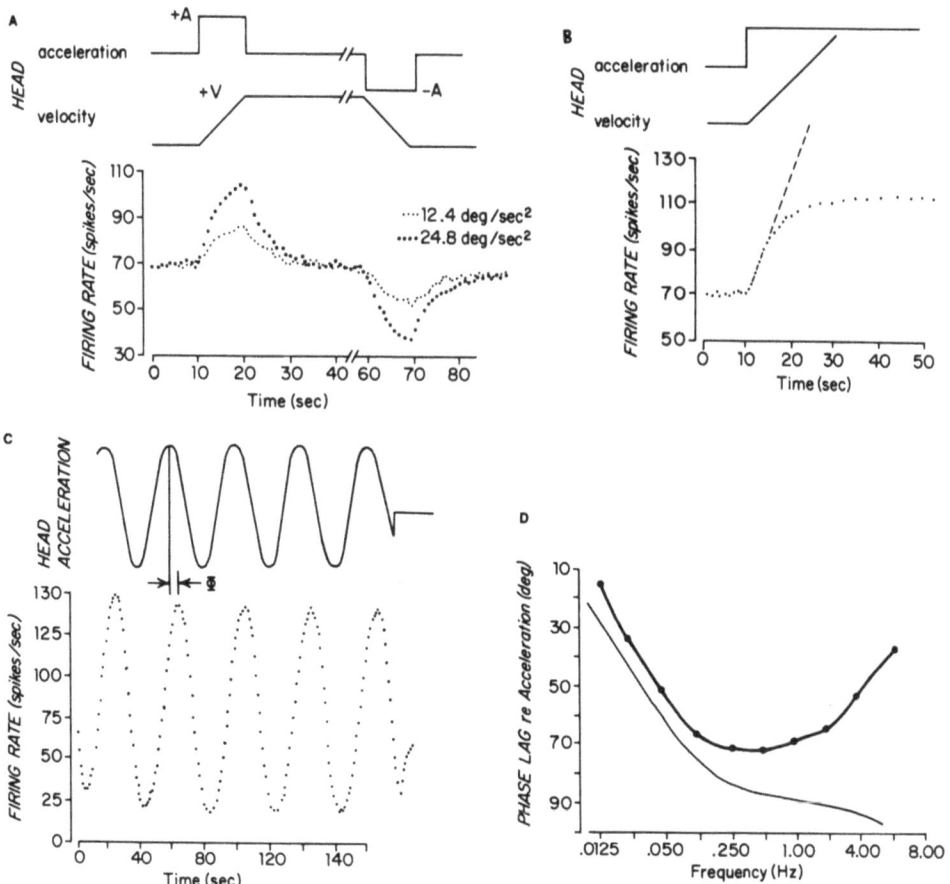

Fig. 8. Response of canal afferents to angular acceleration of the head. Instantaneous firing rate (dotted curves) of an 8th nerve fiber during a brief pulse of two different angular accelerations (A) and a step of angular acceleration (B). For $t < 5$ sec, the firing rate in B (dotted curve) can be approximated by a ramp increase in firing rate (dashed line). (C) Instantaneous firing rate of an 8th nerve fiber during sinusoidal angular accelerations at 0.025 Hz. Phase lag Φ of unit activity relative to peak angular acceleration as a function of frequency is plotted in D ($\bullet$—$\bullet$) for many 8th nerve afferents (Fernandez and Goldberg, 1971; Goldberg and Fernandez, 1969). Thin curve in D represents the phase lag expected from a simple torsion pendulum with $T_1 = 6$ sec, $T_2 = 3$ msec (see discussion in text of hydromechanical properties of the vestibular end organ).

by only 40 deg. However, between about 0.1 and 2 Hz, most units (Fig. 8D) lag head acceleration by at least 70 deg so that, over this frequency range, unit activity is nearly in phase with head velocity.

The discharge characteristics of canal afferents reflect the hydromechanical properties of the peripheral end organ. After measuring the motion of the cupula of the pike in response to pressure applied to the endolymph, Steinhausen (1933) suggested that the cupula-endolymph system could be described as a heavily damped torsion pendulum. The angular displacement of the endolymph (Φ) was related to the angular acceleration of the head (d^2h/dt^2) by the differential equation

$$m \frac{d^2\Phi}{dt^2} + r \frac{d\Phi}{dt} + k\Phi = H \frac{d^2h}{dt^2} \qquad (1)$$

where m represents the moment of inertia of the endolymph-cupula, r the viscous damping of the endolymph due to the canal walls, and k the cupula elasticity returning it to its resting position. Although more recently it has been shown that the cupula may behave like an elastic diaphragm (Hillman, 1972), a similar equation will still describe cupular motion. Since the system is heavily damped, the solution to equation (1) is made up of the sum of two exponential functions. If $(r/m) \gg (k/m)$, the solution to (1) in response to a step change of acceleration (H) would be:

$$\Phi(t) = H [1 - A_1 e^{-t/T_1} - A_2 e^{-t/T_2}] \qquad (2)$$

where $T_2 = (m/r)$ and $T_1 = (r/k)$. T_2, which can be deduced from hydrodynamic principles, has a value of about 0.003 sec in the monkey (Fernandez and Goldberg, 1971), and about 0.005 sec in man. On the other hand, T_1 (particularly k) is difficult to calculate directly. In the past, T_1 has been estimated as the time during which humans either experience the sensation of turning (van Egmond, Groen, Jongkees, 1949) or undergo nystagmus after an applied acceleration (Groen, 1957); by these measures, the value of T_1 ranged from about 10 to 15 sec. However, these measures reflect the dynamics of the entire vestibulo-ocular system, and the recent data on canal afferents presented above suggest that 6 sec may be a more reasonable figure for T_1. In any event, $T_1 \gg T_2$ and, consequently, T_1 dominates any response describing endolymph (and hence cupula and cilia) movement. A step change in angular acceleration, then, causes the cilia to undergo an exponential increase in bending with a time constant on the same order of magnitude as that reported for canal afferents.

Therefore, for brief accelerations, the cupula-endolymph hydromechanics performs like an integrating accelerometer to provide a neural signal proportional to head velocity. Indeed, when subjected to sinusoidal angular accelerations, the canals should function as reasonable velocity transducers over the frequency range

$$\frac{1}{2\pi T_1} < f < \frac{1}{2\pi T_2}$$

that is, from 0.016 ($T_1 = 10$ sec) to 53 ($T_2 = 0.003$ sec) Hz, which includes the bandwidth of normal head movements (Melvill Jones and Milsum, 1965). In fact, the torsion pendulum model provides a reasonable prediction of 8th nerve discharge from 0.025 to about 0.25 Hz; at higher frequencies, 8th nerve activity exhibits a gain enhancement and corresponding phase lead which is not predicted by an overdamped torsion pendulum model (see Fig. 8D). This high-frequency behavior suggests that firing rate is proportional not only to cupular displacement but to cupular velocity as well. In addition to this deviation from the simple velocity transducer model, many 8th nerve fibers, especially those with irregular resting rates, exhibit sensory adaptation to prolonged stimuli with a time constant of 30 to well over 80 sec (Fernandez and Goldberg, 1971).

AFFERENTS FROM THE OTOLITH ORGANS. First-order otolith afferents in the monkey have a somewhat lower resting discharge (about 60 spikes/sec) than do canal afferents. The discharge of each otolith afferent can be characterized by a functional polarization vector whose direction defines the amount of head tilt in pitch and roll that produces the greatest change in unit activity. Consistent with the anatomy, most units in the superior division of the vestibular nerve have vectors near the plane of the utricular macula (i.e., nearly horizontal) whereas those in the inferior division have vectors near the plane of the saccular macula (i.e., nearly vertical) (Fernandez and Goldberg, 1976a; Fernandez, Goldberg, and Abend, 1972). As might be expected from their anatomical orientation, units innervating either the utricle or saccule respond to *both* static pitch and roll, with firing rate being more or less trigonometrically related to the sine of the pitch and roll angle. Unit discharge of otolith afferents in both monkey (Fernandez and Goldberg, 1976b) and cat (Loe, Tomko, and Werner, 1972) is an approximately linear function of the force of gravity (g) for stimuli between ± 1 g. For the monkey, the average sensitivity is on the order of 40 spikes/sec/g. Since plots of firing rate versus force pass smoothly through zero force (Fernandez *et al.*, 1972), otolith organs apparently do not possess a sensory threshold. Humans can barely detect linear accelerations of less than 3×10^{-4} g (Graybiel and Patterson, 1955). Since otolith afferents from the superior and inferior division of Scarpa's ganglion have similar responses to static linear accelerations, the sacculus, long thought to be a vibration sensor, should, like the utricle, now be considered as an equilibrium organ. Saccular afferents in the cat have a lower gain than utricular afferents (Anderson, Blanks, and Precht, 1978) although the lower resting rate and sensitivity of saccular afferents relative to utricular afferents originally reported for the monkey (Fernandez *et al.*, 1972) may be less dramatic than first reported (Fernandez and Goldberg, 1976a).

Both otolith organs also respond to dynamic tilts or changes in linear acceleration (like those created by suddenly assuming a squat from a standing position, or by jogging). In monkeys, centrifugal rotations were used to produce linear accelerations (Fernandez and Goldberg, 1976c), whereas, in the cat, the responses to otolith (largely utricular) stimulation were determined by rolling the animal about a horizontal axis (Anderson *et al.*, 1978). In addition to exhibiting adaptation, otolith primary afferents in the monkey had a significant response assymetry to linear accelerations in opposite directions. The majority of simian otolith afferents had regular resting rates and exhibited a rather flat gain characteristic from dc to 2 Hz. The phase lag relative to the gravity vector or the angle of head tilt (for small angles) was essentially zero from dc to about 0.5 Hz, after which it increased somewhat (less than 25 deg) at higher frequencies (2 Hz).

Most feline otolith afferents show very similar characteristics. However, a small population of units from each species (units with irregular resting rates in the monkey and phasic-tonic units in the cat) exhibit a gain enhancement and slight phase lead at higher frequencies which cannot (like the similar phenomena displayed by canal afferents) be accounted for by a simple velocity sensitivity in the peripheral otolith receptor (see Fernandez and Goldberg, 1976c for details). Although the response of the otolith organs was thought to be governed by a second-order linear differential equation (not unlike equation 1) relating otolith displacement to linear acceleration (Goldberg and Fernandez, 1975b), the data described above indicate that the firing patterns cannot be described by so simple a model. Fitting gain and phase curves of either regular or irregular units revealed a median high frequency time constant of about 15 msec (ca. 60 Hz), which suggests that the otolith receptors are quite capable of responding to rapid head movements.

Afferents from either the canals or the otolith organs can be divided into different populations according to the regularity of their resting discharge rates. Irregular units from the cat or monkey have discharge patterns that are more phasic, show more adaptation, and have a higher sensitivity to adequate stimuli than regular units. Relatively fewer irregular units are found among otolith afferents than among canal afferents (Fernandez et al., 1972). Since the maculae also are innervated by relatively fewer thick fibers (Gacek, 1968), Fernandez et al. (1972) suggested that thick fibers have irregular firing rates and innervate Type I hair cells. This suggestion receives some support from studies on ampullary receptors. Larger fibers contact Type I hair cells and are preferentially distributed to the crest of the cristae (Wersäll, 1956; Wersäll and Bagger-Sjöbäck, 1974). Hair cells on the crest of the cristae would be most sensitive to mechanical stimuli and consequently yield the higher sensitivity seen in irregular units.

EFFERENTS TO THE VESTIBULAR END ORGANS. Of the more than 10,000 fibers in the feline 8th nerve, only 200 to 300 are efferent fibers (about equally crossed and uncrossed) with cell bodies lying in small areas just lateral to both abducens nuclei and ventral to MVN (Gacek and Lyon, 1974; Warr, 1975). Efferent fibers innervate Type II hair cells directly but terminate only on the afferent terminals of Type I hair cells (Gacek, 1974). Stimulation of the nuclei of origin of efferent fibers produces only a modest inhibition in 8% of afferent units (Dieringer, Blanks, and Precht, 1977). In the goldfish and rabbit, about 10% of 8th nerve fibers discharge in relation to eye movements (see Klinke and Galley, 1974 for review). However, even though the monkey also has a population of efferent neurons (Goldberg and Fernandez, 1977), its 8th nerve discharges are apparently unrelated to saccadic, smooth pursuit or optokinetically elicited eye movements (Keller, 1976; Louie and Kimm, 1976).

In summary, afferents from the canals have a discharge pattern related to the angular velocity of the head over the frequency range of normal head movements. Otolith afferents from both the utricle and saccule report the static and dynamic changes of linear accelerations, such as the force of gravity. Most first-order vestibular neurons apparently respond to either linear or angular accelerations (Gold-berg and Fernandez, 1975a; but see Estes *et al.*, 1975) but not to both. Efferents in the vestibular portion of the 8th nerve serve an as yet unknown role.

The Vestibulo-Ocular Reflex

Since the semicircular canal end organs are excited by head *acceleration* and the output of the vestibulo-ocular reflex (VOR) is a compensatory eye *position,* the pathways from the end organ to the extraocular muscles can be thought of as per-forming a double integration in time. As seen above, the information reaching the brain stem and cerebellum over the 8th nerve already reflects one integration since it is essentially a function of head velocity over the range of normal head move-ments. Where and how does the remaining integration take place?

Properties of the Vestibular Nuclei in the Decerebrate Cat. Although Adrian, in 1943, was the first to record from a mammalian (cat) ves-tibular nucleus, it was not until the mid-1960s and early 1970s that discharge pat-terns were analyzed quantitatively. The large majority of those studies employed the decerebrate cat preparation, used horizontal angular acceleration as an ade-quate stimulus, and recorded primarily from the MVN. Contrary to units in the 8th nerve, which exhibit an increase in activity for head rotations *only* toward the recording site (i.e., when the head is viewed from the top, a clockwise rotation causes only increased activity in the right 8th nerve), neurons in the vestibular nuclei can exhibit either an increase in activity (Type I, a designation first used by Gernandt (1949) for vestibular afferents and later by Duensing and Schaefer (1958) for the vestibular nuclei) or a decrease in activity (Type II). As first men-tioned by Adrian (1943), Type I neurons represent about two-thirds of the units in the decerebrate cat (Gernandt, 1949; Melvill Jones and Milsum, 1970; Shimazu and Precht, 1965). In either the alert (Fuchs and Kimm, 1975) or anesthetized monkey (Abend, 1977), both Type I and II neurons occur with almost equal fre-quency. Units that exhibit either an increase (Type III) or decrease (Type IV) in activity for both clockwise *and* counterclockwise rotations occur much less fre-quently and, at least in the monkey, may reflect a latent macular input (Abend, 1978).

As in primary afferents, a step of ipsilateral acceleration causes a roughly exponential increase in firing rate (*FR*) for most Type I feline neurons, that is

$$FR = S_2(1 - e^{-t/T_3})$$

Unlike primary vestibular afferents, neurons in the vestibular nuclei have relatively low resting rates (15.3 ± 13.9 spikes/sec; Shinoda and Yoshida, 1974) and up to 26% of *all* cells in the vestibular nuclei (Melvill Jones and Milsum, 1971) are silent with the head at rest. Type I units with zero resting rates (called *kinetic* neurons by Shimazu and Precht, 1965) have time constants (T_3) of 3.7 ± 0.8 sec, whereas those with finite resting rates (called *tonic* neurons) have time constants more than twice as large (8.1 ± 1.6 sec). Separate populations of short time con-stant (2.8 ± 0.7 sec) and long time constant (7.0 ± 2.5 sec) units have also been reported by others (Shinoda and Yoshida, 1974). Kinetic units can be monosynaptically activated by 8th nerve stimulation, whereas tonic units are activated polysynaptically and are very sensitive to anesthesia (Precht and Shimazu, 1965). Because of their low resting rates, a step of contralateral acceleration drives many units to zero firing rate (cutoff). However, if accelerations are kept less than 4 deg/sec^2, the responses are symmetrical and linear so that a sensitivity factor (S_2) can be estimated at 6.1 $\pm$ 3.3 spikes/sec/deg/sec^2 (Shinoda and Yoshida, 1974).

Discharge properties of Type I neurons in the decerebrate cat have also been studied by applying sinusoidal angular accelerations as adequate stimuli. Unlike 8th nerve discharges, vestibular nucleus discharge patterns have at least two nonlinearities. First, even for small peak accelerations, the increase in unit activity with an ipsilateral sinusoidal acceleration is steeper than the decrease for ipsilateral deceleration (i.e., the response is skewed owing to a third harmonic component). Second, because of the low resting rates in most neurons, even modest increases of peak acceleration drive the unit to zero firing rate during part of the stimulus cycle. By assuming that the cutoff response is part of a sine wave and by correcting the phase shift for the asymmetric rising and falling phases of neural discharge, it has been possible to construct Bode diagrams for Type I neurons (Shinoda and Yoshida, 1974). As with the data obtained earlier for acceleration steps, the neurons fall into two populations with different time constants of about 3 and 7 sec; see Shinoda and Yoshida (1974) for the detailed analysis. A similar time constant (about 4 sec) was previously reported for a smaller population of vestibular nucleus neurons consisting of both Type I and II neurons (Melvill Jones and Milsum, 1971). No quantitative differences between discharge patterns of Type I and II cells were detected (Melvill Jones and Milsum, 1970). Figure 9 shows the averaged phase lag of *all* Type I neurons in the vestibular nucleus recorded by Shinoda and Yoshida (1974).

The Neural Integrator

The Decerebrate Cat. A comparison of the phase shifts of neurons in the vestibular nuclei with the phase shifts of canal afferents (Anderson *et al.,* 1978) reveals that very little additional phase lag is added at the vestibular nuclei (Fig. 9A). The eye movements of the overall VOR in the decerebrate cat lag activity in the vestibular nuclei at all frequencies of head rotation (Fig. 9A). At low frequencies (0.03 Hz), the signal in the vestibular nuclei must be further delayed by 60 deg, but at higher frequencies an additional lag of about 100 deg is required. Since at least some of these vestibular nucleus axons must ascend in the MLF, the final common pathway (extraocular motoneurons and muscle) is the last component in the three-neuron arc that can contribute a delay. Recordings from the abducens nerve reveal that, although the final common path does provide additional phase lag, especially at higher frequencies (Fig. 9A), a phase shift still remains between the 8th nucleus and motor nerve discharge, which cannot be provided by the shortest direct VOR pathways. Therefore, Robinson (1971) suggested that an indirect pathway, involving the reticular formation, delays (or integrates in time) the signal in the vestibular nuclei before sending it to the motoneurons. The phase-shifting properties of this hypothetical reticular network can be obtained by subtracting the phase lag of abducens from vestibular nucleus neurons (data points of Fig. 9B). To a reasonable approximation, the phase lags can be produced by a "lossy" integrator with a time constant of about 2 sec (solid line "fit," Fig. 9B after Shinoda and Yoshida, 1974).

Two specific structures, other than the reticular formation, have been implicated in the integration. First, Type II neurons in and around the nucleus prepositus hypoglossi (NPH) have phase lags similar to those recorded in the abducens nerve (Fukushima, Igusa, and Yoshida, 1977; Fig. 9A), suggesting that these neurons may be part of the "integrator" or receive an integrated signal. Since NPH neurons excite ipsilateral motoneurons, Type II activity would be appropriate to subserve a pathway parallel to the three-neuron route through the vestibular nucleus. On the other hand, in cats anesthetized with ketamine hydrochloride, Blanks, Volkind, Precht, and Baker (1977) found the phase lags of Type II NPH neurons to be only about 20 deg larger than those of vestibular nucleus neurons

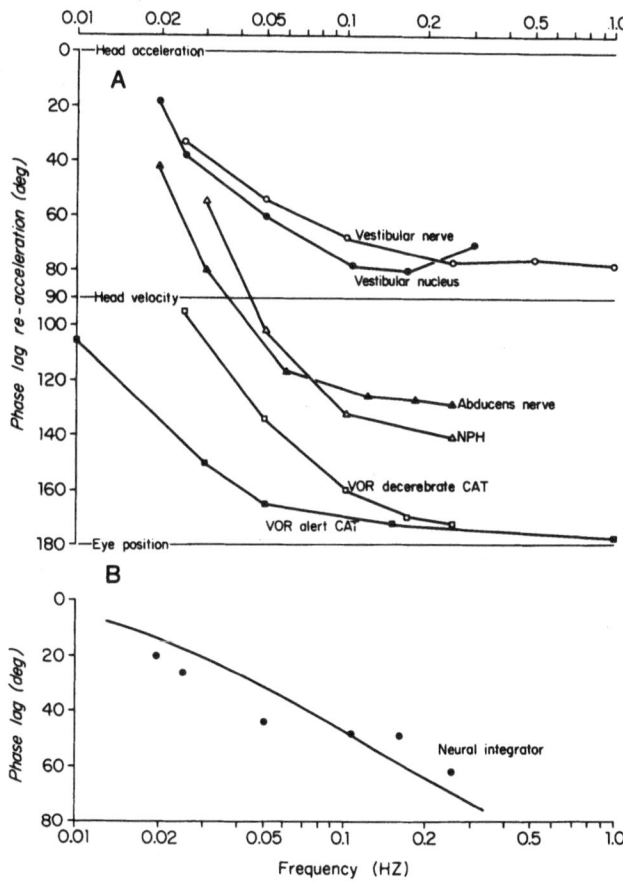

Fig. 9. The phase of averaged neural activity relative to a sinusoidal head acceleration at various stations of the VOR. (A) Vestibular nucleus, abducens nucleus, and VOR decerebrate cat data after Shinoda and Yoshida (1974); vestibular nerve data after Anderson *et al.* (1978); NPH data after Fukushima *et al.* (1977); VOR alert cat data after Robinson (1976). (B) Neural integrator data points obtained by subtracting vestibular nucleus from abducens nerve characteristics; curve represents phase lag of a "lossy integrator" with a time constant of about 2 sec.

and, therefore, too small to match phase lags in the abducens nucleus. Second, cerebellectomy or reversible cooling of the cerebellum causes a tenfold reduction in VOR gain and a phase lead between 45 and 90 deg at low (about 0.05 Hz) frequencies (Carpenter, 1972). By comparing VOR transfer functions before and after cerebellectomy and assuming the cerebellum acts in parallel with MLF and reticular pathways, the Bode plot for the gain of the cerebellum was calculated. Based largely on the gain characteristic, Carpenter (1972) concluded that over a nearly hundredfold (0.05–5.0 Hz) range of frequency, the cerebellar transfer function shows "no significant deviation from a perfect integrator" and suggested that the cerebellum performs the last integration in the VOR. Because of its extensive projections to the brain stem, another possibility is that the cerebellum might provide parametric modulation of the time constant of an integrator residing in the reticular formation. With the feline cerebellum intact, Robinson (1974) estimated the integrator time constant at 20 sec, whereas cerebellectomy reduced it to 1 sec, the apparent time constant of the pontine integrator alone. Unfortunately, these two cerebellectomy studies did not agree on important quantitative details.

The Alert Monkey. There is also strong evidence that a neural integrator is required in the VOR of alert primates. It has already been demonstrated that, over the normal range of head velocities, firing rate in the 8th nerve is proportional to head velocity. The phase of the overall VOR shows a small constant phase advance (less than 5 deg) over the frequency range 0.05 to 1.5 Hz (Skavenski and Robinson, 1973). Only a modest amount of additional phase delay is provided by the viscoelastic properties of the globe and extraocular muscles. (See Robinson, 1964, for details on the mechanics of the oculomotor periphery.)

Simian motoneurons have a very stereotyped discharge pattern for all eye movements. For a lateral saccade to an eccentric target, abducens motoneurons emit an intense burst of activity that begins about 6 to 8 msec before the saccade and lasts nearly for its entire duration. If the eye comes to rest beyond a certain threshold position, the motoneuron discharges at a very regular tonic rate that increases with increasing lateral deviation; individual neurons have different thresholds for tonic firing and different slopes relating firing rate and eye position (Fuchs and Luschei, 1970; Keller and Robinson, 1972). Before and during a medial (i.e., nasalward) saccade, abducens motoneurons often exhibit a complete cessation of activity. Motoneurons in the trochlear and oculomotor nuclei have similar firing patterns (Robinson, 1970; Schiller, 1970), but the direction of eye movement for which they exhibit their most intense increase of activity (the "on" direction) depends on which muscle they innervate. As might be expected from the burst of firing associated with the high-velocity saccade, motoneurons also change their firing rates with eye velocity. In animals trained to follow a slowly moving target, firing rate increases linearly with lateral eye velocity. Although motoneuron firing rate also shows a modest sensitivity to eye acceleration (Keller, 1973), firing rate (FR) can be described, to a good first approximation, by the equation

$$FR = K(\theta - \theta_\mathrm{T}) + R \frac{d\theta}{dt} \tag{3}$$

where θ is eye position and θ_T is the position threshold for steady firing. The value of K is distributed from 1.1 to 14.5 with a clear peak at 3.5 spikes/sec/deg ($n =$

88 from Keller and Robinson, 1972; and Fuchs and Luschei, 1970). R exhibits a skewed distribution ranging from 0.25 to 3.7 spikes/sec/deg/sec with a peak near 0.6 ($n = 20$; Keller and Robinson, 1972); for oculomotor nucleus neurons (Robinson, 1970), R ranges from 0.3 to 5.0. Therefore, an average time constant (R/K) for equation 3 is about 0.17 sec so that the oculomotor periphery (i.e., the muscle and mechanical properties of the moving eyeball) contributes a phase lag of more than 45 deg only at frequencies greater than 1 Hz. Moreover, at higher frequencies, the increasing phase lag in the oculomotor periphery is at least partially compensated for by the decreasing phase lag of 8th nerve firing relative to acceleration (Fernandez and Goldberg, 1971) so that even at high frequencies a neural integrator is probably required. When the discharge patterns of abducens motoneurons were compared during smooth pursuit and the VOR, the time constants were similar in both conditions (11 of 14 neurons), suggesting that equation 3 applies no matter how the eye movement is generated (Skavenski and Robinson, 1973).

The Vestibulo-Collic Integrator. An integrator is apparently also required for the descending vestibular projections to the spinal cord. Like ascending vestibulo-oculomotor projections, vestibulospinal projections involve direct three-neuron pathways via the vestibular nuclei and the lateral and medial vestibulospinal tracts and indirect routes via the reticular formation (see Chapter 4). When EMG activity is recorded from the triceps brachii muscles of the forelimbs during horizontal rotations, the phase shift in EMG activity lags that recorded in the 8th nerve by about 40 deg at 0.15 Hz and 85 deg at 1 Hz (Anderson, Soechting, and Terzuolo, 1977). Similar data with somewhat larger phase lags have also been recorded from neck extensor muscles (Ezure and Sasaki, 1978). Such phase lags suggest that, like the VOR pathway, additional processing in the form of a neural integration is required.

A Model Integrator. Since neurons in the reticular formation are well endowed with axon branches that could end locally (Scheibel and Scheibel, 1958), one possible way to construct an integrator would be by interconnecting a net of neurons with recurrent collaterals. A very elementary net of only six neurons whose connections, one to the other (indicated by the matrix), were determined by the flip of a coin, are shown in Fig. 10. Let us assume that an input volley at $t = 0$ (an impulse of activity) causes every neuron to fire, but in order for each

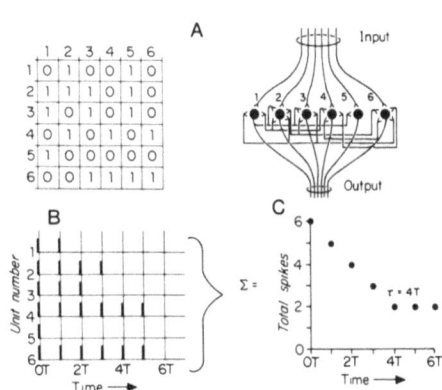

Fig. 10. A hypothetical neural integrator constructed from six neurons with randomly connected recurrent collaterals. (A) Connectivity matrix and circuit diagram generated by a coin flip; a "1" (heads) in a column indicates connections of that neuron (e.g., #2) with other neurons (#1 and #4) or itself (row 2). Firing patterns of each of the six neurons in the net (B) and the total activity in the net (C) as a function of time.

neuron to fire again at least two spikes must impinge on the neuron simultaneously. If T represents the conduction time in every collateral plus the time required to raise the target neurons to threshold, then every cell except No. 5, which receives only one collateral, will fire a second spike at $t = T$ sec (Fig. 10B). At $t = 2T$, four of the units will still be receiving enough synaptic input from their neighbors to fire again, and so on; at $t = 4T$ and thereafter, only units Nos. 4 and 6 will be discharging. If the total activity in the pool, taken as the total number of spikes, is considered as its output, this reticular net has retained the effect of the brief input at $t = 0$ for $4T$ sec. Since a perfect integrator would cause the net to respond indefinitely with 6 spikes (i.e., the integral of an impulse is a step), this net actually performs an imperfect integration. The time constant of this "lossy integrator" would be about $3T$, where T might equal 1 or 2 msec. Clearly, this time constant (on the order of milliseconds) is much too small to provide the 1- or 2-sec time constant actually required. However, by adding more neurons, staggering the values of T, altering the threshold requirements to activate a neuron, etc., it would be possible to construct a suitable neural integrator.

If such an integrator indeed exists, the discharge patterns of different neurons in the reticular formation should exhibit a variety of time constants and hence different phase shifts relative to head acceleration. In fact, the histogram of phase shifts characteristic of neurons in the simian medullary reticular formation reveals that at least 25% of the neurons have lags greater than 110 deg, whereas only 6% of the neurons in the vestibular nuclei have such phase lags (Kimm and Fuchs, 1975). On the other hand, since electrical stimulation of the pons at constant frequencies produces ocular deviations of constant velocity, Cohen and Komatsuzaki (1972) suggest that the integrator may reside in the paramedian pontine reticular formation.

PROPERTIES OF THE VESTIBULAR NUCLEI IN THE ALERT MONKEY. We have seen that the role of the vestibular nuclei as simple relay stations for head velocity information resulted largely from experiments on the anesthetized or decerebrate cat. However, decerebration increases the phase lead of the VOR at low frequencies [compare the VOR in alert (Landers and Taylor, 1975; Robinson, 1976) and decerebrate cats (Fig. 9)], and decerebration also suppresses saccadic and smooth pursuit eye movements, although decerebrate cats pretreated with reserpine do emit saccades (Mergner and Pompeiano, 1978). A parallel series of experiments on alert monkeys has revealed that neurons in the simian vestibular nuclei (and probably the alert feline vestibular nuclei as well) perform roles other than simply forwarding the velocity signal to motoneurons and the integrator.

Single-Unit Activity. Based on their responses to horizontal sinusoidal head rotation on the one hand and eye movement on the other, neurons in the simian vestibular nucleus (mostly MVN) can be divided into four categories (Fuchs and Kimm, 1975; Keller and Daniels, 1975; Keller and Kamath, 1975; Miles, 1974).

The majority of the units are related to rotation of the head *per se.* At frequencies of head oscillation between 0.5 and 0.9 Hz, *vestibular only* neurons always discharge roughly in phase with either ipsilateral (Type I) or contralateral (Type II) head velocity, irrespective of whether the eyes are moving (Fig. 11A). *Vestibular plus saccade* neurons also discharge roughly in phase with head velocity but, in addition, also exhibit a transient change in activity (more often a pause rather than a burst of spikes) for saccades in the same direction as the adequate head rotation, and often in all other directions as well (Fig. 11B). These neurons (and all others with *any* eye-movement sensitivity) discharge at significantly higher resting rates than *vestibular only* neurons (about 87 spikes/sec vs. about 42 spikes/sec) and display higher sensitivities to a standard (0.9 Hz, 680 deg/sec^2) acceleration (about 81 spikes/sec vs. about 45 spikes/sec). Furthermore, Type I neurons in either category have generally higher resting rates and lower sensitivities than Type II neurons. Finally, Type I neurons in either category can be activated at monosynaptic latencies by stimulating electrodes placed on the ipsilateral 8th nerve (Keller and Kamath, 1975). The connectivity of simian Type II neurons has not been studied.

By oscillating animals at very low frequencies, Buettner, Büttner, and Henn (1978) demonstrated that the time constants of both Type I and II vestibular nucleus neurons (*vestibular only* and *vestibular plus saccade*) ranged from 9.5 to 24.5 sec for different monkeys. When the animals were anesthetized, the neural time constants fell to between 5 and 7 sec, values similar to those in the 8th nerve. Therefore, as stimulus frequency decreases (from 0.05–0.007 Hz), 8th nerve activity shows an increasing tendency to lead head velocity (Fig. 8D), whereas the activity in vestibular nucleus neurons remains more nearly in phase with head velocity. These data suggest that 8th nucleus activity represents a signal that has been integrated relative to that on the 8th nerve, and that the effect of this integration is to make the discharge of vestibular nucleus neurons more nearly in phase with head velocity over a wider (low) frequency range. Remember, however, that a second integrator is still required to transform the head velocity signal to eye position.

Another, smaller population of neurons in the vestibular nuclei of monkeys has discharge properties that are quantitatively similar to abducens motoneurons and, hence, these neurons discharge with eye movement *per se* (*eye movement only*, Fig. 11D). In particular, when a rotating animal is required to suppress the VOR by fixating a target rotating with him (suppression) so that the eyes remain fixed in the head, the unit activity is not modulated. During the VOR elicited by sinusoidal head rotation, the peak in firing rate for different units leads eye position by 21 to 36 deg just as with motoneurons (Fuchs and Kimm, 1975; Keller and Kamath, 1975), suggesting that these neurons receive a signal that has already been integrated or that they, themselves, are part of the integrator. Seventy percent of *eye movement only* neurons are activated by 8th nerve stimulation, but at latencies that are at least disynaptic or longer (Keller and Kamath, 1975). Finally, *vestibular plus position* neurons exhibit changes in activity with both vestibular stimulation *and* eye position (Fig. 11C). When the head is stationary, these units show a monotonic increase in firing rate with horizontal eye position but no burst with saccades. The slope relating firing rate and eye position is less than for motoneurons ($K = 1.35 \pm 0.4$; $n = 21$). During suppression of the VOR, these units (like the units already described with vestibular sensitivity) exhibit a deep modulation in activity that is in phase with head velocity; the modulation is superimposed on a steady firing rate established by the fixed-eye position. Unfortunately, neither the connec-

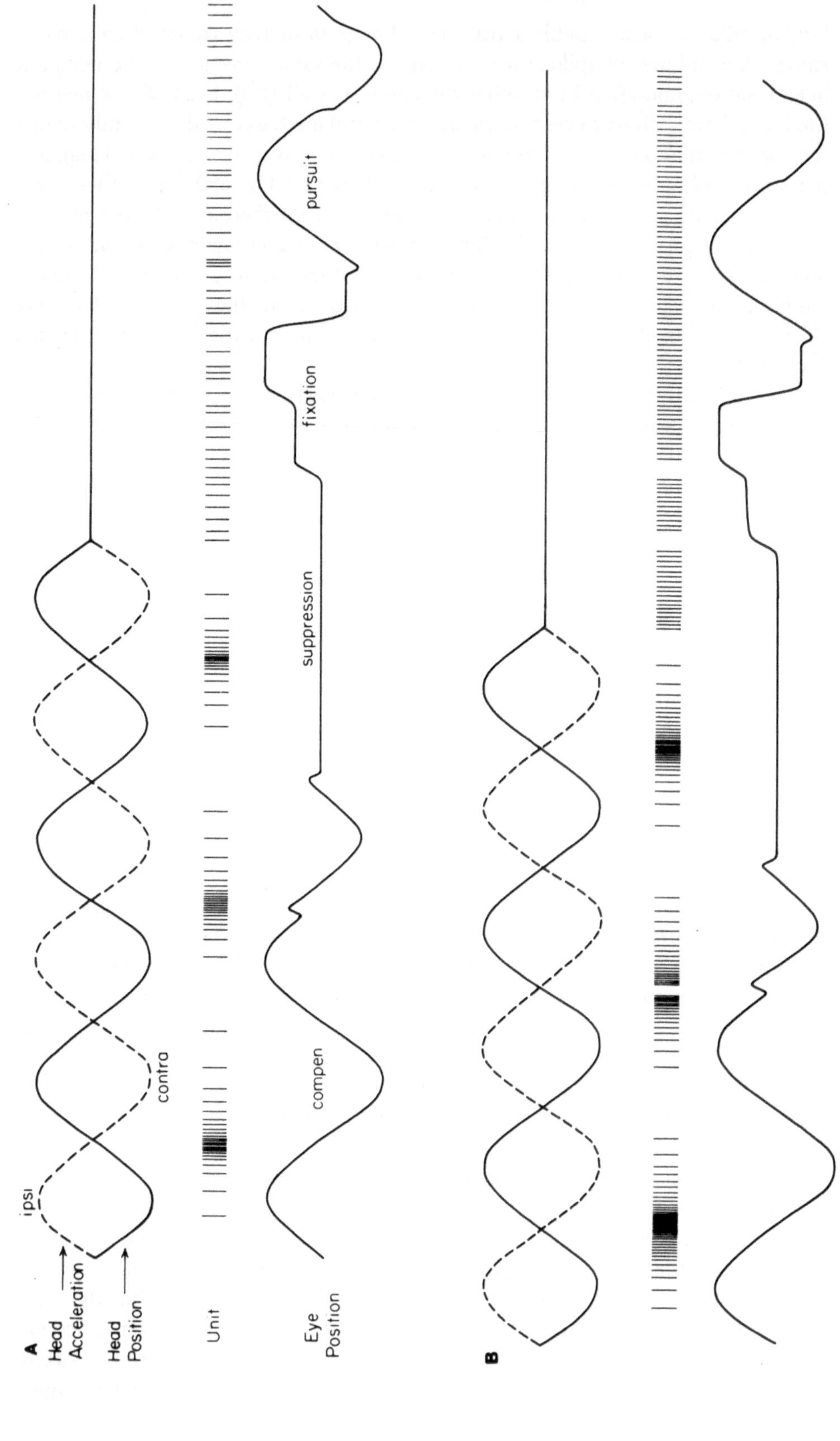
A
Head
Acceleration
Head
Position
ipsi
contra
Unit
Eye
Position
compen
suppression
fixation
pursuit
B

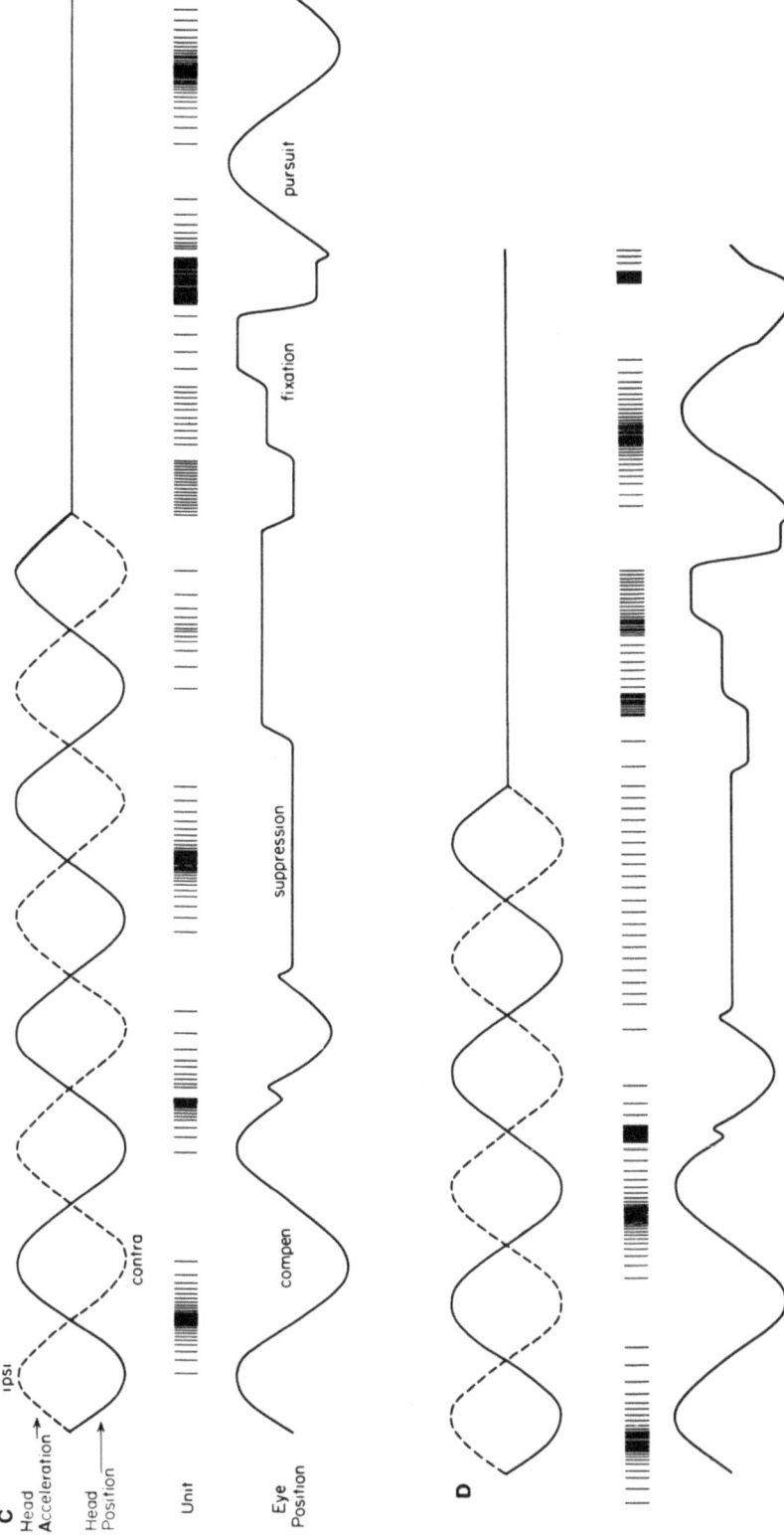

Fig. 11. Schematic discharge patterns of the four categories of neurons in the simian vestibular nucleus classified according to their response to horizontal head rotation and a variety of eye-movement conditions. (A) *Vestibular only* neurons discharge only with head velocity. (B) *Vestibular plus saccade* neurons discharge with head velocity and also pause for saccades. (C) *Vestibular plus position* (TVP) neurons discharge with both head velocity and eye position. (D) *Eye movement only* neurons discharge with eye movements *per se.*

tivity nor the response of these units during the VOR or smooth pursuit eye movements has been studied. Nevertheless, these units are clearly a hybrid whose firing rate is related to both eye position and head velocity *per se* and, therefore, must receive a direct 8th nerve input and an input from the integrator.

Effects of Lesions. Despite the wealth of eye-movement related activity in the vestibular nuclei, lesions there cause rather modest eye-movement defects in the monkey. In fact, lesions in the rostral portions of the main nuclei apparently have no effect on optokinetic nystagmus, optokinetic afternystagmus, or the fast phases of vestibular or optokinetic nystagmus (Uemura and Cohen, 1973); nor do they result in gaze nystagmus or any clear abnormality of voluntary eye movement. One remarkable consequence of unilateral lesions of both SVN and rostral MVN is to cause a marked change in the direction of nystagmus elicited by caloric irrigation contralateral to the lesion ("perverted nystagmus"). In contrast, lesions medial to MVN that possibly involved NPH and the medullary reticular formation have more severe effects on voluntary eye movements than on eye movements induced by caloric stimulation. These findings in the monkey are at odds with those in the cat. Lesions of the cat vestibular nuclei abolish eye movements generated by stimulation of the visual cortex, a finding which led Spiegel (1933) to conclude that commands for voluntary eye movements are funneled through the vestibular complex.

OTHER FEATURES OF THE VESTIBULAR NUCLEI

Visual Responses. In the goldfish, 8th nerve canal afferents respond not only to adequate vestibular stimuli, but also to movement of the entire visual field generated by placing the animal within a large, rotating, striped cylinder (Klinke and Schmidt, 1970). Although similar optokinetic stimuli do not produce modulation of 8th nerve discharge in the monkey (Keller, 1976), all neurons in the vestibular nucleus that respond to head rotation also respond to optokinetic stimuli rotating in the opposite direction (Henn, Young, and Finley, 1974; Waespe and Henn, 1977). In response to a step change of the optokinetic stimulus alone, units increase their firing rates exponentially with a time constant of about 9 sec; Type I neurons have shorter time constants than Type II units. Maximum firing rate increases monotonically with cylinder rotation up to velocities of 60 deg/sec. On the average, the maximum firing rate associated with a given optokinetic velocity was about 70% of the maximum firing rate reached during head rotation at the same velocity. Therefore, when a monkey undergoes a rapid acceleration to a constant velocity within a fixed optokinetic cylinder, a vestibular nucleus neuron first exhibits a rapid exponential rise in activity owing to its short-time-constant vestibular response. After constant velocity has been reached, the neuron's firing rate would begin to decrease if the animal were in the dark. However, by now the contribution of the somewhat longer time constant response to the visual stimulus has become significant so that the net firing rate decreases to a steady discharge of only about 70% of maximum. Consequently, the effect of the visual input is to increase the time constant of the vestibular neuron so that its discharge becomes a better replica of head velocity. Since, in man, the illusion of self rotation (called circular vection) can be induced by a similar full field optokinetic stimulus (Dichgans and Brandt, 1972), the neural data suggest that structures as peripheral as the vestibular nuclei may provide signals related not only to real body motion but apparent body motion as well. Neurons in the feline vestibular nuclei also exhibit an optokinetic response (Keller and Precht, 1978).

Responses to Neck Stimulation: Neck-Ocular Connections. Although other sensory inputs impinge on vestibular nucleus neurons at large, there is relatively little evidence of sensory convergence on vestibular nucleus neurons that are destined for the oculomotor complex. Neck afferents driven by electrical stimulation of the cervical dorsal roots (C2 and C3) or vertebral joints evoke long latency (greater than 3 msec) contralateral IPSPs and ipsilateral EPSPs in cat abducens motoneurons (Hikosaka and Maeda, 1973). This synaptic pattern would indeed produce compensatory eye movements in response to body rotation since a rightward trunk rotation would excite the right cervical joint receptors, which would produce a conjugate rightward eye movement by contraction of the ipsilateral lateral rectus and inhibition of the contralateral lateral rectus. Because vestibular nucleus axons recorded in the VIth nucleus can be affected by neck stimulation and because the spontaneous firing of Type I neurons in the vestibular nuclei can be facilitated by contralateral and inhibited by ipsilateral neck stimulation (see also Fredrickson, Schwarz, and Kornhuber, 1966), it seems likely that the VOR and neck-ocular pathways share common secondary vestibular neurons. However, because noticeable effects owing to neck stimulation often require multiple shocks, it is likely that the neck influences on extraocular motoneurons are usually subordinate to vestibular inputs unless the vestibular input has been eliminated (as in the labyrinthectomized monkeys discussed earlier). About 35% of secondary vestibular neurons whose axons ascend in the MLF exhibit a convergence of labyrinthine and somatosensory input (Rubin, Liedgren, Odkvist, Milne, and Frederickson, 1978). Apparently, convergence of labyrinthine and somatosensory input is more common on spinal-destined neurons in the LVN (Wilson and Peterson, 1978). The functional significance of somatosensory inputs to the extraocular motor nuclei remains to be investigated.

Commissural Connections. It is clear from the existence of Type II neurons in the vestibular nuclei that information originating in the ipsilateral semicircular canals must cross to the contralateral brain stem. The bulk of the crossing information is probably relayed by commissural fibers linking the vestibular nuclei on opposite sides of the midline. Both the SVN and MVN in the cat provide fibers that terminate mostly in the corresponding contralateral vestibular nucleus (Gacek, 1978; Ladpli and Brodal, 1968; Pompeiano *et al.*, 1978). In fact, apparently more fibers from the SVN cross the midline than ascend in the MLF (Ladpli and Brodal, 1968). In the cat, NPH also provides fibers to the contralateral vestibular nuclei (Pompeiano *et al.*, 1978). Commissural connections, at least between the bilateral SVN, have also been demonstrated in the monkey (McMasters *et al.*, 1966).

For vestibular stimuli that activate the horizontal canals, Shimazu and Precht (1966) found evidence suggesting that the commissural system provides inhibition that is mediated by the Type II neurons. First, electrical stimulation of the contralateral 8th nerve caused an excitation of Type II neurons (shortest latency = 3.2 msec) and suppression of spontaneous discharges in Type I neurons (longest latency = 4 msec). Second, transection of commissural fibers running on the floor of the 4th ventricle eliminated both responses. Third, after destruction of one labyrinth, ipsilateral Type I units could not be found, although ipsilateral Type II units were still present. Therefore, they suggested that stimulation of the ipsilateral horizontal canal excites at least two ipsilateral Type I neurons in succession; the axon of the last Type I neuron would cross the midline to excite a contralateral Type II neuron that would, in turn, inhibit a contralateral Type I neuron (Fig. 7C). Recording from vestibular nucleus neurons (not identified as Type I or II), Kasahara and Uchino (1974) demonstrated EPSPs in the monosynaptic range (0.9 ± 0.2 msec) from stimulation of the ipsilateral 8th nerve and IPSPs in the polysynaptic range (2.5 ± 0.5 msec) from contralateral 8th nerve stimulation. These data support the connections suggested by Shimazu and Precht (1966), although shorter inhibitory pathways have also been reported (Kasahara, Mano, Oshima, Ozawa, and Shimazu, 1968; Mano, Oshima, and Shimazu, 1968). Finally, when ampullary nerves from each of the three canals are stimulated individually, most vestibular nucleus neurons exhibit a contralateral inhibition from that canal lying in a plane parallel with the ipsilateral canal whose stimulation produces excitation (Kasahara and Uchino, 1974).

In contrast to crossed inhibition of second-order canal neurons, second-order otolith neurons apparently receive a crossed facilitation. Both LVN neurons affected by lateral tilt (mainly owing to utricular stimulation) and units activated from the saccular nerve are only facilitated by stimulation of the contralateral vestibular nerve (Shimazu and Smith, 1971; Wilson et al., 1978). Furthermore, the crossed utricular effects are not mediated by the commissural system.

The commissural inhibition in the canal system might serve three possible functions: (1) enhance the sensitivity to angular acceleration (i.e., improve the gain); (2) synchronize the inputs from both ears (i.e., affect the phase), or (3) reduce the noise (e.g., common mode rejection). In the acute cat, sensitivities of vestibular nucleus neurons are two to three times greater than those in the 8th nerve (Anderson, Blanks, and Precht, 1978; Shinoda and Yoshida, 1974). Although a twofold increase in sensitivity in the vestibular nucleus could mean that ipsilateral excitatory and contralateral inhibitory inputs of equal size impinge on the neuron, it is also possible that the doubled sensitivity merely reflects different synaptic efficacies of inputs from each side (e.g., the ipsilateral input could have a synaptic gain of 1.5 and the contralateral input a gain of 0.5). In fact, the canal inputs from each side *are* equal because transection of one 8th nerve causes a twofold decrease in the mean sensitivity of horizontal canal neurons in the contralateral MVN (Markham, Yagi, and Curthoys, 1977). Similarly, SVN neurons recorded in an anesthetized squirrel monkey with canals plugged on one side have sensitivities about one-half that of normal animals (Abend, 1978). Because second-order vestibular neurons in the frog have sensitivities equal to those in the 8th nerve, and the frog apparently has no commissural inhibition (Anderson et al., 1978), the role of the commissural system in providing a functionally equal but opposite input from the contralateral canal seems settled. Unfortunately, in both the alert rhesus macaque (Fuchs and Kimm, 1975; Keller, 1976; Louie and Kimm, 1976) and the gerbil (Schneider and Anderson, 1976), the sensitivities in the 8th nerve and nucleus are more nearly equal. It is uncertain whether these differences are a consequence of the variety of species used or the variety of anesthetic states.

Although in the horizontal canal system there are relatively equal numbers of Type I and II neurons (Abend, 1977; Buettner *et al.*, 1978; Fuchs and Kimm, 1975), neurons in the SVN driven by the anterior and posterior canals include only 6% Type II neurons. Because (1) the anatomy reveals a large commissural connection between the SVNs, (2) the SVN seems specifically concerned with the vertical canals, and (3) the ipsilateral anterior and contralateral posterior canals operate in synergism, the small contingent of putative inhibitory Type II neurons is surprising.

In summary, four neuron types have been demonstrated on functional grounds in the rostral vestibular nuclei of the alert monkey. Two types discharge essentially in relation to head velocity over intermediate frequencies of head rotation and thus respond to vestibular stimuli like 8th nerve fibers; indeed, most receive monosynaptic input from the vestibular nerve. The remaining two types discharge with eye position, suggesting that their firing patterns reflect a temporal integration of the velocity signal that reaches the vestibular nucleus; indeed, those cells that can be activated from the 8th nerve respond only at polysynaptic latencies. One type of eye movement cell also discharges with head rotation to create a hybrid cell responding to *both* eye position *and* head velocity. All the types are driven by optokinetic stimuli, which produce the sensation of self-rotation in man, and a few may be driven by neck afferents and somatosensory stimuli as well. However, under normal viewing conditions, the vestibular and eye-movement sensitivities of these neurons clearly predominate over the other sensory inputs.

Now the next question to be addressed is, which of these neurons ascend to the oculomotor nuclei to complete the VOR?

THE ROLE OF THE MLF

Discharge Patterns in Alert Monkeys. The discharge patterns of single MLF fibers recorded in the vicinity of the trochlear nucleus, where the MLF is a compact fiber bundle, can be divided into two populations. The first population exhibits a burst-tonic discharge pattern for contralateral eye movements that is qualitatively and quantitatively similar to that of medial rectus motoneurons (King, Lisberger, and Fuchs, 1976; Pola and Robinson, 1978) and can therefore be described by equation 3. Values of K range from 1.5 to 7.1 with a mean of 3.5 spikes/sec/deg, whereas values of R range from 0.4 to 1.5 with a mean of 0.87. Therefore, the time constant for the neural discharge of these neurons (0.25 sec) can be considered to be essentially identical to that previously determined for abducens motoneurons (0.17 sec). During suppression of the VOR, these fibers discharge at a constant rate proportional to eye position. Most of these neurons probably represent axons of interneurons lying in the abducens nucleus although neurons in both NPH and MVN also have similar burst-tonic discharge properties. Other studies have shown that identified interneurons in the feline abducens nucleus do have discharge patterns like ipsilateral abducens (or contralateral medial rectus) motoneurons (Delgado-Garcia, Baker, and Highstein, 1977; Nakao and Sasaki, 1978).

Neurons in the second population resemble the *vestibular plus eye position* neurons recorded in the vestibular nuclei (Fig. 11C) except that they discharge for *vertical* head rotation and eye position rather than horizontal (King *et al.*, 1976; Pola and Robinson, 1978). There are several quantitative differences between the discharge of these fibers [renamed tonic-vestibular-pause (TVP) neurons by Pola and Robinson, 1978] and vertical oculomotoneurons. First, the slope of the characteristic relating firing rate and eye position (i.e., K_{TVP}) is less (mean 1.2 to 2.5 spikes/sec/deg) than that of vertical motoneurons (3.5 spikes/sec/deg). Second, the eye velocity sensitivity (R_{TVP}) is less (mean 0.47 spikes/sec/deg/sec) than for motoneurons (0.60 spikes/sec/deg/sec). Third, TVP fibers exhibit a pause in activity for saccades in all directions rather than the burst seen in a vertical oculomotoneuron for saccades in its "on" direction. Finally, during suppression of the VOR, these fibers continue to respond vigorously with head velocity (sensitivity of about 1 spike/sec/deg/sec).

Pola and Robinson (1978) suggested that the TVP neuron represents the interneuron in the three-neuron vestibulo-ocular pathway. Using the data cited above, an equation (Fig. 12) can be written describing the firing rate of TVP neurons (FR_{TVP}) as a function of vertical eye position (Φ) and velocity ($\dot{\Phi}$) and vertical head velocity ($\dot{V}$). To produce this discharge pattern, a vertical TVP neuron in the medial vestibular nucleus (MVN) would add an 8th nerve input (FR_{8N}) related to vertical head velocity, an excitatory signal proportional to vertical eye position and velocity, and an inhibitory signal from a burst generator for all saccades. Axons from vertical TVP neurons would ascend in the MLF to impinge on the appropriate vertical motoneurons. Since vertical motoneurons discharge according to equation 3 (FR_{IR} in Fig. 12), they must receive, in addition, an eye-position signal of about 1.7 Φ and an eye velocity signal of about 0.13 $\dot{\Phi}$, an excitatory signal from the saccadic burst generator during "on-direction" vertical saccades, and a head velocity signal ($\dot{V}$) with a sensitivity of 1.0 spikes/sec/deg/sec to cancel the unwanted head rotation signal during suppression of the VOR.

Pola and Robinson (1978) suggest a similar functional synaptology for horizontal VOR connections to the abducens nucleus. However, because the abducens nucleus is so close to the vestibular complex, horizontal TVP fibers would only infrequently be encountered, although their cell bodies do exist in the vestibular nuclei (Fuchs and Kimm, 1975). The horizontal burst-tonic eye-movement fibers in the MLF rostral to abducens (FR_{BT}, Fig. 12) would then represent "surrogate medial rectus motoneurons" (Pola and Robinson, 1978) whose sole purpose is to forward the already complete horizontal signal to medial rectus motoneurons (Fig. 12).

Such a model (see Pola and Robinson, 1978, for more details) predicts that a lesion placed in the rostral MLF would have quite different effects on horizontal and vertical eye movements. Indeed, such lesions cause a complete abolition of the vertical VOR, an inability to maintain eccentric vertical deviations, and impairment of vertical smooth pursuit, but spare vertical saccades (Evinger, Fuchs, and Baker, 1977). On the other hand, the same lesion produces essentially complete paralysis of all medial eye movements while sparing lateral movements.

Discharge Patterns in the Encéphale Isolé Cat. The discharge patterns of axons that are monosynaptically activated from the 8th nerve and therefore presumably represent the middle neuron of the VOR reflex arc have been recorded within the confines of both the abducens and trochlear nuclei in the encéphale isolé cat (Baker and Berthoz, 1974; Hikosaka, Maeda, Nakao, Shimazu, and Shinoda, 1977; Maeda, Shimazu, and Shinoda, 1971). During vestibular nystagmus induced by 8th nerve stimulation, one group of fibers exhibited rhythmic discharge patterns that were in phase with ipsilateral motoneuron discharge. In particular, tonic stimulation of the contralateral 8th nerve caused a gradual increase in fiber discharge frequency, which paralleled that in the motoneurons; during an ipsilateral fast phase, both fiber and motoneuron pool activity went silent. When nystagmus was reversed to cause a contralateral fast phase, the fibers were silent or exhibited spo-

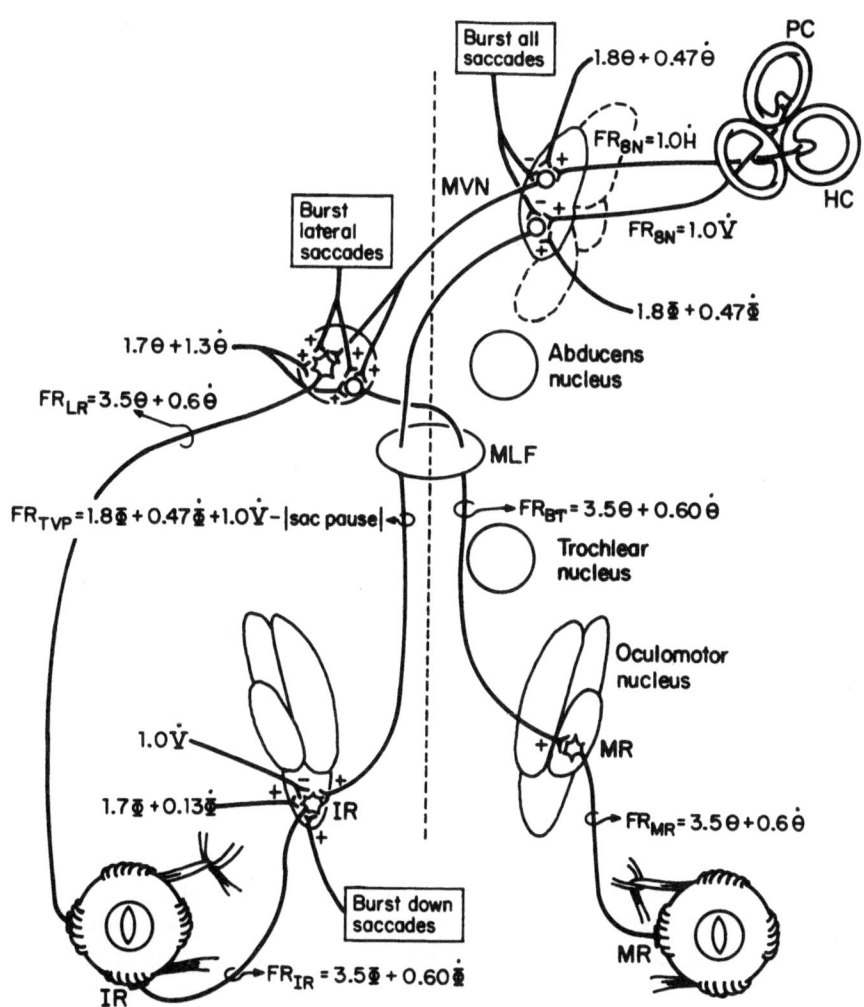

Fig. 12. Highly schematic circuit diagram of possible functional connections of the excitatory VOR pathways (adapted from Pola and Robinson, 1978). Nuclei, canals, and pathways are identified in Fig. 7. Equations relating firing rates to eye movement or head movement parameters are described in the text.

radic bursts unrelated to the fast phase (Hikosaka *et al.*, 1977). Therefore, these neurons behave like TVP units in the monkey. A second group of fibers was activated at monosynaptic latencies by stimulation of the ipsilateral 8th nerve and also displayed the characteristics of a TVP neuron. This latter group of axons could represent the inhibitory VOR interneurons revealed by intracellular techniques, whereas the former group could represent excitatory interneurons. Therefore, data on the cat support the notion of a TVP interneuron in the VOR.

In addition to TVP neurons, inhibitory axons that have burst-tonic discharge patterns and can be driven monosynaptically from the ipsilateral 8th nerve have also been recorded (Baker and Berthoz, 1974; Hikosaka *et al.*, 1977; Maeda *et al.*, 1971), suggesting that VOR interneurons with other discharge patterns may exist. Curiously, other authors recording from secondary vestibular neurons projecting monosynaptically to contralateral abducens motoneurons in the encéphale isolé cat found *no* interneurons that discharged rhythmically with vestibular nystagmus (Grant, Gueritand, Horcholle-Bossavit, and Tyc-Dumond, 1976). The same group of investigators (Horcholle-Bossavit and Tyc-Dumont, 1971) has described a small population of MLF fibers that can apparently be activated directly at latencies of only 0.6 msec by 8th nerve stimulation.

Conclusions. Although some experiments in both cat and monkey are consistent with TVP units serving as the VOR interneuron, the model proposed by Pola and Robinson (1978) is obviously still speculative. As of this writing, there have been no published examples of TVP neurons in the vestibular nuclei whose discharge patterns are related to vertical eye movements. Furthermore, both anatomical evidence (mentioned in the section "Pathways through the Reticular Formation") and neurophysiological evidence (Büttner, Büttner-Ennever, and Henn, 1977; King, 1976) is accumulating that premotor areas for vertical eye movement reside in the mesencephalon. Consequently, information related to vertical eye position and velocity must descend to putative TVP neurons in the vestibular nuclei. Although the MLF carries descending inputs to the vestibular nuclei from midbrain nuclei involved in eye movement (e.g., INC, Markham, 1968; Markham, Precht, and Shimazu, 1966), the studies of MLF discharge patterns described above revealed no neurons with the appropriate discharge patterns. Furthermore, a pure vertical head-velocity signal must ascend to the mesencephalon to cancel the unwanted TVP activity during VOR suppression. Again, there is no evidence of such a signal in the MLF, nor has it been looked for in the mesencephalon.

In summary, despite these reservations, Fig. 12 demonstrates that it may soon be possible to describe quantitatively the behavior of each element in the three-neuron arc. The firing rates of both primary afferents and motoneurons are related in a very stereotypical way to head movement and eye movement, respectively, and therefore their discharge patterns have been amenable to quantitative analysis. Although the firing rate of the interneuron will depend on a variety of inputs, some of these inputs have been looked at in isolation by using an alert, behaving animal. Consequently, more is now known about the operation of this "simple" bulbar reflex than is known about the more widely studied spinal reflexes. Now that the quantitative behavior of the VOR is becoming better understood, it can serve as a model reflex system to study the effects of "descending" connections, like those from the cerebellum, which will be described later.

The Otolith-Ocular Reflex

Behavioral Considerations. In addition to the rotational movements dealt with in detail above, the head also undergoes translational movements such as the vertical displacements that occur during locomotion. Since we have no trouble looking at a stationary object even while jogging at a steady pace, it seems reasonable to expect that compensatory eye movements are elicited by linear as well as angular accelerations. Unfortunately, systematic quantitative studies on eye movement responses to linear acceleration (the otolith-ocular reflex, OOR) are rare, probably because of the difficulty in presenting linear accelerations without potential concomitant angular acceleration.

In the rabbit, linear accelerations produce vertical compensatory eye movements (Baarsma and Collewijn, 1975) with a very low gain varying from 0.1 at 0.3 Hz to 0.01 for frequencies greater than 1 Hz (for comparison, the rabbit's VOR gain at frequencies greater than 0.2 Hz ranges from 0.5 to 0.7). At very low frequencies (0.06 Hz), the gains of the OOR and VOR are nearly equal, although the VOR exhibits a significant phase lead, whereas the OOR lags over the entire frequency range. Since, in addition, eye deviation after a step of linear acceleration takes many seconds to develop, the OOR (at least in the rabbit) seems to represent a slow system that is most effective at low frequencies (Baarsma and Collewijn, 1974) where, in monkey and man, the visual system alone could provide adequate compensation. Comparable data are not yet available for primates, although horizontal but not vertical OORs have been demonstrated in humans (Niven, Hixson, and Correia, 1966).

Another eye-movement response thought to be mediated by the otolith transducers is ocular counterrolling, which results when the head undergoes static rotations around the line of sight. In humans, measurements of the gain of this reflex range from 0.06 (Diamond, Markham, Simpson, and Curthoys, 1978) to almost 0.2 (Kanzaki and Ouchi, 1978). Contrary to the VOR, the gain of the ocular counterrolling reflex is not improved by vision (Krejcova, Highstein, and Cohen, 1971).

In summary, it is possible to demonstrate that the eyes undergo partial compensation for linear accelerations and changes in static head position. However, these compensatory eye movements are very small and it seems doubtful that they assume much significance in normal life. Perhaps they are left over from a time when our ancestors had laterally placed eyes.

Pathways. Information from the utricle and saccule can reach ocular motoneurons over a pathway containing as few as three neurons. As mentioned above, primary afferents from both otolith organs project to the LVN in the cat, and some authors (Gacek, 1971) suggest that the LVN projects to the extraocular motor nuclei. This anatomy has been substantiated by intracellular recordings that have revealed disynaptic EPSPs in contralateral trochlear motoneurons (Baker, Precht, and Berthoz, 1973) and disynaptic EPSPs in ipsilateral abducens motoneurons (Schwindt, Richter, and Precht, 1973) in response to electrical stimulation of the utricular nerve. In the cat, electrical stimulation of the utricular nerve produces EMG changes in extraocular muscles within 5 msec, a figure that is very similar to EMG latencies to canal stimulation (Suzuki, 1972). Electrical stimulation of the saccular nerve causes monosynaptic activation of cells in the ipsilateral LVN, which, in turn, project to the ipsilateral superior rectus subdivision of the oculomotor nucleus (Hwang and Poon, 1975).

In addition to this three-neuron sacculo-ocular pathway [and a longer route originating from the contralateral saccular nerve (Chan, Hwang, and Cheung, 1977)], the cat also possesses another potentially short pathway via the "Y" group, one of the minor vestibular nuclei. Saccular afferents apparently provide an "outstanding termination" to the Y group (Gacek, 1969) and the rostral projection of Y group axons to the IIIrd nucleus is via the brachium conjunctivum (Graybiel and Hartwieg, 1974). The Y group also projects to the oculomotor complex in the monkey (Graybiel, 1977b; Steiger and Büttner-Ennever, 1979).

As with selective stimulation of the canal nerves, eye movements can be elicited by electrical stimulation of the saccular or utricular nerves in the cat or by stimulating the utricular afferents with air puffs delivered to an isolated utricle in cat or monkey (Gernandt, 1970; Hwang and Poon, 1975; Suzuki, 1972). Electrical stimulation of individual maculae elicits eye movements that are appropriate to the orientation of hair cells in the part of the end organ stimulated (Fluur and Mellström, 1970a,b). For example, hair cells in the superior portion of vertically oriented saccules are polarized upward (see Fig. 6C); therefore, an increasing activity in fibers subserving this portion of the maculae would be elicited by an upward deflection of hairs owing to downward head tilt. Downward head tilt would require an upward compensatory eye movement, which is precisely the movement elicited by electrical stimulation in this region of the sacculus. Functionally appropriate compensatory eye movements also accompany focal utricular stimulation.

NEUROPHYSIOLOGY

Primary Afferents. Information about the functional role of the various elements in the OOR has been obtained by subjecting the anesthetized cat to natural linear accelerations. As described earlier, saccular and utricular primary afferents exhibit a firing pattern in response to sinusoidal linear acceleration that usually yields a flat and occasionally an increasing gain characteristic over the frequency range from dc to 1.0 Hz (Anderson *et al.*, 1978; Fernandez and Goldberg, 1976c). Over the same frequency range, the phase varies by no more than 25 deg and is essentially in phase with the angle of head tilt.

Ocular Motoneurons. At the other end of the reflex arc, the vast majority (24/27) of trochlear motoneurons (Blanks, Anderson, and Precht, 1978) and some motoneurons innervating the inferior oblique muscle (Berthoz, Baker, and Precht, 1973) respond to both canal and otolith stimulation created by a sinusoidal roll about a horizontal axis. If the canal contribution is subtracted from the total response to leave the discharge attributed to the otoliths alone, the phase of motoneuron firing rate lags angular head position by 10 deg at 0.025 Hz and roughly 90 deg at 0.5 Hz; the gain is roughly constant over the same frequency range. Since motoneuron discharge is shifted in phase relative to primary afferent discharge, some further neural processing must take place, and Blanks *et al.* (1978) suggest that, as in the VOR, a lossy integrator is required. Based on their phase data, the integrator time constant should be about 1.2 sec. Consistent with the utricular pathway described above, abducens motoneurons also exhibit similar firing patterns (and time constants) to roll stimuli (Anderson, Precht, and Blanks, 1977).

Neurons in the Vestibular Nucleus. Neurons in the vestibular nuclei that have been investigated with linear accelerations have usually been subjected only to steady (constant linear accelerations) tilt stimuli. In the anesthetized cat, step changes in head position (often about a roll axis) usually produce an initial transient response followed by a sustained response that increases monotonically with head tilt (Adrian, 1943; Fujita, Rosenberg, and Segundo, 1968; Peterson, 1970). In response to lateral tilt, the average sensitivity of 21 neurons in the LVN activated monosynaptically from the 8th nerve was about 0.4 spikes/sec/deg; neurons that could not be activated at all or only at polysynaptic latencies were less than half as sensitive (Fig. 9; see also Peterson, 1970).

In one of only two studies of vestibular nucleus neurons during dynamic linear accelerations, Melvill Jones and Milsum (1972) concluded that the phase lag of firing rate relative to linear acceleration was about 0 deg at 0.1 Hz but increased rapidly at higher frequencies to values in excess of 180 deg at 3 Hz. These data imply that the discharge patterns of some OOR interneurons in the vestibular nuclei could already reflect at least one temporal integration of 8th nerve activity. On the other hand, Schor (1974) recorded tilt responses that yielded a Bode plot for gain that was relatively flat to 0.1 Hz and showed a rapid increase in gain at higher frequencies, a characteristic very reminiscent of irregular otolith afferents in the monkey (Fernandez and Goldberg, 1976c) and phasic-tonic otolith afferents in the cat (Anderson *et al.*, 1978). Clearly, more work is needed on the discharge patterns of the OOR interneuron.

INTERACTIONS BETWEEN CANAL AND OTOLITH RESPONSES. As reported above, almost every trochlear motoneuron can be driven by both linear and angular acceleration of the head. The interaction of utricular and canal influences already appears in at least half of the secondary vestibular axons that can be recorded within the confines of the motor nucleus (Blanks *et al.*, 1978); none of these axons conveyed only otolith information. Although disynaptic EPSPs were produced in contralateral trochlear motoneurons following stimulation of either the posterior canal or utricle, the few axons within the IV nucleus that responded to posterior canal stimulation could not also be excited by utricular routes (Baker *et al.*, 1973).

Within the vestibular nuclei themselves, considerable controversy also exists as to whether convergence occurs either between the canals themselves or between otolith and canal inputs. In the anesthetized squirrel monkey tested with natural stimuli, only 9.5% of the canal units in the SVN, 4% of canal neurons in the MVN, and no canal units in the rostral pole of the LVN exhibited a convergence from orthogonal canals (Abend, 1977). In the cat tested with electrical stimuli, there was no positive evidence "for the existence of neural convergence on single vestibular neurons from different ampullary nerves on the same side" (Kasahara and Uchino, 1974, p. 285). On the other hand, in anesthetized cats tested with natural stimuli, only one-third of the units (mostly in MVN) responding to horizontal canal stimulation were *not* affected by stimulation of other vestibular receptors; in addition, many (25%) responded not only to a stimulus effective for one of the vertical canals but for stimuli that excited the otolith organs as well (Curthoys and Markham, 1971). About half of all the units responded to otolithic stimulation. Curthoys and Markham (1971) concluded that these data reflect neural interactions in the vestibular nuclei. On the other hand, Benson, Guedry, and Melvill Jones (1970) also recorded canal sensitive neurons that responded to linear acceleration and concluded that linear acceleration itself could mechanically excite a canal directly. This suggestion is supported by 8th nerve recordings since first-order canal afferents in both the cat (Estes *et al.*, 1975) and the monkey (Goldberg and Fernandez, 1975a) respond as a function of the animal's orientation with respect to the gravity vector.

In summary, although eye-movement responses to linear accelerations are difficult to elicit in an alert animal, the neural connections from the otolith end organs to the extraocular muscles can be as rapid and apparently as effective as those delivered over VOR pathways. Furthermore, the operation of the OOR may also rely on a neural integration before 8th nerve activity is appropriate for motoneurons. In comparison with the VOR, very little is known about the neurophysiology of the OOR, especially in alert animals.

MODIFICATION OF THE VESTIBULO-OCULAR REFLEX

We would be at a great disadvantage watching a tennis match if the VOR were "hardwired" to produce a compensatory eye movement every time we moved our heads. As described earlier, the VOR can be completely suppressed if a rotating subject fixates an object rotating with him. Therefore, it is possible to modify the basic reflex connections described above in order to accommodate to changing visual tracking demands.

ROLE OF THE CEREBELLUM

The flocculus would seem to have all the input and output connections required to effect such a control. First, as shown above, information about head rotation reaches the flocculus directly via collaterals from the 8th nerve (Baker *et al.*, 1972) and indirectly via secondary vestibular neurons (Shinoda and Yoshida, 1975; Wilson, Maeda, Franck, and Shimazu, 1976). Stimulation of the 8th nerve causes largely monosynaptic activation of axons in the ipsilateral flocculus and largely polysynaptic activation in the contralateral flocculus. The crossed polysynaptic input occurs via Type I neurons located in the contralateral rostral MVN (Shinoda and Yoshida, 1975). Second, based mostly on work in the rabbit, the flocculus also receives a visual input. Electrical stimulation of the optic disk evokes both climbing fiber (Maekawa and Simpson, 1973) and mossy fiber (Maekawa and Takeda, 1975) responses in the flocculus. The climbing fiber response (CFR) has a latency of about 12 to 13 msec and is subserved by a three-neuron pathway involving the nucleus of the optic tract and the contralateral dorsal cap of Kooy (a subdivision of the medial accessory olive). The best visual stimulus for the CFR is a contrast-rich, quasi-random pattern moving at very slow velocities of less than 1 deg/sec (Simpson and Alley, 1974); similar stimuli are effective at driving dorsal cap neurons (Barmack, 1977). Mossy fiber responses to adequate visual stimuli can be obtained as modulation of simple-spike activity in some Purkinje cells (P-cells) (Simpson and Hess, 1977). Visual modulation of both mossy fibers (Miles and Fuller, 1975) and simple spikes in P-cells (Noda and Suzuki, 1978) has also been reported in the monkey. Third, according to the anatomy described earlier, the flocculus could receive eye-movement information from one of the oculomotor cell types described in the vestibular nucleus or NPH (Gresty and Baker, 1976). In fact, the discharge patterns of various mossy fibers in the alert monkey are related to every type of eye movement (Lisberger and Fuchs, 1978b). Fourth, stimulation of the C2 dorsal root, which would simulate activation of neck afferents, monosynaptically excites cells in one of the minor vestibular nuclei (the x group). These cells, in turn, project to the flocculus and therefore provide another input, separate from the labyrinthine pathway, to the flocculus (Wilson *et al.*, 1976). Finally, in both the cat and the rabbit, intracellular recordings have demonstrated that flocculus P-cells inhibit neurons in the vestibular nuclei that participate in the VOR. In the cat, stimulation of the ipsilateral 8th nerve evoked a disynaptic IPSP in trochlear motoneurons that was depressed at short latencies by flocculus conditioning stimuli (Baker *et al.*, 1972). Similar depression of the ipsilateral inhibitory pathway owing to flocculus stimulation has been demonstrated in the rabbit abducens (Highstein, 1973) and oculomotor nuclei (Fukuda, Highstein, and Ito, 1972). In the rabbit, the flocculus also apparently inhibits ipsilateral excitatory pathways projecting to the medial and superior rectus muscles (Ito, Nisimaru, and Yamamoto, 1973).

SHORT-TERM MODIFICATION OF THE VOR

All this evidence suggests that the flocculus is strategically placed to compare vestibular, eye-movement, neck, and visual afferent information and produce an output appropriate to modify transmission through VOR pathways. Lesion studies in both monkey and rabbit (Ito, Shiida, Yagi, and Yamamoto, 1974a,b) support this view. In normal monkeys the velocity of the slow phase of nystagmus generated by caloric irrigation can be reduced by 50% if the animal is placed in a stationary visual environment (Takemori and Cohen, 1974a); however, bilateral flocculus destruction virtually eliminates the visual suppression of caloric nystagmus although caloric nystagmus in the dark is unaffected (Takemori and Cohen, 1974b). These data suggest that the flocculus is important only when the VOR must be altered to conform to changing visual conditions.

To test this suggestion, Lisberger and Fuchs (1974) recorded from flocculus P-cells in monkeys trained to suppress the VOR while undergoing sinusoidal head rotations. Although P-cells showed little modulation during head rotation in the dark when the usual compensatory VOR eye movements occurred (Fig. 13, *compensation*), their simple spike activity was deeply modulated during suppression of the VOR when the eyes remained stationary in the head (Fig. 13, *suppression*).

The modulation of firing rate during suppression was related to ipsilateral head velocity (i.e., a Type I neural response) and was quantitatively identical to that recorded in the ipsilateral 8th nerve. Therefore, during suppression of the VOR (the solid cycles in the schematic, Fig. 13), the averaged firing rate of the P-cell (fl) and the vestibular nerve activity (A) impinging on the VOR interneuron would be essentially in phase but of opposite sign since the flocculus synapse is inhibitory. Consequently, interneuron activity ascending to the abducens motoneurons (A–fl) would be essentially unmodulated and the eye (E) would remain stationary. However, during the VOR in the dark (dashed cycles in the schematic, Fig. 13), the flocculus inhibition (fl) is constant and the modulation in the VOR interneuron owing to head rotation is unchecked and ascends to motoneurons to help produce the compensatory eye movements of the VOR.

The lack of P-cell modulation during the VOR could be explained if, in addition to the 8th nerve input, the P-cell was also receiving an input related to eye velocity. Indeed, the same flocculus P-cells that are modulated during ipsilateral head velocities also exhibit roughly equal sensitivities to smooth ipsilateral eye

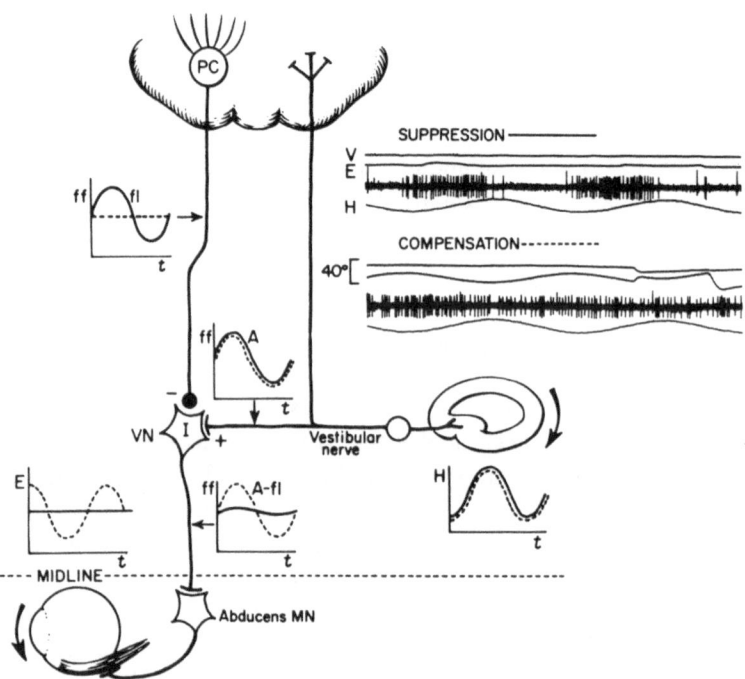

Fig. 13. Schematic diagram of the horizontal VOR pathway and its inhibitory input via the flocculus. The single-unit activity shows the behavior of a flocculus Purkinje cell (PC) while a monkey is undergoing sinusoidal head rotations (0.9 Hz) in the dark *(compensation)* and while he is suppressing the VOR by fixating a target rotating with him *(suppression)*. V and E represent vertical and horizontal eye positions, respectively, and H represents horizontal head position. In the circuit diagram, the head rotation, H(t), causes eye movements, E(t), appropriate to either the suppression (—) or compensation (---) condition. A single cycle of averaged neural activity is traced through the primary afferents (A) and the flocculus (fl) to show a hypothetical discharge (A–fl) in the vestibular nucleus (VN) interneuron.

movements elicited by sinusoidal target movements with the head fixed. Therefore, during the VOR or while a rotating monkey fixates a target stationary in space, the P-cell receives roughly equal but opposite influences owing to head and eye velocity, so its net discharge exhibits no modulation at all (Lisberger and Fuchs, 1978a). In fact, for a large variety of artificial combinations of visual and vestibular interaction created by moving a target spot either in or out of phase with the head at different amplitudes, the net discharge of a P-cell was always accurately predicted by the vector sum of its sensitivity to head and eye velocity alone. In all cases, the net discharge of the P-cell was always appropriate to facilitate the required eye movements. Since these discharge patterns were established within fractions of a second after a new tracking condition was imposed, the flocculus could participate in short-term modification of the VOR.

Lesion studies also suggest that the flocculus plays a role in smooth pursuit eye movements *per se* since flocculus damage causes deficits in optokinetic nystagmus in monkeys (Takemori and Cohen, 1974b) and man (von Reutern and Dichgans, 1977) and abolition of the smooth tracking of food objects in untrained monkeys (Westheimer and Blair, 1974).

LONG-TERM MODIFICATION OF THE VOR

In contrast to the short-term exposures to conflicting visual and vestibular inputs just described, several investigators have outfitted subjects with goggles containing lenses to provide long-term abnormal visual inputs. In experiments begun in 1971 and amplified later, Gonshor and Melvill Jones (1971, 1976) tested the effects of an optical image reversal on the VOR by requiring human subjects to wear spectacles containing "dove" prisms continuously for periods of up to 27 days. In this condition, a head movement to the left, rather than creating movement of the outside world to the right, creates an apparent movement to the left. During the first day, there was a rapid attenuation of VOR gain (measured with 0.16 Hz, ± 60 deg/sec stimuli) to half control values with a subsequent continued decline to about one-quarter control values after four days (Fig. 14). Thereafter, the gain showed a slight gradual increase but the phase underwent large fluctuating changes in a lagging direction. After about two weeks, a new steady state was reached at a gain of about 0.4 and a phase lag of 120 deg (see 14-day record, Fig. 14). Occasionally, eye-movement records displayed a phase lag of 180 deg so that the VOR had actually reversed its direction. These gain and phase changes seemed purposeful since they served to help stabilize the visual world during head movements. Furthermore, the changes could not be attributed to some generalized depression since the gain reduction occurred only in the horizontal and not the sagittal plane. After the prisms were removed, the phase lag was restored to normal in a matter of hours, whereas the restoration of gain, after a marked transient attenuation, proceeded along a time course similar to the original adaptation. Although somewhat different in detail, similar VOR gain and phase changes can be obtained from a prism-adapted cat (Melvill Jones and Davies, 1976).

To test whether these plastic changes in the VOR were indeed adaptive as suggested above, Miles and Fuller (1974) attempted to both decrease and increase the VOR gain by fitting monkeys with telescopic glasses to magnify or minify their "visual world." After wearing ×0.5 diminishing spectacles (the visual world moved half as much as normal during head rotation), the gain in all three monkeys tested decreased to about 0.65 within three days. Similarly, after wearing ×2.0 magnifying spectacles (the world moved twice as much as normal during head rotation), the gain increased to maximum values of about 1.8, again within three days. Prolonged periods of exposure did not further increase or decrease the gain. Recovery back to normal gain also took about two to three days, but immobilizing the head slowed recovery considerably. Similar increases in VOR gain by long-term adaptation to ×2.0 telescopic spectacles have also been reported in humans (Gauthier and Robinson, 1975).

Largely because of its neural connections, Ito (1972) was the first to suggest that the flocculus could be involved in VOR plasticity. The flocculus would be part of a "side path" from the vestibular end organ to VOR interneurons (Fig. 15A) so that by adjusting the sensitivity of P-cells to vestibular input, it would be possible to adjust the gain of the VOR (i.e., in Fig. 15A, the gain of the VOR ($\dot{E}/\dot{H}$) equals $G - H$, where H is the gain of the flocculus "side path"). The adequacy of the adjustment or adaptation would be reflected by the residual slip of the image over the retina as monitored by the visual climbing fiber input (Maekawa and Simpson, 1973). Drawing on Marr's theory of cerebellar cortex (Marr, 1969), Ito further suggested that a persistent error signal arriving over the climbing fibers could serve

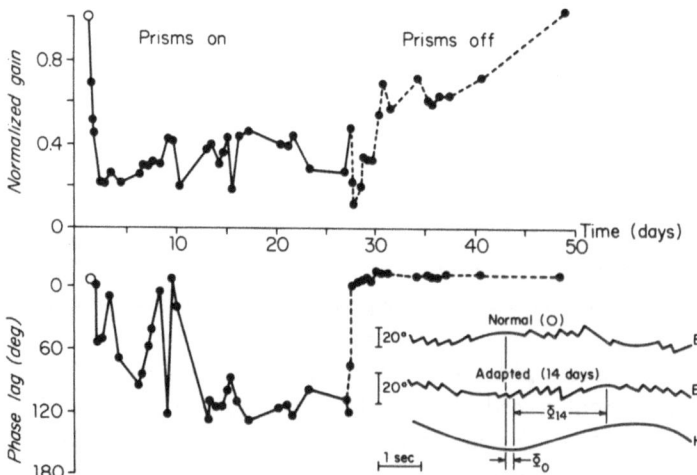

Fig. 14. Effects of long-term reversal of the visual image on the gain and phase of the VOR in humans. Inset shows that, in a normal subject, the VOR compensates quite well for head rotation with a gain near 1 and a slight phase lead (Φ_0). After the subject had been wearing reversing prisms for 14 days, the VOR gain was significantly reduced (note change in eye movement, E, calibration) and the phase (Φ_{14}) lagged head position by about 120 deg. The graph provides a detailed look at the adaptation resulting from putting the prisms on (●) and the recovery upon their removal (●– – – –●). The gains have been normalized to the preadapted value (i.e., the normal VOR gain was taken as 1) and the phase lag was measured relative to a perfect VOR (i.e., a phase lag of zero corresponded to eye movements that were precisely out of phase with head rotation). Assembled from Melvill Jones (1977).

to modify slowly the efficacy of vestibular mossy fiber synapses on flocculus P-cells to promote a "learning process." The modification of the synapses would continue until images no longer slipped on the retina.

To test its role in long-term VOR adaptation, Robinson (1976) removed the flocculus (actually the clearest effects occurred upon removal of the entire vestibulocerebellum) of cats and demonstrated that they no longer exhibited gain changes when outfitted with reversing prisms. Furthermore, the low VOR gain of animals already adapted to reversing prisms returned to control values immediately after flocculus removal. Increases in VOR gain produced by rotating an awake rabbit in front of a single, fixed, vertical slit and decreases in VOR gain produced by rotating the slit with the rabbit for 12 hours were also abolished by flocculectomy (Ito *et al.*, 1974a). Although Ito *et al.* (1974a) produced significant modification of VOR gain in normal rabbits after 12 hours, Collewijn and Kleinschmidt (1975), using full-field optokinetic stimuli for 24 hours, could not reproduce their results.

Most of the evidence presented thus far supports Ito's attractive proposal that the flocculus plays a role in long-term modification (i.e., plasticity) of the VOR. However, at least two more recent experiments suggest that the model in Fig. 15A may be too simple. First, Llinás, Walton, Hillman, and Sotelo (1975) have shown that destruction of the inferior olive prevents recuperation from the motor abnormalities generated by a unilateral labyrinthine lesion in a rat. Furthermore, since

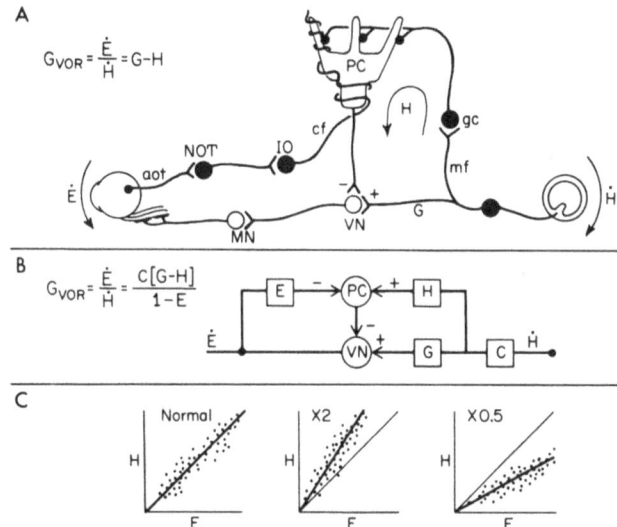

Fig. 15. Circuitry involved in long-term (plastic) modification of the VOR by the flocculus. (A) Ito model (1972) suggests that visual information via the accessory optic system [accessory optic track (*aot*), nucleus of the optic track *(NOT)* and the inferior olive *(IO)*] provides climbing fiber (*cf*) input to Purkinje cells (*PC*), which "educates" the granule cell (*gc*) synapses to alter transmission (*H*) through the mossy fiber (*mf*) pathway. The overall gain of the VOR (G_{VOR}) is altered by varying *H*. (B) More detailed block diagram (Miles and Braitman, 1978) demonstrating four sites where gain changes could affect G_{VOR}. (C) Hypothetical data showing the relation between *H* and *E* from single P-cells in normal animals and animals adapted to ×2 and ×0.5 telescopic goggles.

inferior olive lesions in recuperated animals cause a return of the motor abnormalities, Llinás *et al.* (1975) suggest that the integrity of the olivo-cerebellar system rather than the cerebellum *per se* seems necessary for the acquisition and retention of this form of adaptation. However, the effect of an 8th nerve lesion that removes the tonic drive to one vestibular nucleus and therefore upsets the "operating point" of the system might require a different adaptation than that required to adjust the VOR gain to compensate for the effects of reversing prisms. Second, to test whether the adaptive gain changes that occur during abnormal visual and vestibular interactions can be localized to the flocculus, Miles and Braitman (1978) recorded from populations of P-cells in animals that had been adapted to either $\times 2$ (hi gain) or $\times 0.5$ (low gain) telescopic goggles. In their schematic diagram of Fig. 15B, both pathways G and H are similar to those in the Ito model; the E pathway reflects the P-cell sensitivity to smooth pursuit eye movement, whereas C represents possible changes of 8th nerve sensitivity via an efferent system. In this configuration, the gain $G_{\mathrm{VOR}} = C(G - H)/1 - E$. In normal animals, the sensitivity to head rotation, H, is about equal to the sensitivity to eye rotation, E (Lisberger and Fuchs, 1978a). In high-gain animals, in which adaptation produces a higher VOR gain, E should be greater than H; likewise, in low-gain animals requiring a decreased VOR gain, E should be less than H. Unfortunately, just the opposite results were found (Fig. 15C), suggesting that the adaptation of the VOR might occur elsewhere or that the simple schematic diagram of Fig. 15B is inadequate to explain the adaptive process.

In summary, the long- and short-term effects of altered visual input on the VOR provide a model system to study the potential role of the cerebellum in both immediate and plastic modification of a motor act. A testable model has been proposed, and the next few years should produce experiments that will not only reveal the role of the cerebellum in visual modification of the VOR but also perhaps reveal more about the role of the cerebellum in movement in general.

Appendix

The direction of gaze or the eye position in the head has been measured by three major techniques (Carpenter, 1977). The most widely used technique, the electrooculogram (EOG), measures the orientation of a corneo-retinal dipole potential thought to be generated across one of the layers of the retina. If surface recording electrodes (Ag-AgCl pellets) are placed at the temples, the direction of the dipole and hence the horizontal eye position can be measured with a resolution of about 1 deg of visual angle. The ease of electrode application has made this the method of choice for most human studies; however, its poor sensitivity, its susceptibility to muscle artifacts like orbicularis oculi contractions, its sensitivity to ambient light level, and its unsuitability for vertical eye movements present severe limitations for quantitative studies. A second method determines eye position optically by tracking the position of a light spot reflected or scattered from one of the light-dark boundaries on the eye (iris–sclera or pupil–iris interfaces) or within the eye (a retinal blood vessel or the various reflected Purkinje images). Most of the photoelectric techniques are very sensitive and can resolve movements of a few minutes of arc but usually have a rather limited linear operating range (about ±15 deg), especially in the vertical direction. Finally, a technique that is becoming increasingly popular for the measurement of animal eye movements is an electromagnetic technique first described by Robinson (1963). Briefly, a uniform alternating magnetic field (generated by two large Helmholtz coils on either side of the head) induces a voltage in an eye coil that is created by laying several turns of fine wire under the extraocular muscles (Fuchs and Robinson, 1966). The amplitude of the coil voltage depends on the orientation of the coil within the field and consequently upon the position of the eye in space. By using two magnetic fields, one can measure both horizontal and vertical eye position to within less than 1 min of arc.

Although cooperative human subjects can simply be asked to follow a moving target, animals must be trained to track by one of two basic techniques. With the first technique, monkeys deprived of liquid are trained to shift their direction of gaze by rewarding them with a small amount of liquid for a correct discrimination between the orientations of two fine, small lines that can appear at various random target eccentricities. Alternatively, a monkey can be rewarded for pressing a manipulandum within 0.5 sec of the slight dimming of a moving-target spot. With both these paradigms, accurate discrimination indirectly produces accurate eye tracking. With the second technique, eye position, measured very accurately by the electromagnetic technique, is electronically compared to a signal proportional to target position, and when the difference is less than, say, 1 deg for several seconds, the monkey (or cat) receives a small food reward. In this paradigm, accurate tracking is rewarded directly.

REFERENCES

Abend, W. Functional organization of the superior vestibular nucleus of the squirrel monkey. *Brain Research*, 1977, *132*, 65–84.

Abend, W. Response to constant angular accelerations of neurons in the monkey superior vestibular nucleus. *Experimental Brain Research*, 1978, *31*, 459–473.

Abzug, C., and Peterson, B. Antidromic stimulation in the ponto-medullary reticular formation of local axon branches of contralateral vestibular neurons. *Brain Research*, 1973, *64*, 407–413.

Adrian, E. Discharges from vestibular receptors in the cat. *Journal of Physiology (London)*, 1943, *101*, 389–407.

Anderson, J., Precht, W., and Blanks, R. Central processing in otolith-ocular reflex pathways. In R. Baker and A. Berthoz (Eds.), *Control of Gaze by Brain Stem Neurons* (Vol. 1). New York: Elsevier, 1977.

Anderson, J., Soechting, J., and Terzuolo, C. Dynamic relations between natural vestibular inputs and activity of forelimb extensor muscles in the decerebrate cat. II. Motor output during rotations in the horizontal plane. *Brain Research*, 1977, *120*, 17–33.

Anderson, J., Blanks, R., and Precht, W. Response characteristics of semicircular canal and otolith systems in the cat. I. Dynamic responses of primary vestibular fibers. *Experimental Brain Research*, 1978, *32*, 491–507.

Angaut, P., and Brodal, A. The projection of the "vestibulo cerebellum" onto the vestibular nuclei in the cat. *Archivio Italiano di Biologia*, 1967, *105*, 441–479.

Baarsma, E., and Collewijn, H. Vestibulo-ocular and optokinetic reactions to rotation and their interaction in the rabbit. *Journal of Physiology (London)*, 1974, *238*, 603–625.

Baarsma, E., and Collewijn, H. Eye movements due to linear accelerations in the rabbit. *Journal of Physiology (London)*, 1975, *245*, 227–247.

Baker, R., and Berthoz, A. Organization of vestibular nystagmus in oblique oculomotor system. *Journal of Neurophysiology*, 1974, *37*, 195–217.

Baker, R., and Berthoz, A. Is the prepositus hypoglossi nucleus the source of another vestibulo-ocular pathway? *Brain Research*, 1975, *86*, 121–127.

Baker, R., and Highstein, S. Physiological identification of interneurons and motoneurons in the abducens nucleus. *Brain Research*, 1975, *91*, 292–298.

Baker, R., and Highstein, S. Vestibular projections to medial rectus subdivision of oculomotor nucleus. *Journal of Neurophysiology*, 1978, *41*, 1629–1646.

Baker, R., Mano, N., and Shimazu, H. Postsynaptic potentials in abducens motoneurons produced by vestibular stimulation. *Brain Research*, 1969, *15*, 577–580.

Baker, R., Precht, W., and Llinás, R. Cerebellar modulatory action on the vestibulo-trochlear pathway in the cat. *Experimental Brain Research*, 1972, *15*, 364–385.

Baker, R., Precht, W., and Berthoz, A. Synaptic connections to trochlear motoneurons determined by individual vestibular nerve branch stimulation in the cat. *Brain Research*, 1973, *64*, 402–406.

Baker, R., Berthoz, A., and Delgado-Garcia, J. Monosynaptic excitation of trochlear motoneurons following electrical stimulation of the prepositus hypoglossi nucleus. *Brain Research*, 1977, *121*, 157–161.

Barlow, D., and Freedman, W. The cervico-ocular reflex in the normal adult. *Acta Oto-Laryngologica*, 1980, *89*, 487–496.

Barmack, N. Dynamic visual acuity as an index of eye movement control. *Vision Research*, 1970, *10*, 1377–1391.

Barmack, N. Visually evoked activity of neurons in the dorsal cap of the inferior olive and its relationship to the control of eye movements. In R. Baker and A. Berthoz (Eds.), *Control of Gaze by Brain Stem Neurons*. New York: Elsevier, 1977.

Barr, C., Shultheis, L., and Robinson, D. Voluntary, non-visual control of the human vestibulo-ocular reflex. *Acta Oto-laryngologica*, 1976, *8*, 365–375.

Benson, A., and Barnes, G. Vision during angular oscillation: the dynamic interaction of visual and vestibular mechanisms. *Aviation, Space and Environmental Medicine*, 1978, *49*, 340–345.

Benson, A., Guedry, F., and Melvill Jones, G. Response of semicircular canal dependent units in vestibular nuclei to rotation of a linear acceleration vector without angular acceleration. *Journal of Physiology (London)*, 1970, *210*, 475–494.

Berthoz, A., Baker, R., and Precht, W. Labyrinthine control of inferior oblique motoneurons. *Experimental Brain Research*, 1973, *18*, 225–241.

Bienfang, D. The course of direct projections from the abducens nucleus to the contralateral medial rectus subdivision of the oculomotor nucleus in the cat. *Brain Research*, 1978, *145*, 277–289.

Bizzi, E. The coordination of eye-head movements. *Scientific American*, 1974, *231*, 100–106.

Bizzi, E., Kalil, R., and Tagliasco, V. Eye-head coordination in monkeys: evidence for centrally patterned organization. *Science*, 1971, *173*, 452–454.

Bizzi, E., Kalil, R., Morasso, P., and Tagliasco, V. Central programming and peripheral feedback during eye-head coordination in monkeys. *Bibliotheca Ophthalmologica*, 1972, *82*, 220–232.

Blanks, R., Curthoys, I., and Markham, C. Planar relationships of semicircular canals in the cat. *American Journal of Physiology*, 1972, *223*, 55–62.

Blanks, R., Curthoys, I., and Markham, C. H. Planar relationships of the semicircular canals in man. *Acta Oto-laryngologica*, 1975, *88*, 185–196.

Blanks, R., Estes, M., and Markham, C. Physiologic characteristics of vestibular first-order canal neurons in the cat. II. Response to constant angular acceleration. *Journal of Neurophysiology*, 1975, *38*, 1250–1268.

Blanks, R., Volkind, R., Precht, W., and Baker, R. Responses of cat prepositus hypoglossi neurons to horizontal angular acceleration. *Neuroscience*, 1977, *2*, 391–403.

Blanks, R., Anderson, J., and Precht, W. Response characteristics of semicircular canal and otolith systems in cat. II. Responses of trochlear motoneurons. *Experimental Brain Research*, 1978, *32*, 509–528.

Brodal, A. Anatomy of the vestibular nuclei and their connections. In H. H. Kornhuber (Ed.), *Handbook of Sensory Physiology* (Vol. 6). New York: Springer-Verlag, 1974.

Brodal, A., and Hoivik, B. Site and mode of termination of primary vestibular cerebellar fibers in the cat. An experimental study with silver impregnation methods. *Archivios Italiano di Biologia*, 1964, *102*, 1–21.

Buettner, U. W., Büttner, U., and Henn, V. Transfer characteristics of neurons in vestibular nuclei of the alert monkey. *Journal of Neurophysiology*, 1978, *41*, 1614–1628.

Büttner-Ennever, J. Pathways from the pontine reticular formation to structures controlling horizontal and vertical eye movements in the monkey. In R. Baker and A. Berthoz (Eds.), *Control of Gaze by Brain Stem Neurons* (Vol. 1). Amsterdam: Elsevier, 1977.

Büttner-Ennever, J., and Büttner, U. A cell group associated with vertical eye movements in the rostral mesencephalic reticular formation of the monkey. *Brain Research*, 1978, *151*, 31–47.

Büttner-Ennever, J., and Henn, V. An autoradiographic study of the pathways from the pontine reticular formation involved in horizontal eye movements. *Brain Research*, 1976, *108*, 155–164.

Büttner, U., Büttner-Ennever, J., and Henn, V. Vertical eye movement related unit activity in the rostral mesencephalic reticular formation of the alert monkey. *Brain Research*, 1977, *130*, 239–252.

Carpenter, M., and Nova, H. Descending division of the brachium conjunctivum in the cat: A cerebello-reticular system. *Journal of Comparative Neurology*, 1960, *114*, 295–305.

Carpenter, M., and Strominger, N. Cerebello-oculomotor fibers in the rhesus monkey. *Journal of Comparative Neurology*, 1964, *123*, 211–230.

Carpenter, M., Harbison, J., and Peter, P. Accessory oculomotor nuclei in the monkey: projections and effects of discrete lesions. *Journal of Comparative Neurology*, 1970, *140*, 131–154.

Carpenter, M., Stein, B., and Peter, P. Primary vestibulocerebellar fibers in the monkey: Distribution of fibers arising from distinctive cell groups of the vestibular ganglia. *American Journal of Anatomy*, 1972, *135*, 221–250.

Carpenter, R. Cerebellectomy and the transfer function of the vestibulo-ocular reflex in the decerebrate cat. *Proceedings of the Royal Society (Series B)*, 1972, *181*, 353–374.

Carpenter, R. Appendix 1. Methods of measuring eye movements. In *Movements of the Eyes*. London: Pion Ltd., 1977.

Chan, Y., Hwang, J., and Cheung, Y. Crossed sacculo-ocular pathway via the Deiter's nucleus in cats. *Brain Research Bulletin*, 1977, *2*, 1–6.

Chan-Palay, V. *Cerebellar Dentate Nucleus: Organization, Cytology and Transmitters*. Berlin: Springer-Verlag, 1977.

Clark, B., and Stewart, J. Comparison of sensitivity for the perception of bodily rotation and the oculogyral illusion. *Perceptual Psychophysiology*, 1968, *3*, 253–256.

Cohen, B. The vestibulo-ocular reflex arc. In H. H. Kornhuber (Ed.), *Handbook of Sensory Physiology* (Vol. 6). New York: Springer-Verlag, 1974.

Cohen, B., and Komatsuzaki, A. Eye movements induced by stimulation of the pontine reticular formation: Evidence for integration in oculomotor pathways. *Experimental Neurology*, 1972, *36*, 101–117.

Cohen, B., and Suzuki, J. Eye movements induced by ampullary nerve stimulation. *American Journal of Physiology*, 1963, *204*, 347–351.

Collewijn, H., and Kleinschmidt, H. Vestibulo-ocular and optokinetic reactions in the rabbit: Changes during 24 hours of normal and abnormal interaction. In G. Lennerstrand and P. Bach-y-Rita (Eds.), *Basic Mechanisms of Ocular Motility and their Clinical Implications*. Oxford: Pergamon Press, 1975.

Curthoys, I., and Markham, C. Convergence of labyrinthine influences on units in the vestibular nuclei of the cat. I. Natural stimulation. *Brain Research*, 1971, *35*, 469–490.

Delgado-Garcia, J., Baker, R., and Highstein, S. The activity of internuclear neurons identified within the abducens nucleus of the cat. In R. Baker and A. Berthoz (Eds.), *Control of Gaze by Brain Stem Neurons* (Vol. 1). Amsterdam: Elsevier, 1977.

Diamond, S., Markham, C., Simpson, N., and Curthoys, I. Asymmetries in binocular counterrolling in humans during dynamic rotation. *Society for Neuroscience Abstracts*, 1978, *4*, 162.

Dichgans, J., and Brandt, Th. Visual-vestibular interaction and motion perception. *Bibliotheca Ophthalmologica*, 1972, *82*, 327–338.

Dichgans, J., Bizzi, E., Morasso, P., and Tagliasco, V. Mechanisms underlying recovery of eye-head coordination following bilateral labyrinthectomy in monkeys. *Experimental Brain Research*, 1973, *18*, 548–562.

Dieringer, N., Blanks, R., and Precht, W. Cat efferent vestibular system: Weak suppression of primary afferent activity. *Neuroscience Letters*, 1977, *5*, 285–290.

Dow, R. Efferent connections of the flocculonodular lobe in *Macaca mulatta. Journal of Comparative Neurology*, 1938, *68*, 297–305.

Duensing, F., and Schaefer, K.-P. Die Aktivität einzelner Neurone im Bereich der Vestibuläriskerne bei Horizontalbeschleunigungen unter besonderer Berücksichtigung des vestibulären Nystagmus. *Archiv für Psychiatrie und Nervenkrankheiten*, 1958, *189*, 225–252.

Egmond, A. van, Groen, J., and Jongkees, G. The mechanics of the semicircular canal. *Journal of Physiology (London)*, 1949, *110*, 1–17.

Estes, M., Blanks, R., and Markham, C. Physiologic characteristics of vestibular first-order canal neurons in the cat. I. Response plane determination and resting discharge characteristics. *Journal of Neurophysiology*, 1975, *38*, 1232–1249.

Evinger, L. C., Fuchs, A., and Baker, R. Bilateral lesions of the medial longitudinal fasciculus in monkeys: Effects on the horizontal and vertical components of voluntary and vestibular induced eye movements. *Experimental Brain Research*, 1977, *28*, 1–20.

Ezure, K., and Sasaki, S. Frequency-response analysis of vestibular-induced neck reflex in cat. I. Characteristics of neural transmission from horizontal semicircular canal to neck motoneurons. *Journal of Neurophysiology*, 1978, *41*, 445–458.

Fernandez, C., and Goldberg, J. Physiology of peripheral neurons innervating semicircular canals of the squirrel monkey. II. Response to sinusoidal stimulation and dynamics of peripheral vestibular system. *Journal of Neurophysiology*, 1971, *34*, 661–675.

Fernandez, C., and Goldberg, J. Physiology of peripheral neurons innervating otolith organs of the squirrel monkey. I. Response to static tilts and to long-duration centrifugal force. *Journal of Neurophysiology*, 1976a, *39*, 970–984.

Fernandez, C., and Goldberg, J. Physiology of peripheral neurons innervating otolith organs of the squirrel monkey. II. Directional selectivity and force-response relations. *Journal of Neurophysiology*, 1976b, *39*, 985–995.

Fernandez, C., and Goldberg, J. Physiology of peripheral neurons innervating otolith organs of the squirrel monkey. III. Response dynamics. *Journal of Neurophysiology*, 1976c, *39*, 996–1108.

Fernandez, C., Goldberg, J., and Abend, W. Response to static tilts of peripheral neurons innervating otolith organs of the squirrel monkey. *Journal of Neurophysiology*, 1972, *35*, 978–997.

Fluur, E., and Mellström, A. Utricular stimulation and oculomotor reactions. *Laryngoscope*, 1970a, *80*, 1701–1712.

Fluur, E., and Mellström, A. Saccular stimulation and oculomotor reactions. *Laryngoscope*, 1970b, *80*, 1713–1721.

Frederickson, J., Schwarz, D., and Kornhuber, H. Convergence and interaction of vestibular and deep somatic afferents upon neurons in the vestibular nuclei of the cat. *Acta Oto-laryngologica*, 1966, *61*, 168–188.

Fuchs, A. Saccadic and smooth pursuit eye movements in the monkey. *Journal of Physiology (London)*, 1967, *191*, 609–631.

Fuchs, A. The neurophysiology of saccades. In R. Monty and J. Senders (Eds.), *Eye Movements and Psychological Processes.* Hillsdale, N.J.: Lawrence Erlbaum, 1976.

Fuchs, A., and Kimm, J. Unit activity in vestibular nucleus of the alert monkey during horizontal angular acceleration and eye movement. *Journal of Neurophysiology*, 1975, *38*, 1140–1161.

Fuchs, A., and Luschei, E. Firing patterns of abducens neurons of alert monkeys in relationship to horizontal eye movement. *Journal of Neurophysiology*, 1970, *33*, 382–392.

Fuchs, A., and Robinson, D. A method for measuring horizontal and vertical eye movements chronically in the monkey. *Journal of Applied Physiology*, 1966, *21*, 1068–1070.

Fujita, Y., Rosenberg, J., and Segundo, J. Activity of cells in the lateral vestibular nucleus as a function of head position. *Journal of Physiology (London)*, 1968, *196*, 1–18.

Fukuda, J., Highstein, S., and Ito, M. Cerebellar inhibitory control of the vestibulo-ocular reflex investigated in rabbit IIIrd nucleus. *Experimental Brain Research*, 1972, *14*, 511–526.

Fukushima, Y., Igusa, Y., and Yoshida, K. Characteristics of responses of medial brain stem neurons to horizontal head angular acceleration and electrical stimulation of the labyrinth in the cat. *Brain Research*, 1977, *120*, 564–570.

Furuya, N., Kawano, K., and Shimazu, H. Functional organization of vestibulofastigial projection in the horizontal semicircular canal system of the cat. *Experimental Brain Research*, 1975, *24*, 75–87.

Gacek, R. The innervation of the vestibular labyrinth. *Annals of Otology, Rhinology and Laryngology*, 1968, *77*, 675–685.

Gacek, R. The course and central termination of first order neurons supplying vestibular end organs in the cat. *Acta Oto-laryngologica*, 1969, Suppl. 254, 1–66.

Gacek, R. Anatomical demonstration of the vestibulo-ocular projections in the cat. *Laryngoscope*, 1971, *81*, 1559–1595.

Gacek, R. Morphological aspects of the efferent vestibular system. In H. H. Kornhuber (Ed.), *Handbook of Sensory Physiology* (Vol. 6). New York: Springer-Verlag, 1974.

Gacek, R. Location of brain stem neurons projecting to the oculomotor nucleus in the cat. *Experimental Neurology*, 1977, *57*, 725–749.

Gacek, R. Location of commissural neurons in the vestibular nuclei of the cat. *Experimental Neurology*, 1978, *59*, 479–491.

Gacek, R., and Lyon, M. The localization of vestibular efferent neurons in the kitten with horseradish peroxidase. *Acta Oto-laryngologica*, 1974, *77*, 92–101.

Gacek, R., and Rasmussen, G. Fiber analysis of the stato-acoustic nerve of guinea pig, cat, and monkey. *Anatomical Record*, 1961, *139*, 455–463.

Gauthier, G., and Robinson, D. Adaptation of the human vestibuloocular reflex to magnifying lenses. *Brain Research*, 1975, *92*, 331–335.

Gernandt, B. Response of mammalian vestibular neurons to horizontal rotation and caloric stimulation. *Journal of Neurophysiology*, 1949, *12*, 173–184.

Gernandt, B. Nystagmus evoked by utricular stimulation. *Experimental Neurology*, 1970, *27*, 90–100.

Goldberg, J., and Fernandez, C. Responses of first-order vestibular afferents of the squirrel monkey to angular accelerations. In C. Terzuolo (Ed.), *Systems Analysis in Neurophysiology*. Minneapolis: University of Minnesota Press, 1969.

Goldberg, J., and Fernandez, C. Physiology of peripheral neurons innervating semicircular canals of the squirrel monkey. I. Resting discharge and response to constant angular accelerations. *Journal of Neurophysiology*, 1971, *34*, 635–660.

Goldberg, J., and Fernandez, C. Responses of peripheral vestibular neurons to angular and linear acceleration in the squirrel monkey. *Acta Oto-laryngologica*, 1975a, *80*, 101–110.

Goldberg, J., and Fernandez, C. Vestibular mechanisms. *Annual Review of Physiology*, 1975b, *37*, 129–162.

Goldberg, J., and Fernandez, C. Efferent vestibular system in the squirrel monkey. *Society for Neuroscience Abstracts*, 1977, *3*, 543.

Gonshor, A., and Melvill Jones, G. Plasticity in the adult human vestibulo-ocular reflex arc. *Proceedings of Canadian Federation of Biological Sciences*, 1971, *14*, 11.

Gonshor, A., and Melvill Jones, G. Extreme vestibulo-ocular adaptation induced by prolonged optical reversal of vision. *Journal of Physiology (London)*, 1976, *256*, 381–414.

Grant, K., Gueritand, J., Horcholle-Bossavit, G., and Tyc-Dumont, S. Horizontal vestibular nystagmus. II. Activity patterns of medial vestibular neurons during nystagmus. *Experimental Brain Research*, 1976, *26*, 287–405.

Graybiel, A. Direct and indirect preoculomotor pathways of the brainstem: An autoradiographic study of the pontine reticular formation in the cat. *Journal of Comparative Neurology*, 1977a, *175*, 37–78.

Graybiel, A. Organization of oculomotor pathways in the cat and rhesus monkey. In R. Baker and A. Berthoz (Eds.), *Control of Gaze by Brain Stem Neurons* (Vol. 1.). New York: Elsevier, 1977b.

Graybiel, A., and Hartwieg, E. Some afferent connections of the oculomotor complex in the cat: An experimental study with tracer techniques. *Brain Research*, 1974, *81*, 543–551.

Graybiel, A., and Patterson, J. Thresholds of stimulation of the otolith organs as indicated by the oculogravic illusion. *Journal of Applied Physiology*, 1955, *7*, 666–670.

Gresty, M., and Baker, R. Neurons with visual receptive field, eye movement and neck displacement sensitivity within and around the nucleus prepositus hypoglossi in the alert cat. *Experimental Brain Research*, 1976, *24*, 429–433.

Groen, J. Cupulometry. *Laryngoscope*, 1957, *67*, 894–905.

Haines, D. Cerebellar corticovestibular fibers of the posterior lobe in a prosimian primate, the lesser bushbaby *(Galago senegalensis)*. *Journal of Comparative Neurology*, 1976, *160*, 363–398.

Haines, D. Cerebellar corticonuclear and corticovestibular fibers of the flocculonodular lobe in a prosimian primate *(Galago senegalensis)*. *Journal of Comparative Neurology*, 1977, *174*, 607–630.

Hardy, M. Observations on the innervation of the macula sacculi in man. *Anatomical Record*, 1934, *59*, 403–418.

Henn, V., Young, L., and Finley, C. Vestibular nucleus units in alert monkeys are also influenced by moving visual fields. *Brain Research*, 1974, *71*, 144–149.

Highstein, S. Synaptic linkage in the vestibulo-ocular and cerebellovestibular pathways to the VI nucleus in the rabbit. *Experimental Brain Research*, 1973, *17*, 301–344.

Highstein, S., and Baker, R. Excitatory termination of abducens internuclear neurons on medial rectus motoneurons: Relationship to syndrome of internuclear ophthalmoplegia. *Journal of Neurophysiology*, 1978, *41*, 1647–1661.

Highstein, S., Cohen, B., and Matsunami, K. Monosynaptic projections from the pontine reticular formation to the IIIrd nucleus in the cat. *Brain Research*, 1974, *75*, 340–344.

Hikosaka, O., and Maeda, M. Cervical effects on abducens motoneurons and their interaction with vestibulo-ocular reflex. *Experimental Brain Research*, 1973, *18*, 512–530.

Hikosaka, O., Maeda, M., Nakao, S., Shimazu, H., and Shinoda, Y. Presynaptic impulses in the abducens nucleus and their relation to postsynaptic potentials in motoneurons during vestibular nystagmus. *Experimental Brain Research*, 1977, *27*, 355–376.

Hillman, R. Observations on morphological features and mechanical properties of the peripheral vestibular receptor system in the frog. *Progress in Brain Research*, 1972, *37*, 69–75.

Hirai, N., Uchino, Y., and Watanabe, S. Neuronal organization of the fastigiotrochlear pathway in the cat. *Brain Research*, 1977, *131*, 362–366.

Hoddevik, G., Brodal, A., and Walberg, E. The reticulovestibular projection in the cat. An experimental study with silver impregnation methods. *Brain Research*, 1975, *94*, 383–399.

Horcholle-Bossavit, G., and Tyc-Dumont, S. Evidence for a rapid transmission in the cat vestibulo-ocular pathway. *Experimental Brain Research*, 1971, *13*, 327–338.

Hoyt, W., and Daroff, R. Supranuclear disorders of ocular control systems in man: Clinical, anatomical and physiological correlations. In P. Bach-y-Rita, C. Collins, and J. Hyde (Eds.), *The Control of Eye Movements*. New York: Academic Press, 1971.

Hudspeth, A., and Corey, D. Sensitivity, polarity and conductance change in the response of vertebrate hair cells to controlled mechanical stimuli. *Proceedings of the National Academy of Science*, 1977, *74*, 2407–2411.

Hwang, J., and Poon, W. An electrophysiological study of the sacculo-ocular pathways in cats. *Japanese Journal of Physiology*, 1975, *25*, 241–251.

Ito, M. Neural design of the cerebellar motor control system. *Brain Research*, 1972, *40*, 81-84.

Ito, M., Nisimaru, N., and Yamamoto, M. Specific neural connections for the cerebellar control of vestibulo-ocular reflexes. *Brain Research*, 1973, *60*, 238-243.

Ito, M., Shiida, T., Yagi, N., and Yamamoto, M. The cerebellar modification of rabbit's horizontal vestibulo-ocular reflex induced by sustained head rotation combined with visual stimulation. *Proceedings of the Japan Academy*, 1974a, *50*, 85-89.

Ito, M., Shiida, T., Yagi, N., and Yamamoto, M. Visual influence on rabbit horizontal vestibulo-ocular reflex presumably effected via the cerebellar flocculus. *Brain Research*, 1974b, *65*, 170-174.

Kaneko, C., Steinacker, A., Cohen, B., Maciewicz, R., and Highstein, S. Synaptic linkage of the reticulo-ocular pathway in cat. *Society for Neuroscience Abstracts*, 1975, *1*, 225.

Kanzaki, J., and Ouchi, T. Measurement of ocular countertorsion reflex with fundoscopic camera in normal subjects and in patients with inner ear lesions. *Archives of Oto-Rhino-Laryngology*, 1978, *218*, 191-201.

Kasahara, M., and Uchino, Y. Bilateral semicircular canal inputs to neurons in cat vestibular nuclei. *Experimental Brain Research*, 1974, *20*, 285-296.

Kasahara, M., Mano, N., Oshima, T., Ozawa, S., and Shimazu, H. Contralateral short latency inhibition of central vestibular neurons in the horizontal canal system. *Brain Research*, 1968, *8*, 376-378.

Keller, E. Accommodative vergence in the alert monkey. Motor unit analysis. *Vision Research*, 1973, *13*, 1565-1575.

Keller, E. Behavior of horizontal semicircular canal afferents in alert monkey during vestibular and optokinetic stimulation. *Experimental Brain Research*, 1976, *24*, 459-471.

Keller, E. Gain of the vestibulo-ocular reflex in monkey at high rotational frequencies. *Vision Research*, 1978, *18*, 311-315.

Keller, E., and Daniels, P. Oculomotor related interaction of vestibular and visual stimulation in vestibular nucleus cells in alert monkey. *Experimental Neurology*, 1975, *46*, 187-198.

Keller, E., and Kamath, B. Characteristics of head rotation and eye movement-related neurons in alert monkey vestibular nucleus. *Brain Research*, 1975, *100*, 182-187.

Keller, E., and Precht, W. Persistence of visual response in vestibular nucleus neurons in cerebellectomized cat. *Experimental Brain Research*, 1978, *32*, 591-594.

Keller, E., and Robinson, D. Abducens unit behavior in the monkey during vergence movements. *Vision Research*, 1972, *12*, 369-382.

Kimm, J., and Fuchs, A. Response characteristics of reticular formation neurons to vestibular stimulation and/or eye movements. *Society for Neuroscience Abstracts*, 1975, *1*, 226.

King, W. M. *Quantitative analysis of the activity of neurons in the accessory oculomotor nuclei and the mesencephalic reticular formation of alert monkeys in relation to vertical eye movements induced by visual and vestibular stimulation*. Doctoral Dissertation, University of Washington, Seattle, 1976.

King, W. M., Lisberger, S., and Fuchs, A. Responses of fibers in medial longitudinal fasciculus (MLF) of alert monkeys during horizontal and vertical conjugate eye movements evoked by vestibular or visual stimuli. *Journal of Neurophysiology*, 1976, *39*, 1135-1149.

Klinke, R., and Galley, N. Efferent innervation of vestibular and auditory receptors. *Physiological Reviews*, 1974, *51*, 316-357.

Klinke, R., and Schmidt, C. Efferent influence on the vestibular organ during active movements of the body. *Pflügers Archiv*, 1970, *318*, 325-332.

Kotchabhakdi, N., and Walberg, F. Cerebellar afferent projections from the vestibular nuclei in the cat: An experimental study with the method of retrograde axonal transport of horseradish peroxidase. *Experimental Brain Research*, 1978, *31*, 591-604.

Krejcova, H., Highstein, S., and Cohen, B. Labyrinthine and extra-labyrinthine effects on ocular counter-rolling. *Acta Oto-laryngologica*, 1971, *72*, 165-171.

Ladpli, R., and Brodal, A. Experimental studies of commissural and reticular formation projections from the vestibular nuclei in the cat. *Brain Research*, 1968, *8*, 65-96.

Landers, P., and Taylor, A. Transfer function analysis of the vestibulo-ocular reflex in the conscious cat. In G. Lennerstrand and P. Bach-y-Rita (Eds.), *Basic Mechanisms of Ocular Motility and their Clinical Implications.* New York: Pergamon Press, 1975.

Landolt, E. Nouvelles recherches sur la physiologie des mouvements des yeux. *Archives d'Ophthalmologie,* 1891, *11,* 385–395.

Lindeman, H. H. Studies on the morphology of the sensory regions of the vestibular apparatus. *Ergebnisse der Anatomie und Entwicklungsgeschichte,* 1969, *42,* 1–113.

Lisberger, S., and Fuchs, A. Response of flocculus Purkinje cells to adequate vestibular stimulation in the alert monkey: Fixation vs. compensatory eye movements. *Brain Research,* 1974, *69,* 347–353.

Lisberger, S., and Fuchs, A. Role of primate flocculus during rapid behavioral modification of vestibuloocular reflex. I. Purkinje cell activity during visually guided horizontal smooth-pursuit eye movements and passive head rotation. *Journal of Neurophysiology,* 1978a, *41,* 733–763.

Lisberger, S., and Fuchs, A. Role of primate flocculus during rapid behavioral modification of vestibuloocular reflex. II. Mossy fiber firing patterns during horizontal head rotation and eye movement. *Journal of Neurophysiology,* 1978b, *41,* 764–777.

Llinás, R., Walton, K., Hillman, D., and Sotelo, C. Inferior olive: Its role in motor learning. *Science,* 1975, *190,* 1230–1231.

Loe, P., Tomko, D., and Werner, G. The neural signal of angular head position in primary afferent vestibular nerve axons. *Journal of Physiology (London),* 1972, *230,* 29–50.

Louie, A., and Kimm, J. The response of 8th nerve fibers to horizontal sinusoidal oscillation in the alert monkey. *Experimental Brain Research,* 1976, *24,* 447–457.

Lowenstein, O. E. Comparative morphology and physiology. In H. Kornhuber (Ed.), *Handbook of Sensory Physiology* (Vol. 6). New York: Springer-Verlag, 1974.

Maciewicz, R., and Spencer, R. Oculomotor and abducens internuclear pathways in the cat. In R. Baker and A. Berthoz (Eds.), *Control of Gaze by Brain Stem Neurons* (Vol. 1). New York: Elsevier, 1977.

Maciewicz, R., Eagen, K., Kaneko, C., and Highstein, S. Vestibular and medullary brain stem afferents to the abducens nucleus in the cat. *Brain Research,* 1977, *123,* 229–240.

Maeda, M., Shimazu, H., and Shinoda, Y. Rhythmic activities of secondary vestibular efferent fibers recorded within the abducens nucleus during nystagmus. *Brain Research,* 1971, *34,* 361–365.

Maekawa, K., and Simpson, J. Climbing fiber responses evoked in vestibulocerebellum of rabbit from visual system. *Journal of Neurophysiology,* 1973, *36,* 649–666.

Maekawa, K., and Takeda, T. Mossy fiber responses evoked in the cerebellar flocculus of rabbits by stimulation of the optic pathway. *Brain Research,* 1975, *98,* 590–595.

Mano, N., Oshima, T., and Shimazu, H. Inhibitory commissural fibers interconnecting the bilateral vestibular nuclei. *Brain Research,* 1968, *8,* 378–382.

Markham, C. Midbrain and contralateral labyrinth influences on brain stem vestibular neurons in the cat. *Brain Research,* 1968, *9,* 312–333.

Markham, C., Precht, W., and Shimazu, H. Effect of stimulation of interstitial nucleus of Cajal on vestibular unit activity in the cat. *Journal of Neurophysiology,* 1966, *29,* 493–507.

Markham, C., Yagi, T., and Curthoys, I. The contribution of the contralateral labyrinth to second order vestibular neuronal activity in the cat. *Brain Research,* 1977, *138,* 99–109.

Marr, D. A theory of cerebellar cortex. *Journal of Physiology (London),* 1969, *202,* 437–470.

McMasters, R., Weiss, A., and Carpenter, M. Vestibular projections to the nuclei of the extraocular muscles. Degeneration resulting from discrete partial lesions of the vestibular nuclei in the monkey. *American Journal of Anatomy,* 1966, *118,* 163–194.

Meiry, J. Vestibular and proprioceptive stabilization of eye movement. In P. Bach-y-Rita, C. Collins, and J. Hyde (Eds.), *The Control of Eye Movements.* New York: Academic Press, 1971.

Melvill Jones, G. Plasticity in the adult vestibulo-ocular reflex arc. *Philosophical Transactions of the Royal Society (London), Series B,* 1977, *278,* 319–334.

Melvill Jones, G., and Davies, P. Adaptation of cat vestibulo-ocular reflex to 200 days of optically reversed vision. *Brain Research,* 1976, *103,* 551–554.

Melvill Jones, G., and Milsum, J. Spatial and dynamic aspects of visual fixation. *IEEE Transactions of Bio-medical Engineering*, 1965, *12*, 54–62.

Melvill Jones, G., and Milsum, J. Characteristics of neural transmission from the semicircular canal to the vestibular nuclei of cats. *Journal of Physiology (London)*, 1970, *209*, 295–316.

Melvill Jones, G., and Milsum, J. Frequency-response analysis of central vestibular unit activity resulting from rotational stimulation of the semicircular canals. *Journal of Physiology (London)*, 1971, *219*, 191–215.

Melvill Jones, G., and Milsum, J. Neural response of the vestibular system to translational acceleration. *DRB Aviation Medical Research Unit Reports, 1968–1971*, 1972, *2*, 183–190.

Mergner, T., and Pompeiano, O. Single unit firing patterns in the vestibular nuclei related to saccadic eye movements in the decerebrate cat. *Archivio Italiano di Biologia*, 1978, *116*, 91–119.

Miles, F. Single unit firing patterns in the vestibular nuclei related to voluntary eye movements and passive body rotation in conscious monkeys. *Brain Research*, 1974, *71*, 215–224.

Miles, F., and Braitman, D. Effects of prolonged optic reversal of vision on the vestibuloocular reflex: Some neurophysiological observations. *Society for Neuroscience Abstracts*, 1978, *4*, 167.

Miles, F., and Fuller, J. Adaptive plasticity in the vestibulo-ocular responses of the rhesus monkey. *Brain Research*, 1974, *80*, 512–516.

Miles, F., and Fuller, J. Visual tracking and the primate flocculus. *Science*, 1975, *189*, 1000–1002.

Morasso, P., Bizzi, E., and Dichgans, J. Adjustment of saccade characteristics during head movements. *Experimental Brain Research*, 1973, *16*, 492–500.

Naito, H., Tanimura, K., Taga, N., and Hosoya, Y. Microelectrode study on the subnuclei of the oculomotor nucleus in the cat. *Brain Research*, 1974, *81*, 215–231.

Nakao, S., and Sasaki, S. Firing patterns of interneurons in the abducens nucleus related to vestibular nystagmus in the cat. *Brain Research*, 1978, *144*, 389–394.

Niven, J., Hixson, W., and Correia, M. Elicitation of horizontal nystagmus by periodic linear acceleration. *Acta Oto-laryngologica*, 1966, *62*, 429–441.

Noda, H., and Suzuki, D. Purkinje cell activity in the monkey flocculus during smooth pursuit eye movements. *Society for Neuroscience Abstracts*, 1978, *4*, 167.

Peterson, B. Effect of tilting on the activity of neurons in the vestibular nuclei of the cat. *Brain Research*, 1967, *6*, 606–609.

Peterson, B. Distribution of neural responses to tilting within vestibular nuclei of the cat. *Journal of Neurophysiology*, 1970, *33*, 750–767.

Peterson, B., and Abzug, C. Properties of projections from vestibular nuclei to medial reticular formation in the cat. *Journal of Neurophysiology*, 1975, *38*, 1421–1435.

Pola, J., and Robinson, D. Oculomotor signals in medial longitudinal fasciculus of the monkey. *Journal of Neurophysiology*, 1978, *41*, 245–259.

Pompeiano, D., Mergner, T., and Corvaja, N. Commissural, perihypoglossal and reticular afferent projections to the vestibular nuclei in the cat. An experimental anatomical study with the method of the retrograde transport of horseradish peroxidase. *Archivio Italiano di Biologia*, 1978, *116*, 130–172.

Precht, W. Vestibular system. In A. Guyton and C. Hunt Butterworths (Eds.), *MTP International Review of Sciences, Neurophysiology* (Vol. 3). Baltimore: University Park Press, 1975.

Precht, W. The functional synaptology of brainstem oculomotor pathways. In R. Baker and A. Berthoz (Eds.), *Control of Gaze by Brain Stem Neurons* (Vol. 1). New York: Elsevier, 1977.

Precht, W., and Llinás, R. Functional organization of the vestibular afferents to the cerebellar cortex of frog and cat. *Experimental Brain Research*, 1969, *9*, 30–52.

Precht, W., and Shimazu, H. Functional connections of tonic and kinetic vestibular neurons with primary vestibular afferents. *Journal of Neurophysiology*, 1965, *28*, 1014–1028.

Precht, W., Baker, R., and Okada, Y. Evidence for GABA as the synaptic transmitter of the inhibitory vestibulo-ocular pathway. *Experimental Brain Research*, 1973, *18*, 415–428.

Precht, W., Volkind, R., Maeda, M., and Giretti, M. The effects of stimulating the cerebellar nodulus in the cat on the responses of vestibular neurons. *Neuroscience*, 1976, *1*, 301–312.

Rashbass, C. The relationship between saccadic and smooth tracking eye movements. *Journal of Physiology (London)*, 1961, *159*, 326–338.

Rashbass, C., and Russell, G. Action of a barbiturate drug (amylobarbitone sodium) on the vestibulo-ocular reflex. *Brain*, 1961, *84*, 329–335.

Remmel, R., Pola, J., and Skinner, R. Pontomedullary reticular projections into the region of the ascending medial longitudinal fasciculus in cat. *Experimental Brain Research*, 1978, *32*, 31–37.

Reutern, G. von, and Dichgans, J. Augenbewegungsstörungen als cerebelläre Symptome bei Kleinhirnbrückenwinkeltumoren. *Archiv für Psychiatrie und Nervenkrankheiten*, 1977, *223*, 117–130.

Robinson, D. A method of measuring eye movements using a search coil in a magnetic field. *IEEE Transactions, Bio-Med Electronics*, 1963, *10*, 137–145.

Robinson, D. The mechanics of human saccadic eye movement. *Journal of Physiology (London)*, 1964, *174*, 245–264.

Robinson, D. Oculomotor unit behavior in the monkey. *Journal of Neurophysiolgy*, 1970, *33*, 393–404.

Robinson, D. Models of oculomotor neural organization. In P. Bach-y-Rita and C. Collins (Eds.), *The Control of Eye Movement*. New York: Academic Press, 1971.

Robinson, D. The effect of cerebellectomy on the cat's vestibulo-ocular integrator. *Brain Research*, 1974, *71*, 195–207.

Robinson, D. A quantitative analysis of extraocular muscle cooperation and squint. *Investigative Ophthalmology*, 1975, *14*, 801–825.

Robinson, D. Adaptive gain control of vestibuloocular reflex by the cerebellum. *Journal of Neurophysiology*, 1976, *39*, 954–969.

Roffler-Tarlov, S., and Tarlov, E. Studies of suspected neurotransmitters in the vestibuloocular pathways. *Brain Research*, 1975, *95*, 383–394.

Rubin, A., Liedgren, S., Odkvist, L., Milne, A., and Fredrickson, J. Labyrinthine and somatosensory convergence upon vestibulo-ocular units. *Acta Oto-laryngologica*, 1978, *85*, 54–62.

Scheibel, M., and Scheibel, A. Structural substrates for integrative patterns in the brain stem reticular core. In H. H. Jasper, L. D. Proctor, R. S. Knighton, W. C. Noshay, and R. T. Costello (Eds.), *Reticular Formation of the Brain*. Boston: Little, Brown, 1958.

Schiller, P. The discharge characteristics of single units in the oculomotor and abducens nuclei of the unanesthetised monkey. *Experimental Brain Research*, 1970, *10*, 347–362.

Schneider, L., and Anderson, D. Transfer characteristics of first and second order lateral canal vestibular neurons in gerbil. *Brain Research*, 1976, *112*, 61–76.

Schor, R. Responses of cat vestibular neurons to sinusoidal roll tilt. *Experimental Brain Research*, 1974, *20*, 347–362.

Schwindt, P., Richter, A., and Precht, W. Short latency utricular and canal input to ipsilateral abducens motoneurons. *Brain Research*, 1973, *60*, 259–262.

Shimazu, H., and Precht, W. Tonic and kinetic responses of cat's vestibular neurons to horizontal angular acceleration. *Journal of Neurophysiology*, 1965, *28*, 991–1013.

Shimazu, H., and Precht, W. Inhibition of central vestibular neurons from the contralateral labyrinth and its mediating pathway. *Journal of Neurophysiology*, 1966, *29*, 467–492.

Shimazu, H., and Smith, C. Cerebellar and labyrinthine influences on single vestibular neurons identified by natural stimuli. *Journal of Neurophysiology*, 1971, *34*, 493–508.

Shinoda, Y., and Yoshida, K. Dynamic characteristics of responses to horizontal head acceleration in vestibuloocular pathway in the cat. *Journal of Neurophysiology*, 1974, *37*, 653–673.

Shinoda, Y., and Yoshida, K. Neural pathways from the vestibular labyrinths to the flocculus in the cat. *Experimental Brain Research*, 1975, *22*, 97–111.

Simpson, J., and Alley, K. Visual climbing fiber input to rabbit vestibulocerebellum: A source of direction-specific information. *Brain Research*, 1974, *82*, 302–308.

Simpson, J., and Hess, R. Complex and simple visual messages in the flocculus. In R. Baker and A. Berthoz (Eds.), *Control of Gaze by Brain Stem Neurons* (Vol. 1). New York: Elsevier, 1977.

Skavenski, A., and Robinson, D. Role of abducens neurons in vestibuloocular reflex. *Journal of Neurophysiology*, 1973, *36*, 724–738.

Spiegel, E. Role of the vestibular nuclei in the cortical innervation of the eye muscles. *Archives of Neurology and Psychiatry*, 1933, *29*, 1084–1097.

Spoendlin, H. Ultrastructure of the vestibular sense organ. In R. Wolfson (Ed.), *The Vestibular System and Its Diseases*. Philadelphia: University of Pennsylvania Press, 1966.

Steiger, H.-J., and Büttner-Ennever, J. Relationship between motoneurons and internuclear neurons in the abducens nucleus: A double retrograde tracer study in the cat. *Brain Research* 1978, *148*, 181–188.

Steiger, H.-J., and Büttner-Ennever, J. Oculomotor nucleus afferents in the monkey demonstrated with horseradish peroxidase. *Brain Research*, 1979, *160*, 1–15.

Stein, B., and Carpenter, M. Central projection of portions of the vestibular ganglion innervating specific parts of the labyrinth in the rhesus monkey. *American Journal of Anatomy*, 1967, *120*, 281–318.

Steinhausen, W. Über die Beobachtung der Cupula in den Bogengangsampullen des Labyrinths des lebenden Hechts. *Pflügers Archiv*, 1933, *232*, 500–512.

Suzuki, J. Vestibulo-oculomotor relations: Static responses. *Progress in Brain Research*, 1972, *37*, 507–514.

Suzuki, J., Cohen, B., and Bender, M. Compensatory eye movements induced by vertical semicircular canal stimulation. *Experimental Neurology*, 1964, *9*, 137–160.

Takemori, S., and Cohen, B. Visual suppression of vestibular nystagmus in rhesus monkeys. *Brain Research*, 1974a, *72*, 203–212.

Takemori, S., and Cohen, B. Loss of visual suppression of vestibular nystagmus after flocculus lesions. *Brain Research*, 1974b, *72*, 213–224.

Takemori, S., and Suzuki, J. Eye deviations from neck torsion in humans. *Annals of Otology, Rhinology and Laryngology*, 1971, *80*, 441–444.

Tarlov, E. Organization of vestibulo-oculomotor projections in the cat. *Brain Research*, 1970, *20*, 159–179.

Uchino, Y., Hirai, N., and Watanabe, S. Vestibulo-ocular reflex from the posterior canal nerve to extraocular motoneurons in the cat. *Experimental Brain Research*, 1978, *32*, 377–388.

Uemura, T., and Cohen, B. Effects of vestibular nuclei lesions on vestibulo-ocular reflexes and posture in monkeys. *Acta Oto-laryngologica*, 1973, Suppl. 315, 1–71.

Waespe, W., and Henn, V. Neuronal activity in the vestibular nuclei of the alert monkey during vestibular and optokinetic stimulation. *Experimental Brain Research*, 1977, *27*, 523–538.

Walberg, E. Descending and reticular relations to the vestibular nuclei: Anatomy. *Progress in Brain Research*, 1972, *37*, 585–588.

Walberg, F., Pompeiano, O., Brodal, A., and Jansen, J. The fastigiovestibular projection in the cat. An experimental study with silver impregnation methods. *Journal of Comparative Neurology*, 1962, *118*, 49–76.

Warr, W. Olivocochlear and vestibular efferent neurons of the feline brain stem: Their location, morphology and number determined by retrograde axonal transport and acetylcholinesterase histochemistry. *Journal of Comparative Neurology*, 1975, *161*, 159–181.

Warwick, R. Oculomotor organization. In M. Bender (Ed.), *The Oculomotor System*. New York: Harper and Row, 1964.

Wersäll, J. Studies on the structures and innervation of the sensory epithelium of the cristae ampullares in the guinea pig. A light and electron microscopic investigation. *Acta Oto-laryngologica*, 1956, Suppl. 126, 1–85.

Wersäll, J., and Bagger-Sjöbäck, D. Morphology of the vestibular sense organ. In H. Kornhuber (Ed.), *Handbook of Sensory Physiology* (Vol. 6). New York: Springer-Verlag, 1974.

Westheimer, G. Eye movement responses to a horizontally moving visual stimulus. *Archives of Ophthalmology*, 1954, *52*, 932–941.

Westheimer, G., and Blair, S. Functional organization of primate oculomotor system revealed by cerebellectomy. *Experimental Brain Research*, 1974, *21*, 463–472.

Wilson, V. Physiological pathways through the vestibular nuclei. *International Review of Neurobiology,* 1972, *15,* 27–81.

Wilson, V., Gacek, R., Uchino, Y., and Susswein, A. Properties of central vestibular neurons fired by stimulation of the saccular nerve. *Brain Research,* 1978, *143,* 251–261.

Wilson, V., Maeda, M., Franck, J., and Shimazu, H. Mossy fiber neck and second-order labyrinthine projections to cat flocculus. *Journal of Neurophysiology,* 1976, *39,* 303–310.

Wilson, V., and Peterson, B. Peripheral and central substrates of vestibulospinal reflexes. *Physiological Reviews,* 1978, *58,* 80–105.

Wilson, V., Wylie, R., and Marco, L. Organization of the medial vestibular nucleus. *Journal of Neurophysiology,* 1968a, *31,* 166–175.

Wilson, V., Wylie, R., and Marco, L. Synaptic inputs to cells in the medial vestibular nucleus. *Journal of Neurophysiology,* 1968b, *31,* 176–185.

7

The Basal Ganglia and Movement

MARJORIE E. ANDERSON

INTRODUCTION

Early clinical reports, such as James Parkinson's "An Essay on the Shaking Palsy" (Parkinson, 1817), gave vivid descriptions of the complex motor symptoms that now are associated with pathological changes in the basal ganglia, and it is primarily from the subsequent clinical literature that information has originated regarding possible roles of the basal ganglia in the coordination of movement. Degeneration of neurons in various nuclei of the basal ganglia has been found in the brains of individuals who had marked slowness or apparent absence of voluntary movement under certain conditions (bradykinesia or akinesia), but these same individuals also may have had an involuntary phasic activity of motor units that produced an incessant resting tremor. Other individuals with pathological changes in the basal ganglia exhibit the more dramatic involuntary movements of athetosis, chorea, or ballismus. As experimental studies were added to clinical-pathological correlations, a common theme appeared, and the basal ganglia were referred to as "centers for the automatic or subvoluntary integration of the various motor centers" (Ferrier, 1876), "supravestibular systems" (Muskens, 1922), "a group . . . concerned with posture other than the support of the body against gravity" (Martin, 1967). Denny-Brown (1962) even referred to the globus pallidus, from which many of the basal ganglia output fibers originate, as "the 'head ganglion' of the motor system of primates, forming the essential link between the environment and the reflex organization."

Explicit or implied in all the above descriptions is the idea that the basal

MARJORIE E. ANDERSON   Departments of Rehabilitation Medicine and Physiology and Biophysics, University of Washington School of Medicine, Seattle, Washington, 98195. Supported by Rehabilitation Services Grant 16-P-56818 and Public Health Service Grants NS 10804 and 15017.

ganglia play some role in coordinating the various reflex mechanisms that support and participate in "voluntary" movement. Martin (1967), after an extensive examination of clinical cases, has concluded that all the symptomatology associated with basal ganglia lesions can be viewed as either depressed or released (exaggerated) postural reflexes concerned with stabilizing body parts relative to each other, providing adjustments of the center of gravity over an appropriate base of support (equilibrium), or bringing the body to an upright posture (righting reactions).

To assign to the basal ganglia a responsibility for postural adjustment probably is an oversimplification. We will, however, examine the clinical and experimental data regarding the role of the basal ganglia in motor coordination, with special attention to the fact that coordinated active movement is achieved both by causing rotational and translational movement of joints (primary movement) and by stabilizing other body parts and shifting the center of gravity so that movement, particularly of the extremities, may occur. In attempting to shed light on the complexity of movement disturbances associated with basal ganglia lesions, it also will be useful to examine the anatomically and pharmacologically complex synaptic interconnections between nuclei of the basal ganglia, the sources of basal ganglia efferent fibers, and the destination and action of efferent axons that finally leave the basal ganglia.

ANATOMICAL COMPONENTS OF THE BASAL GANGLIA

The basal ganglia system is composed of a group of nuclei situated at the base of the telencephalon and mesencephalon; in common usage, this includes the *putamen, caudate nucleus, globus pallidus, subthalamic nucleus,* and *substantia nigra* (see Fig. 1). The most rostral components, the caudate nucleus and putamen, collectively are referred to as the (neo)striatum. These two nuclei are separated by the internal capsule in some species such as primates, but they are cytologically similar, have similar connections with other nuclei, and are the sites at which most or all of the well-documented afferent fibers terminate in the basal ganglia (Fig. 2 and see the anatomical review by Carpenter, 1976).

The globus pallidus, located ventromedial to the putamen, is divided by the medial medullary lamina into external (lateral) and internal (medial) segments. The putamen and two pallidal segments together form a purely structural grouping, the lenticular nucleus. Axons from the striatum terminate in both pallidal segments, but according to current data, output from the two pallidal nuclei goes to different sites. Axons from the external globus pallidus (GPe) terminate primarily in the subthalamic nucleus, which in turn sends axons to the internal globus pallidus (GPi) and the substantia nigra. GPi, on the other hand, is one of the two basal ganglia nuclei from which axons originate that connect with other portions of the nervous system.

The most caudal component of the basal ganglia is the substantia nigra, a cellular sheet overlying the cerebral peduncle in the mesencephalon. It, too, has two divisions, the pars reticulata (SNr) and pars compacta (SNc). SNr is the other source of basal ganglia efferents, whereas axons from cells in SNc stay within the

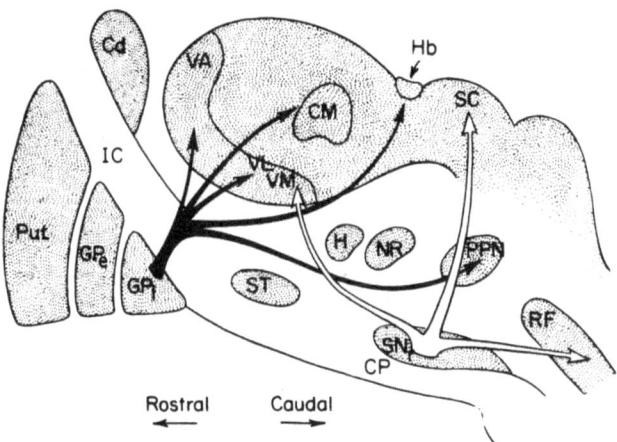

Fig. 1. Basal ganglia efferent systems. Axons leaving the basal ganglia originate from cells in the internal globus pallidus (GPi) and the substantia nigra, especially the pars reticulata (SN_r). Pallidal axons terminate in the ventral anterior (VA), ventrolateral (VL), and centre median (CM) thalamic nuclei, in the habenula (Hb), and in the pedunculopontine nucleus of the mesencephalic-pontine tegmentum (PPN). Nigral axons terminate more medially in the thalamic ventromedial nucleus (VM), in the superior colliculus (SC) and in the pontomedullary reticular formation (RF). Put-putamen; GPe-external globus pallidus; Cd-caudate nucleus; IC-internal capsule; ST-subthalamic nucleus; H-Forel's field; NR-red nucleus; CP-cerebral peduncle. From Anderson, M., and Crill, W. E. The basal ganglia and cerebellum. In T. C. Ruch and H. D. Patton (Eds.), *Physiology and Biophysics,* 20th ed., Vol. 1. Philadelphia. Saunders, 1979. Reprinted by permission.

basal ganglia system and terminate in the striatum, forming the nigrostriatal system (Fig. 3).

Basal ganglia efferent axons from cells in SNr and GPi (Fig. 1) terminate in several thalamic nuclei, the superior colliculus, the pedunculopontine nucleus (PPN) of the dorsal mesencephalic-pontine tegmentum, and perhaps the ponto-medullary reticular formation (Carpenter, 1976; Nauta and Mehler, 1966).

Fig. 2. Basal ganglia afferent systems. Demonstrated afferents originate from ipsilateral and contralateral cerebral cortex and from ipsilateral thalamic intralaminar nuclei and dorsal raphe nuclei of the mibrain. From Anderson, M., and Crill, W. E. The basal ganglia and cerebellum. In T. C. Ruch and H. D. Patton (Eds.), *Physiology and Biophysics,* 20th ed., Vol. 1. Philadelphia. Saunders, 1979. Reprinted by permission.

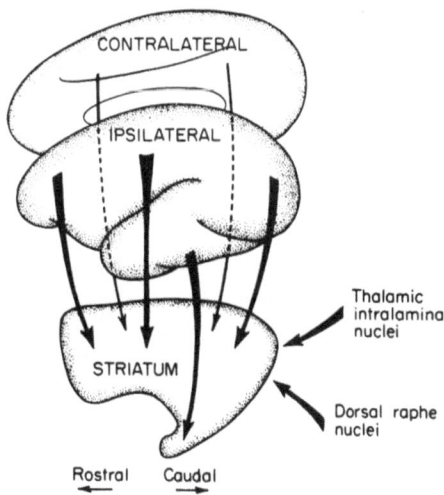

LESIONS OF THE BASAL GANGLIA AND DISORDERED MOTOR
COORDINATION

Most of the information regarding the role(s) of the basal ganglia in the control of posture and movement is derived from the study of defects in motor control that immediately or permanently follow lesions in one or more of the nuclei. In humans, damage to the nervous system results from trauma, infectious processes, vascular lesions, masses such as tumors or abcesses, or biochemically oriented processes that probably are the bases for most degenerative diseases. Consequently, the damage observed in human brains is generally not confined to a particular anatomical structure. This is particularly true for nuclei of the basal ganglia. They are located deep (selective traumatic damage is therefore very unlikely) and in close proximity to each other and to structures, such as the internal capsule, thalamus, optic tract, medial lemniscus, oculomotor nerve, and numerous other fiber systems in the diencephalon and midbrain. Furthermore, the biochemical processes critical to basal ganglia function are also critical to neurons in some other parts of the central or peripheral nervous system. Therefore, care must be taken to evaluate damage to other neural systems in attempts to determine motor defects due to basal ganglia dysfunction in biochemically-based diseases.

In experimental animals, it is possible to produce permanent or reversible lesions that are anatomically relatively discrete. When well-restricted lesions are produced in particular nuclei of the basal ganglia of experimental animals, however, there often are no motor deficits detected by the clinical examination techniques that usually have been used for evaluation. There have been few attempts

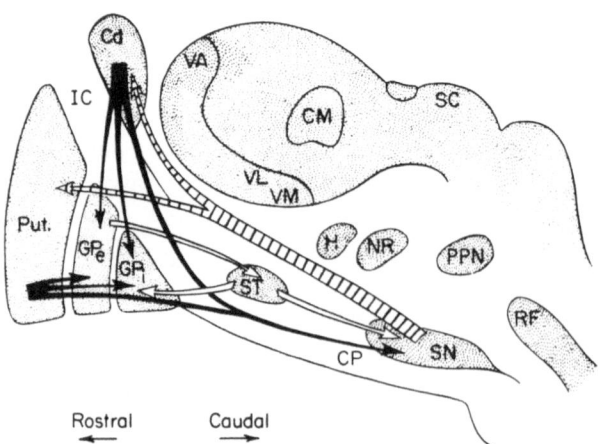

Fig. 3. Interconnections between nuclei of the basal ganglia. Both striatal nuclei (putamen and caudate nucleus) send axons to the substantia nigra and both pallidal segments (black arrows). Axons of GPe cells terminate in the subthalamic nucleus, which, in turn, sends axons to the two sources of basal ganglia output, GPi and SN (hollow arrows). Axons from cells in the pars compacta region of the substantia nigra go to the putamen and caudate nucleus as the nigrostriatal system (hatched arrows). Abbreviations as in Fig. 1. From Anderson M., and Crill, W. E. The basal ganglia and cerebellum. In T. C. Ruch and H. D. Patton (Eds.), *Physiology and Biophysics,* 20th ed., Vol. 1. Philadelphia. Saunders, 1979. Reprinted by permission.

to evaluate quantitatively the effect of basal ganglia lesions on motor function using any techniques other than the observation of ongoing gross motor function or the evaluation of resistance to passive movement.

THE BASAL GANGLIA AND MOTOR DYSFUNCTION IN HUMANS

PARKINSONISM. A trilogy of signs—hypokinesia, rigidity, and tremor—are the common features of the syndrome commonly referred to as *parkinsonism*. These signs, with some modifications, may occur in the aftermath of encephalitis (postencephalitic parkinsonism), after treatment with reserpine or with phenothiazine or butyrophenone tranquilizers used in psychiatric practice (tardive dyskinesia), after manganese or carbon monoxide poisioning, or most frequently, for unknown causes (idiopathic Parkinson's disease or paralysis agitans).

Patients with parkinsonism initiate most voluntary movements very slowly (bradykinesia) and often maintain fixed postures for long periods of time (akinesia). Other manifestations of their general hypokinesia are fixed facial expressions (masked faces) or an absence of armswing during walking. In addition, they do not exhibit the normal postural reflex adjustment to perturbations such as tilt. Although they can assume erect postures on command, they frequently have difficulty maintaining appropriate postural fixation and may assume a flexed or "somersault" posture (Martin, 1967). Grimby and Hannerz (1974) have used electromyographic recording to compare the recruitment of motor units in normal subjects and in individuals with parkinsonian bradykinesia. Parkinsonian patients usually initiated activity in individual motor units only after a delay of several seconds, and this is in contrast to normal subjects, who could recruit motor units within a few hundred milliseconds. Parkinsonian patients also could not terminate the activity in tonically firing motor units as rapidly as could normal individuals. In both groups of individuals, the initial motor units recruited had a regular tonic discharge. With intense effort, parkinsonian patients could reduce the recruitment delay toward the normal value, but when they did this, many motor units were recruited rather synchronously in a burstlike pattern more similar to the pattern of activity seen in normal individuals when they make rapid, brief contractions (Desmedt and Godaux, 1977). Wiesendanger and Rüegg (1978) also saw evidence for a delayed build-up of muscle activity in parkinsonian patients performing in simple- or complex-reaction time tasks. Neither the EMG onset nor the reaction time were significantly prolonged for parkinsonian patients in the simple-reaction task. However, if a visual discrimination was added (complex reaction time), the build-up of muscle activity was markedly slowed in parkinsonian patients, so that both the movement onset and target contact were significantly delayed, compared to those of age-matched control subjects.

Rigidity, as well as bradykinesia, may limit the motor activity of individuals with parkinsonism, although both are not always present. Rigidity is increased motor-unit activity that may be present even when the subject tries to relax completely. The motor-unit activity in a rigid muscle is increased markedly by passive stretch of the muscle, and the resulting resistance to movement is present throughout the range of movement, irrespective of the rate of stretch. Wallin, Hongell, and

Hagbarth (1973) recorded from Ia muscle afferent fibers of individuals with parkinsonian rigidity and found tonic Ia activity even when the subject was asked to relax completely. The presence of tonic EMG and Ia activity with the resting muscle at different lengths was in contrast to the relative electrical silence in muscle and nerve of relaxed normal individuals, and it indicates that both alpha and gamma motoneurons are hyperactive in rigidity. Jansen (1967) suggested that rigidity, in which the muscle resistance is not dependent on the rate of stretch, might involve the "release" of static fusimotor neurons, in contrast to spasticity, in which the resistance has a marked rate dependence and which Jansen proposed might be the result of lesions causing a "release" of dynamic fusimotor neurons. Although this could be tested to some extent by determining the quantitative characteristics of Ia fiber responses to different rates of muscle stretch in normal individuals and in those with rigidity or spasticity, such studies have not yet been done.

The tremor characteristically seen in parkinsonism is due to an alternating rhythmic activity of antagonistic muscle groups at a rate of about 3–6 beats/sec. It usually appears first in the thumb and fingers and may have a "pill-rolling" appearance. The tremor is present when the individual is at rest and may disappear when a voluntary movement is initiated with the tremorous limb; hence, it has been called a "tremor at rest" or "static tremor" to distinguish it from tremorous activity characteristically associated with other lesions (e.g., cerebellum).

Pathological changes in brains of individuals with parkinsonism are complex. Most consistent is a degeneration of catecholamine-containing neurons, including the dopaminergic cells in the pars compacta portion of the substantia nigra (Turner, 1968). Biochemically, there is a decreased dopamine concentration in the substantia nigra and in the striatum, where axons of dopaminergic nigral neurons terminate (see below).

ATHETOSIS, CHOREA, AND BALLISMUS. Athetosis, chorea, and ballismus all are phasic involuntary movements, but with varying characteristics. Athetosis appears as a slow, ceaseless, writhing movement, often present in the hands, lips, and tongue, and less frequently in the neck and foot. Denny-Brown (1962) has suggested that this is an instability of posture, with the hand, for example, going from flexion of the fingers, with flexion and supination of the wrist, to hyperextension of the fingers and extension and pronation of the wrist. Spiegel and Baird (1968) emphasize that these movements start from a state of muscle hypertonia.

The movements of chorea start from a state of normal or decreased muscle tone and are more rapid and jerky. They also primarily involve the hands and face, and at times they may be subtle and appear purposeful. In fact, phasic involuntary movements often have characteristics of both athetosis and chorea, and the term *choreoathetosis* is applied.

Involuntary activity in proximal musculature produces ballismus, a violent, flinging movement of the limbs. This frequently is confined to the side contralateral to the lesion (hemiballismus).

The involuntary movements of athetosis, chorea, and ballismus all tend to disappear when the individual is sleeping, and all are associated with pathological changes in the basal ganglia. In athetosis and chorea, there is atrophy and cell loss in the striatum. Hemiballismus most frequently follows damage to the subthalamic nucleus or its fiber connections, although striatal lesions also may be accompanied by movements that are sufficiently severe to be called ballistic.

DYSTONIA. Dystonia is a fixed, rigid posture produced by tonic activity of motor units. Electromyographically, the muscle shows the tonic activity characteristic of rigidity, but this is so severe as to maintain a fixed position, usually in one of the abnormal positions seen during the phasic involuntary movements of athetosis.

LESIONS OF THE BASAL GANGLIA IN EXPERIMENTAL ANIMALS

Investigators have made restricted lesions of individual nuclei of the basal ganglia in experimental animals in attempts to identify the structures whose dysfunction was responsible for any or all of the clinical signs seen in patients with more complex pathological changes. In general, lesions of individual nuclei have not replicated the clinical signs, and early investigators reported that basal ganglia lesions produced surgically or electrolytically did not produce signs of motor dysfunction that clearly could be distinguished from the signs believed to be due to the accompanying damage to overlying cortical structures (Kennard, 1944; Laursen, 1955; Ranson and Berry, 1941). However, several significant defects in motor activity have been reported following basal ganglia lesions, and these should provide a basis for any comprehensive understanding of how basal ganglia operate to ensure coordinated posture and movement.

INTERRUPTION OF BASAL GANGLIA OUTPUT. In light of current anatomical information which indicates that all basal ganglia output is mediated by neurons in the globus pallidus and SNr, it is logical to begin with an examination of motor performance following lesions of these systems. The most complete, selective lesions of both the globus pallidus (GPe and GPi) and SNr are those reported by Richter (1945). He exposed four monkeys chronically to carbon disulfide, which was used at the time as an industrial solvent and was known to produce parkinsonian symptoms. In all four animals, this produced remarkably selective lesions of the globus pallidus and substantia nigra, especially the pars reticulata; damage to other structures was inconsistent. The consistent findings in all animals were

> motor inactivity and slowness, with great loss of spontaneous and reactive movements but without true paralysis; incoordination in the automatic acts of walking, climbing and jumping; disturbances of posture resulting in exaggerated flexor positions of the trunk and flexor attitudes of the extremities; rigidity of the skeletal muscles, plastic in type and associated with a cog-wheel effect and finally, tremor. (Richter, 1945, p. 351)

It should be noted, however, that the tremor usually appeared when a movement was attempted, and, in most cases, not when the animal was at rest.

Other investigators have attempted to destroy either the globus pallidus or the substantia nigra, but not both. Mettler (1945) observed the symptons that were added when bilateral globus pallidus lesions were superimposed on bilateral cortical and striatal lesions, and he concluded that poverty of movement and the retention of impressed postures were the result of the pallidal lesions. Denny-Brown also made lesions that damaged the globus pallidus in monkeys and reported an akinesia and flexed posture that he attributed to a loss of labyrinthine and optic righting reactions (Denny-Brown, 1962; Denny-Brown and Yanagisawa, 1976). More recently, Hore, Meyer-Lohmann, and Brooks (1977) have used a cooling probe to reversibly interrupt pallidal function unilaterally in monkeys trained to make flexion–extension movements of the elbow. If the animals had no visual cues, the normal rhythmic flexion–extension of the elbow contralateral to the cooled pallidum was interrupted, and the elbow and wrist assumed predominantly flexed postures that were interrupted by brief extensions of restricted amplitude. EMG data showed an increase in the activity of both flexors and extensors, with periods of cocontraction. If visual analogues of the target and the arm movement were presented, there was compensation, the motor defects being markedly reduced.

Stern (1966) used a lateral surgical approach through the temporal lobe to make electrolytic lesions of the nigra in monkeys. All animals with bilateral nigral damage (not necessarily restricted to the pars reticulata) had hypokinesia, manifested by general poverty of movement and a tendency to assume flexed, immobile postures. The hypokinesia was accentuated if the lesions accidentally encroached on the globus pallidus. Stern's findings supported an earlier report of Carpenter and McMasters (1964), although the latter authors saw little deficit if the lesions were only unilateral. Although tremor appears following some nigral lesions, there seems to be general agreement that to produce tremor, additional structures dorsal to the nigra must be damaged by the lesion (Péchadre, Larochelle, and Poirier, 1976; Poirier, 1960; Ward, McCulloch, and Magoun, 1948).

The bulk of the experimental literature would indicate that interruption of output from the globus pallidus and/or the substantia nigra results in hypokinesia. Thus, it is not surprising that surgical lesions of the globus pallidus or ventrolateral thalamus (where many basal ganglia efferent fibers terminate) may reduce rigidity and tremor in parkinsonian patients, but they seldom improve hypokinesia that was not a result of rigidity (Selby, 1967).

STRIATAL LESIONS. Motor symptoms produced by lesions of the putamen and caudate nucleus are quite the opposite of the hypokinesia following pallidal and nigral lesions. Several studies have reported at least a transient circling toward the side of an extensive unilateral caudate lesion and more generalized hyperactivity or hyperreactivity to stimuli following extensive bilateral electrolytic lesions (Mettler, 1945; Villablanca, Marcus, and Olmstead, 1976).

Extensive chemically induced lesions of the striatum have been produced recently by intrastriatal injection of kainic acid (Coyle and Schwartz, 1976). Kainic acid is a rigid glutamate analogue with potent excitatory effects on neuronal soma, but not on axons. In small doses, it causes CNS neuronal degeneration in the area in which it is injected locally, leaving axons passing through the injection site intact. In rats with intrastriatal injections of kainic acid, there is an initial circling away from the side of the injection, perhaps produced by irritant excitation of caudate neurons. Later, animals circle toward the side of the injection and resulting lesion, consistent with the effects of caudate lesions produced electrolytically.

LESIONS OF THE SUBTHALAMIC NUCLEUS. Carpenter, Whittier, and Mettler (1950) reported that a contralateral hyperkinesia, which they called *choreoid hyperkinesia,* was produced by lesioning as little as 20% of the subthalamic nucleus if the lesion did not damage the globus pallidus or its output fibers. This syndrome involved primarily proximal appendicular and axial musculature and closely resembled the hemiballismus associated with lesions of the subthalamic nucleus in humans (Martin, 1927; Whittier, 1947). Hemiballismus (hemichorea) following subthalamic lesions seems to be the most consistent motor symptom following a basal ganglia lesion, although similar signs also may result from lesions at other sites.

BIOCHEMICALLY INDUCED LESIONS OF THE NIGROSTRIATAL SYSTEM. As noted above, the substantia nigra includes two portions, the pars compacta and the pars reticulata. In brains from patients with parkinsonism, nigral cell loss most consistently is in the dopamine-containing neurons of the pars compacta. Experimental lesions produced electrolytically or surgically seldom are so restrictive. In contrast, relatively selective destruction of the dopaminergic nigrostriatal system can be produced by localized intracerebral injection of restricted amounts of 6-hydroxydopamine, a neurotoxic isomer of norepinephrine that is taken up by catecholamine-containing neurons (Kostrzewa and Jacobowitz, 1974; Ungerstedt, 1974). Rats with bilateral 6-OH dopamine-induced nigral lesions are markedly hypokinetic as well as adipsic and aphagic (Ungerstedt, 1974). Those with unilateral lesions have an asymmetric body posture that is concave on the side of the lesion, and they may circle toward the side of the lesion if aroused sufficiently. Spontaneous exploratory behavior is decreased, however, and the animals fail to orient to stimuli presented from the lesioned side.

Depletion of the dopaminergic nigrostriatal system with 6-OH dopamine or electrolytic lesions does not produce the full trilogy of parkinsonian motor signs: hypokinesia, rigidity, and tremor. These are produced more completely by systemic administration of haloperidol or chlorpromazine (antagonists of the catecholamines, dopamine and norepinephrine) or of reserpine, which depletes all the monoamines, including serotonin (Duvoisin, 1976). Electrolytic lesions of the rostral ventromedial tegmentum also may produce combined rigidity, tremor, and hypokinesia, but some structure dorsomedial to the substantia nigra must be destroyed in addition to the nigrostriatal system (Poirier, 1960; Péchadre *et al.*, 1976; Ward *et al.*, 1948). Péchadre *et al.* (1976) have concluded that the effective lesions must interrupt the rubro-olivary-cerebellar loop. The effectiveness of these lesions could be due, however, to interruption of fibers from monoaminergic systems originating more caudally in the brain stem and traveling in the same region.

HYPOTHESES REGARDING ROLES OF THE BASAL GANGLIA IN MOTOR COORDINATION

The motor signs described above have been categorized by Martin (1967) as *negative* or deficiency phenomena (the various manifestations of hypokinesia) and *positive* or release phenomena (athetosis, chorea, ballismus, tremor, and rigidity). The important question is, "Do these signs support the idea that the basal ganglia are involved in the control of a basic aspect of motor coordination?"

Martin (1967) and Kornhuber (1971) both would answer this question affirmatively with different, but perhaps not mutually exclusive, answers. Martin, after studying postural fixation, equilibrium reactions, and locomotion in patients with various types of basal ganglia diseases, has concluded that "the basal ganglia and their associated pathways form a neural system that is devoted to the control of postural reflexes, other than those of the antigravity group." (Martin, 1967, p. 123). Kornhuber (1971), on the basis of less direct inferences from clinical observations, has proposed that "the function of the basal ganglia is to generate slow smooth movements of voluntary speed, i.e., from the technical point of view, the basal ganglia function as a ramp generator." (Kornhuber, 1971, p. 158). Both of these proposals require that we have some understanding of the mechanisms involved in postural control and the changes in postural adjustment demanded in association with movements of different speeds.

The Basal Ganglia and Postural Control

Martin (1967) points out that if posture denotes the position of body parts relative to each other and to gravity, then several mechanisms are necessary to maintain this posture in equilibrium positions. This includes mechanisms to support the body parts against gravity, to stabilize or fixate body parts relative to one another, and to maintain or restore the center of gravity over an appropriate support. Active movement, with or without external loads, requires that all these postural adjustment mechanisms be temporally coordinated with the primary movement, whether that movement is a so-called *voluntary* movement or an *automatic* movement, such as the stepping movement of locomotion.

Martin's examination of patients with postencephalitic or idiopathic parkinsonism showed a significant number who failed to stabilize their head, trunk, or limbs, especially with vision occluded. Consequently, their heads drooped, their trunks became stooped, or their outstretched limbs sagged, in spite of the apparently normal muscular power demonstrated when they were asked to straighten or extend the flexed joints. An even larger number could not maintain their equilibrium when standing, sitting, or, in some cases, when on all fours. Many could not right themselves, and almost none showed the usual reactions to tilting, either when seated on an unstable base or when in the all-fours position.

Of particular interest are the parkinsonian patients' problems with locomotion. Locomotion requires a continual adjustment of the center of gravity in the lateral, vertical, and anterior-posterior directions so that the legs and feet can be lifted in an alternating manner. In Martin's experience, many patients were almost immobile, had a shuffling gait without toe-off, or had uncontrolled retropulsion or forward acceleration in unaided locomotion. These same patients had a relatively normal gait if the examiner rocked their shoulders from side-to-side or forward and backward, thus providing the appropriate shift of the center of gravity. Other stimuli, such as the visual input provided by lines or stairs oriented transversely to the path of progress, also improved locomotion for some patients.

Martin considers all these signs as evidence that the basal ganglia play a critical role in organizing the motor activity necessary for maintaining static postures or for providing the postural adjustment necessary to insure that the center of gravity stays within the limits of the vertical projection of the base. Moreover, he points out that the involuntary movements of hemiballismus are minimized when the patient sits on a firm base and are accentuated when he has an unstable base of support. In the latter case, the involuntary movements can be interpreted as exaggerated reactions to tilting, the converse of the depressed reactions characteristic of patients with parkinsonism.

The Basal Ganglia and Rate of Movement

Kornhuber (1971) has concluded that the basal ganglia play an essential role in controlling "ramp" movements, which may be made at variable speeds. These ramp movements may be contrasted with "ballistic" movements, which are steplike and presumably occur too rapidly to be modified or controlled by sensory feedback produced by the ongoing movement.

Kornhuber's hypothesis was based primarily on saccadic eye movements, which he thought were intact in parkinsonism, but dysmetric in patients with cerebellar degeneration. He interpreted this evidence to indicate that the cerebellum generated ballistic movements and the basal ganglia generated ramp movements. However, others have reported that parkinsonian patients also tend to make hypometric, successive, saccadic eye movements to a target, instead of the normal large saccade, followed, if necessary, by a corrective adjustment.

Movements of the extremities do not fall into two distinct groups that are as clearly ballistic and nonballistic as are saccadic and nonsaccadic eye movements. However, Flowers (1975, 1976) has examined visuomotor tracking movements made with the arm by normal and parkinsonian subjects and subjects with tremor, sometimes of the cerebellar type. Subjects moved a joystick that controlled a visible spot to be moved to a target, and if the movement was completed in less than the mean reaction time for that individual, it was considered to be ballistic. Normal and tremorous subjects often made ballistic movements to the target position, and for movements of 128 mm or less, the speed of the movement (ballistic or nonballistic) had little influence on its accuracy. Parkinsonian subjects usually made the movements much more slowly, but with accuracy comparable to normals. When they did make ballistic movements, however, the error of the movement increased markedly. Moreover, if the visual target or joystick-controlled dot display was blanked out during the movement, the error made by parkinsonian patients was increased markedly and resembled the error they made in ballistic movements, which presumably would have occurred too rapidly to allow control by visual feedback. Several investigators (Kay, 1979; Woodworth, 1899) have reported that in the development of skilled movement, movements become more rapid and reach a ballistic speed, at which accurate preprogrammed motor commands must be issued by the nervous system. Flowers, in contrast to Kornhuber, has interpreted his data to indicate that parkinsonian subjects cannot generate these appropriate "preprogrammed" motor commands for ballistic movement and operate, instead, in a mode where movements are slowed sufficiently to allow ongoing control by sensory feedback (Flowers, 1976).

The pattern of EMG activity at both moving and stabilizing joints normally varies as a function of the rate of movement. When a rapid movement is made around a joint such as the elbow, a characteristic triphasic pattern of EMG activity normally occurs in the muscles acting across that joint (Hallet, Shahani, and Young, 1975; Stetson and Bouman, 1935; Wacholder, 1928). The agonist shows a brief burst of activity that is terminated before maximum displacement. During the subsequent pause in agonist activity, there is a burst in the antagonistic muscles, and this is followed by a second period of activity in the agonists. During slower movements, the agonist is recruited gradually and does not show a silent period (Desmedt and Godaux, 1977), and there also may be continuously adjusted cocontraction of the antagonistic muscle groups (Stetson and Bouman, 1935). Wiesendanger and Rüegg (1978) have reported briefly that in parkinsonian subjects, the triphasic reciprocal pattern normally seen with rapid movements is absent and is replaced by simultaneous activity that persists throughout the movement in both the agonist and antagonist. For some reason, the parkinsonian subject does not generate the pattern of activity that, in normals, appears to be preprogrammed and produces a rapid movement (Desmedt and Godaux, 1978; Hallet *et al.*, 1975).

In addition to the different patterns of activity in agonists and antagonists during movements of different speeds, the activity of other muscles, some of which may be important for postural stabilization, also is a function of the speed of the movement. For example, Basmajian (1978) has reported that the activity of brachioradialis is greatest during rapid flexion of the elbow, when a significant centrifugal force is generated. Since this muscle acts along the long axis of the forearm, it provides centripetal force to stabilize the elbow joint. In monkeys, the activity of proximal shoulder and paraspinal muscles precedes that of more distal arm muscles during push-pull and side-to-side movements of the arm, and DeLong and Strick (1974) reported that deep paraspinal muscles were activated earlier and more intensely during slow movements than they were during rapid movements. It should be noted, however, that the monkey's heads were stabilized mechanically and the two types of movements differed in other respects than just speed: rapid movements were terminated by a mechanical stop, whereas the monkey had to terminate slow movements in a trigger zone. Thus, the slow movements may have required more accurate postural stabilization than did the rapid ones. In a standing human subject who raises his arm rapidly, muscles acting across the hip and along the spine become active much earlier than do shoulder muscles (Belen'kii, Gurfinkel', and Pal'tsev, 1967; Horak, Lynch, and Anderson, unpublished observations), but early activity in the hip muscles does not occur if the arm is raised more slowly.

Thus, as the velocity of a movement changes, the required postural stabilization and adjustment also change, and the basal ganglia could play a role that relates to both postural stabilization and speed of movement. At first glance, it might appear that this could be tested easily by quantitatively comparing small amplitude, distal movement that presumably would not require much postural control with larger amplitude movements that require obvious postural adjustment. However, EMG evaluation would have to be done to overcome the investigator's common error: the assumption that changes in EMG activity are restricted to or exemplified by the few muscles he chooses to monitor.

The basal ganglia, like other portions of the nervous system, could participate in the production of coordinated movement in several ways. One way, often overlooked, is that they could provide some sort of "bias" or "set point" mechanism. For example, the basal ganglia could provide a tonic excitatory or inhibitory input to the thalamo-corticospinal system or to various brain stem systems to which pallidal or nigral axons project. In this way, the basal ganglia could determine the level at which these systems would be activated by other transient mechanisms, resulting in movement and including postural adjustment. Note that a constant biasing or "enabling" function would not necessarily require that basal ganglia neurons have phasic changes in activity that are temporally correlated with the movement.

Another possibility is that the basal ganglia initiate certain aspects of some or all movements. In this case, transient changes in the firing of basal ganglia neurons would precede these aspects of motor activity. The usual experimental problem is that a motor act includes many sequential components, and a neuron's activity may covary with all of them, even though the neurons might play a role in controlling only a part of the sequence.

Finally, the basal ganglia could be involved in some "feedback" aspect of the motor coordination, in which input from peripheral sensory sources or from "corollary discharge" mechanisms modifies the activity of basal ganglia neurons and, eventually, some aspect of the motor output.

DISCHARGE OF BASAL GANGLIA NEURONS DURING VARIOUS TYPES OF MOTOR ACTIVITY

The activity of neurons in the caudate nucleus, putamen, and external and internal globus pallidus has been recorded in awake monkeys during stable posture and during phasic motor activity. In the absence of overt movement, most neurons that have been studied in the striatum, both the putamen (Anderson, 1977; DeLong, 1973; DeLong and Strick, 1974) and the caudate nucleus (Buser, Ponderous, and Mereaux, 1974; Hull, Levine, Buchwald, Heller, and Browning, 1974; Matsunami and Cohen, 1975), fire irregularly at very low discharge rates, usually less than 10 impulses/sec. Most neurons in both segments of the globus pallidus show more regular, higher-frequency discharge, with mean discharge rates as high as 100/sec (Anderson, 1977; DeLong, 1971; Matsunami and Cohen, 1975). Those neurons in the external segment may have pauses in their discharge if the animal's head is mechanically stabilized and the animal does not make overt movements (DeLong, 1971), but these pauses are much less common if the animal must actively maintain a restricted head position (Anderson, 1977).

Some neurons in the putamen and both segments of the globus pallidus show phasic changes in discharge rate that are temporally correlated with movement. DeLong (1971) examined the activity of pallidal neurons in monkeys making push-pull or side-to-side movements of the contralateral or ipsilateral arm or leg (see Fig. 4A). Of the units studied in GPe or GPi of 2 monkeys, 19% showed a clear increase or decrease in discharge during particular phases of arm movement, usually of the contralateral arm. Arm-related cells formed a slightly smaller percentage of the sample studied in another animal that also was trained to make leg movements. In this monkey, less than 5% of the neurons studied showed a relationship to leg movement. Anderson (1977) trained monkeys to make the postural adjustments necessary to maintain a restricted head position when their primate chair was tilted. The activity of pallidal neurons was recorded during sinusoidal chair tilt (Fig. 4B), and approximately 70% of the cells studied showed changes in firing rate that were temporally correlated with particular phases of the chair-tilt cycle when the chair was tilted ± 10–$15°$ at frequencies of 0.1 to 0.5 Hz. Neurons with moderate tonic discharge rates (20–50/sec) when the chair was horizontal had phase lags with respect to maximum posterior tilt when the chair was tilted sinusoidally. This phase shift changed to a slight phase lead for neurons whose tonic discharge rate was 50/sec or higher when the chair was stable in the horizontal position.

The movement-correlated phasic changes in basal ganglia activity refute any notion that the basal ganglia provide only some constant tonic bias to other systems related to motor control. However, this does not negate some type of "changing set point" function of the basal ganglia, with other systems supplying the "trigger mechanism" or the generating signal to initiate the movement or determine its temporal or spatial form (Buchwald, Hull, Levine, and Villablanca, 1975).

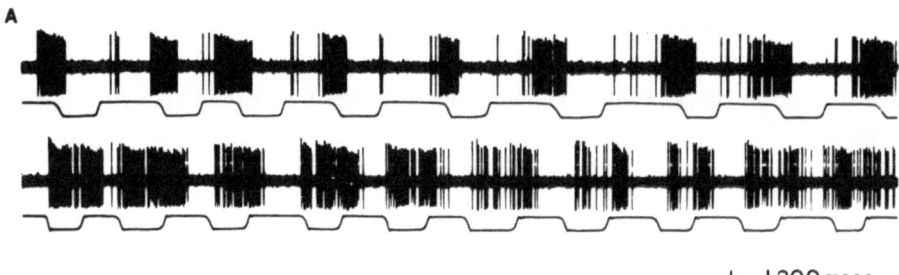

A

H—H 200 msec

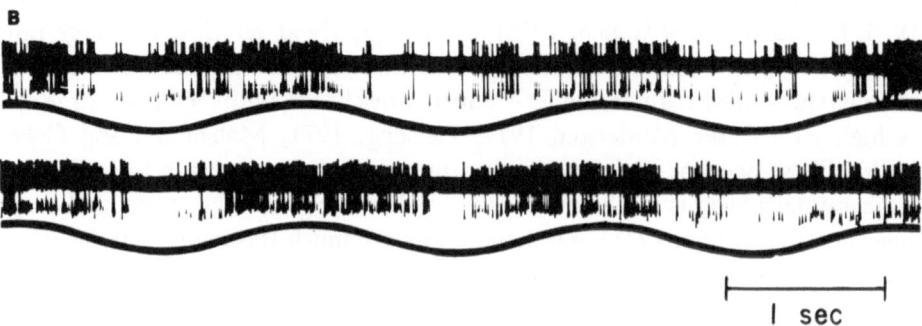

B

H———H
I sec

Fig. 4. (A) Activity of a GPE neuron in a monkey during contralateral arm movement in a push-pull *(top)* or side-to-side *(bottom)* direction. The lever position is shown below the unit activity, and an upward deflection indicates a pull movement *(upper records)* or an extension movement *(lower records)*. From DeLong (1971). (B) Activity of a GPe neuron in another monkey during anterior-posterior chair tilt. The monkey was required to make postural adjustments actively to maintain a restricted head position. The lower traces of each pair indicate chair position, from 10° posterior tilt *(upward deflection)* to 10° tilt *(downward deflection)*. From Anderson (1977).

It is the temporal-spatial form of slow movements in particular that Korn-huber suggested was generated by the basal ganglia. DeLong (1973) examined the discharge pattern of neurons in the putamen during slow and rapid movements, and DeLong and Strick (1974) compared the activity of putamen neurons to the activity of cells in the globus pallidus, cerebellum, and precentral cerebral cortex during similar movements. Of the putamen cells that showed a movement-corre-lated change in firing rate, about 40 to 50% showed more intense or longer duration changes in discharge during slow movements than they did during rapid move-ments. The slow movements, however, also were of longer duration, and it is not clear whether the discharge of putamen neurons was related to the duration or to the velocity of the movement. Nevertheless, it was of interest that only 17% of movement-related cells in the globus pallidus were similarly "ramp-related," and only 3% of the cerebellar neurons and one of the cells in the precentral cerebral cortex showed similar preferential relationships. Again, it should be pointed out that there were differences other than rate in the ramp versus ballistic paradigms used by DeLong and Strick; in particular, the rapid movement was ended by a mechanical stop and the slow movement was not.

If basal ganglia activity played a role in determining some aspect of motor activity that was particularly associated with slow movements, then this ultimately must be mediated by basal ganglia output neurons, located in the internal globus pallidus or pars reticula portion of the substantia nigra (Carpenter, 1976). It is interesting, then, that so few movement-related pallidal neurons in the experiments of DeLong and Strick (1974) showed a strong change in firing in association with slow movements.

In summary, data from awake animals performing phasic motor tasks indicate that some basal ganglia neurons show phasic changes in discharge rate and that such phasic changes can be temporally correlated with the motor activity. These data do not prove that the basal ganglia initiate motor activity, and they provide little information as to any particular properties of motor function to which the activity of basal ganglia neurons is related.

EXPERIMENTAL ANALYSES OF BASAL GANGLIA SYNAPTIC RELATIONSHIPS

Whatever the role of the basal ganglia in motor coordination, this complex network of nuclei deals with input from several sources and, after processing, must emit the appropriate signals via cells in the globus pallidus and substantia nigra. Many studies during recent years have been directed at determining the anatomi-cal, physiological, and pharmacological characteristics of basal ganglia synaptic connections.

AFFERENT INFORMATION

In studies on animals anesthetized with general or local anesthesia, basal gan-glia neurons in the caudate nucleus and globus pallidus have been reported to change their firing in response to various modalities of stimuli: cutaneous, auditory, vestibular, and visual (Albe-Fessard, Oswalds-Cruz, and Rocha-Miranda, 1960; Levine, Hull, and Buchwald, 1974; Sedgwick and Williams, 1967). In general, the latency of these responses is relatively long; for example, about 15–30 msec following stimulation of cutaneous nerves in the contralateral forelimb. By comparison, cells in somatosensory regions of the cerebral cortex are activated at about 10 to 15 msec by stimulation of the contralateral forepaw (Morse, Adkins, and Towe, 1965).

Most anatomical studies have indicated that all afferent information enters the basal ganglia by way of the striatum (see Carpenter, 1976, for review). Fibers from all the ipsilateral cerebral cortex and the contralateral precentral cortex, from the central lateral, center median, and parafascicular thalamic nuclei, and from the dorsal raphe nuclei of the rostral brain stem central grey terminate in both the putamen and the caudate nucleus (see Fig. 2). Those from the cortex, in particular, connect with the striatum in a topographically organized manner (Kemp and Powell, 1970).

Stimulation of the cerebral cortex or thalamus elicits initial EPSPs in striatal neurons, indicating that the fastest, and perhaps all corticostriatal and thalamo-striatal fibers make excitatory connections (Buchwald, Price, Vernon, and Hull, 1973; Kitai, Kocsis, and Wood, 1976a). Data from Kitai and his colleagues (Kitai *et al.*, 1976a) suggest that corticostriatal neurons are distinct from cortical cells that send their axons into the pyramidal tract. Using electrophysiological techniques, these investigators observed that neurons in the sensorimotor cortex of cats could be activated antidromically either from the caudate nucleus or the medullary pyramids, but not from both sites. They also used horseradish peroxidase (HRP) retrograde transport techniques to show with anatomical methods that the cortical neurons labeled by HRP injected into the caudate nucleus were located in both cortical layers III and V, whereas others have reported that corticospinal and corticobulbar neurons labeled by HRP injected into the spinal cord or brain stem of cat or monkey were located exclusively in layer V (Berrevoets and Kuypers, 1975; Coulter, Ewing, and Carter, 1976). However, Hedreen (1977), who used HRP techniques to identify corticostriatal neurons in the rat, concluded that layer V cortical neurons were the ones labeled by HRP injected in the caudate-putamen, and they, as well as others (Kitai *et al.*, 1976; Nauta, Pritz, and Lasek, 1974), have emphasized that corticostriatal neurons are difficult to identify using HRP retrograde labeling techniques.

Little is known about the influence of the dorsal raphe nuclei on the striatum. The raphe nuclei contain large amounts of serotonin (5-hydroxytryptamine), and preliminary evidence indicates that this may be the neurotransmitter for raphe-evoked inhibition of neurons in the caudate nucleus and putamen (Olpe and Koella, 1977).

Recent data indicate that some afferents terminate directly in basal ganglia nuclei other than the putamen and caudate. Clausing, Anderson, and DeVito (1977) put HRP into the globus pallidus of monkeys and found retrogradely labeled neurons in the pedunculopontine nucleus, dorsal raphe nuclei, and intralaminar thalamus, as well as the substantia nigra and striatum. Stimulation of the brain stem tegmentum in the region of the pedunculopontine nucleus (PPN) causes excitation of neurons in the entopeduncular nucleus (ENTO, feline homologue of GPi), and this excitatory input may be important in producing the high-frequency discharge that is characteristic of globus pallidus neurons in awake animals (Gonya and Anderson, unpublished observations).

Output from the Globus Pallidus and Substantia Nigra

Axons of neurons in the internal segment of the globus pallidus (GPi) and the pars reticulata of the substantia nigra (SNr) leave the basal ganglia and terminate in several areas via which they might reach other parts of the brain involved in motor control (see Fig. 1). Pallidal axons have been shown to terminate in the ventroanterior (VA) and ventrolateral (VL) thalamic nuclei; nigral axons terminate slightly more medially, in an area called the ventromedial nucleus (VM) in the cat and the medial portion of the ventrolateral nucleus (VLm) and magnocellular region of the ventroanterior nucleus (VAmc) in the monkey (see review by Carpenter, 1976). Both VA and VL are known to influence the precentral cerebral cortex (Carmel, 1970; Strick, 1976), but there are some data that indicate that basal ganglia output may inhibit thalamic neurons that project to cortical area 6, and not to area 4, which receives input from thalamic neurons excited by cerebellar efferents (Uno, Ozawa, and Yamamoto, 1978). Other thalamic destinations of pallidal or nigral axons are the centromedian and dorsomedial nuclei, which also are reported to project to the cerebral cortex (Macchi, Bentiroglio, D'Atena, Rossini, and Tempesta, 1977; Scollo-Lavizzari and Akert, 1963).

In addition to the thalamic projections, both pallidal and nigral axons terminate in other brain stem areas that may influence motor control. A major pallidal output is to the pedunculopontine nucleus (PPN), which is located near the brachium conjunctivum in the caudal mesencephalon (Nauta and Mehler, 1966) and which, as described above, also appears to send afferent fibers to the globus pallidus. It is not known whether PPN could provide access to other spinal-destined neurons, such as those of the reticulospinal system, but it is of interest that it also receives afferent fibers from the motor cortex (Hartmann-von Monakow, Akert, and Künzle, 1979) and the red nucleus (Strominger, Truscott, Miller, and Royce, 1979).

Nigral axons project to two brain stem areas that are known to influence motor coordination, in addition to their projection to the thalamus. One is the pontomedullary reticular formation (Hopkins and Niessen, 1976; Rinvik, Grofová, and Ottersen, 1976), and the other is the superior colliculus (Faull and Mehler, 1976; Graybiel and Sciascia, 1975; Hopkins and Niessen, 1976; Jayaraman, Batton, and Carpenter, 1977; and Rinvik *et al.*, 1976). Many nigral neurons have branching axons, with one branch going toward the thalamus and the other branch going toward the superior colliculus; this can be shown by activating the same neuron antidromically by low-intensity stimulation of both the thalamus and the superior colliculus (Anderson and Yoshida, 1977). Figure 5 shows antidromic activation of a nigral neuron when the superior colliculus (C) or the thalamus (T) was stimulated. In each case, the response was blocked if a spontaneously occurring spike (spon) or one evoked by stimulation of the other branch had not had time to pass the test stimulus position. Branching axons to more than one brain stem destination also have been reported for ENTO neurons (Filion, Harnois, and Guano, 1976). If each branch carries the same information, then this raises interesting questions with respect to the specificity of basal ganglia output. For example, is this a mechanism by which nigral neurons could play some role in coordinating motor activity of the head and trunk?

There are conflicting opinions as to the synaptic actions of pallidal and nigral axons on their target neurons. At the thalamus, investigations on cats anesthetized with sodium pentobarbital have shown that monosynaptically evoked responses were only inhibitory (IPSPs) following stimulation of either the entopeduncular nucleus (Uno, Ozawa, and Yoshida, 1978) or the substantia nigra (Ueki, Uno, Anderson, and Yoshida, 1977). Deniau, Lackner, and Feger (1978) also found nigral inhibition of VA and VL neurons in cats rendered anesthetic with local

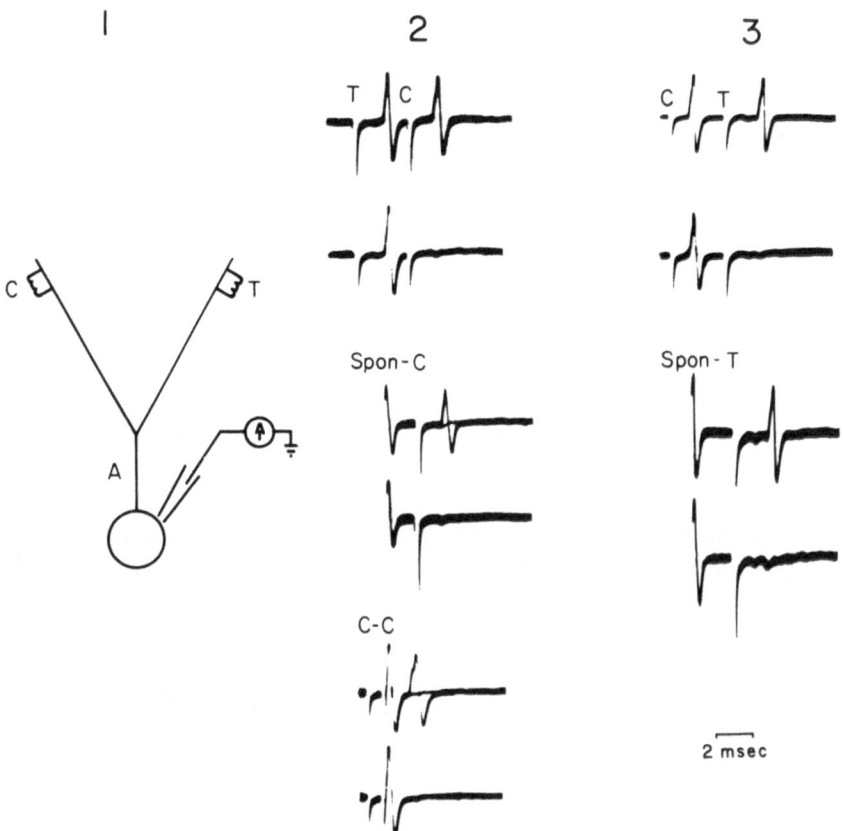

Fig. 5. Demonstration of nigral axon branching using collision blockade. (1) Experimental arrangement. C-collicular branch; T-thalamic branch; A-unbranched axonal segment. (2) A test stimulus to the colliculus is preceded by a thalamic-evoked response *(upper traces)*, a spontaneously occurring spike (spon—*middle traces*), or an earlier collicular-evoked response *(bottom traces)*. In the lower record of each pair of traces, the collicular stimulus followed the conditioning spike at an interval that was too short to allow the second spike to reach the soma. (3) Same as in 2, but the test stimulus is applied to the thalamic branch. From Anderson and Yoshida (1979).

anesthesia and spinal transection. Dormont and Ohye (1971), who stimulated the entopeduncular nucleus and recorded extracellularly from neurons in VL of cats anesthetized with chloralose, found that only 3 of 200 VL neurons were excited at latencies short enough to be produced by a direct action of entopeduncular axons, and in 31 cells studied under local analgesia, only another 3 showed an initial excitation following entopeduncular stimulation. Most cells studied by Dormont and Ohye had very little ongoing activity, and inhibition probably would not have been detected reliably.

Uno and Yoshida (1975) and Ueki et al. (1977) found almost no convergence on individual thalamic neurons of output from the basal ganglia and the cerebellum. Entopeduncular or nigral-evoked IPSPs occurred in thalamic neurons ventromedial and, in some cases, slightly rostral to the neurons in which EPSPs were elicited by stimulation of the brachium conjunctivum (BC). Deniau et al. (1978) have reported, however, that convergence of nigral inhibition and cerebellar (BC) excitation does occur on individual thalamic neurons. Deniau and co-workers recorded from cats without pentobarbital, and when histological figures from the papers by Ueki et al. (1977) and Deniau et al. (1978) are compared, it appears that the latter authors recorded at more caudal locations than did Ueki et al. Perhaps the difference in results is because of one or both of these factors.

The action of basal ganglia efferents at the superior colliculus recently has been examined by Deniau, Chevalier, and Feger (1978), and they reported that nigral axons also inhibit collicular neurons. Thus, most recent data indicate that basal ganglia output is exclusively inhibitory.

Other investigators reported, however, that EPSPs were evoked at short latencies (less than 1.0 msec) in some VL neurons following stimulation of ENTO efferent axons in the ansa lenticularis of cats anesthetized with sodium pentobarbital (Desiraju and Purpura, 1969), and Frigyesi and Machek (1970) reported that stimulation of the substantia nigra produced EPSPs at short latencies in some thalamic neurons.

One possible source of confusion could be spread of the stimulus to fibers other than the axons of nigral or entopeduncular neurons. High-intensity stimulation of the substantia nigra can produce enough current spread to excite the fibers of the brachium conjunctivum (BC) that pass between the substantia nigra and the red nucleus on their way to the thalamus. Proof that such spread occurs can be obtained by showing the occlusive interaction between this strong "SN" stimulus and a subsequent stimulus to the brachium conjunctivum at a more caudal brain stem site. At relatively short interstimulus intervals, the EPSP expected following BC stimulation fails to occur, owing to occlusion in the common axons excited at different points by both SN and BC stimulation (Yoshida, Ueki, and Anderson, unpublished observations). Deniau, Lackner, and Feger (1978) also found short-latency excitation of some VL neurons when they stimulated SN, but this was not present in animals that had undergone ablation of the pericruciate cortex three weeks before the acute experiments. Thus, this excitation must have been because of stimulus spread to the cerebral peduncles or to excitation of cortical efferent fibers coursing through SN en route to other structures. Yoshida and his colleagues (Uno and Yoshida, 1975; Ueki et al., 1977) tried to evaluate stimulus spread by determining the stimulus positions from which synaptic potentials were evoked at lowest stimulus intensities; these low-threshold points were restricted to the entopeduncular nucleus or the substantia nigra. Desiraju and Purpura (1969) and Frigyesi and Machek (1970) used only the presence of potentials evoked in the sensorimotor cortex as an indicator of stimulus spread to the nearby internal capsule, cerebral peduncle, brachium conjunctivum, or medial lemniscus.

If the synaptic actions of both sets of basal ganglia output neurons, those in the globus pallidus and those in the substantia nigra, are exclusively inhibitory, both at their thalamic terminations and at other brain stem destinations, then it means that all basal ganglia influence on motor coordination is exerted via an inhibitory modulation of subsequent brain stem systems. At the thalamus, there may be some convergence with information from the cerebellum, but we know little about other sources of information with which basal ganglia inhibition may interact.

INTERCONNECTIONS BETWEEN NUCLEI OF THE BASAL GANGLIA

As illustrated in Fig. 3, there are at least three systems interconnecting various nuclei of the basal ganglia: (1) an output from the external globus pallidus to the subthalamic nucleus, which in turn sends axons to GPi and SN; (2) a projection of caudate and putamen axons to neurons of GPi and SN (the striatopallidal and striatonigral systems); and (3) a projection of axons from the pars compacta portion of the substantia nigra (SNc) to the putamen and caudate nucleus (the nigrostriatal system).

CONNECTIONS OF THE SUBTHALAMIC NUCLEUS. Little has been done physiologically to characterize the subthalamic nucleus or its interaction with the pallidal segments. Frigyesi and Rabin (1971) reported that stimulation of external globus pallidus (GPe) produced EPSPs in most subthalamic neurons studied, with longer duration IPSPs following.

STRIATAL ACTIONS ON THE GLOBUS PALLIDUS AND SUBSTANTIA NIGRA. Projections to the globus pallidus and substantia nigra originate from neurons in both the putamen and the caudate nucleus. Earlier Golgi studies (Kemp and Powell, 1971) suggested that few striatal neurons had long axons extending out of the nucleus, but horseradish peroxidase studies have shown that a large number of striatal efferent axons originate from medium spiny striatal neurons (Grofová, 1975). Stimulation of the caudate nucleus in barbiturate-anesthetized animals monosynaptically inhibits entopeduncular or nigral neurons via slowly conducting caudate axons (Dray, Gonye, and Oakley, 1976; McNair, Sutin, and Tsubokawa, 1972; Obata and Yoshida, 1973; Yoshida and Precht, 1971). Inhibition of the tonic activity of neurons in both segments of the globus pallidus also can be produced by stimulation of the putamen or caudate nucleus in awake monkeys (Ohye, LeGuyader, and Feger, 1976), although a "postinhibitory rebound" often follows.

In contrast, it has been reported in other intracellular studies in unanesthetized encéphale isolé cats that caudate stimulation produces an initial excitation, followed by inhibition (EPSP–IPSP sequences) in both entopeduncular (Malliani and Purpura, 1967; Levine *et al.*, 1974) and nigral neurons (Frigyesi and Purpura, 1967; Frigyesi and Szabo, 1975). In studies of striatal actions on *nigral* neurons in unanesthetized monkeys (Feger and Ohye, 1975), about 50% had a caudate-evoked increase in discharge rate prior to a prominent inhibition. Again, anesthetics may be responsible for some differences in the data, but all authors seem to agree that at least some of the striatal fibers directly inhibit pallidal and nigral neurons. Whether or not some striatal efferent axons are excitatory is less certain.

Some nigral neurons inhibited by striatal stimulation are part of the basal ganglia efferent system, since they can be antidromically activated from the thalamus and/or the superior colliculus (Anderson and Yoshida, unpublished observations; Deniau, Feger, and LeGuyader, 1976). Guyenet and Aghajanian (1978) have reported that antidromically identified nigrostriatal neurons also are inhibited by stimulation of the striatum, and this is consistent with biochemical evidence that striatal output influences the activity of the nigral dopaminergic system (Racagni, Bruno, Cattabeni, Maggi, DiGulio, Parenti, and Groppetti, 1977).

THE NIGROSTRIATAL SYSTEM. Nigrostriatal axons originate primarily from neurons in the pars compacta region of the substantia nigra, and there is abundant evidence that at least a portion of these contain dopamine, both in the cell bodies and in their striatal axon terminals (Andén, Dahlström, Fuxe, and Larsson, 1964; Moore, Bhatnager, and Heller, 1971; Ungerstedt, 1971). Stimulation of the substantia nigra exerts synaptic actions on caudate neurons at rather long latencies (12–18 msec in the cat; Connor, 1970; Feltz and MacKenzie, 1969; Kitai, Wagner, Precht, and Ono, 1975; McLennan and York, 1967). The latency and configuration of PSPs recorded intracellularly remain constant in individual neurons, however, and at least some of these probably reflect the direct action of slowly conducting nigrostriatal axons (Kitai *et al.*, 1975).

There has been a considerable amount of disagreement in the literature regarding the action of the nigrostriatal system. All the evidence for an inhibitory nigrostriatal action has come from extracellular studies. Connors (1970) reported that 74% of the caudate neurons affected by nigral stimulation in his sample had an initial inhibition and 26%, an initial excitation. Richardson, Miller, and McLennan (1977) distinguished two groups of caudate neurons that had action potentials of different amplitudes in extracellular records from the rat and presented evidence that these two groups respond differently to nigral stimulation. The hypothesis of Richardson *et al.* (1977) is that one type of neuron is an output neuron and the other is an inhibitory interneuron. In contrast, Norcross and Spehlmann (1977, 1978) reported that nigral stimulation only facilitated and never depressed the firing rates of caudate neurons that they studied extracellularly in encéphale isolé cats, and investigators who have recorded intracellularly consistently have reported that nigral stimulation produced an initial EPSP in most caudate neurons, although a few cells had an initial inhibition (Fuller, Hull, and Buchwald, 1975; Kitai *et al.*, 1976b). The controversy over the action of the nigrostriatal system is paralleled by a controversy over the striatal action of dopamine, the presumed neurotransmitter of this system.

MARJORIE E.
ANDERSON

It is beyond the scope of this chapter to review extensively the large body of literature concerning potential transmitter substances and their actions in the basal ganglia. However, a brief discussion of the evidence that indicates where these transmitters might act in the systems discussed above is appropriate.

DOPAMINE

Dopamine (DA) is present in higher concentrations in the substantia nigra and the striatum than in any other part of the nervous system (Hornykiewicz, 1966). With the use of histochemical techniques (Andén, Carlsson, Dahlström, Fuxe, Hillarp, and Larsson, 1964; Fuxe, 1965), it has been shown that in the substantia nigra, DA is located in the cell bodies of neurons in the pars compacta, whereas in the striatum, it is restricted to axon terminals that disappear if ascending nigrostriatal fibers are destroyed (Hökfelt and Ungerstedt, 1969). If DA is the transmitter substance released by nigrostriatal fibers, then it would be expected that the effect of DA applied iontophoretically to striatal neurons on which nigrostriatal fibers terminate would be similar (excitatory or inhibitory) to the synaptic action produced by stimulation of the substantia nigra. In many extracellular studies, the actions of iontophoretically applied DA and nigral-evoked responses were not examined in the same neurons, but most investigators reported a predominantly inhibitory action of DA on caudate cells (Bloom *et al.,* 1965; Connor, 1970; Feltz and DeChamplain, 1972; and McLennan and York, 1967). Of these investigators, only Connor consistently examined nigral-evoked responses in the same neurons, and, as indicated earlier, nigral-evoked inhibitory responses in this study outnumbered excitatory responses by about 3 to 1. Of the caudate neurons examined with both nigral stimulation and iontophoretic application of DA, neurons depressed by nigral stimulation were consistently depressed by DA, but those facilitated by nigral stimulation showed mixed responses to iontophoretically applied DA (Connor, 1970). Norcross and Spehlmann (1977, 1978), the investigators who found that stimulation of SN only produced excitation of caudate cells, reported that DA applied iontophoretically only facilitated and never depressed the excitatory nigrostriatal action. They did find that DA sometimes depressed caudate excitatory responses following stimulation at other locations in the caudate nucleus, but the iontophoretic current necessary to produce the depression was high, and the dose-response curves for facilitation and depression produced by iontophoretically applied DA differed markedly. Norcross and Spehlmann (1978) have proposed that the two responses to DA may be the result of its actions on two different types of receptor sites. The presence of high- and low-affinity DA receptor sites has been suggested on the basis of biochemical studies, and other investigators have proposed that there are distinct excitatory and inhibitory striatal DA receptors (Cools, Struyker Boudier, and Van Rossum, 1976).

A major technical problem in interpreting the results of studies in which multibarreled extracellular electrodes are used for both recording and iontophoresis is that there is no guarantee that the drug-induced response is exerted directly on the neuron whose activity is being recorded. In fact, action potentials of many neurons can be recorded over an electrode excursion of 100 μm or more, and the caudate nucleus appears to have many interneurons via which the drug-evoked response could be mediated synaptically. The effects of iontophoretically applied drugs often appear only after many *seconds* of application (Bloom, Costa, and Salmoiraghi, 1965; Norcross and Spehlmann, 1978), and the drugs must diffuse considerable distances in this amount of time. Kitai *et al.* (1976b) attempted to reduce the probability of indirect action of iontophoretically applied DA by recording intracellularly from one barrel of a multibarreled pipette and applying pharmacological agents through other extracellularly located barrels. Using this technique, they found that iontophoretic application of DA caused a depolarization of caudate neurons, and with sufficient DA application, action potentials occurred. Thus, they concluded that the dopaminergic nigrostriatal action on caudate neurons was excitatory, and they also showed that SN-evoked EPSPs could be blocked by the iontophoretic application of chlorpromazine, a DA blocker. Investigators who record extracellularly, however, point out that adequate current neutralization is difficult to maintain, and in intracellular studies, this may result in current-induced direct actions on the neuron (Siggins, Hoffer, Bloom, and Ungerstedt, 1976). Further careful studies are needed to sort out the actions of dopamine and the nigrostriatal pathway. However, the marked loss of dopaminergic nigral neurons in the brains of individuals with parkinsonism and the alleviation of their hypokinetic signs by L-dopa therapy makes it apparent that this system must play an important role in achieving appropriate mobility. Dopamine, which is believed to act as the active synaptic transmitter substance, does not cross the blood-brain barrier, and hence its precursor, L-dopa, is administered systemically, usually in combination with peripheral decarboxylase inhibitors that reduce the conversion of dopamine to norepinephrine (Papavasilious, Cotzias, Duby, Steck, Fehling, and Bell, 1972).

ACETYLCHOLINE

Acetylcholine (ACh) also is a potential neurotransmitter in the basal ganglia, and its synthesizing enzyme, choline acetyltransferase (CAT) is present in particularly high concentrations in the caudate nucleus and putamen (Hebb and Silver, 1956). There is considerable evidence to indicate that a large portion of the striatal acetylcholine is associated with striatal interneurons (Butcher and Butcher, 1974; McGeer, McGeer, Fibiger, and Wickson, 1971), and McGeer, McGeer, Grewall, and Singh (1975) have presented evidence derived from combined immunohistochemical and 6-OH dopamine-induced degeneration techniques to show that at least some of the dopaminergic nerve endings in the striatum terminate on cholinergic neurons. It has been known for some time that anticholinergic drugs alleviate some of the symptomatology of parkinsonism, and it has been proposed (Barbeau, 1962) that normal basal ganglia function requires a balance between DA and ACh; if DA deficiency results in ACh overbalance, parkinsonian signs appear, and if DA overbalances ACh, signs of chorea appear.

Iontophoretically-applied ACh also had a mixed action on caudate neurons. In the study of Bloom *et al.* (1965) the response was largely dependent on the anesthetic state. In brain stem-transected animals without anesthesia, the large majority of neurons was facilitated after ACh application; in animals anesthetized with sodium pentobarbital, a smaller number of cells showed any response, but those that did were mostly depressed. The number of neurons whose activity was depressed in unanesthetized decerebrated animals increased markedly when glutamate was released iontophoretically and otherwise "silent" neurons were tested with ACh. McLennan and York (1967) also saw a mixed action of ACh, but they reported that its effect on caudate neurons varied in different regions of the head of the caudate nucleus. Again, the latency of response to iontophoretically released ACh is very long (as much as 10–15 sec), and it is not known whether the ACh acted directly on the cells studied or whether these cells were caudate efferent neurons, interneurons, or both.

Gamma-Aminobutyric Acid (GABA)

A third important potential neurotransmitter in the basal ganglia is GABA. GABA is present in the striatum, globus pallidus, and substantia nigra. In the nigra, its synthesizing enzyme, glutamate decarboxylase (GAD), is located in axon terminals that form axodendritic and axosomatic synapses (Ribak, Vaughn, Saito, and Barber, 1976). Since interruption of the striatonigral pathway markedly reduces GAD or GABA levels in the substantia nigra (Fonnum, Grofová, Rinvik, Storm-Mathisen, and Walberg, 1974; Kim, Bak, Hassler, and Okada, 1971), it is possible that GABA is the neurotransmitter for the striatonigral pathway.

Further evidence that GABA is the neurotransmitter mediating striatal-evoked inhibition of pallidal and nigral neurons has been obtained in electrophysiological experiments. Precht and Yoshida (1971) showed that striatal-evoked depression of nigral neuron firing could be blocked by picrotoxin, a drug known to block GABA actions in other systems (Robbins and VanDerKloot, 1958). Feltz (1971) also showed an inhibitory action of iontophoretically applied GABA, and both striatal and GABA-evoked inhibition were blocked by picrotoxin and by bicuculline, another GABA antagonist.

There also is evidence (Hattori, McGeer, Fibiger, and McGeer, 1973) that some of the GABA containing terminals in the substantia nigra are terminals of axons that originate in the globus pallidus and that some pallidal neurons send projections to the substantia nigra (Bunney and Aghajanian, 1976; Kanazawa, Marshall, and Kelly, 1976). Fonnum, et al. (1974), however, concluded that little, if any of the nigral GABA really is from axons of pallidal neurons.

Other Potential Neurotransmitters in the Basal Ganglia

Serotonin (5-hydroxytryptamine) and the polypeptides, substance P and enkephalin, are other potential neurotransmitters in the basal ganglia. It has been suggested in brief reports that serotonin may be an inhibitory transmitter for dorsal raphe axons terminating in the striatum (Olpe and Koella, 1977) and that substance P may be an excitatory transmitter released by axons (perhaps striatal) terminating in the substantia nigra (Walker, Kemp, Yajima, Kitagawa, and Woodruff, 1976). At present, a better understanding of the role of these substances in the basal ganglia awaits further experimental studies.

MODELS OF BASAL GANGLIA FUNCTION

Present models of basal ganglia synaptic interconnections and their relationship to the symptoms of motor dysfunction must be tenuous and incomplete, owing to the incomplete and often conflicting data outlined above. Nevertheless, attempts to model the system may highlight deficiencies or conflicts in the data, and it is useful to attempt to combine electrophysiological, pharmacological, pathological, and experimental lesion findings into a whole.

Several facts that seem clear, although certainly incomplete, are shown schematically in Fig. 6A. First, a major part of the afferent information enters the basal

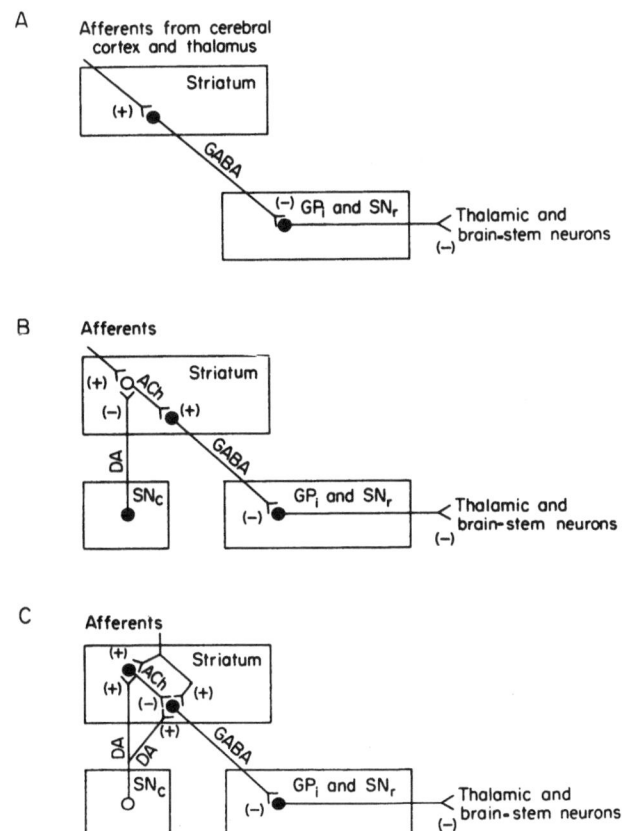

Fig. 6. Three simplified models of basal ganglia synaptic relationships. (A) Oversimplified model to emphasize basal ganglia afferents, efferents, and striatonigral system. (B) Model with ACh interneuron and proposed inhibitory nigrostriatal system added. (C) Model with excitatory nigrostriatal system, + represents excitatory synaptic action; − represents inhibitory action. Putative neurotransmitter substances (DA, GABA, ACh) are indicated for systems in which they have been studied.

ganglia via the striatum, and basal ganglia output leaves via pallidal and nigral neurons that have a high basic firing rate in awake animals. At least part of the striatal efferents, which have low rates of tonic activity in awake animals, inhibit pallidal and nigral neurons, perhaps via branching striatal axons that liberate GABA as a neurotransmitter (Feltz, 1971; Obata and Yoshida, 1973; Precht and Yoshida, 1971; Yoshida, Rabin, and Anderson, 1971). The high tonic firing of pallidal and nigral neurons would provide an appropriate background against which to detect increases in striatal inhibition of these basal ganglia "output" cells, and this would result in a disinhibition of target cells in the thalamus or other brain stem areas. Striatal inhibitory control of pallidal and nigral output would be consistent with the opposing signs of abnormal motor control seen in animals or humans with lesions of the striatum (chorea and other hyperkinetic signs) versus those with lesions of the pallidum and substantia nigra (the hypokinesia seen in animal studies).

The model of Fig. 6A is clearly incomplete, however, if for no other reason than that it leaves out the many known striatal interneurons, the nigrostriatal system(s), and interactions with the subthalamic nucleus.

The model of Fig. 6B adds elements that would be consistent with the following data: (1) ACh seems largely associated with striatal interneurons, (2) at least some dopaminergic nigrostriatal axons terminate directly on cholinergic striatal interneurons, and (3) some signs of parkinsonism can be at least partially reduced by either an *increase* in dopamine (L-dopa therapy) or a *decrease* of ACh (anticholinergic therapy). Figure 6B, however, uses the extracellularly derived iontophoretic findings that dopamine inhibits striatal neurons, and this is not consistent with the finding from intracellular studies that dopamine excites striatal cells.

The model of Fig. 6C would better suit those who contend that dopamine in the nigrostriatal dopaminergic system is excitatory, and it perhaps could account for the fact that L-dopa and anticholinergic therapy both can reduce parkinsonian signs, but L-dopa is more effective. However, it is difficult to see how this model could account for hypokinesia following *either* loss of the dopaminergic system *or* of the pallidal and nigral output systems. Furthermore, it does not account for the fact that ACh applied iontophoretically produces facilitation of discharge in some caudate neurons.

Thus, it is not possible with current data to describe a model that is consistent with all the reported electrophysiological, pharmacological, and lesion data. Subsequent studies must clarify some areas of conflict (synaptic actions of nigrostriatal fibers; excitatory or inhibitory action of dopamine in the nigrostriatal system) and must add data identifying the actions of systems not yet studied in detail (pallido-subthalamic-pallidal, raphe-striatal).

CONCLUDING REMARKS

Ultimately, it would be desirable to know both the role(s) played by the basal ganglia in achieving coordinated movement and the synaptic mechanisms by which such control is mediated. Experiments should proceed in both areas, but useful experiments necessary to clarify functional roles are probably more difficult to design than are those aimed at determining synaptic relationships.

One clear need is for quantitative descriptions of motor activity following basal ganglia lesions. Modern psychophysical techniques should be used to quantify various parameters of movement (amplitude, velocity, accuracy of end point) and, in combination with EMG recordings from prime movers, antagonists, and postural support muscles, to determine the temporal and spatial characteristics of movements performed with or without lesions of different components of the basal ganglia. These studies should include movements made without the mechanical postural support (such as elbow stabilization) usually used in experiments to date.

Ideally, one would like to be able to establish a cause and effect relationship between the discharge pattern of basal ganglia neurons and particular components or characteristics of motor activity. Any motor act, however, involves a complex series of neuromuscular events, any one or combination of which might be controlled by basal ganglia activity. Thus, when we examine basal ganglia unit activity during repeated, stereotyped, motor behavior, we have little chance of determining whether the activity of basal ganglia neurons is related to particular aspects of the complex motor pattern. If we are to determine whether there are particular aspects of motor activity in which basal ganglia neurons play an essential role, behavioral techniques must be used to elicit varying motor behaviors that occur at different rates, from different postural starting points, and in tasks that require cocontraction and stabilization, as well as reciprocal activation. By examining the discharge of basal ganglia neurons during varying patterns of movement we may be able to determine particular aspects of motor control in which the basal ganglia play an essential role.

REFERENCES

Albe-Fessard, D., Oswalds-Cruz, E., and Rocha-Miranda, C. Activités évoquées dans le noyau caudé du chat en résponse à des types divers d'afférences. I. Etude Macrophysiologique. *Electroencephalography and Clinical Neurophysiology*, 1960, *12*, 405–420.

Andén, N. E., Carlsson, A., Dahlström, A., Fuxe, K., Hillarp, N. A., and Larsson, K. Demonstration and mapping out of nigro-neostriatal dopamine neurons. *Life Sciences*, 1964, *3*, 523, 530.

Andén, N. E., Dahlström, A., Fuxe, K., and Larsson, K. Further evidence for the presence of nigro-neostriatal dopamine neurons in the rat. *American Journal of Anatomy*. 1965, *116*, 329–333.

Anderson, M. E. Discharge properties of basal ganglia neurons during active maintenance of postural stability and adjustment to chair tilt. *Brain Research*, 1977, *143*, 325–338.

Anderson, M. and Yoshida, M. Electrophysiological evidence for branching nigral projections to the thalamus and the superior colliculus. *Brain Research*, 1977, *137*, 361–364.

Barbeau, A. The pathogenesis of Parkinson's disease: A new hypothesis. *Canadian Medical Association Journal*, 1962, *87*, 802–807.

Basmajian, J. *Muscles Alive*. Baltimore: Williams & Wilkins, 1978.

Belen'kii, V., Gurfinkel', V., and Pal'tsev, Y. Elements of control of voluntary movements. *Biophysics*, 1967, *12*, 135–141.

Berrevoets, C., and Kuypers, H. Pericruciate cortical neurons projecting to brain stem reticular formation, dorsal column nuclei, and spinal cord in the cat. *Neuroscience Letters*, 1975, *1*, 257–262.

Bloom, F., Costa, E., and Salmoiraghi, G. Anaesthesia and the responsiveness of individual neurons of the caudate nucleus to acetylcholine, norepinephrine, and dopamine administered by microelectrophoresis. *Journal of Pharmacology and Experimental Therapeutics*, 1965, *150*, 244-252.

Buchwald, N., Price, D., Vernon, L., and Hull, C. Caudate intracellular response to thalamic and cortical inputs. *Experimental Neurology*, 1973, *38*, 311-323.

Buchwald, N., Hull, C., Levine, M., and Villablanca, J. The basal ganglia and the recognition of response and cognitive sets. In M. Brazier (Ed.), *Growth and Development of the Brain*, New York: Raven Press, 1975.

Bunney, B., and Aghajanian, G. The precise localization of nigral afferents in the rat as determined by a retrograde tracing technique. *Brain Research*, 1976, *117*, 423-435.

Buser, P., Ponderous, G., and Mereaux, J. Single unit recording in the caudate nucleus during sessions with elaborate movements in the awake monkey. *Brain Research*, 1974, *21*, 337-344.

Butcher, S. G., and Butcher, L. L. Origin and modulation of acetylcholine activity in the neostriatum. *Brain Research*, 1974, *71*, 167-171.

Carmel, P. W. Efferent projections of the ventral anterior nucleus of the thalamus in the monkey. *American Journal of Anatomy*, 1970, *128*, 159-184.

Carpenter, M. B. Anatomical organization of the corpus striatum and related nuclei. In M. Yahr (Ed.), *The Basal Ganglia, Research Publications: Association for Research in Nervous and Mental Disease* (Vol. 55). New York: Raven Press, 1976.

Carpenter, M. B., and McMasters, R. E. Lesions of the substantia-nigra in the Rhesus monkey. Efferent fiber degeneration and behavioral observations. *American Journal of Anatomy*, 1964, *114*, 293-320.

Carpenter, M. B., Whittier, J., and Mettler, F. A. Analysis of choreoid hyperkinesia in the Rhesus monkey. Surgical and pharmacological analysis of hyperkinesia resulting from lesions of the subthalamic nucleus of Luys. *Journal of Comparative Neurology*, 1950, *92*, 293-331.

Clausing, K. W., Anderson, M. E., and DeVito, J. L. Afferent connections of the globus pallidus of *M. mulatta* as determined by retrograde transport of horseradish peroxidase. *Society for Neuroscience Abstracts*, 1977, *3*, 36.

Connor, J. Caudate nucleus neurons: Correlations of the effects of substantia nigra stimulation with iontophoretic dopamine. *Journal of Physiology (London)*, 1970, *208*, 691-703.

Cools, A., Struyker Boudier, H., and Van Rossum, J. Dopamine receptors: Selective agonists and antagonists of functionally distinct types within the feline brain. *European Journal of Pharmacology*, 1976, *37*, 283-293.

Coulter, J. D., Ewing, L., and Carter, C. Origin of primary sensorimotor cortical projections to lumbar spinal cord of cat and monkey. *Brain Research*, 1976, *103*, 366-373.

Coyle, J. T., and Schwarz, R. Lesion of striatal neurons with kainic acid provides model for Huntington's chorea. *Nature*, 1976, *263*, 244-246.

DeLong, M. Activity of pallidal neurons during movement. *Journal of Neurophysiology*, 1971, *34*, 414-427.

DeLong, M. Putamen: Activity of single units during slow and rapid arm movements. *Science*, 1973, *179*, 1240-1242.

DeLong, M., and Strick, P. Relation of basal ganglia, cerebellum, and motor cortex units to ramp and ballistic limb movements. *Brain Research*, 1974, *71*, 327-335.

Deniau, J. M., Feger, J., and LeGuyader, C. Striatal evoked inhibition of identified nigrothalamic neurons. *Brain Research*, 1976, *104*, 152-156.

Deniau, J. M., Chevalier, G., and Feger, J. Electrophysiological study of the nigro-tectal pathway in the rat. *Neuroscience Letters*, 1978, *10*, 215-220.

Deniau, J. M., Lackner, D., and Feger, J. Effect of substantia nigra stimulation on identified neurons in the VL-VA thalamic complex: Comparison between intact and chronically decorticated cats. *Brain Research*, 1978, *145*, 27-35.

Denny-Brown, D. *The Basal Ganglia and Their Relation to Disorders of Movement*. London: Oxford University Press, 1962.

Denny-Brown, D., and Yanagisawa, N. The role of the basal ganglia in the initiation of movement.

In M. Yahr (Ed.), *The Basal Ganglia, Research Publications: Association for Research in Nervous and Mental Disease* (Vol. 55). New York: Raven Press, 1976.

Desmedt, J., and Godaux, E. Ballistic contractions in man: Characteristic recruitment pattern of single motor units of the tibialis anterior muscle. *Journal of Physiology (London)*, 1977, *264*, 673–693.

Dormont, S. F., and Ohye, C. Entopeduncular projections to the thalamic ventrolateral nucleus of the cat. *Experimental Brain Research*, 1971, *12*, 254–264.

Desiraju, T., and Purpura, D. Synaptic convergence of cerebellar and lenticular projections to thalamus. *Brain Research*, 1969, *15*, 544–547.

Dray, A., Gonye, T. J., and Oakley, N. R. Caudate stimulation and substantia nigra activity in the rat. *Journal of Physiology (London)*, 1976, *259*, 825–849.

Duvoisin, R. Parkinsonism: Animal analogues of the human disorder. In M. Yahr (Ed.), *The Basal Ganglia, Research Publications: Association for Research in Nervous and Mental Disease* (Vol. 55). New York: Raven Press, 1976.

Faull, R. L. M., and Mehler, W. R. Studies of the fiber connections of the substantia nigra in the rat using the method of retrograde transport of horseradish peroxidase. *Society for Neuroscience Abstracts*, 1976, *1*, 62.

Feger, J., and Ohye, C. The unitary activity of the substantia nigra following stimulation of the striatum in the awake monkey. *Brain Research*, 1975, *89*, 115–159.

Feltz, P. Gamma-aminobutyric acid and caudato-nigral inhibition. *Canadian Journal of Physiology and Pharmacology*, 1971, *48*, 113–115.

Feltz, P., and DeChamplain, J. Persistence of caudate unitary responses to nigral stimulation after destruction and functional impairment of the striatal dopaminergic terminals. *Brain Research*, 1972, *43*, 595–600.

Feltz, P., and MacKenzie, J. S. Properties of caudate unitary responses to repetitive nigral stimulation. *Brain Research*, 1969, *13*, 612–616.

Ferrier, D. *The Functions of the Brain*. London: Smith, Elder, 1876.

Filion, M., Harnois, C., and Guano, G. Electrophysiological study of the distribution of axonal branches of individual entopeduncular neurons in the cat. *Society for Neuroscience Abstracts*, 1976, *2*, 63.

Flowers, K. Ballistic and corrective movements on an aiming task. *Neurology*, 1975, *25*, 413–421.

Flowers, K. Visual "closed-loop" and "open-loop" characteristics of voluntary movement in patients with parkinsonism and intention tremor. *Brain*, 1976, *99*, 260–310.

Fonnum, F., Grofová, I., Rinvik, E., Storm-Mathisen, S., and Walberg, F. Origin and distribution of glutamate decarboxylase in substantia nigra of the cat. *Brain Research*, 1974, *71*, 77–92.

Frigyesi, T., and Machek, J. Basal ganglia-diencephalon synaptic relations in the cat. I. An intracellular study of dorsal thalamic neurons during capsular and basal ganglia stimulation. *Brain Research*, 1970, *20*, 201–217.

Frigyesi, T., and Purpura, D. Electrophysiological analysis of reciprocal caudatonigral relations. *Brain Research*, 1967, *6*, 440–456.

Frigyesi, T. L., and Rabin, A. Basal ganglia-diencephalon synaptic relations in cat. III. An intracellular study of ansa lenticularis, lenticular fasciculus, and pallido-subthalamic projection activities. *Brain Research*, 1971, *35*, 67–87.

Frigyesi, T., and Szabo, J. Caudate-evoked synaptic activities in nigral neurons. *Experimental Neurology*, 1975, *49*, 123–139.

Fuller, D. R. G., Hull, C. D., and Buchwald, N. A. Intracellular responses of caudate output neurons to orthodromic stimulation. *Brain Research*, 1975, *96*, 337–341.

Fuxe, K. Evidence for the existence of monoamine neurons in the central nervous system. *Acta Physiologica Scandinavica*, 1965, *64*, Supplementum 247.

Graybiel, A. M., and Sciascia, T. R. Origin and distribution of nigrotectal fibers in the cat. *Society for Neuroscience Abstracts*, 1975, *1*, 174.

Grimby, L., and Hannerz, J. Disturbances in the voluntary recruitment order of anterior tibial motor units in bradykinesia of parkinsonism. *Journal of Neurology, Neurosurgery, and Psychiatry*, 1974, *37*, 47–54.

Grofová, I. The identification of striatal and pallidal neurons projecting to substantia nigra. An experimental study by means of retrograde axonal transport of horseradish peroxidase. *Brain Research*, 1975, *91*, 286–291.

Guyenet, P. G., and Aghajanian, G. K. Antidromic identification of dopaminergic and other output neurons of the rat substantia nigra. *Brain Research*, 1978, *150*, 69–84.

Hallett, M., Shahani, B. T., and Young, R. R. EMG analysis of stereotyped voluntary movements in man. *Journal of Neurology, Neurosurgery, and Psychiatry*, 1975, *38*, 1154–1162.

Hartmann-von Monakow, K., Akert, K., and Künzle, H. Projections of precentral and premotor cortex to the red nucleus and other midbrain areas in *Macaca fascicularis*. *Experimental Brain Research*, 1979, *34*, 91–105.

Hattori, T., McGeer, P. L., Fibiger, H. C., and McGeer, E. G. On the source of GABA-containing terminals in the substantia nigra. Electron microscopic, autoradiographic and biochemical studies. *Brain Research*, 1973, *54*, 103–114.

Hebb, C. O., and Silver, A. Choline acetylase in the central nervous system of man and some other mammals. *Journal of Physiology (London)*, 1956, *134*, 718–728.

Hedreen, J. C. Corticostriatal cells identified by the peroxidase method. *Neuroscience Letters*, 1977, *4*, 1–7.

Hökfelt, T., and Ungerstedt, U. Electron and fluorescence microscopical studies on the nucleus caudatus putamen of the rat after unilateral lesions of ascending nigroneostriatal dopamine neurons. *Acta Physiologica Scandinavica*, 1969, *76*, 415–426.

Hopkins, D., and Niessen, L. W. Substantia nigra projections to the reticular formation, superior colliculus, and central gray in the rat, cat, and monkey. *Neuroscience Letters*, 1976, *2*, 253–259.

Hore, J., Meyer-Lohmann, J., and Brooks, V. B. Basal ganglia cooling disables learned arm movements of monkeys in the absence of visual guidance. *Science*, 1977, *195*, 584–586.

Hornykiewicz, O. Dopamine (3-hydroxytyramine) and brain function. *Pharmacological Reviews*, 1966, *18*, 925–964.

Hull, C. D., Levine, M. S., Buchwald, N. A., Heller, A., and Browning, R. A. The spontaneous firing pattern of forebrain neurons. I. The effects of dopamine and non-dopamine depleting lesions on caudate unit firing patterns. *Brain Research*, 1974, *73*, 241–262.

Jansen, J. S. On the functional properties of stretch receptors of mammalian skeletal muscles. In A.V.S. de Reuck and J. Knight (Eds.), *Myotatic, Kinesthetic, and Vestibular Mechanisms*, Ciba Foundation Symposium. Boston: Little, Brown, 1967.

Jayaraman, A., Batton, R., and Carpenter, M. B. Nigrotectal projections in the monkey: An autoradiographic study. *Brain Research*, 1977, *135*, 147–152.

Kanazawa, I., Marshall, G. R., and Kelly, J. S. Afferents to the rat substantia nigra studied with horseradish peroxidase, with special reference to fibers from the subthalamic nucleus. *Brain Research*, 1976, *115*, 485–491.

Kay, H. Analyzing motor skill performance. In K. Connolly (Ed.), *Mechanisms of Motor Skill Development*. New York: Academic Press, 1979.

Kemp, J., and Powell, T. The corticostriate projection in the monkey. *Brain*, 1970, *93*, 525–546.

Kemp, J. M., and Powell, T. P. S. The structure of the caudate nucleus of the cat: Light and electron microscopy. *Philosophical Transactions of the Royal Society (London), Series B*, 1971, *262*, 383–401.

Kennard, M. Experimental analysis of functions of the basal ganglia in monkeys and chimpanzees. *Journal of Neurophysiology*, 1944, *7*, 127–148.

Kim, J. S., Bak, I. J., Hassler, R., and Okada, Y. Role of γ-aminobutyric acid (GABA) in the extra-pyramidal motor system. 2. Some evidence for the existence of a type of GABA-rich strio-nigral neurons. *Experimental Brain Research*, 1971, *14*, 95–104.

Kitai, S. T., Wagner, A., Precht, W., Ohno, T. Nigro-caudate and caudate-nigral relationship: An electrophysiological study. *Brain Research*, 1975, *85*, 44–48.

Kitai, S. T., Kocsis, J. D., and Wood, J. Origin and characteristics of the cortico-caudate afferents: An anatomical and electrophysiological study. *Brain Research*, 1976, *118*, 137–141.

Kitai, S., Sugimori, M., and Kocsis, J. Excitatory nature of dopamine in the nigro-caudate pathway. *Experimental Brain Research*, 1976, *24*, 351–363.

Kornhuber, H. H. Motor functions of cerebellum and basal ganglia: The cerebellocortical saccadic (ballistic) clock, the cerebellonuclear hold regulator, and the basal ganglia ramp (voluntary speed smooth movement) generator. *Kybernetik*, 1971, *8*, 157–162.

Kostrzewa, R., and Jacobowitz, D. Pharmacological actions of 6-hydroxydopamine. *Pharmacological Reviews*, 1974, *26*, 199–287.

Laursen, A. Experimental study of pathways from the basal ganglia. *Journal of Comparative Neurology*, 1955, *102*, 1–25.

Levine, M., Hull, C. D., and Buchwald, N. A. Pallidal and entopeduncular intracellular responses to striatal, cortical, thalamic, and sensory inputs. *Experimental Neurology*, 1974, *44*, 448–460.

Macchi, G., Bentivoglio, M., D'Atena, C., Rossini, P., and Tempesta, E. The cortical projections of the thalamic intralaminar nuclei restudied by means of HRP retrograde axonal transport. *Neuroscience Letters*, 1977, *4*, 121–126.

Malliani, A., and Purpura, D. P. Intracellular studies of the corpus striatum. II. Patterns of synaptic activities in lenticular and entopeduncular neurons. *Brain Research*, 1976, *6*, 341–354.

Martin, J. P. Hemichorea resulting from a local lesion of the brain. *Brain*, 1927, *50*, 637–650.

Martin, J. P. *The Basal Ganglia and Posture*. Philadelphia: Lippincott, 1967.

Matsunami, K., and Cohen, B. Afferent modulation of unit activity in globus pallidus and caudate nucleus: Changes induced by vestibular nucleus and pyramidal tract stimulation. *Brain Research*, 1975, *91*, 140–146.

McGeer, E. G., McGeer, P. L., Grewall, D. S., and Singh, V. K. Cholinergic interneurons and their relation to dopaminergic nerve endings. *Journal of Pharmacology*, 1975, *2*, 143–152.

McGeer, P. L., McGeer, E. G., Fibiger, H. C., and Wickson, V. Neostriatal cholineacetylase and cholinesterase following selective brain lesions. *Brain Research*, 1971, *35*, 308–314.

McLennan, H., and York, D. The action of dopamine on neurons of the caudate nucleus. *Journal of Physiology (London)*, 1967, *189*, 393–402.

McNair, J., Sutin, J., and Tsubokawa, T. Suppression of cell firing in the substantia nigra by caudate nucleus stimulation. *Experimental Neurology*, 1972, *37*, 395–411.

Mettler, F. A. Effects of bilateral simultaneous subcortical lesions in the primate. *Journal of Neuropathology and Experimental Neurology*, 1945, *4*, 99–122.

Moore, R., Bhatnagcr, R., and Heller, A. Anatomical and chemical studies of a nigro-neostriatal projection in the cat. *Brain Research*, 1971, *30*, 119–135.

Morse, R. W., Adkins, R. J., and Towe, A. L. Population and modality characteristics of neurons in the coronal region of somatosensory area I of the cat. *Experimental Neurology*, 1965, *11*, 419–440.

Muskens, L. The central connection of the vestibular nuclei with the corpus striatum and their significance for ocular movements and for locomotion. *Brain*, 1922, *45*, 452–478.

Nauta, H. J. W., Pritz, B. B., and Lasek, R. J. Afferents to the rat caudatoputamen studied with horseradish peroxidase. An evaluation of a retrograde neuroanatomical research method. *Brain Research*, 1974, *67*, 219–238.

Nauta, J. H., and Mehler, W. R. Projections of the lentiform nucleus in the monkey. *Brain Research*, 1966, *1*, 3–42.

Norcross, K., and Spehlmann, R. Selective blockade of excitatory caudate responses to nigral stimulation by microiontophoretic application of dopamine antagonists. *Neuroscience Letters*, 1977, *6*, 323–328.

Norcross, K., and Spehlmann, R. A quantitative analysis of the excitatory and depressant effects of dopamine on the firing of caudatal neurons: Electrophysiological support for the existence of two distinct dopamine-sensitive receptors. *Brain Research*, 1978, *156*, 168–174.

Obata, K., and Yoshida, M. Caudate-evoked inhibition and actions of GABA and other substances on cat pallidal neurons. *Brain Research*, 1973, *64*, 455–459.

Ohye, C., LeGuyader, C., and Feger, J. Responses of subthalamic and pallidal neurons to striatal stimulation: An extracellular study on awake monkeys. *Brain Research*, 1976, *111*, 241–252.

Olpe, H. R., and Koella, W. P. The response of striatal cells upon stimulation of the dorsal and median raphe nuclei. *Brain Research*, 1977, *122*, 357–360.

Papavasiliou, P. S., Cotzias, G. C., Duby, S. E., Steck, A. J., Fehling, C., and Bell, M. A. Levodopa in parkinsonism: Potentiation of central effects with a peripheral inhibitor. *New England Journal of Medicine*, 1972, *286*, 8–14.

Parkinson, J. *An Essay on the Shaking Palsy.* London: Sherwood, Neely and Jones, 1817.

Péchadre, J., Larochelle, L., and Poirier, L. J. Parkinsonian akinesia, rigidity, and tremor in the monkey. *Journal of Neurological Sciences*, 1976, *28*, 147–157.

Poirier, L. J. Experimental and histological study of midbrain dyskinesias. *Journal of Neurophysiology*, 1960, *23*, 534–551.

Precht, W., and Yoshida, M. Blockage of caudate-evoked inhibition of neurons in the substantia nigra by picrotoxin. *Brain Research,* 1971, *32*, 229–233.

Racagni, G., Bruno, F., Cattabeni, F., Maggi, A., DiGulio, A. M., Parenti, A., and Groppetti, A. Functional interaction between rat substantia nigra and striatum: GABA and dopamine interrelation. *Brain Research*, 1977, *134*, 353–358.

Ranson, S. W., and Berry, C. Observations on monkeys with bilateral lesions of the globus pallidus. *Archives of Neurology and Psychiatry*, 1941, *46*, 504–508.

Ribak, C., Vaughn, J., Saito, K., and Barber, R. Immunocytochemical localization of glutamate decarboxylase in the substantia nigra of the rat. In M. Yahr (Ed.), *The Basal Ganglia, Research Publications: Association for Research in Nervous and Mental Disease* (Vol. 55). New York: Raven Press, 1976.

Richardson, T. L., Miller, J. J., and McLennan, H. Mechanisms of excitation and inhibition in the nigrostriatal system. *Brain Research,* 1977, *127*, 219–234.

Richter, R. Degeneration of the basal ganglia in monkeys from chronic carbon-disulfide poisoning. *Journal of Neuropathology and Experimental Neurology*, 1945, *4*, 324–353.

Rinvik, E., Grofová, I., and Ottersen, O. Demonstration of nigrotectal and nigroreticular projections in the cat by axonal transport of proteins. *Brain Research*, 1976, *112*, 388–394.

Robbins, J., and VanDerKloot, W. G. The effect of pictotoxin on peripheral inhibition in the crayfish. *Journal of Physiology (London)*, 1958, *143*, 541–552.

Scollo-Lavizzari, G., and Akert, K. Cortical area 8 and its thalamic projection in *Macaca mulatta. Journal of Comparative Neurology*, 1963, *121*, 259–269.

Sedgwick, E. M., and Williams, T. D. The response of single units in the caudate nucleus to peripheral stimulation. *Journal of Physiology (London)*, 1967, *189*, 281–298.

Selby, G. Stereotaxic surgery for the relief of parkinson's disease. Part 1. A critical review. *Journal of Neurological Science*, 1967, *5*, 315–342.

Siggins, G. R., Hoffer, B. J., Bloom, F. E., and Ungerstedt, U. Cytochemical and electrophysiological studies of dopamine in the caudate nucleus. In M. Yahr (Ed.), *The Basal Ganglia, Research Publications: Association for Research in Nervous and Mental Disease* (Vol. 55). New York: Raven Press, 1976.

Spiegel, E. A., and Baird, H. W. Athetotic syndromes. In P. Vinken and G. Bruyn (Eds.), *Handbook of Clinical Neurology*, Vol. 6: *Diseases of the Basal Ganglia.* New York: Wiley, 1968.

Stern, G. The effects of lesions in the substantia nigra. *Brain*, 1966, *89*, 449–478.

Stetson, R., and Bouman, H. The coordination of simple skilled movements. *Archives de Physiologie Neerlandia*, 1935, *20*, 177–254.

Strick, P. Anatomical analysis of ventrolateral thalamic input to primate motor cortex. *Journal of Neurophysiology*, 1976, *39*, 1020–1031.

Strominger, N. L., Truscott, T. C., Miller, R. A., and Royce, G. J. An autoradiographic study of the rubroolivary tract in the rhesus monkey. *Journal of Comparative Neurology*, 1979, *183*, 33–46.

Turner, Brian. Pathology of paralysis agitans. In P. Vinken and G. Bruyn (Eds.), *Handbook of Clinical Neurology*, Vol. 6: *Diseases of the Basal Ganglia.* New York: Wiley, 1968.

Ueki, A., Uno, M., Anderson, M., and Yoshida, M. Monosynaptic inhibition of thalamic neurons produced by stimulation of substantia nigra. *Experientia*, 1977, *33*, 1480–1481.

Ungerstedt, U. Brain dopamine and behavior. In F. Schmitt and F. Worden (Eds.), *The Neurosciences: Third Study Program.* Cambridge: The M.I.T. Press, 1974.

Ungerstedt, U. Stereotaxic mapping of the monoamine pathways in the rat brain. *Acta Physiologica Scandinavica*, 1971, *82* (Supplementum 367), 1–48.

Uno, M., Ozawa, N., and Yamamoto, K. Antidromic responses of thalamic VL neurons to cortical stimulation in cats. In M. Ito, N. Tsukahara, K. Kubota, and K. Yagi (Eds.), *Integrative Functions of the Brain* (Vol. 1). Tokyo: Kodansha, 1978.

Uno, M., Ozawa, N., and Yoshida, M. The mode of pallido-thalamic transmission investigated with intracellular recording from cat thalamus. *Experimental Brain Research*, 1978, *33*, 493–507.

Uno, M., and Yoshida, M. Monosynaptic inhibition of thalamic neurons produced by stimulation of the pallidal nucleus in cats. *Brain Research*, 1975, *99*, 377–380.

Villablanca, J., Marcus, R., and Olmstead, C. Effects of caudate nuclei or frontal cortical ablations in cats. I. Neurology and gross behavior. *Experimental Neurology*, 1976, *52*, 389–420.

Wacholder, K. Wilkürliche haltung und bewegung. *Ergebnisse Physiologie*, 1928, *26*, 568–775.

Walker, R. J., Kemp, J. A., Yajima, H., Kitagawa, K., and Woodruff, G. N. The action of substance P on mesencephalic reticular and substantia nigral neurons of the cat. *Experientia*, 1976, *32*, 214–215.

Wallin, B. G., Hongell, A., and Hagbarth, K. E. Recordings from muscle afferents in parkinsonian rigidity. In J. Desmedt (Ed.), *New Developments in Electromyography and Clinical Neurophysiology* (Vol. 3). Basel: Karger, 1973.

Ward, A. A., McCulloch, W. S., and Magoun, H. W. Production of an alternating tremor at rest in monkeys. *Journal of Neurophysiology*, 1948, *11*, 317–330.

Whittier, J. R. Ballism and the subthalamic nucleus. *Archives of Neurology and Psychiatry*, 1947, *58*, 672–692.

Wiesendanger, M., and Rüegg, D. G. Electromyographic assessment of central motor disorders. *Muscle and Nerve*, Sept/Oct. 1978, 407–412.

Woodworth, R. S. The accuracy of voluntary movement. *Psychological Review Monographs* (Suppl.), 1899, *3*, 1–114.

Yoshida, M., and Precht, W. Monosynaptic inhibition of neurons of the substantia nigra by caudato-nigral fibers. *Brain Research*, 1971, *32*, 225–228.

Yoshida, M., Rabin, A., and Anderson, M. Two types of monosynaptic inhibition of pallidal neurons produced by stimulation of the diencephalon and substantia nigra. *Brain Research*, 1971, *30*, 235–239.

8

The Pyramidal Tract

Its Structure and Function

MARIO WIESENDANGER

> *Peu de questions ont suscité autant de travaux que celles qui se rattachent à l'anatomie et la physiologie des pyramides antérieures.**
>
> Magendie, 1839

HISTORICAL DEVELOPMENT OF CONCEPTS ON PYRAMIDAL TRACT FUNCTION

Throughout the short history of experimental brain research, the pyramidal tract has attracted much interest from brain scientists (for a detailed account about the history of the pyramidal tract, see Lassek, 1954). Early research on the structure and function of the brain in the nineteenth century was most often performed in a clinical neurological setting; accordingly, the questions asked were mostly of a clinical nature. Hemiplegia or hemiparesis following a stroke, one of the most frequently encountered motor disorders, was commonly caused by a local impairment of the cerebral blood flow that affected volitional movements and often also speech and was thus a most dramatic illness. It was known since antiquity (Hippocrates, 460–377 B.C.) and rediscovered by Wepfer (1658; see Lassek, 1954) that hemiplegia was due to contralateral brain damage. The discovery by Pourfour du Petit (1710; see Lassek, 1954) of crossing fibers on the ventral side of the spinomedullary junction, the pyramid decussation, had far-reaching consequences; but 100 years

---

*No other questions generated so much work as those concerned with the anatomy and physiology of the anterior (bulbar) pyramids.

MARIO WIESENDANGER   Department of Physiology, University of Fribourg, Pérolles CH-1700 Fribourg, Switzerland. The author's work was supported by the Swiss National Science Foundation.

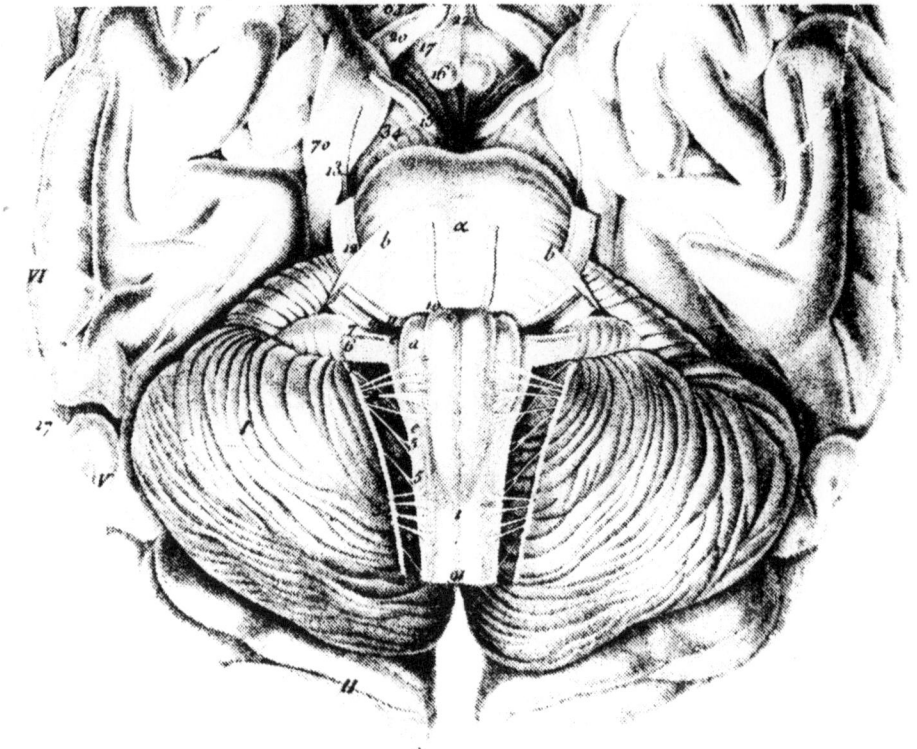

Fig. 1. Early picture of the basal aspect of the human brain by Gall and Spurzheim, published in "Anatomie et Physiologie du système nerveux en général, et celui du cerveau en particulier" (1810–1819) and reproduced in Clarke and Dewhurst (1974). The bulbar pyramids on the ventral aspect of the medulla are clearly seen.

went by before Gall and Spurzheim in 1810 (i.e., before their venture into phrenology earned them a bad reputation, see Clark and Dewhurst, 1974) dissected the fibers from the decussation upward to the cerebral cortex (Fig. 1). The further course of the fibers down to the spinal cord was documented by Türck (1851; see Lassek, 1954), who also introduced the term *Pyramidenstrang* (pyramidal fascile). Did Magendie (1834), who was the first to cut the medullary pyramid in living rabbits, expect to reproduce the symptomatology of hemiplegia? He described his completely negative result as follows: "Je n'ai point remarqué de lésion sensible dans les mouvements, et surtout je n'ai aperçu aucune paralysis, soit du côté lésé, soit du côté opposé."* Schiff (1858–1859), who operated with an improved technique (ventral approach), came to the same negative conclusion; namely that "diese Operation weder eine vorübergehende noch eine bleibende bemerkliche Lähmung notwendig nach sich zieht."†

However, the neurologist Cruveilhier (1853) first noted a correlation of hemiparesis with contralateral damage in the brain and a shrinkage of the medullary

*I did not observe a gross motor disturbance and in particular no paralysis contralaterally or ipsilaterally to the lesion.
†This operation produces neither a transient nor a permanent paralysis.

pyramid ipsilateral to the lesion. This report undoubtedly had an immense influence on the formation of concepts about pyramidal tract function. But it also led to a discrepancy between the views of clinicians and the views of investigators involved in lesion experiments. On the one hand, the neurologist had proof that atrophy of the bulbar pyramid followed long-standing cortical or capsular lesions (Fig. 2), and it was natural to conclude that the pyramidal tract constitutes the fiber connection between the higher motor centers, soon to be discovered by Fritsch and Hitzig (1870) and by Hughlings Jackson (1873/1958), and the spinal centers. The pyramidal tract was becoming synonymous for the pathway mediating volitional movements. Ultimately it was discovered that the "giant" pyramidal cells of Betz contributed to the pyramidal tract, and after Holmes and May's paper, published in 1909, it was universally believed that Betz cells were the sole origin of the pyramidal tract. On the other hand, physiologists continued to be disappointed by their lesion experiments. Following the pioneer experiments on rabbits, the bulbar pyramids were sectioned in rats (Barron, 1934; Castro, 1972), cats (Laursen and Wiesendanger, 1966; Liddell and Phillips, 1944; Marshall, 1934; Ranson, 1932; Tower, 1935), dogs (Morin, Donnet, and Zwirn, 1949; Schüller, 1906; Starlinger, 1897), hamster (Kalil and Schneider, 1975), brush-tailed possum (Hore, Phillips, and Porter, 1973) and monkeys (Schüller, 1906; Rothmann, 1907; Schäfer, 1910). In all these experiments, it became clear that voluntary movements were still performed by the lesioned animals. Even in monkeys, the motor deficits appeared to be minor when compared to those seen in patients suffering from a stroke.

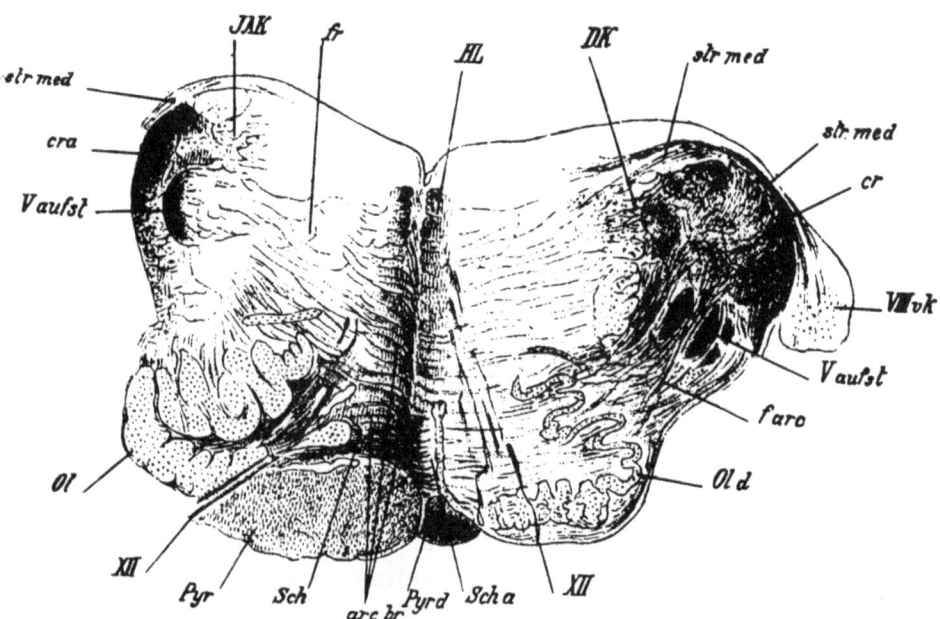

Fig. 2. Total degeneration of the right pyramid (pyr d) because of a large, long-standing lesion in the pericentral region of the ipsilateral cerebral hemisphere. There is also an atrophy (retrograde degeneration) of the right olivary nucleus (ol d) as a result of another lesion in the contralateral cerebellar hemisphere. From von Monakow (1897).

Tower (1940) was the first to describe the subtle but nevertheless important deficits of monkeys with pyramidal lesions. From her careful observations, she concluded that the main role of the pyramidal tract is to govern *discrete digital skill*. Although digital skill is undoubtedly an important, but not the sole, expression of voluntary action, the pyramidal tract continued to retain its attribute of the pathway for voluntary actions in many textbooks. The classification into *pyramidal* and *extrapyramidal* motor control originated from animal lesion experiments (Prus, 1898). The terms were soon adopted by clinicians for two broad categories of motor disabilities: those most commonly caused by a lesion in the internal capsule and those involving the basal ganglia and related structures—the *pyramidal* and the *extrapyramidal* syndrome, respectively. The belief that the extrapyramidal system has to do with involuntary movements was perhaps strengthened by the frequent occurrence of involuntary movements, such as tremor, choreiform, and ballistic movements.

Considerable progress was made in the last 20 and 25 years in unravelling the course and connections of fibers issuing from the electrically excitable motor cortex, and their input–output relations. These morphological investigations paralleled microelectrophysiological studies on pyramidal tract cells, the modifiability of their discharge rate by various afferent systems and effects as revealed by electrical stimulation. In the 1950s, with the introduction of the influential new method of recording from single cells in the brain of awake monkeys (Jasper, Ricci, and Doane, 1958), it was unequivocally shown that discharges of motor cortical cells were associated with conditioned movements. Moreover, it was convincingly shown by Evarts (1966) that activity in identified pyramidal tract neurons could precede the earliest detectable muscle activity in a reaction time task, as must be expected from central command neurons. The views on the role of the pyramidal tract found in current textbooks of medical physiology are that this tract has indeed to do with voluntary movements, expecially with digital skill, but that it is not the *only* tract transmitting the "command signals."

The aim of this chapter is to evaluate critically our present knowledge of the structure and function of the pyramidal tract, and integrate this detailed knowledge in a broader analysis of pyramidal tract function in the behaving animal or in man. The term *pyramidal tract* is thus defined as the fiber bundle traversing the medullary pyramid longitudinally.

It might be useful to start the review by a description of the well-known *pyramidal syndrome,* as understood by the clinician. Although it is now widely accepted that the signs and symptoms observed in patients with a typical pyramidal syndrome are not caused by interruption of only the pyramidal fibers, the clinical terminology is now so deeply established that an attempt to change it appears hopeless. The reality of the clinical syndrome has to be taken seriously by those engaged in basic research. The clinical picture demonstrates clearly that functions, especially such complex ones as the control of voluntary movements, are distributed in many components of the central nervous system (Edelman and Mountcastle, 1978). For instance, a lesion that destroys fibers in the internal capsule unavoidably will interrupt fiber systems of internal feedback loops engaging the basal ganglia, precerebellar nuclei, cerebellum, and thalamic relays, which are all likely to play an important role in the programming and execution of voluntary movements. Conceptually, "multicomponent lesions" may be as revealing as "single-component lesions." It is true that the pyramidal tract attains its highest development in humans, together with the cerebral cortex, but the same could be said for the basal ganglia, some of the thalamic nuclei, the inferior olive, the pontine nuclei, and the lateral cerebellum.

Although some physiologist think that lesion experiments should be abandoned altogether, because the rationale for deducing functions from deficits is weak, it is the reviewer's opinion that careful quantitative studies of motor deficits in lesion experiments are still valuable for giving the physiologist and the behavioral neurobiologist clues for formulating new hypotheses about motor control mechanisms, hypotheses that preferably may be tested in intact animals or perhaps also in healthy human subjects. A historical account of the role of stimulation and ablation, the most widely used methods in cerebral physiology up to the 1950s, was presented by Walker (1957).

> *The pyramidal system is such a human feature.*
> Sherrington, in Walshe, 1947

The "Pyramidal" Syndrome of Clinical Neurology

The Typical Deficits

No attempt will be made here to review the large body of literature on this subject, and the reader is referred to standard textbooks of neurology for details (see also special articles by Walshe, 1947; Brodal, 1973; Zülch, 1975, and the review by Lassek, 1954). As will be discussed briefly in the next section, the deficits largely depend on location and extent of the lesion and also on the time course of the disease. Essentially, the syndrome consists of a *paralysis* (total inability to perform volitional movements) or a *paresis* (partial inability to perform volitional movements). Most often, the distal muscles are more affected than the proximal ones ("predilection type" of Wernicke-Mann; Fig. 3), but the reverse has also been observed in cases of pontine tumors (Zülch, 1975). Volitional movements of the lower facial muscles (perioral) are more affected than the upper ones, and emotional expressions (believed to be controlled by the thalamus and the pallidum) may be preserved. Recently, Jung and Dietz (1975) found increased reaction times on the paretic side as compared to the normal side in hemiparetic patients. Weakness of the muscles is not associated with signs of denervation, as is the case in lower motor neuron disease, and wasting is usually not prominent (except in cases of severe disuse). *Exaggerated stretch reflexes* constitute a major symptom of the pyramidal syndrome. *Tendon reflexes,* elicited with the reflex hammer, are not only abnormally brisk but also involve more muscles than in healthy subjects. For instance, the patellar reflex may include not only the quadriceps muscle but also the adductor muscles. A rapid and then maintained dorsiflexion of the foot can produce repetitive stretch reflexes known as ankle *clonus.* Rapid stretching of any

MARIO
WIESENDANGER

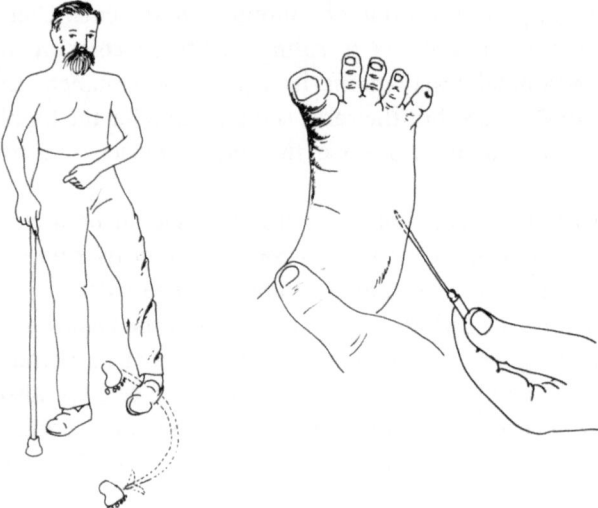

Fig. 3. Spastic posture and gait in a patient with a left-sided hemiparesis of the type described by Wernicke and Mann. Note circumduction of the leg and flexion posture of the arm. Stroking the foot sole produces a positive sign of Babinski, that is, fanning of the toes and dorsiflexion of the big toe. After photographs, drawn by R. Wiesendanger.

muscle may be felt as muscle resistance or muscle *tone*. As already described by Sherrington (1909) for the decerebrate cat, the resistance suddenly yields when a certain tension is produced (clasp-knife phenomenon). All these phenomena are components of the general term *spasticity*. In hemiplegic patients, spasticity is most pronounced in the arm flexors and the leg extensors. With spinal cord lesions,

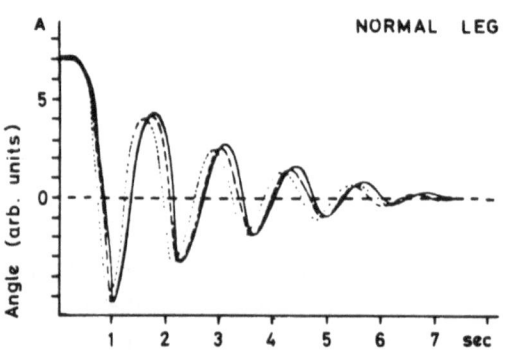

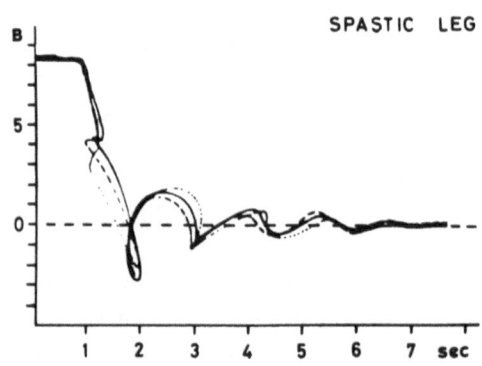

Fig. 4. Pendulousness test of Wartenberg. Mechanographic registration of toe movement by means of a camera which rotates at constant speed orthogonally to the swinging leg with a light source fixed on the big toe. The first downward deflection corresponds to the first downward swing of the leg after sudden release of the foot support. (A) Damped curve of a normal person. Total time of oscillations was about 7 sec. (B) Curve from a patient with spasticity of spinal origin. Note small jerk on the first downward swing of the leg which is caused by an exaggerated phasic stretch reflex (patellar tendon jerk). Irradiation of the stretch reflex to adductors results also in lateral movements of the leg which are seen as loops on the tracing. Increased muscle tone causes an abnormal damping of the swinging leg. Camera angle velocity is the same in A and B. Three superimposed consecutive trials. Redrawn after Wiesendanger and Mumenthaler (1959).

spasticity of flexors often dominates that of extensors. Spasticity can be evaluated semiquantitatively and in a simple way by means of the pendulousness test of Wartenberg (Fig. 4). Clinically, one relies mainly on asymmetries of tendon reflexes (left-right in hemispasticity, upper-limb and lower-limb in paraspasticity) and on muscle tone estimated by rapid stretching of muscle groups. Responses to exteroceptive, nociceptive, or innocuous stimuli may also be exaggerated in a pyramidal syndrome. This is often the case in patients with spinal cord lesions. A light stimulus, such as contact with a bed sheet, may trigger a complex flexion synkinesia, sometimes accompanied by a crossed extension or even a few cycles of alternate flexions and extensions. Light stroking of the abdominal wall normally elicits a reflex contraction of the abdominal muscles (abdominal reflexes), and stroking the medial aspect of the thigh a reflex lifting of the testicles (cremasteric reflex); these exteroceptive, polysynaptic reflexes are typically depressed or absent in patients with a pyramidal syndrome. An important sign is the appearance of the positive *Babinski reflex:* scratching the lateral part of the foot sole elicits dorsal extension of the big toe with fanning out of the toes (Fig. 3). Normally, the same stimulus results in a plantar flexion of the toes, that is, in the *plantar reflex.*

Volitional control of bowel function, micturition, sexual activity all depend on supraspinal control and may thus be disturbed in patients with a pyramidal syndrome.

TIME COURSE

Sudden onset of a *pyramidal syndrome* because of a traumatic lesion or a cerebrovascular insult is characterized by an initial flaccid paralysis. Gradually, over several weeks or months, some voluntary movements of proximal muscles may again be restored, accompanied by a gradually increasing spasticity. Acute spinal transection is followed by a complete loss of excitability of the cord, with a lack of segmental reflexes. This period, called the *spinal shock* phase, also gradually develops into a stage of spastic paraplegia. Such patients are especially prone to severe trophic changes if appropriate treatment and care are not provided.

In the presence of a slowly growing brain or spinal cord tumor, the deficits may be compensated for over a relatively long period and the signs and symptoms, such as spasticity, may be discrete even if the lesion is large.

ANATOMO-CLINICAL CORRELATION

Basically, the same deficits may be seen with lesions in the crebral cortex, in the internal capsule (the most frequent site of a vascular insult), in the brain stem, or in the spinal cord. Collectively, one uses the term *upper motor neuron disease.* The task of the neurologist is to determine exactly the location of the lesion by considering the combination of symptoms. Thus, a *lower motor neuron* paralysis of a cranial nerve (with signs of muscle denervation) combined with a contralateral *upper motor neuron* hemiplegia is indicative of a brain stem lesion (at the level of the affected cranial nerve). Pure motor deficits are relatively rare but may be present in cases of motor cortical lesions or medullary pyramid lesions (see Chokroverty, Rubino, and Haller, 1977; Fischer and Curry, 1965; Leestma and Noronha, 1976). Paraplegia is typical for a spinal lesion, but may also develop in cases of tumors in the midsagittal plane near the motor cortex.

The most important question to be asked in this chapter is whether or not the deficits in motor control are correlated with a lesion of the pyramidal tract, as has been traditionally assumed. Lassek (1954) showed that the degree of cerebral palsy does not correlate well with the amount of secondary degeneration of fibers in the medullary pyramids. In 11 postmortem cases presenting a total loss of pyramidal tract fibers, the reported motor deficits varied from a mild to a more severe paralysis. Furthermore, Lassek analyzed 331 human cases and came to the conclusion that "the majority of paralytic cases occur with little or no destruction of pyramidal tract fibers." In cases of space-occupying, progressive diseases in which motor deficits insidiously increase, degeneration of the pyramidal tract fibers was particularly rare. With respect to a positive sign of Babinski, Lassek noted that it can develop in the absence of destruction of pyramidal tract fibers. These statements cannot be adopted without caution: Lassek used an indirect method (counting "surviving" fibers); van Crevel (1958) and Russel and DeMyer (1961) showed, however, that axolysis of small fibers takes up to one year following a lesion. Thus, pyramidal tract fibers, considered to be normal by Lassek, may not, in fact, have been functional.

THE QUESTION OF PURE PYRAMIDAL TRACT LESIONS IN HUMANS

As noted before, lesions in the human neuropathological material are usually complex. The pyramidal tract fibers "are hopelessly contaminated with other functional systems" (Patton and Amassian, 1960), except at the level of the bulbar pyramid. Ischemic lesions of the medullary pyramids are rare, but cases have recently been described by Meyer and Herndon (1962), Chokroverty, Rubino, and Haller (1975), and by Leestma and Noronha (1976). In the first case, the infarct was bilateral and resulted in quadriplegia, first flaccid, then spastic. Although described as a "pure pyramidal lesion," reconstruction of the lesion, as published by Meyer and Herndon (1962), revealed a considerable involvement of the medial reticular formation of the medulla. The case of Leestma and Noronha appears to be purer, although some lemniscal and reticular involvement cannot be excluded. These cases all had a severe contralateral palsy which was described as flaccid in two cases and slightly spastic in one case.

Complementary to these clinical observations is a report of a case representing virtually a "pyramidal preparation" (Wiesendanger, 1973): a malignant glioma had destroyed a large part of the medulla and the caudal pons, sparing the pyramids. The patient developed multiple signs of cranial nerve involvement. The day before his death, the patient suffered respiratory troubles and a slight, transient hemiparesis. That same evening, the patient was still conscious and when asked, was able to move his limbs symmetrically.

These latter observations are difficult to interpret. At first sight, they emphasize the important role played by the pyramidal tract in the mediation of the voluntary commands. Unfortunately, the deficits in these few patients have been described conventionally and not in very much detail. They do not agree with observations made in patients subjected to section of the middle third of fibers in the cerebral peduncle with the aim to abbolish hemiballistic movements (Bucy, Keplinger, and Siqueira, 1964; Walker, 1949). Bucy and co-workers found that their patients not only lost their involuntary movements but postoperatively regained a remarkably good control of voluntary movements. A postmortem investigation of the brain stem of one patient, 2½ years after the operation, revealed that only about 17% of the fibers in the pyramid were not degenerated. Surprisingly, Bucy *et al.* (1964) claimed that this patient had "an almost normal volitional control" of limbs contralateral to the peduncular lesion.

One has to emphasize that these patients had a severe motor disability *before* the operation and that the description of the improvement in motor control was naturally the basic issue of this paper, rather than a refined analysis of what the patient could not do.

Finally, there are a few reports on unilateral (Balthasar and Schlagenhauff, 1966) or bilateral (Bubis and Landau, 1964) congenital agenesis of the pyramidal tract. These cases were investigated in detail postmortem, but the clinical description is fragmentary. It appears that the paresis was slight in all cases.

To summarize the main results of human neuropathology one can assert that

1. The pyramidal tract is involved in the mediation of voluntary movements as evidenced from a patient who was able to perform voluntary movements after a tumor had virtually interrupted all descending fibers except those in the bulbar pyramids. The evidence comes also from patients who suffered a severe paralysis after a sudden brain-stem lesion which was mainly restricted to the bulbar pyramid.
2. It also appears that systems other than the pyramidal tract are endowed with similar functions as the pyramidal tract. These systems seem to be able to perform a large amount of voluntary control, especially in those cases with long-standing or congenital lesions of the pyramids.
3. There is no clear-cut correlation between the severity of paralysis and degeneration of the pyramids.

> *Though we are well aware that the method of ablation studies has many obvious pitfalls we are convinced that this is still the only method to seek the general principles of control of movement by the brain.*
>
> Denny-Brown, 1966

EFFECTS OF PYRAMIDAL TRACT LESIONS IN ANIMAL EXPERIMENTS

VERBAL DESCRIPTION OF DEFICITS

Gross observations reveal few deficits following pyramidal tract section in nonprimate mammals. Therefore, only experiments performed on monkeys will be considered. The first lesion experiment in a monkey was reported early in this century. Rothmann (1907), who sectioned the pyramidal tract fibers at the decussation, reviewed his own results and that of others. Even in cases with complete section of the pyramidal tract fibers, motor behavior was reported to be almost normal after a few days of postoperative recovery. However, a histological documentation was not given by the author. Rothmann claimed that the finest finger movements were possible, but that the movements were considerably slower than before the operation. Because of this slowing of movements and because a more serious deficit was apparently seen immediately after surgery, the author did not dismiss the current view that the pyramidal tract plays a role in motor control, but he assumed that "extrapyramidal" pathways (Monakow's rubrospinal tract and a fiber bundle in the ventral funiculus) would take over voluntary control of movements. Schäfer (1910) destroyed one bulbar pyramid completely in three monkeys be electrocautery (fine curved platinum wire placed on the ventral surface of the pyramid). The symptoms following this lesion seemed to have been more serious than those described by Rothmann (1907): grasping of currants was not possible any longer with the affected hand, or was performed "feebly and clumisily." In a detailed study, Tower (1940, 1944) also emphasized the role of the pyramidal tract in "the discrete usage of the musculature, especially of the digits, which is characteristic of the order." Apart from the "phasic or episodic," roles of the pyramidal tract, essential for rapid initiation and performance of movements, the pyramidal tract was also considered to exert a tonic function providing "for smooth, continuous, efficient action." Tonic functions of the pyramidal tract were deduced from the following observations in operated animals: hypotonus in flexor muscles, difficulty in relaxing a grasp, lack of placing, and hopping reactions. In chimpanzees, "the paresis is outstanding," and, a bit further on, Tower (1944) describes the unilaterally lesioned chimpanzee as follows:

> Again, as the animals move about a large cage, swinging from bar to bar, grasping alternately with each hand or foot, the paretic extremity swings like a flail, unchecked by tonic innervation, while a normal extremity is always visibly in tone. (p. 166)

Tower's observations and her interpretation of the results are widely cited in the literature (Fulton, 1949). More recent "clinical" descriptions of monkeys with pyramidal lesions have added but little to that of Tower (Denny-Brown, 1966; Gilman and Marco, 1971; Goldberger, 1969; Woolsey, Gorska, Wetzel, Erickson, Earls, and Allman, 1972). Muscle tone and weakness of particular muscle groups are extremely difficult to assess in untrained animals; conflicting reports are therefore not surprising.

In recent years, an effort was made to quantify deficits (or remaining capacities) of pyramidotomized animals in view of the unavoidable subjective bias when interpreting gross observations. A simple test, like "shaking hands" of operated dogs (Starlinger, 1897), or a more complex task like picking up currants from the floor by monkeys (Schäfer, 1910) had been used a long time ago. Although testing of motor deficits with standardized conditioning techniques was in use already at the turn of the century (see Bechterew, 1923), pyramidal lesions were only evaluated by means of operant conditioning techniques much later.

Stretch-evoked electromyographic activity was investigated in various muscles of lesioned animals. In accordance with the clinical observation of hypotonia (rather than spasticity), the electromyographic responses to stretch were weaker on the side contralateral to the lesion, especially in flexor muscles (Gilman and Marco, 1971; Wiesendanger, 1973). Depressed myotatic reflexes were probably caused by a reduced gamma bias exerted on muscle spindles (Gilman *et al.*, 1971). The myotatic reflex responses were elicited by manual stretching of muscles (passive joint movements). A more quantitative assessment of stretch-evoked responses in pyramidotomized monkeys remains to be done.

SEMIQUANTITATIVE EVALUATION OF DIGITAL SKILL BY MEANS OF KINEMATOGRAPHY

The capability of monkeys to pick up food morsels from holes of varying diameter (Klüver board; see Glees, 1961) was observed and filmed in a large series of animals by Lawrence and Kuypers (1968). The most important outcome of this study was that the monkeys, although using their hands for grasping food, did so in a less delicate way than normal monkeys. The precise picking up of a food morsel from small holes with the index finger and the thumb ("precision grip," Napier, 1962) was no longer performed. The animals were more successful in retrieving food from the larger holes of the Klüver board, the fingers then "closing in concert" ("power grip," Napier, 1962). These observations have been confirmed (Hepp-Reymond and Wiesendanger, unpublished).

OPERANT CONDITIONING TECHNIQUES FOR THE EVALUATION OF MOTOR DEFICITS IN ANIMALS WITH PYRAMIDAL TRACT LESIONS

Conditioning techniques have been widely used to evaluate the performance of a motor task, the instrumental response. Experiments designed to test animals with pyramidal tract lesions were usually based on alimentary (food) reinforcement, whereas classical (Pavlovian) conditioning has not been used.

INVESTIGATION IN RATS. Castro (1972) counted successful retrievals of small food pellets. To obtain the food, the animals had to extend their forepaw into a slot to grasp the food. Reaching through the slot and touching the food was counted as an attempt; a success was scored when the rat grasped the pellet and placed it into its mouth. After bilateral and total lesions involving not only the pyramids but also virtually all of the lemniscal fibers (3 animals), the animals showed a large drop in performance in both attempts and percentage of success. All other animals tested had incomplete pyramid lesions and a variable amount of additional lesion of lemniscal fibers. This complication makes it difficult to evaluate the role played by the pyramidal tract in the observed motor deficits in this study.

INVESTIGATION IN DOGS. Gorska (1967) reported the effects of pyramidotomy in dogs subjected to Type II conditioning (Konorski, 1967). If a conditioning auditory stimulus was followed by a conditioned response, the dogs were reinforced with food. Alternatively, the dogs performed the conditioned movement to avoid electric stimulation of skin nerves. The conditioned (instrumental) responses were simple motor acts: lifting either the forelimb or the hind limb or rubbing the cheek with the forelimb. All these motor responses were either preserved after pyramidotomy or were soon restored. In naive pyramidotomized dogs, all motor responses could be established, indicating that the pyramidal tract is not essential for learning such simple motor acts. A consistent effect of pyramidotomy was, however, a delay in the execution of the response (the animals were not especially trained to respond quickly). Ioffe (1973a) used a similar experimental paradigm. In response to presentation of the food, the dog had to learn to lift the forepaw and to maintain it in a flexed position. With this maintained postural task, a food tray was lifted to various heights with respect to the dog's head. It was found that the leg was maintained in the flexed position also when snatching the food, but only if the tray was at a critical height so that the head had not to be lowered. Thus, it appears that the pyramidal tract was important for maintaining a complex postural adjustment in this study, but not for the execution of limb flexion *per se*. In another study (Ioffe, 1973b), the EMG amplitude of a "reflex response" (an index of motoneuronal excitability) was measured when it was applied in the interval between a warning stimulus (conditioning stimulus) and an avoidance or escape response. The latter consisted in lifting of the hind limb to avoid or to escape a noxious electrical stimulus (unconditioned stimulus) occurring 500 msec after the warning stimulus. It was found that, in normal dogs, the reflex response increased in magnitude the closer the reflex response occurred with respect to the conditioned hind-limb movement. After pyramidotomy, the conditioned response, hind-limb flexion, was again present but occurred at a longer mean response latency (measured from the warning stimulus). The premovement excitability test (amplitude plots of the EMG reflex response) was also altered after pyramidotomy: either the gradual increase of excitability before the movement was less pronounced (two dogs) or, first increased and then decreased again, before movement onset (two dogs). The latter result was also obtained in two dogs with combined lesions of the pyramidal tract and the red nucleus. The author concluded that during the period immediately preceding the conditioned movement (60–80 msec), the pyramidal tract and the rubrospinal tract are "presetting" the excitability of motoneurons involved in the motor task. It is noteworthy that the behavioral response latency again was increased in this experimental paradigm.

INVESTIGATION IN CATS. Normally occurring automatic movement sequences, such as the scratch reflex, the cleaning reaction, and rubbing the cheek, all requiring postural adjustments, were reinforced with food and could then be elicited by a conditioning stimulus alone, the sight of food (Gorska, Jankowska, and Mossakowski, 1966b). During the process of "instrumentalization," the motor task steadily became simpler (symbolic) in comparison to the unconditioned reaction. Transection of the bulbar pyramid was reported to have a deteriorating effect on these simple motor reactions. This is most surprising, since the same authors found little effect of pyramidotomy in dogs subjected to a similar task and since, in another report (Gorska, Jankowska, and Mossakowski, 1966a), it was noted that fine manipulatory movements in cats suffered very little when the medullary pyramids were lesioned. Similar experiments in which the conditioned scratch reflex and cleaning reaction (extension of the hind limb) were tested, revealed no loss of this type of motor task (Laursen and Wiesendanger, 1966b; Wiesendanger and Tarnecki, 1966). In another study (Voneida, 1967) involving the withdrawal of a forelimb to avoid an electric shock delivered to the grid of the experimental chamber, lesioning of the pyramidal tract only slightly and transiently affected the motor response.

In view of repeated observations of postoperative slowing of movements in older reports (see Marshall, 1934), an attempt was made to quantify the speed of a conditioned motor task (Laursen and Wiesendanger, 1967). This appeared necessary, since no detectable slowing of movements was seen in cats when observed in their free behavior (Laursen and Wiesendanger, 1966b). It was found that repetitive lever pressing tested with a fixed ratio schedule was not altered; that is they were performed at the same rate, after unilateral or bilateral pyramidotomy. However, it was found that lever pressing in response to a visual two-choice discrimination was significantly slowed. The increase in response latencies was related to the amount of pyramidal tract lesioned. Even more significantly, differential reinforcement of short response latencies failed to improve the performance after pyramidotomy, whereas preoperatively this training procedure was quite effective in reducing the median response latency (Fig. 5).

The aforementioned results of Ioffe (1973a) indicated that the pyramidal tract may be important in adapting posture to movements. It is clear that, especially in quadrupeds, each manipulatory movement of one limb must be accompanied by a subtle adjustment of posture. Massion and collaborators (Regis, Trouche, and Massion, 1976) devised an experimental test which allowed measurements of the redistribution of body weight to be taken during unconditioned paw lifting (evoked by a brief train of cortical stimulation). The cats were standing on a platform. Changes in force under each paw were measured separately. The decrease in force exerted by the paw which was lifted off the ground occurred often simultaneously or even following weight shifts measured under the three other paws. (Gahéry and Nieoullon, 1978). This was taken to indicate that a cortical mechanism was set in action which may simultaneously control the movement and adjust the posture.

Fig. 5. Distributions of response latencies (lever pressing) in a brightness discrimination task performed by a cat before and after bulbar pyramidotomy (Laursen adn Wiesendanger, 1967). The histograms in (B) and (D) are obtained wtih selective reinforcement for short latencies. In the normal cat, this procedure decreases markedly the median value (hatched column). After pyramidotomy, the distribution was flatter (C) and the cat could no longer decrease the response time when short latencies were selectively reinforced (D).

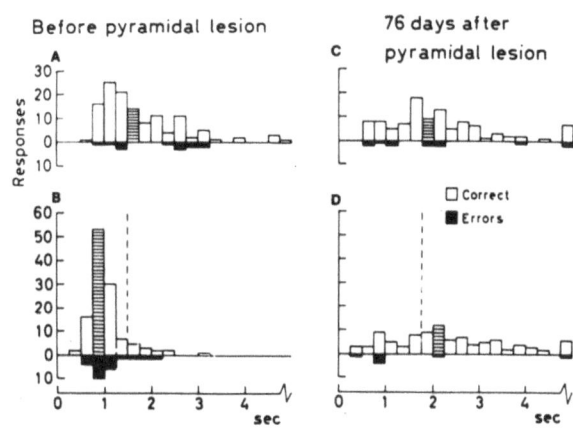

This hypothesis was tested more directly in cats subjected to the bulbar pyramidotomy (Nieoullon and Gahéry, 1978). Postoperatively, stimulation of the motor cortex still elicited a flexion movement in contralateral fore- or hind limbs at about the same thresholds. Obviously, a redistribution of body weight has then also to take place. The main change was an increase in weight-shift latency (with an increased scatter in the latency histograms) and a reduction in amplitude and speed of the weight shifts. This change was more marked for forelimbs as compared to hind limbs.

INVESTIGATIONS IN MONKEYS. The failure of pyramid-sectioned monkeys to use the index finger in isolation to retrieve food from small holes (Lawrence and Kuypers, 1968), was studied in a more formal way by Beck and Chambers (1970) by measuring the capability of their lesioned monkeys to reach a key placed at varying distance from a small hole, thus requiring the animal to extend the index finger and to flex the other fingers. The monkeys had particular difficulty in flexing these fingers, but whether this is because of a general impairment in flexion or an inability of the fingers to act independently from the index fingers was not clear. It was noted in this study that those animals with the largest involvement of the medial lemniscus also had the most serious deficit "in controlling accuracy of movement and in opening the paw or bringing food to the mouth." The same monkeys had increased choice reaction time. Slowing of both choice and simple reactions was noted by Laursen (1971), even if the lesion of the bulbar pyramid comprised as little as 25% of the fibers. Hepp-Reymond and Wiesendanger (1972) and Hepp-Reymond, Trouche, and Wiesendanger (1974) trained monkeys to perform a precision grip with the index and the thumb, the most "pyramidal" motor task so far tested under rigid operant-conditioning criteria. In one type of experiment, the rise time was measured as well as the duration of phasic EMG activity necessary to attain various force levels. In another type of experiment, the threshold force level was kept constant and the EMG and mechanical response latencies to a visual signal (preceded by a warning auditory stimulus at random intervals) were mea-

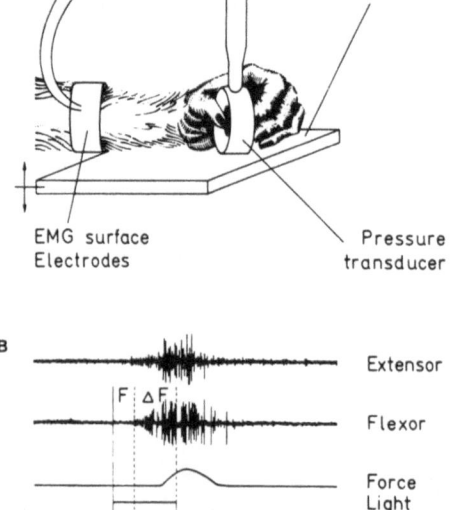

Fig. 6. Evaluation of motor skill (precision grip) in monkeys subjected to bulbar pyramidotomy. (A) Motor task consists of compressing a small wafer mounted in a manipulandum containing a pressure transducer. Surface EMG activity is recorded from hand extensors and flexors. (B) Typical recording of one trial showing phasic EMG bursts (co-contraction) and force curve. Reinforcement occurs when a prescribed force threshold is attained. The first vertical bar indicates occurrence of the visual "go" signal. Measured parameters are the response latency (RL), the EMG onset latency (F) and the EMG duration to force threshold (ΔF). In other experiments, the build-up time to force threshold was measured with various threshold levels. From Hepp-Reymond *et al.* (1974).

sured. Surprisingly, the rather precise manipulation (Fig. 6) was performed again by all monkeys, even by the animal with a total pyramidal lesion. Certainly, the deficits were serious in the first postoperative days, and the animals did not succeed in activating the manipulandum, but recovery took place rather quickly, that is, within 1 to 3 weeks of systematic postoperative retraining. The persistent and measurable deficit was an increased rise time of the force curve. This was particularly prominent when monkeys had to squeeze up to 700 g (the highest force thresholds for consistent successful responses in normal monkeys); the "build-up" time of the EMG activity was likewise increased (Fig. 7).

Animals trained to perform the precision grip as quickly as possible in response to a light signal (using a differential reinforcing procedure for rapid

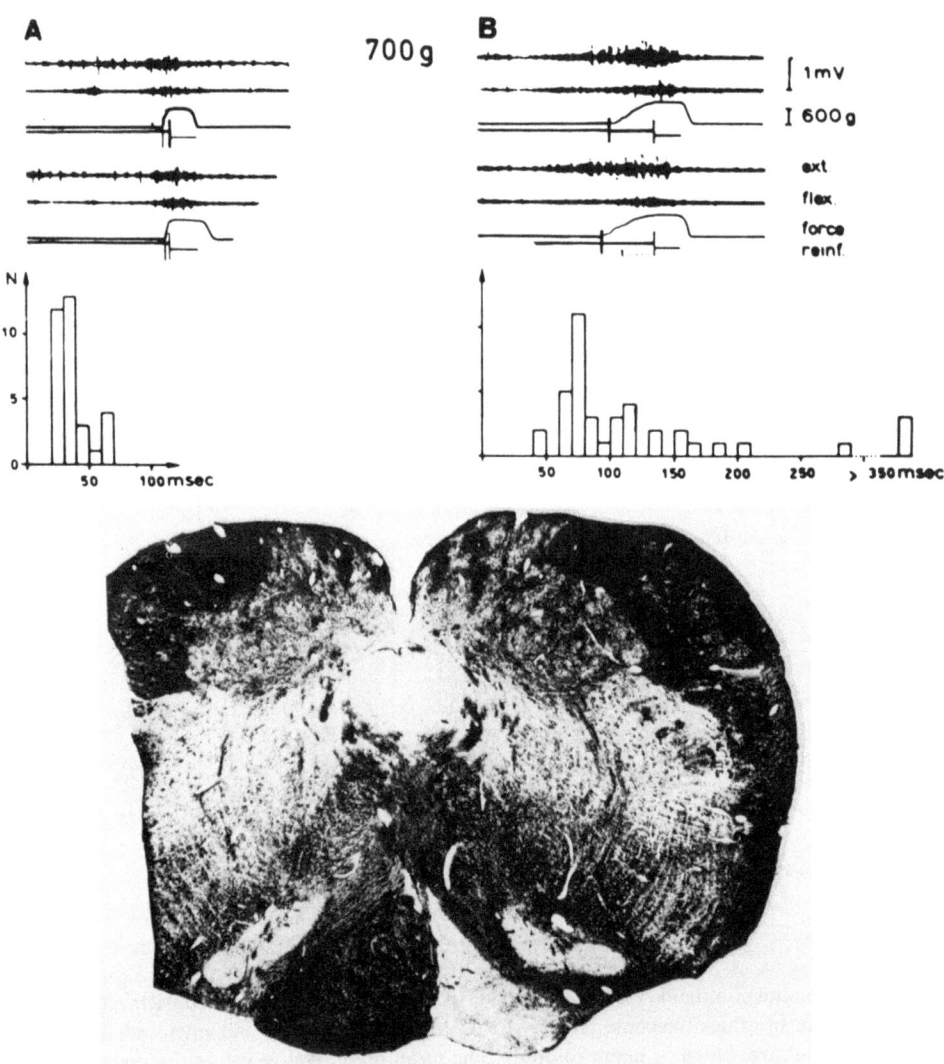

Fig. 7. Distribution of force build-up time (interval between vertical bars) of a monkey before (A) and after (B) a unilateral subtotal bulbar pyramidal lesion. The ensuing degeneration, caudal to the lesion, is shown in a Weil-stained micrograph. Note delayed build-up of force in the lesioned animal. The force threshold was 700 g. From Hepp-Reymond and Wiesendanger (1972).

MARIO
WIESENDANGER

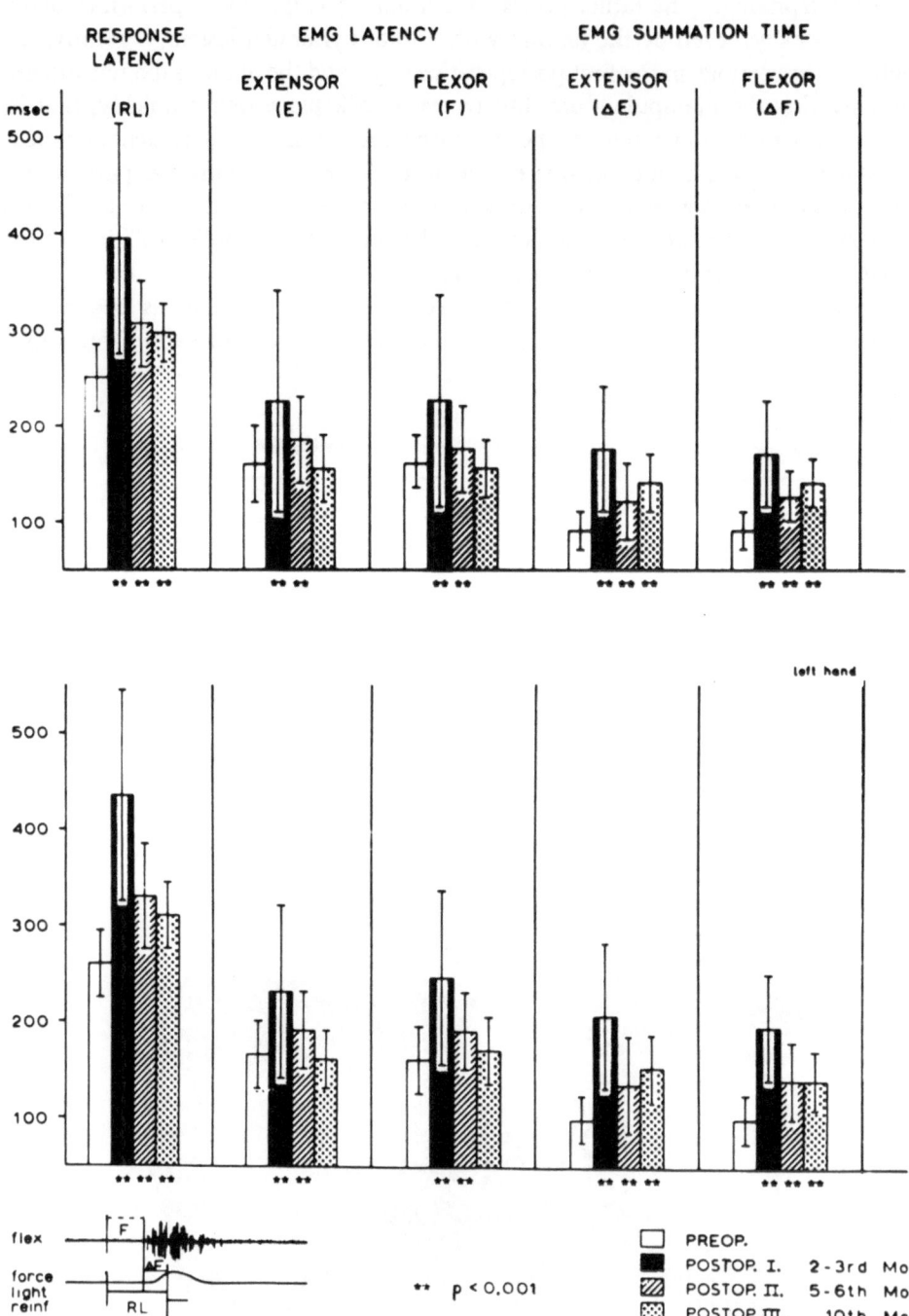

Fig. 8. Performance of a monkey on a precision grip task shown in Fig. 6 (Hepp-Reymond *et al.*, 1974). The control values (response latency, EMG latencies for flexors and extensors, and EMG summation time) are shown as blank columns. The mean postoperative values are represented as black and hatched columns. Bilateral incomplete pyramidotomy. Note the almost total recovery 10 months postoperatively (III).

responses; Miller, Glickstein, and Stebbins, 1966) were clearly slower after pyramidotomy. A mean increase was seen for the response latency and for the duration of the EMG burst to reach force threshold. This indicates that motor-unit recruiting was delayed in the absence of the pyramidal tract (EMG summation time), and this may have contributed to the increased response latency. However, delayed execution of the task was not the only factor, since a delayed onset was also measured in the electromyographic recordings. Pre- and postoperative results from a monkey with a bilateral, but incomplete lesion, are illustrated in Fig. 8. This graph also demonstrates the gradual recovery in performance over the postoperative testing period lasting 10 months.

It appears from results of other experiments performed on monkeys subjected to bulbar pyramidotomy that no simple solution to the question of pyramidal tract function is yet available: Laursen (1977) reported that highly trained monkeys performing a visual multidiscrimination task were not slower postoperatively than preoperatively. However, an increase in the *motor* performance time might be completely masked in view of the long choice reaction time (increasing with each additional choice, Hick, 1952) and the parallel increase of variability. This is particularly likely to occur if the motor task is simple (bar pressing in Laursen's study). The same monkeys were slower when forced to perform quickly on a fixed ratio schedule: the mean interresponse interval increased by 20 to 30%. Although one can agree with Laursen's statement that the contribution of the pyramidal tract in the control of movements depends on the behavioral context, it is the author's opinion that it depends much more on the type of motor task.

COMPARISON OF MOTOR CORTEX LESIONS WITH PYRAMIDAL TRACT LESIONS

No attempt is made here to review the voluminous literature on experimental lesions of the motor cortex in animals and of the equivalent lesions seen in neurology (for area 4 and area 6 lesions see recent reinvestigation by Gilman, Lieberman, and Marco, 1974). The most instructive experiments in monkeys were done a long time ago by Trendelenburg (1911). He was the first to use the elegant technique of reversibly cooling cerebral tissue in awake animals. Although remarkable conceptually and technically, these experiments have been forgotten and the technique of blocking nervous tissue reversibly was rediscovered and used with great success in recent years by Bénita and Condé (1972) and by Brooks and co-workers (see Brooks, 1975). Trendelenburg (1911), working on freely moving monkeys, observed that the arm contralateral to the cortical area being cooled became limp and was not used when the cooling fluid entered the cooling chamber overlying the cortex. The arm was slightly flexed (similar to the hemiplegic posture) and was used neither for isolated manipulative movements nor for climbing and locomotion on the floor. Occasional proximal movements were seen only if the animal became agitated. This massive deficit disappeared immediately upon rewarming the cooling probe. This experiment leaves no doubt that, in primates, the motor cortex is the chief executive not only of fine manipulative movements but also of gross, more automatic movements. These immediate effects, seen on cooling, do not of course invalidate many other observations made on animals with long-standing motor cortical lesions. As is so often seen in chronic lesion experiments, recovery of function is astounding, although fine digital skill will never return in primates with large motor cortex lesions. Unfortunately, reversible cooling of the pyramidal tract is not feasible for technical reasons. But all reports on the effects of pyramidal tract section seem to indicate that the immediate effects (in the first few postoperative days) are much more severe than several months later.

Synopsis of Lesion Experiments

The interpretation of neural function based on lesion experiments is hampered by any one of the following complicating factors: incomplete destruction of the structure, involvement of unwanted structures in the lesions (medial lemniscus in case of pyramidal lesions!), postoperative edema and infections, and postoperative recovery of function. This last factor represents a major challenge for neurobiologists. What are the mechanisms of recovery? It is interesting that the question had been asked a long time ago. Von Monakow (1914), in his theoretical consideration of *diaschisis,* a term he introduced to denote the depression of neural function caused by a sudden lesion and its gradual recovery, conceived the possibility of reorganization on the basis of newly formed connections (synapses) which is a rather modern concept.

Recognizing the limiting factors of lesioning experiments, and taking into account the important results of reversible cooling in monkeys (which might be compared to the reversible postictal paresis in epileptic patients), one comes to these conclusions: (1) the motor cortex (including the posterior part of Brodmann's area 6) of primates is the main motor command center of the brain. It is *execution,* rather than programming or initiation, of the movement which is governed by the motor cortex. (2) The pyramidal tract is only partially responsible for the deficit seen after cortical lesions. Nonpyramidal pathways are equally endowed with the capability of controlling "voluntary" movements. On purely anatomical grounds, one can assume that the share of the pyramidal tract is higher in man than in monkeys. Pure pyramidal lesions in man have been observed in only a very few cases. Precise digital movements suffer most in the presence of both pyramidal tract and cortical lesions. Pyramidal tract lesions affect not only highly skilled digital movements but also postural adjustments, especially contact placing. The often-invoked dichotomy of posture and movement is artificial; postural adjustments are often as precise and often more complex than the movements, and are tightly linked to, or even precede, the movements, and therefore also require fast-conducting neural control systems.

> The pyramidal tract has been evolving in the primates
> pari passu with the increase in fine control of the hand.
> Phillips, 1971

Developmental and Structural Aspects of the Pyramidal Tract

Methodological Remarks

A large amount of useful data about the origin of the pyramidal tract in the cerebral cortex, about the number and spectrum of its fibers and about their destination has been collected and presented in several reviews. The reader interested in data from older literature is referred to reviews of Schäfer (1900), Tower (1944), Nyberg-Hansen and Rinvik (1963), Verhaart (1970), Nathan and Smith (1955), Blinkov and Glezer (1968), Lassek (1954), Petras and Towe (1937b). These data will be summarized briefly in essentially following an earlier review (Wiesendanger, 1969). In addition, some new results will be reported which have been obtained with the powerful axoplasmic flow techniques.

> Minute amounts of an exogenous peroxydase enzyme are injected in presumed targets of pyramidal fiber endings. The enzyme, horseradish peroxydase (HRP), appears to be captured by the endings (possibly also by lesioned parent fibers) and transported backwards by axoplasmic flow to the cell bodies. After treatment of the histological section with adequate reagents (for details see LaVail, 1975), HRP appears as brownish granules in the cell bodies. Labeled cells appear most impressively in dark-field microscopy. Alternatively, the tritiated amino acids leucin or prolin may be injected in the known field of origin. Such amino acids were found to be rapidly incorporated into protein molecules and transported in an anterograde direction. The radioactive substance accumulates in synaptic endings and can be visualized by using autoradiographic techniques (for technical details, see Rogers, 1975). Again, silver grains appear most clearly in dark-field micrographs of the projection fields. These two "axoplasmic flow techniques" have almost supplanted the classical degeneration methods.

PHYLOGENESIS AND ONTOGENESIS

The pyramidal tract first occurs in mammals and its development seems to parallel that of the cerebral cortex. In view of Schäfer's (1910) and Tower's (1940) assertion that the pyramidal tract governs skilled movements, one may ask whether available data on its structure support this claim. In particular, it would be interesting to know which of the following features of the pyramidal tract is correlated with motor skill: origin, extent, termination, and fiber spectrum. Useful information bearing on this question has been collected by Nyberg-Hansen and Rinvik (1963), Noback and Shriver (1966), Petras (1969), Phillips (1971), Towe (1973c), and, most comprehensively, by Heffner and Masterton (1975). The latter authors ranked the dexterity of 69 different mammals from 1 to 7 according to an algorithm for the use of forelimbs, especially the hands (Fig. 9). The scaling was based on the work of Napier and Napier (1967), who took into account the ability of animals to manipulate objects, but not other types of motor skills seen, for instance in the seal and the porpoise. These animals ranked lowest in Heffner and Masterton's study in spite of their well-known and remarkable motor skill. Thus, the correlation found by the authors might be somewhat distorted. Nevertheless, the main outcome of the analysis is interesting in that *extent* and mode of *termination* most closely corresponded with dexterity. Thus, animals with a pyramidal tract penetrating the spinal cord down to the lowest segments and with terminals in the region of motoneurons also ranked high in dexterity. The authors also found that the two structural criteria have an independent relationship with dexterity. A collateral conclusion was that body weight alone accounts for more than 65% of the variation in pyramidal tract size and for the number of its fibers. Towe (1973b,c) came essentially to the same conclusion. This, of course, does not preclude dexterity as a contributing factor for the variation in the number of fibers. Surprisingly, the diameter of the largest fibers found for a given species was not significantly related to dexterity. However, it is noteworthy that the seal, although deprived of hands, displays an exquisitely well developed motor skill and ranked high in both number of pyramidal fibers and diameter of largest fibers. The review of Heffner and Masterton (1973) as well as that by Towe (1973b) bring out clearly the fact that in many species, the pyramidal tract disappears at cervical levels, before the cervical enlargement (insectivores, chiropterons, artiodactyls), or within the enlargement (marsupials, proboscidians, edentates). The question is, then, whether the pyramidal tract is composed of different subsystems with quite different functions. Pyramidal fibers of small diameter ending in the brain stem are numerous and found in all mammalian species.

Cortico-motoneuronal connections appear as a new addition in primates (Hoff and Hoff, 1934). Phillips (1971) emphasized the role played by the cortico-motoneuronal system in the evolution of manual skill and especially of the precision grip requiring an opposable thumb. Shapovalov (1975) has recently reviewed comparative studies of supraspinal synaptic effects on spinal neurons in electrical stimulation experiments.

Ontogenetically, the pyramidal tract matures at a late stage. Humphrey (1960) found that, in human fetuses of 9 to 29 weeks (menstrual age), the pyr-

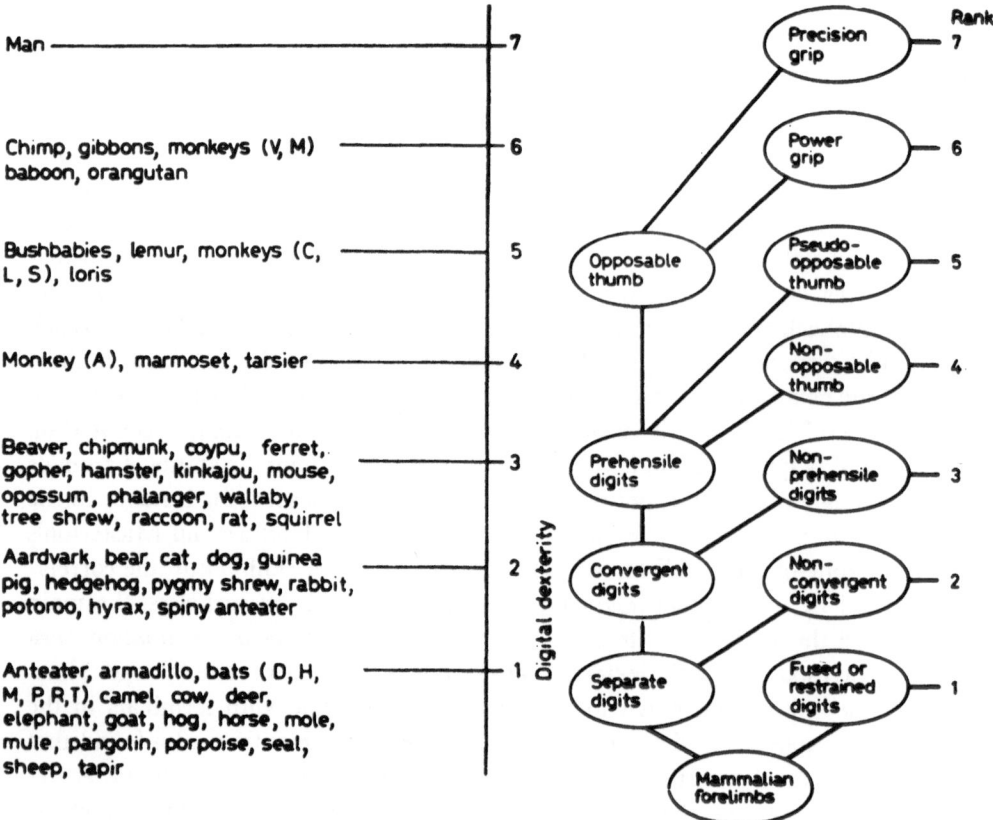

Fig. 9. Algorithm (left) for ranking dexterity of mammals according to manual skill (right; Heffner and Masterton, 1975).

amidal tract fibers had grown down to the level of the upper medulla within 10 weeks and that decussation occurred between the sixteenth and seventeenth week. By 29 weeks the fibers reached the most caudal segments of the cord (Fig. 10).

In man (Flechsig, 1876) and monkeys (Artom, 1925), myelin formation in fibers of spinal tracts occurs at or shortly after birth, with the exception of pyramidal tract fibers. The larger fibers of the pyramidal tract myelinate first, and this occurs approximately at the time walking is first attempted. According to Langworthy (1933), myelination is completed at the age of about 2 years. The pyramidal fibers, being of uniformally small diameter at birth, continue to increase up to somatic maturity. The establishment of synapses also continues after birth. In monkeys, Kuypers (1962) found that cortico-motoneuronal synapses were formed only several months postnatally. In infant monkeys, excitability of motoneurons by motor cortex stimulation and dexterity were found to resemble that of adult monkeys with pyramidal lesions (Felix and Wiesendanger, 1971; Lawrence and Hopkins, 1976).

In kittens, which already display the contact placing reaction requiring the thalamocortical system, the pyramidal tract fibers were shown to have very low conduction velocities, not exceeding 3 m/sec (Amassian and Ross, 1977). Forelimb movements could be elicited by motor cortical stimulation, and motor cortical outflow could be engaged by thalamic stimulation, but the latencies were very long and synaptic efficacy seemed low. Thus, the behavioral transcortical response was established, even when the neuronal system was still immature.

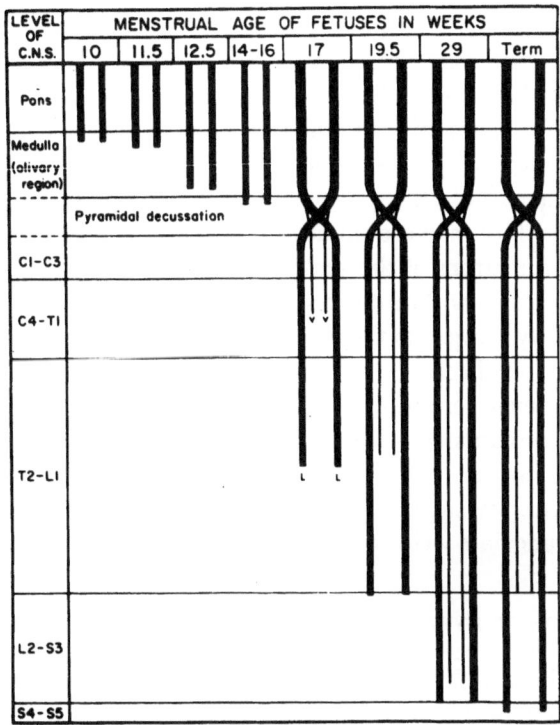

Fig. 10. Progress of growth of the pyramidal tract of human fetus. From Humphrey (1960).

MARIO
WIESENDANGER

In *cats* (Chambers and Liu, 1967; Nyberg-Hansen and Brodal, 1963), the origin was found to be restricted to the rostral third of the cortex, including the sigmoid, coronal, and anterior ectosylvian gyri, with a few fibers from the proreate gyrus. Subsequently, Nyberg-Hansen (1969a) found also some fibers originating from the mesial wall of the anterior sigmoid gyrus and the anterior cingulate gyrus, an area which may correspond to the supplementary motor area of primates. A few fibers were also seen to originate from the middle suprasylvian gyrus and the lateral gyrus, corresponding to the parietal association areas 5 and 7 (Nyberg-Hansen, 1969b). A renewed detailed investigation done with the retrograde labeling technique by means of HRP confirmed most of the above results (Berrevoets and Kuypers, 1975; Groos, Ewing, Carter, and Coulter 1978). After injection of the enzyme at various spinal levels, labeled cells were found in areas 4, 3a, 1, 2, 5, and in the second somatosensory area (SII). Interestingly, SII neurons appeared to be labeled only after cervical injections, and almost no cells were labeled, by any of the injections, in area 6 (see Hassler and Muhs-Clement, 1964, for boundaries of the various areas of the sensorimotor cortex in the cat).

In *monkeys,* area 4 of Brodmann gave rise to 31%, area 6 to 29%, and the parietal cortex to 40% of pyramidal fibers (Russel and DeMyer, 1961). The large "giganto-pyramidal neurons," typical for area 4, account for only about 3% of all fibers of the bulbar pyramid. Most of them disappear in monkeys with long-standing lesions of the bulbar pyramid (Pernet and Hepp-Reymond, 1975).

HRP experiments in monkeys revealed labeled large and small pyramidal-shaped cells in both pre- and postcentral gyri of the macaque monkey, including area 1, 2, 3b, 3a, and 4 (Catsman-Berrevoets and Kuypers, 1976; Coulter, Ewing, and Carter, 1976), and some also in the lateral aspect of area 6 and in the parietal association area 5 (Biber, Kneisley, LaVail, 1978; Jones and Wise, 1977; Murray and Coulter, 1976). A nest of medium-sized neurons in the depth of the cingulate sulcus was also detected by Biber *et al.* (1978) and Murray and Coulter (1977). This area might be part of the supplementary motor area. All efferent corticospinal cells, large and small ones, were situated in the fifth cortical layer.

The origin of neurons of the sensorimotor areas projecting to various nuclei of subcortical structures will be discussed below. Some of these neurons, but probably only a minority, project to these structures *via* collaterals of fibers descending further down the spinal cord.

In *man,* the exact origin of the pyramidal tract is not known. It appears, however, that approximately 60% of fibers originate from the precentral gyrus (Jane, Yashon, DeMyer, and Bucy, 1967). Other evidence from long-standing cases of cortical atrophy suggest that the remaining 40% of pyramidal fibers originate from the immediate vicinity of area 4 (see Wiesendanger, 1969).

THE DESTINATION OF PYRAMIDAL TRACT AXONS

In *cats,* the bulk of fibers terminate in Rexed's laminae V and VI at the base of the dorsal horn (Chamber and Liu, 1967; Nyberg-Hansen, 1966; Nyberg-Hansen and Brodal, 1963). It is therefore assumed that contacts are established exclusively with interneurons in this species. Of course, one cannot exclude the possibility that contacts are also established on dorsally oriented dendrites of motoneurons (Nyberg-Hansen, 1969c), but these direct connections would not be very effective.

In *monkeys,* the terminals of the pyramidal tract have a more widespread distribution on the cord cross-section. Fibers from the hand and foot areas near the central sulcus terminate chiefly in the dorsolateral part of the intermediate zone and to flexor motoneurons, whereas fibers from the remaining fields of areas 4 and from caudal area 6 terminate more in ventromedial parts of the intermediate zone and to extensor motoneurons (Kuypers and Brinkman, 1970). More and more direct connections seem to be established with motoneurons in the ventral horn on moving from *young* and *adult monkeys,* to *chimpanzee* (Kuypers, 1964), and to *man* (Schoen, 1964). The importance of the direct cortico-motoneuronal component in primate evolution and in the development of manual skill has already been mentioned and was especially emphasized by Petras (1969) and Phillips (1971). With the autoradiographic tracing techniques, Coulter and Jones (1977) found that subdivisions of the somatosensory cortex project to the dorsal horn, area 4 and 3a to the intermediate zone. The anterior wall of the precentral sulcus was found to be the exclusive source of fibers projecting to the motor nucleus.

Number and Fiber Spectrum of the Pyramidal Tract

The *macaque's* pyramid contains about 400,000 fibers (five times more than that of cats). About 38% of these fibers were found to be unmyelinated (DeMyer and Russel, 1958).

In *man,* the bulbar pyramid has about 1 million fibers (749,900 to 1,391,100), but, in contrast to the monkey, almost all (94%) pyramidal fibers were classified as myelinated (DeMyer, 1959). In quantitative terms, the pyramidal tract, also in man, is essentially a fine fiber tract. Medium-sized and large (3–20 μm) fibers do not exceed 10% of the tract (Lassek, 1954).

Ipsilateral Connections and Their Possible Role in the Recovery After Unilateral Lesions

It is well known that the bulk of pyramidal tract fibers project to the contralateral half of the spinal cord. Observations of a smaller contingent of uncrossed pyramidal fibers descending the spinal cord in the ventral funiculus led to the assumption that these fibers may play a role in the recovery of function in hemiplegic patients. Fibers in the ventral tract, however, seem to cross mostly at segmental levels (Liu and Chambers, 1964). A considerable amount of ipsilateral degeneration has been found by Liu and Chambers (1964) and by Kuypers and Brinkman (1970). The terminals appeared to be derived mostly from the dorsolateral tracts (Chambers and Liu, 1967; Liu and Chambers, 1964) and to project mainly to medial segments of the intermediate zone and of the ventral horn; these neurons govern proximal and axial muscles (Kuypers and Brinkman, 1970). Therefore, ipsilateral connections might be important for adjusting posture of the ipsilateral muscles during the performance of fine movements with the contralateral hand. Ipsilateral corticospinal connections were presumed to exist also because of cortical stimulation experiments. Thus, Bucy and Fulton (1933) were able to evoke ipsilateral movements, especially of the hind limb, which were still obtained after ablation of the opposite cortex. These authors attributed the ipsilateral movements to uncrossed pyramidal tract fibers, known to arise in area 6. Most recently it was shown that neurons of area 6 of the mesial cortex (the supplementary motor area), when studied in awake monkeys, were often bilaterally related to arm movements (Brinkman and Porter, 1978). Even more surprising was the observation of Matsunami and Hamada (1978), who found that 63 out of 197 neurons of the motor cortex were related to the ipsilateral as well as to the contralateral hand movements, a phenomenon to which passing reference has been made by Evarts (1966). However, ipsilateral cortical control is achieved not only via the pyramidal tract; it is known, and it has again been confirmed by Catsman-Berrevoets and Kuypers (1976), that area 6 has a substantial bilateral projection to subcortical targets, such as the medial bulbar reticular formation. It is thus tentative to suggest that the rather astounding clinical recovery in volitional control by patients with extensive lesions of the motor cortex or its outflow is because of he ipsilateral, direct and indirect, connections of the motor cortex with the spinal cord. This problem is possibly related to the next one which will be discussed.

The Problem of Compensatory Hypertrophy of the Intact Pyramid in Cases with Long-Standing Lesions of a Pyramidal Tract

Hypertrophy of a bulbar pyramid has been observed by many neuropathologists. The issue was discussed by Walker at a symposium on the pyramidal tract (see Lassek, Woolsey, Walker, and Boshes, 1957). An impressive hypertrophy of the medullary pyramid, with a concomitant (probably hereditary) agenesis of the other pyramid, has been published by Balthasar and Schlagenhauff (1966). The patient, who died at age 45, had only a slight weakness of distal hand musles contralateral to the pyramidal atrophy. It is highly improbable that hypertrophy of the contralateral pyramid was because of an increased number of fibers but rather to an increased percentage of large fibers (see Lassek *et al.*, 1957). What is the cause, the mechanism, and the effect of this hypertrophy? No ready answer seems available, but the question is an intriguing one, a challenge to neurobiologists in search for the general problem of recovery of functions.

> *This is not to say ... that foci for different muscles, or groups of muscles, are discrete entities, as in a mosaic. Rather there is very extensive overlap in the cortical motor patterns ... there are no such sharp lines of demarcation.*
>
> Woolsey, Settlage, Meyer, Sencer, Pinto-Hamuy, and Travis, 1950

The Motor Maps

General Organization of the Motor Cortex as Revealed by Prolonged, Repetitive, Electrical Stimulation

With the discovery of the "excitable cortex" by Fritsch and Hitzig (1870), a new area of brain research had begun. In rapid succession the cortical "pattern of localization" of the motor apparatus was established for various mammals. In Fig. 11 are assembled some classical maps, demonstrating the increasing complexity and fractionation of somatotopic representation. The exact extent of the "excitable cortex" and the "minuteness" of representation vary somewhat in different publications. This is not at all surprising when one considers that evoked movements were protocolled according to visual observations, that the thresholds and repertoire of motor effects vary with the depth of anesthesia, and that different stimulus parameters were used. In spite of these variable factors, the general picture of motor representation is well established. It appears from such maps of primates that the distal muscles are represented near or in the anterior bank of the central fissure, whereas the proximal and axial muscles are represented more anteriorly.

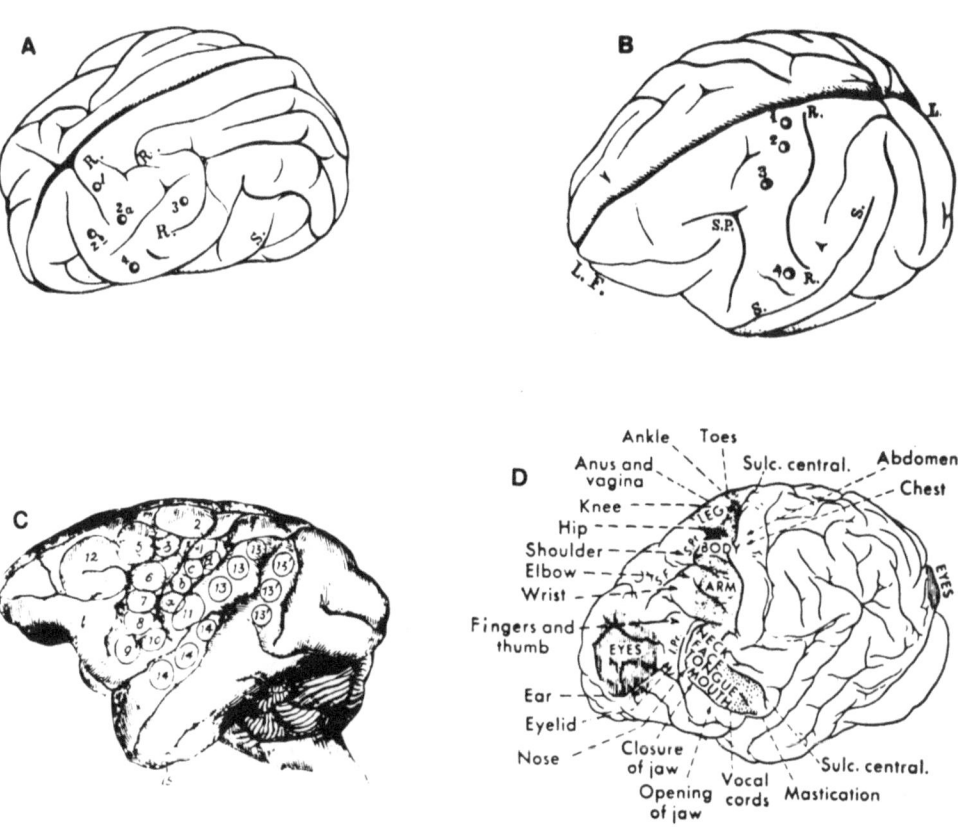

Fig. 11. Historical motor maps: (A) dog and (B) macaque are from Hitzig (1874) and illustrate the representation of the hind-limb muscles (1), the forelimb muscles (2a), the neck muscles (2b), the facial muscles (3) and the masticatory muscles (4). R = Rolandic sulcus. S = Sylvian sulcus. Hitzig investigated one monkey only and found the excitable cortex confined to the precentral cortex. In contrast to this pioneering study, Ferrier (1886) found the excitable cortex in monkeys to be more extensive (C). Some of the movements observed upon electrical stimulation were: (1) protraction of contralateral hind limb (as in locomotion); (2) complex movement of the hind limb including the foot; (3) tail movement; (4) retraction and adduction of the arm; (5) extension of forearm and of the hand; (a,b,c,d) various finger and hand movements; (7,8) facial movements; (12) head and eye movements. (D) The classical map of the excitable cortex in the chimpanzee published by Grünbaum and Sherrington (1903). Like Hitzig but unlike Ferrier, they found the excitable cortex to be limited to the precentral cortex (except for the eye muscles).

MARIO
WIESENDANGER

All who have worked in stimulation experiments of the motor cortex were faced with the difficulty of giving an adequate pictorial representation of their results. The outlines of motor maps shown in Fig. 12 are an oversimplification. The "simiusculi" and "homunculi" give but a crude impression of the relative representation of the various body parts: caricatures with large representation of distal and poor representation of proximal and axial body parts.

Woolsey and collaborators undertook a long series of investigations in various animals under "late pentobarbital anesthesia." They introduced the "figurine" representation of evoked movements which, even if not ideal, has given so far the most detailed information of such systematic mapping experiments with surface stimulation. Woolsey worked with a team of collaborators, each of them assigned with the observation of a particular part of the body. A 60-Hz sinusoidal current of 2 sec-duration and of 0.2 to 2.5 mA threshold intensity was used in these experiments. Figure 13 gives an impression of the "figurines" of the precentral cortex

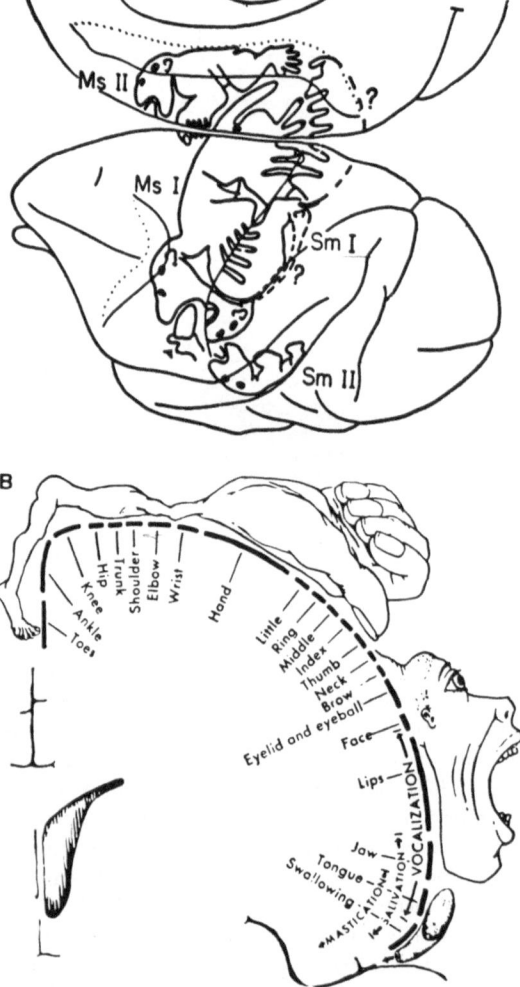

Fig. 12. Motor fields of the monkey brain (A; Woolsey, 1958) and of the human brain (B; Penfield and Rasmussen, 1950). The composite representation of the body parts gives an impression about the importance of cerebral control for the various muscles. The details, of evoked movements and of overlap in representation cannot be visualized. In the monkey, Woolsey described four separate motor maps: motorsensory I, motorsensory II (= supplementary motor area), sensorimotor I, and sensorimotor II. Note large representation of distal muscles. In man, representation of the hand and of the oral region becomes dominant. The map of the human motor cortex was constructed from electrical stimulation in many patients.

of a macaque monkey. The black and shaded areas are an indication of the involvement and strength of evoked movements. It is evident from these maps that there is a considerable degree of overlap in the representation of the motor apparatus. This is an important point, emphasizing the fact that cortical "localization" cannot be interpreted as a simple point-to-point representation of the muscles, but that the periphery is, as it were, re-represented in many combinations.

RESPONSES EVOKED BY REPETITIVE STIMULATION OF THE CORTICAL SURFACE

Ferrier (1886), a pioneer of brain-stimulation experiments, was the first to use repetitive stimulation of the motor cortex, and François-Franck (1887) to combine it with myographic recording. François-Franck undertook his work under

Fig. 13. Motor map (precentral cortex) from one experiment performed on a macaque (Woolsey, 1958). The figurine representation gives a more detailed (albeit still simplified) account of the motor responses obtained by repetitive electrical stimulation of the brain surface. The most vigorously contracting parts are in black. Rotations are indicated by arrows. The surfaces hidden in the central sulcus (c) and the arcuate sulcus are unfolded.

favorable conditions: on the one hand, he was guided and inspired by clinical colleagues and eminent neurologists, Charcot and Pîtres, and, on the other hand, he had at his disposal the technique of graphic recording introduced by Marey, in whose laboratory in the Collège de France the stimulation experiments were done. With weak stimulation, typically one or more seconds elapsed after onset of stimulation until movements were detected. Depending on the frequency of stimulation, the mechanogram revealed an incomplete tetanus (for instance with a frequency of 3/sec in Fig. 14) or a smooth contraction with build-up of force. François-Franck also noted that direct stimulation of efferent fibers (after ablation of the gray matter) was more effective than surface stimulation. Finally, he was the first to document the epileptiform (clonic) after-discharge occurring upon prolonged and strong stimulation. These after-effects (beats of 2–5 sec) were shown to occur only if the gray matter was intact (Fig. 14).

Cooper and Denny-Brown (1928) then went one step further and recorded both tension and electrical activity of the muscle in monkeys subjected to electrical stimulation of the cortex. These experiments fully confirmed the results described earlier by François-Franck in dogs. They termed the time elapsing between the onset of the stimulus and the first appearance of discharge of motor units the "summation" period. Once contraction begins, each stimulus is followed by an EMG burst. The interval between the individual stimuli and the "primary wave" in the EMG, *the latent period,* was short (about 15 msec for hand and finger extensors) and further decreased 8.8 msec) upon local application of strychnine. This fact led the authors to assume that "a monosynaptic relation (with motoneurons) seems most likely," a hypothesis which was proven with intracellular recording techniques some 30 years later (see p. 434).

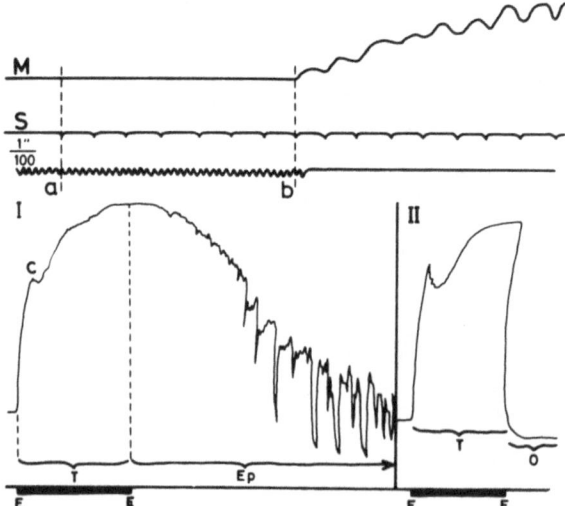

Fig. 14. First mechanographic recordings (M) of motor responses to repetitive stimulation of the motor cortex (indicated in S; François-Franck, 1887). In the upper record, beginning of repetitive stimulation (20 Hz) is marked by (a). At (b) contractions begin to develop. Each stimulus then evokes a twitch which is superposed on the previous twitch (incomplete tetanus). Note that contraction starts only with the eighth stimulus. The time elapsing between (a) and (b) (in this case 350 ms) is called the summation time. The aftereffects of intensive repetitive stimulation (E) are shown in I of the lower recordings. Note the clonic, epileptiform contractions (Ep) which follow stimulation. These epileptiform beats depend on an intact cortex: after lesioning the gray matter (II), stimulation of the white matter produced still a vigorous contraction during stimulation, but no epileptiform aftereffects.

THE SUPPLEMENTARY MOTOR AREA (SMA). Penfield and collaborators (for reviews see Penfield and Rasmussen, 1950; Penfield and Jasper, 1954; Wiesendanger, Séguin, and Künzle, 1973) described another motor representation which is situated "just anterior to the classical representation of the foot." In man, tonic contractions of the arm and adversive movements were elicited upon electrical stimulation of this area. A frequent response was vocalization and (in the locally anesthetized patients) arrest of speech. Penfield and co-workers did not see a clear somatotopic pattern in patients or monkeys, and the motor effects appeared to depend on the precentral motor cortex. Woolsey's systematic mapping led him to a different conclusion; namely, that there is, on the mesial surface, a second complete representation of the body which is independent of the precentral cortex, as evidenced by motor cortical ablation. The problem of somatotopic representation in the SMA has been taken up again by Talairach and Bancaud (1966) during systematic explorations with multiple intracortical electrodes in epileptic patients. Like Penfield, the French group was unable to detect a somatotopic representation within the SMA; the evoked movements were always complex and were often accompanied by vocalization ("le cri de l'aire motrice supplémentaire"*) or arrest of speech. Wiesendanger et al. (1973) concluded that the motor effects seen in monkeys were mediated by indirect paths. No corticospinal cells were found in this study, a finding which was in agreement with degeneration studies and with anterograde axoplasmic flow studies (Künzle, 1978; Wiesendanger et al., 1973). The problem is, however, not yet settled: as mentioned on p. 422, HRP-labeled cells were detected in the depth of the cingulate sulcus, anterior to the hind-limb area after spinal injection. This area of labeled cells is much smaller than the SMA as described by Woolsey. At present, it is safe to conclude that the SMA exerts its influence on the motor apparatus mainly via indirect routes (pontine nuclei and cerebellum, red nucleus, basal ganglia) and also via cortico-cortical connections with the precentral cortex, but probably only to a minor degree directly via corticospinal fibers. Techniques other than electrical stimulations will have to be used in order to get a better understanding of this cortical area.

THE POSTCENTRAL GYRUS. Whether motor effects evoked by electrical stimulation of the postcentral gyrus (the first somatosensory area) are because of current spread or are mediated by a motor pathway arising postcentrally has been debated since Leyton and Sherrington's mapping experiments in apes published in 1917. Woolsey, Travis, Barnard, and Ostenso (1953) reinvestigated this problem by using monkeys with chronic precentral and SMA lesions. This work has never been published in detail, but in a symposium Woolsey (1958) reported that "a well-organized postcentral motor outflow" had been demonstrated in animals with "complete degeneration of the motor pathways from both frontal lobes." Figure 15 illustrates the similarity of sensory and motor representation in the postcentral gyrus. It is noteworthy that the postcentral representation includes almost exclusively distal muscles which are represented in the depth of the central sulcus. As

*The cry of the supplementary motor area.

long as no precise histology from such lesion experiments is available, one has to be cautious with the notion of a postcentral motor area.

The postcentral gyrus is intimately and reciprocally connected with the precentral cortex (Jones and Powell, 1969a). Furthermore, many neurons were found in the postcentral gyrus which were activated in awake monkeys during the performance of a motor task (Evarts, 1972). A small proportion of these neurons increased their discharge rate well *before* movement onset and thus behaved like "command neurons."

THE MOTOR CORTEX IN THE PARIETAL OPERCULUM. Sugar, Chusid, and French (1948) described another motor area that was situated in roughly the same area as the second somatosensory cortex (SII). Welker, Benjamin, Miles, and Woolsey (1957) later confirmed, in the squirrel monkey, the existence of a motor area coinciding somatotopically with SII. Motor effects were also reported from stimulation of the anterior ectosylvian gyrus in cats (Garol, 1942). In a study of the output organization of SII in cats, intracortical microstimulation with high intensities in this area, near the face representation of SI, produced movements of the face only (Atkinson, Séguin, and Wiesendanger, 1973). Very few corticospinal neurons were found in this and other studies (Jones and Wise, 1977; Weisberg and Rustioni, 1976). In summary, it appears doubtful that somatosensory areas are endowed with neurons controlling directly, or via a few synapses, the motoneurons of the spinal cord. The corticospinal neurons, whose existence is fairly

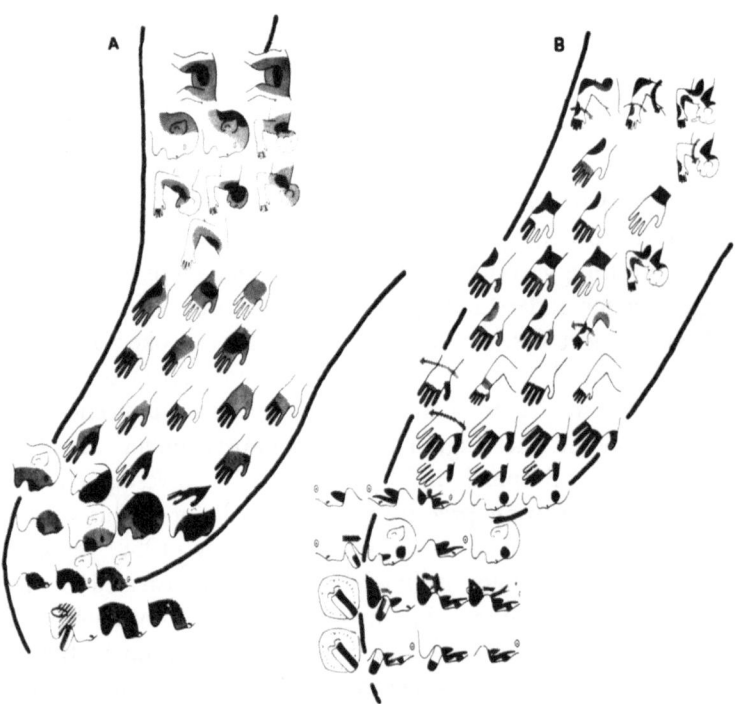

Fig. 15. Figurine representation of the (A) postcentral tactile and (B) postcentral motor representation in the monkey (Woolsey, 1958). Note almost exclusive representation of distal muscles in the postcentral cortex.

well established, both anatomically and electrophysiologically (e.g., Atkinson *et al.*, 1973), may be engaged in modulatory functions exerted on dorsal horn neurons. It is noteworthy that these neurons seem to have a small fiber diameter (Brech, Gordon, and Powell, 1977; Humphrey and Corrie, 1978).

MOTOR EFFECTS PRODUCED BY INTRACORTICAL MICROSTIMULATION

For a long time it has been known that fibers can be excited by passing current through a microelectrode at threshold intensities that are by far lower than those used for cortical surface stimulation (Wall, McCulloch, Lettvin, and Pitts, 1956; for a recent extensive discussion of the method, see Abzug, Maeda, Peterson, and Wilson, 1974; Gustafsson and Jankowska, 1976). Sakata and Miyamoto (1968) first observed the twitch movements in awake cats evoked by intracortical application of trains of stimuli (15–20 cathodal pulses of 0.2 msec duration, separated by 2.5 msec). Characteristically, movements were seen in the forelegs (mainly in distal muscles) and not in the trunk muscles. Thresholds for EMG responses in distal muscles were as low as 4 μA. Similarly, Asanuma and Sakata (1967) found low-threshold zones for facilitation (2 μA) or inhibition of monosynaptic test reflexes of various forelimb muscles innervated by the radial nerve. These microstimulation experiments have proven useful in studying intracortical organization. More will be said about this aspect on pp. 459–470. A number of investigations were performed by Asanuma and colleagues in lightly anesthetized cats (see Asanuma, 1973, 1975). Here, the discussion will be limited to the results obtained in primates. In Cebus monkeys, under sedation with small doses of Nembutal, Asanuma and Rosén (1972) found, as in the cat, small efferent zones in the forelimb area of the precentral cortex. With stimulus intensities of 10 μA, however, the effects were more complex than in cats, in that frequently more than one muscle contracted. With further reduction of the stimulating current, down to 5 μA, 75% of the spots produced "simple" effects (probably in one muscle only). The efferent zones were oriented radially (parallel to the apical dendrites) with a diameter of the cylinder of the order of 1 mm. The fractionation in different efferent zones was more prominent in Cebus monkeys than was previously found in cats. The lowest thresholds were in layer V, reaching values as low as 1.5 μA.*

However, matters appear to be more complicated. Andersen, Hagan, Phillips, and Powell (1975) were not convinced of the "mosaic" model of the motor cortex. They also explored the precentral hand area of the baboon's cortex with stimulating microelectrodes and found that single motor units, or aggregates of motor units (from one muscle), received an input from cortical territories occupying an area of up to 6.0 mm $\times$ 5.5 mm. The various territories overlapped considerably. It will be shown in the next section that such wide and overlapping territories projecting to individual motor units (the "colonies" with monosynaptic connections to motoneurons) were already assumed to exist on the basis of surface stimulation experiments. In response to these conflicting results, Asanuma and Arnold (1975) and

---

*The monkeys were moving spontaneously which might have been a necessary condition of obtaining thresholds as low as 1.5 μA.

Asanuma, Arnold, and Zarzecki (1976) argued that Andersen *et al.* (1975) had used excessive current strengths, damaging the cortex, and trains of stimuli which were too short (six pulses instead of 12). However, these arguments do not solve the problem, because one might reason that the cortical zones would have been even larger, had the Oxford group used longer trains, which stir up more intracortical elements and/or would have produced overt motor responses from a wider cortical area, owing to longer lasting temporal facilitation at the spinal synapses (Phillips and Porter, 1964; see also next section). Most important was the finding in the baboon (and to some extent this was also the case in Cebus monkeys) that the architecture of the efferent zones was "thin-textured and patchy" with "a fine grained internal structure" (Fig. 16). Finally, it must be said that high stimulating current (80 μA or less) were applied only in fringe areas of the cortical efferent zones, and that thresholds fell "with characteristic steepness" as the stimulating microelectrode was moved into the "hot spots" of the efferent zones. Andersen *et al.* (1975) had no difficulty in finding minima of 5 μA in quiescent baboons (their Figs. 6 and 8). These thresholds are surely acceptable. The baboons were under light barbiturate anesthesia and the limbs were not manipulated during threshold determinations, a procedure considered important by Asanuma and Arnold (1975) "so that the animal is kept alert."

Independently, Jankowska, Padel, and Tanaka (1975a), using both minimal

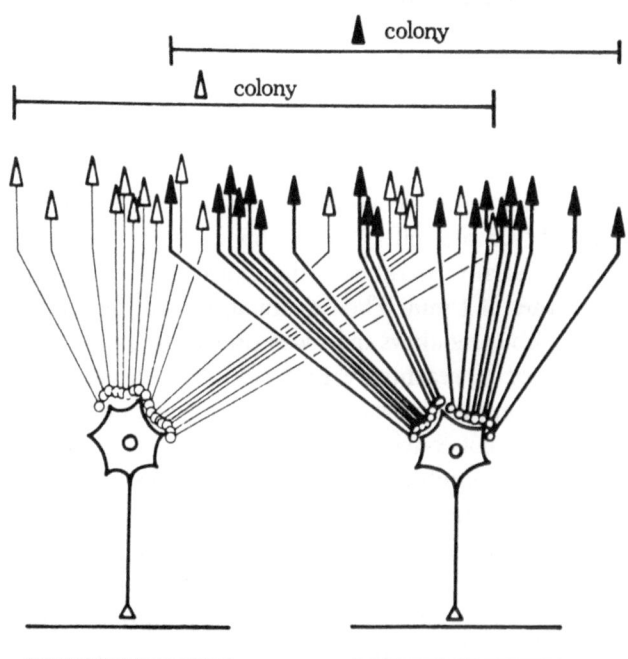

Fig. 16. Schematic representation of the cortico-motoneuronal organization (Andersen *et al.*, 1975). Corticofugal cells projecting to one motoneuron in the spinal cord constitute a "colony." The cortical cells of different colonies are highly intermingled. Clusters of cells (subsets of a colony) are the "hot spots" for a given motoneuron.

surface and intracortical microstimulation of the hind-limb cortex and recording synaptic potentials in spinal neurons, came to the same conclusion as Phillips and co-workers. In their own words, the results "give further evidence for overlapping of areas of cortical projections to motoneurons and speak against a mosaic-like organization of pyramidal tract cells projecting to different motor nuclei" (p. 637).

What Is Represented in the Motor Cortex?

The question of whether individual muscles or motor units, or rather, muscle groups are "represented" in the motor cortex has been around since the early days of electrical stimulation of the brain surface (Chang, Ruch, and Ward, 1947; Ruch, 1951). It is now taken for granted that the artificial method used by neurophysiologists to activate the motor apparatus from the motor cortex is not adequate to call up motor units in a natural spatial and temporal order, as seen during the performance of voluntary movements. Natural patterning of movements can only be elicited by electrical stimulation of cortical areas of the limbic system (for instance of area 24 of Brodmann on the mesial surface; Bancaud, Talairach, Geier, Bonis, Trottier, and Manrique, 1976) or from diencephalic structures (Hess, 1956).

A good account of the historical development of concept-formation about what might be represented in the motor cortex was given by Evarts (1967) and more recently by Phillips (1975) in an opening address at a motor-control symposium. As pointed out by Phillips, it is time that "the hoary ghost of *muscles versus movements*" may be laid to rest. Electrical stimulation experiments, useful and necessary as they are to understand structural relationships, cannot reveal how the clusters of output neurons of the motor cortex are activated (and, perhaps more importantly, inhibited) during the course of natural movements. On the other hand, it is evident that the elegant microstimulation and recording experiments have shed fresh light on how sensory input reaches an efferent zone and how excitation spreads intracortically (findings which will be summarized on p. 467). On the efferent side, the picture emerging is that of small, low-threshold foci projecting to a few motoneurons. However, it seems highly probable that multiple, discontinuous foci converge on one motor unit. Also, it is likely (experimental evidence will be presented on p. 451) that one focus may address more than one motor unit.

One may imagine that central commands (for instance, arriving in the motor cortex from neurons of the thalamic VL nucleus) activate much foci in ever changing combinations. In fact, it was first shown by Evarts (1967) that two motor cortical neurons, close by each other, (i.e., picked up by the same microelectrode) may be called up in concert for one movement or may show a reciprocal pattern of activation and inhibition for another movement. One should therefore not misinterpret the experiments by Asanuma and his co-workers in the sense that in the motor cortex there is a one-to-one, discrete representation of muscles and that each little "stone" (the efferent zone) of a "mosaic" is linked with one particular muscle.

Finally, it must be emphasized that the distal hand musculature is preferentially accessible to intracortical microstimulation, probably by neurons that establish monosynaptic connections with spinal motoneurons in primates. The proof of the monosynaptic excitation from the motor cortex was made with intracellular recordings from motoneurons, as will be discussed in the next section.

> *These colonies are intermingled. Localized action within the cortico-motoneuronal system thus requires a highly selective channeling of activity in the horizontally running intracortical networks which excite and inhibit the corticofugal neurons.*
>
> Phillips, 1966

THE CORTICO-MOTONEURONAL SYSTEM AND THE CONCEPT OF MOTOR CORTICAL COLONIES

THE CORTICAL TERRITORY PROJECTING TO INDIVIDUAL MOTOR UNITS

Liddell and Phillips (1950, 1951), working on the exposed cortex of baboons under hexobarbitone anesthesia and using single rectangular pulses of 5 msec duration for stimulation of the surface, discovered a simpler kind of motor map than those described in the preceding section. Under these conditions, movements were confined to distal limb muscles and face muscles. The authors established three zones, representing, respectively, the "thumb complex," the "toe complex," and the "face complex." The extent varied considerably according to the depth of the narcosis and the strength of stimulation. The overlap of the zones is illustrated in Fig. 17. The "thumb complex" zone always had the lowest threshold (1–1.5 mA). This remarkable preferential representation of distal muscles was an experimental result matching well the clinical observation of Jacksonian seizures, which

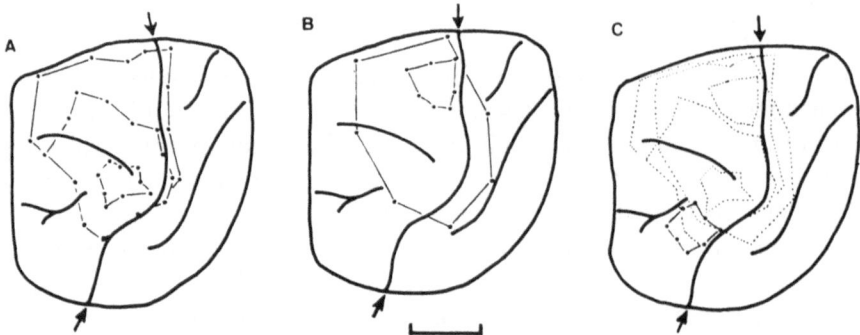

Fig. 17. Motor map established by means of long (5 ms) anodal stimulating pulses applied to the brain surface of the baboon (Liddell and Phillips, 1950). (A) "thumb complex"; (B) "toe complex"; (C) "face complex". Rolandic fissure marked with arrows. The smallest zones in (A), (B), and (C) correspond to the extent from which threshold stimuli (about 1.5 mA) evoked a response. As the stimulus intensity was raised to several mA, the same response could be elicited from a much wider zone. (C) is also a composite of the three complexes. The most important outcome of this study was that threshold currents preferentially activated distal muscles. It also suggested that a particular group of muscles may be controlled from a relatively wide cortical territory.

have their characteristic onset in distal muscles. Liddell and Phillips (1950) referred to Walshe's hypothetical comment which now had found a sound experimental basis: "It is suggested that Jacksonian fits have their characteristic form of onset because the movements concerned are those that have their widest fields of low threshold excitability" (p. 134). Single motor units of the first interosseus muscle were activated from several points of the cerebral cortex. The latencies and the order of firing were variable over time (Liddell and Phillips, 1951).

These experiments indicated that corticofugal cells projecting to a given motor nucleus may have a wide cortical origin. This question was systematically investigated in a series of experiments on baboons with intracellular recording techniques (Hern, Landgren, Phillips, and Porter, 1962; Kernell and Wu, 1967; Landgren, Phillips, and Porter, 1962a,b; Phillips and Porter, 1962). A summary of these experiments can also be found in Phillips and Porter (1964) and in Phillips (1966, 1967, 1969, 1973, 1975). First, one has to know how far the current spreads when applied to the cortical surface. This was estimated by plotting the stimulus intensity as a function of the distance from the optimum point from which a single corticospinal axon could be excited. Fiber spikes were recorded in the cervical spinal cord (Fig. 18). The parabolic threshold curves indicate, for instance, that current spread for pulses of 3 mA involves a cortical surface area with a radius of more than 5 mm. When motoneurons of the cervical enlargement were impaled, EPSPs were found to grow in amplitude as the stimulus intensity was increased from, say, 0.8 mA to 3.0 mA. This indicates that with increasing current spread, more and more corticofugal cells were activated which projected to that particular motoneuron. In other words, the maximum EPSP amplitude was a measure of number of corticofugal neurons contributing to that EPSP. This population of cortical cells was called a *colony* by Phillips and co-workers. As already suggested by the previous experiments, these intracellular recordings clearly revealed that the colonies had overlapping territories.

SOME FEATURES OF THE CORTICO-MOTONEURONAL PROJECTION FROM LOW-THRESHOLD ZONES OF THE MOTOR CORTEX TO MOTONEURONS

That monosynaptic contacts are established by corticofugal neurons and motoneurons was suggested by the stimulation experiments of Cooper and Denny-Brown (1928) and by the anatomical work of Hoff and Hoff (1934), who used the "bouton method." Ventral root and peripheral nerve recordings by Bernhard, Bohm, and Petersen (1953), Bernhard and Bohm (1954a,b) and by Preston and Whitlock (1960) strongly supported the notion of monosynaptic connections to both hind-limb and forelimb motoneurons. The final proof was then provided with intracellular recordings from forelimb (see Phillips and Porter, 1964) and hind-limb motoneurons (Jankowska *et al,* 1975a; Muir and Porter, 1973; Porter, 1970; Porter and Hore, 1969; Preston and Whitlock, 1961; Stewart and Preston, 1967; Tamarova, Shapovalov, Karamjan, and Kurchavyi, 1972). The monosynaptic nature of the EPSPs was evident from the short delay between arrival of the descending volley at segmental level and onset of the synaptic potential. The smooth, synchronous potential was similar to the monosynaptic EPSPs produced

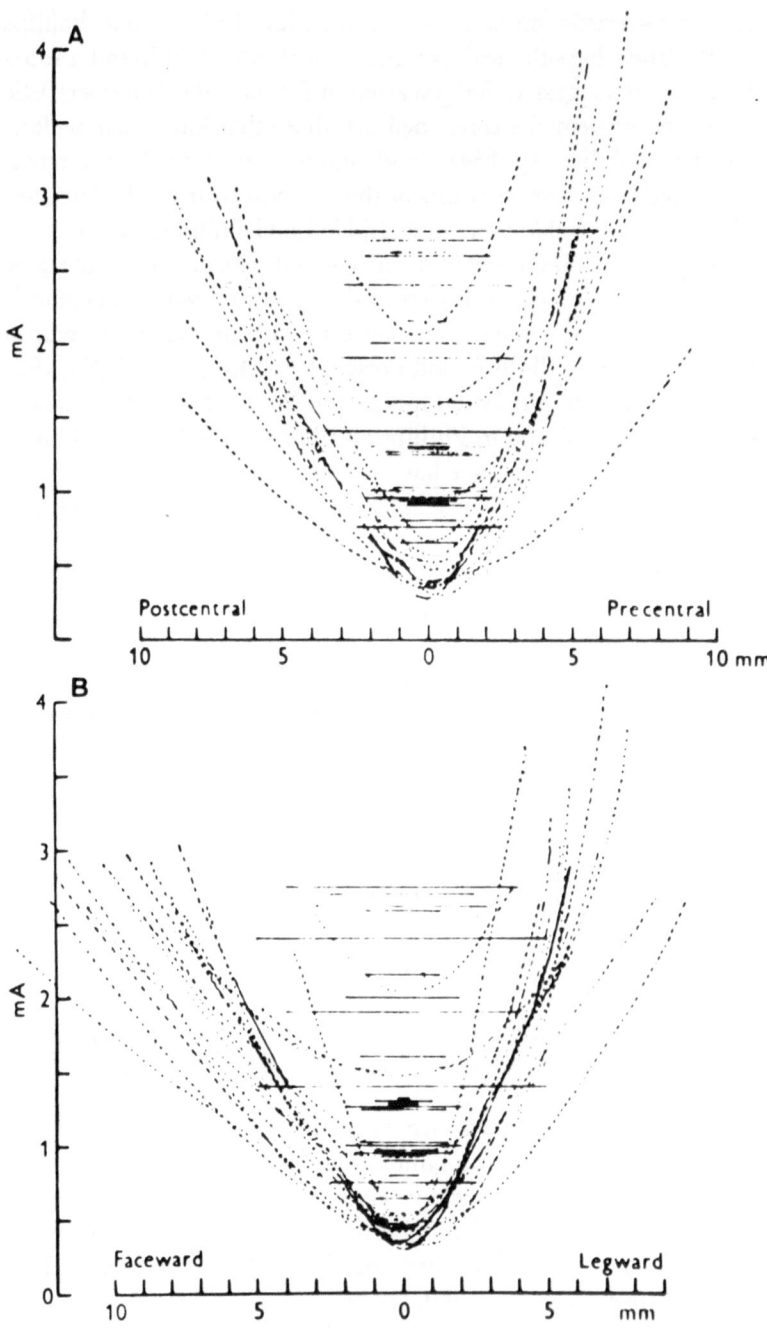

Fig. 18. Family of parabolic threshold curves for direct activation of an individual pyramidal tract neuron (Phillips and Porter, 1964). Action potentials of corticospinal axons were recorded in the dorsolateral funiculus of the cervical spinal cord. Threshold intensity (surface anodal stimulation) is plotted as a function of the distance from the "best" point (lowest threshold) in anteroposterior (A) and mediolateral (B) direction. These curves allow to estimate the amount of physical current spread for a given stimulus intensity. Since, in other experiments, synaptic potentials recorded from many cervical motoneurons were growing in amplitude as the stimulus intensity was raised to 3 or more mA, it can be assumed that the cortical colonies of these motoneurons extended over a territory of several mm diameter (experiments in baboons under barbiturate anesthesia).

by volleys from group Ia afferent fibers. It was also shown by Phillips and co-workers that distal motor units receive considerably more monosynaptic excitation than proximal muscles. Furthermore, repetitive discharges of pyramidal tract axons (set up for instance by a 5 msec cathodal pulse) result in remarkable build-up of depolarization, which contrasts with the lack of frequency potentiation of Ia-motoneuronal synapses (Fig. 19).

From the functional point of view, the existence of direct synaptic connections between pyramidal tract fibers (only a fraction of the fibers) and motoneurons in primates was taken as structural basis for the developing manual skill in primates (see p. 419 and p. 420). One may ask, however, what skill means: what does the cortico-motoneuronal component—still lacking in cats (Lloyd, 1941; Szentagothai-Schiemert, 1941)—add to motor performance in terms of skills? As was pointed out, lesion experiments indicate that it is the fractionation in movements involving only a small number of motor units of hand muscles which is lost in pyramid-lesioned and new-born monkeys (Lawrence and Hopkins, 1976). It is tempting to assume that this particular resolution for discrete usage of motor units, necessary for instance for the precision grip, is based on the monosynaptic control of motor units by the motor cortex. It appears that primates without monosynaptic connections, such as squirrel monkeys (Harting and Noback, 1970) and marmoset monkeys (Shriver and Matzke, 1965), are not capable of performing the precision grip (Fig. 20). Noteworthy also is the recent electrophysiological discovery of Elger, Speckmann, Caspers, and Janzen (1977) that rats, which use their forelimbs with considerable skill, also possess monosynaptic connections with cervical motoneurons.

It was suggested by Phillips and Porter (1964) that temporal summation with rapid build-up of motoneuron-membrane depolarization could be the physiological basis of rapid movement initiation, another functional attribute of the pyramidal tract. Although maximal facilitation occurred at stimulus frequencies of 500 Hz, a firing frequency not observed in pyramidal tract neurons of awake monkeys performing motor task (Evarts, 1968), significant cortico-motoneuronal facilitation was still observed when the frequency was reduced to a more physiological range of about 100 Hz (Porter, 1970). Under natural conditions, pyramidal tract neurons do not discharge at a fixed frequency. Porter, Lewis, and Horne (1971) analyzed

Fig. 19. Monosynaptic excitatory postsynaptic potentials recorded in the same cervical motoneuron of the baboon (Phillips and Porter, 1964), evoked by Ia volleys (A) and pyramidal volleys (B). The stimulus intensity was adjusted so to produce about the same initial EPSPs. Note the pronounced build-up of depolarization upon repetitive bombardment by pyramidal volleys, not present upon bombardment with Ia volleys. This "frequency potentiation" is characteristic of synapses of the pyramidal system and is likely to reflect a progressively augmenting transmitter release with successive stimuli (presynaptic facilitation).

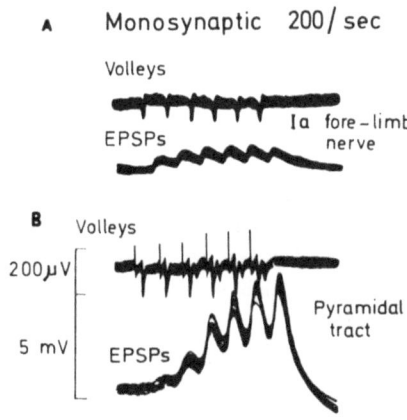

this point in alert monkeys and found that phasic discharges preceding movements (see Chapter 9) may contain interspike intervals as short as 3 msec. It was furthermore shown by Porter and Muir (1971) that the intraburst pattern of discharge was decisive for the amount of frequency potentiation at the cortico-motoneuronal synapses. Thus, the authors found that "for a given mean frequency of 100/sec, 11 corticospinal volleys with increasing intervals were more effective than other patterns" (p. 139).

As expected, responses of distal forelimb muscles to single anodal pulses were lost in monkeys with pyramidal lesions (Felix and Wiesendanger, 1971). In baboons, the familiar motor map, as revealed by repetitive 60/sec cathodal stimulation (which brings into action a large number of intracortical elements impinging on output neurons of the motor cortex) was preserved to a large extent (Lewis and Brindley, 1965). In macaque monkeys, Woolsey and colleagues (Woolsey, Gorska, Wetzel, Erickson, Earls, and Allman, 1972), however, found the pattern of movements elicited by cortical stimulation in animals with chronic pyramidal tract lesions much simpler and essentially restricted to proximal muscles. Clearly, the thresholds were higher on the cortical side lacking a pyramidal projection (Fig. 21). It is beyond the scope of this chapter to review the nonpyramidal pathways that may be responsible for transmission of cortical excitation to the spinal cord in

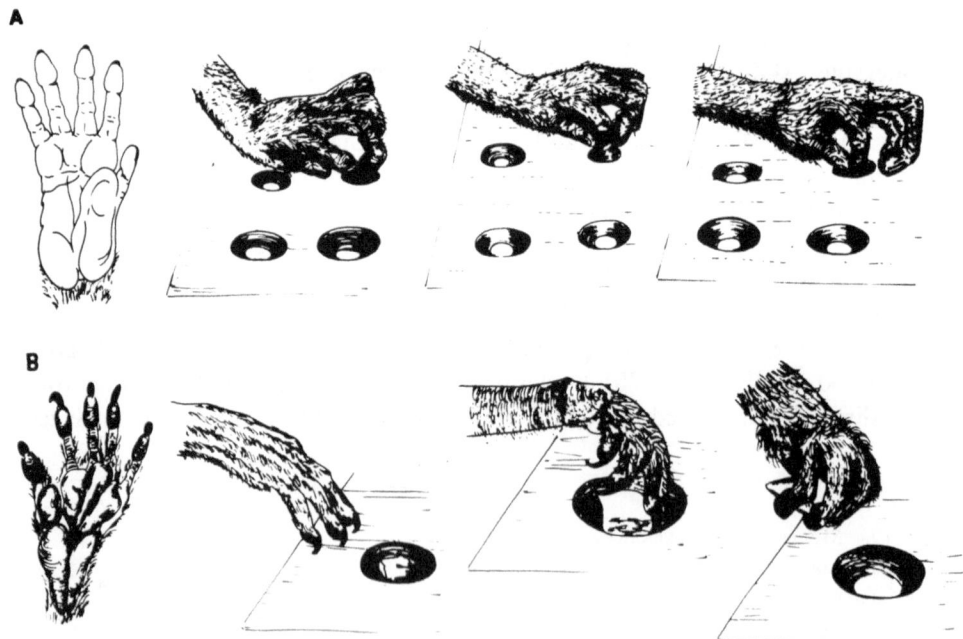

Fig. 20. (A) Hand of macaque *(Macaca fascicularis)*. Food morsels are taken out from small holes of a Klüver board by means of the precision grip. Note independent movement of index finger and opposition of thumb. (B) Hand of a marmoset monkey with its sharp claws adapted to arboreal climbing. The light weight of marmoset monkeys and the claws make vertical clinging possible. As a prehensile instrument the marmoset's hand lacks opposability for a precision grip; the fingers close together. The capability to perform a precision grip is believed to depend on the presence of a direct cortico-motoneuronal system. After photographs, drawn by R. Wiesendanger.

pyramidal-lesioned animals (but see Chapter 4 and the recent review by Shapovalov, 1975). It appears that, in monkeys, the rubrospinal tract also has monosynaptic access to distal motoneurons and that repetitive volleys arriving via rubrospinal fibers bring about a similar frequency potentiation in motoneurons as corticospinal volleys. Reticulospinal and vestibulospinal fibers were also found to be monosynaptically connected with motoneurons, but more to those innervating proximal muscles.

THE CONVERGENT ACTIONS OF MUSCLE SPINDLE FEEDBACK AND CORTICO-MOTONEURONAL COMMANDS

Current views about servo mechanisms imply the involvement of muscle spindles in motor control (see Chapter 3); accordingly, the interplay of muscle spindle afferents and descending pyramidal tract fibers on motoneurons has been extensively studied in the baboon by Phillips and colleagues (Clough and Sheridan, 1968; Clough, Kernell, and Phillips, 1968; Clough, Phillips and Sheridan, 1971; Koeze, 1968; Koeze, Phillips, and Sheridan, 1968: for review see Phillips, 1969). Conforming with the hypothesis of "servo-assistance of movements" (Matthews, 1964), it was found that motor cortical stimulation could activate γ-motoneurons together with α-motoneurons. In normal functioning, cortical command signals would coactivate α- and γ-motoneurons. This is the concept of α–γ linkage, introduced by Granit (1955), and which was considered important for opposing the

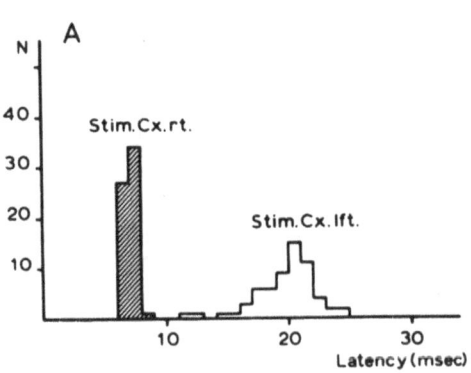

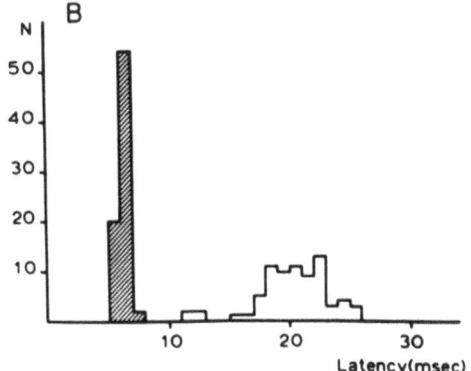

Fig. 21. Latency histograms of EMG responses recorded in hand extensors (A) and hand flexors (B) of a monkey with a chronic left-sided partial lesion of the bulbar pyramid. Hatched columns are response latencies recorded on the left side upon stimulation of the right motor cortex with an intact pyramidal projection. Blank columns are responses evoked from the cortex whose connection with the spinal cord was lesioned. Responses were obtained during repetive stimulation (30 Hz) at threshold intensity on both sides. From Felix and Wiesendanger, (1971).

unloading effect on muscle spindles when the muscle shortens during contraction. If the movement progresses as programmed, the γ-drive would exactly balance the unloading effect; unexpected load changes would increase or decrease the muscle spindle feedback. This change in muscle spindle firing, the "error signal," would tend to compensate automatically for load changes. In this context, it is important that the intrinsic hand muscles were found to be richly supplied by muscle spindles and the γ-motoneurons innervating them are under powerful control from the motor cortex, partly also via monosynaptic connections (Clough *et al.,* 1971). The instructive diagram of Fig. 22 shows the mean relative inputs from the cortico-motoneuronal source and from the muscle spindle source on motoneurons of different muscles. The largest segmental inputs converge essentially to the same motoneurons that receive the richest cortical inputs. In Phillips's words, "the output of the CM-selected motoneurons could be governed automatically by marginal changes in a cortically sustained spindle feedback, reinforcing or withdrawing support from the cortico-motoneuronal input in response to changes in peripheral load" (1963, p. 167).

The cortical control of hind-limb γ-motoneurons has also been studied by several authors (Grigg and Preston, 1971; Koeze, 1973; Mortimer and Akert, 1961). However, the contribution of motoneuron excitation via the γ-fiber-spindle loop in cortical stimulation experiments was considered to be small for hind-limb muscles (Lewis and Porter, 1971).

The deteriorating effect of total deafferentation of a forelimb on motor control was first described by Mott and Sherrington (1895) and Hering (1897), and the dramatic lack of spontaneous use of the affected hand has since been noted repeatedly (Gilman, Carr, and Hollenberg, 1976; Lassek, 1953; Twitchell, 1954; Wiesendanger, unpublished observations). The remarkable recovery of function reported by Taub and Goldberg (1974) when monkeys were trained in an operant-conditioning situation does not invalidate the notion that segmental feedback is of great importance for the execution of fine digital movements. Since manual skill suffers most in deafferented monkeys, it was speculated that "an important part of the disability is owing to the loss of the spindle feedback" (Phillips, 1969).

A new dimension of the problem of servo-assistance by muscle spindles was

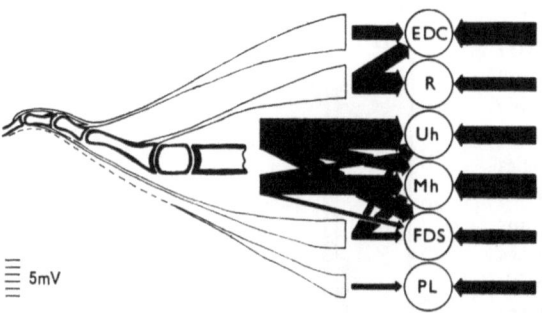

Fig. 22. Diagrammatic representation of convergence of pyramidal tract fibers and muscle spindle (Ia) afferents on various motoneurons of the baboon's hand (Phillips, 1969). EDC = m. extensor digitorum communis; R = remaing dorsiflexors of the wrist; Uh = intrinsic hand muscles supplied by the ulnar nerve; Mh = intrinsic hand muscles supplied by the median nerve; FDS = m. flexor digitorum sublimis; PL = m. palmaris longus. The size of the arrows are proportional to the average maximum size of the respective monosynaptic EPSPs recorded in these motoneurons (calibration on the lower left).

introduced by Phillips's hypothesis (1969) of a transcortical servo loop, and the discovery that signals from muscle spindle afferent fibers have access to the cerebral cortex (Lucier, Rüegg, and Wiesendanger, 1975; Phillips, Powell, and Wiesendanger, 1971). This aspect will be discussed on p. 468.

There is now extensive evidence showing convergence of descending pathways and primary afferents on common interneurons projecting to motoneurons.

Lundberg, 1975

Pyramidal Actions on Spinal Internuncial Cells

General Comment

The brain controls motoneurons, in large measure, via indirect routes. It has already been mentioned that the corticospinal tract is only one of several descending pathways. Thus, the motor cortex is capable of influencing the motor apparatus via subcortical motor centers. Furthermore, the descending volleys exert their effects chiefly via spinal interneurons. This was first clearly established by Lloyd (1941) for the pyramidal tract in cats. From degeneration and axoplasmic flow studies (see p. 423), it is now well established that spinal neurons at the base of the dorsal horn and in the intermediate zone receive the bulk of terminals of the corticospinal tract. There is now a large body of evidence suggesting that the brain motor centers use neurons of the propriospinal apparatus and of segmental reflexes in their control of the "final common path," the motoneurons. Figuratively, one might think in terms of an operator calling up subroutines. To unravel the complex actions and the integration taking place at the spinal level is, for the neurophysiologist with his microelectrophysiological equipment, a formidable task. Lundberg and his colleagues (see Lundberg, 1975) probably contributed most to our present knowledge. The picture of the spinal circuitry with its descending and segmental inputs is becoming progressively more complex. In spite of intensive research, with new discoveries every year, only part of the spinal machinery is so far known. A special chapter by Schwindt (see Chapter 4) is devoted to the problem. In the present context, the accent will be on pyramidal tract effects on different types of spinal neurons. That the pyramidal tract exerts its influence on segmental reflexes, monosynaptic and polysynaptic, was confirmed many times after Lloyd's pioneering study, and the literature is extensive (see Shapovalov, 1975; Wiesendanger, 1969). More recently, the emphasis has been to record intracellularly from identified interneurons which are tested for their afferent and descending inputs and which, by means of intracellular staining, may be localized with respect to Rexed's laminae (Jankowska, 1975).

It cannot be emphasized enough that the mechanism of inhibition is a prerequisite for the organization of skilled movements, or, put in Eccles's (1973, p. 89) words: "The inhibition, as it were, chisels away at the diffuse and rather amorphous mass of excitatory action and gives a more specific form to the neuronal performance at every stage of synaptic relay." Not surprising then was the observation that cortical stimulation evoked inhibitory actions on motoneurons in primates (Hern *et al.*, 1962; Preston and Whitlock, 1960, 1961). The latency difference between the monosynaptic EPSPs and the earliest IPSP was of the order of 1.2 msec, and was even less in the recent experiments of Jankowska, Padel, and Tanaka, (1976), suggesting one interposed inhibitory neuron in the pathway. In a natural situation, such as the reaction-time situation, suppression of preexisting activity of motor units may be the first detectable response to a "go" signal (Evarts, 1974). The question is, whether this turning off of motor units is achieved by an intracortical or by a spinal mechanism. This question was raised long ago by Sherrington (1906), who favored the spinal mechanism, since reciprocal effects on antagonistic muscles were seen not only upon electrical stimulation of the cerebral cortex but also of the internal capsule (Sherrington and Hering, 1897). This observation, of course, does not preclude the possibility of inhibitory actions organized at the cortical level (Evarts, 1967). Some known intracortical inhibitory mechanisms will be discussed on pp. 463–464.

Focusing now on spinal mechanisms, the question is: Which neurons are capable of inverting pyramidal tract excitation into inhibition? Theoretically, the candidates could be (1) interneurons common to the descending and the segmental reflex path, (2) propriospinal neurons, or (3) "private" neurons (see Lundberg, 1975). In cats, most authors have found that interneurons influenced by pyramidal tract volleys also react to some sort of peripheral afferent input (Asanuma, Stoney, and Thompson, 1971; Fetz, 1968; Lloyd, 1968; Lundberg, 1967). The possibility remains, however, that some interneurons, "private" to the pyramidal tract, exist (Kostyuk and Vasilenko, 1978). Common to all experiments to be described was the observation that temporal summation in interneurons was needed to fire the motoneurons (Lloyd, 1941).

PYRAMIDAL TRACT ACTIONS ON Ia-INTERNEURONS

Intensive research by the Gothenburg school revealed that Ia-interneurons have a key position in the spinal circuitry necessary for motor control. These interneurons are monosynaptically activated by Ia afferent fibers from a given muscle and inhibit motoneurons of antagonistic muscles. The degree of reciprocal linkage must be critically controlled. Thus, reciprocal inhibition must be avoided when antagonistic muscles are called up for cocontraction. The full story of the control of Ia-interneurons is given elsewhere in this volume (see Chapter 4; see also Hultborn, 1972, and Lindström, 1973). It may suffice here to mention that the pyramidal tract was found to facilitate cervical Ia-interneurons disynaptically in cats (Illert, Lundberg, and Tanaka, 1976b) and monosynaptically in monkeys (Jankowska *et al.*, 1976). The effects were found to be rather weak, however, and it was proposed that convergent excitation from other sources and linked excitation of α- and γ-motoneurons could substantially increase the excitability of these interneurons ("α-γ-linked reciprocal inhibition"). Indirect evidence obtained with H-reflex testing in the soleus muscle of man seems to indicate that the Ia inhibitory path is facilitated during a rapid dorsiflexion of the foot (Tanaka, 1974).

In monkeys, pure pyramidal tract volleys produced predominantly short-latency inhibition in soleus motoneurons, whereas facilitation prevailed in gastrocnemius and flexor motoneurons (Preston and Whitlock, 1963). It was concluded that the output from the motor cortex may be organized to curtail the activity of tonic postural muscles during cortically initiated movement. Because of the short latency, this inhibition may have been mediated by Ia interneurons.

PYRAMIDAL TRACT ACTIONS ON RENSHAW INTERNEURONS

The Renshaw cell is another important interneuron which, activated by motoneuron collaterals, mediates inhibition to motoneurons especially of the tonic type. These interneurons were also found to be under segmental and supraspinal control (Wilson, Talbot, and Kato, 1964). The recent discovery that Renshaw interneurons are crucial in their inhibitory actions on Ia interneurons puts the Renshaw neuron in a key position for the regulation of reciprocal inhibition from Ia afferents (Hultborn, 1972). Therefore, it would be of greatest importance to know whether or not the pyramidal tract exerts a control on Renshaw cells. So far, there is only indirect evidence that the motor cortex inhibits, via the pyramidal tract, Renshaw cells (MacLean and Leffman, 1967); this question should, however, be studied in more detail.

PYRAMIDAL TRACT EFFECTS ON INTERNEURONS OF THE IB INHIBITORY PATH

These afferent fibers, connected to Golgi tendon organs, come into play when muscle tension increases. Again, it was shown (so far only in cats) that these interneurons receive convergent actions from a variety of peripheral afferent fibers (Lundberg et al., 1975) and from descending tracts, including the pyramidal tract (Lundberg and Voorhoeve, 1962).

PYRAMIDAL TRACT EFFECTS ON INTERNEURONS OF THE FLEXOR REFLEX PATHWAY

That reflexes evoked by excitation of cutaneous afferent fibers are under powerful pyramidal tract control has been found by several investigators (see Shapovalov, 1975). It appears that supraspinal control of many interneurons is exerted tonically (Wall, 1967). In unilaterally pyramid-sectioned cats, it was shown that the resulting flexor hypotonia was accompanied by a diminished tonic activity of both α- and γ-motoneurons in the nerve to the tenuissimus muscle, which responded only weakly to air jets applied to the paws (Wiesendanger and Tarnecki, 1966). The net tonic effect of descending control on flexor reflexes must, however, be inhibitory, since patients with spinal lesions often have greatly enhanced flexor reflexes, with irradiation to many muscles, resulting in almost continuous flexor spasms.

PYRAMIDAL TRACT EFFECTS ON PROPRIOSPINAL NEURONS

Propriospinal neurons are cells interconnecting several spinal segments, for instance cervical and lumbosacral segments. A class of propriospinal neurons was identified in the segments C3–C4 of the cat's spinal cord which send their axons to the forelimb motor nuclei (Illert and Tanaka, 1978; Illert *et al.*, 1976a,b). These integrative cells receive short-latency inputs from cutaneous afferent fibers and descending inputs from the dorsal mesencephalic tegmentum (probably the colliculi). Furthermore, evidence has been presented that these neurons also receive a monosynaptic input from the pyramidal tract and from the rubrospinal tract (Illert, Lundberg, and Tanaka, 1977). Lundberg (1975) suggested "that a command from the motor cortex for a forelimb movement can be influenced *en route* by on-going visual activity and exteroceptive activity from the forelimb" (p. 263).

Kostyuk and Maisky (1972) described another population of propriospinal neurons in upper lumbar segments with axons traveling over several segments in the dorsolateral funiculus. Their termination was in the lumbosacral cord just dorsal to motor nuclei. In a recent symposium, Kostyuk and Vasilenko (1978) presented evidence that these propriospinal neurons receive convergent actions from corticospinal, rubrospinal, and reticulospinal fibers, but that they are only weakly influenced by primary afferent fibers.

When one is pitching a baseball, the lower extremities must be effectively manoeuvred to give the right upper extremity its greatest effectiveness. For this reason fibres passing to the foot area are commingled with those governing the movements of the arm.

Fulton, 1949

Branching of Corticofugal Fibers from the Sensorimotor Cortex

General Comments

The collateralization of the bulbar pyramid has struck many investigators of the pyramidal tract. Figure 23 is a classical Golgi picture of the cat's brain stem, cut sagittaly. It demonstrates unequivocally the presence of pyramidal tract collateral fibers in the brain stem. These fine fibers can be seen to fan toward the retic-

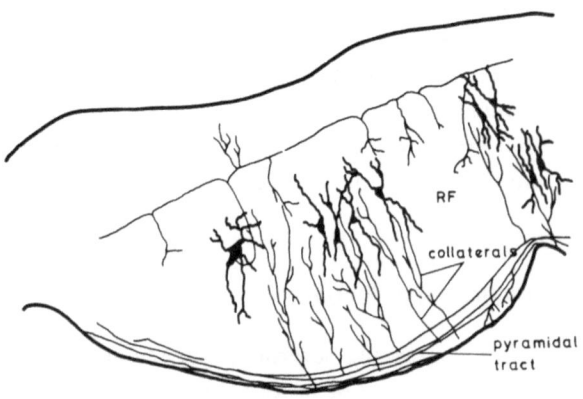

Fig. 23. Parasagittal section of the brain stem of the rat. Redrawn following Scheibel and Scheibel (1958). Note collaterals of the pyramidal tract fanning toward neurons of the reticular formation (RF). Similar contacts are also established with specific brain stem nuclei (not shown).

ular formation and the somatosensory relay nuclei. With the advent of degeneration methods, which allowed tracing the degenerating preterminal and terminals after placing lesions in areas of the sensorimotor cortex, the striking number of corticofugal fibers to subcortical centers became more and more evident. Of course, the presence of "feedback-fibers" from the cortex to deeper relays were known to the masters of neurohistology. Thus, Cajal (1903), impressed by the two-way traffic between cortex and thalamus, commented about the possible significance of these fibers in controlling sensory inflow to the brain by either facilitation or inhibition. This functional aspect of corticofugal control on sensory transmission will be discussed in the next section.

The question can be raised as to whether all corticofugal fibers are derived from collaterals of the pyramidal tract. This is a very unlikely proposition, because corticofugal fibers to the basal ganglia, thalamus, the pontine nuclei, and to other structures can be found to originate from a much wider field than the corticospinal tract. The discussion will be limited to those corticofugal fibers originating from the sensorimotor areas. For much of the older studies (mainly done in cats and rats), see the reviews by Wiesendanger (1969) and Towe (1973a,b). It is now well established that collateralization is a characteristic feature of pyramidal tract fibers, from the cortex down to the spinal cord. The diagram of Fig. 24 is an up-dated version of a previously published one. However, an important qualification is necessary to this statement: whether an ending is from a collateral or not cannot be determined in most anatomical studies. The proof of branching can be gained from the laborious antidromic stimulation technique: cell of origin must be fired antidromically by stimulation of collaterals within the collateral target, as well as by stimulation of the parent fibers or targets further down the neuraxis. Alternatively, cells of presumed collateral targets can be tested for descending monosynaptic and ascending monosynaptic excitation, preferably by demonstration of impulse collision in the common stem fibers. From such studies, it has become apparent that many fibers from sensorimotor areas end in subcortical centers of the thalamus, basal ganglia, pons, and medulla and do not project further down to the spinal cord.

The collateralization within the spinal cord is a matter of much current interest. The presence of branching corticospinal fibers may have been anticipated by the observation that circumscribed lesions of the precentral cortex produced degeneration over a considerable longitudinal extent of the spinal cord (Glees and Cole, 1950; Sherrington, 1889; Tyner, 1974).

As a last general comment, it must be emphasized that only very few electrophysiological studies have been made so far in which particular attention has been paid to the question whether the cortical cells have multiple targets (Atkinson, Séguin, and Wiesendanger, 1974; Endo, Araki, and Yagi, 1973; Humphrey and Corrie, 1978; Steriade and Yossif, 1977). One can anticipate that much future research in cortical physiology will be devoted to this important question. Also, a newly developed technique for double labeling by retrograde axonal transport after injections of fluorescent substances in two presumed targets of a neuronal population (Kuypers, Catsman-Berrevoets, and Padt, 1977) will certainly clarify the question of collaterals for many structures in the near future.

MARIO
WIESENDANGER

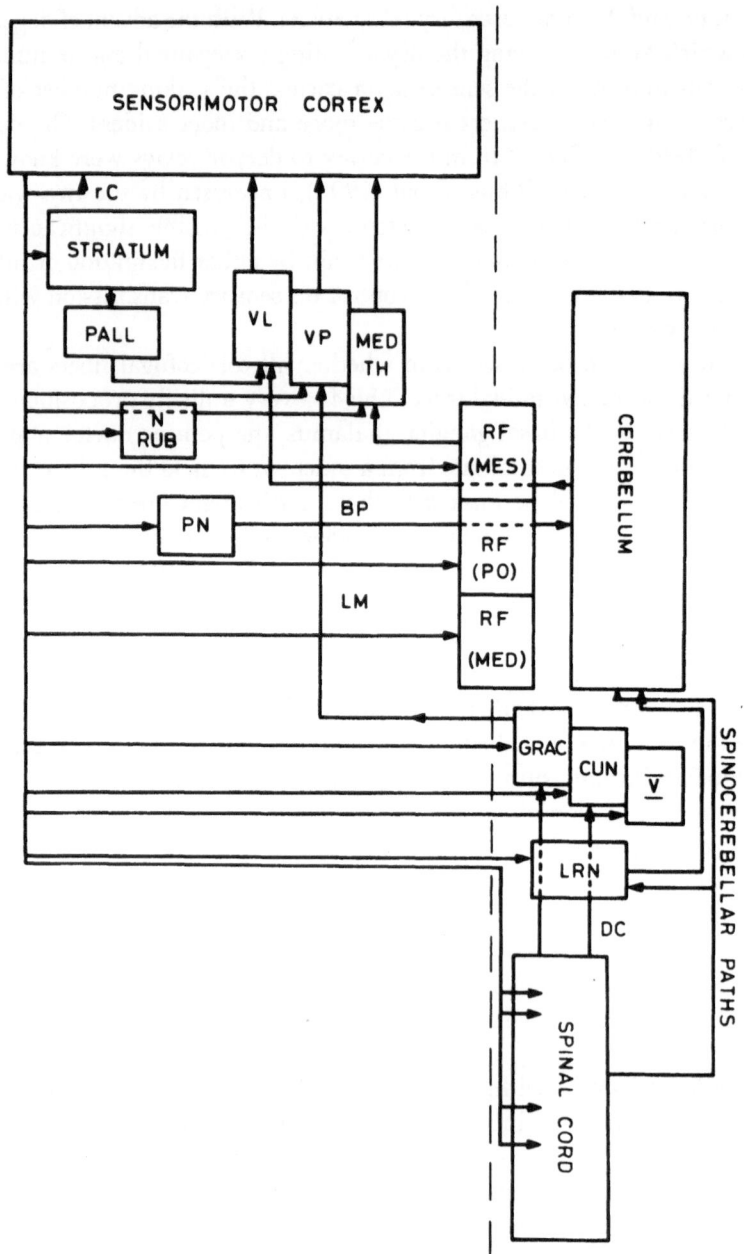

Fig. 24. Schematic representation of the collateralization of the pyramidal tract. Collaterals are well known to occur within the cerebral cortex (rC = recurrent collaterals). Further targets of pyramidal tract collaterals are the basal ganglia (striatum), the specific (VP) and unspecific (MED) thalamic nuclei, the red nucleus (N RUB), the pontine nuclei (PN), the mesencephalic (MES), pontine (PO), and the medullary (MED) reticular formation (RF), the dorsal column and trigeminal nuclei (GRAC, CUN, $\overline{V}$), the lateral reticular nucleus (LRN). Collaterals are also seen to distribute to various spinal levels. The targets of pyramidal tract collaterals are in their turn relays of internal feedback loops. The main reentrance loops are via the lemniscal system (LM)—specific thalamus (VP), via the cerebellum—ventrolateral nucleus of the thalamus (VL), and via the pallidum (PALL)—VL system.

Recurrent collaterals of pyramidal-shaped cells in layers III and V are well known. They probably form a considerable proportion of intra-gray horizontal connections, reaching to a maximum width of 5 to 6 mm (for a recent analysis in the visual cortex of monkeys see Fisken, Garey, and Powell, 1975). Physiologically, both collateral inhibition (Kameda, Nagel, and Brooks, 1969) and collateral facilitation (Takahashi, Kubota, and Umo, 1967) have been observed (see pp. 464). The former mechanism is probably of great importance in sharpening the excited focus or efferent zone of pyramidal tract cells, since weak excitation of efferent neurons is much more attenuated by collateral inhibition than strong excitation. In analogy to the situation in the spinal cord, one may infer that collaterals play a role not only in the elaboration of spatial, but also of temporal, contrast.

Since most cortico-cortical neurons of the monkey's sensorimotor cortex were located to supragranular layers II and III, and since most descending neurons are in layers V and VI (Jones and Wise, 1977) the number of pyramidal tract fibers with long cortico-cortical collaterals must be small. Deschenes (1977) found a small subpopulation of neurons in the motor cortex of cats which were antidromically invaded by stimulation of both the first somatosensory area and of the cerebral peduncle.

Pyramidal Collaterals to the Basal Ganglia

The massive projections from the peri-Rolandic cortex to the caudate nucleus and the putamen are well established (Kemp and Powell, 1970) and have recently been reexamined with axoplasmic flow studies (Goldman and Nauta, 1977; Jones, Coulter, Burton, and Porter, 1977; Künzle, 1975). These connections are highly developed in primates (and especially in man) and undoubtedly play an important role in motor control. The current trend is to consider the basal ganglia as important structures for programming movements, in that their place in the hierarchy of motor control is seen "above" the motor cortex through the return pathway via ventral thalamus and motor cortex (Brooks, 1975; Kemp and Powell, 1971). However, one cannot gloss over the fact that the motor cortex itself projects heavily onto the striatum. Thus, a pertinent question is whether this is achieved, at least partly, via collaterals of the pyramidal tract.

Cajal (1955) saw collaterals of cortico-fugal fibers intruding the striatum, but whether there were pyramidal tract collaterals is not clear. Injections of HRP into the caudate and putamen of squirrel monkeys revealed labeled cells in the motor cortex (ipsi- and contralaterally) which were located in layer V, particularly in the superficial, smaller celled part of layer V (Jones et al., 1977; Jones and Wise, 1977). The corticostriate cells were among the smallest pyramidal-shaped cells. Although corticospinal cells tend to be in the deeper part of layer V, and corticostriate in the most superficial part, one cannot exclude from these anatomical experiments the idea that some of the corticostriate connections are established via collaterals of the pyramidal tract. In fact, Endo et al. (1973) found, while stimulating various subcortical sites, that about 6% of identified pyramidal tract fibers had collaterals to the basal ganglia. Whether the parent axon went further down from the bulbar pyramid is not known.

PYRAMIDAL TRACT CONNECTIONS TO THE THALAMUS

As already mentioned, thalamic nuclei, specific and nonspecific, receive a massive corticofugal projection. Known thalamic targets of the sensorimotor cortex are the ventral nuclear group (including the nuclei ventralis anterior, ventralis lateralis, ventralis posterior lateralis and medialis, and area x) and some of the "nonspecific" nuclei, especially the center median-parafascicular complex (see Burton and Jones, 1976; Jones and Burton, 1976; Jones and Powell, 1970; Künzle, 1976; Rinvik, 1972). Congruent with these anatomical results are electrophysiological observations that stimulation of sensorimotor areas may influence synaptic transmission in these nuclei (for discussion of these results and their functional implication see pp. 453). The motor-cortical input to ventralis-lateralis (Dormont and Massion, 1970) was claimed to be, at least partly, via collaterals of the pyramidal tract (Clare, Landau, and Bishop 1964). Similarly, Tsumoto, Nakamura, and Iwama (1975) presented evidence that the cortical control exerted on cells of the ventrobasal complex is via the pyramidal tract. Endo *et al.* (1973) backfired identified pyramidal tract neurons from the VA nucleus (4.6% of the sample of pyramidal tract neurons), from the VPL nucleus (16.4%), and from the CM nucleus (8.2%). Again, it is not known whether these "PT-neurons" had fibers reaching the spinal cord. Steriade and Yossif (1977) were unable to backfire corticothalamic neurons of SI by peduncular or pyramidal tract stimulation (only 18 cells tested).

These electrophysiological observations thus suggest that part (but hardly all) of the cortical control on thalamic targets is via pyramidal tract, possibly corticospinal, collaterals. According to retrograde labeling experiments with HRP (Jones and Wise, 1977), it appears that a subpopulation of corticothalamic neurons have a common location with corticospinal neurons in layer V. The majority of corticothalamic neurons, however, were found to be in layer VI, and thus represent an independent population.

PYRAMIDAL TRACT CONNECTIONS TO THE RED NUCLEUS

The red nucleus is another target of neurons of sensorimotor areas (see Massion, 1967). With the retrograde labeling technique, Jones and Wise (1977) located the corticorubral neurons mainly in the motor and premotor fields of the monkey brain. There they occupied the same layer as corticospinal neurons, that is, layer V, and are thus potential candidates of corticospinal neurons with collaterals to the ipsilateral red nucleus.

Electrophysiological evidence for pyramidal collaterals to the red nucleus comes from intracellular recording experiments of Tsukahara, Fuller, and Brooks (1968). These authors concluded that large pyramidal tract cells inhibit and small pyramidal tract cells excite neurons of the red nucleus (interaction was also found to occur at cortical level via recurrent collaterals to corticorubral cells in these experiments). Whether these "pyramidal tract" neurons had axons going further down the spinal cord was not investigated. Recently, Humphrey and Corrie (1978) reported that 16% of identified pyramidal tract cells of the precentral efferent zone for wrist extensors in the monkey had collaterals to the red nucleus. Some of these (number not reported) were corticospinal neurons. This percentage compares well with the 18% reported by Endo *et al.* (1973) for the cat.

PYRAMIDAL TRACT CONNECTIONS TO PRECEREBELLAR NUCLEI

As is well known, the sensorimotor areas of the cerebral cortex are reciprocally linked with the cerebellar cortex. Activity of the cerebral cortex is transmitted to the cerebellum by mossy fibers, mainly from the pontine nuclei, and by climbing fibers from the inferior olive (Allen and Tsukahara, 1974). Minor relays (for mossy fibers) are the nucleus reticularis lateralis and the nucleus reticularis tegmenti pontis. After injections of HRP in the inferior olive (Bishop, McCrea, and Kitai, 1976) and the pontine nuclei (Jones and Wise, 1977), labeled cells were located in layer V of the cortical fields of origin. The neurons of the pontine nuclei outnumber by far those of the pyramidal tract (in man by about 20 to 1!). It thus seems unlikely that all connections are derived from pyramidal tract fibers.

Nevertheless, there is positive electrophysiological evidence that a subpopulation of corticospinal cells gives off collaterals to precerebellar nuclei. The contribution of corticopontine neurons from the peri-Rolandic cortex is particularly powerful in monkeys (Dhanarajan, Rüegg, and Wiesendanger, 1977; Wiesendanger, Wiesendanger, and Rüegg, 1979). Of those pontocerebellar neurons influenced from sensorimotor cortex, the contingent activated by corticospinal collaterals was found to be small (about 10%; Rüegg and Wiesendanger, 1977). A similar proportion was found in cat experiments, and the pyramidal tract neurons were found to be of the fast and slow type (Allen, Korn, and Oshima, 1975; Allen, Korn, Oshima, and Toyama, 1975; Endo *et al.*, 1973; Rüegg and Wiesendanger, 1975). In monkeys, corticopontine neurons were found in layer V, usually superficial to the giant Betz cells.

Kitai, Oshima, Provini, and Tsukahara (1969) reported that slowly conducting pyramidal tract fibers terminate in the inferior olive. The corticofugal neurons were identified by stimulation of the bulbar pyramid, and it is therefore not known whether corticospinal fibers provide collaterals as well.

Finally, it was shown that the cortical projection to the contralateral lateral reticular nucleus (Brodal, Maršala, and Brodal, 1967; Bruckmoser, Hepp-Reymond, and Wiesendanger, 1970) is partly mediated by collaterals of corticospinal fibers (Zangger and Wiesendanger, 1973).

PYRAMIDAL TRACT CONNECTIONS TO THE SOMATOSENSORY NUCLEI OF THE MEDULLA

The dorsal column nuclei were among the first relays shown to receive cortical modulation, inhibitory (Towe and Jabbur, 1961), as well as excitatory (Jabbur and Towe, 1961). A large body of experimental data from experiments in rats and cats accumulated in the last 15 years confirming the early observations (see Towe,

1973a,b; Wiesendanger, 1969). As emphasized by Towe (1973b), the fiber fascicle which takes off from the bulbar pyramid is probably the phylogenetically oldest component of the pyramidal tract.

In cats, it was shown that transection of the bulbar pyramid did not abolish cortically induced inhibition in the dorsal column nuclei (Towe and Jabbur, 1961) and in the trigeminal nuclear complex (Hepp-Reymond and Wiesendanger, 1969; Wiesendanger, Hammer, and Tarnecki, 1967), but did abolish cortically induced excitation of trigeminothalamic neurons (Wiesendanger and Felix, 1969).

Again, the question arises what proportion of the invoked pyramidal tract neurons influencing the bulbar nuclei had parent axons running further down the spinal cord? Several authors addressed themselves to this problem: Endo et al. (1973) found that 16 out of 100 identified corticospinal tract cells projected via collaterals to the dorsal column nuclei. Brech et al. (1977), in a very small sample of corticobulbar cells, tested for spinal projection and obtained negative results. The cortical origin of corticonuclear cells in the cat is mainly in area 3a (Brech et al., 1977; Gordon and Miller, 1969). Somatosensory area II appears to contribute little to corticonuclear control (Atkinson et al., 1974): out of 39 SII neurons activated antidromically by a peduncular stimulating electrode, only 4 cells were also driven antidromically by dorsal-column nuclei stimulation, and two of these were further identified as corticospinal neurons. The small contribution of SII corticonuclear neurons was confirmed by Weisberg and Rustioni (1976), Brech et al. (1977) and by Jones and Wise (1977).

In a large sample of neurons of the monkey's precentral cortex (area for wrist extensors), Humphrey and Corrie (1978) found that 7.6% of identified corticospinal neurons provided collaterals to the cuneate nucleus. With the retrograde labeling technique (HRP), corticonuclear cells were found mostly in areas 4 and 3a, then progressively diminishing in number, in areas 3b, 1, and 2 (Weisberg and Rustioni, 1976). In cats, the cells were situated in layer V (Weisberg and Rustioni, 1976), and were not of the giant Betz type (Berrevoets and Kuypers, 1975). The same was reported for monkeys (Catsman-Berrevoets and Kuypers, 1976), with the additional information that the corticonuclear neurons were of rather uniform size in all cortical fields (unlike the corticospinal neurons), and that they were on the whole smaller than corticospinal neurons (Jones and Wise, 1977). Weisberg and Rustioni (1977) found labeled cells also in area 5 and in the supplementary motor area. As one might expect, there is a somatotopical relationship between hind-limb cortex and gracile nucleus, forelimb area and cuneate nucleus, and face area and trigeminal nucleus (Kuypers, Fleming, and Farinholt, 1962).

PYRAMIDAL TRACT CONNECTIONS WITH THE PONTOMEDULLARY RETICULAR FORMATION

The presence of a corticoreticular projection is generally accepted and has been documented in numerous anatomical and electrophysiological studies (see Wiesendanger, 1969). The cortical origin of corticoreticular fibers is, however, not coexistent with that of corticospinal neurons. With degeneration techniques, it was shown, in cats (Kuypers, 1958) and in monkeys (Kuypers and Lawrence, 1967), that area 6 is the main origin of corticoreticular neurons. Also, the corticoreticular system has a larger ipsilateral component than the corticospinal system. The results of degeneration studies were recently confirmed with the retrograde labeling method in cats (Berrevoets and Kuypers, 1975) and in monkeys (Catsman-Berrevoets and Kuypers, 1976). Since the corticoreticular neurons were again located in layer V, and since caudal area 6 also contributes to the corticospinal tract, the possibility of a collateral projection to the reticular formation via corticospinal fibers exists, especially since some corticoreticular neurons are also found in area 4. In electrophysiological experiments, using the antidromic stimulation technique, Humphrey and Corrie (1978) indeed found that 11% of identified corticospinal neurons of the precentral cortex were also backfired by stimulation of the medullary reticular formation. When stimulating the bulbar pyramid in cats for identification of PT cells, Endo *et al.* (1973) found a slightly higher proportion, that is, 16%.

INTRASPINAL BRANCHING PATTERN OF PYRAMIDAL TRACT FIBERS

Abzug *et al.* (1974) were the first to describe the extensive branching pattern of a descending tract of the spinal cord. With microstimulation of local branches and recording of antidromic potentials from cell bodies, these authors found that 50% of vestibulospinal axons giving off local branches at low cervical or upper thoracic levels projected further down to lumbar levels. In a further study (Peterson, Maunz, Pitts, and Mackel, 1975), branching was also established for reticulospinal neurons. For this tract, it was found that 85% of the axons with branches to the cervical ventral horn continued their course to lower levels. It was suggested that this might be a general pattern of descending tracts, and that branching may represent the anatomical basis "for extensive 'hard-wired' coordination of spinal activity at multiple levels." Spinal branching of corticospinal axons has now also been established in cats (Shinoda, Arnold, and Asanuma, 1976); 30% of the fibers giving off collaterals to cervical segments were shown to project to lower levels. Of the 70% corticospinal axons, whose projection was limited to the cervical cord, some had multiple branches within as many as five segments of the cervical cord. This longitudinal extent of branching is longer than the longitudinal extent of a motor nucleus (Romanes, 1951). Thus, it can be concluded that single corticospinal neurons are likely to influence several motor nuclei. With the obligatory interneurons or propriospinal neurons in cats, a further divergence at the spinal level is likely to occur. These most intriguing new discoveries seem to indicate that supraspinal neurons may call up *sets of spinal neurons* in varying combinations (depending on the background excitability), rather than single spinal neurons. Corticospinal neurons with branches to both cervical and lumbar cord were located mainly in between more specific zones controlling the hind-limb and the forelimb, respectively (Armand and Aurenty, 1977).

Investigations in monkeys bearing on this problem are at a preliminary stage (Asanuma, Hongo, Jankowska, Marcus, Shinoda, and Zarzecki, 1978; Asanuma, Zarzecki, Jankowska, Hongo, and Marcus, 1979; Shinoda, Zarzecki, and Asanuma, 1979). Nevertheless, it appears that the same principle of branching holds also for primates. An equal percentage (30%) of corticospinal axons was shown to form collaterals to cervical levels and projecting to lumbar segments. Again it was found that several motor nuclei were likely to receive collaterals from one parent fiber. In the cortex, nearby somata of pyramidal tract fibers often sent collaterals to the same motor nucleus.

Support for the concept of spinal branching of pyramidal tract neurons comes also from a quite different methodological approach: Fetz, Cheney, and German (1976) used correlation techniques to establish connections of cells in the motor cortex to motor units. In awake monkeys performing a motor task, action potentials of the motor cortex were utilized as "trigger signals" for averaging rectified EMG activity of 5 to 6 synergistic wrist and finger muscles. By averaging several-hundred times, the authors showed that about 50% of cells exhibited postspike facilitation in one, two, or even more muscles. Similarly, microstimulation in the "hot spots" during the execution of a conditioned movement revealed a poststimulus facilitation pattern similar to that seen for postspike facilitation (Fetz and Cheney, 1978). This is thus a remarkable demonstration of subtle effects on motor units produced by a single or a few motor cortical cells. It also shows that even miniature cortical outputs diverge to several muscles. Fetz and Cheney (1978) coined the term "muscle field" as an attribute of motor cortical cells.

Taken together with the results of intracortical microstimulation, one can now conclude that "hot spots" of the motor cortex are connected with "hot spots" in spinal motor nuclei. This model explains how motor units can be activated artificially from the motor cortex, and also volitionally (Basmajian, 1967), with relative discreteness. However, it also suggests that, depending on the degree of preexisting modulation of motoneurons, complex sets of spinal neurons, and with it, motor patterns, including several muscles that may be related to one another for a given motor task, may be thrown into action.

CONCLUDING REMARKS ABOUT THE PHENOMENON OF PYRAMIDAL TRACT BRANCHING

That collaterals of the pyramidal tract (and other descending tracts) exist has been known to neuroanatomists for a long time. It is only recently, with the introduction of new techniques, that it has become possible to learn more about the properties of these collaterals. It is clear that a small, but nevertheless measurable, contingent of corticofugal fibers that synapse in the major subcortical relays—thalamic, striatal, precerebellar, and bulbar—are collaterals from pyramidal tract fibers. In the spinal cord, the branching of parent axons seems to be rather extensive, and common innervation of cervical and lumbar segments seems to occur frequently. It would be interesting to know whether this common input to cervical and lumbosacral centers is linked to quadrupedal locomotion, or whether the same pattern exists also in man.

It is evident that much remains to be done in order to understand the details of branching. It might be that a subpopulation of pyramidal tract cells will be found which shows particular combinations of branching, although multiple branching to subcortical centers was seen but rarely (Endo *et al.*, 1973). A virtually unexplored field of research is ahead of us: to understand under which behavioral conditions activation in the various collateral targets occurs.

Hence we accept that there is also a centrifugal side in the process of sensation, of vision, of hearing, and so on; I believe that a further analysis of these descending tracts to pure sensory centres will help physiologists and psychologists to understand some of their experience.

Brouwer, 1933

453

THE PYRAMIDAL
TRACT

Pyramidal Modulation of Sensory Input

General Comments

The identification of collateral inputs from the pyramidal tract to various subcortical targets, which are themselves interposed in return loops to the cortex and to sensory ascending systems, makes it increasingly clear that the separate discussion of "sensory" and "motor" functions is artificial. Downstream traffic of impulses has necessarily, by way of collaterals, an influence on ascending traffic, as ascending impulses have the potency to alter the motor commands. The notion of how it all works in the behavioral situation is unfortunately denied us. The idea of centrifugal control, particularly on sensory transmission, is so intriguing that it has stirred up a great deal of discussion and speculation. Apart from electrophysiological evidence that synchronous descending volleys decrease or increase transmission postsynaptically in somatosensory relays or presynaptically, mainly at the endings of primary afferent fibers, little has been done experimentally to demonstrate, in a behavioral context, the functional significance of the modulatory mechanisms exerted by the pyramidal tract. Centrifugal modulation exerted on primary afferent fiber terminals, spinal tract cells, and medullary and thalamic somatosensory relays has been most intensively investigated. We shall therefore restrict the discussion to the "lemniscal system" and make an attempt to give an overview rather than a detailed review of the subject.

Sites of Pyramidal Modulation

That the *dorsal horn of the spinal cord* and the *dorsal column nuclei* are sites of corticofugal modulation was seen already with surface potential recordings (see Wiesendanger, 1969). In 1961, the now classical papers of Towe and Jabbur and Jabbur and Towe appeared, reporting, in detail, corticofugal inhibition as well as excitation of single neurons in these relays. These observations in cats were extended and confirmed in the main lines in experiments on monkeys (Felix and Wiesendanger, 1970; Harris, Jabbur, Morse, and Towe, 1965). The inhibitory effects were probably not mediated by the pyramidal tract, at least not directly via collaterals in the lower brain stem since inhibition persisted after bulbar pyramidotomy, but rather via the reticular formation (Cesa-Bianchi and Sotgiu, 1969; Sotgiu and Margnelli, 1976).

Similar findings were made in the *trigeminal system* (Darian-Smith, 1969; Towe, 1973a). In this nucleus it was shown that the pyramidal tract excited neurons projecting to the thalamus. Pyramid section abolished the short-latency responses evoked by cortical stimulation but left cortically induced inhibition unchanged (Wiesendanger and Felix, 1969). Inhibition was, however, sensitive to barbiturates (Hepp-Reymond and Wiesendanger, 1969) suggesting a multisynaptic pathway from the cortex to the trigeminal nucleus (possibly via the reticular formation).

Spinal neurons, part of them identified as ascending tract cells, were also shown to be modulated (Fetz, 1968; Kasprzak, Mann, and Tapper, 1970); the pyramidal tract as the mediating pathway was ascertained in these animals by working in "pyramidal preparations" (bulb transected except pyramids). Peripherally evoked discharges were either depressed (predominantly in ventral cells) or enhanced (predominantly in dorsal cells). The pyramidal fibers involved in the control of dorsal cells and of ventral cells, respectively, may represent separate populations since, according to Nyberg-Hansen and Brodal (1963), the neurons projecting to dorsal layers originate mainly from the posterior sigmoid gyrus, those to the ventral layers mainly from the anterior sigmoid gyrus.

Wall (1967) made the interesting observation that the descending control is, at least partly, of tonic nature, in that spontaneous and evoked discharges of spinal neurons in the dorsal horn, and also the size of their receptive fields, changed when all descending influence was reversibly blocked by cooling. Similar observations were also made when "pyramidal" cats were used (Wiesendanger, unpublished). The presence of considerable spontaneous activity of pyramidal tract neurons was shown by Adrian and Moruzzi (1939). Since modulation of synaptic transmission occurs in the dorsal horn and in more ventral layers of the spinal cord, one can expect that *neurons of ascending tracts* will be influenced equally by descending tracts. This has indeed been shown for most ascending tracts (see Wiesendanger, 1969), recently also for the spinocervical tract (Brown, Coulter, Rose, Short, and Snow, 1977) and for the spinothalamic tract (Coulter, Maunz, and Willis, 1974). Undoubtedly, descending tracts other than the pyramidal tract contribute to the central control (Willis, Haber, and Martin, 1977). Future experiments will have to assess the relative roles played by the various descending tracts.

Finally, corticofugal modulation of somatosensory transmission *in the ventrobasal complex of the thalamus* was also reinvestigated recently. Short-latency (monosynaptic) excitations of ventrobasal neurons projecting to the cortex were found by Tsumoto *et al.* (1975). Interestingly, the thalamic neurons receiving pyramidal excitation were mostly those that reacted to passive joint movements. "Hair-units" received predominantly inhibitory effects. Corticofugal facilitation of thalamic transmission was also observed by Andersen, Junge, and Sveen (1972). Again, there is evidence that descending inhibitory control of thalamic transmission has a tonic component: during cooling of the first or second somatosensory areas, Burchfiel and Duffy (1974) noted a remarkably increased transmission in thalamocortical projection fibers.

POSSIBLE SIGNIFICANCE OF CORTICOFUGAL MODULATION OF SOMATOSENSORY INPUT

In his review about descending influences on ascending systems, Towe (1973a) remarked that "the brain may not behave simply as a passive recipient of sensory information" and concluded his chapter as follows:

The old view of the mammal as a passive receiver and responder has long since been supplanted by our current view of the mammal as an active experiencer. In interacting with its environment, the mammal seeks input and indeed 'expects' specific inputs, especially as a consequence of specific output. (p. 715)

Obviously, the brain is capable of modifying its own input according to the particular context in which the sensory information reaches the sense organs. We will summarize views that have been proposed as to how this central control of sensory transmission should be interpreted.

SELECTIVE ATTENTION. Walsh (1957) made the following comment relevant to this proposed mechanism:

probably there is available at the periphery much more sensory data than can be transmitted centrally. It is almost to be expected that according to the conditions at any given time the sensory systems can be readjusted by means of activity in descending pathways *to listen in to specific aspects of the information* that is available. (italics added)

Selective attention implies two things: (1) suppression of unimportant incoming data and (2) enhancement of specific (relevant, important) information. We may speculate that descending (tonic?) inhibitory control, which appears to include the reticular formation, may have such a "filtering" function. On the other hand, the pyramidal tract may contribute to sharpen, in time and in space, the passage of an important piece of information. In neurophysiological terms, the pyramidal tract would selectively increase synaptic transmission in strongly excited lines and, by increasing collateral inhibition, depress weak surround excitation. If it turns out that the same subsets of pyramidal tract neurons that command a given group of motoneurons simultaneously, via collaterals, influence sensory transmission in related channels, the particular role of the pyramidal tract neurons could be to "sensitize" the moving body part. It is known from behavioral experiments in man that tactile sensibility of the hand (Kesten, 1956) and position sense of the arm are more accurate when the subject is moving the limb actively. It is still an open question whether "pyramidal sensitization" plays a role in the phenomenon. Modifications of evoked potentials in the lemniscal path were noted when recordings were made in alert animals trained to perform certain movements. These changes occurred usually *before* the movement started (Coquéry, 1972; Coulter, 1974; Ghez and Pisa, 1972; Gottlieb and Agarwal, 1972). The observations that spinothalamic neurons are also influenced by corticofugal impulses may be of considerable importance to the understanding of selective attention to painful stimuli, a well-known behavioral phenomenon. A special case is the pyramidal control of muscle spindle feedback via the γ-efferents. (The possible role of "γ-loop" in motor control has already been alluded to and is discussed in more detail in Chapter 3). Could it be that pyramidal tract neurons also exert an influence on central structures receiving an input from muscle spindles? If muscle spindles are used by the cerebral cortex and the cerebellum for evaluating the length of muscles, the centers would have to compute the length by correcting the "error" introduced in spindle discharges by the "γ-bias." It is interesting in this respect that slow pyramidal tract neurons and neurons of cortical area 3a (known to receive muscle spindle feedback) discharge intensively when monkeys perform small but precisely controlled movements (Fromm and Evarts, 1977; Tanji, 1976).

These latter examples lead us to the next hypothetical function of pyramidal tract collaterals.

COROLLARY DISCHARGE, EFFERENCE COPY, INTERNAL (ABRIDGED) FEEDBACK LOOPS. These are more or less synonymous concepts for the same mechanism deemed necessary from a purely conjectural point of view. One may say that "internal feedback loops" provide the structural basis for a "corollary discharge" (a term coined by Sperry, 1950, in the sense of collateral discharge). The result would be an "efference copy" computed by the collateral targets, such as sensory nuclei or cerebellar structures. The concept of "efference copy" was introduced by von Holst and Mittelstaedt (1950) to account for constant space perception when one moves the eye actively as opposed to jumping of the surrounding world when the eye is passively moved or when, in paralyzed eyes, the subject intends unsuccessfully to move his eye. "Exafference" signals (i.e., external feedback) and "reafference" signals (i.e., internal feedback) are thought to cancel each other centrally. The reader is encouraged to read the summarizing account on the problem of feedback control of movements discussed at a meeting on "Central Control of Movements" (Evarts, Bizzi, Burke, DeLong, and Thach, 1971).

A special case which is, at least to some extent, amenable to experimental testing, is the internal feedback loop formed by last-order interneurons of the spinal cord that signal to higher centers how effectively central commands were transmitted to motoneurons after interneuronal processing (Oscarsson, 1973). It remains to be shown with microstimulation techniques that the proposed ascending paths receive specific inputs from sets of interneurons ("reafferent" signals) rather than from receptors ("exafferent" signals). The discovery that rhythmic activity persisted in some ascending paths (like the ventral spinocerebellar tract) and not in others (like the dorsal spinocerebellar tract) during "fictitious locomotion" (induced locomotor activity in curarized cats) is an indication that there are separate channels for external and internal feedback signals (Orlovsky and Shik, 1976).

> *If the horizontal spread of the above structures is taken into account (collaterals of the descending axons "plumes" of the apical dendrites, etc.), the resulting imaginary cylinders or "modules" overlap to such an extent that they cannot provide a material basis for the sharp borderlines that delimit the functional columns.*
> Ramon-Moliner, 1970

EXTRINSIC AND INTRINSIC ORGANIZATION OF THE SENSORIMOTOR CORTEX: STRUCTURAL CONSIDERATIONS

EXTRINSIC CONNECTIONS OF CORTICAL AREAS FROM WHICH THE PYRAMIDAL TRACT TAKES ITS ORIGIN

As previously mentioned, the principal areas containing pyramidal tract (PT) neurons are areas 4, 3a, 3b, 1, and 2. To a lesser degree PT neurons were also found in areas 5 and 6 and SII. The main connections of these areas will now be considered.

IPSILATERAL CORTICO-CORTICAL CONNECTIONS. *Area 4* relates somatotopically and reciprocally to areas of the postcentral gyrus. In addition, area 4 has also a forward projection to area 6 on the lateral hemisphere, the so-called *premotor* cortex, and to area 6 on the mesial surface, the so-called *supplementary motor area* (SMA). *Area 6,* including the SMA, has again a backward projection to area 4 and a forward projection to *frontal association areas* (including the "prefrontal cortex"). The *SMA* has an afferent input from pre- and postcentral cortices and projects back to area 4 of the motor cortex, but not to the postcentral cortex. *SII* appears to project more heavily to anterior areas (area 4 and SMA) than posteriorly to *parietal association cortex. SI,* on the other hand, has an equally strong projection in the anterior direction, that is, to the motor cortex and to the SMA, and in the posterior direction, that is, to *area 5* of the parietal association cortex, also termed *supplementary sensory area* by Penfield and Jasper (1954). The two "supplementary areas", motor and sensory, are again reciprocally interconnected. Figure 25, based on the extensive work of Pandya and Kuypers (1969) and of Jones and Powell (1969a), gives a summarizing diagram of the main relations. In such a simplified diagram it is not possible to give a realistic picture of the precise, somatotopically arranged patterns of interconnections that have been shown to exist at least for the primary areas. What the diagram does bring out, however, is the central position of area 6, the *premotor cortex,* in that it interconnects high-order association cortex (frontal and temporal) with the chief area of motor cortical outflow, area 4. Also, one has to consider that, in addition to its connection with area 4, the premotor cortex has a more direct descending control, predominantly

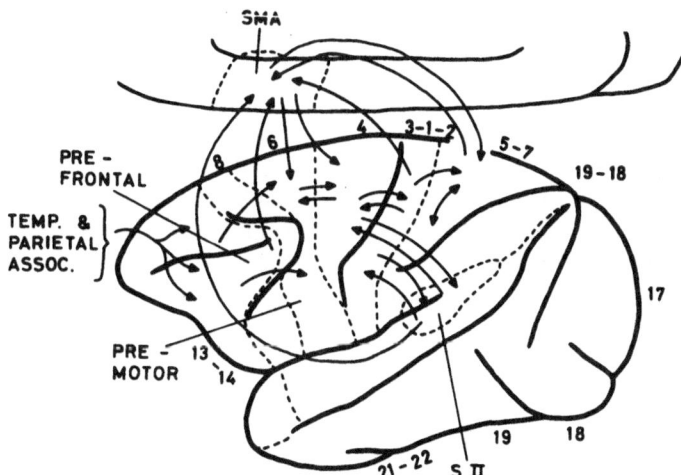

Fig. 25. Cortico-cortical connections in relation to the areas of Brodmann as evidenced by degeneration experiments. An important outcome of these studies is that temporal and parietal association cortex converge in the high order association cortex in the prefrontal area. This area, in turn, projects successively to area 6 and area 4. It thus appears that area 6 on the convexity (and to some extent also the supplementary motor area, SMA) is an area of convergence transmitting complex sensory information to the motor cortex. Somatosensory areas project directly to the precentral motor and supplementary motor areas. From Pandya and Kuypers (1969) and Jones and Powell (1969).

directed to trunk and proximal limb muscles, via brain-stem motor centers. From association areas, the motor cortex may receive processed sensory information. Recognizing this particular position of area 6, Kuypers and his collaborators (Haaxma and Kuypers, 1974; Moll and Kuypers, 1975) devised experiments to test the hypothesis that area 6 may be important in visually guided reaching movements. Preliminary results of Kuypers and his associates and also of Kubota and Hamada (1978), who recorded from single cells in area 6 in monkeys performing a visual tracking task, indeed support Kuypers's hypothesis.

CALLOSAL CONNECTIONS. Homotopic points of the sensorimotor areas, that is, points with the same somatotopic relationship with the periphery, were found to be interconnected precisely via corpus callosum (Jones and Powell, 1969b; Karol and Pandya, 1971; Pandya and Vignolo, 1969) without much divergence and overlap. Interestingly, however, the hand and foot representations were said to lack interhemispheric connections. It would be important to reinvestigate this problem in detail with modern axoplasmic flow methods (Jones, Burton, and Porter, 1975; Künzle, 1976a) in view of recent observations indicating that, in the SI area, representation of distal limb parts appear to be connected callosally, though weakly, by neurons in the subfield 3a (Shanks, Rockel, and Powell, 1975). Also, it would be interesting to follow the interhemispheric connections with the ontogenetic development since it was shown that the callosal connections of kittens gradually decrease with maturation (Innocenti, Fiore, and Caminiti, 1977). Neurons providing callosal fibers are a separate population of cortical cells located predominantly in layer III (Jones and Wise, 1977).

ASCENDING CONNECTIONS. The thalamus provides the most powerful ascending input to the cerebral cortex. It is only recently, with the development of fluorescence histochemical methods, that other sources were discovered which project to wide areas of the cerebral cortex: noradrenergic fibers originating in the pontine nucleus coeruleus (Ungerstedt, 1971) and dopaminergic fibers from the mesencephalon (Hökfelt, Ljungdahl, Fuxe, and Johansson, 1974; Thierry, Stinus, Blanc, and Glowinski, 1973). The catecholaminergic fibers terminate in the most superficial (molecular) layer of the cerebral cortex. The functional significance of this "vegetative" innervation of the cerebral cortex is unknown at present.

It is well established that neurons of the thalamic VPL and VPM nuclei, also collectively termed *ventrobasal complex,* send their axons in a precise somatotopic order to SI and SII (see Jones and Powell, 1973). The VL–VA complex on the other hand projects to precentral areas, VL to area 4 and the caudal part of area 6, VA to the anterior part of area 6 (Strick, 1976).

> The question whether the motor cortex (area 4) receives an additional "lemniscal" projection from the ventrobasal complex has been debated for some time and answered in the affirmative by Amassian and Weiner (1966) and by Rosén and Asanuma (1972). These latter authors recorded in the primate's motor cortex from cells that could be activated at short latencies from peripheral sources and that displayed small receptive fields. A reversible block of the postcentral cortex by cooling did not abolish these responses, suggesting that the motor cortex receives a precise, lemniscal-type input independent of the postcentral cortex. However, recent neuroanatomical investigations failed to confirm any substantial projection from thalamic nuclei receiving lemniscal inputs (Strick, 1976; Kalil, 1976). For the time being, it may be safe not to

exclude entirely the possibility of inputs from VPL neurons to the motor cortex (perhaps by thalamic neurons with axons bifurcating to post- and precentral cortex?); but in view of the negative anatomical results, one would rather favor the idea that topographically organized sensory input reaches the motor cortex via U-fibers from corresponding points of the postcentral cortex that are known to exist (see also Steriade, Wyzinski, and Apostol, 1973). In surface cooling experiments and ablation studies, area 3, which is part of the buried cortex and which receives the bulk of thalamocortical fibers (Jones and Powell, 1969c), may have escaped the lesion.

Other fibers reaching sensorimotor areas originate from the center medianparafascicular complex of the thalamus. These "unspecific" afferent fibers were found to reach superficial layers, whereas "specific" afferent fibers terminated predominantly in layer IV, and to a lesser extent in adjacent layers (Jones, 1975). In the motor cortex, the "specific" thalamocortical terminals are split by the inner band of Baillarger in layer V (Powell, 1973).

Area 5 and area 6 appear to have a common thalamic input from the VA nucleus. Further thalamic sources for area 5 in the monkey are the lateroposterior nucleus, rostral pulvinar, and the ventral part of the laterodorsal nucleus (Robertson, 1977).

Since the VA–VL complex of the thalamus is the target zone for pallidal and cerebellonuclear fibers (see Kemp and Powell, 1971), it may be concluded that the motor cortex, the premotor cortex, and area 5 of the parietal association cortex are dominated by inputs from two large return paths—from basal ganglia and from cerebellum—whereas the somatosensory areas SI and SII are dominated by inputs from specific somatosensory thalamic nuclei.

INTRINSIC CIRCUITRY OF SENSORIMOTOR AREAS

The morphological and microphysiological investigation of the intrinsic network is still at an early stage. However, fast progress has been made in this field in recent years, and available data are already too complex to be discussed in detail in the present account. The reader may get an idea of the new developments and the status of knowledge in 1975 by consulting the proceedings of the 7th International Neurobiology Meeting devoted to problems of "Afferent and Intrinsic Organization of Laminated Structures in the Brain" (Creutzfeldt, 1976). Gradually, one begins to see the picture of functional "modules," defined by their common inputs and outputs. Already in ordinary Nissl-stained material of the motor cortex, one can recognize an orderly arrangement of the largest Betz cells, grouped together in clusters of 3 or more cells (von Economo and Koskinas, 1925), occurring at regular intervals of 380 ± 50 μm (Kemper, Caveness, and Yakovlev, 1973). With the use of retrograde labeling techniques, it became apparent that, in cats, corticospinal neurons were arranged in nests of 5 to 10 cells in areas of 300 to 500 μm (Groos et al., 1978). Similar clusters of labeled cells in the monkey cortex were described by Jones and Wise (1977) upon injections in various subcortical sites. The counterpart of disjunctive organization on the site of origin is found also in the projection fields: ipsi- and contralateral cortico-cortical projections were organized in alternating zones of high- and low-axon terminal density (Grant, Landgren, and

Silfvenius, 1975; Jones *et al.,* 1975; Künzle, 1976a). This indicates that various afferent inputs to a given slab of cortex have interdigitating fields of termination. Finally, a segregation in radially oriented columns appears also when looking at apical dendrites and at thalamo-cortical fibers, both of which are organized in radially oriented bundles (Colonnier, 1966; Kemper *et al.,* 1973; von Bonin and Mehler, 1971) which reinforce the impression of radial cell columns. Thus, one may conclude that the *horizontal* lamination of the cortex is determined by the targets to which the cells project; the *radial* organization of the cortex would then be more an expression of the afferent input. *Within* a "module," elements are present which, by their structure, lend themselves to the role of transmitting incoming thalamocortical impulses in layer IV and of spreading the excitation radially. Figure 26 illustrates a morphological model of a cortical unit, based on the work of Jones (1975). The width of the column would be given by the terminal spread of one or a few thalamocortical fibers. According to Jones, impulses would be relayed to small spiny interneurons (his type 7 cells) which, in turn, with their ascending and descending axonal branches, would spread the excitation over the entire radial extent by making multiple articulations with the spiny apical dendrites of pyramidal cells. The Golgi material also revealed clearly that the radial modules are not isolated from each other, but that there is ample opportunity for horizontal "cross talk." Jones suggests that some of these horizontally oriented cells, particularly his large type I cells, which are similar to the "basket" cells of the human motor cortex described by Marin-Padilla (1969, 1972), might be inhib-

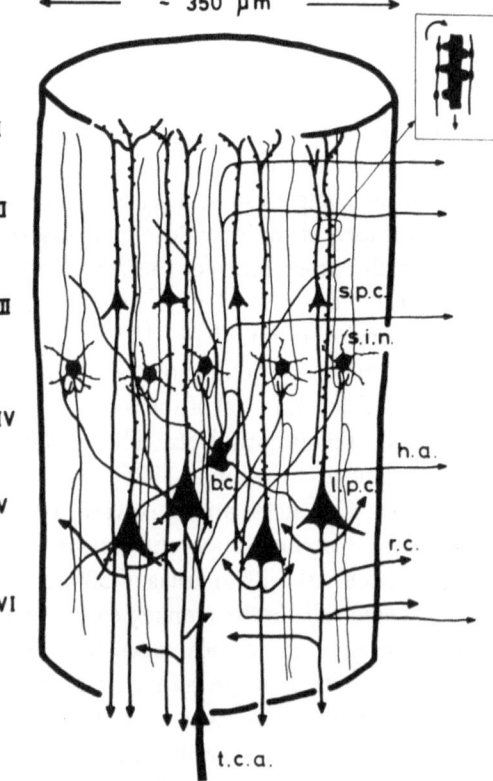

Fig. 26. Model of a cortical column (somatosensory cortex) based on anatomical work of Jones (1975). Width of column is determined by the branching of a thalamocortical afferent fiber (t.c.a.). These fibers synapse with spiny interneurons (s.i.n.) in the IV layer. These elements distribute the excitation radially and establish contacts with dendrites of small (s.p.c.) and large (l.p.c.) pyramidal cells (see inset). Some cells (b.c.) were shown to exert their influence horizontally (h.a.). It was suggested that these cells (similar to "basket cells") may be inhibitory. This model may represent at best a simplified version of the basic cortical circuit.

itory. This would provide for an "inhibitory fringe." Recurrent collaterals of pyramidal cells have also a considerable lateral spread.

This morphological model, which was also proposed in similar form by Szentágothai (1978), is likely to be much too simple; but it is a useful starting point for micro-electrophysiological studies in the various cortical fields (see also quantitative fine-structural studies by Sloper and Powell, 1979a,b,c and by Sloper, Hiorns, and Powell, 1979). Probably, the functional modules will turn out to be quite different from area to area. A first attempt was made by Mountcastle and Powell in a series of four papers that appeared in 1959 (Mountcastle and Powell, 1959a,b; Powell and Mountcastle, 1959a,b). In these classical papers, the "column" concept for the postcentral gyrus of the macaque's brain was developed on the basis of cytoarchitectonic and micro-electrophysiological considerations. The common denominator of a column was its topological relation with the periphery and the submodality of its afferent input. The size of the proposed columns approximates the size of those proposed by Jones (1975), determined by the terminal arbor of one or a few thalamocortical fibers. The "columnar" size is also in the same order of magnitude as the territory occupied by a cluster of the large pyramidal cells in layer V. However, the disjunctive patches of terminals in degeneration pictures or autoradiographs would probably represent larger functional units of cerebral cortex (for a critical discussion of the "column" concept, as originally proposed by Mountcastle, see Towe, 1975). The discussion that follows will treat available micro-electrophysiological data which may or may not corroborate the concept of cortical columns in the motor cortex.

It seems that the machinery of the brain will perform properly and provide the required information if its parts are correctly assembled, just as a radio will emit music or noise, depending on how its circuits are wired.
Kuffler and Nicholls, 1976

MICRO-ELECTROPHYSIOLOGY OF THE MOTOR CORTEX

The large neurons in layer V are easiest to investigate, and many of them can be identified with antidromic stimuli as pyramidal tract (PT) neurons. Accordingly, most data are from this class of neurons. As emphasized repeatedly in the past, one has to realize that this population of neurons is small if compared with the much higher proportion of small neurons. Gradually, some information about "slow" PT neurons is coming forth; virtually nothing is yet known about intrinsic, inhibitory or excitatory, interneurons. Sorting out the various elements and studying their interactions are more difficult in this area than in the allocortex or in the cerebellar cortex, in which the neuronal elements have a much more regular spatial arrangement.

"FAST" AND "SLOW" PT NEURONS

Identified PT neurons (whether antidromic stimulation was effected at bulbar or spinal level is of no importance in this respect) may be divided into fast-conducting or slow-conducting neurons, the dividing line usually being taken at 20 m/sec (Takahashi, 1965). Both types of cells were located in layer V (Naito, Nakamura, Kurosaki, and Tamura, 1969). Readers interested in the biophysical properties of the two classes of neurons may find it useful to consult some of the recent papers on the subject (Calvin and Sypert, 1976; Koike, Okada, Oshima, Takahashi, 1968; Koike, Mano, Okada, and Oshima, 1970; Koike, Mano, Okada, and Oshima, 1972). To study the membrane properties, it is necessary to impale the cells (first achieved by Phillips in 1956). Cardiovascular pulsations and the smallness of some of the PT cells make it very difficult to make measurements over a prolonged time. Intracellular injection of current steps reveals some of the membrane properties that may be of functional significance. Thus, "fast" PT neurons could be fired at relatively high initial frequency, soon followed by adaptation. The phenomenon was interpreted in terms of a posttetanic hyperpolarization linked to an electrogenic sodium pump. The capability of producing short, high-frequency bursts makes the fast PT cell suitable for producing presynaptic facilitation and thus a rapid build-up of depolarization at target neurons (Phillips and Porter, 1964; Porter, 1970). It was also noted that fast PT neurons, driven by transmembrane currents, tend to fire in doublets or triplets with very short interspike intervals. This "intraburst" structure may be of great significance for the frequency potentiation at target synapses as shown by Porter and Muir (1971). The fast PT cells associated with rapid arm movements, such as those described by Evarts (1966), typically display phasic properties: they have a burst discharge preceding and during the movement and do not fire when the limb does not move.

"Slow" PT neurons are different. In biophysical experiments, they tended to fire more regularly under the various conditions of injected current. This property makes them more suitable for tonic, sustained discharges. As first shown by Adrian and Moruzzi (1939), many PT neurons fire "spontaneously" in the absence of any movement. Tonic features of PT cells are typical, but their functional significance is not well understood. One can only speculate that tonic PT neurons are most likely involved in the control of posture via γ-motoneurons and tonic α-motoneurons. As an alternative or additional function, one has to consider the modulatory effects exerted on ascending sensory system (see p. 453).

Fresh light was shed in the discussion about possible roles of "slow" PT neurons by Fromm and Evarts (1977), who discovered that a relatively large proportion of "slow" PT neurons of the monkey's motor cortex discharges intensively in relation to visual tracking movements that require precise and small displacements of a handle. The intense and rather prolonged firing of this category of PT cells were in sharp contrast to the brief bursts of phasic PT cells associated with ballistic movements. It is also remarkable that the intense firing of "tonic" PT neurons was associated with minimal EMG activity. Furthermore, Evarts and Fromm (1977) found that those PT neurons that fired intensely with precisely controlled movements were the most sensitive to small load perturbations added to the handle. These fine experiments add weight to the proposition that the "small" PT neurons may be used for sensory guidance of precise movements. Interestingly, the background ("spontaneous") activity of "slow" PT neurons increased when the monkey was in an aroused state, whereas "fast" PT neurons tended to stop firing spontaneously when the animals were aroused (Steriade, Deschênes, and Oakson, 1974).

These observations emphasize anew the need for further investigation of the very large population (about 90%!) of "slow" PT neurons that have been neglected mainly because of technical difficulties. As pointed out by Humphrey and Corrie (1978), the "viewing" distance of microelectrodes is much shorter for "slow" PT neurons (60–70 μm in those experiments) than for "fast" PT neurons (up to 300 μm). Experiments therefore require appropriately fine microelectrodes.

THE QUESTION OF INTERNEURONS OF THE MOTOR CORTEX

In many structures of the central nervous system, the interneurons (i.e., intrinsic neurons) by far outnumber projection neurons. To identify interneurons of the cerebral cortex electrophysiologically is hardly possible in view of the many projection sites. The lack of an antidromic response (from the pyramid or from the cerebral peduncle) is thus not more than a hint that the neuron under investigation is an interneuron. Additional criteria may help in classifying a neuron as a putative interneuron, but the evidence has remained circumstantial. Thus, Biscoe and Curtis (1967) recorded from nearby cortical neurons by means of multibarrelled micropipettes and found that an increased activity of a "non-PT" neuron could be coupled by a depression of the neighboring PT neuron, suggesting a causal relation of the PT neuron inhibition by the interneuron. More recently, Renaud and Kelly (1974a,b) and Kelly and Renaud (1974) extended these experiments and improved the analysis of data by statistical online computation (joint scatter diagrams). In this way it was possible to see negative cross-correlation functions when recording from neuron pairs. The effect was *unidirectional* from the "non-PT" cell to the PT cell, which makes the assumption of a causal relationship (inhibitory interneuron → output neuron) more likely, without proving it of course. In addition, these histograms made it possible to estimate the latencies of the synaptic effect and its duration. Characteristically, the presumed inhibitory interneurons displayed high-frequency bursts of spikes (up to 600 impulses per sec) which could last 40 msec or longer. Such prolonged bursts, induced by afferent or recurrent actions, did not occur in PT neurons. In the nonanesthetized monkey, Steriade *et al.* (1974) found similar high-frequency bursts in presumed interneurons of the precentral cortex (i.e., neurons which could not be backfired by peduncular stimulation). It is to be hoped that micropharmacological tests and staining of individual, functionally investigated cells, by means of intracellular labeling techniques, will further aid in the understanding of the basic cortical circuits.

INHIBITION AND EXCITATION IN THE MOTOR CORTEX AND ITS POSSIBLE PHARMACOLOGICAL BASIS

INHIBITION. In the first successful impalements of PT neurons, Phillips (1956) noted a hyperpolarization that followed the antidromically evoked spike. It is now generally taken for granted that this hyperpolarization is mediated synaptically by a recurrent collateral pathway, much like the well-known Renshaw inhibitory circuit of the spinal cord. Inhibition appears to be a prominent phenomenon in the motor cortex, which can be elicited not only by antidromic stimulation of PT axons but perhaps most powerfully by cortical surface stimulation (Krnjević,

Randic, and Straughan, 1966a,b). Inhibition is also prominent when afferent paths to the cortex are electrically stimulated. Whenever excitation precedes, inhibition can be because of a recurrent collateral path. Augmenting responses and recruiting responses recorded from the cortical surface upon stimulation of specific and non-specific thalamic relays were shown to be associated with EPSP–IPSP sequences in neurons of the motor cortex. It was also shown that weak tetanic stimulation of the reticular formation could produce a sustained hyperpolarization of motor cortical neurons (see Klee, 1966). Numerous experiments have been made to study inhibitory cortical mechanisms by using electrical stimulation techniques. A large volume of results has been assembled, but at present it would be impossible to subsume all the available data under a unifying general principle of cortical inhibition under "natural" conditions. Even for the best-known type of intracortical inhibition, recurrent inhibition, one can but speculate about its role in motor control. Recent developments of thought about the possible role of collateral inhibition, which appears to be such a prominent phenomenon, have been reviewed by Brooks and Stoney (1971). Briefly, the most popular view is that recurrent inhibition may shape the output signals of the motor cortex by suppressing fringe excitation to peripheral volleys. The most important experimental evidence to support this hypothesis was the finding that receptive fields of cortical cells could be modified by repetitive stimulation of the pyramidal tract (Asanuma and Brooks, 1965; Brooks and Asanuma, 1965).

At a more global and pathological level, one cannot escape the conclusion that proper inhibitory control is a prerequisite for not having constant seizures: convulsive substances, like strychnine, bicucculin, and tetanus toxin, are known to remove inhibitory actions (including some of the cerebral cortex). To learn more about the transmitters involved in mediating inhibitory effects in various pathways has been the aim of several research groups using micro-iontophoretic techniques, notably of Curtis, Krnjević, Phillips, and Stone and their colleagues (see Krnjević, 1974). Most recently, Stone (1973, 1976) suggested that glutamic acid might be the transmitter of pyramidal tract neurons. This hypothesis certainly requires further corroboration and confirmation.

EXCITATION. One of the basic problems in understanding the functional organization in presumed cortical "columns" or "modules" is that of spread of excitation as a result of a small afferent volley. The question was studied by utilizing natural volleys set up by discrete somatosensory stimuli. The results of these studies will be considered below. The degree of spread can also be studied by stimulating directly with a microelectrode. In experiments requiring a remarkable technical skill, Asanuma and co-workers were able to adjust the position of a second recording microelectrode in varying distances from the stimulating microelectrode to measure the spread. Stoney, Thompson, and Asanuma (1968) measured the distance over which the PT cell could be excited directly. It was concluded that, with currents of about 10 μA or less (0.2 ms duration), direct excitation of PT neurons occurred when the tip of the stimulating microcathode was within a radius of 80 to 90 μm from the cell. In other experiments, Asanuma and Rosén (1973) recorded synaptic potentials intracellularly or extracellular spikes in lower layers of the cortex when stimulating in upper layers, and vice versa. It appeared that excitation in layers II, III, and IV spread to the deeper layers only (mono- and disynaptically). Polysynaptic spread was limited to 1 mm (horizontally and vertically) when stimulus intensities of 4 μA were used. Stimulation of deeper layers produced predominantly inhibitory effects in upper layers. Characteristically, inhibition was seen to spread further. These results seemed to fit nicely with the concept of radially organized narrow-cell columns, about 1 mm in diameter, with excitation spreading radially, and bounded laterally by inhibition. Excitation within a column would be followed by feedback inhibition, securing contrast also in the time domain. As recently pointed out by Jankowska *et al.* (1975a,b), the estimates of spread of excitation depends on whether one is using single or repetitive stimuli. It was already found by Stoney *et al.* (1968) that direct activation of PT cells by intracortical stimulation rarely occurred. This was confirmed by Jankowska and co-workers. Furthermore, these authors found that repetitive intracortical stimulation resulted in a tremendous build-up of transsynaptically induced descending waves in the pyramidal tract ("I-waves" of Patton and Amassian, 1960). Weak surface stimulation, on the contrary, activated PT cells more directly without much contamination of indirect (transsynaptic) effects. At present, there is still no precise notion as to how far a focal excitation can spread in the horizontal direction upon *repetitive* intracortical stimulation. According to Jankowska and her colleagues, the spread may be far greater with repetitive intracortical stimulation than with single-pulse surface stimulation. This was concluded from experiments in which the cortical territory was mapped from where particular motoneurons could be influenced by the two methods of cortical stimulation.

It thus appears that much more has to be learned about horizontal connections in the motor cortex and about their functional significance to further support or to refute the concept of efferent columns in the motor cortex.

Sensory Input to the Motor Cortex

Types of Inputs to the Motor Cortex. Although in most primates the central fissure provides a clear-cut boundary between precentral-motor and postcentral-somatosensory cortex, it has always been noted that one may record sensory evoked potentials also from the precentral area as one can elicit movements by stimulation of postcentral areas. This has led some authors to propose the term *motor-sensory* for the precentral and *sensorimotor* for the postcentral cortex. For reviews of the early literature, including evoked potential studies, consult Patton and Amassian (1960), Buser and Ascher (1960), Buser and Imbert (1961), Buser (1966), Buser, Ascher, Bruner, Jassik-Gerschenfeld, and Sindberg (1963), Towe, Patton, and Kennedy (1964), and Wiesendanger (1969).

Recordings from single units of the "peri-cruciate" cortex in cats revealed in essence 3 types of cell responses to sensory stimuli. The "lemniscal"-type neurons displayed small receptive fields on the body surface and reacted usually to one modality only. These cells were typically found in the somatosensory cortex ("s-neurons"). Other cells had quite different properties, discharging to cutaneous stimulation over wide fields. These neurons were typically found in the motor cortex ("m-neurons"). In between were cells which were termed "polyvalent" (Buser and Imbert, 1961) because they reacted to somatosensory stimuli of different submodalities. It has repeatedly been found that the responses of neurons in the motor cortex to sensory stimuli vary "spontaneously" and are greatly influenced by the depth and the kind of anesthesia. Thus, barbiturates strongly depress the reactivity of neurons, whereas chloralose rather enhances the responses. Many of the conflicts in the literature can probably be explained by this high degree of variability and susceptibility to anesthetics. Further difficulties were that, in cats, the complicated boundaries between functional areas are difficult to establish, that neurons were not always identified as PT-cells or non-PT cells, and that the type of sensory stimulation (electrical stimulation of peripheral nerves, bright flashes, or clicks) were often not meaningful in terms of natural functioning. Nevertheless, these experiments were necessary in providing the evidence that sensory information has access to the motor cortex. As expected, the latencies for somatosensory inputs were always much shorter than for visual or auditory input.

> That visual signals have access to the motor cortex has been denied by Kornhuber (1974). He and other authors believe that responses to visual stimuli in the motor cortex are "artefacts" due to chloralose anesthesia. Although it is correct that chloralose enhances the probability to obtain visually evoked responses, they may also be recorded in unanesthetized cats (Buser and Imbert, 1961) and in patients who are not under general anesthesia (Cooper, Crow, and Papakostopoulos, 1975). Under chloralose anesthesia, the type of visual stimulation seemed to be important. Moving dots appeared to be quite effective to excite motor cortical cells (Garcia-Rill and Dubrovsky, 1974). These authors (Garcia-Rill and Dubrovsky, 1973) proposed that visual input to the motor cortex might provide "the definition of the spatial context in which the movements take place" (p. 161).

It is quite evident that signals from distance receptors in the eye and ear are essential in the guidance of movements. Therefore, one might conceive that visual and auditory cues are "built" into a central program, which, in turn, singles out the efferent zones to be activated. We do not know which one of the anatomical connections from the visual system (or auditory system) to the motor cortex is operative and under which conditions. The long time lag between activation of visual cortex and motor cortex in a visual-reaction time experiment suggested a rather complex pathway. Glickstein (1972) proposed that a cortico-pontocerebellar loop may be involved. As was already mentioned, however, area 6 of the motor cortex may receive a more direct (cortico-cortical) input from visual cortex.

Information about the orientation of the body in space, as well as about linear and angular acceleration, would be useful for the guidance of movements (see Chapter 6). In fact, it appears that vestibulomotor regulation may involve not only brain-stem centers and cerebellum but also the motor cortex. Kornhuber and Aschoff (1964) were the first to demonstrate a vestibular input to single cells of the motor cortex in cats. As yet, no analysis of responses to controlled vestibular stimulation has been made in the motor cortex of primates. Interestingly, however, it was reported that area 3a, which receives an input from primary muscle spindle afferents, also has a short-latency input from vestibular afferents (Ödkvist, Schwarz, and Fredrickson, 1974; Sans, Raymond, and Marty, 1970). Since, in cats, connections between area 3a and area 4 have been found, both anatomically

(Grant *et al.*, 1975) and electrophysiologically (Zarzecki, Shinoda, and Asanuma, 1978), it is likely that motor cortex neurons are subject to the influence from vestibular and proprioceptive signals from area 3a.

THE CONCEPTS OF SENSORY COLUMNS AND THE PROBLEM OF INPUT-OUTPUT LINKAGE IN THE MOTOR CORTEX. Brooks, Rudomin, and Slayman (1961a,b) noticed that, under their experimental condition (light anesthesia), a surprisingly high proportion of cells in the motor cortex had small receptive fields. In these papers, a first indication for columnar arrangement was presented. Thus, *clusters* of cortical neurons with common field type were found. The width of these clusters was 0.3 to 1.0 mm, which indicated "that the clustered appearance of neurons with common field type, may reflect anatomical afferent connections" (Brooks, 1963, p. 24). However, the possibility of "cross-talk" between various clusters was also considered in view of previous findings of Smith and Burns (1959) that weak electrical stimulation of the cortex could produce subtle changes in the firing of cortical cells at a long distance from the stimulating electrode.

In subsequent studies, Brooks and colleagues corroborated their findings and now proposed the term *column* for the functional units of motor cortex with a common input (Welt, Aschoff, Kameda, and Brooks, 1967). In the meantime, it was asserted that microstimulation of the motor cortex revealed a similar "columnar" type of organization (Asanuma and Sakata, 1967). It was logical to combine the experiments and to look specifically for the afferent–efferent linkage in the motor cortex (Sakata and Miyamoto, 1968; Asanuma, Stoney, and Abzug, 1968). These and several subsequent papers by Asanuma and co-workers supported the concept of a tight input–output organization in terms of radially oriented cell columns (see Asanuma, 1973, 1975). In awake Cebus monkeys, the columnar organization turned out to be even more pronounced and clear-cut than in cats. The essentials of the impressive results of Asanuma and Rosén (1972a) and Rosén and Asanuma (1972) are now to be found in textbooks. The input–output columns were considered to be "the neural substrates of cortical reflex for motor responses to tactile stimulation of the paw, as for instance, the tactile placing reaction" (Welt *et al.*, 1967, p. 276). The possible role of the tight input–output columns for the grasp reflex was then also discussed by Brooks (1969), a suggestion which found support by the results obtained in the hand area of the primate's precentral cortex (see Asanuma and Rosén, 1972b).

As we already cautioned before, when discussing the experiments of microstimulation, it would be wrong to assume that these "columnar" units function in isolation. The branching of thalamocortical fibers over wide areas of the motor cortex and the presence of horizontally connecting fibers suggest that, under natural conditions, "columns" may be activated simultaneously over a large territory. Furthermore, the findings of both convergence and divergence from spots of the motor cortex to motoneuron pools, the presence of somatosensory "wide field" neurons, and of polysensory convergence in the motor cortex suggest that the model of discrete input–output columns that operate independently is too restrictive. It appears also too simple to view the motor cortex only as "a specialized device for making tactile and proprioceptive adjustments of the motor patterns generated by the cerebellum and basal ganglia" (Kornhuber, 1974, p. 610).

PROPRIOCEPTIVE INPUT TO THE MOTOR CORTEX. It is well established that proprioceptive afferent fibers may influence cells of the somatosensory cortex (Mountcastle and Powell, 1959a). Although it was first thought that signals from primary muscle spindle endings have no access to the cerebral cortex (Mountcastle, Covian, and Harrison, 1952), positive evidence for a group I projection to the sensorimotor cortex of cats was first provided by Amassian and Berlin (1958). The cortical projection area for group I afferent fibers of various forelimb and hindlimb nerves of cats was systematically investigated by two Scandinavian schools (see Rosén, 1972, and Silfvenius, 1972). A correlation of field potential recordings with cytoarchitectonic definition of the recording sites (performed in the same animals) localized the projection area for low-threshold muscle afferents in the baboon's forelimb area of the cerebral cortex (Phillips *et al.*, 1971). Thus, it was found that "positive" points were located predominantly in the posterior portion of area 3a (i.e., near the 3a/3b border), in the depth of the central fissure. This projection system was characterized by fast and secure synaptic transmission, involving the dorsal columns, and by precise somatotopic relationships. None of the cells responding to group I volleys could be backfired from the spinal cord. Controlled stretches of a single forelimb muscle were capable of activating 3a neurons at low amplitude thresholds. Such an example is shown in Fig. 27.

In subsequent studies, convergent inputs from cutaneous (Heath, Hore, and Phillips, 1976) and from vestibular afferent fibers (Schwarz, Deecke, and Fredrickson, 1973) were described to occur in the same cytoarchitectonic field 3a.

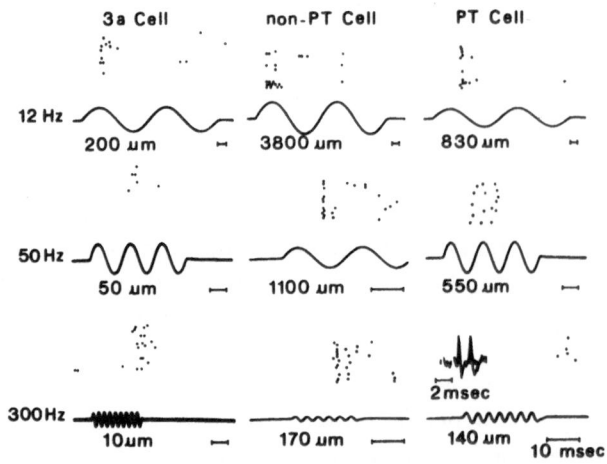

Fig. 27. Responses of cortical neurons to controlled stretches of m. extensor digitorum communis. Frequencies of sinusoidal displacement were varied between 6 and 300 Hz and the threshold amplitudes determined. Selected responses to 12, 50, and 300 Hz are shown for a neuron in area 3a, for a non-PT neuron in the precentral hand area, and for a PT neuron in the same area (inset: antidromic response). Action potentials are transformed and displayed as dots. Each horizontal row of dots corresponds to one of 20 stimulus presentations. The minimal amplitude is noted below each displacement signal. Note that the 3a neurons has the lowest threshold which was typical for the sample. The threshold amplitude typically decreased for all cell types when the displacement frequency was raised indicating that these cells are more responsive to dynamic stimuli. From experiments by Lucier *et al.* (1975).

It was concluded (Phillips *et al.*, 1971) that area 3a, intermediate between area 3b, receiving a predominant cutaneous input, and area 4, the major output region to distal muscles, constitute a subdivision of the first somatosensory cortex dealing primarily with inputs from muscle spindle primaries. However, the modality specificity is not strictly bound to cytoarchitectonic subdivisions. Analysis of the distribution of the various responses reveals gradients rather than sharp boundaries (Dreyer, Schneider, Metz, and Whitsel, 1974). There is electrophysiological and anatomical evidence that afferent inputs arriving at the first cortical projection zones may be further distributed to neighboring zones via intracortical connections (Heath *et al.*, 1976).

The immediate question then arises whether or not these proprioceptive inputs also reach the adjacent motor cortex. Degeneration studies in cats (Grant *et al.*, 1975) provided positive evidence for connections between 3a and area 4. In locally anesthetized cats, Murphy, Wong, and Kwan (1975) were able to activate cortical neurons with stretch stimuli that probably excited muscle spindle primaries exclusively. One focus was situated in area 3a (predimple area), the other in area 4γ. Interestingly, microstimulation of the motor cortex receiving an input from the stretched muscle also produced a contraction in that muscle. This result indicates a tight input–output linkage with respect to proprioceptive signals. In monkeys, Lucier *et al.* (1975) found similar stretch-evoked responses in the motor cortex. Characteristically, the responses were seen much more often among neurons that did not project to the spinal cord. Amplitude-threshold determinations for various frequencies of sinusoidal and ramp stretches led the authors to the conclusion that the responses were elicited not only by activation of secondary spindle endings (as originally proposed on the basis of electrical stimulation experiments, Wiesendanger, 1973) but also of primary spindle endings. Hore, Preston, Durkovic, and Cheney (1976) confirmed that ramp displacements of variable amplitude and slopes, imposed on hind-limb muscles, effectively excited neurons in area 3a as well as in area 4 in the baboon.

In further experiments, it was shown that stretch-evoked discharges of the output cells of the motor cortex are strongly inhibited by conditioning stimuli applied to the supplementary motor area on the mesial surface (Wiesendanger *et al.*, 1975). The hypothesis was discussed that inputs from muscle spindles (and perhaps also from other receptors) are directed primarily to non-PT cells of the motor cortex. The difficulty in activating the large PT cells may reflect the presence of an inhibitory barrier preventing the feedback from muscles that necessarily would influence the output cells. In other words, the supplementary motor region was considered to be one of the control systems that might adjust the gain of synaptic transmission within the motor cortex. Tanji and Taniguchi (1978) tested this hypothesis in monkeys trained to push or pull a handle in response to a somatosensory stimulus. The choice between the two responses had to be made according to a preceding instruction. A group of cells located in the supplementary motor area was selectively correlated with the instruction rather than with the movement *per se*. The EMG response, occurring at a latency compatible with a transcortical loop, was modified in parallel with the changes occurring in the cells of the supplementary motor area. It was thus concluded that this area may indeed contribute in the "gating" of sensory inflow to the motor cortex. These experiments are preliminary and require further investigation. However, they were discussed in this context because it illustrates the type of questions one may now ask, knowing that somesthetic signals can modify motor cortical output. The functional significance of inputs from muscle spindle afferents to the motor cortex are discussed in Chapter 3. It may suffice to mention here that a hypothesis formulated by Phillips in 1969 was influential in the type of research just discussed (see Phillips and Porter, 1977). Phillips proposed that signals from muscle spindle afferents could be utilized by the motor cortex to adjust its own output for a given movement according to the demand imposed by the load.

> *The scientist's trials and errors consist of hypotheses. He formulates them in words, and often in writing. He can then try to find flaws in any one of these hypotheses, by criticizing it, and by testing it experimentally, helped by his fellow scientists who will be delighted if they can find a flaw in it. If the hypothesis does not stand up to these criticisms and to these tests at least as well as its competitors, it will be eliminated.*
>
> Popper, 1972

A Synthetic View and Conclusion

Classical Concept of Pyramidal Tract Function

The description of pyramidal tract function in textbooks is largely based on lesion experiments. Monkeys with chronically severed bulbar pyramids lack digital skill. This is evident when monkeys are tested by means of a "Klüver board." The animals, although far from being deprived of voluntary movements, have difficulties grasping food morsels with the precision grip. They also show a slowing of movements when tested in a reaction-time situation. Rapid movements and the fractionation of movements have been ascribed to the development of cortico-motoneuronal connections in primates. Electrophysiologically, it was shown that monosynaptic cortical influence is directed mainly to distal motoneurons and that repetitive activation results in a conspicious frequency potentiation in synaptic transmission. This phenomenom, investigated by means of intracellular recordings from single pyramidal tract cells revealed, furthermore, that their activity can be modified by sensory stimuli, especially by excitation of cutaneous and deep receptors. It was discovered that this sensory input activates clusters of cells in the motor cortex in a somatotopical order. Artificial activation of a small cortical segment by means of microstimulation revealed, in turn, a somatotopic relationship on the efferent side, which is analogous to the somatotopic relationship on the afferent side. Thus emerged the concept of a tight input–output linkage in radially arranged "columns" of the motor cortex. These columns were considered to be operational units put to use for some motor reactions depending critically on the motor cortex: the placing reaction, the grasp reflex, and, perhaps also, load compensation. In a broader sense, sensory-feedback information was considered to merge with central command signals and to adjust these commands according to the need imposed by the ever changing environment in which movements are performed.

Lastly, single-cell recordings from identified pyramidal tract neurons revealed a class of neurons tightly associated in their discharge rate with a given movement that the monkey is trained to perform. Typically, a relatively short series of action potentials precedes and accompanies a movement, as one would expect of central neurons with the attribute of motor command functions. The functional investigation of the pyramidal tract has thus been brought to a more appropriate and "natural" level of inquiry. The importance of such investigations for understanding the role of the pyramidal tract in motor control is discussed in Chapter 9.

This brief summary of the most salient features of pyramidal tract function raises a number of questions. Some of these questions as well as some apparent inconsistencies will be discussed in this final section. It is hoped that the reader will become aware that nothing should be taken absolutely for granted. Pointing out inconsistencies in current views may lead to new insights.

INCONSISTENCIES AND OPEN QUESTIONS

The association of direct cortico-motoneuronal connections with skilled movements probably needs some qualifications. Thus, the direct connections are made up of *fast-conducting fibers*. So far, the activity of these neurons has been shown to be associated with fast ballistic, but usually very simple, movements. On the other hand, it was recently shown that when monkeys are trained to steer their hands in a narrow target, a task requiring great precision in motor output, *slow-conducting* pyramidal tract neurons discharge intensively and in a larger proportion than for ballistic movements. It is unlikely that a significant proportion of slow-conducting neurons establish monosynaptic connections with motoneurons. This leads to the next problem of defining skills. To move rapidly may be considered as some sort of skill. But many mammals without cortico-motoneuronal connections can move and react as quickly as monkeys or man. Also, the investigation of other descending tracts reveals that they may all show the phenomenon of frequency potentiation at target synapses in the spinal cord (see Shapovalov, 1975).

Are complicated postural adjustments of proximal and trunk muscles, needed in the performance of acrobatics and of dance, less skillful than independent finger movements? Postural adaptations and manipulation may be both equally skillful, and, in fact, the two are not independent. To what extent the pyramidal tract contributes to postural skills has hardly been investigated. But some new experiments by Massion and his co-workers (see p. 413) give an indication that the pyramidal tract plays a role in postural adaptation. It was also emphasized by Goldberger (1974) and by Wiesendanger (1973) that deficits of pyramidotomized monkeys are not restricted to distal muscles. These animals also have clear-cut postural deficits and changes in muscle tone (flexor hypotonia).

It is as if the fast-conducting cortico-motoneuronal component of the pyramidal tract could account exclusively for the classical functions ascribed to the pyramidal tract. What about the 90% or more small-fibered, pyramidal tract cells? Or the neurons that establish collateral connections with sensory relays? Or the neurons that branch extensively at several segments of the spinal cord? What about the neurons that establish ipsilateral connections with the spinal cord? What is the functional significance of the wide territories of some motor cortical cells projecting to a given motor unit? These are all important yet practically unanswered questions. It would be tempting to assume that those neurons that have a wider field of action are used especially for postural adjustments; on the other hand, neurons with more restricted actions on motor units would be used for fractionation of distal finger movements. Surely, this dichotomy in a "specific" and a "diffuse" component of the pyramidal tract is again too simple but may serve as a basic hypothesis. In fact, recent anatomical studies of the cat's motor cortex revealed a "common" zone with neurons projecting to both cervical and lumbosacral segments of the spinal cord and two specific zones with neurons projecting to cervical or lumbosacral segments only (Armand and Aurenty, 1977).

Another complex problem that needs clarification is the role played by the ipsilateral projection in the amazing capacity of mammals to recover from motor deficits following lesions of descending tracts. Could it be that voluntary control (as for instance in stroke patients) is regained, at least to some extent, via ipsilateral pyramidal connections? These fibers may sprout and form new synapses at locations of degenerated synapses.

Conclusion

In summary, the pyramidal tract should be considered as a *multicomponent system*. Its functions have to be understood not in isolation but in conjunction with inputs and targets. There is experimental evidence that the pyramidal tract is important for the control of independent finger movements and postural adjustments (controls that are mutually dependent) and for active "focusing" on somatosensory inputs. The controls are determined by central commands that often appear to blend with peripheral, tactual, and proprioceptive, feedback signals.

Readers will find it useful to consult the recently published monograph on the pyramidal tract by Phillips and Porter (1977), which was published after completion of this chapter.

References

Abzug, C., Maeda, M., Peterson, B. W., and Wilson, V. J. Cervical branching of lumbar vestibulospinal axons. *Journal of Physiology (London)*, 1974, *243*, 499–522.

Adrian, E. D., and Moruzzi, G. Impulses in the pyramidal tract. *Journal of Physiology (London)*, 1939, *97*, 153–199.

Allen, G. I., Korn, H., and Oshima, T. The mode of synaptic linkage in the cerebro-ponto-cerebellar pathway of the cat. I. Responses in the brachium pontis. *Experimental Brain Research*, 1975, *24*, 1–14.

Allen, G. I., Korn, H., Oshima, T., and Toyama, K. The mode of synaptic linkage in the cerebro-ponto-cerebellar pathway of the cat. II. Responses of single cells in the pontine nuclei. *Experimental Brain Research*, 1975, *24*, 15–36.

Allen, G. I., and Tsukahara, N. Cerebrocerebellar communication systems. *Physiological Review*, 1974, *54*, 957–1006.

Amassian, V. E., and Berlin, L. Early cortical projection of group I afferents in the forelimb muscle nerves of cat. *Journal of Physiology (London)*, 1958, *143*, 61P.

Amassian, V. E., and Ross, R. Developing role of sensori-motor cortex and pyramidal tract neurons in contact placing kittens. *Journal de Physiologie* (Paris) 1978, *74*, 165–184.

Amassian, V. E., and Weiner, H. Monosynaptic and polysynaptic activation of pyramidal tract neurons by thalamic stimulation. In D. P. Purpura and M. D. Yahr (Eds.), *The Thalamus*. New York. Columbia University Press, 1966.

Andersen, P., Junge, K., and Sveen, O. Cortico-fugal facilitation of thalamic transmission. *Brain, Behavior and Evolution*, 1972, *6*, 170–184.

Andersen, P., Hagan, P. J., Phillips, G. C., and Powell, T. P. S. Mapping by microstimulation of overlapping projections from area 4 to motor units of the baboon's hand. *Proceedings of the Royal Society, Series B*, 1975, *188*, 31–60.

Armand, J., and Aurenty, R. Dual organization of motor cortico-spinal tract in the cat. *Neuroscience Letters*, 1977, *6*, 1–7.

Artom, G. Untersuchungen über die Myelogenese des Nerven-systems der Affen. *Archiv für Psychiatrie und Nervenkrankheiten*, 1925, *75*, 169–234.

Asanuma, H. Cerebral cortical control of movement. *Physiologist (Wash.)*, 1973, *16*, 143–166.

Asanuma, H. Recent developments in the study of the columnar arrangement of neurons within the motor cortex. *Physiological Reviews*, 1975, *55*, 143–156.

Asanuma, H., and Arnold, A. P. Noxious effects of excessive currents used for intracortical microstimulation. *Brain Research*, 1975, *96*, 103–107.

Asanuma, H., and Brooks, V. B. Recurrent cortical effects following stimulation of internal capsule. *Archives Italiennes de Biologie*, 1965, *103*, 220–246.

Asanuma, H., and Rosén, I. Topographical organization of cortical efferent zones projecting to distal forelimb muscles in the monkey. *Experimental Brain Research*, 1972a, *14*, 243–256.

Asanuma, H., and Rosén I. Functional role of afferent inputs to the monkey motor cortex. *Brain Research*, 1972b, *40*, 3–5.

Asanuma, H., and Rosén, I. Spread of mono- and polysynaptic connections within cat's motor cortex. *Experimental Brain Research*, 1973, *16*, 507–520.

Asanuma, H., and Sakata, H. Functional organization of a cortical efferent system examined with focal depth stimulation in cats. *Journal of Neurophysiology*, 1967, *30*, 35–54.

Asanuma, H., Stoney, S.D., and Abzug, C. Relationship between afferent input and motor outflow in cat motor sensory cortex. *Journal of Neurophysiology*, 1968, *31*, 670–681.

Asanuma, H., Stoney, S. D., and Thompson, W. D. Characteristics of cervical interneurones which mediate cortical motor outflow to distal forelimb muscles of cats. *Brain Research*, 1971, *27*, 79–95.

Asanuma, H., Arnold, A., and Zarzecki, P. Further study on the excitation of pyramidal tract cells by intracortical microstimulation. *Experimental Brain Research*, 1976, *26*, 443–461.

Asanuma, H., Hongo, T., Jankowska, E., Marcus, S., Shinoda, Y., and Zarzecki, P. Pattern of projections of individual pyramidal tract neurons to the spinal cord of the monkey. *Journal de Physiologie (Paris)*, 1978, *74*, 235–236.

Asanuma, H., Zarzecki, P., Jankowska, E., Hongo, T., and Marcus, S. Projection of individual pyramidal tract neurons to lumbar motor nuclei of the monkey. *Experimental Brain Research*, 1979, *34*, 73–89.

Atkinson, D. H., Séguin, J. J., and Wiesendanger, M. Organization of corticofugal neurons in somatosensory area II of the cat. *Journal of Physiology (London)*, 1974, *236*, 663–679.

Balthasar, K., and Schlagenhauff, R. Unilateral agenesis of pyramidal system. *Proceedings, Vth International Congress of Neuropathology*. Excerpta Medica International Congress Series, *100*, 1966, 881.

Bancaud, J., Talairach, J., Geier, S., Bonis, A., Trottier, S., and Manrique, M. Manifestations comportementales induites par la stimulation électrique du gyrus cingulaire antérieur chez l'homme. *Revue Neurologique (Paris)*, 1976, *132*, 705–724.

Barron, D. H. The results of unilateral pyramidal section of rat. *Journal of Comparative Neurology*, 1934, *60*, 45–55.

Basmajian, J. V. Control of individual motor units. *American Journal of Physical Medicine*, 1967, *46*, 480–486.

Bechterew, W. Studium der Funktionen der Präfrontal- und anderer Gebiete der Hirnrinde vermittelst der assoziativ-motorischen Reflexe. *Schweizer Archiv für Neurologie und Psychiatrie*, 1923, *13*, 61–76.

Beck, Ch. H., and Chambers, W. W. Speed, accuracy, and strength of forelimb movement after unilateral pyramidotomy in rhesus monkeys. *Journal of Comparative and Physiological Psychology* (Monograph), 1970, *70*, No. 2, Part 2, 1–22.

Bénita, M., and Condé, H. Effects of local cooling upon conduction and synaptic transmission. *Brain Research*, 1972, *36*, 133–151.

Bernhard, C. G., and Bohm, E. Cortical representation and functional significance of the cortico-motoneuronal system. *Archives of Neurology and Psychiatry* (Chicago), 1954, *72*, 473–502.

Bernhard, C. G., and Bohm, E. Monosynaptic corticospinal activation of forelimb motoneurones in monkeys (macaca mulatta). *Acta Physiologica Scandinavica*, 1954b, *31*, 104–112.

Bernhard, C. G., Bohm, E., and Petersen, J. Investigations on the organization of the cortico-spinal system in monkeys (macaca mulatta). *Acta Physiologica Scandinavica*, 1953, *29*, Supplemention 106, 79–103.

Berrevoets, C. E., and Kuypers, H. G. J. M. Pericruciate cortical neurons projecting to brain stem reticular formation, dorsal column nuclei and spinal cord in the cat. *Neuroscience Letters*, 1975, *I*, 157–262.

Biber, M. P., Kneisley, L. W., LaVail, J. H. Cortical neurons projecting to the cervical and lumbar enlargements of the spinal cord in young and adult Rhesus monkeys. *Experimental Neurology*, 1978, *59*, 492–508.

Biscoe, T. J., and Curtis, D. R. Strychnine and cortical inhibition. *Nature (London)*, 1967, *214*, 914–915.

Bishop, G. A., McCrea, R. A., and Kitai, S. T. A horseradish peroxydase study of the cortico-olivary projection in the cat. *Brain Research*, 1976, *116*, 306–311.

Blinkov, S. M., and Glezer, I. I. *The Human Brain in Figures and Tables. A Quantitative Handbook*. New York: Basic Books, 1968.

Bonin, G. von, and Mehler, W. R. On the columnar arrangements on nerve cells in cerebral cortex. *Brain Research*, 1971, *27*, 1–9.

Brech, A. J., Gordon, G., and Powell, T. P. S. Corticofugal cells responding antidromically to stimulation of the cuneate or gracile nuclei of the cat. *Brain Research*, 1977, *128*, 39–52.

Brinkman, J., and Porter, R. Supplementary motor area of the monkey: Activity of neurones during performance of a learned motor task. *Journal de Physiologie (Paris)*, 1978, *74*, 313–316.

Brodal, A. Self-observations and neuro-anatomical considerations after a stroke. *Brain*, 1973, *96*, 675–694.

Brodal, P., Maršala, Y., and Brodal, A. The cerebral cortical projection to the lateral reticular nucleus in the cat, with special reference to the sensorimotor cortical areas. *Brain Research*, 1967, *6*, 252–274.

Brooks, V. B. Variability and redundancy in the cerebral cortex. *Electroencephalography and Clinical Neurophysiology*, 1963, Supplementum 24, 13–32.

Brooks, V. B. Information processing in the motorsensory cortex. In K. N. Leibovic (Ed.), *Information Processing in the Nervous System*. Berlin: Springer, 1969.

Brooks, V. B. Roles of cerebellum and basal ganglia in initiation and control of movements. *Canadian Journal of Neurological Sciences*, 1975, *2*, 265–277.

Brooks, V. B., and Asanuma, H. Recurrent cortical effects following stimulation of medullary pyramid. *Archives Italiennes de Biologie*, 1965, *103*, 247–278.

Brooks, V. B., and Stoney, S. D. Motor mechanisms: The role of the pyramidal system in motor control. *Annual Review of Physiology*, 1971, *33*, 337–392.

Brooks, V. B., Rudomin, P., and Slayman, C. L. Sensory activation of neurons in the cat's cerebral cortex. *Journal of Neurophysiology*, 1961a, *24*, 286–301.

Brooks, V. B., Rudomin, P., and Slayman, C. L. Peripheral receptive fields of neurons in the cat's cerebral cortex. *Journal of Neurophysiology*, 1961b, *24*, 302–325.

Brouwer, B. Centrifugal influence on centripetal systems in the brain. *Journal of Nervous and Mental Diseases*, 1933, *77*, 621–627.

Brown, A. G., Coulter, J. D., Rose, P. K., Short, A. D., and Snow, P. J. Inhibition of spinocervical tract discharges from localized areas of the sensorimotor cortex in the cat. *Journal of Physiology (London)*, 1977, *264*, 1–16.

Bruckmoser, P., Hepp-Reymond, M. C., and Wiesendanger, M. Cortical influence on single neurons of the lateral reticular nucleus of the cat. *Experimental Neurology*, 1970, *26*, 239–252.

Bubis, J. J., and Landau, W. M. Agenesis of the pyramidal tracts associated with schizencephalic clefts in Rolandic cortex. *Neurology (Minneapolis)*, 1964, *14*, 821–824.

Bucy, P. C., and Fulton, J. F. Ipsilateral representation in the motor and premotor cortex of monkeys. *Brain*, 1933, *56*, 318–342.

Bucy, P. C., Keplinger, J. E., and Siqueira, E. B. Destruction of the "pyramidal tract" in man. *Journal of Neurosurgery*, 1964, *21*, 385–398.

Burchfiel, J. E., and Duffy, F. H. Corticofugal influence upon cat thalamic ventrobasal complex. *Brain Research*, 1974, *70*. 395–411.

Burton, H., and Jones, E. G. The posterior thalamic region and its cortical projection in new world and old world monkeys. *Journal of Comparative Neurology*, 1976, *168*, 249–302.

Buser, P. Subcortical controls of pyramidal activity. In D. P. Purpura and M. D. Yahr (Eds.), *The Thalamus*. New York: Columbia University Press, 1966.

Buser, P., and Ascher, P. Mise en jeu réflexe du système pyramidal chez le chat. *Archives Italiennes de Biologie*, 1960, *98*, 123–164.

Buser, P., and Imbert, M. Sensory projections to the motor cortex in cats: A microelectrode study. In W. A. Rosenblith (Ed.), *Sensory Communication*. New York: Wiley, 1961.

Buser, P., Ascher, P., Bruner, J., Jassik-Gerschenfeld, D., and Sindberg, R. Aspects of sensorimotor reverberation to acoustic and visual stimuli. The role of primary specific cortical areas. In G. Moruzzi, A. Fessard, and H. H. Jasper (Eds.), *Progress in Brain Research*, Vol. 1: *Brain Mechanisms*. Amsterdam: Elsevier, 1963.

Cajal, S. Ramón y. Las fibras nerviosas de origen cerebral del tuberculo cuadrigémino anterior y tálamo óptico. *Trabajos del Laboratorio de Investigaciones Biologia Universita Madrid*, 1903, Vol. *2*, 5–21.

Cajal, S. Ramón y. *Histologie du système nerveux de l'homme et des vertébrés* (Vol. II). Madrid. Instituto Ramón y Cajal, 1955.

Calvin, A., and Sypert, W. Fast and slow pyramidal tract neurons: An intracellular analysis of their contrasting repetitive firing properties in the cat. *Journal of Neurophysiology*, 1976, *39*, 420–434.

Castro, A. J. Motor performance in rats. The effects of pyramidal tract section. *Brain Research*, 1972, 313–323.

Catsman-Berrevoets, C. E., and Kuypers, H. G. J. M. Cells of origin of cortical projections to dorsal column nuclei, spinal cord and bulbar medial reticular formation in the Rhesus monkey. *Neuroscience Letters*, 1976, *3*, 245–252.

Cesa-Bianchi, M. G., and Sotgiu, M. L. Control by brain stem reticular formation of sensory transmission in Burdach nucleus. Analysis of single units. *Brain Research*, 1969, *13*, 129–139.

Chambers, W. W., and Liu, C. N. Cortico-spinal tract of the cat. An attempt to correlate the pattern of degeneration with deficits in reflex activity following neocortical lesions. *Journal of Comparative Neurology*, 1967, *108*, 23–55.

Chang, H.-T., Ruch, T. C., and Ward, A. A. Topographical representation of muscles in motor cortex of monkeys. *Journal of Neurophysiology*, 1947, *10*, 39–56.

Chokroverty, S., Rubino, F. A., and Haller, C. Pure motor hemiplegia due to pyramidal infarction. *Archives of Neurology (Chicago)*, 1975, *32*, 647–648.

Chokroverty, S., Rubino, F. A., and Haller, C. Pure motor hemiplegia due to cerebral cortical infarction. *Archives of Neurology*, 1977, *34*, 93–95.

Clare, M. H., Landau, W. M., and Bishop, G. H. Electrophysiological evidence of a collateral pathway from the pyramidal tract to the thalamus in the cat. *Experimental Neurology*, 1964, *9*, 262–267.

Clark, E., and Dewhurst, K. *An Illustrated History of Brain Function*. Berkeley. University of California Press, 1974.

Clough, J. F. M., and Sheridan, J. D. A fast pathway for cortical influence on cervical gamma motoneurones in the baboon. *Journal of Physiology, (London)*, 1968, *195*, 26-27P.

Clough, J. F. M., Kernell, D., and Phillips, C. G. The distribution of monosynaptic excitation from the pyramidal tract and from primary spindle afferents to motoneurons of the baboon's hand and forearm. *Journal of Physiology (London)*, 1968, *198*, 145-166.

Clough, J. F. M., Phillips, C. G., and Sheridan, J. D. The short latency projection from the baboon's motor cortex to fusimotor neurones of the forearm and hand. *Journal of Physiology (London)*, 1971, *216*, 257-279.

Colonnier, M. The structural design of the neocortex. In J. C. Eccles (Ed.), *Brain and Conscious Experience*, Berlin: Springer, 1966.

Coquéry, J. M. Fonctions motrices et contrôle des messages sensoriels d'origine somatique. *Journal de Physiologie (Paris)*, 1972, *64*, 533-560.

Cooper, R., Crow, H. J., and Papakostopoulos, D. Response of the motor cortex to sensory input in man. *Journal of Physiology (London)*, 1975, *245*, 70-72P.

Cooper, S., and Denny-Brown, D.E. Responses to stimulation of the motor area of the cerebral cortex. *Proceedings of the Royal Society, Series B*, 1928, *102*, 222-236.

Coulter, J. D. Sensory transmission through lemniscal pathway during voluntary movement in the cat. *Journal of Neurophysiology*, 1974, *37*, 831-845.

Coulter, J. D., and Jones, E. G. Differential distribution of corticospinal projections from individual cytoarchitectonic fields in the monkey. *Brain Research*, 1977, *129*, 335-340.

Coulter, J. D., Maunz, R. A., and Willis, W. D. Effects of stimulation of sensorimotor cortex on primate spinothalamic neurons. *Brain Research*, 1974, 351-356.

Coulter, J. D., Ewing, L., and Carter, C. Origin of primary sensorimotor cortical projections to lumbar spinal cord of cat and monkey. *Brain Research*, 1976, *103*, 366-372.

Creutzfeldt, O. (Ed.) Afferent and intrinsic organization of laminated structures in the brain. *Experimental Brain Research, Supplementum 1*, 1976.

Crevel, H. van *The Rate of Secondary Degeneration in the Central Nervous System. An Experimental Study in the Pyramid and Optic Nerve of the Cat*. Leiden: Ijdo, 1958.

Cruveilhier, J. *The Anatomy of the Human Body*. New York: Harper & Brothers, 1853.

Darian-Smith, I. Somatic Sensation. *Annual Review of Physiology*, 1969, *31*, 417-450.

DeMyer, W. Number of axons and myelin sheaths in the adult human medullary pyramids. Study with silver impregnation and iron hematoxylin staining methods. *Neurology (Minneap.)*, 1959, *9*, 42-47.

DeMyer, W., and Russel, J. The number of axons in the right and left medullary pyramids of macaca rhesus and the ratio of axons to myelin sheaths. *Acta Morphologica Neerlandica et Scandinavica*, 1958, *2*, 134-139.

Denny-Brown, D. *The Cerebral Control of Movements*. Liverpool. Liverpool University Press, 1966.

Deschenes, M. Dual origin of fibers projecting from motor cortex to SI in cat. *Brain Research*, 1977, *132*, 159-162.

Dhanarajan, P., Rüegg, D. G., and Wiesendanger, M. An anatomical investigation of the corticopontine projection in the primate (Saimiri sciureus). The projection from motor and somatosensory areas. *Neuroscience*, 1977, *2*, 913-922.

Dormont, J. F., and Massion, J. Duality of cortical control on ventrolateral thalamic activity. *Experimental Brain Research*, 1970, *10*, 205-218.

Dreyer, D. A., Schneider, R. J., Metz, C. B., and Whitsel, B. L. Differential contributions of spinal pathways to body representation in postcentral gyrus of macaca mulatta. *Journal of Neurophysiology*, 1974, *37*, 119-145.

Eccles, J. C. *The Understanding of the Brain*. New York: McGraw-Hill, 1973.

Economo, C. von, and Koskinas, G. N. *Die Cytoarchitektonik der Hirnrinde des erwachsenen Menschen*. Berlin. Springer, 1925.

Edelman, G. M., and Mountcastle, V. B. *The Mindful Brain. Cortical Organization and the Group-Selective Theory of Higher Brain Function.* Cambridge. M.I.T. Press, 1978.

Elger, C. E., Speckmann, E. J., Caspers, H., and Janzen, R. W. C. Cortico-spinal connections in the rat. I. Monosynaptic and polysynaptic responses of cervical motoneurons to epicortical stimulation. *Experimental Brain Research,* 1977, *28,* 385–404.

Endo, K., Araki, T., and Yagi, N. The distribution and pattern of axon branching of pyramidal tract cells. *Brain Research,* 1973, *57,* 484–491.

Evarts, E. V. Pyramidal tract activity associated with a conditioned hand movement in the monkey. *Journal of Neurophysiology,* 1966, *29,* 1011–1027.

Evarts, E. V. Representation of movements and muscles by pyramidal tract neurons of the precentral motor cortex. In M. D. Yahr and D. P. Purpura (Eds.), *Neurophysiological Basis of Normal and Abnormal Motor Activities.* New York: Raven Press, 1967.

Evarts, E. V. Relation of pyramidal tract activity to force exerted during voluntary movement. *Journal of Neurophysiology,* 1968, *31,* 14–27.

Evarts, E. V. Contrasts between activity of precentral and postcentral neurons of cerebral cortex during movement in the monkey. *Brain Research,* 1972, *40,* 25–31.

Evarts, E. V. Sensorimotor cortex activity associated with movements triggered by visual as compared to somesthetic inputs. In F. O. Schmitt and F. G. Worden (Eds.), *The Neurosciences,* Vol. III, Cambridge. M.I.T. Press, 1974.

Evarts, E. V., and Fromm, Ch. Sensory responses in motor cortex neurons during precise motor control. *Neuroscience Letters,* 1977, *5,* 267–272.

Evarts, E. V., Bizzi, E., Burke, R. E., DeLong, M., and Thach, W. T. Central control of movement. *Neuroscience Research Program Bulletin,* 1971, *9,* 170 pp.

Felix, D., and Wiesendanger, M. Cortically induced inhibition in the dorsal column nuclei of monkeys. *Pflügers Archiv,* 1970, *320,* 285–288.

Felix, D., and Wiesendanger, M. Pyramidal and nonpyramidal motor cortical effects on distal forelimb muscles of monkeys. *Experimental Brain Research,* 1971, *12,* 81–91.

Ferrier, D., *The functions of the brain.* London: Smith, Elder, 1886.

Fetz, E. E. Pyramidal tract effects on interneurons in the cat lumbar dorsal horn. *Journal of Neurophysiology,* 1968, *31,* 69–80.

Fetz, E. E., and Cheney, P. D. Muscle fields of primate corticomotoneuronal cells. *Journal de Physiologie (Paris),* 1978, *74,* 239–245.

Fetz, E. E., Cheney, P. D., German, D. C. Corticomotoneuronal connections of precentral cells detected by postspike averages of EMG activity in behaving monkeys. *Brain Research,* 1976, *114,* 505–510.

Fisher, C. M., and Curry, H. B. Pure motor hemiplegia of vascular origin. *Archives of Neurology (Chicago),* 1965, *13,* 30–44.

Fisken, R. A., and Garey, L. J., and Powell, T. P. S. The intrinsic, association and commissural connections of area 17 of the visual cortex. *Philosphical Transactions of the Royal Society (London),* Series B, 1975, *272,* 487–536.

Flechsig, P. *Die Leitungsbahnen im Gehirn und Rückenmark des Menschen auf Grund entwicklungsgeschichtlicher Untersuchungen.* Leipzig: Engelmann, 1876.

François-Franck, C. E. *Leçons sur les fonctions motrices du cerveau.* Paris: Doin, 1887.

Fritsch, G., and Hitzig, E. Ueber die elektrische Erregbarkeit des Grosshirns. *Archiv für Anatomie und Physiologie,* 1870, 300–332.

Fromm, Ch., and Evarts, E. V. Relation of motor cortex neurons to precisely controlled and ballistic movements. *Neuroscience Letters,* 1977, *5,* 259–265.

Fulton, J. F. *Physiology of the nervous system.* New York. Oxford University Press, 1949.

Fulton, J. F. *Physiology of the nervous system.* New York: Oxford University Press, 1951.

Gahéry, Y., and Nieoullon, A. Postural and kinetic co-ordination following cortical stimuli which induce flexion movements in the cat's limbs. *Brain Research,* 1978, *149,* 25–37.

Garcia-Rill, E., and Dubrovsky, B. Topographical organization of visual input to pericruciate cortex of cat. *Brain Research,* 1973, *56,* 151–163.

Garcia-Rill, E., and Dubrovsky, B. Responses of motor cortex cells to visual stimuli. *Brain Research*, 1974, *82*, 185–194.

Garol, H. W. The "motor" cortex of the cat. *Journal of Neuropathology and Experimental Neurology*, 1942, *1*, 139–145.

Ghez, C., and Pisa, M. Inhibition of afferent transmission in cuneate nucleus during voluntary movement in the cat. *Brain Reserach*, 1972, *40*, 145–161.

Gilman, S., and Marco, L. A. Effects of medullary pyramidotomy in the monkey. I. Clinical and electromyographic abnormalities. *Brain*, 1971, *94*, 495–514.

Gilman, S., Marco, L. A., and Ebel, H. C. Effects of medullary pyramidotomy in the monkey. II. Abnormalities of spindle afferent responses. *Brain*, 1971, *94*, 515–530.

Gilman, S., Lieberman, J. S., and Marco, L. A. Spinal mechanisms underlying the effects of unilateral ablation of areas 4 and 6 in monkeys. *Brain*, 1974, *97*, 49–64.

Gilman, S., Carr, D., and Hollenberg, J. Kinematic effects of deafferentation and cerebellar ablation. *Brain*, 1976, *99*, 311–330.

Glees, P. *Experimental Neurology*. Oxford: Oxford University Press, 1961.

Glees, P., and Cole, J. Recovery of skilled motor functions after small repeated lesions of motor cortex in macaque. *Journal of Neurophysiology*, 1950, *13*, 137–148.

Glickstein, M. Brain mechanisms in reaction time. *Brain Research*, 1972, *40*, 33–37.

Goldberger, M. E. The extra-pyramidal systems of the spinal cord. II. Results of combined pyramidal and extra-pyramidal lesions in the macaque. *Journal of Comparative Neurology*, 1969, *135*, 1–26.

Goldberger, M. E. Recovery of movement after CNS lesions in monkeys. In D. G. Stein *et al.* (Eds.), *Plasticity and Recovery of Function in the Central Nervous System*. New York: Academic Press, 1974.

Goldman, P. S., and Nauta, W. J. H. An intricately patterned prefronto-caudate projection in the Rhesus monkey. *Journal of Comparative Neurology*, 1977, *171*, 369–386.

Gordon, G., and Miller, R. Identification of cortical cells projecting to the dorsal column nuclei of the cat. *Quarterly Journal of Experimental Physiology*, 1969, *54*, 85–98.

Gorska, T. Instrumental conditioned reflexes after pyramidotomy in dogs. *Acta Biologica Experimentales (Warsaw)*, 1967, *27*, 103–121.

Gorska, T., Jankowska, E., and Mossakowski, M. Effects of pyramidotomy on instrumental conditioned reflexes in cats. I. Manipulatory reflexes. *Acta Biologica Experimentales (Warsaw)*, 1966a, *26*, 441–450.

Gorska, T., Jankowska, E., and Mossakowski, M. Effects of pyramidotomy on instrumental conditioned reflexes in cats. II. Reflexes derived from unconditioned reactions. *Acta Biologica Experimentales (Warsaw)*, 1966b, *26*, 451–462.

Gottlieb, G. L., and Agarwal, G. C. The role of the myotatic reflex in the voluntary control of movements. *Brain Research*, 1972, *40*, 139–143.

Granit, R. *Receptors and Sensory Perception*. New Haven: Yale University Press, 1955.

Grant, G., Landgren, S., and Silfvenius, H. Columnar distribution of U-fibres from the postcruciate cerebral projection area of the cat's group I muscle afferents. *Experimental Brain Research*, 1975, *24*, 57–74.

Grigg, P., and Preston, J. B. Baboon flexor and extensor fusimotor neurons and their modulation by motor cortex. *Journal of Neurophysiology*, 1971, *34*, 428–436.

Groos, W. P., Ewing, L. K., Carter, C. M., and Coulter, J. D. Organization of corticospinal neurons in the cat. *Brain Research*, 1978, *143*, 393–419.

Grünbaum, A. S. F., and Sherrington, C. S. Observation on the physiology of the cerebral cortex of the anthropoid apes. *Proceedings of the Royal Society*, 1903, *72*, 152–155.

Gustafsson, B., and Jankowska, E. Direct and indirect activation of nerve cells by electrical pulses applied extracellularly. *Journal of Physiology (London)*, 1976, *258*, 33–61.

Hassler, R., and Muhs-Clement, K. Architektonischer Aufbau des sensomotorischen und parietalen Cortex der Katze. *Journal für Hirnforschung*, 1964, *6*, 377–420.

Haaxma, R., and Kuypers, H. G. J. M. Role of occipitofrontal cortico-cortical connections in visual guidance of relatively independent hand and finger movements in rhesus monkeys. *Brain Research*, 1974, *71*, 361–366.

Harris, F., Jabbur, S. J., Morse, R. W., and Towe, A. L. Influence of the cerebral cortex on the cuneate nucleus of the monkey. *Nature (London)*, 1965, *208*, 1215–1216.

Harting, J. K., and Noback, Ch. R. Corticospinal projections from the pre- and postcentral gyri in the squirrel monkey *(Saimiri sciureus)*. *Brain Research*, 1970, *24*, 322–328.

Heath, C. J., Hore, J., and Phillips, C. G. Inputs from low threshold muscle and cutaneous afferents of hand and forearm to areas 3a and 3b of Baboon's cerebral cortex. *Journal of Physiology (London)*, 1976, *257*, 199–227.

Heffner, R., and Masterton, B. Variation in form of the pyramidal tract and its relationship to digital dexterity. *Brain, Behavior and Evolution*, 1975, *12*, 161–200.

Hepp-Reymond, M.-C., and Wiesendanger, M. Pyramidal influence on the spinal trigeminal nucleus of the cat. *Archives Italiennes de Biologie*, 1969, *107*, 54–66.

Hepp-Reymond, M.-C., and Wiesendanger, M. Unilateral pyramidotomy in monkeys: Effect on force and speed of a conditioned precision grip. *Brain Research*, 1972, *36*, 117–131.

Hepp-Reymond, M.-C. Trouche, E., and Wiesendanger, M. Effects of unilateral and bilateral pyramidotomy on a conditioned rapid precision grip in monkeys *(Macaca fascicularis)*. *Experimental Brain Research*, 1974, *21*, 519–527.

Hering, H. E. Ueber centripetale Ataxie beim Menschen und beim Affen. *Neurologisches Centralblatt*, 1897, *16*, 1077–1094.

Hern, J. E. C., Landgren, S., Phillips, C. G., and Porter, R. Selective excitation of corticofugal neurones by surface-anodal stimulation of the baboon's motor cortex. *Journal of Physiology (London)*, 1962, *161*, 73–90.

Hess, W.R. *Hypothalamus and Thalamus*. Documentary pictures. Stuttgart: Thieme, 1956.

Hick, W. E. On the rate of gain of information. *Quarterly Journal of Experimental Psychology*, 1952, *4*, 11–26.

Hitzig, E. *Untersuchungen über das Gehirn. Abhandlungen physiologischen und pathologischen Inhalts*. Berlin: Hirschwald, 1874.

Hoff, E. C., and Hoff, H. E. Spinal terminations of the projection fibres from the motor cortex of primates. *Brain*, 1934, *57*, 454–474.

Hökfelt, T., Ljungdahl, A., Fuxe, K., and Johansson, O. Dopamine nerve terminals in the rat limbic cortex: Aspects of the dopamine hypothesis of schizophrenia. *Science*, 1974, *184*, 177–179.

Holmes, G. L., and May, W. S. On the exact origin of the pyramidal tracts in man and other mammals. *Brain*, 1909, *32*, 1–42.

Holst, E. von, and Mittelstaedt, H. Das Reafferenzprinzip (Wechselwirkungen zwischen Zentralnervensystem und Peripherie). *Naturwissenschaften*, 1950, *37*, 464–476.

Hore, J., Phillips, C.G., and Porter, R. The effects of pyramidotomy on motor performance in the brushtailed possum *(Trichosurus vulpecula)*. *Brain Research*, 1973, *49*, 181–184.

Hore, J., Preston, J. B., Durkovic, R. G., and Cheney, P. D. Responses of cortical neurons (areas 3a and 4) to ramp stretch of hindlimb muscles in the baboon. *Journal of Neurophysiology*, 1976, *39*, 484–500.

Hultborn, H. Convergence on interneurones in the reciprocal Ia inhibitory pathway to motoneurones. *Acta Physiologica Scandinavica*, 1972, *85*, Supplementum 375.

Humphrey, D. R., and Corrie, W. S. Properties of the pyramidal tract neuron system within a functionally defined subregion of primate motor cortex. *Journal of Neurophysiology*, 1978, *41*, 216–243.

Humphrey, T. The development of the pyramidal tracts in human fetuses, correlated with cortical differentiation. In D. B. Tower and J. P. Schadé (Eds.), *Structure and Function of the Cerebral Cortex*. Amsterdam: Elsevier, 1960.

Illert, M., and Tanaka, R. Integration in descending motor pathways controlling the forelimb in the cat. 4. Corticospinal inhibition of forelimb motoneurones mediated by short propriospinal neurones. *Experimental Brain Research*, 1978, *31*, 131–141.

Illert, M., Lundberg, A., and Tanaka, R. Integration in descending motor pathways controlling the forelimb in the cat. 1. Pyramidal effects on motoneurones. *Experimental Brain Research*, 1976a, *26*, 509–519.

Illert, M., Lundberg, A., and Tanaka, R. Integration in descending motor pathways controlling the forelimb in the cat. 2. Convergence on neurones mediating disynaptic cortico-motoneuronal excitation. *Experimental Brain Research*, 1976b, *26*, 521–540.

Illert, M., Lundberg, A., and Tanaka, R. Integration in descending motor pathways controlling the forelimb in the cat. 3. Convergence on propriospinal neurones transmitting disynaptic excitation from the corticospinal tract and other descending tracts. *Experimental Brain Research*, 1977, *29*, 323–346.

Innocenti, G. M., Fiore, L., and Caminiti, R. Exuberant projection into the corpus callosum from the visual cortex of newborn cats. *Neuroscience Letters*, 1977, *4*, 237–242.

Ioffe, M. E. Pyramidal influences in establishment of new motor coordinations in dogs. *Physiology and Behavior*, 1973a, *11*, 145–153.

Ioffe, M. E. Supraspinal influences on spinal mechanisms activated prior to learned movement. *Acta Neurobiologica Experimentales (Warsaw)*, 1973b, *33*, 729–741.

Jabbur, S. J., and Towe, A. L. Cortical excitation of neurons in dorsal column nuclei of the cat, including an analysis of pathways. *Journal of Neurophysiology*, 1961, *24*, 499–509.

Jackson, Hughlings, J. On the anatomical and physiological localisation of movements in the brain. In J. Taylor (Ed.), *Selected Writings of John Hughlings Jackson*. New York: Basic Books, 1958.

Jane, J. A., Yashon, D., DeMyer, W., and Bucy, P. C. The contribution of the precentral gyrus to the pyramidal tract of man. *Journal of Neurosurgery*, 1967, *26*, 244–248.

Jankowska, E. Identification of interneurons interposed in different spinal reflex pathways. In M. Santini (Ed.), *Golgi Centennial Symposium: Perspectives in Neurobiology*. New York: Raven Press, 1975.

Jankowska, E., Padel, Y., and Tanaka, R. Projections of pyramidal tract cells to α-motoneurones innervating hindlimb muscles in the monkey. *Journal of Physiology (London)*, 1975a, *249*, 637–667.

Jankowska, E., Padel, Y., and Tanaka, R. The mode of activation of pyramidal tract cells by intra-cortical stimuli. *Journal of Physiology (London)*, 1975b, *249*, 617–636.

Jankowska, E., Padel, Y., and Tanaka, R. Disynaptic inhibition of spinal motoneurones from the motor cortex in the monkey. *Journal of Physiology (London)*, 1976, *258*, 467–487.

Jasper, H., Ricci, G. F., and Doane, B. Patterns of cortical neuronal discharge during conditioned responses in monkeys. In G. E. W. Wolstenholme and C. M. O'Connor (Eds.), *Neurological Basis of Behavior*. London: Churchill, 1958.

Jones, E. G. Varieties and distribution of non-pyramidal cells in the somatic sensory cortex of the squirrel monkey. *Journal of Comparative Neurology*, 1975, *160*, 205–268.

Jones, E. G., and Burton, H. Areal differences in the laminar distribution of thalamic afferents in cortical fields of the insular, parietal and temporal regions of primates. *Journal of Comparative Neurology*, 1976, *168*, 197–247.

Jones, E. G., and Powell, T. P. S. Connexions of the somatic sensory cortex of the rhesus monkey. I. Ipsilateral cortical connexions. *Brain*, 1969a, *92*, 477–502.

Jones, E. G., and Powell, T. P. S. Connexions of the somatic cortex of the rhesus monkey. II. Contralateral cortical connexions. *Brain*, 1969b, *92*, 717–730.

Jones, E. G., and Powell, T. P. S. The cortical projection of the ventroposterior nucleus of the thalamus in the cat. *Brain Research*, 1969c, *13*, 298–318.

Jones, E. G., and Powell, T. P. S. Connexions of the somatic sensory cortex of the rhesus monkey. III. Thalamic connexions. *Brain*, 1970, *93*, 37–56.

Jones, E. G., and Powell, T. P. S. Anatomical organization of the somatosensory cortex. In A. Iggo (Ed.), *Handbook of Sensory Physiology*, (Vol. II). Berlin: Springer, 1973.

Jones, E. G., and Wise, S. P. Size, laminar and columnar distribution of efferent cells in the sensory-motor cortex of monkeys. *Journal of Comparative Neurology*, 1977, *175*, 391–438.

Jones, E. G., Burton, H., and Porter, R. Commissural and corticocortical "columns" in the somatic sensory cortex of primates. *Science*, 1975, *190*, 572–574.

Jones, E. G., Coulter, J. D., Burton, H., and Porter, R. Cells of origin and terminal distribution of corticostriatal fibers arising in the sensory-motor cortex of monkeys. *Journal of Comparative Neurology*, 1977, *173*, 53–80.

Jung, R., and Dietz, V. Verzögerter Start der Willkürbewegung bei Pyramidenläsionen des Menschen. *Archiv für Psychiatrie und Nervenkrankheiten*, 1975, *221*, 87–109.

Kalil, K. Motor and sensory regions of the rhesus monkey ventral thalamic nuclei defined by their afferent and efferent connections. *Abstracts, Society of Neuroscience*, 1976, *2*, 546.

Kalil, K., and Schneider, G. E. Motor performance following unilateral pyramidal tract lesions in the hamster. *Brain Research*, 1975, *100*, 170–174.

Kameda, K., Nagel, R., and Brooks, V. B. Some quantitative aspects of pyramidal collateral inhibition. *Journal of Neurophysiology*, 1969, *32*, 540–553.

Karol, E. A., and Pandya, D. N. The distribution of the corpus callosum in the rhesus monkey. *Brain*, 1971, *94*, 471–486.

Kasprzak, H., Mann, M. D., and Tapper, D. N. Pyramidal modulation of responses of spinal neurons to natural stimulation of cutaneous receptors. *Brain Research*, 1970, *24*, 121–124.

Kelly, J. S., and Renaud, L. P. Physiological identification of inhibitory interneurones in the feline pericruciate cortex. *Neuropharmacology*, 1974, *13*, 463–474.

Kemp, J. M., and Powell, T. P. S. The corticostriate projection in the monkey. *Brain*, 1970, *93*, 525–546.

Kemp, J. M., and Powell, T. P. S. The connexions of the striatum and globus pallidus: Synthesis and speculation. *Philosophical Transactions of the Royal Society (London), Series B*, 1971, *262*, 441–457.

Kemper, T. L., Caveness, W. F., and Yakovlev, P. I. The neuronographic and metric study of the dendritic arbours of neurons in the motor cortex of macaca mulatta at birth and at 24 months of age. *Brain*, 1973, *96*, 765–782.

Kernell, D., and Wu, C.-P. Post-synaptic effects of cortical stimulation of forelimb motoneurones in the baboon. *Journal of Physiology (London)*, 1967, *191*, 673–690.

Kesten, W. Die Grenzen der haptischen Leitungsfähigkeit. *Zeitschrift für Biologie*, 1956, *109*, 24–40.

Kitai, S. T., Oshima, T., Provini, N., and Tsukahara, N. Cerebro-cerebellar connections mediated by fast and slow conducting pyramidal tract fibers of the cat. *Brain Research*, 1969, *15*, 267–271.

Klee, M. R. Different effects on the membrane potential of motor cortex units after thalamic and reticular stimulation. In D. P. Purpura and M. D. Yahr (Eds.), *The Thalamus*, New York: Columbia University Press, 1966.

Koeze, T. H. The independence of cortico-motoneuronal and fusimotor pathways in the production of muscle contraction by motor cortex stimulation. *Journal of Physiology (London)*, 1968, *197*, 87–105.

Koeze, T. H. Thresholds of cortical activation of baboon α- and γ-motoneurones during halothane anaesthesia. *Journal of Physiology (London)*, 1973, *229*, 319–337.

Koeze, T. H., Phillips, C. G., and Sheridan J. D. Thresholds of cortical activation of muscle spindles and α-motoneurones of the baboon's hand. *Journal of Physiology (London)*, 1968, *195*, 419–449.

Koike, H., Okada, Y., Oshima, T., and Takahasi, K. Accomodative behavior of cat pyramidal tract cells investigated with intracellular injection of currents. *Experimental Brain Research*, 1968, *5*, 173–188.

Koike, H., Mano, Okada, Y., and Oshima, T. Repetitive impulses generated in fast and slow pyramidal tract cells by intracellularly applied current steps. *Experimental Brain Research*, 1970, *11*, 263–281.

Koike, H., Mano, N., Okada, Y., and Oshima, T. Activities of the sodium pump in cat pyramidal tract cells investigated with intracellular injection of sodium ions. *Experimental Brain Research*, 1972, *14*, 449–462.

Konorski, J. *Integrative activity of the brain*. An interdisciplinary approach. Chicago: University of Chicago Press, 1967.

Kornhuber, H. H. The vestibular system and the general motor system. In H. H. Kornhuber (Ed.), *Handbook of Sensory Physiology*. Vol. VI/2: *Vestibular System*. Berlin: Springer, 1974.

Kornhuber, H. H., and Aschoff, J. C. Somatisch-vestibuläre Integration an Neuronen des motorischen Cortex. *Naturwissenschaften*, 1964, *51*, 62–63.

Kostyuk, P. G., and Maisky, V. A. Propiospinal projections in the lumbar spinal cord of the cat. *Brain Research*, 1972, *39*, 530–535.

Kostyuk, P. G., and Vasilenko, D. A. Propriospinal neurones as a relay system for transmission of corticospinal influences. *Journal de Physiologie (Paris)*, 1978, *74*, 247–250.

Krnjević, K. Chemical nature of synaptic transmission in vertebrates. *Physiological Reviews*, 1974, *54*, 418–540.

Krnjević, K., Randic, M., and Straughan, D. W. An inhibitory process in the cerebral cortex. *Journal of Physiology (London)*, 1966a, *184*, 16–48.

Krnjević, K., Randic, M., and Straughan, D. W. Nature of a cortical inhibitory process. *Journal of Physiology (London)*, 1966b, *184*, 49–77.

Kubota, K., and Hamada, I. Visual tracking and neuron activity in the postarcuate area in monkeys. *Journal de Physiologie (Paris)*, 1978, *74*, 297–312.

Kuffler, S. W., and Nicholls, J. G. *From Neuron to Brain*. Sunderland, Mass.: Sinauer Associates, 1976.

Künzle, H. Bilateral projections from precentral motor cortex to the putamen and other parts of the basal ganglia. An autoradiographic study in *Macaca fascicularis*. *Brain Research*, 1975, *88*, 195–210.

Künzle, H. Alternating afferent zones of high and low axon terminal density within the macaque motor cortex. *Brain Research*, 1976a, *106*, 365–370.

Künzle, H. Thalamic projections from the precentral motor cortex in *Macaca fascicularis*. *Brain Research*, 1976b, *105*, 253–267.

Künzle, H. An autoradiographic analysis of the efferent connections from "premotor" and adjacent prefrontal regions (areas 6 and 9) in *Macaca fascicularis*. *Brain, Behavior and Evolution*, 1978, *15*, 185–234.

Kuypers, H. G. J. M. Corticobulbar connexions from the pericentral cortex to the pons and lower brain stem in monkey and chimpanzee. *Journal of Comparative Neurology*, 1958, *110*, 221–256.

Kuypers, H. G. J. M. Corticospinal connections: Postnatal development in the rhesus monkey. *Science*, 1962, *138*, 678–680.

Kuypers, H. G. J. M. The descending pathways to the spinal cord, their anatomy and function. *Progress in Brain Research*, 1964, *11*, 178–202.

Kuypers, H. G. J. M., Catsman-Berrevoets, C. E., and Padt, R. E. Retrograde axonal transport of fluorescent substances in the rat's forebrain, *Neuroscience Letters*, 1977, *6*, 127–135.

Kuypers, H. G. J. M., and Brinkman, J. Precentral projections to different parts of the spinal intermedial zone in the rhesus monkey. *Brain Research*, 1970, *24*, 29–48.

Kuypers, H. G. J. M., and Lawrence, D. G. Cortical projections to the red nucleus and the brain stem in the rhesus monkey. *Brain Research*, 1967, *4*, 151–188.

Kuypers, H. G. J. M., Fleming, W. R., and Farinholt, J. W. Subcorticospinal projections in the rhesus monkey. *Journal of Comparative Neurology*, 1962, *118*, 107–137.

Landgren, S., Phillips, C. G., and Porter, R. Minimal synaptic actions of pyramidal impulses on some alpha-motoneurones of baboon's hand and forearm. *Journal of Physiology (London)*, 1962a, *161*, 91–111.

Landgren, S., Phillips, C. G., and Porter, R. Cortical fields of origin of the monosynaptic pyramidal pathways to some alpha-motoneurones of the baboon's hand and forearm. *Journal of Physiology (London)*, 1962b, *161*, 112–125.

Langworthy, O. R. Development of behavior patterns and myelinization of the nervous system in the human fetus and infant. *Contributions to Embryology of the Carnegie Institution*, 1933, *24*, 1–58.

Lassek, A. M. Inactivation of voluntary motor function following rhizotomy. *Journal of Neuropathology and Experimental Neurology*, 1953, *12*, 83–87.

Lassek, A. M. *The pyramidal tract. Its status in medicine.* Springfield, Ill.. Charles C Thomas, 1954.

Lassek, A. M., Woolsey, C. N., Walker, A. E., Boshes, B. The pyramidal tract. *Neurology (Minneapolis)*, 1957, *7*, 496–509.

Laursen, A. M. A kinesthetic deficit after partial transection of a pyramidal tract in monkeys. *Brain Research*, 1971, *31*, 263–274.

Laursen, A. M. Task dependence of slowing after pyramidal lesions in monkeys. *Journal of Comparative and Physiological Psychology*, 1977, *91*, 897–906.

Laursen, A. M., and Wiesendanger, M. Pyramidal effect on alpha and gamma motoneurones. *Acta Physiologica Scandinavica*, 1966a, *67*, 165–172.

Laursen, A. M., and Wiesendanger, M. Motor deficits after transection of a bulbar pyramid in the cat. *Acta Physiologica Scandinavica*, 1966b, *68*, 118–126.

Laursen, A. M., and Wiesendanger, M. The effect of pyramidal lesions on response latency in cats. *Brain Research*, 1967, *5*, 207–220.

LaVail, J. H. Retrograde cell degeneration and retrograde transport techniques. In W. M. Cowan and M. Cuénod (Eds.), *The Use of Axonal Transport for Studies of Neural Connectivity*, Amsterdam: Elsevier, 1975.

Lawrence, D. G., and Kuypers, H. G. J. M. The functional organization of the motor system in the monkey. I. The effects of bilateral pyramidal lesions. *Brain*, 1968, *91*, 1–14.

Lawrence, D. G., and Hopkins, D. A. The development of motor control in the rhesus monkey. Evidence concerning the role of corticomotoneuronal connections. *Brain*, 1976, *99*, 235–254.

Leestma, J. E., and Noronha, A. Pure motor hemiplegia, medullary pyramid lesion, and olivary hypertrophy. *Journal of Neurology, Neurosurgery and Psychiatry*, 1976, *39*, 877–884.

Lewis, R. P., and Brindley, G. S. The extrapyramidal cortical motor map. *Brain*, 1965, *88*, 397–406.

Lewis, M. McD., and Porter, R. Lack of involvement of fusimotor activation in movements of the foot produced by electrical stimulation of monkey cerebral cortex. *Journal of Physiology (London)*, 1971, *212*, 707–717.

Leyton, A. S. F., and Sherrington, C. S. Observations on the excitable cortex of the chimpanzee, orang-utan and gorilla. *Quarterly Journal of Experimental Physiology*, 1917, *11*, 135–222.

Liddell, E. G. T., and Phillips, C. G. Pyramidal section in the cat. *Brain*, 1944, *67*, 1–9.

Liddell, E. G. T., and Phillips, C. G. Thresholds of cortical representation. *Brain*, 1950, *73*, 125–140.

Liddell, E. G. T., and Phillips, C. G. Overlapping areas in the motor cortex of the baboon. *Journal of Physiology (London)*, 1951, *112*, 392–399.

Lindström, S. Recurrent control from motor axon collaterals of Ia inhibitory pathways in the spinal cord of the cat. *Acta Physiologica Scandinavica*, 1973, Supplementum 392.

Liu, C. N., and Chambers, W. W. An experimental study of the cortico-spinal system in the monkey (macaca mulatta). The spinal pathway and preterminal distribution of degenerating fibers following discrete lesions of the pre- and postcentral gyri and bulbar pyramid. *Journal of Comparative Neurology*, 1964, *123*, 257–284.

Lloyd, D. P. C. The spinal mechanism of the pyramidal system in cats. *Journal of Neurophysiology*, 1941, *4*, 525–546.

Lloyd, D. P. C. Note on convergence of pyramidal and primary afferent impulses in the spinal cord of the cat. *Proceedings of the National Academy of Sciences*, 1968, *59*, 381–384.

Lucier, G. E., Rüegg, D. G., and Wiesendanger, M. Responses of neurones in motor cortex and in area 3a to controlled stretches of forelimb muscles in Cebus monkeys. *Journal of Physiology (London)*, 1975, *251*, 833–853.

Lundberg, A. The supraspinal control of transmission in spinal reflex pathways. *Electroencephalography and Clinical Neurophysiology*, 1967, Supplementum *25*, 35–46.

Lundberg, A. Control of spinal mechanisms from the brain. In D. B. Tower (Ed.), *The Nervous System*. Vol 1: *The Basic Neurosciences*. New York: Raven Press, 1975.

Lundberg, A., and Voorhoeve, P. Effects from pyramidal tract on spinal reflex arcs. *Acta Physiologica Scandinavica*, 1962, *56*, 201–219.

Lundberg, A., Malmgren, K., and Schomburg, E. D. Convergence from Ib, cutaneous and joint afferents in reflex pathways to motoneurones. *Brain Research*, 1975, *87*, 81–84.

MacLean, J. B., and Leffman, H. Supraspinal control of Renshaw cells. *Experimental Neurology*, 1967, *18*, 94–104.

Magendie, F. *Précis élémentaire de physiologie*. 4ᵉ éd. Bruxelles: H. Dumont, 1834.

Magendie, F. *Leçons sur les fonctions et les maladies du système nerveux*. Paris: Ebrard, 1839.

Marin-Padilla, M. Origin of the pericellular baskets of the pyramidal cells of the human motor cortex: a Golgi study. *Brain Research*, 1969, *14*, 633–646.

Marin-Padilla, M. Double origin of the pericellular baskets of the pyramidal cells of the human motor cortex: A Golgi study. *Brain Research*, 1972, *38*, 1–12.

Marshall, C. Experimental lesions of the pyramidal tract. *Archives of Neurology and Psychiatry (Chicago)*, 1934, *32*, 778–796.

Massion, J. The mammalian red nucleus. *Physiological Reviews*, 1967, *47*, 383–436.

Matsunami, K., and Hamada, I. Precentral neuron activity associated with ipsilateral forelimb movements in monkeys. *Journal de Physiologie (Paris)*, 1978, *74*, 319–322.

Matthews, P. C. Muscle spindles and their motor control. *Physiological Reviews*, 1964, *44*, 219–288.

Meyer, J. C., and Herndon, R. M. Bilateral infarctions of the pyramidal tracts in man. *Neurology (Minneapolis)*, 1962, *12*, 637–642.

Miller, J. M., Glickstein, M., and Stebbins, W. C. Reduction of response latency in monkeys by a procedure of differential reinforcement. *Psychonomic Science*, 1966, *5*, 177–178.

Moll, L., and Kuypers, H. Role of premotor cortical areas and VL nucleus in visual guidance of relatively independent hand and finger movements in monkeys. *Abstract Collection Vol. 1, European Neuroscience Meeting, Munich*, 1975, *142* (Abstract no. 280).

Monakow, C. von *Die Gehirnpathologie*. Wien: Hölder, 1897.

Monakow, C. von *Die Lokalisation im Grosshirn und der Abbau der Funktion durch kortikale Herde*. Wiesbaden: Bergmann, 1914.

Morin, G., Donnet, V., and Zwirn, P. Nature et évolution des troubles consécutifs à la section d'une pyramide bulbaire chez le chien. *Comptes Rendus des Séances de la Société de Biologie*, 1949, *143*, 710–712.

Mortimer, E. M., and Akert, K. Cortical control and representation of fusimotor neurones. *American Journal of Physical Medicine*, 1961, *40*, 228–248.

Mott, V. W., and Sherrington, C. S. Experiments upon the influence of sensory nerves upon movement and nutrition of the limbs. *Proceedings of the Royal Society*, 1895, *57*, 481–488.

Mountcastle, V. B., and Powell, T. P. S. Central nervous mechanisms subserving position sense and kinesthesis. *Bulletin of the Johns Hopkins Hospital*, 1959a, 173–200.

Mountcastle, V. B., and Powell, T. P. S. Neural mechanisms subserving cutaneous sensibility, with special reference to the role of afferent inhibition in sensory perception and discrimination. *Bulletin of the Johns Hopkins, Hospital*, 1959b, *105*, 201–232.

Mountcastle, V. B., Covian, M. R., and Harrison, C. R. The central representation of some forms of deep sensibility. *Research Publications: Association for Research in Nervous and Mental Disease*, 1952, *30*, 339–370.

Muir, R. B., and Porter, R. The effect of a preceding stimulus on temporal facilitation at corticomotoneuronal synapses. *Journal of Physiology (London)*, 1973, *228*, 749–763.

Murphy, J. T., Wong, Y. C., and Kwan, H. C. Afferent-efferent linkages in motor cortex for single forelimb muscles. *Journal of Neurophysiology*, 1975, *38*, 990–1014.

Murray, E. A., and Coulter, J. D. Origins of cortical projections to cervical and lumbar spinal cord in monkey. *Abstracts, Society of Neuroscience*, 1976, *2*, 917.

Murray, E. A., and Coulter, J. D. Corticospinal projections from the medial cerebral hemisphere in monkey. *Abstracts, Society of Neuroscience*, 1977, *3*, 275.

Naito, H., Nakamura, K., Kurosaki, T., and Tamura, Y. Precise location of fast and slow pyramidal tract cells in cat sensorimotor cortex. *Brain Research*, 1969, *14*, 237–239.

Napier, J. The evolution of the hand. *Scientific American*, 1962, *207*, 56–62.

Napier, J. R., and Napier, P. H. *A Handbook of Living Primates*. New York: Academic Press, 1967.

Nathan, P. W., and Smith, M. C. Long descending tracts in man. I. Review of present knowledge. *Brain*, 1955, *78*, 248–303.

Nieoullon, A., and Gahéry, Y. Influence of pyramidotomy on limb flexion movements induced by cortical stimulation and on associated postural adjustment in the cat. *Brain Research*, 1978, *149*, 39–52.

Noback, Ch. R., and Shriver, J. E. Phylogenetic and ontogenetic aspects of the lemniscal systems and the pyramidal system. In R. Hassler and H. Stephan (Eds.), *Evolution of the Forebrain*, Stuttgart: Thieme, 1966.

Nyberg-Hansen, R. Functional organization of descending supraspinal fibre systems to the spinal cord. Anatomical observations and physiological correlations. *Reviews of Anatomy, Embryology and Cell Biology*, 1966, *39*, 1–48.

Nyberg-Hansen, R. Corticospinal fibres from the medial aspect of the cerebral hemisphere in the cat. An experimental study with the Nauta method. *Experimental Brain Research*, 1969a, *7*, 120–132.

Nyberg-Hansen, R. Further studies on the origin of corticospinal fibres in the cat. An experimental study with the Nauta method. *Brain Research*, 1969b, 39–54.

Nyberg-Hansen, R. Do cat spinal motoneurones receive direct supraspinal fibre connections? A supplementary silver study. *Archives Italiennes de Biologie*, 1969c, *107*, 67–78.

Nyberg-Hansen, R., and Brodal, A. Sites of termination of corticospinal fibers in the cat. An experiental study with silver impregnation methods. *Journal of Comparative Neurology*, 1963, *120*, 369–391.

Nyberg-Hansen, R., and Rinvik, E. Some comments on the pyramidal tract, with special reference to its individual variations in man. *Acta Neurologica Scandinavica*, 1963, *39*, 1–30.

Ödkvist, L. M., Schwarz, D. W. F., Fredrickson, J. M., and Hassler, R. Projection of the vestibular nerve to the area 3a arm field in the squirrel monkey *(Saimiri sciureus)*. *Experimental Brain Research*, 1974, *21*, 97–105.

Orlovsky, G. N., and Shik, M. L. Control of locomotion: a neurophysiological analysis of the cat locomotor system. In R. Porter (Ed.) *International Review of Physiology*, Vol. 10. *Neurophysiology II*. Baltimore: University Park Press, 1976.

Oscarsson, O. Functional organization of spinocerebellar paths. In A. Iggo (Ed.), *Handbook of Sensory Physiology*. Vol. II. *Somatosensory System*. Berlin: Springer, 1973.

Pandya, D. N., and Kuypers, H. G. J. M. Cortico-cortical connections in the rhesus monkey. *Brain Research*, 1969, *13*, 13–36.

Pandya, D. N., and Vignolo, L. A. Interhemispheric projections of the parietal lobe in the rhesus monkey. *Brain Research*, 1969, *15*, 49–65.

Patton, H. D., and Amassian, V. E. The pyramidal tract. Its excitation and function. In *Handbook of Physiology, Neurophysiology* (Vol. 2). Washington: American Physiological Society, 1960.

Penfield, W., and Jasper, H. *Epilepsy and the Functional Anatomy of the Human Brain*. Boston: Little, Brown, 1954.

Penfield, W., and Rasmussen, T. *The Cerebral Cortex of Man*. New York: McMillan, 1950.

Pernet, U., and Hepp-Reymond, M.-C. Retrograde Degeneration der Pyramidenbahnzellen im motorischen Kortex beim Affen *(Macaca fascicularis)*, *Acta Anatomica (Basel)*, 1975, 552–561.

Peterson, B. W., Maunz, R. A., Pitts, N. G., and Mackel, R. G. Patterns of projection and branching of reticulospinal neurons. *Experimental Brain Research*, 1975, *23*, 333–351.

Petras, J. M. Some efferent connections of the motor and somatosensory cortex of simian primates and felid, canid and procynid carnivores. *Annals of the New York Academy of Sciences*, 1969, *167*, 469-505.

Phillips, C. G. Intracellular records from Betz cells in the cat. *Quarterly Journal of Experimental Physiology*, 1956, *41*, 58-68.

Phillips, C.G. Changing concepts of the precentral motor area. In J. C. Eccles (Ed.), *Brain and Conscious Experience*. Berlin: Springer, 1966.

Phillips, C. G. Corticomotoneural organization: Projection from the arm area of the baboon's motor cortex. *Archives of Neurology*, 1967, *17*, 188-195.

Phillips, C. G. Motor apparatus of the baboon's hand. *Proceedings of the Royal Society, Series B*, 1969, *173*, 141-174.

Phillips, C. G. Evolution of the corticospinal tract in primates with special reference to the hand. *Proceedings of the Third International Congress of Primatology* (Vol. 2). Basel: Karger, 1971.

Phillips, C. G. Cortical localization and "sensorimotor processes" at the "middle level" in primates (Hughlings Jackson Lecture 1973), *Proceedings of the Royal Society of Medicine*, 1973, *66*, 987-1002.

Phillips, C. G. Laying the ghost of "muscles versus movements." *Canadian Journal of Neurological Sciences*, 1975, *2*, 209-218.

Phillips, C. G., and Porter, R. Unifocal and bifocal stimulation of the motor cortex. *Journal of Physiology (London)*, 1962, *162*, 532-538.

Phillips, C. G., and Porter, R. The pyramidal projection to motoneurones of some muscle groups of the baboon's forelimb. *Progress in Brain Research*, 1964, *12*, 222-242.

Phillips, C. G., and Porter, R. *Corticospinal Neurones. Their Role in Movement*. Monographs of the Physiological Society No. 34. New York. Academic Press, 1977.

Phillips, C. G., Powell, T. P. S., and Wiesendanger, M. Projection from low-threshold muscle afferents of hand and forearm to area 3a of baboon's cortex. *Journal of Physiology (London)*, 1971, *217*, 419-446.

Popper, K. R. *Objective Knowledge. An Evolutionary Approach*. Oxford: Clarendon Press, 1972.

Porter, R. Early facilitation at corticomotoneuronal synapses. *Journal of Physiology (London)*, 1970, *207*, 733-745.

Porter, R., and Hore, J. Time course of minimal cortico-motoneuronal excitatory postsynaptic potentials in lumbar motoneurons of the monkey. *Journal of Neurophysiology*, 1969, *32*, 443-451.

Porter, R., Lewis, M. McD., and Horne, M. Analysis of patterns of natural activity of neurones in the precentral gyrus of conscious monkeys. *Brain Research*, 1971, *34*, 99-113.

Porter, R., and Muir, R. B. The meaning for motoneurones of the temporal pattern of natural activity in pyramidal tract neurones of conscious monkeys. *Brain Research*, 1971, *34*, 127-142.

Powell, T. P. S. The organization of the major functional areas of the cerebral cortex. *Symposia of the Zoological Society, London*, 1973, *33*, 235-252.

Powell, T. P. S., and Mountcastle, V. B. The cytoarchitecture of the postcentral gyrus of the monkey macaca mulatta. *Bulletin of the Johns Hopkins Hospital*, 1959a, *105*, 108-131.

Powell, T. P. S., and Mountcastle, V. B. Some aspects of the functional organization of the cortex of the postcentral gyrus of the monkey: A correlation of findings obtained in a single unit analysis with cytoarchitecture. *Bulletin of the Johns Hopkins Hospital*, 1959b, *105*, 133-162.

Preston, J. B., and Whitlock, D. G. Precentral facilitation and inhibition of spinal motoneurones. *Journal of Neurophysiology*, 1960, *23*, 154-170.

Preston, J. B., and Whitlock, D. G. Intracellular potentials recorded from motoneurones following precentral gyrus stimulation in primate. *Journal of Neurophysiology*, 1961, *24*, 91-100.

Preston, J. B., and Whitlock, D. G. A comparison of motor cortex effects on slow and fast muscle innervations in the monkey. *Experimental Neurology*, 1963, *7*, 327-341.

Prus, J. Ueber die Leitungsbahnen und Pathogenese der Rindenepilepsie. *Wiener Klinische Wochenschrift*, 1898, *11*, 857-863.

Ramón-Moliner, E. Discussion in "Aspects of the structural and functional organization of the neocortex." *Neurosciences Research Progress Bulletin*, 1970, *8*, 157-217.

Ranson, S. W. Rigidity caused by pyramidal lesions in the cat. *Journal of Comparative Neurology*, 1932, *55*, 91–97.

Regis, H., Trouche, E., and Massion, J. Effet de l'ablation du cortex moteur ou du cervelet sur la coordination posturo-cinétique chez le chat. *Electroencephalography and Clinical Neurophysiology*, 1976, *41*, 348–356.

Renaud, L. P., and Kelly, J. S. Identification of possible inhibitory neurons in the pericruciate cortex of the cat. *Brain Research*, 1974a, *79*, 9–28.

Renaud, L. P., and Kelly, J. S. Simultaneous recordings from pericruciate pyramidal tract and non-pyramidal tract neurons. Response to stimulation of inhibitory pathways. *Brain Research*, 1974b, *79*, 29–44.

Rinvik, E. Organization of thalamic connections from motor and somatosensory cortical areas in the cat. In T. L. Frigyesi, E. Rinvik, and M. D. Yahr (Eds.), *Corticothalamic Projections and Sensorimotor Activities*. New York: Raven Press, 1972.

Robertson, R. T. Thalamic projections to parietal cortex. *Brain, Behavior and Evolution*, 1977, *14*, 161–184.

Rogers, A. W. Autoradiography and the study of the central nervous system. In P. B. Bradley (Ed.), *Methods in Brain Research*. New York. Wiley, 1975.

Romanes, G. J. The motor cell columns of the lumbosacral spinal cord of the cat. *Journal of Comparative Neurology*, 1951, *94*, 313–358.

Rosén, I. Projection of forelimb group I muscle afferents to the cat cerebral cortex. *International Review of Neurobiology*, 1972, *15*, 1–25.

Rosén, I., and Asanuma, H. Peripheral afferent inputs to the forelimb area of the monkey motor cortex: Input-output relations. *Experimental Brain Research*, 1972, *14*, 257–273.

Rothmann, M. Ueber die physiologische Wertung der corticospinalen (Pyramiden-) bahn. *Archiv für Physiologie*, 1907, 217–275.

Ruch, T. C. Motor system. In S. S. Stevens (Ed.), *Handbook of Experimental Psychology*. New York: Wiley, 1951.

Rüegg, D. G., and Wiesendanger, M. Corticofugal effects from sensorimotor area I and somatosensory area II on neurons of the pontine nuclei in the cat. *Journal of Physiology (London)*, 1975, *247*, 745–757.

Rüegg, D. G., Séguin, J. J., and Wiesendanger, M. Effects of electrical stimulation of somatosensory and motor areas of the cerebral cortex on neurones of the pontine nuclei in squirrel monkeys. *Neuroscience*, 1978, *2*, 923–927.

Russel, J. R., and DeMyer, W. The quantitative cortical origin of pyramidal axons of macaca rhesus. With some remarks on the slow rate of axolysis. *Neurology (Minneapolis)*, 1961, *11*, 96–108.

Sakata, H., and Miyamoto, J. Topographic relationship between the receptive fields of neurons in the motor cortex and the movements elicited by focal stimulation in freely moving cats. *Japanese Journal of Physiology*, 1968, *18*, 489–507.

Sans, A., Raymond, J., and Marty, R. Réponses thalamiques et corticales à la stimulation électrique du nerf vestibulaire chez le chat. *Experimental Brain Research*, 1970, *10*, 265–275.

Schäfer, E. A. Experiments on the paths taken by volitional impulses passing from the cerebral cortex to the cord: The pyramids and the ventrolateral descending tracts. *Quarterly Journal of Experimental Physiology*, 1910, *2*, 356–373.

Schäfer, E. A. *Text-Book of Physiology*. Edinburgh: Young J. Pentland, 1900.

Scheibel, M. E., and Scheibel, A. B. Structural substrates for integrative patterns in the brain stem reticular core. In H. H. Jasper *et al.* (Eds.), *Reticular Formation of the Brain*. Boston: Little, Brown, 1958.

Schiff, J. M. *Lehrbuch der Physiologie des Menschen. Muskel- und Nervenphysiologie* (Vol. 1). Lahr: Schauenburg, 1858 to 1859.

Schoen, J. H. R. Comparative aspects of the descending fibre systems in the spinal cord. *Progress in Brain Research*, 1964, *11*, 203–222.

Schüller, A. Experimentelle Pyramidendurchschneidung beim Hunde und Affen. *Wiener Klinische Wochenschrift*, 1906, *19*, 57–62.

Schwarz, D. W. F., Deecke, L., and Fredrickson, J. M. Cortical projection of group I muscle afferents to area 2, 3a, and the vestibular field in the rhesus monkey. *Experimental Brain Research*, 1973, *17*, 516–526.

Shanks, M. F., Rockel, A. J., and Powell, T. P. S. The commissural fibre connections of the primary somatic sensory cortex. *Brain Research*, 1975, *98*, 166–171.

Shapovalov, A. I. Neuronal organization and synaptic mechanisms of supraspinal motor control in vertebrates. *Reviews of Physiology, Biochemistry and Pharmacology*, 1975, *72*, 1–54.

Sherrington, C. S. On nerve tracts degenerating secondarily to lesions of the cortex cerebri. *Journal of Physiology (London)*, 1889, *10*, 429–432.

Sherrington, C. S. *The Integrative Action of the Nervous System.* Cambridge: Cambridge University Press, 1906.

Sherrington, C. S. On plastic tonus and proprioceptive reflexes. *Quarterly Journal of Experimental Physiology*, 1909, *2*, 109–156.

Sherrington, C. S., and Hering, H. H. Antagonistic muscles and reciprocal innervation. Fourth note. *Proceedings of the Royal Society, Series B*, 1897, 62, 183–187.

Shinoda, Y., Arnold, A. P., and Asanuma, H. Spinal branching of corticospinal axons in the cat. *Experimental Brain Research*, 1976, *26*, 215–234.

Shinoda, Y., Zarzecki, P., and Asanuma, H. Spinal branching of pyramidal tract neurons in the monkey. *Experimental Brain Research*, 1979, *34*, 59–72.

Shriver, J., and Matzke, H. A. Corticobulbar and corticospinal tracts in the marmoset monkey *(Oedipomidas oedipus)*. *Anatomical Record*, 1965, *151*, 416.

Silfvenius, H. Projections to the cat cerebral cortex from fore- and hindlimb group I muscle afferents. *Umea University Medical Dissertations*, 1972, *4*, 1–46.

Sloper, J. J., and Powell T. P. S. Ultrastructural features of the sensori-motor cortex of the primate. *Philosophical Transactions of the Royal Society (London), Series B*, 1979a, *295*, 123–139.

Sloper, J. J., and Powell, T. P. S. A study of the axon initial segment and proximal axon of neurons in the primate motor and somatic sensory cortices. *Philosophical Transactions of the Royal Society (London), Series B*, 1979b, *285*, 173–197.

Sloper, J. J., and Powell, T. P. S. An experimental electron microscopic study of afferent connections to the primate motor and somatic sensory cortices. *Philosophical Transactions of the Royal Society (London), Series B*, 1979c, *285*, 199–226.

Sloper, J. J., Hiorns, R. W., and Powell, T. P. S. A qualitative and quantitative electron microscopic study of the neurons in the primate motor and somatic sensory cortices. *Philosophical Transactions of the Royal Society (London), Series B*, 1979, *285*, 141–171.

Smith, G., and Burns, G. D. Probability analysis of unit activity in the cerebral cortex. *Proceedings of the Canadian Federation of Biological Societies*, 1959, *2*.

Sotgiu, M. L., and Margnelli, M. Electrophysiological identification of pontomedullary reticular neurons directly projecting into dorsal column nuclei. *Brain Research*, 1976, *103*, 443–453.

Sperry, R. W. Neural basis of the spontaneous optokinetic response produced by visual inversion. *Journal of Comparative Physiology and Psychology*, 1950, *43*, 482–489.

Starlinger, J. Die Durchschneidung beider Pyramiden beim Hunde. *Jahrbücher für Psychiatrie und Neurologie*, 1897, *15*, 1–42.

Steriade, M., and Yossif, G. Afferent and recurrent collateral influences on cortical somatosensory neurons. *Experimental Neurology*, 1977, *56*, 334–360.

Steriade, M., Wyzinski, P., and Apostol, V. Differential synaptic reactivity of simple and complex pyramidal tract neurons at various levels of vigilance. *Experimental Brain Research*, 1973, *17*, 87–110.

Steriade, M., Deschênes, M., and Oakson, G. Inhibitory processes and interneuronal apparatus in motor cortex during sleep and waking. I. Background firing and responsiveness of pyramidal tract neurons and interneurons. *Journal of Neurophysiology*, 1974, *37*, 1065–1092.

Stewart, D. H., and Preston, J. B. Functional coupling between the pyramidal tract and segmental motoneurons in cat and primate. *Journal of Neurophysiology*, 1967, *30*, 453–465.

Stone, T. W. Cortical pyramidal tract interneurones and their sensitivity to L-glutamic acid. *Journal of Physiology (London)*, 1973, *233*, 211–225.

Stone, T. W. Blockade by amino acid antagonists of neuronal excitation mediated by the pyramidal tract. *Journal of Physiology (London)*, 1976, *257*, 187–198.

Stoney, S. D., Thompson, W. D., and Asanuma, H. Excitation of pyramidal tract cells by intracortical microstimulation: Effective extent of stimulating current. *Journal of Neurophysiology*, 1968, *31*, 659–669.

Strick, P. Anatomical analysis of ventrolateral thalamic input to primate motor cortex. *Journal of Neurophysiology*, 1976, *39*, 1020–1031.

Sugar, O., Chusid, J. G., and French, J. D. A second motor cortex in the monkey (macaca mulatta). *Journal of Neuropathology and Experimental Neurology*, 1948, *7*, 182–189.

Szentágothai, J. The Ferrier Lecture 1977. The neuron network of the cerebral cortex: A functional interpretation. *Proceedings of the Royal Society (London)*, Series B, 1978, *201*, 219–248.

Szentágothai-Schiemert, J. Die Endigungsweise der absteigenden Rückenmarksbahnen. *Zeitschrift für Anatomie und Entwicklungsgeschichte*, 1941, *111*, 322–330.

Takahashi, K. Slow and fast groups of pyramidal cells and their respective membrane properties. *Journal of Neurophysiology*, 1965, *28*, 908–924.

Takahashi, K., Kubota, K., and Uno, M. Recurrent facilitation in cat pyramidal tract cells. *Journal of Neurophysiology*, 1967, *30*, 22–34.

Talairach, J., and Bancaud, J. The supplementary motor area in man. *International Journal of Neurology*, 1966, *5*, 330–347.

Tamarova, Z. A., Shapovalov, A. I., Karamjan, O. A., and Kurchavyi, G. G. Cortico-pyramidal and cortico-extrapyramidal synaptic effects on the monkey lumbar motoneurons. *Neurofiziologia USSR*, 1972, *4*, 587–596.

Tanaka, R. Reciprocal Ia inhibition during voluntary movements in man. *Experimental Brain Research*, 1974, *21*, 529–540.

Tanji, J. Selective activation of neurons in cortical area 3a associated with accurate maintenances of limb positions. *Brain Research*, 1976, *115*, 328–333.

Tanji, J., and Taniguchi, K. Does the supplementary motor area play a part in modifying motor cortex reflexes? *Journal de Physiologie (Paris)*, 1978, *74*, 317–318.

Taub, E., and Goldberg, I. A. Use of sensory recombination and somatosensory deafferentation techniques in the investigation of sensory-motor integration. *Perception*, 1974, *3*, 393–408.

Thierry, A. M., Stinus, L., Blanc, G., and Glowinski, J. Dopamine terminals in the rat cortex. *Science*, 1973, *182*, 499–501.

Towe, A. L. Somatosensory cortex: Descending influences on ascending systems. In A. Iggo (Ed.), *Handbook of Sensory Physiology*. Vol. II: *Somatosensory System*. Berlin: Springer, 1973a.

Towe, A. L. Motor cortex and the pyramidal system. In J. Maser (Ed.), *Efferent Organization and the Integration of Behavior*. New York: Academic Press, 1973b.

Towe, A. L. Relative numbers of pyramidal tract neurons in mammals of different sizes. *Brain, Behavior and Evolution*, 1973c, *7*, 1–17.

Towe, A. L. Notes on the hypothesis of columnar organization in somatosensory cerebral cortex. *Brain, Behavior and Evolution*, 1975, *11*, 16–47.

Towe, A. L., and Jabbur, S. J. Cortical inhibition of neurons in dorsal column nuclei. *Journal of Neurophysiology*, 1961, *24*, 488–498.

Towe, A. L., Patton, H. D., and Kennedy, T. T. Response properties of neurons in the pericruciate cortex of the cat following electrical stimulation of the appendages. *Experimental Neurology*, 1964, *10*, 325–344.

Tower, S. S. The dissociation of cortical excitation from cortical inhibition by pyramid section and the syndrome of that lesion in the cat. *Brain*, 1935, *58*, 238–255.

Tower, S. S. Pyramidal lesion in the monkey. *Brain*, 1940, *63*, 36–90.

Tower, S. S. The pyramidal tract. In P. C. Bucy (Ed.), *The Precentral Motor Cortex*. Urbana. University of Illinois Press, 1944.

Trendelenburg, W. Untersuchungen über reizlose verübergehende Ausschaltung am Zentralnervensystem. III. Mitteilung. Die Extremitätenregion der Grosshirnrinde. *Pflüger's Archiv für die gesamte Physiologie,* 1911, *137,* 515–544.

Tsukahara, N., Fuller, D. R. G., and Brooks, V. B. Collateral pyramidal influences on the cortico-rubrospinal system. *Journal of Neurophysiology,* 1968, *31,* 467–484.

Tsumoto, T., Nakamura, S., and Iwama, K. Pyramidal tract control over cutaneous and kinesthetic sensory transmission in the cat thalamus. *Experimental Brain Research,* 1975, *22,* 281–294.

Twitchell, T. E. Sensory factors in purposive movement. *Journal of Neurophysiology,* 1954, *17,* 239–252.

Tyner, C. F. Anatomic specificity in the feline corticospinal system. *Brain Research,* 1974, *69,* 336–340.

Ungerstedt, U. Stereotaxic mapping of the monoamine pathways in the rat brain. *Acta Physiologica Scandinavica,* Supplementum, 1971, *367,* 1–48.

Verhaart, W. J. C. The pyramidal tract in the primates. In C. R. Noback and W. Montagna (Eds.), *The Primate Brain,* Vol. 1. *Advances in Primatology.* New York: Appleton-Century-Crofts, 1970.

Voneida, T. J. The effect of pyramidal lesions on the performance of a conditioned avoidance response in cats. *Experimental Neurology,* 1967, *19,* 483–493.

Walker, A. E. Cerebral pedunculotomy for the relief of involuntary movements: hemiballismus. *Acta Psychiatrica Scandinavica,* 1949, *24,* 723–726.

Walker, A. E. Stimulation and ablation. Their role in the history of cerebral physiology. *Journal of Neurophysiology,* 1957, *20,* 435–449.

Wall, P. D. The laminar organization of dorsal horn and effects of descending impulses. *Journal of Physiology (London),* 1967, *188,* 403–423.

Wall, P. D., McCulloch, W. S., Lettvin, J. Y., Pitts, W. H. The terminal arborization of the cat's pyramidal tract determined by a new technique. *Yale Journal of Biological Medicine,* 1956, *28,* 457–464.

Walsh, E. G. *Physiology of the Nervous System.* London: Longmans and Green, 1957.

Walshe, F. M. R. On the role of the pyramidal system in willed movements. *Brain,* 1947, *70,* 329–354.

Weisberg, J. A., and Rustioni, A. Cortical cells projecting to the dorsal column nuclei of cats. An anatomical study with the horseradish peroxydase technique. *Journal of Comparative Neurology,* 1976, *168,* 425–438.

Weisberg, J. A., and Rustioni, A. Cortical cells projecting to the dorsal column nuclei of rhesus monkeys. *Experimental Brain Research,* 1977, *28,* 521–528.

Welker, W. I., Benjamin, R. M., Miles, R. C., and Woolsey, C. N. Motor effects of cortical stimulation in squirrel monkey (Saimiri sciureus). *Journal of Neurophysiology,* 1957, *20,* 347–364.

Welt, C., Aschoff, J. C., Kameda, K., Brooks, V. B. Intracortical organization of cat's motor-sensory neurons. In M. D. Yahr and D. P. Purpura (Eds.), *Neurophysiological Basis of Normal and Abnormal Motor Activities.* New York: Raven Press, 1967.

Wiesendanger, M. The pyramidal tract. Recent investigation on its morphology and function. *Ergebnisse der Physiologie,* 1969, *61,* 71–136.

Wiesendanger, M. Input from muscle and cutaneous nerves of the hand and forearm to neurones of the precentral gyrus of baboons and monkeys. *Journal of Physiology (London),* 1973, *228,* 203–219.

Wiesendanger, M. Some aspects of pyramidal tract functions in primates. In J. E. Desmedt (Ed.), *New Developments in Electromyography and Clinical Neurophysiology* (Vol. 3). Basel: Karger, 1973.

Wiesendanger, M., and Felix, D. Pyramidal excitation of lemniscal neurons and facilitation of sensory transmission in the spinal trigeminal nucleus of the cat. *Experimental Neurology,* 1969, *25,* 1–17.

Wiesendanger, M., and Mumenthaler, M. Der Pendeltest. Ein Beitrag zur Objektivierung von muskulären Hypertonien. *Schweizerische Medizinische Wochenschrift,* 1959, *89,* 1301–1305.

Wiesendanger, M., and Tarnecki, R. Die Rolle des pyramidalen Systems bei der sensomotorischen

Integration. *Bulletin der Schweizerischen Akademie der Medizinischen Wissenschaften*, 1966, 22, 306–328.

Wiesendanger, M., Hammer, B., and Tarnecki, R. Corticofugal control of the presynaptic inhibition in the spinal trigeminal nucleus of the cat. The effect of pyramidotomy and barbiturates. *Schweizer Archiv für Neurologie, Neurochirurgie und Psychiatrie*, 1967, *100*, 255–276.

Wiesendanger, M., Séguin, J. J., and Künzle, H. The supplementary motor area—a control system for posture? In R. B. Stein, K. B. Pearson, R. S. Smith, and J. B. Redford (Eds.), *Control of Posture and Locomotion*. New York: Plenum Press, 1973.

Wiesendanger, M., Rüegg, D. G., and Lucier, G. E. Why transcortical reflexes? *Canadian Journal of Neurological Sciences*, 1975, *2*, 295–301.

Wiesendanger, R., Wiesendanger, M., and Rüegg, D. G. An anatomical investigation of the corticopontine projection in the primate (*Macaca fascicularis* and *Saimiri sciureus*). II. The projection from frontal and parietal association areas. *Neuroscience*, 1979, *4*, 747–765.

Willis, W. D., Haber, L. H., and Martin, R. F. Inhibition of spinothalamic tract cells and interneurons by brain stem stimulation in the monkey. *Journal of Neurophysiology*, 1977, *40*, 968–981.

Wilson, V. J., Talbot, W. H., and Kato, M. Inhibitory convergence upon Renshaw cells. *Journal of Neurophysiology*, 1964, *27*, 1064–1080.

Woolsey, C. N. Organization of somatic sensory and motor areas of the cerebral cortex. In H. F. Harlow and C. N. Woolsey (Eds.), *Biological and Biochemical Bases of Behavior*. Madison: University of Wisconsin Press, 1958.

Woolsey, C. N., Settlage, P. H., Meyer, D. R., Sencer, W., Pinto-Hamuy, T., and Travis, A. M. Patterns of localization in precentral and "supplementary" motor areas and their relation to the concept of a premotor area. *Proceedings of the Association for Research in Nervous and Mental Diseases*, 1950, *30*, 238–264.

Woolsey, C. N., Travis, A. M., Barnard, J. W., and Ostenso, R. S. Motor representation in the postcentral gyrus after chronic ablation of precentral and supplementary motor areas. *Federation Proceedings*, 1953, *12*, 160.

Woolsey, C. N., Gorska, T., Wetzel, A., Erickson, T. C., Earls, F. J., and Allman, J. M. Complete unilateral section of the pyramidal tract at the medullary level in macaca mulatta. *Brain Research*, 1972, *40*, 119–123.

Zangger, P., and Wiesendanger, M. Excitation of lateral reticular nucleus neurones by collaterals of the pyramidal tract. *Experimental Brain Research*, 1973, *17*, 144–151.

Zarzecki, P., Shinoda, Y., and Asanuma, H. Projection from area 3a to the motor cortex by neurons activated from group I afferents. *Experimental Brain Research*, 1978, *33*, 269–282.

Zülch, K. J. Pyramidal and parapyramidal motor systems in man. In K. J. Zülch, O. Creutzfeldt, and G. C. Galbraith (Eds.), *Cerebral Localization. An Otfried Foerster Symposium*. Berlin. Springer, 1975.

Acknowledgments

I am greatly indebted to E. Wild for secretarial assistance and to R. Wiesendanger for bibliographical help. I also thank Drs. C. G. Phillips, E. Jankowska, and J. Büttner-Ennever who critically read parts of this chapter.

9

Neuronal Activity Associated with Conditioned Limb Movements

EBERHARD E. FETZ

INTRODUCTION

The development of techniques for recording activity of single neurons in the central nervous system of conscious, behaving animals was optimistically hailed two decades ago as a major advance toward the analysis of neuronal mechanisms of behavior. Previously, information concerning the functions of particular regions of the central nervous system had been obtained primarily by observing the behavioral consequences of experimental and clinical lesions, or by analyzing the responses evoked by electrical stimulation. Ablation experiments can provide important clues as to those functions that depend on the integrity of specific areas and that cannot be subserved or compensated by other regions. Likewise, "electroanatomy," that is, investigation of pathways by electrical stimulation, can elucidate the neural connections between various centers. Neither method, however, can provide any information on the normal activity of cells in the relevant regions. Until recently, activity of single neurons had been recorded only in anesthetized or surgically reduced preparations; although such preparations were useful to document sensory responses to peripheral stimulation, they were of limited value in analyzing normal motor mechanisms.

    With the advent of "chronic" unit recording, it has become possible to observe the activity of single neurons under normal physiological conditions, and even more significantly, to document neuronal responses during relevant behavioral events.

EBERHARD E. FETZ   Department of Physiology and Biophysics and Regional Primate Research Center, University of Washington, Seattle, Washington 98195. Supported in part by NIH grant RR 00166 and NS 12542.

This means that particular hypotheses about the function of cells in movement can be tested by observing their activity in animals trained to make responses specifically designed to resolve alternative hypotheses. In fact, these hypotheses often derive from the behavioral deficits produced by lesions. Moreover, the source and consequences of the activity recorded in single cells can be adequately understood only in light of their input and output connections. Thus, information from both lesion and acute electrophysiological studies forms an essential prerequisite context in which neuronal activity observed in behaving animals can be meaningfully interpreted. This chapter will summarize some contributions of chronic unit experiments to our understanding of the neural mechanisms underlying voluntary limb movement, assess the limitations of this approach, and suggest promising directions for future research.

To record neuronal activity in alert animals, experimenters typically employ extracellular microelectrodes advanced through the brain with a remotely controlled microdrive. A given brain region can be systematically explored over many recording sessions, and the relative locations of different cell types determined. Recording sites of particular interest can be marked by electrolytic lesions, typically made by passing about 10 μA of current briefly through the electrode; the marked sites can be subsequently identified by gliosis or metal deposits in histologic sections of the brain. Since the cell's target sites are of considerable interest, the projection of the axons of recorded neurons is sometimes determined during recording by electrical stimulation; for example, an important class of cortical cells that project to the spinal cord and medulla can be identified by their antidromic response to stimulating the medullary pyramidal tract.

During recording periods the animal is usually restrained, with the head immobilized for recording stability. Animals are trained to perform the relevant behavioral responses by operant conditioning techniques; their behavior is "shaped" by differentially reinforcing (rewarding) those responses that are closest to the desired final behavior. Depending on the goal of the experiment, the behavioral task can be designed to answer particular questions concerning the possible functions of the recorded cells. For example, a rapid and repeatable movement in response to a sensory signal, called a "reaction–time" response, is useful for investigating the relative timing of neuronal activity in different regions during generation of a simple movement. On the other hand, to test whether neuronal activity is related to particular movement parameters, such as limb displacement or active force, animals can be trained to move different loads through the same displacement. As a third example, to investigate the functional relations between cells and muscles, it is possible to condition operantly the response patterns of both directly. Finally, primates performing delayed alternation or delayed match-to-sample tasks can be used to study the neural coding of information that must be remembered in the absence of either stimuli or movements.

The following review of such experiments will illustrate the power of combining neural recording with appropriate behavioral strategies. (See also Phillips, 1973; Phillips and Porter, 1977.) Although this chapter deals mainly with investigations of the neuronal mechanisms of somatic movement, similar strategies have been used to elucidate neuronal activity underlying perception, operant and classical conditioning, sleep, and a variety of natural, unconditioned behaviors (e.g., locomotion, mastication, and postural adjustments). Neural activity during real and "fictive" locomotion is beyond the scope of this chapter, but is discussed by Schwindt (Chapter 4) and by Wetzel and Howell (Chapter 11) in this volume.

NEURAL ACTIVITY DURING CONDITIONED AVOIDANCE RESPONSE

The first comprehensive study of single unit activity in relation to conditioned limb movement in conscious animals was that of Jasper, Ricci, and Doane (1958, 1960). To investigate neural mechanisms mediating a conditioned avoidance response, they recorded activity of cells in different cortical areas in monkeys trained to withdraw their forearms on cue to avoid shock. Figure 1, from these pioneering experiments, illustrates activity of representative neurons recorded in frontal, motor, sensory, and parietal cortex. The conditioned withdrawal response (CR) is shown by the step in trace D, which indicates when the monkey's arm withdrawal opened the switch to the stimulator; this trace also includes EMG activity recorded from the responding limb. The conditioning stimulus (CS) was a train of stroboscopic light flashes at 5/sec, indicated by the evoked potentials recorded from visual cortex, shown in trace C. Unless the monkey withdrew his arm, this CS was followed after 6 sec by a shock. Direct responses of cells to light were tested before conditioning and also with a differential photic stimulus (DS) of higher frequency that was not paired with the shock, and did not produce arm withdrawal.

Under these conditions, Jasper *et al.* found that cells in each cortical area tended to be related to different aspects of the task. Activity of the motor cortex cell illustrated in Fig. 1 increased after onset of the CS and well before the occurrence of the conditioned response. The activity of this unit was clearly related to the

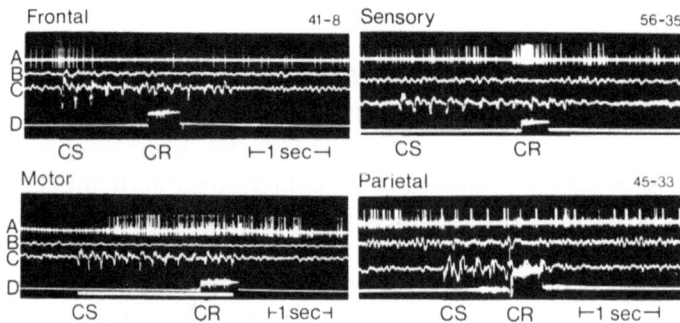

Fig. 1. Responses of single cortical neurons in an awake monkey during conditioned avoidance response. Action potentials of cortical cell (trace A) and overlying electrocorticogram (trace B) were recorded from four cortical areas: frontal (Brodmann area 8), precentral motor, postcentral sensory, and parietal. Trace C is the electrocorticogram recorded in occipital cortex, showing visual evoked potentials in response to the conditioned stimulus (CS), a train of light flashes at 5/sec. The conditioned arm withdrawal response (CR) is represented in trace D by the step, marking the opening of a switch, and by the superimposed EMG activity of the responding arm. From Jasper *et al.* (1960).

subsequent movement rather than being a direct sensory response to photic stimulation, because the unit activity did not change on trials in which there was no movement (e.g., for the DS). Jasper *et al.* found that more than half of the units in motor cortex increased their discharge in relation to the response. Units in postcentral sensory cortex also tended to fire in relation to limb movement, but their increased discharge seemed more closely related, in time, to the muscle contraction. The sensory cortex cell in Fig. 1 fired with onset of the CR, perhaps in response to input from peripheral receptors.

In contrast, neurons in frontal cortex (area 8 of Brodmann) were less consistently related to the conditioned withdrawal. The frontal cell illustrated in Fig. 1 was typical of others related to the task in being activated at onset of the CS and then suppressed throughout the subsequent conditioned response. In posterior parietal cortex, Jasper *et al.* found units whose activity tended to be more related to the conditioning stimuli than the motor response, as shown by the discharge of the cell in Fig. 1. After the monkey had learned the conditioned avoidance response, the CS activated some parietal cells in synchrony with the photic flash. The conditioning procedure had established a differential responsiveness to the CS, linked to the frequency of the stimulus. The higher frequency photic stimulation of the differential stimulus produced no arm withdrawal and produced general inhibition of such parietal cells.

Thus, Jasper *et al.* found a general distinction between neurons of motor and sensory cortex, on the one hand, which tended to be predominantly excited with the CR, and the frontal and parietal cells, which tended more often to be inhibited with the CS. In each cortical area these experimenters actually found a variety of cell types in relation to the conditioned avoidance task. Following these pioneering studies, the relations of cells in each of these regions, as well as in other subcortical centers, to conditioned limb movement have been investigated in more detail. Surprisingly few studies since then, however, have followed the example of Jasper *et al.* in comparing activity of cells in diverse regions to the same conditioned behavior.

Neural Response Patterns during Active and Passive Limb Movement: Central versus Peripheral Input

In alert, behaving animals, the activity of many neurons in the central nervous system can be affected by two separable sources of input, namely, *peripheral* and *central*. Peripheral input originates from activation of sensory receptors and can be characterized by the neuron's responses to adequate natural stimulation. Cells involved in limb movement often receive input from somatic receptors, particularly in muscle and skin; the nature and location of such receptors can be determined in awake animals that have been trained to remain passive during manipulation of their limbs. Sensory input from peripheral receptors often provides useful clues about the region of the body to which the cell may be related during active movements. In addition to such peripheral input, neural activity may also be modulated during active movement by inputs originating centrally. Changes in cell discharge that precede onset of any muscle activity can be attributed to central mechanisms.

The relation between the central and peripheral inputs to different types of cells involved in generating active movements is relevant to understanding the functional organization of the motor system.

To illustrate the distinction between central and peripheral input, Fig. 2 shows the responses of a neuron in the precentral "motor" cortex of a macaque monkey during active and passive forearm movements. With his forearm restrained in a cast that was movable about the elbow joint, the monkey was trained to flex and extend his forearm alternately between two stops; each movement involved a phasic change of position followed by a static hold (called a *ramp-and-hold* movement), making it possible to elucidate the dynamic and static components of the response patterns. The typical response of this precentral unit during active movements (upper records) was a burst of activity during active flexion of the forearm about the elbow. The repeatability of the unit's response pattern is illustrated by the dot raster of cell activity during successive flexion movements (Fig. 2, left). By comparison, when the monkey actively *extended* the arm, this cell exhibited no phasic response. These firing patterns can be compared with the responses evoked when the monkey's arm was passively moved by the experimenter (lower records). With the monkey inactive, as shown by the absence of EMG activity, this neuron responded primarily during passive elbow flexion; in addition to its phasic response, some tonic discharge occurred during maintained flexion.

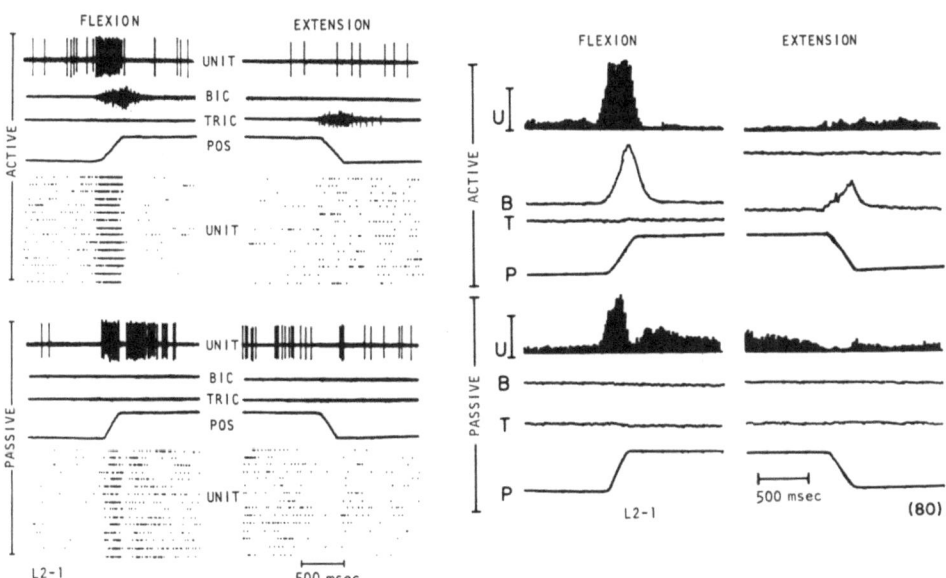

Fig. 2. Responses of monkey motor cortex cell during active and passive flexions and extensions of the forearm about the elbow. Also shown are activity of biceps (B) and triceps (T) muscles, and elbow position (P); elbow angle changed between 90° *(lower level)* and 30° *(upper level)*. Single trials are shown in the top four traces *(left)*; the dot raster below shows unit activity during successive movements to illustrate repeatability of responses. Response averages *(right)* were aligned at onset of movement and show average activity of unit (U), muscles, and position during 80 successive responses. This cell fired phasically with passive and active elbow flexion. From Fetz *et al.* (1980).

The "response averages" at the right in Fig. 2 quantify the average profile of unit and muscle activity and help to identify the onset time of unit activity—namely, the time at which the firing rate exceeds premovement levels. During active flexion, this neuron began firing before the onset of biceps activity, suggesting that its initial activity was determined by input from central structures. This unit also received sensory input from peripheral receptors stimulated by passive arm movements, possibly stretch receptors in the triceps muscles or elbow joint receptors. Many motor cortex cells are driven by passive joint movement, and these neurons typically also fire during active movement of the same joint(s). For motor cortex cells in general, the effective directions of active and passive movement are not always the same, as in Fig. 2; at least as often they are in the opposite direction, as shown in Fig. 3 for a neuron recorded in the vicinity of the cell in Fig. 2. Like its neighbors, this neuron was also driven by passive flexion, both phasically and tonically; during active movement, however, it began to fire before active extension. Because this early activity preceded any peripheral changes, it can be attributed to central sources.

Once the response is under way, peripheral and central input may affect the cell simultaneously, and the two sources of input become difficult to distinguish from one another. The presence of peripheral input during active movement may be demonstrated by sudden perturbations applied to the responding limb. Figure 4 shows the responses of a motor cortex pyramidal tract neuron (PTN) recorded by Conrad, Meyer-Lohmann, Matsunami, and Brooks (1975) during active elbow flexion and illustrates the effect of suddenly changing the resistance to movement.

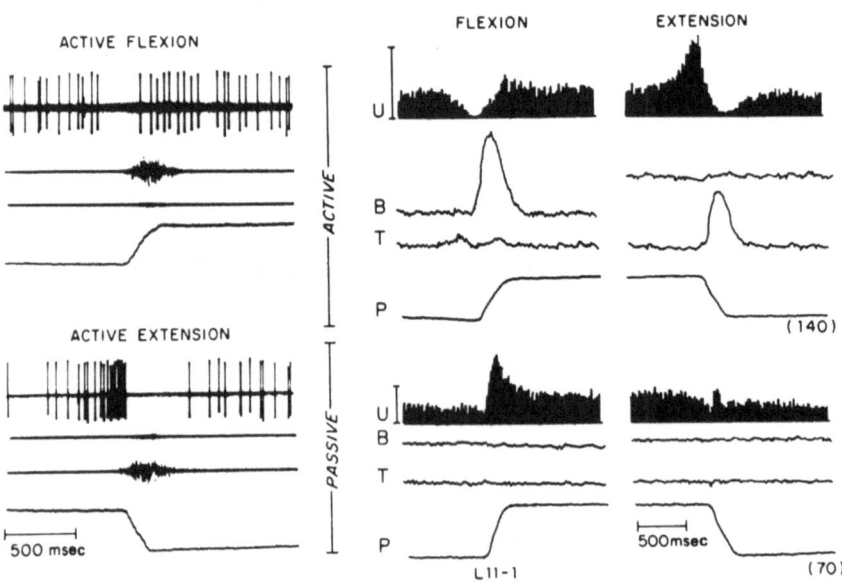

Fig. 3. Motor cortex neuron that fired with active extension and passive flexion of the forearm about the elbow. The phasic and tonic responses to passive elbow flexion could be from stretch receptors in triceps or elbow joint receptors. During active extension, unit activity peaked as the agonist muscle (triceps) became active; during active flexion, cell activity was phasically suppressed. Trace arrangement and identification as for Fig. 2. From Fetz *et al.* (1980).

This PTN discharged during flexion movements and the associated biceps activity (Fig. 4A). When a transient torque pulse that increased the load opposing flexion was introduced with a servomotor at the beginning of flexion (Fig. 4B), the movement was delayed and the stimulated receptors initiated reflex activity in the biceps muscle as well as a burst of activity in the cortical unit. If this PTN facilitated biceps motoneurons, such a response could contribute to the immediately subsequent EMG response in the biceps. Introducing a transient *decrease* in the flexion load at the beginning of the flexion movement (Fig. 4C) produced a pause in unit activity, which Conrad *et al.* related to a subsequent pause in agonist muscle activity. Assuming this cell had an effect on the flexor muscles, these observations are consistent with the concept of a cortical reflex appropriate to compensate for load perturbations (Phillips, 1969).

Motor cortex cells are obviously not the only ones firing before active movements. Some "somatosensory" cortex cells also have central as well as peripheral input. Figure 5 illustrates a neuron recorded in postcentral area 2 that responded to passive forearm flexion. The phasic burst of firing during the passive flexion movement and tonic discharge in the flexed position are typical of input from joint receptors. When the monkey actively flexed and extended the elbow (top), the fir-

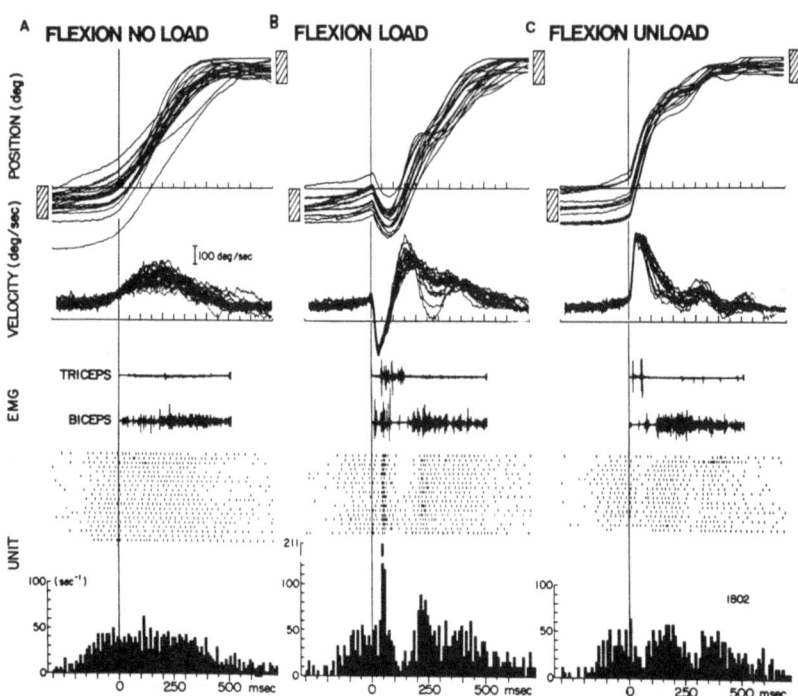

Fig. 4. Response of a flexion-related pyramidal tract neuron to transient load changes. (A) From top to bottom, elbow position trajectories during flexion movements between target zones (hatched rectangles); computed velocity traces; superimposed traces of triceps and biceps activity, dot raster, and time histogram of unit responses during flexion movements. (B) Effect of transient increase in load opposing flexion, triggered at onset of movement: a reflex response in arm muscles and an evoked response in the PTN. (C) Effect of transient decrease in flexor load. Origin of time scale is onset of torque pulse. From Conrad *et al.* (1975).

ing rate of this unit clearly increased well before activity of agonist muscles. It would be important to determine whether such early activity in a sensory cortex cell might be because receptors in other parts of the arm are stimulated prior to agonist muscle activity. Peripheral input to the cell illustrated in Fig. 5 was carefully determined using natural stimulation, and found to derive only from the elbow joint; furthermore, other muscles in the forearm were shown not to become active prior to the agonist muscles during these movements (Soso and Fetz, 1980). This is one example of other postcentral cells that also changed their activity before the onset of agonist-muscle activity, suggesting that postcentral cells may also be affected by central input prior to active movement. On the average, postcentral cells begin firing later than precentral cells during active movements (Evarts, 1974; Fetz, Finocchio, Baker, and Soso, 1980).

The fact that cells may receive both peripheral and central input during movement of a limb has now been documented by recording in diverse areas of the brain, including motor cortex (Brinkman, Bush, and Porter, 1978; Evarts, 1973; Evarts and Fromm, 1977, 1978; Evarts and Tanji, 1974, 1976; Fetz and Baker, 1973; Fetz and Finocchio, 1975; Fetz *et al.*, 1980; Goldring and Ratcheson, 1972; Lamarre, Bioulac, and Jacks, 1978; Lemon and Porter, 1976; Lemon, Hanby, and Porter, 1976; Lewis, Porter, and Horne, 1971; Luschei, Garthwaite, and Armstrong, 1971; Matsumura, 1979; Porter and Rack, 1976; Wong, Kwan, and Murphy, 1979; Wyler and Burchiel, 1978a,b), supplementary motor cortex (Brinkman and Porter, 1979), postcentral sensory cortex (Bioulac and Lamarre, 1979; Soso and Fetz, 1980; Yumiya, Kubota, and Asanuma, 1974; MacKay, Kwan, Murphy, and Wong, 1978), thalamus (Joffroy and Lamarre, 1974; Strick, 1976), cerebellum (Burton and Onoda, 1978; Harvey, Porter, and Rawson, 1977; Strick, 1978; Soechting, Burton, and Onoda, 1978), red nucleus (Burton and Onoda, 1978; Cheney, 1980; Ghez and Vicario, 1978), and spinal cord (Bromberg and Fetz, 1977).

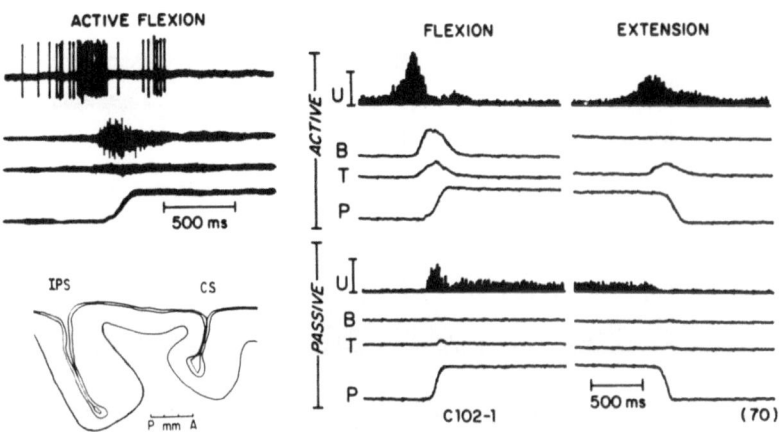

Fig. 5. Response of postcentral neuron during active and passive forearm movements. This area 2 cell responded phasically and tonically to passive flexion, and fired prior to active flexion and extension. Trace arrangement and identification as for Figs. 2 and 3. Insert shows location of cell in postcentral cortex, relative to central sulcus (CS) and intraparietal sulcus (IPS). From Soso and Fetz (1980).

Although a given cell often responds during both active and passive movements of the same joint, neurons in a given region of the brain seem to exhibit many possible relations between the effective direction of active and passive movement. A more consistent relation may exist, however, for cells with specific output projections; for example, precentral PTNs that fire with active limb movement in one direction commonly respond to passive movement in the opposite direction (Evarts and Fromm, 1978; Wolpaw, 1980).

Although the types of *input* that may activate neurons can often be adequately characterized, the *output* "target" sites of the neurons are usually more difficult to determine. The inputs may reveal the sources of a neuron's activity, but its output connections determine the functional consequences of that activity. One method of confirming a causal linkage between activity of a neuron and potential target cells is by cross-correlation, or spike-triggered averaging (STA). If action potentials of the recorded neuron are followed by a clear statistical increase (or decrease) of activity in a target cell, one can infer the presence of a sufficiently potent anatomical connection between the two cells to generate a correlational link. For example, STAs of rectified EMG activity have been used to identify output neurons in motor cortex (Fetz, Cheney, and German, 1976; Fetz and Cheney, 1977) and red nucleus (Cheney, 1980) that generate a postspike facilitation (PSF) of motor-unit firing.

The example illustrated in Fig. 6 is a motor cortex cell that fired with active

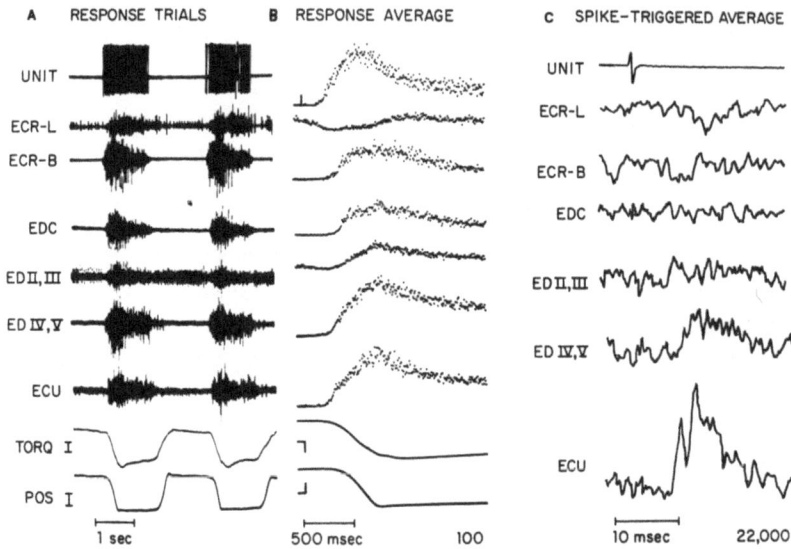

Fig. 6. (A) Responses of precentral cortex cell that produced postspike facilitation of EMG activity of forelimb muscles. Monkey was alternately flexing and extending the wrist against an elastic load (requiring active torque proportional to wrist displacement). Records show activity of unit and six extensors of the wrist (ECR-L, ECR-B, ECU) and digits (EDC, ED II, III, and ED IV, V). (B) Response averages synchronized at onset of extension indicate that unit exhibited phasic peak at movement onset and tonic discharge during static hold; this phasic-tonic pattern is typical of two-thirds of presumed CM cells. (C) Spike-triggered average shows mean rectified EMG activity in the coactivated muscles associated with action potentials of the cell. Analysis interval includes 5 ms before and 25 ms after spike. STAs show transient postspike facilitation in two of the six coactivated muscles: ED IV, V, and ECU. From Fetz and Cheney (1980).

wrist extension; it was coactivated with the six recorded forelimb extensor muscles, and the STA (right) indicates that it facilitated two of the six muscles. The strength and latency of the PSF in the wrist extensor, ECU, suggest a direct cortico-moto-neuronal (CM) connection from this cell to motoneurons of ECU. When such clear output effects of single action potentials can be established, it is reasonable to assume that the neuron exerts an effect on its target cells that is proportional to its activity. The peripheral input to cells with confirmed output, such as the cell shown in Fig. 6, is of interest because it establishes a functional input/output loop. Motor cortex cells that clearly facilitate activity of forelimb muscles tend to respond to passive joint movements that stretch their target muscles (Fetz *et al.*, 1976; Cheney and Fetz, 1980), that is, they fire with active and passive movements in opposite directions. Thus, they would participate in a cortical reflex that could function to subserve load compensation (Phillips, 1969).

RELATIVE TIMING OF CELL ACTIVITY: REACTION–TIME RESPONSES

In the above experiments, the monkey generated limb movements at his own rate; such "free operant" responses were not initiated by any external signals (except perhaps the successful completion and reinforcement of the preceding response). To better control the onset of the motor response and to document the relative timing of neural activity underlying initiation of a well-timed movement, it is often useful to study an animal performing a reaction–time response (Glickstein, 1972). In such experiments the subject is trained to detect a brief stimulus and to make a particular motor response, such as releasing a key, as rapidly as possible. The time interval between the stimulus and the response, typically on the order of 125–180 msec, depends on the stimulus modality and intensity as well as on the species. Minimal reaction times are longer for visual stimuli than for auditory stimuli, perhaps because of the relatively slow photochemical process in the retina. Comparing reaction times to auditory and visual stimuli in humans and monkeys, Luschei, Saslow, and Glickstein (1967) found that monkeys took slightly longer than humans to activate agonist muscles under comparable conditions. An EMG change that usually precedes onset of the agonist muscle activity is suppression of the antagonist muscle (Hufschmidt and Hufschmidt, 1954). In addition to response-related changes in EMG, Luschei *et al.* (1967) observed an early stimulus-related muscle potential in the monkey occurring 25–50 msec after the visual stimulus. In monkeys trained to delay the key release, these early EMG responses remained correlated with the stimulus.

Reaction–time responses are particularly useful for defining the temporal sequence of activation of cells in different regions of the nervous system because they recur repeatedly with a relatively brief interval between stimulus and response. Just as the neural pathways underlying short-latency reflexes, such as the knee jerk or the withdrawal reflex, have been elucidated, it may be hoped that the sequence of neural responses occurring in the interval between the reaction-time stimulus and response might also be understood. For a visually triggered key release, one can imagine a sequence of neural activity beginning with stimulation of retinal cells; activity would then be propagated to diverse cortical and subcortical centers, including cortical association areas. In an animal prepared to make the response, some activity would converge again in proper combination to activate the agonist motoneurons that produce the movement. Although the peripheral links at the input and output stages of such a sequence are beginning to be elucidated, the central steps remain relatively obscure.

In the first of a series of systematic studies of the relation of motor cortex cells to limb movements, Evarts (1966) recorded activity of pyramidal tract neurons in monkeys trained to release a key after a visual stimulus. He found that many precentral cells began to change their activity 10 to 100 msec before the onset of activity in agonist muscles; the onset of discharge of a given PTN covaried more closely with initiation of muscle activity than with the occurrence of the light stimulus. In this study, the earliest responses of precentral cortex cells began about 100 msec after the visual stimulus. This latency is considerably longer than the minimal latency of PTNs to photic stimulation (30 msec) in animals anesthetized with chloralose (Wall, Remond, and Dobson, 1953).

To investigate the relative onset times of other cells that might precede activation of motor cortex neurons, Thach (1970, 1975, 1978) recorded unit activity in cerebellar nuclei and motor cortex during the same responses. To a large extent, the recruitment times of units in cerebellar nuclei overlapped those of precentral motor cortex cells (Fig. 7). Recruitment times of cells in both cerebellum and cortex were distributed over hundreds of milliseconds, and the difference in their mean onset times was relatively slight. Comparable overlapping distributions of recruit-

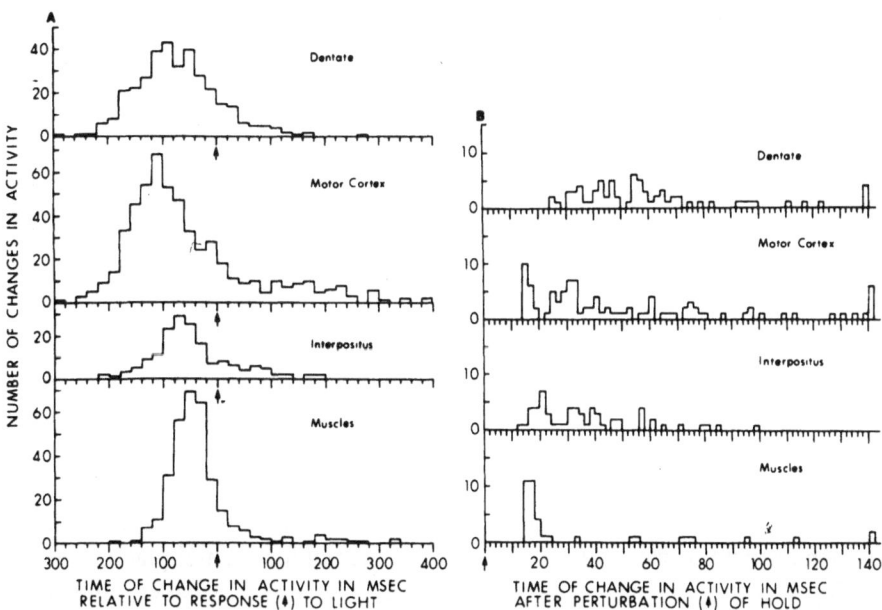

Fig. 7. (A) Distribution of times of change of neural activity relative to a light-triggered wrist movement (arrow). Histograms from bottom up show relative onset times in arm and trunk muscles, and units in interpositus nucleus, motor cortex, and dentate nucleus. (B) Histogram plots time of change of activity following a torque pulse perturbation. From Thach (1978).

ment times of motor cortex and red nucleus neurons were found by Otero (1976). Similar results were also obtained by Neafsey, Hull, and Buchwald (1978) in comparing relative onset time of units in basal ganglia, thalamus, and motor cortex of cats trained to release a bar by lifting their forepaws. In this self-paced task, the distribution of onset times of units in all areas extended to more than a second before initiation of agonist EMG. The neurons with early onset of activity were suggested to be involved in a "response set," in contrast to those neurons recruited later in closer relation to paw movement. The possibility of early postural adjustments of axial muscles was also noted.

A fundamental problem in attempting to demonstrate serial activation of different motor centers is that any particular region, including motoneuron pools, contains cells that are recruited over diverse times, making it difficult to interpret the relative onset times of particular cells in different regions. Moreover, since the duration of most movements greatly exceeds the conduction time between centers, recurrent loops could be "traversed" many times during a single response; thus, the conceptually appealing notion that initiation of movement involves sequential activation of cells in hierarchically related centers seems difficult to prove.

Since any particular region includes cells projecting to diverse areas, it might be hoped that subsets of neurons projecting to particular sites would have a more restricted range of recruitment times. This has recently been tested for those precentral cortex cells that have sufficiently strong correlational linkages to motoneurons to produce clear postspike facilitation (PSF) of forelimb muscle activity (see Fig. 6). During ramp-and-hold wrist movements, such CM cells exhibited a wide range of onset times relative to onset of their facilitated target muscles, extending from 300 msec before to 100 msec after their target muscles became active (Cheney and Fetz, 1980). Thus, even that subset of precentral cortex cells that generate spike-correlated effects within 10 msec is recruited over a wide range of times relative to the beginning of activity in their target muscles. The conclusion would be that generation of a movement involves a progressive recruitment of cells in many regions more or less in parallel, rather than a sequential activation of cells in spatially separate regions. If the location of cells and even their output connections have so little relation to their relative recruitment order, other factors, such as size (see Chapter 2), may be more related to their onset times.

In contrast to a visual or auditory reaction–time response, in which the stimulus and response can be arbitrarily chosen, and their association operantly reinforced, a reflex response, such as a placing reaction, can also be used to analyze the sequence of neural activity between stimulus and response. A contact placing reaction is initiated by cutaneous stimulation of a free limb and involves placing the limb onto a higher surface. Such placing reactions are repeatable and easily elicited, and provide a useful model for analysis of involvement of higher supraspinal centers, such as motor cortex, whose integrity is essential for normal placing reactions. Studying contact placing in cats, Amassian, Ross, Wertenbaker, and Weiner (1972) have analyzed the sequence of neural activity from cerebellum to thalamus to cortex involved in this automatic response, and related these to development of the response in kittens. In cats standing on four force transducers, Padel and Steinberg (1978) documented activity of red nucleus cells during a forepaw placing reaction. They found that most of the red nucleus cells related to the placing response were associated with the phasic muscle activity during movement, but relatively unrelated to static postural forces. Documenting the activity of ventral lateral thalamic nucleus during placing movement, Smith, Massion, Gaherty, and Romieu (1978) found that most were responsive with contralateral limb movement; relatively few responded directly to sensory input.

LONG-LOOP REFLEXES: EFFECT OF MOTOR PREPARATION

A particularly interesting reaction–time response is one initiated by a perturbation of the responding limb. In this case, the same limb perturbation may be used to initiate both a short-latency reflex and a long-latency reaction–time response. In one of the earliest such experiments, Hammond (1956) asked human subjects to resist a sudden, externally imposed extension of their flexed forearm, and found two distinct components in the biceps EMG response. An early component, with a latency of 18 msec, was apparently produced via a segmental reflex initiated by the perturbation, which stretched the biceps. A later component, beginning at a latency of 50 msec, appeared when the subjects were asked to resist the perturbation (by activating biceps), but did not appear when they were instructed to "let go" when they felt the perturbation. The neural pathways mediating this late response may involve "long-loop" circuits via supraspinal structures, a hypothesis that can be investigated in chronic animal experiments.

Figure 8 schematically illustrates the EMG responses evoked by muscle stretch in a subject prepared to contract that muscle after detecting the perturbation, and shows possible mediating reflex pathways. The shortest latency "M1" response is probably mediated by segmental reflexes, including the monosynaptic stretch reflex. The longer latency "M2" response seems to involve a long-loop reflex, rather than slowly conducting afferent fibers or repetitive firing of the same motoneurons. As illustrated, one contributing pathway involves motor cortex CM cells, which do respond to externally applied movements that stretch their target

Fig. 8. Schematic diagram of sequence of responses initiated by muscle stretch in a subject prepared to make a voluntary reaction–time contraction of that muscle. Arrows indicate conduction of neural activity over segmental loop (mediating M1 response) and transcortical loop (contributing to M2 response). Although motor cortex is illustrated as a representative supraspinal center involved in the long latency responses, many other regions could be similarly involved.

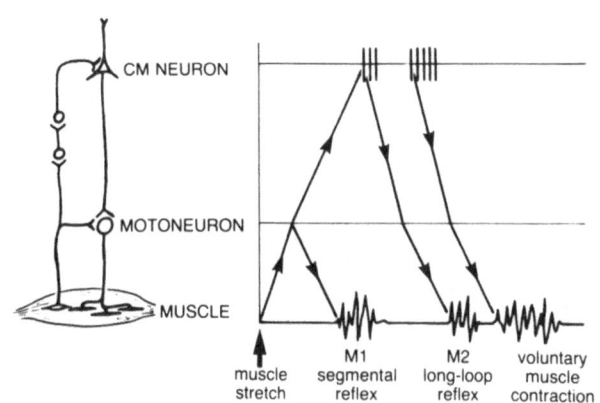

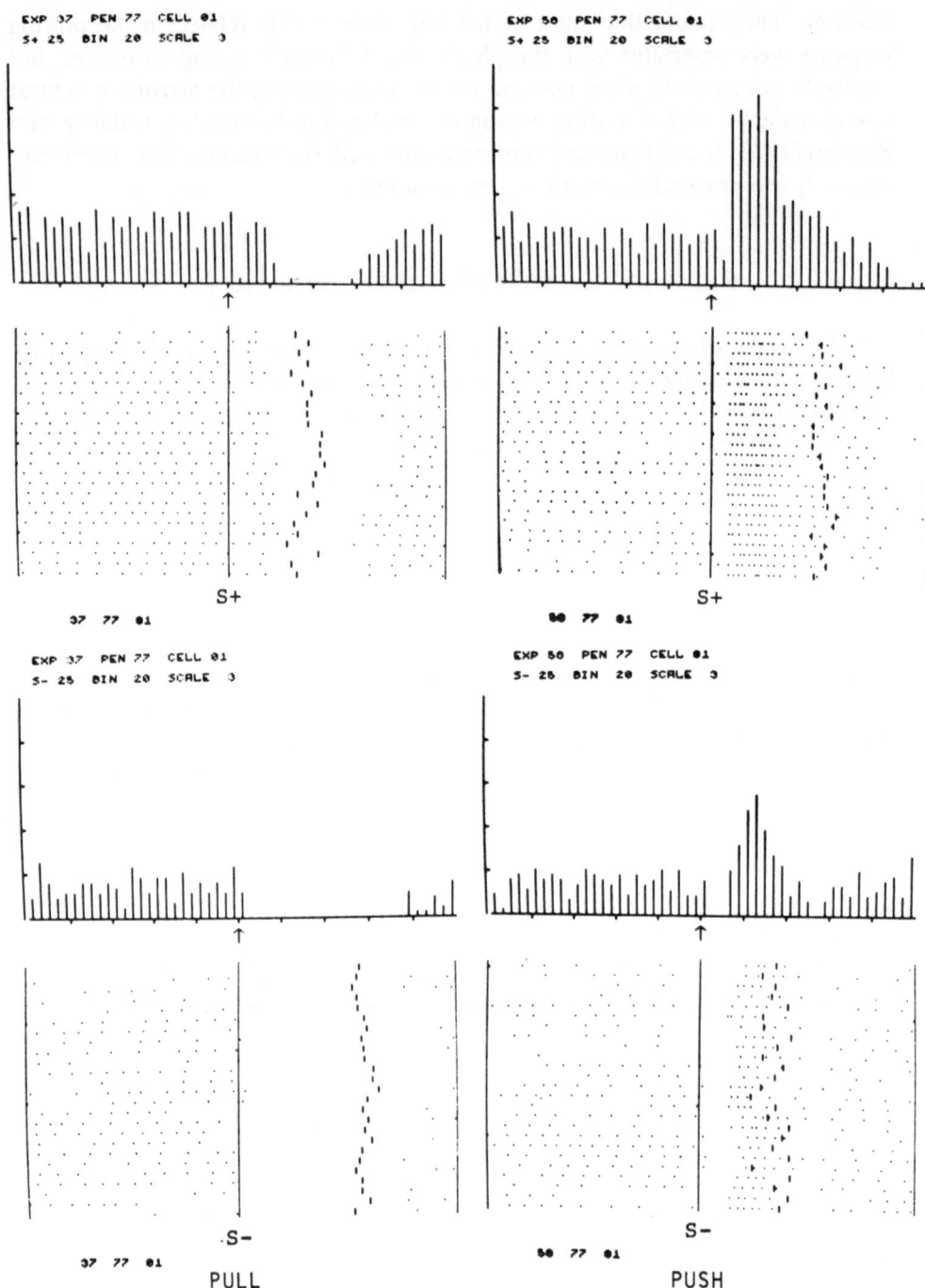

Fig. 9. Responses of motor cortex PTN during two arm movements (push or pull) initiated by either of two perturbations (S+ or S−). The monkey was cued by a prior instruction light as to which arm movement to make when the handle was perturbed. This PTN fired with push movements *(trials on right)* and was silenced with pull movements *(trials on left)*. When the perturbation opposed push movements (S+, *upper trials*) this PTN exhibited a short-latency response; in contrast, when the perturbation assisted the push movement (S−, *lower trials*) the PTN activity paused. Comparison of the upper right with upper left indicates that the prior instruction to push

muscles (Cheney and Fetz, 1980); other subcortical reflex pathways could also contribute as well.

Evidence that precentral motor cortex cells may participate in the later EMG responses was obtained by Evarts and Tanji (1974, 1976). In an experimental paradigm comparable to Hammond's (1956), they trained monkeys to either push or pull on a handle after it had been perturbed. When the monkey was instructed to pull the handle (a movement requiring contraction of biceps), a perturbation of the handle that stretched biceps evoked two EMG responses: a short-latency (M1) response at 12 msec and a longer latency (M2) response at 30–40 msec. A still later phase of EMG activity seemed to be related to the voluntary reaction–time pull response. Recording activity of motor cortex cells, Evarts and Tanji (1974) found many precentral neurons responding at latencies of 32 to 34 msec, appropriately timed to contribute to the M2 EMG activity. A motor cortex PTN whose response pattern is consistent with such a transcortical reflex loop is illustrated in Fig. 9. Before the perturbation, the monkey had been cued with a light indicating whether he should push or pull the handle when it was perturbed. This PTN was activated when the monkey pushed the handle. The trials are illustrated separately for different stimulus and response combinations. The responses in the upper right were push movements initiated by a perturbation that opposed the push movement, that is, that stretched the agonist muscle; this stimulus evoked a strong early response in the PTN as shown by the broad histogram peak. This broad peak includes the neuron's response to the perturbation as well as the subsequent activity involved in making the appropriate push movement. At the bottom right, a perturbation that released the muscle resulted in cessation of activity of the PTN, followed by a period of discharge associated with the push movement. When the monkey made the opposite arm movement, that is, pulled the handle, this PTN was silent, and the perturbation that stretched the biceps evoked a smaller excitatory reflex response in the cell (upper left). Thus, the magnitude of the cell's direct response to muscle stretch depended on which movement the monkey was prepared to make.

Another dramatic example of the influence of behavioral "set" on the sensory responses of motor system cells was obtained by Strick (1978) in the dentate nucleus (Fig. 10). In a behavioral paradigm similar to that used by Hammond (1956) and Evarts and Tanji (1974, 1976), the monkey's movement was triggered by a perturbation that moved the handle away from the monkey. When the monkey was set to push the handle (Fig. 10A), the perturbation activated this dentate cell; when he was prepared to pull (Fig. 10B), the *same* perturbation produced short latency suppression of activity. One might wonder whether such differences could be due to changes in the way the monkey grasped the handle in preparation for each response. This seems unlikely because under the same behavioral conditions

---

(right) enhanced the magnitude of the direct response to S+. The perturbation occurred at the center line of the dot rasters and arrow under the histogram (bin width: 20 msec); activity is shown for 500 msec before and after the perturbation. Dark bars following perturbation indicate completion of required arm movement. From Evarts and Tanji (1974).

neighboring cells in the interpositus nucleus showed no such modification of their sensory response as a function of set (Strick, 1978).

A similar modulation of sensory input has been found in the oculomotor system, in certain neurons related to visually triggered eye movements. Presenting a visual stimulus in the receptive field of such cells evokes a more intense short-latency response in those trials in which the subject subsequently makes a saccade to the stimulus. Such movement-contingent enhancement of a sensory response has been documented in superior colliculus, posterior parietal cortex, and frontal eye fields (Goldberg and Wurtz, 1972; Wurtz and Mohler, 1976a,b).

These experiments suggest that even in the absence of overt movements, the preparation, or "set," to perform a response involves detectable changes in activity and responsiveness of cells in diverse regions of the brain. In monkeys cued to make a limb movement to a subsequent stimulus, the cue signal alone has been found to generate changes in firing of cells in motor cortex (Jasper *et al.*, 1960; Evarts and Tanji, 1974, 1976; Tanji and Evarts, 1976), basal ganglia (Soltysik, Hull, Buchwald, and Feheti, 1975; Neafsey *et al.*, 1978), prefrontal cortex (Fuster, 1973; Niki and Watanabe, 1976a,b) and supplementary motor cortex (Tanji, Taniguchi, and Saga, 1980). Thus, the subsequent reaction–time stimulus initiates the overt response in a nervous system preset to generate the appropriate movement.

RELATION OF CELLULAR ACTIVITY TO PARAMETERS OF MOVEMENT

Knowing that cells in particular brain regions are activated during a motor response, one can next investigate more specifically which parameters of the response might be coded in cellular activity. Just as neurons in sensory regions code

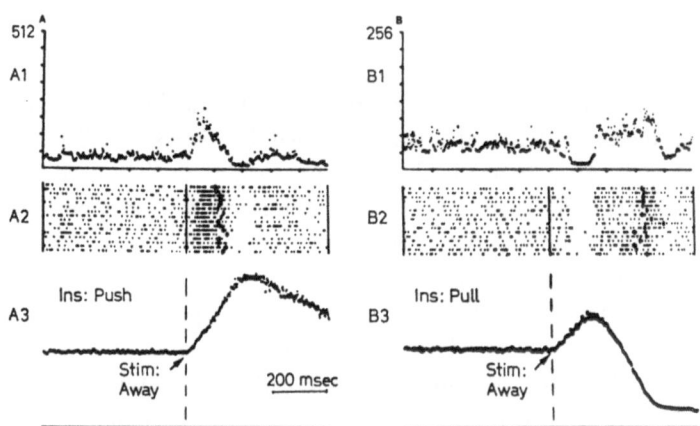

Fig. 10. Effect of motor preparation on short-latency response of a dentate cell to the same load perturbation. Trials in which the monkey had to push the handle (A) are contrasted with trials in which he pulled (B). The average handle trajectories (A3 and B3) indicate that the perturbation produced the same initial deflection in both cases. Short-latency unit response, shown by histograms (A1 and B1) and dot rasters (A2, B2), was excitation when monkey was prepared to push and suppression when prepared to pull. From Strick (1978).

stimulus features, such as size, location, and motion, it seems reasonable that cells in motor regions could code particular movement parameters, such as position of the limb or the force required to move it. To determine whether motor cortex PTN activity is more strongly related to limb position or active force, Evarts (1967, 1968) trained monkeys to move a handle through the same displacement but against different loads. In these studies most motor cortex cells were more consistently related to the force exerted or to changes in force than to the displacement of the wrist. Similarly, when the monkey was required to hold the handle steady in opposition to different loads, motor cortex cell activity again correlated with the direction and degree of isometric force that the muscles generated (Evarts, 1969). Humphrey, Schmidt, and Thompson (1970) further analyzed these relations by simultaneously recording the activity of a group of motor cortex cells during alternating wrist movements. They found that a weighted average of the firing rates of several cells could be used to predict the subsequent time course of force and position, as well as their temporal derivatives. The degree of correlation between the predicted and the observed movement parameters improved with the number of nonredundant, task-related motor cortex cells included in the weighted average. The correlations were somewhat better for force and velocity trajectories than for position or rate of change of force. An important contribution of this study was to point out that the ability to predict accurately the movement parameters improved with the number of cells taken into account.

Others have challenged the hypothesis that motor cortex cell activity exclusively codes force. Schmidt, Jost, and Davis (1975) documented activity of motor cortex cells during wrist movements against elastic loads and found that a fourfold difference in active force produced, on the average, a one-and-one-half-fold difference in firing rate for all task-related cells. Differences in firing rate were considerably greater when flexion and extension movements were compared, so Schmidt *et al.* concluded that the direction of force, or the agonist muscles, might be the more relevant variable encoded. Whereas previous experimenters (Evarts, 1967, 1968; Humphrey *et al.*, 1970) had used isotonic loads (force constant in magnitude and direction, independent of displacement), Schmidt *et al.* used elastic loads (force proportional to and opposing displacement), which probably accentuated the contrast between neural activity in the two directions of movement.

Another variable affecting the relation of neural activity to force generated during a movement is the speed of the movement. Evidence is accumulating to suggest that rapid, ballistic movements may involve different motor mechanisms than slower, finely controlled movements. Certain cells in basal ganglia are more active with slow than rapid limb movement (DeLong, 1971, 1973; DeLong and Strick, 1974). Some motor cortex cells also fire more strongly with controlled hand movement than rapid ballistic responses (Cheney and Fetz, 1980; Fromm and Evarts, 1977). Certain jaw-related motor cortex cells also appear less active with chewing than with a reaction–time bite (Luschei *et al.*, 1971) or a controlled force bite (Hoffman and Luschei, 1980). Neurons in cortical area 3a have also been found to be strongly activated during fine control of limb position (Tanji, 1975, 1976). The differences between ballistic and controlled movements may involve different motor units as well as different degrees of sensory feedback.

To quantify the neural activity associated with the dynamic and static components of limb movement, many investigators have studied tasks involving ramp-and-hold responses. An example, already illustrated in Fig. 6, involved alternating wrist movements against an elastic load. The response pattern of this neuron, a phasic burst of activity during the ramp of movement followed by tonic discharge during the static hold period, is typical of many motor cortex cells observed during different types of ramp-and-hold responses, including isometric precision grip on a force transducer with the fingers (Hepp-Reymond, Wyss, and Anner, 1978; Smith, Hepp-Reymond, and Wyss, 1975), ramp-and-hold force bites with the jaw (Hoffman and Luschei, 1980) isometric wrist torque trajectories (Cheney and Fetz, 1980), and forearm movements (Conrad, Wiesendanger, Matsunami, and Brooks, 1977). In addition to such phasic-tonic patterns, all these investigators also observed other cortical cells that fired only during the static hold; still other neurons fired only during the phasic ramp and many cells had more complex or variable relation to the response. In all these studies the cells that discharged during the hold period typically fired tonically at rates that increased with the amount of static force exerted.

The quantitative relation of discharge with force depends critically on which cells are examined, so it becomes relevant to identify those cells that have a direct output effect on the agonist muscles. Precentral CM neurons with sufficiently

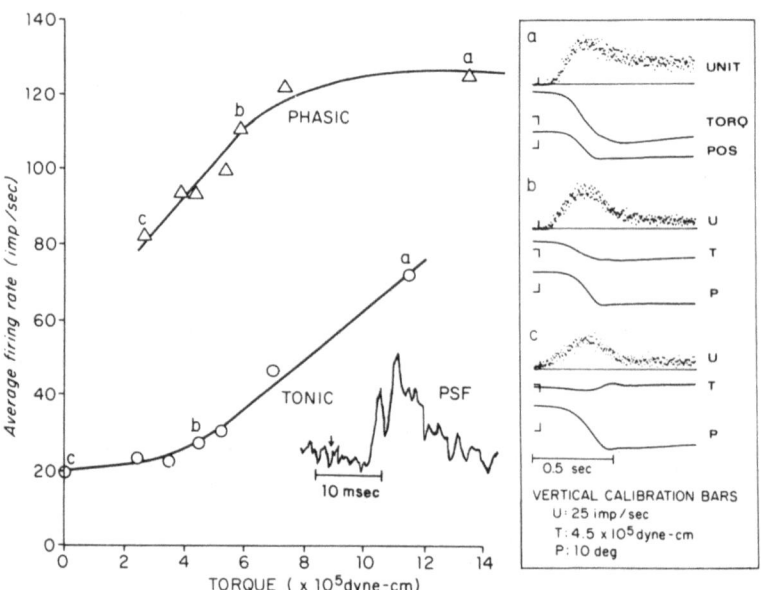

Fig. 11. Relation between activity of CM neuron and active torque produced by its target muscles. Response averages *(right)* of unit activity (U), torque, (T), and position (P) were compiled separately for ramp-and-hold wrist movements into the same hold zone but against elastic loads of different stiffness. The graph plots firing rate of cell during phasic peak at onset of movement and tonic discharge during the static hold period. For higher loads, tonic discharge increased with static torque. Since ramp movements were completed in about the same time, the phasic activity increased with rate of change of torque in lower end of range but saturated at upper end. PSF of a target extensor muscle is illustrated in inset. From Cheney and Fetz (1980).

potent linkages to motoneurons to facilitate muscle activity have a tonic firing rate during the hold period that increases with active force (Cheney and Fetz, 1980). Figure 11 illustrates the responses of the CM cell in Fig. 6 during ramp-and-hold wrist movements; by changing the stiffness of the load, different levels of active torque were required to move into the same hold zone, as indicated by the response averages at the right. When the movements required substantial active torque (a), the tonic activity of this cell during the static hold period was relatively high. When wrist displacement was made with no external load (c), the tonic activity during the hold period was lower. Plotting this tonic discharge rate as a function of external load (left) shows that the average firing rate increased as a function of static torque. This cell is typical of a larger number of such CM cells that produced postspike facilitation; their tonic firing rate increased with static torque, and increased linearly over much of the range. This was also true for isometric responses, in which the wrist was not displaced, indicating that the active torque, not the displacement, was the primary variable to which cell activity was related. The fact that action potentials of these cells facilitate the activity of agonist muscles, which generate the torque, establishes a causal link between CM cell activity and force output.

Studies documenting the relation of cell activity to the mechanical parameters of movement have so far been done primarily in motor cortex, whose cells have a fairly direct output relation to agonist muscles. In contrast to the motor cortex cells firing tonically with static force, activity of cells in supplementary motor cortex (Smith, 1979) and red nucleus (Cheney, 1980; Ghez and Vicario, 1978; Padel and Steinberg, 1978) appeared less related to the degree of static force (but see Ghez and Kubota, 1977). Clearly, similar experiments for cells in other regions, such as thalamus and cerebellum, would help define the regions involved in maintenance of active force.

OPERANTLY CONDITIONED UNIT ACTIVITY

In the above experiments, designed to test the relation of cell activity to movement parameters, the animals were trained to perform relevant responses and task-related neurons were sought. Usually, a small proportion of the recorded cells in any region are clearly related to a particular pretrained movement. Whether the remaining cells are related to different movements or, perhaps, unrelated to any, could, in principle, be determined by having the animal perform an exhaustive repertoire of possible responses. A more practical approach, suggested by Olds' (1965) original experiments on operant conditioning of a single unit activity, is to train the subject to generate activity in the recorded unit and observe the correlated responses.

In his first unit-conditioning study with awake rats, Olds made food reinforcement contingent on increased activity of hippocampal and midbrain tegmentum cells. To preclude reinforcement of movement artifacts, Olds devised a circuit to withhold food delivery when movement artifacts were detected. Nevertheless, conditioned burst activity in pontine cells was often seen to be associated with slight eye or head movements, and activity of some hippocampal cells was associated with nose or whisker movements. Operantly conditioning activity of pontine units in unrestrained cats (using lateral hypothalamic stimulation reward), Breedlove, McGinty, and Siegel (1979) noted that increased discharge was often associated with increased levels of motor activity. Whether the units conditioned in these experiments were involved in generating the movement or were responding to input from the movement remains an open question.

Many studies on operant conditioning of single unit activity have involved motor cortex cells in awake primates. With contingent food reward and visual feedback, monkeys readily learned to increase or decrease precentral cell activity (Fetz, 1969; Fetz and Baker, 1973; Wyler and Burchiel, 1978a,b), and to sustain tonic levels of unit discharge (Schmidt, Bak, McIntosh, and Thomas, 1977; Wyler and Finch, 1978). Operantly reinforced bursts of precentral cell activity were often accompanied by movements of the contralateral limb. For some cells, these movements were relatively specific and repeatable with each operant burst, such as flexion of a distal joint. For many other cells, the movements were complex and variable from one burst to the next; in some of these cases the movements became more restricted and repeatable with continued reinforcement of the same cell, perhaps because unrelated components of the response dropped out. Still other motor cortex cells could be activated with no observable concomitant motor response (Fetz and Baker, 1973).

With the arm mechanically immobilized, the set of muscles that were coactivated with operant unit bursts was more repeatable than with free movements, and typically included several different arm muscles (Fetz and Finocchio, 1975). Neighboring cells in the same region of motor cortex were associated with different degrees of coactivation in the same set of forearm muscles, suggesting that each neuron has its preferential relation to different muscles. Wyler and colleagues have reported greater success in operantly conditioning cells related to distal forearm muscles than those related to proximal muscles (Wyler and Burchiel, 1978a; Wyler and Finch, 1978).

As a means of eliciting movements associated with individual neurons, operant conditioning of neural activity might fruitfully be applied with cells in other regions of the nervous system as well, such as cerebellum and basal ganglia; such a strategy may be useful when the relevant movements cannot be anticipated in advance, or may differ for neighboring cells. A major question in such experiments is whether the neuron's activity contributes to the movement or is a consequence of the movement. Since many cells have peripheral input as well as central input, this issue may be only partially resolvable. If the cell has demonstrable responses to adequate natural stimulation and fires during movements that stimulate its peripheral receptors, its activity may be largely a consequence of sensory feedback. If it fires before active movement and has no sensory input, it may be under more direct central control.

One practical problem in using unit conditioning to elicit related movements is that bursts of many cells are associated with such variable and complex movements that it becomes difficult to determine which component(s) of the responses

may be related to the cell activity. In some cases, continued reinforcement of the
unit eventually leads to more specific correlated responses. It is never certain, however, that even a specific correlated response is the only one in which that cell plays
a role, because some cells may be involved in several different movements.

513

NEURONAL
ACTIVITY AND
LIMB MOVEMENTS

RELATIONS BETWEEN SINGLE CELLS AND MUSCLE CONTRACTION

Most behavioral tasks involve movements produced by groups of coactivated
muscles; such movements presumably result from the activity of a population of
central cells that may each be related in different ways to particular subsets of the
activated muscles. For example, a given cell may be related to one muscle, another
to several; moreover, their activity may be related to particular features of muscle
activity, such as onset or termination. Such questions concerning the functional
relations of cells to specific muscles can be further resolved by training the animal
to control the activity of particular muscles directly.

In a study designed to determine whether individual motor cortex cells were
related to one or to multiple arm muscles, monkeys were trained to contract each
of four representative forelimb muscles in relative isolation (Fetz and Finocchio,
1975). With its arm held in a cast, the monkey was differentially rewarded for
generating bursts of isometric EMG activity in a given muscle and maintaining
relative inactivity in the other three. In addition to reinforcement for the correct
responses, the monkey had continuous visual feedback in the form of a meter whose
needle deflection indicated how close his response patterns approached criterion for
reinforcement. Figure 12 illustrates responses of one motor cortex cell during different operantly reinforced response patterns. The activity of this cell was documented in relation to isolated contractions of a wrist flexor muscle (A), a wrist
extensor (B), elbow flexor (C), and elbow extensor (D). This neuron became active
with both the wrist flexor and the elbow flexor muscle under isometric conditions;
this relation was consistent with a clear response when the forearm was actively
flexed about the elbow (not shown). This result is typical of other motor cortex
cells observed during isolated muscle contractions; most were activated with several
sets of forearm muscles, and some fired with both flexors and extensors of the same
joint. Certain precentral cells were coactivated in the same way with all four sets
of muscles. Such observations suggested that many motor cortex cells may have a
"higher order" relation to several muscles, in contrast to the simple one-to-one
relation of motoneurons to a single muscle.

To elucidate further the relation between the unit in Fig. 12 and the forelimb
muscles, the monkey was also rewarded for activating the cell in "operant bursts,"
and was allowed to coactivate any of the muscles. Under these conditions the monkey coactivated biceps and both wrist muscles (Fig. 12E). This result was also
typical of other motor cortex cells whose activity was operantly rewarded; usually
the monkey coactivated several muscles with operant bursts. However, the coactivated muscles and their relative amplitude were usually quite different for adjacent
cells in the same region of motor cortex. In fact, some precentral cells were activated

without any observed muscle activity or movements. Although most cells responded to peripheral stimulation, their activity usually increased and peaked prior to onset and peak of EMG activity, suggesting their major drive originated centrally.

Another useful application of operantly conditioning cell and muscle activity is to test the stability of observed correlations by rewarding their dissociation. For example, the cell in Fig. 12 was consistently coactivated with the biceps and wrist flexor muscle under several different behavioral conditions: (1) when isometric muscle activity was rewarded (A, C), (2) when unit activity was rewarded (E), and (3) when the monkey actively flexed the elbow. However, when the monkey was rewarded specifically for activating the cell and suppressing muscle activity, he quickly learned to do so (F), demonstrating an unexpected degree of plasticity in these unit-muscle correlations. Thus, by operantly rewarding the dissociation of correlated events (in this case, unit and muscle activity), one may directly test the flexibility of such relations.

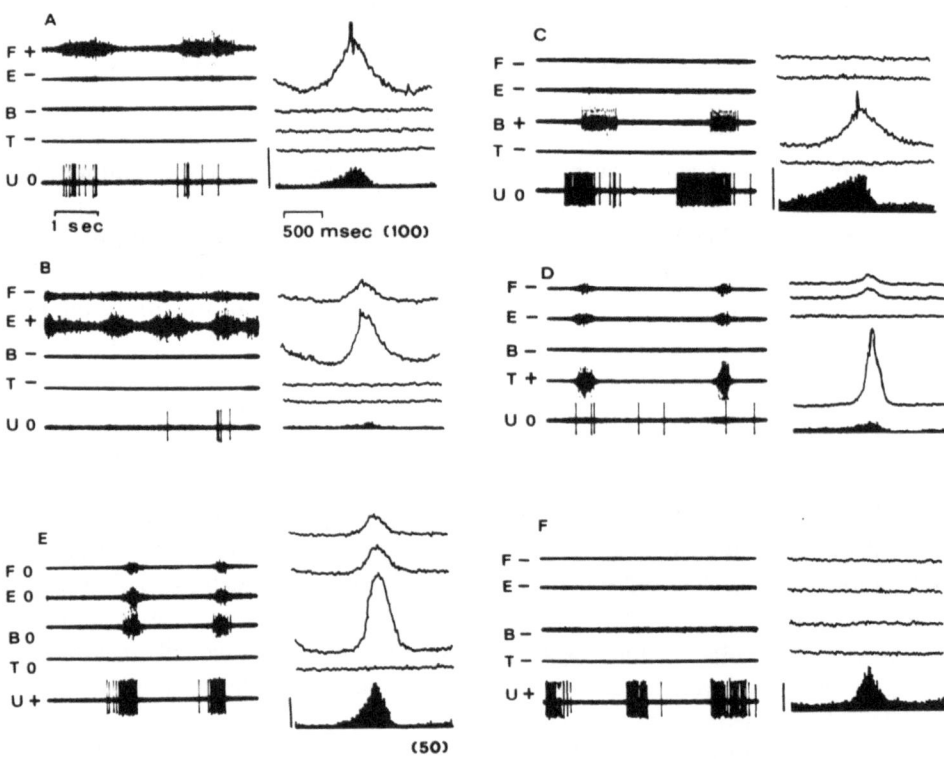

Fig. 12. Relation of motor cortex cell to isolated contraction of different arm muscles (A–D). Records illustrate activity of flexor carpi radialis (F), extensor carpi radialis (E), biceps (B), triceps (T), and the cortical unit (U). The arm was held fixed while patterns of isometric muscle activity were operantly rewarded; symbol after each letter indicates whether the reinforced pattern included activation (+) or suppression (−) of that element, or did not include its activity (0). Each set of records shows sample trials *(left)* and response averages *(right)* compiled for successive reinforced responses. Lower sets show response patterns when operant bursts of unit activity were reinforced with no contingency on muscles (E) and with simultaneous suppression of all recorded muscle activity (F). From Fetz and Finocchio (1975).

The results of this study suggested some caution in interpreting coactivation of cell and muscles as evidence for a causal relation. Single motor cortex cells were typically coactivated with many muscles, and the coactivation of a given unit-muscle pair could be quite flexible, depending on the behavioral conditions. Even "strong" unit-muscle correlations that appeared during a variety of different response patterns (e.g., the unit-biceps correlation of the cell in Fig. 12) could be dissociated by differential reinforcement of unit activity and muscle suppression.

In contrast to the widespread and variable coactivation patterns, more reliable and definitive evidence of causal relation to muscle activity can be provided by their cross-correlation. The short-latency PSF in STAs of EMG reveals which of the many coactivated muscles, if any, are actually facilitated. An extremely small proportion of the STAs of coactivated EMG show any features indicating a causal linkage (Fig. 6). Interestingly, the firing patterns of some CM cells, which do generate PSF, are quite different from the average profile of EMG activity of their target muscles; they may show distinct bursts at onset of EMG activity (Fetz and Cheney, 1977; Cheney and Fetz, 1980).

While STAs are useful for identifying cells with direct linkages to motoneurons, such as CM cells or rubromotoneuronal cells (Cheney, 1980), their reliability in detecting polysynaptic connections is dubious (see Soechting *et al.,* 1978). Most CM cells were found to facilitate a set of forelimb muscles, called their "muscle field," indicating that their activity has divergent effects on a group of synergistic muscles involved in a movement. Such muscle fields were usually a subset of all the coactivated muscles, suggesting that any movement involves many CM cells (and other descending cells) with overlapping muscle fields.

UNIT ACTIVITY RELATED TO DELAYED RESPONSES

Most of the experiments cited above have investigated activity of neurons related in a relatively direct way to initiation of movement. The relation of neural activity to more subtle aspects of behavior can be similarly analyzed, as illustrated by experiments in which monkeys were required to remember the appropriate response during a delay period. Ablation studies have shown that lesions in dorsolateral prefrontal cortex permanently impair a monkey's ability to remember the correct response during a delay period between presentation of a cue and initiation of the response. These lesions do not, however, impair movements directly. Kubota and Niki (1971) recorded activity of prefrontal cortex units in monkeys trained to press two levers alternately; during the intervening 5-sec delay periods, a screen blocked the levers, and the monkey had to remember the direction of the next correct lever press. In these studies, some prefrontal cells fired during the delay period and others fired in direct relation to movement, much like precentral motor cortex cells. Using a Wisconsin General Test Apparatus, Fuster (1973) trained monkeys to retrieve food from one of two covered wells after a delay. Each trial sequence involved (a) a cue period in which the monkey observed food being placed in one of the wells and then covered, (b) a delay period of 18 sec in which vision was occluded and access to the food was prevented, and (c) a response period in which the monkey retrieved food from the baited well. Figure 13 illustrates the types of prefrontal cortex cells observed. The most common types of task-related pattern in prefrontal units were types C and D (34% and 17% of all cells, respectively), which exhibited increased activity during the delay period. This increased firing occurred only during those trials which were followed by correct responses; such firing could be disrupted by distracting stimuli that also led to subsequent behavioral errors. This increased prefrontal cell activity was not related to postural changes, nor was it specific to the side on which the response was made. Such activity was clearly the consequence of the behavioral training, because naive animals exhibited no such changes in prefrontal unit activity during the delay periods. Subsequently, similar classes of response patterns were found in cells of the thalamic nucleus medialis dorsalis, which has a close reciprocal interrelation with prefrontal cortex (Alexander and Fuster, 1973; Fuster and Alexander, 1973). These responses appear to be involved in coding short-term memory of the next appropriate movement.

Recently, Niki and Watanabe (1976a) found evidence that activity of some prefrontal cortex cells may even code specific information concerning the location of the spatial cue or the direction of the subsequent behavioral response. They required their monkeys to press, after a delay, one of four choice keys (left, right, up, or down), depending on which key had been illuminated before the delay period (Fig. 14). To differentiate whether cell activity was related to the cue or the target position, the monkey was trained to perform three different cue–choice com-

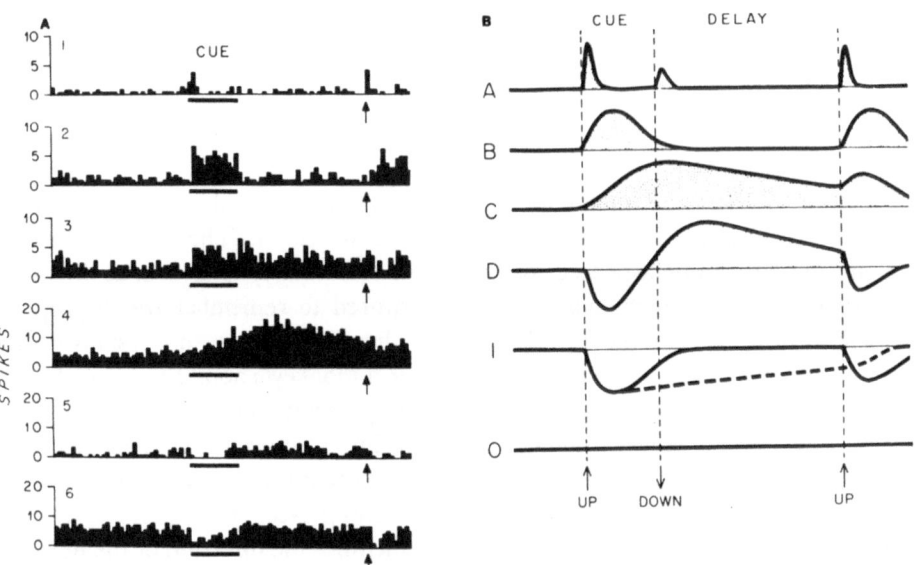

Fig. 13. Response patterns of prefrontal cortex neurons during delayed response trials. (A) Histograms of average unit activity over 5 trials are shown for representative cells. Horizontal bar denotes cue period, followed by an 18-sec delay period before the response period, indicated by the arrow. (B) Schematic representations of the types of response patterns found in prefrontal cortex and medialis dorsalis neurons. Arrows indicate movement of the screen blocking the monkeys' vision. The most common types of cells were C and D, for both prefrontal cortex (Fuster, 1973) and medialis dorsalis (Fuster and Alexander, 1973). From Fuster (1973).

binations. Among the task-related units in prefrontal cortex, almost one fifth were differential delay units; about three-quarters of these were related to the cue location, and the rest were related to the direction of the impending responses. An example of the former is shown in Fig. 14. This neuron was activated during the delay period only when the cue occurred in the lower choice key (D), independently of whether the subsequent response was on the left key (D→L) or the lower key (D→D). Other differential delay units in prefrontal cortex discharged tonically during the delay period only in those trials in which subsequent movement was made toward a specific choice key. When the delay period was prolonged until the monkey made errors, these response-related differential units fired in relation to the direction of the subsequent movement, whether it was correct or not. In contrast, activity of the cue-related differential units was not predictive of subsequent responses. Other types of task-related cells in prefrontal cortex were found to respond differentially just during cue presentation or the choice period (Niki, 1974a,b). Thus, one can already speculate on the sort of relations between these types of cells that might be involved in delayed response behavior. The prefrontal neurons responding differentially when the spatial cue is presented may activate the cue-related differential delay units (like the one in Fig. 14), whose sustained activity during the delay holds information concerning the spatial position of the preceding cue. During a correct performance, these cells may in turn activate the response-related differential delay units, whose activity at the time of choice appears to determine the direction of the subsequent response (Niki and Watanabe, 1976a).

Fig. 14. Response patterns of a prefrontal cortex neuron differentially active during delay periods, depending on cue location. Four keys in a diamond configuration served as cues and choice keys. Dot rasters show unit activity separately for different cue–choice combinations; from left to right, each 13-sec period includes intertrial interval (4.5 sec sample), cue period (1 sec) in which one key is illuminated, delay period (3 sec) in which keys are turned off, choice period in which two choice keys were illuminated. Top two sets show responses on the "up-down delayed response" task, in which upper (U) or lower (D) key was illuminated in the cue period, and the money was reinforced for pressing the same key when both were illuminated in the choice period. Bottom sets show the equivalent "left-right delayed response" task. Middle set shows "conditioned position" task in which upper or lower cue key indicated whether to press right or left choice key, respectively. This unit discharged at higher rate during the delay period only on trials in which the lower choice key ("D") had been illuminated. From Niki and Watanabe (1976a).

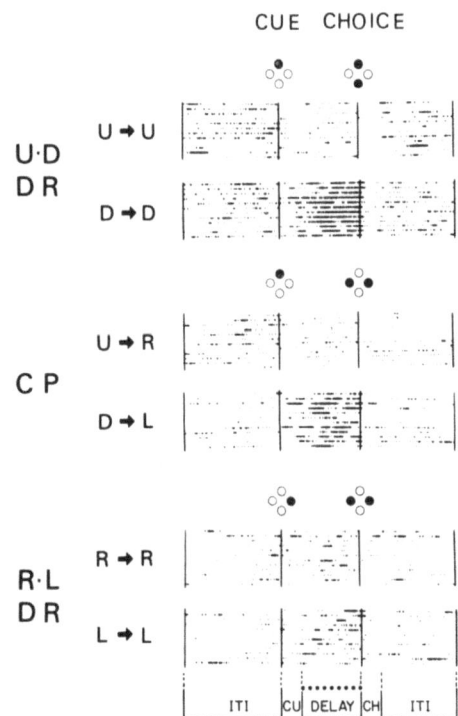

Prefrontal cells can code even more subtle learned information. More recently Watanabe (1980) investigated a three-stage task in which a conditional cue indicated which of a pair of subsequent discriminative cues marked the correct key for a succeeding delayed response; he found prefrontal cells differentially related not only to the spatial cues but also to the context defined by the conditional cues. These intriguing experiments demonstrate neural correlates of specific abstract information whose behavioral relevance had been learned by training.

Motor Functions of Posterior Parietal Cortex

Many of the behavioral deficits produced by lesions in posterior parietal "association" cortex can be summarized as a sensory and motor neglect of the contralateral half of the body. Regarding motor deficits, monkeys with unilateral ablation of posterior parietal cortex (Brodmann's areas 5 and 7) tend to make fewer spontaneous movements with the contralateral limb and make errors in reaching targets with the contralateral limb. Bilateral lesions of this area result in a marked reduction in exploratory behavior, suggesting that this cortical area plays a role in programming of certain goal-directed movements. Documenting the response properties of single units in posterior parietal cortex, experimenters have found many neurons that respond to passive joint rotation and/or cutaneous input; other cells fire during particular active limb or eye movements (Hyvarinen and Poranen, 1974; Mountcastle, Lynch, Georgopoulos, Sakata, and Acuna, 1975; Leinonen, Hyvarinen, Nyman, and Linankoski, 1979). So-called arm-projection neurons in areas 5 and 7 were activated during specific reaching movements, particularly those directed toward objects of interest within reach, such as food. A similar class of

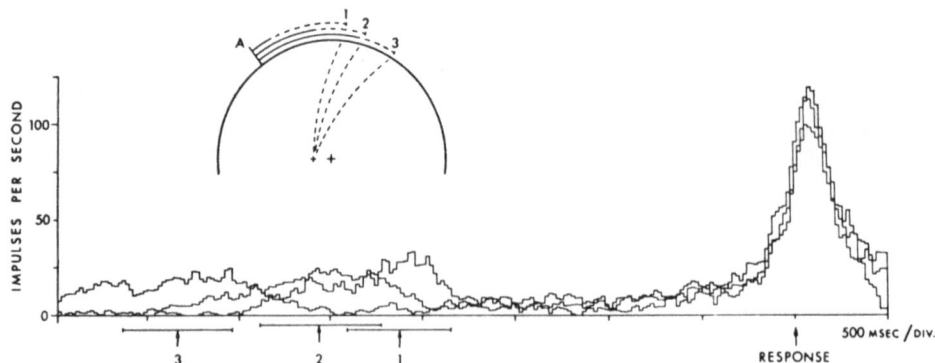

Fig. 15. Activity of an arm projection neuron recorded in posterior parietal cortex, area 7. The task is schematically shown in inset diagram. The monkey sat in the center of a circular track on which an illuminated target moved from left to right, starting at position A. The monkey reached out to touch the target at different positions (1, 2, 3) depending on the duration of a foreperiod delay. The histograms of average unit activity are aligned on the "response," when the target was touched. The brackets below the histogram toward the left show the approximate times at which the target appeared at A for the three response positions. The similarity in histograms, independent of target positions, was taken to suggest that unit activity was more related to a command signal to move than to the pattern of muscle activity. From Mountcastle *et al.* (1975).

"hand manipulation" cells discharged while the monkey manipulated objects with the hand. Such cells did not respond to passive somatosensory, visual or auditory stimuli. Moreover their discharge during limb movements depended on the motivational properties of the object grasped; they were active only in relation to movements aimed at securing an object which the animal desired, such as food when hungry, or reaching a manipulandum to acquire a food reward. Most of these cells were inactive during other limb movements, such as clawing or aversive movements, which employed the same muscles. Thus, these posterior parietal cells differed from motor cortex neurons in not being related in an obligatory fashion to muscle activity; instead, they fired only during those limb movements aimed at securing a desired object. About 80% of these cells were active only with movements of the contralateral limb (Mountcastle *et al.,* 1975).

The activity of "arm-projection" cells appears independent of the direction of reaching, as indicated by the example in Fig. 15; in this study, the monkey reached out to touch a lighted switch that moved around him along a circular track (Mountcastle *et al.,* 1975). The inset diagram illustrates the relative location of the monkey (center of circle) and the trajectory of the moving target, which he touched at three different positions to obtain a food reward. The histogram of unit activity associated with reaching three different target positions shows that activity of this arm-projection neuron occurred around the time of the response, and was independent of the limb trajectory. Another class of posterior parietal neurons fired during hand manipulation and, like the arm-projection cells, were unresponsive to passive stimulation of the skin or deep tissues of the hand, or passive joint movement; instead, they fired specifically during exploratory or manipulative movements of the hand and fingers, such as retrieving a raisin from a small well, or during grooming. Such neurons were not active during other hand movements, for example, pinching or scratching. In area 7, experimenters have also found cells with a relation to specific eye movements; they respond during visual fixation of objects of interest, particularly those within reach, but did not fire during general eye movements, such as gazing around the room. These "visual searching" cells exhibited little spontaneous activity and tended to be unresponsive when the monkey was not alert.

These experiments suggest that certain neurons in posterior parietal cortex are related to generating specific classes of movements directed toward objects of interest. Being contingent on the motivational properties of the object and on the attentional state of the animal, the activity of posterior parietal cells differs from the activity of movement-related neurons in precentral cortex. In view of the complementary deficits in motivated movements produced by parietal lesions, such cells have been suggested to play a role in parietal command functions for manual or visual exploration of surrounding space (Mountcastle *et al.,* 1975).

CONCLUDING COMMENTS

In the two decades since Jasper's first cortical unit recording experiments in trained monkeys, many studies have investigated the relations between activity of single neurons and behavior. Those cited here are only representative examples. As information on neural responses has accumulated, some of the limitations and pitfalls of this technique have also become more apparent. From the millions of neurons involved in any particular behavior, experimenters usually sample one or, at best, several units at a time. The staggering number and complexity of connections between neurons and the diversity of muscle patterns underlying limb movement make it difficult to determine definitely to which aspect of the overall behavior each cell is related. This problem will make it difficult to study more advanced questions about, for example, neural mechanisms involved in programming complex sequences of movements; whether a given cell is involved in coordinating the overall sequence or more directly related to a particular component would require detailed analysis (Grimm and Rushmer, 1974). Similarly, questions about plasticity of the relation between cell activity and movement with learning are of considerable interest (see Gilbert and Thach, 1977); even without learning, however, the relation of cell activity to different movements can vary so drastically (Fetz and Finocchio, 1975; Schmidt, Jost, and Davis, 1974) that it becomes difficult to determine whether observed changes are attributable to changes in movement strategies or learning.

Interpreting the functional significance of the observed activity is often limited by the lack of information on the input and output connections of the recorded cells. Since most regions of the nervous system contain cells with diverse response properties and each neuron may project to a different target site, it becomes crucial to know where the recorded cells project. Such information can be obtained by identifying antidromic responses evoked by stimulating potential target sites, or by cross-correlating activity of the cell and the activity of potential target neurons.

An obvious pitfall in chronic unit recording experiments is the degree to which the pretrained behavior determines the observed neural responses. Documenting the activity of cells in a given region in relation to one particular behavioral response ignores the potential involvement of these same cells in many other behavioral functions. The fact that motor cortex contains cells related to a variety of functions was clearly documented in a recent study by Thach (1978). To determine statistically whether the activity of precentral cells is more strongly related to muscle activity, limb position, or direction of the next intended movement, Thach designed a task in which all three response variables could be tested; surprisingly and significantly, he found that "all the cells that were looked for were found, in nearly equal numbers" (!) In addition to being related to active movements, many precentral cells also discharge repeatedly to passive joint rotation (Lemon et al., 1976; Murphy, Kwan, and Wong, 1979), and could conceivably be involved in perception of passive movements (Fetz et al., 1980). Similarly, posterior parietal cortex contains many different types of neurons; besides those related to directed limb or eye movement, other cells are involved in coding complex visual and proprioceptive stimuli (Hyvarinen and Poranen, 1974; Leinonen et al., 1979; Mountcastle et al., 1975). Thus, it seems probable that many regions are involved in multiple functions, rather than being exclusively devoted to one. Whether different functions are subserved by the same set of cells used in different combinations or by different sets of cells is a matter for further investigation.

When each area is studied under a different behavioral condition, the potential similarities between cells in different regions may escape notice. In fact, such similarities could provide important clues to the interaction between regions. For example, "early onset" cells related to preparation for response have now been observed in diverse cortical areas, including primary motor cortex (Evarts and Tanji, 1974, 1976; Neafsey *et al.,* 1978; Thach, 1978), supplementary motor cortex (Tanji and Taniguchi, 1978), prefrontal cortex (Fuster, 1973; Niki and Watanabe, 1976), and posterior parietal cortex (Mountcastle *et al.,* 1975), as well as subcortical sites, such as basal ganglia and thalamus (Alexander and Fuster, 1973; Neafsey *et al.,* 1978). These observations underscore the fact that preparation for movement involves a large population of cells distributed over diverse areas. The tendency to investigate the role of a given region in a particular behavioral function risks missing the degree to which such functions are actually performed by cells distributed over diverse centers.

These considerations suggest that one promising direction for future work is further analysis of the interaction between different regions during the same behavior. Although cortical areas have been extensively explored, their interrelations with thalamic nuclei remain relatively neglected. The use of cooling probes to eliminate reversibly the contribution of one center while recording from another has provided important clues to their interaction (Alexander and Fuster, 1973; Amassian *et al.,* 1972; Brooks, Adrien, and Dykes, 1972; Jasper, Lamarre, and Joffroy, 1972; Meyer-Lohmann, Conrad, Matsunami, and Brooks, 1975; Meyer-Lohmann, Hore, and Brooks, 1977; Vilis, Hore, Meyer-Lohmann, and Brooks, 1976). Such extensions of chronic unit recording procedures would further elucidate interaction between neurons subserving behavior.

REFERENCES

Alexander, G. E., and Fuster, J. M. Effects of cooling prefrontal cortex on cell firing in the nucleus medialis dorsalis. *Brain Research,* 1973, *61,* 93–105.

Amassian, V. E., Ross, R., Wertenbaker, C., and Weiner, H. Cerebellothalamocortical interrelations in contact placing and other movements in cats. In T. L. Frigyesi, E. Rinvik, and M. D. Yahr, (Eds.), *Corticothalamic Projections and Sensorimotor Activities.* New York: Raven Press, 1972.

Bioulac, B., and Lamarre, Y. Activity of postcentral cortical neurons of the monkey during conditioned movements of a deafferented limb. *Brain Research,* 1979, *172,* 427–437.

Breedlove, S. M., McGinty, D. J., and Siegel, J. M. Operant conditioning of pontine gigantocellular units. *Brain Research Bulletin,* 1979, *4,* 663–668.

Brinkman, C., and Porter, R. Supplementary motor area in the monkey: Activity of neurons during performance of a learned task. *Journal of Neurophysiology,* 1979, *42,* 681–709.

Brinkman, J., Bush, B. M., and Porter, R. Deficient influences of peripheral stimuli on precentral neurones with dorsal column lesions. *Journal of Physiology (London),* 1978, *276,* 27–48.

Bromberg, M. A., and Fetz, E. E. Responses of single units in cervical spinal cord of alert monkeys. *Experimental Neurology,* 1977, *55,* 469–482.

Brooks, V. B., Adrien, J., and Dykes, R. W. Task-related discharge of neurons in motor cortex and effects of dentate cooling. *Brain Research,* 1972, *40,* 85–90.

Burton, J. E., and Onoda, N. Dependence of the activity of interpositus and red nucleus on sensory input data generated by movements. *Brain Research,* 1978, *152,* 41–63.

Cheney, P. D. Response of rubromotoneuronal cells identified by spike-triggered averaging of EMG activity in awake monkeys. *Neuroscience Letters,* 1980, *17,* 137–142.

Cheney, P. D., and Fetz, E. E. Functional classes of primate corticomotoneuronal cells and their relation to active force. *Journal of Neurophysiology*, 1980, *44*, 773–791.

Conrad, B., Meyer-Lohmann, J., Matsunami, K., and Brooks, V. B. Precentral unit activity following torque pulse injections into elbow movements. *Brain Research*, 1975, *94*, 219–236.

Conrad, B., Wiesendanger, M., Matsunami, K., and Brooks, V. B. Precentral unit activity related to control of arm movements. *Experimental Brain Research*, 1977, *29*, 85–95.

DeLong, M. R. Activity of pallidal neurons during movement. *Journal of Neurophysiology*, 1971, *34*, 414–427.

DeLong, M. R. Putamen: Activity of single units during slow and rapid arm movements. *Science*, 1973, *179*, 1240–1242.

DeLong, M. R., and Strick, P. L. Relation of basal ganglia, cerebellum and motor cortex units to ramp and ballistic limb movements. *Brain Research*, 1974, *71*, 327–335.

Evarts, E. V. Pyramidal tract activity associated with a conditioned hand movement in the monkey. *Journal of Neurophysiology*, 1966, *29*, 1011–1027.

Evarts, E. V. Representation of movements and muscles by pyramidal tract neurons of the precentral motor cortex. In M. D. Yahr and D. P. Purpura (Eds.), *Neurophysiological Basis of Normal and Abnormal Motor Activities*. New York: Raven Press, 1967.

Evarts, E. V. Relation of pyramidal tract activity to force exerted during voluntary movement. *Journal of Neurophysiology*, 1968, *31*, 14–27.

Evarts, E. V. Activity of pyramidal tract neurons during postural fixation. *Journal of Neurophysiology*, 1969, *32*, 375–385.

Evarts, E. V. Motor cortex reflexes associated with a learned movement. *Science*, 1973, *179*, 501–503.

Evarts, E. V. Precentral and postcentral cortical activity in association with visually triggered movement. *Journal of Neurophysiology*, 1974, *37*, 373–381.

Evarts, E. V., and Fromm, C. Sensory responses in motor cortex neurons during precise motor control. *Neuroscience Letters*, 1977, *5*, 267–272.

Evarts, E. V., and Fromm, C. The pyramidal tract neuron as summing point in a closed-loop control system in the monkey. *Progress in Clinical Neurophysiology*, 1978, *4*, 56–69.

Evarts, E. V., and Tanji, J. Gating of motor cortex reflexes by prior instruction. *Brain Research*, 1974, *71*, 479–494.

Evarts, E. V., and Tanji, J. Reflex and intended responses in motor cortex pyramidal tract neurons of monkey. *Journal of Neurophysiology*, 1976, *39*, 1069–1080.

Fetz, E. E. Operant conditioning of cortical unit activity. *Science*, 1969, *163*, 955–958.

Fetz, E. E., and Baker, M. A. Operantly conditioned patterns of precentral unit activity and correlated responses in adjacent cells and contralateral muscles. *Journal of Neurophysiology*, 1973, *36*, 179–204.

Fetz, E. E., and Cheney, P. D. Muscle fields of primate corticomotoneuronal cells. *Journal de Physiologie (Paris)*, 1977, *74*, 239–246.

Fetz, E. E., and Cheney, P. D. Post-spike facilitation of forelimb muscle activity by primate corticomotoneuronal cells. *Journal of Neurophysiology*, 1980, *44*, 751–772.

Fetz, E. E., and Finocchio, D. V. Correlations between activity of motor cortex cells and arm muscles during operantly conditioned response patterns. *Experimental Brain Research*, 1975, *23*, 217–240.

Fetz, E. E., Cheney, P. D., and German, P. C. Corticomotoneuronal connections of precentral cells detected by post-spike averages of EMG activity in behaving monkeys. *Brain Research*, 1976, *114*, 505–510.

Fetz, E. E., Finocchio, D. V., Baker, M. A., and Soso, M. J. Sensory and motor responses of precentral cortex cells during comparable passive and active joint movements. *Journal of Neurophysiology*, 1980, *43*, 1070–1089.

Fromm, C., and Evarts, E. V. Relation of motor cortex neurons to precisely controlled and ballistic movements. *Neuroscience Letters*, 1977, *5*, 259–265.

Fuster, J. M. Unit activity in the prefrontal cortex during delayed response performance: Neuronal correlates of short-term memory. *Journal of Neurophysiology*, 1973, *36*, 61–78.

Fuster, J. M., and Alexander, G. E. Firing changes in cells of the nucleus medialis dorsalis associated with delayed response behavior. *Brain Research,* 1973, *61,* 79–91.

Ghez, C., and Kubota, K. Activity of red nucleus neurons associated with a skilled forelimb movement in the cat. *Brain Research,* 1977, *131,* 383–388.

Ghez, C., and Vicario, D. Discharge of red nucleus neurons during voluntary muscle contraction. *Journal de Physiologie (Paris),* 1978, *74,* 283–286.

Gilbert, P. F. C., and Thach, W. T. Purkinje cell activity during motor learning. *Brain Research,* 1977, *128,* 309–328.

Glickstein, M. Brain mechanisms in reaction time. *Brain Research,* 1972, *40,* 33–37.

Goldberg, M. E., and Wurtz, R. H. Activity of superior colliculus in behaving monkeys. *Journal of Neurophysiology,* 1972, *35,* 560–574.

Goldring, S., and Ratcheson, R. Human motor cortex: Sensory input data from single neuron recordings. *Science,* 1972, *175,* 1493–1495.

Grimm, R. J., and Rushmer, D. S. The activity of dentate neurons during an arm movement sequence. *Brain Research,* 1974, *71,* 309–326.

Hammond, P. H. The influence of prior instruction to the subject on an apparently involuntary neuromuscular response. *Journal of Physiology (London),* 1956, *132,* 17P–18P.

Harvey, R. J., Porter, R., and Rawson, J. A. The natural discharges of Purkinje cells in paravermal regions of lobules V and VI of the monkey's cerebellum. *Journal of Physiology (London),* 1977, *271,* 515–536.

Hepp-Reymond, M.-C., Wyss, U. R., and Anner, R. Neuronal coding of static force in the primate motor cortex. *Journal de Physiologie (Paris),* 1978, *74,* 287–292.

Hoffman, D. S., and Luschei, E. S. Responses of monkey precentral cortical cells during a conditioned jaw bite task. *Journal of Neurophysiology,* 1980, *44,* 333–348.

Hufschmidt, H.-J. and Hufschmidt, T. Antagonist inhibition as the earliest sign of a sensory-motor reaction. *Nature,* 1954, *174,* 607.

Humphrey, D. R., Schmidt, E. M., and Thompson, W. D. Predicting measures of motor performance from multiple cortical spike trains. *Science,* 1970, *179,* 758–762.

Hyvarinen, J., and Poranen, A. Function of the parietal association area 7 as revealed from cellular discharges in alert monkeys. *Brain,* 1974, *97,* 673–692.

Jasper, H., Lamarre, Y., and Joffroy, A. The effect of local cooling of the motor cortex upon experimental Parkinson-like tremor, shivering, voluntary movements, and thalamic unit activity in the monkey. In T. L. Frigyesi, E. Rinvik, and M. D. Yahr (Eds.), *Corticothalamic Projections and Sensorimotor Activities.* New York: Raven Press, 1972.

Jasper, H., Ricci, G. F., and Doane, B. Patterns of cortical neuronal discharge during conditioned responses in monkeys. In G. Wolstenholme and C. O'Connor (Eds.), *Neurological Basis of Behavior.* Boston: Little, Brown, 1958.

Jasper, H., Ricci, G., and Doane, B. Microelectrode analysis of cortical cell discharge during avoidance conditioning in the monkey. *Electroencephalography and Clinical Neurophysiology,* 1960, Supplement 13, 137–155.

Joffroy, A. J., and Lamarre, Y. Single cell activity in the ventral lateral thalamus of the unanesthetized monkey. *Experimental Neurology,* 1974, *42,* 1–16.

Kubota, K., and Niki, H. Prefrontal cortical unit activity and delayed alternation performance in monkeys. *Journal of Neurophysiology,* 1971, *34,* 337–347.

Lamarre, Y., Bioulac, B., and Jacks, B. Activity of precentral neurons in conscious monkeys: Effects of deafferentation and cerebellar ablation. *Journal de Physiologie (Paris),* 1978, *74,* 253–264.

Leinonen, L., Hyvarinen, J., Nyman, G., and Linankoski, I. Functional properties of neurons in lateral part of associative area 7 in awake monkeys. *Experimental Brain Research,* 1979, *34,* 299–320.

Lemon, R. N., Hanby, J. A., and Porter, R. Relationship between the activity of precentral neurones during active and passive movements in conscious monkeys. *Proceedings of the Royal Society,* 1976, *B194,* 341–373.

Lemon, R. N., and Porter, R. Afferent input to movement-related precentral neurones in conscious monkeys. *Proceedings of the Royal Society,* 1976, *B194,* 313–339.

Lewis, M. McD., Porter, R., and Horne, M. The effects of impairment of afferent feedback from the moving limb on the natural activities of neurones in the precentral gyrus of conscious monkeys: a preliminary investigation. *Brain Research*, 1971, *32*, 467–473.

Luschei, E. S., Johnson, R. A., and Glickstein, M. Response of neurones in the motor cortex during performance of a simple repetitive arm movement. *Nature*, 1968, *217*, 190–191.

Luschei, E. S., Garthwaite, C. R., and Armstrong, M. E. Relationship of firing patterns of units in face area of monkey precentral cortex to conditioned jaw movements. *Journal of Neurophysiology*, 1971, *34*, 552–562.

Luschei, F., Saslow, C., and Glickstein, M. Muscle potentials in reaction time. *Experimental Neurology*, 1967, *18*, 429–442.

MacKay, W. A., Kwan, M. C., Murphy, J. T., and Wong, Y. C. Responses to active and passive wrist rotation in area 5 of awake monkeys. *Neuroscience Letters*, 1978, *10*, 235–239.

Matsumura, M. Intracellular synaptic potentials of primate motor cortex neurons during voluntary movement. *Brain Research*, 1979, *163*, 33–48.

Meyer-Lohmann, J., Conrad, B., Matsunami, K., and Brooks, V. B. Effects of dentate cooling on precentral unit activity following torque pulse injections into elbow movements. *Brain Research*, 1975, *94*, 237–251.

Meyer-Lohmann, J., Hore, J., and Brooks, V. B. Cerebellar participation in generation of prompt arm movements. *Journal of Neurophysiology*, 1977, *40*, 1038–1050.

Mountcastle, V. B., Lynch, J. C., Georgopoulos, A., Sakata, H., and Acuna, C. Posterior parietal association cortex of the monkey: Command functions for operations within extrapersonal space. *Journal of Neurophysiology*, 1975, *38*, 871–908.

Murphy, J. T., Kwan, H. C., and Wong, Y. C. Differential effects of reciprocal wrist torques on responses of somatotopically identified neurons of precentral cortex in awake primates. *Brain Research*, 1979, *172*, 329–337.

Neafsey, E. J., Hull, C. D., and Buchwald, N. A. Preparation for movement in the cat. I. Unit activity in the cerebral cortex; II. Unit activity in the basal ganglia and thalamus. *Electroencephalography and Clinical Neurophysiology*, 1978, *44*, 706–713; 714–723.

Niki, H. Prefrontal unit activity during delayed alternation in the monkey. I. Relation to direction of response, and II. Relation to absolute versus relative direction of response. *Brain Research*, 1974a, *68*, 185–196; 197–204.

Niki, H. Differential activity of prefrontal units during right and left delayed response trials. *Brain Research*, 1974b, *70*, 346–349.

Niki, H., and Watanabe, M. Prefrontal unit activity and delayed response. Relation to cue location versus direction of response. *Brain Research*, 1976a, *105*, 79–88.

Niki, H., and Watanabe, M. Cingulate unit activity and delayed response. *Brain Research*, 1976b, *110*, 381–386.

Olds, J. Operant conditioning of single unit responses. *Excerpta Medica International Congress Series*, 1965, *87*, 372–380.

Otero, J. B. Comparison between red nucleus and precentral neurons during learned movements in the monkey. *Brain Research*, 1976, *101*, 37–46.

Padel, Y., and Steinberg, R. Red nucleus cell activity in awake cats during a placing reaction. *Journal de Physiologie (Paris)*, 1978, *74*, 265–282.

Phillips, C. G. Motor apparatus of the baboon's hand. *Proceedings of the Royal Society*, 1969, *B173*, 141–174.

Phillips, C. G., and Porter, R. *Corticospinal Neurones: Their Role in Movement.* New York: Academic Press, 1977.

Phillips, M. I. (Ed.). *Brain unit activity during behavior.* Springfield, Ill.: Charles C Thomas, 1973.

Porter, R., and Rack, P. M. H. Timing of the response in the motor cortex of monkeys to an unexpected disturbance of finger position. *Brain Research*, 1976, *103*, 201–213.

Schmidt, E. M., Bak, M. J., McIntosh, J. S., and Thomas, J. S. Operant conditioning of firing patterns in monkey cortical neurons. *Experimental Neurology*, 1977, *54*, 467–477.

Schmidt, E. M., Jost, R. G., and Davis, K. K. Plasticity of cortical cell firing patterns after load changes. *Brain Research*, 1974, *73*, 540–544.

Schmidt, E. M., Jost, R. G., and Davis, K. K. Reexamination of the force relationship of cortical cell discharge patterns with conditioned wrist movements. *Brain Research*, 1975, *83*, 213–223.

Smith, A. M. The activity of supplementary motor area neurons during a maintained precision grip. *Brain Research*, 1979, *172*, 315–327.

Smith, A. M., Hepp-Reymond, M.-C., and Wyss, U. R. Relation of activity in precentral cortical neurons to force and rate of force change during isometric contractions of finger muscles. *Experimental Brain Research*, 1975, *23*, 315–332.

Smith, A. M., Massion, J., Gahery, Y., and Roumieu, J. Unitary activity of ventrolateral nucleus during placing movement and associated postural adjustment. *Brain Research*, 1978, *149*, 329–346.

Soechting, J. F., Burton, J. E., and Onoda, N. Relationships between sensory input, motor output and unit activity in interpositus and red nuclei during intentional movement. *Brain Research*, 1978, *152*, 65–79.

Soltysik, S., Hull, C. D., Buchwald, N. A., and Feketi, T. Single unit activity in basal ganglia of monkeys during performance of a delayed response task. *Electroencephalography and Clinical Neurophysiology*, 1975, *29*, 65–78.

Soso, M. J., and Fetz, E. E. Responses of identified cells in postcentral cortex of awake monkeys during comparable active and passive joint movements. *Journal of Neurophysiology*, 1980, *43*, 1090–1110.

Strick, P. L. Activity of ventrolateral thalamic neurons during arm movement. *Journal of Neurophysiology*, 1976, *39*, 1032–1044.

Strick, P. L. Cerebellar involvement in "volitional" muscle responses to load changes. *Progress in Clinical Neurophysiology*, 1978, *4*, 85–93.

Tanji, J. Activity of neurons in cortical area 3a during maintenance of steady postures by the monkey. *Brain Research*, 1975, *88*, 549–553.

Tanji, J. Selective activation of neurons in cortical area 3a associated with accurate maintenance of limb positions. *Brain Research*, 1976, *115*, 328–333.

Tanji, J., and Evarts, E. V. Anticipatory activity of motor cortex neurons in relation to direction of intended movement. *Journal of Neurophysiology*, 1976, *39*, 1062–1068.

Tanji, J., and Taniguchi, K. Does the supplementary motor area play a part in modifying motor cortex reflexes? *Journal de Physiologie (Paris)*, 1978, *74*, 317–318.

Tanji, J., Taniguchi, K., and Saga, T. Supplementary motor area: Neuronal response to motor instruction. *Journal of Neurophysiology*, 1980, *43*, 60–68.

Thach, W. T. Discharge of cerebellar neurons related to two maintained postures and two prompt movements. I. Nuclear cell output; II. Purkinje cell output and input. *Journal of Neurophysiology*, 1970, *33*, 527–536; 537–547.

Thach, W. T. Timing of activity in cerebellar dentate nucleus and cerebral motor cortex during prompt volitional movement. *Brain Research*, 1975, *88*, 233–241.

Thach, W. T. Correlation of neural discharge with pattern and force of muscular activity, joint position and direction of intended movement in motor cortex and cerebellum. *Journal of Neurophysiology*, 1978, *41*, 650–676.

Vilis, T., Hore, J., Meyer-Lohmann, J., and Brooks, V. B. Dual nature of the precentral responses to limb perturbations revealed by cerebellar cooling. *Brain Research*, 1976, *117*, 336–340.

Wall, P. D., Remond, A. G., and Dobson, R. C. Studies on the mechanism of the action of visual afferents on motor cortex excitability. *Electroencephalography and Clinical Neurophysiology*, 1953, *5*, 385–393.

Watanabe, M. Prefrontal unit activity and delayed conditional discrimination. In Y. Tsukada and B. Agranoff (Eds.), *Neurobiological Basis of Learning and Memory*. New York: Wiley, 1980.

Wolpaw, J. R. Correlations between task-related activity and responses to perturbation in primate sensorimotor cortex. *Journal of Neurophysiology*, 1980, *44*, 1122–1138.

Wong, Y. C., Kwan, H. C., and Murphy, J. T. Activity of precentral neurons during torque triggered hand movements in awake primates. *Canadian Journal of Physiology and Pharmacology*, 1979, *57*, 174–184.

Wurtz, R. H., and Mohler, C. W. Organization of monkey superior colliculus: Enhanced visual response of superficial layer cells. *Journal of Neurophysiology*, 1976a, *39*, 745–765.

Wurtz, R. H., and Mohler, C. W. Enhancement of visual response in monkey striate cortex and frontal eye fields. *Journal of Neurophysiology,* 1976b, *39,* 766–772.

Wyler, A. R., and Burchiel, K. J. Operant control of pyramidal tract neurons: The role of spinal dorsal columns. *Brain Research,* 1978a, *157,* 257–265.

Wyler, A. R., and Burchiel, K. J. Factors influencing accuracy of operant control of pyramidal tract neurons in monkey. *Brain Research,* 1978b, *152,* 418–421.

Wyler, A. R., and Finch, C. A. Operant conditioning of tonic firing patterns from precentral neurons in monkey neocortex. *Brain Research,* 1978, *146,* 51–68.

Yumiya, H., Kubota, K., and Asanuma, H. Activities of neurons in area 3a of the cerebral cortex during voluntary movements in the monkey. *Brain Research,* 1974, *78,* 169–177.

10

Analysis of Stance Posture in Humans

LEWIS M. NASHNER

INTRODUCTION

The generation of a purposeful movement requires that the sensorimotor system regulate and control three kinds of activities: (1) the basic patterns of the movement must be generated in each limb and the patterns of activities among the limbs must be coordinated, (2) the muscular activities generating the basic movement pattern must be adaptively modified to suit the external loading and sensory conditions of the particular task, and (3) the movement patterns must be adjusted to maintain the equilibrium of the body as a whole. The control of upright stance by normal human subjects is one specific example of such a sensorimotor process. Understanding upright posture is made difficult because most of what is known about sensorimotor control has been learned from studies of acute and chronically prepared experimental animals. Apart from the immediate interest in understanding more about ourselves, one might be initially inclined to question how a study in which the experimental paradigms and measurement techniques are severely limited can contribute significantly to the detailed physiological information about the generation and coordination of movements. One aim of this chapter is to show that studies of the performance of awake human subjects are answering important questions about the adaptive and equilibrium controls of the sensorimotor system that have not been addressed in experiments using animal preparations. Also, studies of the upright posture controls of patients with various CNS disorders affecting their balance and walking are contributing new insights about the disease process and new approaches for diagnosis and evaluation of treatment.

LEWIS M. NASHNER   Neurological Sciences Institute, Good Samaritan Hospital and Medical Center, Portland, Oregon 97209. This work was prepared by the author while funded by Grants NS 00148 and NS 12661 from the NINCDS.

In recent years, a predominant concept has been that the sensorimotor system produces purposeful movement, movement adaptation, and equilibrium control utilizing hierarchically organized groups of specialized subsystems. Bernstein (1967) has argued, on theoretical grounds, that hierarchical simplification of these motor activities is essential, because, otherwise, the brain could not independently regulate the activity of the many muscles controlling the motions of the mechanical linkages that collectively compose purposeful movements. Based on these theoretical arguments, Gelfand, Gurfinkel, Fomin, and Tsetlin (1971) proposed that complex locomotor movements are broken down into much simpler, stereotyped patterns, each of which is executed by an autonomous spinal "function generator" upon the issuance of a single command from the brain. Although many questions about the validity of the hierarchical concept remain, animal experiments have produced many results consistent with this concept. For example, two detailed reviews of the literature describing the ways in which acute cat preparations walk upon treadmills (Grillner, 1975; Wetzel and Stuart, 1976) have both cited extensive evidence that the spinal programs alone produce the basic rhythmic stepping movements in the individual limbs and coordinate the movements among the limbs during the various gait patterns. In these acute preparations, the supraspinal descending commands provide the nonspecific excitation necessary to sustain the locomotor activity. It is important to note, however, that these preparations were deprived of normal vestibular and visual inputs and could not freely balance or guide themselves.

Although demonstration of locomotor function generators in the spinal cord has strengthened the hierarchical concept and revealed many of the physiological mechanisms of these generators, studies of animal preparations have shown little about the integrative functions of the brain in adapting the movement activity to external conditions and in maintaining equilibrium. One exception has been a recent study in experimental animals showing the role of the cerebellum in adapting the vestibulo-ocular reflex. The convergence of vestibular and visual information within the cerebellum acts to correct erroneous performance by changing the gain of the vestibulo-ocular reflex (e.g., Ito, 1976).

Subjective observations of human motor performance have suggested that novel movements are performed quite differently once they become well rehearsed and are performed under familiar external conditions. Although somatosensory, vestibular, and visual inputs exert strong and continuous influences upon motor activity during the performance of novel movements or during performance under unknown external conditions (e.g., Lee and Lishman, 1975), well-rehearsed movements are performed as if being played out from a "motor tape." Sensory feedback has a weaker corrective influence on practiced movement. By accepting the detailed knowledge about sensorimotor organization gained from the study of experimental animal preparations as valid for the human motor system, human studies are particularly useful to examine some of the more integrative aspects. Because human subjects are easily incorporated into experimental paradigms that do not place severe constraints on CNS functions or on the predictability of external conditions, the attention in human studies can be focused on what happens during well-rehearsed tasks, while learning new tasks, and during times immediately following unexpected changes in the external conditions.

This chapter will attempt to integrate information about the sensorimotor control system that has been gained through the study of upright stance in human subjects. As a background to detailed observations related in later sections, we will begin by defining the characteristics of body motions during stance and by reviewing some of the many subjective observations which, though lacking in detail, have given a good general description of the task. In the third section, we will examine what has been learned from experimental observations about the transformations of sensory inputs into the commands for postural movement and about how these transformations adapt to the conditions of the task. The fourth section will treat the way in which commands for movement are coordinated among the postural muscles, and the fifth section will introduce some of the theoretical models for hierarchically and distributionally organized control systems and will attempt to formulate from these a speculative, conceptual model of posture control that is consistent with the experimental data presented in the preceding sections. The concluding section will assess the possible applications of platform tests in the diagnosis and evaluation of patients with CNS disorders affecting their postural and walking abilities.

GENERAL FEATURES OF STANCE POSTURE

KINEMATICS OF UPRIGHT STANCE

An understanding of the kinematic relations between motions of the body and muscular, gravitational, and perturbational forces is important, since the performance of stance is ultimately expressed as the combined orientations and motions of many body parts. Although the complete mathematical expressions for defining the movements of the six major body parts (two lower legs, two upper legs, the trunk, and the head) are formidable, very simplified expressions can help define the basic features of postural motions and provide analytical models with which to test ideas about the organization of postural controls. Two very simplified expressions will be examined here.

The first presupposes that the body is rigid and sways only about the ankle joints. This model shows that the body is inherently unstable during stance without the continuous, active support afforded by muscles of the legs. Apart from its mathematical simplicity, this expression is motivated by observations that the major axes of sway rotations during stance are the ankle joints (e.g., Mori, 1973; Nashner, 1970). In the model shown in Fig. 1A, sway motion about the ankles (θ_A) is defined by prescribing parameter values for the mass of the body (m), the height of the center of mass above the ankles (h), the moment of inertia of the body about the ankles (I), and the input torque exerted about the ankle joint [T_A]. Equation (1) relating the body sway angle to the torque exerted by the ankle musculature is

$$
\begin{bmatrix} \dot{\theta}_1 \\ \theta_2 \end{bmatrix} = \begin{bmatrix} 0 & 1 \\ \dfrac{mgh}{I} & 0 \end{bmatrix} \begin{bmatrix} \theta_1 \\ \theta_2 \end{bmatrix} + \begin{bmatrix} 0 \\ \dfrac{1}{I} \end{bmatrix} T_A \qquad (1)
$$

written in a "state variable" format. This format has been adopted because it greatly simplifies the multidimensional equations which are necessary to express intrabody postural motions (to be considered next). For readers unfamiliar with this notation a brief introduction is given in the Appendix.

The inherent instability of the body is indicated by the positive value of term A_{21} within the characteristic matrix. One way to stabilize this system would be to provide a restoring torque proportional to the ankle angle, $T_A = -K_1\theta_A$, where K_1 has the units of stiffness (kg-m/radian). Now term A_{21} becomes $(mgh/I) - K_1$, and a stiffness of sufficient magnitude will render it negative. The lack of a negatively valued term A_{22} also indicates that body sway motions are undamped, and oscillations induced by a torsional correction would continue unabated. Damping is provided by a compensatory torque proportional to the rate of sway: $T_A = -K_2\dot\theta_A$, where K_2 has the units kg-m/radian/sec. Defining the ankle viscosity and stiffness necessary to provide stability and damping is useful in evaluating the viscoelastic and active contractile characteristics of the ankle musculature.

Although it has set requirements for the restorative and damping forces of the ankle musculature, the single-link model has given no insights as to how the relative movements of the ankle, knee, hip, and neck joints are coordinated. A two-link model of body sway allows one degree of intrabody motion along with sway about the ankle joints. The model describes motion about the ankle and hip joints, assuming that the body is composed of two rigid structures, the legs and the trunk (Fig. 1B). Motions of the two links (θ_A and θ_H) are defined by prescribing the height of the center of mass of each link above its pivot point, the mass of each link, the moment of inertia of each link about its pivot point, and the torsional moments exerted by the postural muscles about each joint. Because the parameters of the two-link equation are functionally dependent on the angular variables, the equations of this model are highly nonlinear and impossible to solve analytically. However, the equations can be linearized to describe approximately the small angular motions of the links about a fixed operating joint. Equation (2) omits the complex mathematical prescription for these linearized parameters, but gives some idea about the relative interactions between the two links by showing the sign of each of the parameters.

$$
\begin{bmatrix} \dot\theta_{A_1} \\ \dot\theta_{A_2} \\ \hline \dot\theta_{H_1} \\ \dot\theta_{H_2} \end{bmatrix}
=
\begin{bmatrix} 0 & 1 & 0 & 0 \\ + & 0 & - & 0 \\ \hline 0 & 0 & 0 & 1 \\ - & 0 & + & 0 \end{bmatrix}
\begin{bmatrix} \theta_{A_1} \\ \theta_{A_2} \\ \hline \theta_{H_1} \\ \theta_{H_2} \end{bmatrix}
+
\begin{bmatrix} 0 & 0 \\ + & - \\ \hline 0 & 0 \\ - & + \end{bmatrix}
\begin{bmatrix} T_A \\ T_H \end{bmatrix}
\tag{2}
$$

The partitioned characteristic and input matrices of the two-link system reveal that the dynamical properties of each link (in isolation) are similar to those of the single-link system (upper left and lower right sections of each matrix). Each link is inherently unstable (positive value of A_{23} and A_{43}) and undamped (zero value of A_{22} and A_{44}) without the exertion of stabilizing muscular forces. Thus, the two-link expression refines but does not alter potential requirements for restorative and damping forces defined by the single-link model.

The other two sections of the characteristic and input matrices (lower left and upper right portions) describe the way in which small motions of one link affect the motions of the other. The mutual negativity of the characteristic coupling terms A_{23} and A_{41} indicates that motions about the two links will tend to be of opposite phase. The mutual negativity of the input coupling terms B_{22} and B_{41} indicates that a muscular effort about either one of the joints also tends to create antiphasically coupled movements about the two joints. For example, a positive torque about the ankle joint $(+T_A)$ will tend to extend the ankle but also flex the hip. Thus, the antiphasic interaction of adjacent links is a fundamental characteristic of body motions during stance. Given this property, the exertion of a compensatory torque about the ankle joints in an attempt to stabilize sway would "buckle" the body at the hips. To prevent such "buckling," coordinated muscular forces are necessary. Thus, the two-link expression gives some valuable insights into the need for coordination of proximal, distal, and trunkal muscles.

Although linearized models of this type can only be applied with great caution, the two-link model also predicts which muscles are most critical in controlling ankle and hip joint sway motions. The multilink mathematical expression of body sway offers a second insight about the coordination of posture controls and hip motions. Theoretically, postural sway could be stabilized with the exertion of muscular torques about either one of the two joints alone! This property is determined by performing certain tests on the characteristic and input matrices described in Equation (2) (e.g., Padulo and Arbib, 1974). A computer simulation of the two-link body which retained many of the nonlinear properties of the body not included in the two-link expression also confirmed that controls about either joint alone are potentially sufficient for stabilization. This conclusion is corroborated by the common observation that standing on ones "tiptoes" requires a good deal of added hip movement control (ankle torques cannot be exerted) but does not end by falling. The control of sway motions by torques exerted about either joint alone may appear to be in conflict with the need for coordinated action between these two joints. However, these alternative possibilities only illustrate that there is a *redundancy* among the possible control strategies. This section will shortly argue that a similar redundancy (of information) occurs among the sensory inputs to the system. Redundancy, it will be shown, has a strong influence on the organization of the posture control system.

Except when locked knees align the vertical gravitational forces of the body

Fig. 1. Kinematic models of upright stance posture. (A) Single-link model of sway about the ankle joints (θ_A). (B) Two-link model of sway about the ankle and hip (θ_H) joints. (C) Model of coordinated vertical movements about the ankle, knee (θ_K), and hip joints. (D) Functional anatomy of the gastrocnemius (G), tibialis anterior (T), hamstrings, (H), and quadriceps (Q) muscles.

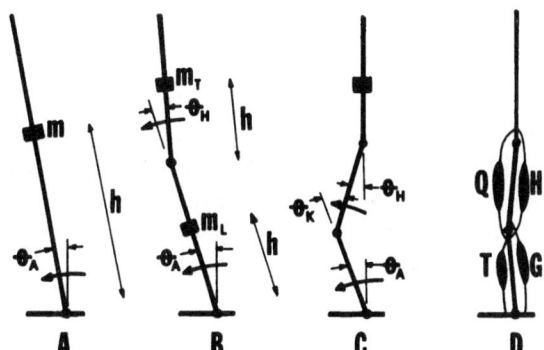

with the skeletal structure of the legs, the downward force of gravity on the body must be resisted by forces within the leg musculature. Figure 1C shows the position of a leg during vertical movements in which the center of body mass remains aligned with the ankle joints so that there is no sway component to the movement. The configuration of the leg is such that the torsional movements created by the force of gravity upon the bent leg are most concentrated at the knee joint, and that the angular rotations of this joint are twice as large as the ankles or hips.

The placement of leg muscles is well suited to perform the dual tasks of suspensory and sway posture controls (Fig. 1D). With the exception of the tibialis anterior, each of the other three major leg muscle groups (gastrocnemius, hamstrings, and quadriceps) performs dual functions about adjacent joints. The gastrocnemius extends the ankles and flexes the knees. The hamstrings muscles extend the hips and flex the knees, whereas the quadriceps muscles extend the knee and flex the hip. During anteroposterior (AP) sway, motions are largest about the ankles and hips. The coordinated activation of gastrocnemius muscle pairs and hamstrings or of tibialis anterior and quadriceps would therefore produce the necessary corrective movements about the ankles and hips. During changes in height, motions are largest about the knee joint. Now, the coordinated activation of either the gastrocnemius and quadriceps or the tibialis anterior and hamstrings muscle pairs would produce the necessary upward or downward thrust of the leg.

OBSERVATIONS OF UPRIGHT STANCE BEHAVIORS

Studies of postural behaviors have helped to define the significant features of performance by observing how subjects respond to various environmental conditions and the imposition of external disturbances. Although these behavioral studies have not quantitatively defined the mechanisms by which the sensory inputs are transformed into the necessary postural adjustments, they have demonstrated in qualitative ways the roles played by visual, vestibular, and somatosensory inputs and the flexibility with which the system can utilize different combinations of these inputs to achieve stability. Behavioral studies have therefore helped to define problems most suitable for detailed study and have suggested some fruitful experimental approaches.

Begbie (1967) devised an unstable platform that rotated in response to foot pressure. By changing somatosensory inputs and the biomechanics of stance, the platform caused corresponding changes in the amplitude of two oscillatory components of postural sway, one a slow oscillatory component at 0.5–1 Hz and the other a fast component between 1.5 and 2.5 Hz. Conditions under which either the entire or the peripheral visual input was disturbed (closing the eyes, rapidly moving the head, or reading a paper) preferentially increased the faster of these two oscillatory components. Presenting vestibular deficit patients with display of their sway movements reduced sway amplitudes in both frequency ranges. Begbie speculated that the fast oscillations occurred within the stretch reflex pathways to enhance the somatosensory inputs whenever visual or vestibular inputs were disrupted, whereas the slow oscillations were corrections based on the use of vestibular information.

The stabilogram is an easily measured variable which is generated by standing a subject upon a force plate and measuring the position of the contact forces exerted by the feet on the plate. However, because the position of the center of force upon the plate is a complex function of gravitational, inertial, and muscular forces, the stabilogram cannot be directly associated with any one variable of posture, such as the sway angle or the muscular forces being generated about the ankles. Nevertheless, several investigators have examined the frequency spectrum of stabilographic records to infer something about the underlying control systems. Scott and Dzendolet (1972) grouped the performance of 74 standing subjects into two categories, suggesting that subjects in each category were relying on a different combination of sensory inputs to achieve stability. Other investigators found that changes in the stabilographic spectrum (frequency range unspecified) were associated with performance of Jendrassik maneuvers and mental arithmetic calculations, two exercises that are known to increase the excitability of stretch-reflex pathways (Gurfinkel, Alexeef, Elner, and Baron, 1972). In another study, fatigue was shown to increase higher frequency sway motions (Bensel and Dzendolet, 1968). Thus, stabilographic measurements have led to some qualitative inferences similar to Begbie's, that frequency bands within the spectrum of stabilographic oscillations are signatures of particular regulatory subsystems.

Eklund (1969) cited the large sway disturbances that followed vibration of various postural muscles of the legs as evidence of the powerful influence of muscle spindle afferents on posture. The twofold increases in the vibratory-induced sway when subjects closed their eyes suggested that the influence of proprioception is significantly enhanced whenever visual inputs are disrupted (Eklund, 1973).

Although the exact relation between motion inputs and electric-current stimuli to the vestibular system has not been physiologically well-defined (e.g., Lowenstein, 1955), normal subjects swayed in relation to small, sinusoidal or intermittent currents imposed between the mastoid bones (Bizzo and Baron, 1972; Coats, 1972; Dzendolet, 1963). Disrupting the somatosensory and visual inputs during electrical vestibular stimulation increased the amplitude of elicited sway by three to five times, suggesting that the vestibular input is more influential when others have been disrupted (Nashner and Wolfson, 1974).

Moving the visual scene surrounding an otherwise unperturbed standing subject caused the subject to sway in the same direction as that of the surrounding medium (Lestienne, Soechting, and Berthoz, 1977). Although these visually induced sway motions were compensatory (attempting to stabilize the head with respect to the visual scene), they were quite small. Lee and Lishman (1975), however, showed that subjects became much more sensitive to visual motion cues whenever the normal somatosensory inputs were disrupted.

The above observations have shown qualitatively that visual, vestibular, and somatosensory inputs each exert an influence. Yet, disrupting or perturbing any one of these inputs alone does not destabilize the normal subject. Combining the perturbation of one sensory modality and the disruption of another had a decidedly greater destabilizing effect. The most logical conclusion is that the posture control system receives a good deal of redundant sensory information and it is able to recognize quickly when one has been disrupted or perturbed, shifting attention to others that are not. A particular input modality is critically important only when other redundant inputs have already been lost. Thus, the redundancy among sensory inputs and controlling subsystems is a feature of posture control system which enables it to perform under a wide variety of different physiological and environmental conditions.

One of the difficult challenges in the study of posture control is to devise experimental techniques that can measure the performance of a highly adaptable system. Techniques that rely on repetitive stimulation and averaging over a large number of trials may yield deceptive results, because the system can reorganize its strategy of control to minimize the effect of a predictable disturbance. Interpretations of such results cannot be readily generalized beyond the specific conditions imposed by the experiment. The second major problem is to devise a language for expressing experimental results that can synthesize a diverse body of physiological, biomechanical, and behavioral data.

SENSORY INPUTS TO POSTURE

Three sensory modalities impart information about the orientation and motions of the standing subject. However, because each sensory modality derives information within a different frame of reference, the presence of three independent modalities is critically important to resolve the frequently conflicting input information. To understand how the system might resolve a conflict between inputs, three frames of reference for sensory input information have been defined (Gibson, 1966). Unfortunately, the functional categories do not coincide with the modal categories. (1) *Proprioception* is the sense of position and movement of one part of the body relative to another. (2) *Exproprioception* imparts information about the position and movement of a part of the body relative to the external environment. (3) *Exteroception* locates objects in the external environment relative to one another. The posture control system uses information from all three of these reference frames: upright stance requires the proper relative positioning of the body links (proprioception). The body must be correctly oriented with respect to gravity (exproprioception). Also, standing requires the support afforded by a firm surface which is stable with respect to gravitational vertical (exproprioception and exteroception).

Each sensory modality may receive information from one or several of the above frames of reference. The rotational position of the ankles and the contact forces between the feet and the supporting surface (somatosensory inputs) include information about the orientation of the ankle joints (proprioceptive), and about the sway orientation of the body (exproprioception). However, somatosensory inputs can be in error if the supporting surface is moved. The vestibular system is a purely exproprioceptive sense that measures the orientation and the motions of the head with respect to the inertial and the gravitational fields. Because the vestibular system is not subject to external perturbations, it is a most useful input to recognize other sensory errors, as when motion of the supporting surface perturbs the somatosensory inputs. Vision is the most complex of the three modalities because it includes proprioceptive, exproprioceptive, and exteroceptive information.

The overall system regulates stance posture by simultaneously transforming many sensations of body and environmental motions into compensatory muscular activities. It is possible that each sensory modality is transformed into appropriate muscular activity by unique neural mechanisms, although it is clear that the processes of transformation interact with one another. The system can select a subset of sensorimotor transformations from among those available to suit the demands of the specific task. Presumably, the system can also adaptively modify the characteristics of individual sensorimotor transformations. The focus of this section is limited to experimental results that have described the transformational characteristics between a single sensory of input and the resulting muscular activities, whereas the other sensory modalities were either eliminated or not experimentally modified. Unfortunately, most of the important experimental questions relating to the interactions among the sensory inputs have yet to be examined, principally because the current experimental and analytical techniques are unable to cope effectively with experimental data in which multiple inputs are used to generate responses.

ANKLE ROTATIONAL INPUTS

Much of our postural stability and a part of our sense of orientation are derived from the sense of contact between the feet and the supporting surface and the orientation of the ankle joints. A person becomes aware of these important inputs whenever, for example, the supporting surface is very slippery or compliant. Subjects experiencing these unusual conditions often report that they can feel changes in muscle tone and orientation of the limbs. These subjective impressions are indications that a new strategy for posture control has been employed.

Before examining the role of ankle rotational inputs, it is necessary to define the viscoelastic properties of tonically active skeletal muscles. The stiffness with which the soleus muscle of the cat resists length changes is exceptionally high when the imposed displacements are no greater than $\pm 0.45\%$ of the total muscle length (Nichols and Houk, 1973). Presupposing that the human gastrocnemius and soleus muscles are also exceptionally stiff for the first $\pm 0.45\%$ change in length, the ankle joint would be exceptionally stiff in resisting angular changes equal to or less than approximately ± 1.0 degrees (El'ner, Popov, and Gurfinkel, 1972; Gurfinkel and Shik, 1973; Nashner, 1976). This is, in fact, the angular range within which most spontaneous sway oscillations about the ankle joints are constrained. By direct measurements the inherent stiffness of the ankle joints during stance was found to vary between 1.3 and 1.7 kg-m/deg, magnitudes which the kinematic model predict are sufficiently large to counteract the destabilizing forces of gravity (Gurfinkel, Lipshits, Mori, and Popov, 1976; Nashner, 1976). However, stance posture cannot be stabilized by the viscoelastic properties of tonically active ankle joint muscles alone, since external forces and loads or motions of the supporting surface would perturb the body beyond the very narrow region of stability. Under these conditions, additional compensatory EMG adjustments are necessary to provide additional stabilizing torques.

The traditional viewpoint, based on experiments with anesthetized animals, was that the myotatic stretch reflexes helped the postural muscles resist displacements by controlling their length. However, a number of more recent studies have challenged this assumption.

Houk (1979) has presented convincing experimental and theoretical evidence that limb perturbations are effectively resisted by the reflex regulation of muscle length–tension (stiffness) properties. Other concepts of movement control suggest that peripheral feedback may not be necessary at all to compensate for occasional load perturbations. Bizzi, Dev, Morasso, and Polit (1978) have tested this concept of positional control by measuring the accuracy of the voluntary arm movements of monkeys deprived of all relevant inputs and subjected to unexpected perturbations. Other investigators have also shown that myotatic reflexes are either totally inhibited or are too weak to effectively resist changes in muscle length when perturbations are imposed on the limbs of actively performing animals and human subjects (e.g., El'ner *et al.*, 1972; Gurfinkel, Lipshits, and Papov, 1974).

Other investigators have suggested that the spinal reflex pathways help organize the activity of muscles prior to the onset of movements (Gottlieb and Agarwal, 1972; Kots, 1969; Kots and Zhukov, 1972). Because there are no conclusive concepts for reflex control function, the most probable conclusion is that reflexes perform a number of different functions, depending on the task. Sometimes the viscoelastic properties of tonically active muscles are sufficient and no reflex activation is necessary. Other times reflexes may enhance the stiffness of muscles or may activate coordinated patterns of limb movement.

When positional disturbances of the limbs are imposed on performing human subjects, the first functionally effective compensatory activity occurs at latencies approximately twice that of the myotatic stretch reflex, about 60 msec in the flexor of the thumb (Marsden, Merton, and Morton, 1973) and 100–110 msec in the gastrocnemius muscle of the leg (Melvill Jones and Watt, 1971; Nashner, 1976). These active responses to stretch probably include inputs from joint and cutaneous afferents as well as muscle proprioceptors, and may possibly involve central brain pathways (e.g., Evarts and Tanji, 1974) as well as segmental and intersegmental ones. For lack of a better name, these responses will be termed *rapid postural adjustments*, although other investigators have adopted such names as *functional stretch reflex responses, long-loop reflexes, transcortical stretch reflexes,* or *long-latency stretch reflexes.*

As the above names might suggest, there is considerable disagreement about the origin and hence the characteristics of these adjustments. Houk (1979) has argued that these rapid postural adjustments are "triggered reactions" having characteristics of voluntary activities; namely, latencies increasing with the number of potential choices, and the occurrence of occasional executional errors based on an incorrect choice. Chan, Kearney, and Melvill Jones (1979) have examined tibialis anterior EMG responses to perturbations of the ankle joint of normal subjects and those with Parkinson's disease. They described a narrow dividing line between automatic "reflex" adjustments that occur at 90–95 msec latency and adjustments with some characteristics of choice that begin as early as 125–130 msec. Hence, any examination of rapid postural adjustments during stance must include careful latency measurements and tests for possible voluntary choice reactions.

Rapid postural adjustments helped subjects stabilize anteroposterior sway, although there were significant intersubject differences. Also, subjects varied depending on their postural "set." When AP sway disturbances are imposed on quietly standing subjects (by horizontally translating the platform upon which they are standing), only about one-half the group regularly evidences vigorous EMG activity commencing in the stretching ankle muscles at 100–110 msec (Nashner, 1976). As shown in Fig. 2, the subjects with rapid postural adjustments arrest sway perturbations significantly faster than subjects with very little or more delayed EMG changes. Based on latencies, these adjustments are probably "automatic."

Whenever the supporting base upon which a subject stands is rotated, rotation of the ankles is not correlated with body sway motion, and rapid postural adjustments utilizing this input would be erroneous. Gurfinkel, Lipshits, Mori, and Popov (1975) employed a platform that rotated sinusoidally about an axis colinear with the ankle joints to show that normal subjects can completely disassociate the EMG activity in ankle muscles from the positional inputs from ankle joint rotation. When the platform was rotated at frequencies of 2–3 Hz, EMG activity of the ankle muscles was still closely related to the sway motion of the body, but became completely disassociated from that of the imposed ankle rotation. The process by which the relation between ankle movements and postural adjustments adapts to the conditions of the task was further explored using a rotating and horizontally translating platform (Nashner, 1976). The mode of perturbation was unexpectedly altered between sequences of horizontally induced sway rotation of the ankles (adjustments destabilized the sway orientation). Subjects who utilized rapid postural adjustments triggered to help stabilize AP sway adaptively modified the gain of these responses during the first three to five trials following each transition in the mode of perturbation (Fig. 3). During the first trials of unexpected direct rotation, large inappropriate responses caused exaggerated swaying. These inappropriate responses were progressively attenuated during subsequent trials. During

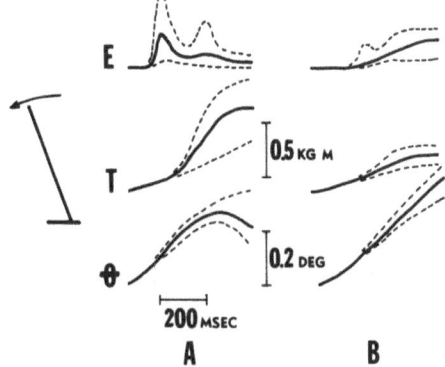

Fig. 2. Response characteristics to induced sway of three subjects utilizing rapid postural adjustments (A) and of three subjects not utilizing these responses (B), showing averaged data ± 1 *SD* of the rectified, integrated EMG response of gastrocnemius muscles (E), the torque exerted by the feet upon the platform (T), and the body sway angle (θ).

the first unexpected sway perturbations (following a sequence of direct ankle rotations), rapid postural adjustments were not seen and the subjects swayed more. However, the responses were facilitated and the amplitude of sway was reduced during subsequent sway perturbations.

The observations on the ankle rotational input to posture control have provided one clear example of the way in which a subject's postural "set" can be characterized by a specific transformational relation between somatosensory inputs from ankle rotation and automatic changes in the EMG activities of the ankle muscles. Subjects modified the relation between ankle rotation and postural adjustments, so that the system could perform during direct rotations, presumably utilizing information from other sensory inputs. This presumption, however, provokes several new questions: How does a subject stand when the ankle rotational inputs are not being employed? How does the system differentiate between useful and erroneous ankle rotational inputs? Some of the answers to these important questions are found by observing the role of vestibular and visual inputs to the posture control system.

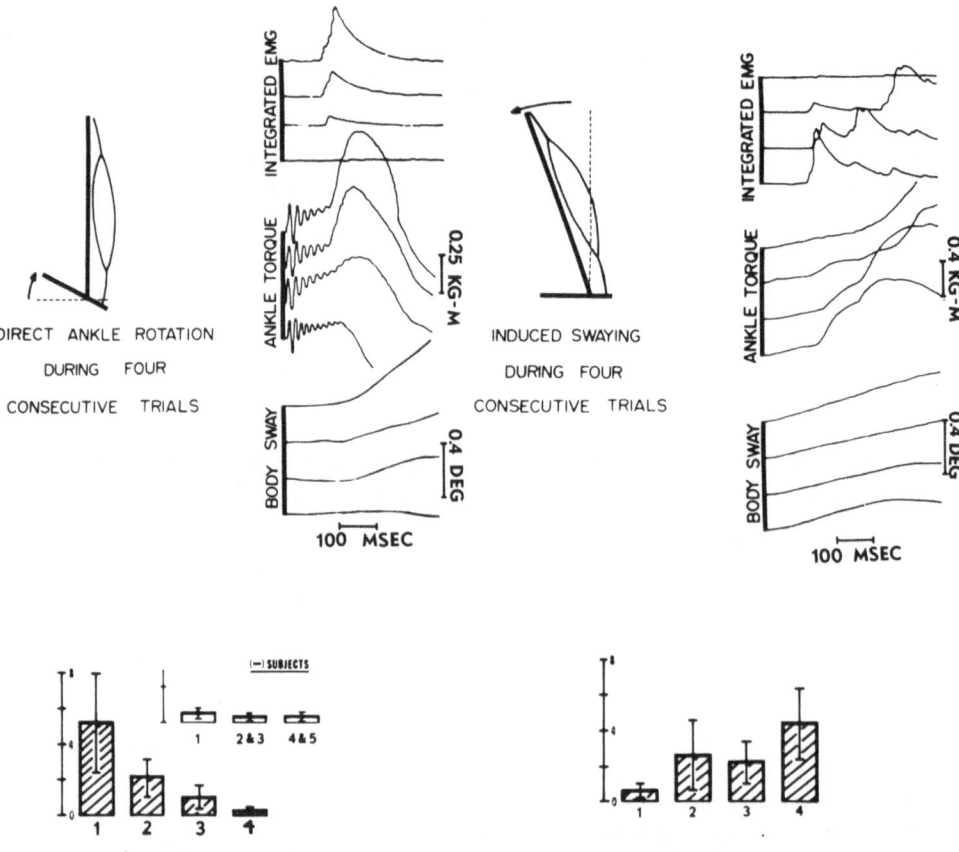

Fig. 3. Adaptive changes in the gains of rapid postural adjustments elicited by ankle rotations during stance following unexpected transition to direct rotations (A) and to sway rotations (B). Bar graphs show data averaged for three subjects. Redrawn from Nashner (1976).

The strong influences of vestibular inputs on posture and the subjective sense of orientation are most dramatically experienced when a person disembarks from a vehicle which has been rapidly accelerating. A simple experiment which can recreate this experience is the act of rotating oneself about the vertical axis, stopping quickly (leaving the horizontal canals in a state of transient activation), and then rotating the head downward in front so that the active horizontal canals are now oriented in the frontal plane. Since the human body is inherently much less stable in sway than it is about the vertical axis, the transient vestibular perturbation (now in the frontal plane) will cause the experimenter to stumble sideways and to experience a brief period of disorientation. In addition to dramatizing the effect of an erroneous vestibular input on posture, this simple experiment also introduces another important aspect of the vestibular system: its signals interact strongly with those from the neck proprioceptors that measure the orientation of the head with respect to the body.

The anatomy and physiology of the vestibular receptors have been studied extensively in experimental animal preparations. Using anesthetized squirrel monkeys, Goldberg and Fernandez (1971; Fernandez and Goldberg, 1971) showed conclusively that single afferent fibers innervating the semicircular canals respond to angular accelerations in a manner consistent with the torsion pendulum model. For the purpose of studying human posture control, however, the most directly relevant studies are those that have developed control system models of the vestibular input–output characteristics. These studies have been based on the subjective reports of motion and the eye movements of subjects experiencing angular and linear accelerations.

Each labyrinth consists of three roughly orthogonal semicircular canals and two orthogonally placed linear acceleration sense organs. By applying circular motions in the plane of a single set of canals, Meiry (1966) ascribed to each the characteristics of an overdamped, second-order torsion pendulum. Because the long time constant of each canal effectively integrates the acceleration input, each measures the *rate* of rotation over the approximate frequency range 0.2–2 Hz. The utricular otoliths, linear accelerometers oriented approximately in the horizontal plane, measure the apparent vertical (vector sum of gravitational and linear accelerations) over the frequency range 0–0.2 Hz.

Because of the low frequencies and relatively small amplitudes characterizing postural sway, the threshold levels for the vestibular receptors are important parameters in the analysis of vestibular posture control. Unfortunately, subjective and objective measures of threshold vary over a wide range, depending on the measurement paradigm used. The lowest values for the semicircular canals were recorded during the oculogyral illusion produced in the laboratory (Clark and Stewart, 1969), in a slowly turning aircraft (Öosterveld, 1970), and during stance posture control (Nashner, 1971). Using subjective reports, Young and Meiry (1968) measured a linear acceleration threshold of 0.005g (about a 0.3 degree tilt with respect to the vertical).

Vestibular projections to the motor system are best understood for the eye movement control system, but have also been described for forelimb and hind-limb areas of the cat spinal cord. Grillner, Hongo, and Lund (1966, 1970, 1971) showed that vestibular projections to lumbar cord exerted a reciprocal influence on the ankle extensor and flexor motoneurons in the anesthetized cat. Evidence also suggested that these vestibular inputs could modulate the strength of spinal stretch reflexes (Gernandt, Iranyi, and Livingstok, 1959; Kim and Partridge, 1969). Presumably, if such mechanisms were also operating in standing human subjects, sway stimulation of the vestibular system could reciprocally modulate the activity in the ankle flexor and extensor muscles.

The early works of von Holst and Mittelstaedt (1950) have argued that vestibular inputs cannot be used for postural stabilization unless the position of the head relative to the body is also measured. Cohen (1961) dramatized this functional relationship by showing that the balance and postural orientation of monkeys were severely disrupted by the removal of neck proprioceptors. Physiological studies in anesthetized cats have since shown that vestibular and neck proprioceptive inputs exert a mutually anatagonistic influence on extensor and flexor motoneurons of the forelimb (Ehrhardt and Wagner, 1970). This mutual influence was described quantitatively by applying combinations of sinusoidal head and neck rotational stimuli (Berthoz and Anderson, 1971). The close interactions between neck proprioceptive and vestibular inputs also subserve important functions during stance posture control. The system must be able to distinguish between head accelerations caused by motions of the body relative to the gravitational vertical and those caused by motions of the head relative to the body. For example, the horizontal canals, normally sensing the rotations of the head (and body) about the vertical axis, will sense rotational motions of the body in the frontal plane whenever the head is tilted forward 90 degrees.

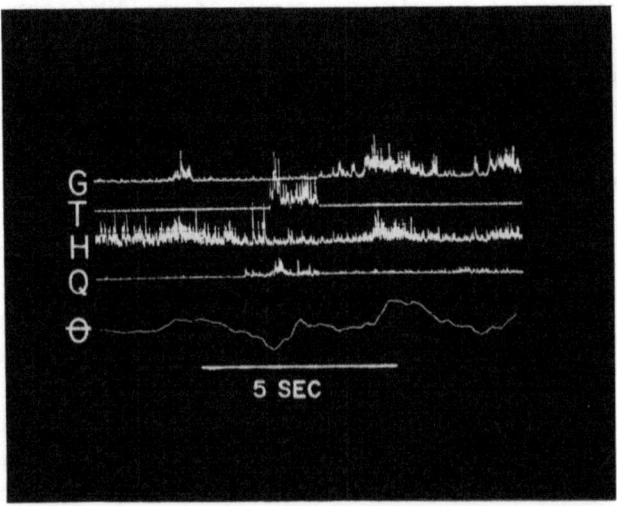

Fig. 4. Representative period of stance under stabilized ankle conditions, showing the raw gastrocnemius (G), tibialis anterior (T), hamstrings (H), and quadriceps (Q) EMG activity and the body sway angle (θ).

It has been difficult to study vestibular postural adjustments using traditional experimental approaches in which isolated stimuli are imposed on a sensor organ. Because the vestibular system responds to accelerations (within the fixed inertial-gravitational frame of reference), any movements of a standing subject influence the somatosensory and visual systems. Any attempt to restrain the motions of the subject would necessarily change the posture control task and therefore also the vestibular influence on the postural muscles. Several techniques have been devised in an attempt to surmount this difficulty. One has been to eliminate almost all somatosensory inputs relating the rotational motions of the ankle joint to the sway motions of the body. Standing the subject upon a movable platform with eyes closed, the rotational position of the ankles was stabilized by rotating the platform (axis colinear with the ankle joints) in direct proportion to the AP sway angle of the body. This was accomplished by measuring the sway angle and using this feedback signal to drive the rotational axis of the platform. Under this condition spontaneous AP sway motions could be observed or sway imposed by horizontally translating the platform. Figure 4 illustrates the performance of normal subjects on the rotating platform. AP sway oscillations approximately ±0.5 to 1.0 degrees in amplitude were large compared to the 0.1 to 0.2 degree amplitudes measured under normal somatosensory and visual conditions. Two factors contributed to the significant increases in the sway: (1) the loss of stabilization supplied by the viscoelastic stiffness of the ankle joints, and (2) the interruption of rapid postural adjustments normally elicited by ankle joint motions.

When the ankle joints were rotationally stabilized and the eyes closed, the earliest EMG responses occurred in the appropriate ankle joint muscles, but at latencies of 180 msec or greater. Figure 5 shows how the latency of EMG responses was related to the rate of the induced sway motion. Because vision did not reduce latency times, it appeared that the vestibular system provides the initial sway corrections when somatosensory inputs are absent. However, the damping of sway oscillation was considerably greater when visual inputs were intact.

Another approach to study of vestibular posture control follows from the knowledge that electrical stimulation of the vestibular system elicits subtle postural sway motions. Electrically elicited postural adjustments have been observed to measure the way in which vestibular inputs are modulated by the position of the head relative to the body during stance (Nashner and Wolfson, 1974). Because subjectively perceptible and sometimes even uncomfortably high levels of current were

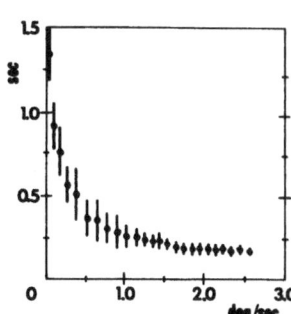

Fig. 5. Latency of EMG responses to induced sway motion during stabilized ankle conditions as a function of induced sway rate. Averages for three subjects ± 1 *SD*.

required to elicit consistently measurable sway motions, changes in the background level of gastrocnemius EMG activity were measured instead so that subjectively imperceptible levels of current (250 μA between the mastoid bones) could be used. The results, summarized in Fig. 6, show that changes in the background level of EMG activities begin 100 msec after the onset of current and last approximately 350 msec. Depending upon the orientation of the head, the same stimulus can either increase EMG activity (extending the ankles) or decrease it (flexing the ankles). The sign of an EMG change is reversed either by reversing the polarity of the current (roughly equivalent to reversing the direction of the vestibular input) or by reorienting the head 180° with respect to the body. This is precisely the spatial transformation necessary to reinterpret accelerational motions of the body perceived in the frame of reference of the head. For example, a given AP sway motion of the body would have the opposite stimulus upon the vestibular system if it occurred while the head was facing 90° over the other shoulder. Unless the vestibular input to posture was also reversed following such a repositioning of the head, the input would be appropriate to stabilize the sway with the head held in one of the positions but would be destabilizing with the head held in the other position.

The predominance of ankle rotational inputs for stabilization of stance on rigid flat surfaces was dramatized by the extremely small amplitude of postural sway associated with the 250 μA vestibular stimuli. Rotationally stabilizing the ankle joints during stance was used to eliminate the somatosensory inputs, thereby maximizing the postural influence of the electrical vestibular stimuli (Nashner, 1973). Somatosensory inputs and the stiffness of the ankle joints varied from a low of zero (platform rotations completely stabilize the ankles) to a high of twice the normal input and stiffness (platform rotations are directionally opposite AP sway, thereby doubling the sway-related motions of the ankles). Figure 7 shows that the amplitude of sway elicited by 250-μA currents sustained for eight sec continued to increase as ankle inputs decreased from twice normal, to normal, and then to zero.

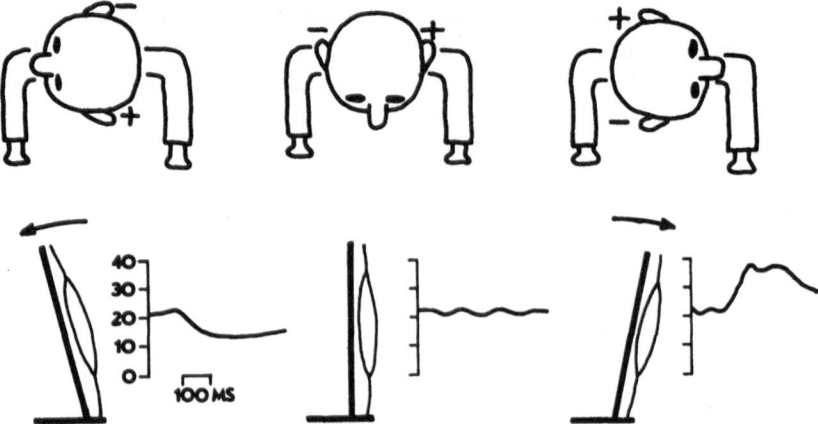

Fig. 6. Rectified and filtered gastrocnemius EMG response (average, 50 trials) and the direction of resulting sway (average, 50 trials) elicited by 250-μA vestibular stimuli delivered between the mastoid bones of three standing subjects with their heads held in each of three positions.

The result was consistent with the interpretation that vestibular and somatosensory inputs are normally combined. Thus, the strength of vestibular input may be unaltered by the presence or absence of somatosensory inputs. Vestibular influences on postural adjustments are normally dominated by the more powerful somatosensory inputs and become more apparent when somatosensory inputs have been disrupted.

The results of the above three vestibular posture control experiments have been synthesized into a systems model of vestibular posture control (Fig. 8). This model provides a good example of one way in which a diverse body of data about the vestibular receptors, the dynamics of muscular contractions, and the kinematics of body sway motions can be mathematically combined to infer something about neural processes within the posture control system. The model was developed in several stages. First, the semicircular canal and the utricular otolith responses to body sway motions were described (Nashner, 1970, 1971). Second, the model of the vestibular system was combined with a simple one-link model of the body and its muscles to describe how the vestibular input might be used to stabilize AP sway motions within a closed-loop control system (Nashner, 1972).

In the first stage of simulation, the time constants for canal dynamics described by Meiry (1966) were adequate, but the threshold of the semicircular canals had to be set significantly lower to sense AP sway motions. The model predicts that the semicircular canals sense the rate of sway for frequencies above 0.1 Hz. The otoliths sense sway motions below this frequency. Higher frequencies are functionally excluded from the otolith because conflicts between gravitational and linear accelerations become increasingly larger and cannot be resolved by this receptor.

In the second stage of the synthesis, the closed-loop model of vestibular posture control predicted that high-frequency inputs from the semicircular canals and low-frequency inputs from the otoliths are both necessary to control the full-frequency spectrum of sway motions. A neural integration of the canal input was necessary in the feedback loop to provide sway angle information above 0.1 Hz. Because the determination of sway angle based on the integration was erroneous for frequencies below 0.1 Hz, the semicircular canals alone could not stabilize the model. Similarly, because the utricular otoliths lacked dynamic information above 0.1 Hz, they alone could not stabilize the model. A spatial transformation of vestibular signals based on the neck proprioceptive inputs was included to compensate for different head positions.

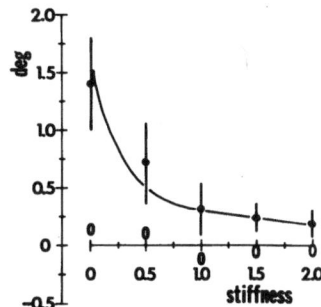

Fig. 7. Average sway amplitude of three subjects ($\pm$ 1 SD) elicited by 250-μA vestibular stimuli delivered between the mastoid bones with the positive electrode facing backward. Ankle stiffness was modified by platform rotations. Open circles show control trials during which no electrical stimuli were imposed.

Based on the performance of subjects who were deprived of normal somato-sensory inputs from ankle rotation and normal visual inputs, the vestibular system acting alone is clearly a less sensitive regulator of posture than are the somatosensory or visual inputs. This observation is in agreement with those of other investigators already mentioned, some of whom have suggested that the vestibular system plays little if any significant role in posture control. However, the experimental results are consistent with the conclusion that the vestibular system can provide critically important inputs under conditions that disrupt somatosensory or visual inputs. Another perhaps more important role of the vestibular system is to resolve conflicts among the three input modalities; for example, when the supporting or a visually perceived surface moves with respect to the body. Experimental studies of adaptive behaviors of this kind have yet to be conducted.

Visual Inputs

Probably everyone has experienced a sense of self-motion and resulting postural sway while standing at the cinema or while a passenger in a stationary vehicle as a neighboring one to the side begins to accelerate forward or backward. These experiences are examples of the strong visual influence on the sense of orientation and on posture.

Under normal conditions, movements of the body cause relative motions of the visual surround in a direction opposite to that of the body movement. The effects of isolated movements of the visual surround have been studied by recording postural adjustments, compensatory eye movements, and subjective reports of self-motion. In all instances, the latencies for the onset of visual-evoked activities were quite long compared to those of normal postural adjustments. Lestienne *et al.* (1977) observed small postural sway motions in the direction of the visual field motion at latencies of 0.5–2 sec. Reports of self-motion commenced approximately

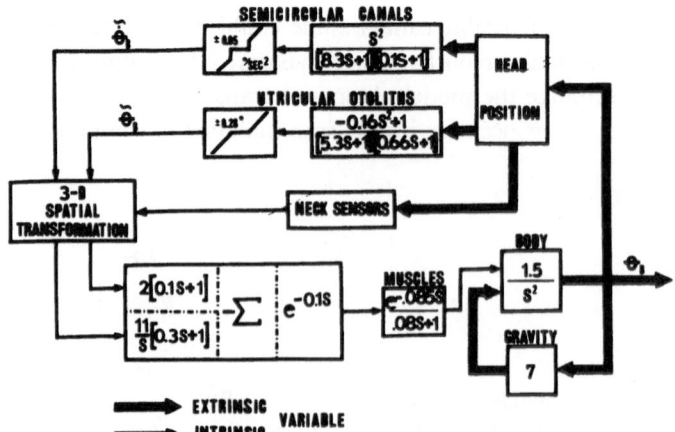

Fig. 8. The vestibular posture control model. $\tilde{\theta}_B$ and $\tilde{\dot{\theta}}_B$ refer to estimates of body sway rate and sway angle sensed by the vestibular system. S is the LaPlace transform operator. From Nashner (1973).

one second following visual stimulation (Berthoz, Pavard, and Young, 1975), whereas compensatory eye movements occurred at 1–14-sec latencies (Brandt, Dichgans, and Koenig, 1973). With such long latencies, visually guided postural adjustments would be too late to prevent frequent falls. However, long delays between visual inputs and motor responses are not consistent with electrophysiological evidence that visual as well as vestibular inputs can evoke changes in activity within the vestibular nuclei within several millisec in goldfish (Dichgans, Schmidt, and Graf, 1973) and monkeys (Henn, Young, and Finley, 1974). Large moving patterns presented at the periphery evoke rapid changes in the activity of neurons in brain-stem nuclei that influence motoneurons of forelimb and neck muscles (Baker, Gibson, Glickstein, and Stein, 1976; Maeda, Magherini, and Precht, 1977). Inconsistencies between behavioral and physiological data suggest that vestibular signals dominate in the short-term perception of motion and that visual signals dominate in long-term perception whenever there are conflicts between visual and vestibular inputs (Young, 1970). Human studies, however, have shown that vestibular sensations influence the perception of motion over the long term when the subject is tilted (Dichgans, Held, Young, and Brandt, 1972). A better explanation may lie in understanding that sensory inputs normally work together rather than in isolation and that experiments that impose isolated stimuli may lead to deceptive results. Perhaps much of our sense of motion is derived from the *relations* between the sensory inputs. Unfortunately, these kinds of issues have yet to be experimentally examined.

The visual input to posture is not an autonomous sense, but its influence is closely connected with the vestibular and probably somatosensory inputs. The long onset latencies of responses to isolated visual stimuli are not inconsistent with close sensory interaction; nor do they rule out the possibility that visual inputs have rapid influences on posture as well. The clinical observation that patients with bilateral vestibular loss cannot stand on narrow surfaces with eyes closed, but can with them opened, is evidence that visual inputs are at least as effective as vestibular ones under the appropriate conditions (Begbie, 1967). Lee and Lishman (1975) have argued that visual inputs are a major influence on the posture of normal subjects. They found that normal subjects often stumbled or fell when the somatosensory inputs were disrupted by having the subject stand upon a narrow beam and when the visual inputs were simultaneously disrupted by translating the visual surround. This result was most interesting because the subjects with normal vestibular inputs could perform similar stance tasks upon narrow beams with eyes opened or closed (e.g., Graybiel and Fregly, 1966). Apparently, with loss of somatosensory inputs, the system could not resolve the conflict between visual and vestibular inputs and, therefore, subjects became unstable. This result suggests a close interaction and similar levels of dominance for visual and vestibular inputs to posture. To explain an apparent lack of visual influence during the performance of simple familiar tasks with normal visual, vestibular, and somatosensory inputs, Lee and Aronson (1974) studied infants learning to walk and adults learning new postural tasks. They found that vision is most important while subjects are learning a new task, but that somatosensory inputs are most important once the task has been learned.

There has been little experimental data that quantitatively describes the properties of visually induced activity in postural muscles, probably because of the difficulty in constraining the interactions between the visual and vestibular inputs during posture control. This author has performed an experiment that shows that visual inputs affect motor activity as rapidly as somatosensory and vestibular inputs (100 msec) whenever visual perturbations are given in conjunction with other motion inputs, rather than in isolation. When a subject unexpectedly experienced a discongruence between visual and other motion inputs during a sway perturbation (the visual scene was unexpectedly moved to eliminate sway-related visual input), the adjustments as early as 100 msec were significantly attenuated, as shown in Fig. 9 (Nashner and Berthoz, 1978). However, visual stabilization had a progressively diminished effect during subsequent trials. Two conclusions follow from this observation. First, vision influences the rapid postural adjustments that were formerly believed to be dependent primarily on somatosensory inputs from the ankles. Second, the visual influence, like the somatosensory, is highly adaptable. The posture control system quickly ceases to use visual inputs that experience has shown to be erroneous.

Although unable to describe quantitatively some transformational relations between visual inputs and postural adjustments, studies of visually guided posture control have confirmed the major, rapid influence of this input. Visual inputs are clearly most effective when they occur in conjunction with other sensory inputs, which is, of course, usually the case. It may thus be concluded that there is a close interrelation between visual and vestibular inputs, with evidence favoring the vestibular reception of more rapid accelerations and visual reception of steady-state orientations and velocities. However, there is a good deal of overlap between these functions, and the system is readily able to adapt the transformational relations between these inputs and postural outputs to suit particular environmental conditions.

ORGANIZATION OF POSTURAL ADJUSTMENTS AMONG LEG MUSCLES

It is oftentimes difficult to learn a simple but unusual combination of movements without concentrated effort. For example, one's first attempt to "circularly

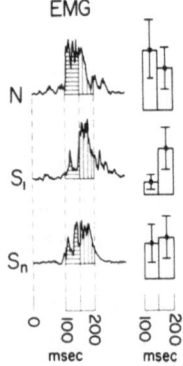

Fig. 9. Gastrocnemius EMG responses elicited by induced forward sway during normal (N) visual conditions, during first instances of unexpected stabilized (S_1) visual conditions, and during repeated stabilized (S_n) visual conditions. Traces at left are representative EMG records, with horizontally shaded portion being the one most influenced by vision. Bars at left show areas under the curve for each of the two intervals (6 subject average $\pm$ 1 SD). Note the strong visual influence, but only during the first interval.

stroke his stomach and pat the head," and then to perform the reverse combination, is amusing because these tasks are deceptively difficult. The common knowledge that humans cannot easily produce two differing rhythmical movements about adjacent joints was in part a motivation for Bernstein's (1967) theoretical concept that the interactions among joints are fixed into stereotyped patterns. He reasoned that the structuring of all movements into fixed patterns of muscle activities was necessary because the independent regulation of each muscle and linkage of the body would be an impossibly difficult task. However, one disadvantage of the structuring of movement activities is that the range of possible behaviors is severely limited to combinations of the available stereotyped patterns, and the number of possible patterns must be kept manageably small. The presence of structural constraints within the posture control system has been substantiated by observations that the range of normal movement behaviors in animals and humans is in fact quite limited when compared to the number of behaviors that are theoretically possible (e.g., Engberg and Lundberg, 1969; Kots, Krinski, Naydin, and Shik, 1971). The reduction of movement behaviors into stereotyped patterns has also been supported by the extensive physiological evidence demonstrating that the isolated cat spinal cord can generate stereotyped stepping and coordination of the hind limbs (Grillner, 1975).

Organization of Postural Motions during Stance

Two studies have observed the organization of postural motions and muscular activities in standing subjects in order to test the hypothesis that patterns of interaction are fixed during the control of stance posture. The absence of stabilographic oscillations correlated with respiration (which rhythmically changes the position of the center of mass of the thorax during upright stance) was ascribed to a synergic compensation for this internal postural disturbance (Gurfinkel, Kots, Paltsev, and Feldman, 1971). Angular measures revealed antiphase movements of the torso and hips that were in precise phase with the respiratory rhythm and in fixed proportion to the tidal volume (Fig. 10). The lack of respiratory sway under normal conditions was shown to be related to these antiphasic movements by experimentally constraining the hips: this interrupted the "respiratory synergy" and produced sway correlated with the respiratory rhythm. A second example involves the motor skills used in marksmanship. Because aiming the arm precisely is an acquired skill, Arutyunyan, Gurfinkel, and Mirskii (1969) examined the way in which it was learned to show that highly trained subjects performed with a fixed pattern of interaction among the links of the body. During aiming, they observed that the relative motions among the body links of skilled marksmen were tightly organized into a fixed pattern, whereas interactions among the links were much less organized in the unpracticed subjects. Although neither of these studies was able to establish directly a neural basis for the observed patterns of motion, the authors used kinematic models to argue that the patterned movements were produced by the organization of muscular activities, a conclusion that is entirely consistent with concepts of hierarchical organization.

LEWIS M. NASHNER

The organization of muscle activities produced by rapid postural adjustments has been observed in order to provide stronger evidence that muscular activities are neurally organized into fixed patterns during stance posture control. However, before introducing these experimental observations, a brief examination of the functional relations between activities of the leg muscles and the resulting motions of the body will be useful. During AP sway motions, the knee joints remain relatively stationary, and the ankle and hip joints tend to rotate in an antiphasic relation. When the knees are stationary, the G and H muscles function as a complementary pair to extend the ankles and hips, while the T and Q muscles flex these two joints. In marked contrast, ankle, knee, and hip joints are all either extended or flexed (with the largest angular changes occurring about the knees) whenever the body moves vertically up or down. Under these conditions the G and Q muscles function as a complementary pair to extend the leg downward (elevating the body), while the T and Q muscles flex the leg (lowering the body). Figure 11 shows the

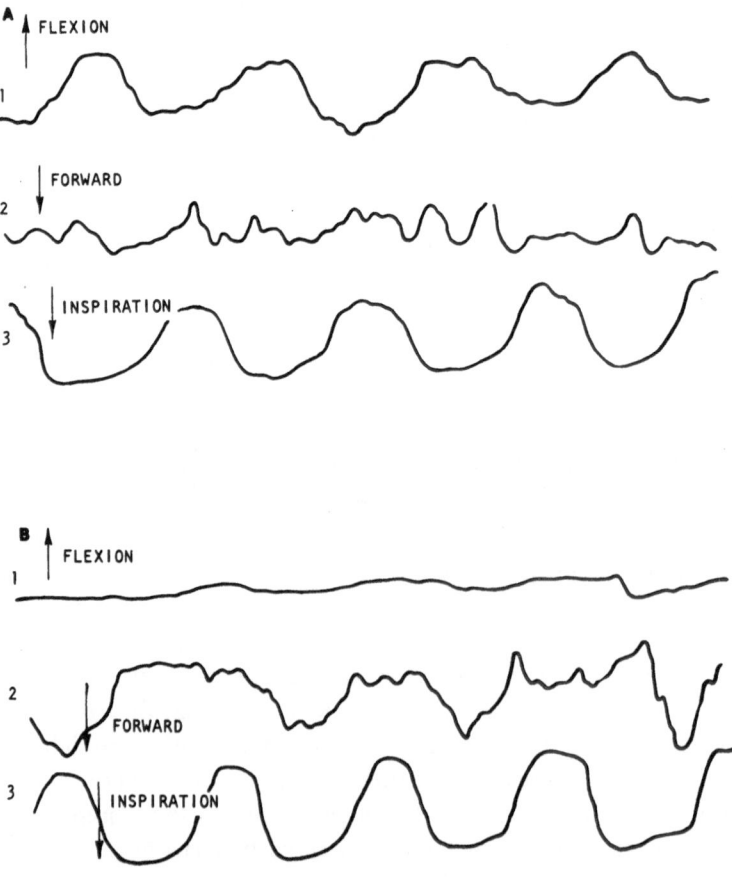

Fig. 10. Changes of a sagittal stabilogram with immobilization of the hip joints. (A) initial recording; (B) during immobilization of the hips by a temporary splint. (1) hip angle; (2) sagittal stabilogram; (3) pneumogram. Reproduced from Gurfinkel *et al.* (1971).

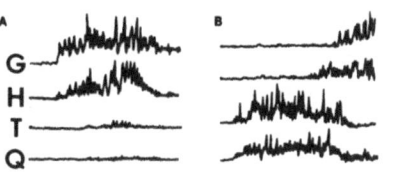

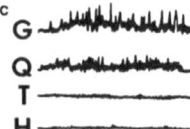

Fig. 11. EMG activity in leg muscles of standing subjects during a voluntarily sustained forward sway attitude (A), backward sway attitude (B), and with knees flexed approximately 10° (C).

muscle activity patterns necessary to resist forward and backward AP sway displacements and to resist the downward force of gravity when the knee joints are flexed.

The stretching ankle muscle and its proximal complement (H extending the hips and G extending the ankles, or Q flexing the hips and T flexing the ankles) contracted in fixed proportion during AP sway disturbances, independent of the magnitude of the sway adjustment (Nashner, 1977). The coupling of EMG activities between ankle and hip muscles was also observed during responses to unexpected, direct rotations of the ankles (Fig. 12A), stimuli which produced a completely different pattern of motion and yet the same coupling ratios between proximal and distal muscles. Because the fixed coupling of G-H and T-Q muscle pairs could be elicited either by sway rotations or direct rotations of the ankle joints, this pattern appeared to be a neurally programmed response to ankle rotations and not the result of independent, but mechanically coupled, somatosensory inputs to the ankle and hip muscles. The proportionate activation of ankle and hip flexors and extensors in response to ankle joint rotation is functionally useful during normal stance to stabilize the sway about the ankles and at the same time to counteract

Fig. 12. Processed EMG responses elicited in the leg muscles of standing subjects by unexpected platform perturbations. (A) Responses to stimuli that rotated the ankles: (1) ankles flexed by forward sway, (2) ankles flexed by direct rotation, (3) ankles extended by backward sway, and (4) ankles extended by direct rotation. (B) Responses to stimuli that flexed or extended the legs: (1) leg extended by downward platform displacement, (2) leg extended by upward platform displacement and ankle flexed by platform rotation, (3) leg flexed by upward displacement, and (4) leg flexed by upward displacement and ankle extended by rotation.

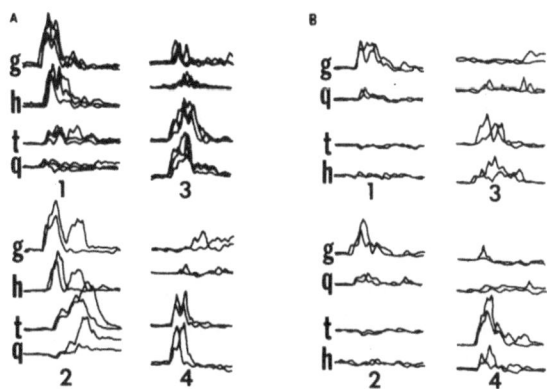

the antiphase motions about the hips. Therefore, this organizational structure has been termed the "sway synergy."

When vertical motions of the platform flexed a leg upward or extended it downward, the opposite relations were observed between the proximal and distal muscles (Nashner, Woollacott, and Tuma, 1979). Muscles that extend the ankle and the knee joints (G and Q) or flexed these joints (T and H) contracted proportionately (Fig. 12B).

Activation of flexor or extensor muscles of the ankle and knee joints will flex the leg upward or extend it downward. These coordinated adjustments are functionally useful in regulating the vertical load carried by the limb and in maintaining the height of the body. Therefore, this organizational structure has been termed the "suspensory" synergy. To demonstrate that this stereotyped pattern of activity is also neurally organized by the system rather than the result of independent sensory inputs to each muscle, the platform was rotated during vertical perturbations to disassociate motions of the ankles from those of the knees. However, because the same suspensory patterns of activity were observed in this condition, ankle rotational inputs cannot act independently whenever the leg moves vertically. This is in marked contrast to the sway adjustments that followed isolated ankle rotational inputs.

The direction of the vertical thrust exerted by a leg differed, depending on the relative movements of the two legs. The active vertical thrust of a leg *resisted* the vertical displacement whenever the two legs were moved in the same vertical direction. In contrast, the thrust of each leg *followed* the direction of the platform motion whenever the two legs were moved in opposite vertical directions. Thrusting each leg in the direction of opposing vertical movements was necessary to maintain the share of the load carried by each leg and to prevent lateral sway. The rapidity with which these adjustments occurred, however, appeared to rule out central vestibular and/or visual influences upon the direction of vertical thrust.

ADAPTIVE CHANGES IN SYNERGIC ORGANIZATION

The observation of three, movement-specific patterns of EMG activity during stance, each subserving a specific function, suggests that some process establishes the appropriate muscle pattern following changes among sway, vertical, and opposing vertical perturbations. Previous observations had shown that several trials were required for the system to alter the gain of sway adjustments following unexpected changes in the relation between ankle rotation and AP sway. The same kind of experimental procedures were reapplied here to determine how many trials were required to change from one synergic pattern to another. Response patterns were measured during the first three trials following unexpected changes among three modes of stimulation: AP swaying, simultaneous vertical changes in the height of both feet, and reciprocal changes in the height with one leg elevated and the other lowered.

Surprisingly, the functionally correct organization of EMG activities was always observed, even during trials immediately following an unexpected transition from one platform movement to another. Sequences in Fig. 13 illustrate EMG activity patterns during the first three trials following an unexpected change from one mode of movement to another. Sequences A, B, and C (simultaneous and reciprocal vertical perturbations) were all imposed following AP sway tests. Sequence D (direct rotational perturbations) was also imposed after AP sway tests, whereas the three E trials (AP sway tests) were imposed after direct rotational perturbations. Activation of the suspensory synergy by vertical changes in the height of both platforms was always in the direction which *resisted* the changes in height, although these responses were adaptively attenuated during repeated trials. Normally, resisting changes in vertical height of the body relative to the floor would be quite useful. However, resisting changes in the height of the platform was functionally inappropriate and tended to exaggerate the vertical perturbation of the body. Probably the adaptive attenuation of inappropriate suspensory responses after several trials was a similar process to that which adaptively attenuated the functionally inappropriate sway responses during direct platform rotations.

Even though it is not possible to assess fully the neuronal mechanisms mediating the coordination of muscular activities without more complete knowledge of the many sensory inputs and of the many other active muscles, the experimental observations suggest two principles: (1) Rapid postural adjustments are organized into a limited number of synergic arrangements, each of which is movement specific.

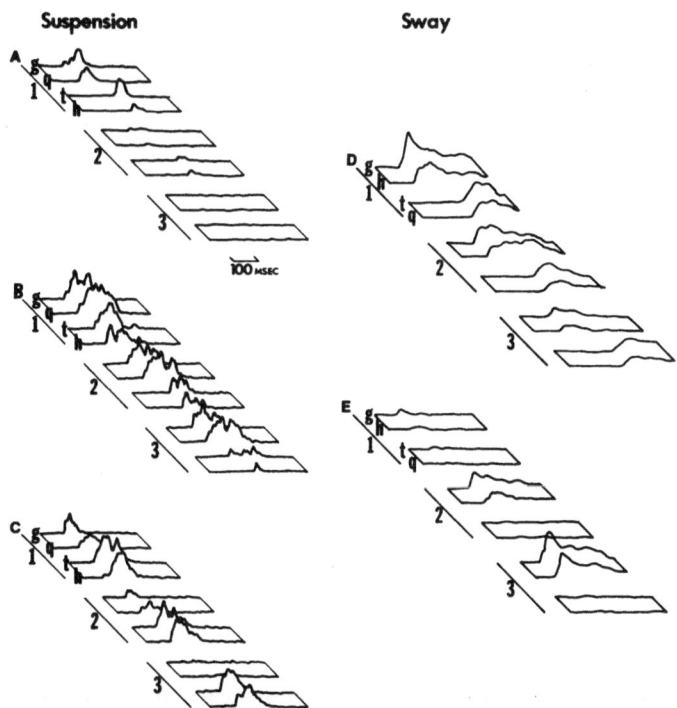

Fig. 13. Processed EMG responses of the leg muscles of standing subjects elicited by sequences of three unexpected platform perturbations. (A) Simultaneous upward movement of both legs. (B) Downward movement of the measured leg during reciprocal stimulation. (C) Upward movement of the measured leg during reciprocal stimulation. (D) Rotational flexion of the ankles. (E) Flexion of ankles by forward sway perturbations.

(2) Synergic organization appears to be performed automatically by local mechanisms, utilizing inputs primarily from the receptors within leg muscles, joints, and skin surfaces. (3) *In contrast,* processes that adapt responses to erroneous inputs are performed more slowly by central mechanisms making use of vestibular and visual as well as somatosensory information. The functional distinction between local and central components of the system is an important one. The evidence suggests that local processes, such as synergic structuring, are automatic and rely principally on somatosensory inputs. Acting solely within the internal "proprioceptive" frame of relative body movements, these mechanisms regulate the basic movement patterns. Adaptive processes modifying the locally organized patterns to suit the external conditions must also utilize the exproprioceptive and exteroceptive information afforded by the vestibular system and vision. These processes require considerably more time to act than processes that generate a rapid postural adjustment.

A Conceptual Model for the Organization of Postural Adjustments

The experimental observations have revealed a number of particularly prominent features that will be the primary focus of the conceptual model. Somatosensory inputs (from rotation of leg joints) were necessary to elicit rapid postural adjustments within the interval of 100–150 msec following a perturbation (although vestibular and visual inputs probably had a modulating influence). A second prominent feature was that the synergic pattern was always appropriate, even following unexpected changes in the mode of perturbation. In marked contrast, the magnitudes of sway adjustments elicited by direct rotations and of suspensory adjustments elicited by direct vertical motions were inappropriate during the first unexpected trials and adapted to more appropriate levels only after three to five trials. The premise used in synthesizing a posture control model from these observations is that automatic processes expressed within the latent period of single adjustments are peripheral and are mediated primarily by somatosensory inputs. In contrast, adaptive changes which require several trials involve the integrative functions of the central nervous system and utilize more complex combinations of somatosensory, vestibular, and visual inputs. If this hypothesis is correct, we can then begin to distinguish between the activities that generate the basic patterns of movement and those that select from among the many appropriate patterns and then regulate the levels of activity within them. Using only somatosensory inputs, local processes can only distinguish movements of one part of the body *relative* to other body parts. However, a local process cannot control a movement that requires a distinction between relative body motions and motions of external objects, because the resolution of body and environmental motions can only be performed with the complex combinations of visual, vestibular, and somatosensory inputs.

The process for conceptually modeling a multilevel control system is somewhat different than that for formulating control systems models quantifying the transformational characteristics between specific inputs and outputs. The conceptual model focuses attention on defining the theoretical constructs that give rise to the characteristic behavioral patterns of the system. It so happens that control-systems engineers have been interested in designing systems with many of the desirable features that have been observed in the sensorimotor system. Therefore, there is a rich literature of control theory and practical application that have already examined hierarchical and adaptive systems. Because a detailed review of this literature would not be appropriate within this chapter, an examination of some of these constructs will provide useful analogies with which to develop a conceptual model of posture control. Whether these analogies have any physiological validity or not, they nevertheless provide useful hypotheses that can be experimentally tested and refined.

HIERARCHICALLY CONTROLLED FUNCTION GENERATORS

Complex patterns of activity within a system can be created by piecing together combinations of much simpler stereotyped behaviors, each of which is generated by a preprogrammed "function generator." A function generator can greatly simplify the task of controlling a complex system by fixing the interrelations among its many inputs and outputs so that only a few inputs are required to produce a complete behavior. Within such a system, an "executive" is necessary to compose the desired behavior by specifying one or a sequence of stereotyped functions. Presumably, other controls would also be necessary to "fine tune" the activity and to smooth the transitions between functions. However, if the repertoire of functions is kept manageably small compared to the number of degrees-of-freedom of the system, the task of controlling a complex system will be much simpler than controlling each part separately. Greene (1972) drew the analogy of a graphic display system that utilized a family of preprogrammed curvilinear function generators to produce any desired complex curve on an oscilloscope. Because generating each curve on a point-by-point basis would have required a large storage capacity, many fewer coordinates were needed to specify a sequence of these curvilinear functions. A simpler example of how this system works is to compare it to a draftsman's French curve, portions of which can be used to reproduce any curved line.

There are a number of benefits and drawbacks to controlling a complex system by constraining its performance to combinations of programmed functions. Under this form of organization, the executive system formulates the activity as a prescription of "functions" (in much the same way as a computer programmer combines subroutines to specify whole operations when writing in other than a machine language). Within this simpler language, the executive can make gross changes in the control strategy by respecifying fewer critical parameters. Presumably, also, new activities can be learned more quickly and are more easily generalized to produce similar behaviors with different parts of the system. The major limitation to this hierarchical arrangement is that the behaviors of the system are constrained to be combinations of the stereotyped functions, and the number of stereotyped functions must be kept small compared to the total range of possible systemic activities. Also, the performance of the system is constrained because function generators cannot interact directly with one another. However, some investigators are of the opinion that sensorimotor activities in man and animals are in fact quite limited compared to that theoretically possible (e.g., Bernstein, 1967; Evarts, Bizzi, Burke, DeLong, and Thach, 1971).

Another school of thought, exemplified by Davis (1976), has emphasized the importance of distributionally organized structures within sensorimotor systems. In contrast to the hierarchical system in which specialized subsystems produce complete stereotyped behaviors, activities in the distributed system are "emergent" properties of the interactions among all subsystems within the group. Individual subsystems can play more than one functional role, depending on the task, and the interactions among subsystems are mutual rather than unidirectional. One important consequence is that the locus of command can vary from one task to another, whereas in the hierarchical system the executive always assumes the role of commander. Much less attention has been paid to examining these forms of organization, probably because the intuitive understanding and the analytic techniques for dealing with them are far less developed than for hierarchical systems. Nevertheless, some behavioral characteristics are wholly consistent with a distributional system and these deserve more attention in future work.

A MODEL REFERENCE SYSTEM THAT ADAPTS

"Model reference" control systems have been realized by engineers to regulate very complex and changeable processes that could not have been successfully stabilized using a fixed set of feedback parameters. Basically, reference model systems incorporate into the feedback loops an approximate model of the actual process to be regulated (Fig. 14). The model is in a simple enough form so that it yields rapid solutions amenable to analysis. The scheme works by issuing to the reference model the same regulatory commands as issued to the actual process, thereby generating an "anticipated" response more rapidly than the actual. The regulator then uses the anticipated outcome to modify its command to the actual process before any undesirable consequences of the initial command have occurred. The parameters of the reference model can also be fine-tuned so that its subsequent predictions more accurately reflect the actual responses. As the reference model is progressively fine tuned, it becomes a more exact predictor of the actual process. The system can now regulate actual behaviors based upon the predicted feedback alone! Model reference control systems can also adapt to changing conditions by changing the parameters in the model. Because the reference model is itself a simplified, deterministic process, the necessary adaptive changes can be more easily and rapidly realized by formulating and testing them within the model. However, in actual practice, the reference model scheme will work only if the model accurately embodies all the important dynamical characteristics of the actual process. Also, the algo-

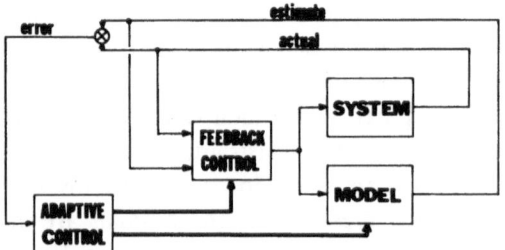

Fig. 14. Schematic arrangement for "model reference" control system. Narrow lines represent the variables of the system and the input control signals. Heavy lines represent operations that modify the parameters of the feedback controller and the model.

rithms for adjusting the parameters of the model based on discrepancies between the predicted and actual outcomes are very complex dynamical processes in themselves that can, if not properly realized, exhibit undesirable or even totally unstable characteristics.

Although the relevance of model reference control systems to some of the multisensory adaptive processes described during stance posture control is questionable, some adaptive motor behaviors perform similarly to those predicted by reference model controllers. For example, if a subject is asked to pick up rapidly an object of unknown weight and bring it to a specific point in space, the speed and accuracy of his movement will improve with repeated trials. If the weight of the object is then unexpectedly changed, the performance of the subject will initially be *poorer* picking up the object of incorrectly judged weight than it was previously picking up the object of unknown weight. This observation is consistent with the interpretation that adaptation to a repetitive task involves the generation of a stereotyped behavior based increasingly on an internal "model" of the task and less on actual sensory feedback. In terms of posture control, the progressive changes in rapid adjustments to ankle rotational inputs (following unexpected changes from sway to direct ankle rotations) could be interpreted in terms of a reference model controller that changes the ankle rotational input to posture based on a model of somatosensory, vestibular, and visual responses to normal body sway.

The Conceptual Model

The conceptual posture control model is shown in Fig. 15. The model is divided into three subsystems: (1) the motor apparatus of each leg, (2) the local spinal circuits comprising the movement generators, and (3) the supraspinal centers. Separate generators produce the fixed proportions of muscular activity characteristic of the "sway" and "suspensory" synergies. A separate input to each generator phasically regulates its level of activity. Spinal organizing machinery selects the appropriate synergic generators and provides the phasic activating inputs. The four-sided diamond represents a modulator in which each output is in phase with the adjacent input when the two inputs are in-phase. When the two inputs are antiphasically related, each output is phased to the diagonally opposite input. The three-sided diamond is a modulator that produces two phase outputs in direct proportion to the common mode (in-phase) portion of the two inputs. Filled circles represent inhibitory action of suspensory inputs on the sway inputs. Heavy lines represent a descending (adaptive) modulation of sway and suspensory input amplitudes.

Although its simplicity renders it far from complete, this conceptual model can be useful to interpret some of the complex experimental observations. It illustrates the importance of local (relative body) inputs in generating and coordinating basic movements and of multimodal sensory inputs and more complex central processes in the adaptive and balance functions. Even though the sway and suspensory patterns do characterize the majority of activities seen during stance, the system is no doubt capable of producing other synergic patterns of muscular activity. For example, the function of the individual muscle changes and other forms of organization might be necessary when a standing subject assumes other postures, carries loads, or performs specific voluntary acts. Also, the model provides a format with which to interpret the behaviors of patients with disorders of posture and movement control. Platform tests and conceptual models have already suggested some ways to differentiate among deficits within the basic peripheral motor programs, the adaptive central motor programs, and the sensory inputs to these systems.

CLINICAL APPLICATIONS OF POSTUROGRAPHIC STUDIES

The recent assurgence of neurophysiological studies with human subjects has motivated a reexamination of some of the traditional models of neuropathology in humans and the methods for their diagnosis and treatment. The capabilities of the neurologist to measure quantitatively the function of individual component parts of the nervous system and to intervene surgically, biochemically, and prosthetically have increased dramatically. However, the neurologist is significantly hampered in his attempts to apply these new techniques fully by an inability to evaluate quantitatively performance of the motor system as whole. This inability forces the neurologist to make a number of assumptions about the relations between the neurological signs exhibited by a patient (elevated tendon reflexes, positive Babinski reflexes, clonus, etc.) and the principal symptoms of the patient (spasticity, rigidity,

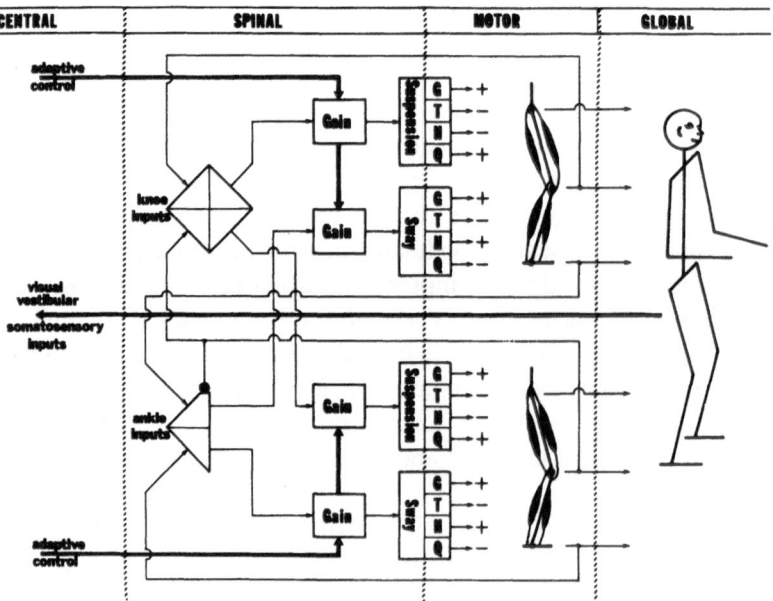

Fig. 15. A conceptual model for the production of rapid postural adjustments by standing subjects. The motions of the body are governed by activity of the four representative leg muscles. The direction of contractile motion of each is indicated by a "+" sign *(extensor)* or a "−" sign *(flexor)*. Note that, depending on the synergy involved, individual muscles can be flexors or extensors. The spinal regulating circuitry is described in the text.

tremor, dyscoordination, postural instability). For example, if the neurologist observes elevated tendon reflexes in a patient with spasticity or tremor, he must question whether these symptoms are caused by the elevated tendon reflexes. In fact, a very different relation might be the case; elevated tendon reflexes could be a useful adaptive change made by the motor system to help compensate for its losses in other subsystems. Without accurate knowledge about function of the system as a whole, the neurologist must make a decision whether to treat the patient's abnormal signs (i.e., administer a drug which is known to depress reflex activity) based on a subjective evaluation of the patient's overall condition. The risk is that treatment of abnormal changes that may in fact be useful compensatory ones will lead to deterioration rather than improvement in the patient's symptoms. A quantitative systems analysis of the patient's performance could decrease significantly this type of risk.

A thorough discussion of sensorimotor physiology and its relationship to various types of clinical pathology would require a review of considerable depth. Therefore, this section will be limited to several concrete examples of how the quantitative evaluation of a patient's posture might be used in a clinical setting to help the neurologist evaluate a patient's sensorimotor disorders in a systematic way. This field is in its infant stages of development at the present time, and relatively little can be found in the literature. Therefore, most of the discussion in this section is focused upon the works of the author and his colleagues.*

Some Techniques for Patient Evaluation

Techniques for measuring the performance of patients standing upon a controlled platform and the systems models of performance derived from the study of normal human subjects together provide the tools for some quantitative measures of the integrative actions of the posture control system during stance. The specific characteristic features that can be assessed at the present time are: (1) the capability of the system to detect postural sway and to generate the appropriate stabilizing adjustments in leg muscles in the absence of normal somatosensory (from ankle joint rotations) and visual inputs, (2) the capability of the system to change adaptively the somatosensory input to posture control when ankle rotational inputs are alternately stabilizing and destabilizing, and (3) the capability of the system to produce activities within the correctly patterned synergy of muscles (including the "sway" and "suspensory" synergies and the simultaneous or reciprocal phasing of the two legs). A synthesis of this author's techniques with those developed by Lishman and Lee (1973) would lead to another very important group of tests measuring the capability of the system to adapt when presented with visual-vestibular conflicts during stance. According to the conceptual model presented in the preceding section, each of these tests would measure the performance of particular subsystems of posture control.

*Unpublished observations have been made in collaboration with Drs. R. J. Grimm and Robert Rafal.

Stance posture control in the absence of somatosensory and visual information about sway orientation is a critical test of vestibular functions and of the neural circuits transmitting this information to the leg muscles. According to the systems model of this task, the semicircular canals sense higher frequency sway-rate information, and the utricular otoliths at low frequencies sense the sway orientation of the body with respect to gravity. Furthermore, sway-rate information must be neurally integrated to determine the body's sway orientation at higher frequencies. If this model is correct, this test might selectively measure the performance of several sensory inputs to posture. Because sway motions are among the smallest accelerations that the vestibular receptors can sense, this test is a sensitive measure of these receptors. Also, because the semicircular canals and the utricular otoliths each perform a specialized function, differences in the characteristics of a patient's instability may be used to evaluate canal and otolith functions differentially. The model predicts that deficits selective to the utricular otolith or to the circuits that neurally integrate the vestibular input would result in appropriate EMG responses to sway perturbations but would permit a "slow drift" instability after several seconds. In contrast, loss of semicircular canal function would result in significantly delayed or absent EMG responses and immediate instability. Tests upon subjects with loss of semicircular canal functions have shown grossly delayed or absent responses and immediate instability in this condition (Nashner, 1970, and unpublished). Patients whose instabilities were of the "slow drift" kind, presented cerebellar deficits or other deficits of clinically unknown etiology. There is physiological evidence to suggest that (mathematical) integration of vestibular rate information may be performed by the cerebellum. Kornhuber (1974), for example, has described the close interrelations between vestibular and cerebellar functions during movement and posture control. More specifically, Carpenter (1972) measured the phase relations between labyrinthine inputs and eye position before and after creating cerebellar lesions in cats. He found phase shifts that were consistent with the interpretation that the cerebellum integrated the canal angular rate input in order to compute an angular position input to the eye-movement control system.

To adapt the rapid postural adjustments elicited by rotation of the ankle joints during stance, the system must be able to detect conflicts between the somatosensory, vestibular, and visual inputs and then appropriately modify the strategy of control in favor of vestibular and visual controls. Principally because normal subjects required several trials to perform these adaptive changes, and because sensory inputs from among all the modalities were necessary, this adaptive process is most likely performed centrally. Again, drawing analogy to physiological studies of the eye-position control system, several investigators have shown that the cerebellum is critically involved in the slow adaptive changes in vestibulo-ocular reflex gains following changes in the relations between vestibular and visual motion cues (Ito, 1976; Robinson, 1976). Perhaps, the quantitative analysis of the patient's adaptation to somatosensory (and visual) conflicts during stance will be another sensitive measure of certain types of cerebellar lesions.

Studies with normal human subjects and evidence from the animal literature have both suggested that muscle activities during stance are organized by local neural circuits of the spinal cord utilizing, primarily, the local somatosensory inputs. Therefore, the organization of EMG activities during various vertical and sway perturbations may be a sensitive measure of peripheral circuits. The critical issue, however, is whether the results of the above tests will give details about a patient's overall performance that can be relevant in answering the important questions for diagnosis and evaluation of treatment. Some preliminary results of studies with selected patients have suggested that quantitative tests of this kind will be of value to the neurologist.

Evaluation of Selected Patients

Platform tests may be of greatest value to those patients who have subtle deficits that affect balance and walking. These patients generally show a good deal of compensatory behavior, which must be systematically evaluated to pinpoint the pathological changes.

Patients with clinically defined cerebellar lesions were evaluated independently with the controlled platform and with a more traditional neurological examination that included a subjective rating of ability to perform normal stance and walking exercises (Nashner and Grimm, 1978, and unpublished observations). To varying degrees, these patients all presented the typical "shuffling" cerebellar gait and difficulty in performing rapid turns on command. Based on results of platform tests evaluating vestibular inputs, ability to adapt, and synergic organization of muscles, there was a strong correlation between the severity of a patient's symptoms (rated on a subjective scale from -1 to -10) and the *number* of test categories in which performance was abnormal. In itself, this general observation relates something about the importance of compensatory mechanisms to the individual patient. A deficit in only one or perhaps two subsystems, no matter how severe, can potentially be compensated by the remaining normal ones, whereas multimodal deficits leave the patient fewer alternatives and he is severely disabled.

Among the cerebellar patients whose symptoms were the least severe, abnormality was limited to a "slow drift" instability during "vestibular" posture control. Those with somewhat greater disabilities, as judged clinically, evidenced abnormalities in the adaptation of rapid postural adjustments as well as those in vestibular controls. However, in all but the most severe cases, adaptive and vestibular abnormalities were accompanied by the normal organization of muscular activities. The cerebellar patients could adjust to postural perturbations in a relatively normal way so long as somatosensory inputs were not disturbed. Since cerebellar patients may present abnormal functions that are similar to those characterizing primary vestibular lesion, an important question for future work will be to show whether these two groups of patients can be separated on the basis of differences in the adaptation or in the utilization of visual inputs.

Our most recent study has been with a selected group of patients with multiple sclerosis whose lesions were mostly peripheral. The major symptoms of these patients included spasticity, elevated tendon reflexes, and leg clonus. However, with only one exception, none of these patients evidenced abnormally large rapid postural adjustments. In the exceptional case, small myotatic responses of the stretched ankle muscles did not appear to be of direct functional significance to the task of stance posture control. However, all these patients suffered a gross disorganization of muscle activities during postural adjustments. Typically, all the muscles were activated, therefore, attempted adjustments were grossly discoordinated. To varying degrees, most of these patients had also lost vestibular controls. When compared to the cerebellar patients, the most significant difference was the degree of muscular disorganization.

In conclusion, the use of controlled platforms and systems models derived from studies of normal subjects is providing some new observations and ideas about the fundamental organization of posture controls and about the roles played by the sensory inputs, the cerebellum, and the spinal cord in the production and regulation of movements. Although the primary intent of these studies has been to study normal motor activities, the information gained by quantitative human studies is beginning to suggest new ways to characterize neurological disorders and new diagnostic techniques that may be a valuable complement to the classical neurological examination.

Acknowledgments

A special acknowledgment is given to Dr. Victor Gurfinkel for his exceptional contributions to the understanding of the human sensorimotor system. The author also appreciates the helpful assistance of Dr. Marjorie Woollacott in the preparation and the editing of the manuscript.

Appendix

State Space Representation of a Linear Differential Equation

The order of a dynamical linear system is determined by the number of derivatives of the dependent variable (x) that, if defined for one point, will completely characterize the response of the system for all succeeding times. The dependent variable and its n derivatives form the characteristic equation for the system:

$$x^{(n)} + a_{n-1} x^{(n-1)} + \ldots + a_0 x = 0 \qquad (A.1)$$

A forcing function can be added to the equation to describe how the system responds to an external input:

$$x^{(n)} + a_{n-1} x^{(n-1)} + \ldots + a_0 x = f \qquad (A.2)$$

The response of the forced nth order system is completely characterized by defining the n derivatives of the system at one point (the initial conditions) and by defining the forcing function for this and all succeeding times.

The state space representation of the system is formulated by designating each derivative of the system as an independent "state," thereby forming a vector of states:

$$x_0 = x$$
$$x_1 = x^{(1)}$$
$$x_2 = x^{(2)}$$
$$\cdot$$
$$\cdot$$
$$\cdot$$
$$x_{n-1} = x^{(n-1)}$$

$$\bar{x} = \begin{bmatrix} x_0 \\ x_1 \\ \cdot \\ \cdot \\ \cdot \\ x_{n-1} \end{bmatrix} \qquad (A.3)$$

A matrix equation can then be formulated quite simply by realizing that the following equalities hold:

$$\dot{x}_0 = x_1$$
$$\dot{x}_1 = x_2$$
$$\dot{x}_2 = x_3$$
$$\cdot$$
$$\cdot$$
$$\cdot$$
$$\dot{x}_{n-1} = x^{(n)} = -a_{n-1}x_{n-1} \ldots - a_0 x_0 + f$$

These equations can now be rewritten in a familiar matrix form:

$$\dot{\bar{x}} = \begin{bmatrix} 0 & 1 & 0 & 0 & 0\ldots & & 0 \\ 0 & 0 & 1 & 0 & 0\ldots & & 0 \\ 0 & 0 & 0 & 1 & 0\ldots & & 0 \\ \cdot & & & & & & \\ \cdot & & & & & & \\ \cdot & & & & & & \\ -a_c & -a_1 & \cdot & \cdot & \cdot & & -a_{n-1} \end{bmatrix} \bar{x} + \begin{bmatrix} 0 \\ 0 \\ 0 \\ \cdot \\ \cdot \\ \cdot \\ 1 \end{bmatrix} f \qquad (A.4)$$

The $n \times n$ matrix is called the *characteristic matrix* because its coefficients completely describe the dynamical properties of the system. Notice that in this form, the characteristic matrix can be written simply by forming the off diagonal of "1" elements and by placing the n coefficients across the bottom. (This is not the only way in which this system can be represented.)

In the second-order system the coefficients a_1 and a_0 are directly related to the natural frequency (w_n) and the damping (ζ) of the system by the relations:

$$a_1 = 2\zeta w_n$$
$$a_0 = w_n^2$$

Thus, inspection of the characteristic matrix can tell much about how the system will behave. For example, positive values for either of the terms a_0 or a_1 indicate that the system is unstable.

References

Arutyunyan, G. A., Gurfinkel, V. S., and Mirskii, M. L. Organization of movements on execution by man of an exact postural task. *Biophysics*, 1969, *14*, 1162–1167.

Begbie, J. V. Some problems of postural sway. In A. V. S. deReuck and J. Knight (Eds.), *CIBA Foundation Symposium on Myotatic, Kinesthetic and Vestibular Mechanisms*. London: Churchill Limited, 1967.

Bensel, C. K., and Dzendolet, E. Power spectral density analysis of the standing of males. *Perception and Psychophysics*, 1968, *4*, 285–287.

Bernstein, N. *Coordination and Regulation of Movements*. New York: Pergamon Press, 1967.

Berthoz, A., and Anderson, J. H. Frequency analysis of vestibular influences on extensor motoneurons. II. Relationship between neck and forelimb extensors. *Brain Research*, 1971, *34*, 376–380.

Berthoz, A., Pavard, B., and Young, L. R. Perception of linear self-motion induced by peripheral vision (linear vection). *Experimental Brain Research*, 1975, *23*, 471–489.

Bizzi, E., Dev, P., Morasso, P., Polit, A. The effect of load disturbances during centrally initiated movements. *Journal of Neurophysiology*, 1978, *41*, 542–556.

Bizzo, G., and Baron, J. B. Aspect cybernétique des déplacements du centre de gravité du corps induits par des stimulations labyrinthiques électriques rectangulaires ou sinusoidales. *Agressologie*, 1972, *13B*, 41–50.

Brandt, Th., Dichgans, J., and Koenig, E. Differential effects of central versus peripheral vision on egocentric and exocentric motion perception. *Experimental Brain Research*, 1973, *16*, 476–491.

Carpenter, R. H. S. Cerebellectomy and the transfer function of the vestibulo-ocular reflex in the decerebrate cat. *Proceedings of the Royal Society, Series B*, 1972, *181*, 353–374.

Chan, C. W. Y., Kearney, R. E., and Melvill Jones, G. Tibialis anterior response to sudden ankle displacements in normal and Parkisonian subjects. *Brain Research*, 1979, *173*, 303–314.

Clark, B., and Stewart, J. D. Effects of angular acceleration on man: Thresholds of the perception of rotation and the oculogyral illusion. *Aerospace Medicine*, 1969, *40*, 952–956.

Coats, A. C. The sinusoidal galvanic body-sway response. *Acta Otolaryngologica*, 1972, *74*, 155–162.

Cohen, L. A. Role of eye and neck proprioceptive mechanisms in body orientation and motor coordination. *Journal of Neurophysiology*, 1961, *24*, 1–11.

Davis, W. J. Organizational concepts in the central motor networks of invertebrates. In R. M. Herman, S. Grillner, R. G. S. Stein, and D. G. Stuart (Eds.), *Neural Control of Locomotion*. New York: Plenum Press, 1976.

Dichgans, J., Held, R., Young, L. R., and Brandt, Th. Moving visual scenes influence the apparent direction of gravity. *Science*, 1972, *178*, 1217–1219.

Dichgans, J., Schmidt, C. L., and Graf, W. Visual input improves the speedometer function of the vestibular nuclei of the goldfish. *Experimental Brain Research*, 1973, *18*, 319–322.

Dzendolet, E. Sinusoidal electrical stimulation of the human vestibular apparatus. *Perceptual Motor Skills*, 1963, *17*, 171–185.

Ehrhardt, K. J., and Wagner, A. Labyrinthine and neck reflexes recorded from spinal single motoneurons in the cat. *Brain Research*, 1970, *19*, 87–104.

Eklund, G. Influence of muscle vibration on balance in man. *Acta Society Medica Upsala*, 1969, *74*, 113–117.

Eklund, G. Further studies of vibration-induced effects on balance. *Upsala Journal of Medical Science*, 1973, *78*, 65–72.

El'ner, A. M., Popov, K. E., and Gurfinkel, V. S. Changes in stretch reflex system concerned with the control of postural activity of human muscle. *Agressologie*, 1972, *13D*, 19–23.

Engberg, I., and Lundberg, A. An electromyographic analysis of muscular activity in the hindlimb of the cat during unrestrained locomotion. *Acta Physiologica Scandinavica*, 1969, *75*, 614–630.

Evarts, E. V., and Tanji, J. Gating of motor cortex reflexes by prior instruction. *Brain Research*, 1974, *71*, 479–494.

Evarts, E. V., Bizzi, E., Burke, R. E., DeLong, M., and Thach, W. T. Jr. Central control of movement. In *NRP Bulletin 9*. Cambridge: M.I.T. Press, 1971.

Fernandez, C., and Goldberg, J. M. Physiology of peripheral neurons innervating semicircular canals of the squirrel monkey. II. Response to sinusoidal stimulation and dynamics of peripheral vestibular system. *Journal of Neurophysiology*, 1971, *34*, 661–675.

Gelfand, I. M., Gurfinkel, V. S., Fomin, S. V., and Tsetlin, M. L. (Eds.). *Models of the Structural-Functional Organization of Certain Biological Systems*. Cambridge: M.I.T. Press, 1971.

Gernandt, B. D., Iranyi, M., and Livingstok, R. B. Vestibular influences on spinal mechanisms. *Experimental Neurology*, 1959, *1*, 248–273.

Gibson, J. J. *The Senses Considered as Perceptual Systems*. Boston: Houghton Mifflin, 1966.

Goldberg, J. M., and Fernandez, C. Physiology of peripheral neurons innervating semicircular canals of the squirrel monkey. I. Resting discharge and response to constant angular accelerations. *Journal of Neurophysiology*, 1971, *34*, 635–660.

Gottlieb, G. L., and Agarwal, G. C. The role of the myotatic reflex in the voluntary control of movements. *Brain Research*, 1972, *40*, 139–143.

Graybiel, A., and Fregly, A. R. A new quantitative ataxia test battery. *Acta Otolaryngological (Stockholm)*, 1966, *61*, 292–312.

Greene, P. H. Problems of organization of motor systems. In F. M. Snell (Ed.), *Progress in Theoretical Biology* (Vol. 2). New York: Academic Press, 1972.

Grillner, S. Locomotion in vertebrates: Central mechanisms and reflex interaction. *Physiological Reviews*, 1975, *55*, 247–304.

Grillner, S., Hongo, T., and Lund, S. Interaction between the inhibitory pathways from the Deiters' nucleus and Ia afferents to flexor motoneurons. *Acta Physiologica Scandinavica*, 1966, *68*, Supplementum 277.

Grillner, S., Hongo, T., and Lund, S. The vestibulospinal tract. Effects on alpha-motoneurons in the lumbosacral spinal cord in the cat. *Experimental Brain Research*, 1970, *10*, 94–120.

Grillner, S., Hongo, T., and Lund, S. Convergent effects on alpha-motoneurons from the vestibulospinal tract and a pathway descending in the medial longitudinal fasciculus. *Experimental Brain Research*, 1971, *12*, 457–479.

Gurfinkel, V. S., and Shik, M. L. The control of posture and locomotion. In A. Gydikov, N. Tankov, and D. Kosarov (Eds.), *Motor Control*. New York: Plenum Press, 1973.

Gurfinkel, V. S., Kots, Y. M., Paltsev, Y. I., and Feldman, A. G. The compensation of respiratory disturbances of the erect posture of man as an example of the organization of interarticular interaction. In I. M. Gelfand, V. S. Gurfinkel, S. V. Fomin, and M. L. Tsetlin (Eds.), *Models of the Structural-Functional Organization of Certain Biological Systems*. Cambridge, Massachusetts: M.I.T. Press, 1971.

Gurfinkel, V. S., Alexeef, M., Elner, G., and Baron, J. B. Variations de l'activité tonique posturale et du réflexe achiléen sous l'influence du calcul mental et d'une manoeuvre dérivée de celle de Jendrassik. *Agressologie*, 1972, *13D*, 63–68.

Gurfinkel, V. S., Lipshits, M. I., and Papov, K. Y. Is the stretch reflex the main mechanism in the system of regulation of the vertical posture of man? *Biophysics*, 1974, *19*, 744–748.

Gurfinkel, V. S., Lipshits, M. I., Mori, S., and Popov, K. E. The state of stretch reflex during quiet standing in man. In S. Homma (Ed.), *Progress in Brain Research* (Vol. 44). Amsterdam: Elsevier, 1975.

Gurfinkel, V. S., Lipshits, M. I., Mori, S., and Popov, K. E. Postural reactions to the controlled sinusoidal displacement of the supporting platform. *Agressologie*, 1976, *17*, 71–76.

Henn, V., Young, L. R., and Finley, C. Vestibular units in alert monkeys are also influenced by moving visual fields. *Brain Research*, 1974, *71*, 144–149.

Holst, E. von, and Mittelstaedt, H. Das Reafferenzprinzip (Wechselurrkungen zwischen Zentralnervensystem und Peripherie). *Naturwissenschaften*, 1950, *37*, 464–475.

Houk, J. C. Regulation of stiffness by skeletomotor reflexes. *Annual Review of Physiology*, 1979, *41*, 99–114.

Ito, M. Cerebellar learning control of vestibulo-ocular mechanisms. In T. Desiraju (Ed.), *Mechanisms in Transmission of Signals for Conscious Behavior*. Amsterdam: Elsevier, 1976.

Kim, J. H., and Partridge, L. D. Observations on types of response to combinations of neck, vestibular and muscle stretch signals. *Journal of Neurophysiology*, 1969, *32*, 239–250.

Kornhuber, H. H. The vestibular system and the general motor system. In H. H. Kornhuber (Ed.), *Handbook of Sensory Physiology* (Vol. 6, Part 2). Berlin: Springer-Verlag, 1974.

Kots, Y. M. Supraspinal control of the segmental centres of muscle-antagonists in man. I. Reflex excitability of the motor neurones of muscle-antagonists in the period of organization of voluntary movement. *Biophysics*, 1969, *14*, 176–183.

Kots, Y. M., and Zhukov, V. I. Supraspinal control of the segmental centres of the muscle-antagonists in man. III. "Tuning" of the spinal apparatus of reciprocal inhibition in the period of organization of voluntary movement. *Biofizica*, 1972, *16*, 1129.

Kots, Y. M., Krinski, V. I., Naydin, V. L., and Shik, M. L. The control of movements of the joints and kinesthetic afferentation. In I. M. Gelfand, V. S. Gurfinkel, S. V. Fomin, and M. L. Tsetlin (Eds.), *Models of the Structural-Functional Organization of Certain Biological Systems*. Cambridge: M.I.T. Press, 1971.

Lee, D. N., and Aronson, E. Visual proprioceptive control of standing in human infants. *Perception and Psychophysics*, 1974, *15*, 529–532.

Lee, D. N., and Lishman, J. R. Visual proprioceptive control of stance. *Journal of Human Movement Studies*, 1975, *1*, 87–95.

Lestienne, F., Soechting, J., and Berthoz, A. Postural readjustments induced by linear motion of visual scenes. *Experimental Brain Research*, 1977, *28*, 363–384.

Lishman, J. R., and Lee, D. N. The autonomy of visual kinaesthesis. *Perception*, 1973, *2*, 287–294.

Lowenstein, O. The effect of galvanic polarization on the impulse discharge from sense endings in the isolated labyrinth of the thornback ray. *Journal of Physiology (London)*, 1955, *127*, 104–117.

Maeda, M., Magherini, P. C., and Precht, W. Functional organization of vestibular and visual inputs to neck and forelimb motoneurons in the frog. *Journal of Neurophysiology*, 1977, *40*, 225–243.

Marsden, C. D., Merton, P. A., and Morton, H. B. Latency measurements compatible with a cortical pathway for the stretch reflex in man. *Journal of Physiology (London)*, 1973, *230*, 58–59.

Meiry, J. L. *The Vestibular System and Human Dynamic Space Orientation*. National Aeronautics Space Administration-Contractor Report-628, 1966.

Melvill Jones, G., and Watt, D. G. D. Observations on the control of stepping and hopping movements in man. *Journal of Physiology (London)*, 1971, *219*, 709–727.

Mori, S. Discharge patterns of soleus motor units with associated changes in force exerted by the foot during quiet stance in man. *Journal of Neurophysiology*, 1973, *36*, 458–477.

Nashner, L. M. *Sensory Feedback in Human Posture Control*. Massachusetts Institute of Technology Report MVT-70-3, 1970.

Nashner, L. M. A model describing vestibular detection of body sway motion. *Acta Otolaryngologica*, 1971, *72*, 429–436.

Nashner, L. M. Vestibular posture control model. *Kybernetik*, 1972, *10*, 106–110.

Nashner, L. M. Vestibular and reflex control of normal standing. In R. B. Stein, K. G. Pearson, R. S. Smith, and J. B. Redford (Eds.), *Control of Posture and Locomotion*. New York: Plenum Press, 1973.

Nashner, L. M. Adapting reflexes controlling the human posture. *Experimental Brain Research*, 1976, *26*, 59–72.

Nashner, L. M. Fixed patterns of rapid postural responses among leg muscles during stance. *Experimental Brain Research*, 1977, *30*, 13–24.

Nashner, L. M., and Berthoz, A. Visual contribution to rapid motor responses during posture control. *Brain Research*, 1978, *150*, 403–407.

Nashner, L. M., and Grimm, R. J. Analysis of multiloop dyscontrols in standing cerebellar patients. In J. E. Desmedt (Ed.), *Progress in Clinical Neurophysiology* (Vol. 4). Basel: S. Karger, 1978.

Nashner, L. M., and Wolfson, P. Influence of head position and proprioceptive dues on short latency postural reflexes evoked by galvanic stimulations of the human labyrinth. *Brain Research*, 1974, *67*, 255–268.

Nashner, L. M., Woollacott, M., and Tuma, G. Organization of rapid responses to postural and locomotor-like perturbations of standing man. *Experimental Brain Research*, 1979, *36*, 463–476.

Nichols, T. R., and Houk, J. C. Reflex compensation for variations in the mechanical properties of a muscle. *Science*, 1973, *181*, 182–184.

Öosterveld, W. J. Threshold value for stimulation of the horizontal semicircular canals. *Aerospace Medicine*, 1970, *41*, 386–391.

Padulo, L., and Arbib, M. A. *Systems Theory*. Washington, D.C.: Hemisphere Publishing, 1974.

Robinson, D. A. Adaptive gain control of vestibulo-ocular reflex by the cerebellum. *Journal of Neurophysiology*, 1976, *39*, 954–969.

Scott, D. E., and Dzendolet, E. Quantification of sway in standing humans. *Agressologie*, 1972, *13D*, 35–40.

Wetzel, M. D., and Stuart, D. G. Ensemble characteristics of cat locomotion and its neural control. In G. A. Kerkut and J. W. Phillis (Eds.), *Progress in Neurobiology* (Vol. 7). London: Pergamon Press, 1976.

Young, L. R. On visual-vestibular interactions. In *Proceedings of the Fifth Symposium on the Role of the Vestibular Organs in Space Exploration*. NASA-SP-314, 1970.

Young, L. R., and Meiry, J. L. A revised dynamic otolith model. *Aerospace Medicine*, 1968, *39*, 606–610.

11

Properties and Mechanisms of Locomotion

Mary C. Wetzel and Leon G. Howell

INTRODUCTION

An account of any behavior must include descriptions of both its "properties" and its "mechanisms." The "properties" of locomotion include its topographical characteristics, the *kinematics,* or movement in space and time, and an analysis of these in terms of masses and forces, *kinetics*. There are two groups of "mechanisms" to be considered. One of these concerns the hardware of which the organism is composed, and the functional relationships between the components. This mechanism has received much attention from neurophysiologists, who have established functional connections between neurons by examining impulses in numerous cells under many different conditions of activation.

The other class of mechanisms is a set of relationships between environments and the behavior of organisms. These functional relationships, or behavioral mechanisms, are usually studied by psychologists. They are concerned with the relationships between behavior and the current circumstances under which it occurs, as well as with the phylogenetic and ontogenetic histories. These histories comprise the recurrent sequences of environmental events that are necessary to establish the current relationships. Again, the biological and the environmental conditions must be considered.

A functional analysis of behavior is difficult because, as we shall see, any given locomotor act is produced by a large number of events, called *stimuli,* in the envi-

MARY C. WETZEL AND LEON G. HOWELL   Department of Psychology, University of Arizona, Tucson, Arizona 85721.

ronment. Conversely, any given stimulus, whether acting from outside the body or within, affects several and often all features of stepping. In addition, the biological conditions (which are to be distinguished from stimuli) may differ upon separate occasions for which responses and controlling stimuli are specified.

To understand how complex response repertoires are produced, it will be profitable first to discuss locomoting machines, in which there are few properties and mechanisms. The appropriate control theories come from physics and engineering. Systems theory, as provided by Wiener (1948) in the West and Bernstein (1947) in Russia, together with a related theory of information processing, as given in Shannon's (1948) communication theory, have been the dominant influences in this area.

The second major topic traces work on spinalized, decerebrated, or anesthetized animals. Although these "reduced" preparations are enormously more complicated than robots, their behaviors are still far more limited than are those of intact, mobile animals. Ramón y Cajal (1911) and Sherrington (1906) established a neuroscience that was long based largely on the concept of the reflex. Only recently has a different theoretical account gained general favor—that of intrinsic pattern generators lying within the central nervous system. The facts that gave rise to both of these control concepts will be interpreted under the assumption that in reduced animals stimuli have only one function: to *elicit* behavior.

The third major topic is locomotion by intact animals, from invertebrates to humans. As Sherrington pointed out, reflex behavior can account for only a small percentage of animal or human movement or experiences (1906; see 1948 "forward" in which the emphasis on reflex action is seen to have paled). We must distinguish between what are often called involuntary (automatic) behaviors and those that are called voluntary (willed) movements. A behavioristic analysis will give us a precise account of both kinds of movement (Skinner, 1953). To show how they are produced, we will add and explain three other stimulus functions, for stimuli not only *elicit* responses but also *evoke* and *discriminate* responses and *reinforce* the control exerted by evoking and discriminating stimuli.

MECHANICAL LOCOMOTION

Understanding machine locomotion can be useful in the study of animal locomotion for several reasons. First, a mechanical system is simple in the sense that the builder determines the properties and the mechanisms; the performance of the machine is completely specified. Second, the approach tells us the physical constraints within which all living, moving organisms evolved. Third, most of the technical terms have been standardized in the scientific community for many years, although some revision continues. The terms evolved in that branch of physics called *mechanics* and also in contemporary *systems theory*. We will discuss the two theories briefly, tracing their origins in the history of science and showing how both theories have continued to be useful to the present day.

In studying automatic walking devices, we will exclude those machines whose operation depends on a human being. McGhee (1976) has described, for example, the General Electric Quadruped Transporter. Although it had a "significant ability to climb over obstacles and to traverse difficult terrain" (p. 239), the human operator was quickly exhausted by trying to coordinate twelve independent joints. Man–machine interactive devices are at least as difficult to study as locomotion by intact people (a topic discussed in a later section). Nor will we be concerned with children's pull toys or other devices that have no independently powered joints. Finally, biped machines are discussed only briefly, because no truly effective ones have been built. McGhee (1976) mentioned only two biped vehicles. One was a true robot, a pneumatically powered exoskeleton designed for paraplegics by Vukobratović, Ciric, Hristic, and Stepanenko in Yugoslavia (1972), that had no control of speed or direction. The other was a series of Japanese machines whose coordination was computer-controlled (Kato and Tsuiki, 1972). Although one of these machines can climb stairs and carry a load, it is too slow (90 sec/step) to be practical.

A two-legged creature is easily tipped over, whether at rest or moving, especially if the feet are close together. The problem of building a machine that will respond effectively to a push is a problem of *stability,* a basic term in every account of posture and locomotion. We cannot discuss stability independently of the concept of *force,* which is derived, in turn, from the fundamental concepts of mechanics: length (space), time, and mass (matter). The necessary concepts emerged generations ago, when Newton derived force in three laws of motion that related time and mass. Ziwet and Field (1912, pp. 139–140) provided a literal translation of the three laws of motion contained in Newton's *Principia.* We repeat them here and emphasize their importance in banishing fictional explanations from science. The first law clearly stated that matter has no intrinsic tendency to change its condition of rest or motion. There is no effect without cause, in this precise sense, and the causal agent is force, as defined in the second law.

1. "Every body persists in its state of rest or of uniform motion along a straight line, except in so far as it is compelled by impressed (i.e., external) forces to change that state." *Particle* has more recently been substituted for body, with a particle being a geometrical point to which a definite mass is assigned. Since this finite mass is considered to exist in a finite space whose dimensions are equal to zero, the concept of particle is abstract. It is also useful because sometimes the motion can be determined for a rigid body (one of invariable size and shape) if its mass is considered to be concentrated at a single point called the *center of mass.* Today's locomoting machines are almost always conceptualized as sets of rigid bodies in different configurations.

2. "Change of motion is proportional to the impressed moving force and takes place along the straight line in which that force acts." Change of motion is acceleration in contemporary terms. The law may be stated as $F = ma$, where F, m, and a represent, respectively, the resultant force acting on the particle, the mass of the particle, and the acceleration of the particle expressed in consistent units (see Beer and Johnston, 1962). Any number of forces, whose actions Newton tacitly assumed to be independent, may act on a particle. They are equivalent to a single force—their *resultant.*

3. "To every action there is an equal and contrary reaction; or, the mutual

MARY C. WETZEL
AND LEON G.
HOWELL

actions of two bodies on one another are always equal and directed in contrary senses." When a car moves forward, we can say that the wheels exert a force backward and the ground exerts a forward force.

The second and third laws are especially important in understanding locomotion because they permit precise accounts of balance and equilibrium. In the condition of equilibrium of a system of forces acting on the same particle, the resultant is equal to zero. There is no unbalanced force, and no acceleration. The particle may be at rest or in motion. Consider first the case of a rigid body at rest. If the weight of the body (i.e., the force with which the earth pulls the body) is downward, then a single upward force can support the body. Only at one point will it be true that a single force equal to the weight of the body will keep the body in equilibrium. This point is called the *center of gravity*, which coincides with the center of mass. Note that the conditions for equilibrium of a rigid body are more complicated than for a particle, since the body may rotate. If the body is supported at any other point than the center of gravity, it will rotate until it comes to rest. A rigid body is in equilibrium when its linear acceleration is zero and its angular acceleration is zero.

Equilibrium may be *stable* or *unstable,* as can be tested by displacing the body slightly from the equilibrium position. Observation indicates that if it is stable, the body returns to the original position when released; if it is unstable, it does not return. In the first case, the disturbance raises the center of gravity, and when released, the body falls back to the original position. In the second case, the disturbance lowers the center of gravity, and upon release, the body continues to move down until the center of gravity has reached the lowest possible position. In *neutral* equilibrium, the body is supported such that slight displacement does not change the height of the center of gravity, as is true for a cylinder lying on its side.

If a resting body is to move, with or without disturbances (temporary unbalanced forces), then an additional force must act upon it. Newton's laws of motion apply in the general sense that the internal forces associated with the masses of a walking device's elements must counter the external forces of gravity, friction at the interface with the surface, and resistance of the air or water medium. In a machine, mechanical members are substituted for bones, and power elements, or *actuators,* are substituted for living muscles. Although the laws of motion are conceptually simple, the contemporary science of mechanics is complicated by additional statements. The number of equations required to describe even simple machine locomotion is large because there are many moving parts of the "body" and "legs," each with a different mass. It is unnecessary to review the mathematics here, but we may note that the simplifying assumptions for computer-modeling of a biped machine have resulted in forms unlike the human one. The "body" is usually a rigid framework, the "legs" have zero mass and are limited to a few joints (perhaps only a hip and knee), and the "foot" is usually part of the leg (Hemami, Weimer, and Koozekanini, 1973; Vukobratović, Frank, and Juričić 1970). Sometimes the foot has a rocker configuration, and sometimes it is entirely absent. The idealized biped moves with frictionless pivots in a straight line, across a smooth plane without slippage.

In a biped machine, there will always be numerical errors in the generation of various angles by the controlling mechanism, and the feet may slip on the surface. Nor will the surface be perfectly smooth. Even if small, these variations will perturb the locomotion. Vukobratović *et al.* (1970) discussed three kinds of robot stability: *body, body path,* and *gait stability.* Trunk or body stability concerned only the body's angular position, with respect to earth, and its altitude. Body-path stability was that of the trajectory of the center of gravity in its forward motion and depended on the return of an average velocity vector to its original position after a disturbance. Note that this concept of equilibrium was not limited to the case of constant velocity during locomotion.

Gait stability concerned the ways by which the leg movements returned to their original values after some number of steps. This kind of "stability" is quite far removed from the classically simple descriptions of stability, although no new principles of mechanics are involved. Several kinematic variables were introduced to define a stationary gait. These definitions will be summarized here, because they are commonly accepted conventions in specifying both animal and machine locomotion. A *stationary gait* has a constant mean velocity, constant *stride length* (the length of step for all individual limbs is equal, ideally), constant *phasing* (timings within and between limbs), constant *step cycle time* (period of a complete step for one limb, which is ideally the same for all limbs), and constant *duty factor* (proportion of *stance,* or down time, in relation to total cycle time). The up time of the leg is called its *swing.*

The concepts of stability apply to machines with more than two legs, and when there are at least four, stable locomoting machines can be built. Quadrupeds are emphasized here, but the treatments are general for any number of legs. Since control of the machines comes from applications of systems theory, cybernetics, and information (communication) theory, their basic concepts will be discussed briefly. As was true for the review of classical Newtonian mechanics, the intent is not to teach physics or engineering but to point out some useful assumptions and laws and their historical development. It is important to know what the system theories can and cannot do in understanding locomotion, for they have also been applied widely to human affairs. Wiener (1948) explicitly suggested the term *cybernetics* for the study of "control" and "communication" in the animal and the machine. Our concern here, however, is to account for machine locomotion, with extension of the theories to biological creatures considered in later sections of the chapter.

Control is brought about in Wiener's theory by feedback, such that the operation of a machine at any given time provides a way to modify its future operation. The communication aspect of cybernetics concerns *information,* in which concept a quantity (physically, it may be an electrical pulse) that enters a machine can reduce uncertainty among a set of available alternatives. A common example is that of an electronic relay with two possible positions. A pulse that puts the relay into one of the two positions has conveyed to the machine one *bit* (binary digit) of information, by means of a "choice" of zero or one. There is, of course, no implication of volition on the part of the machine. The general concepts appeared almost simultaneously in many fields. They were appropriate for telephone systems (see Shannon's communication theory, 1948), computers, and automatic machines.

The terms in systems theory have not been completely standardized, but we have drawn major concepts from several sources (Arbib, 1964, 1972; Carlson, 1968; Chestnut and Mayer, 1951; McFarland, 1971). The theory accounts for relationships between variables, and often a block diagram is convenient to display these relations, as in Fig. 1. Usually a further distinction is made between a *variable*—a measurable property of the system that changes value during the period of observation—and a parameter—a system property that is assumed not to change during this period. The causes, or independent variables, are called *input variables* and are shown by arrows pointing into each block or box. The effects, or dependent variables, are called *output variables* and appear as arrows pointing out of each box. Alongside each arrow is the label for that variable. Each box is a subsystem, and though these boxes are often labeled as if they were just machine components, the components should be conceptualized together with their associated mathematical operations. To emphasize the distinction, Fig. 1 has two sets of labels for each box: the parameters and the physical components (in parentheses). The history of the system is described by a set of variables that are called *state variables*. There will be no state variables (no history) if there are no storage components within the system. The output of the system is completely determined by the state variables and the input variables. The *transfer function* is a mathematical function that describes the relation between the input variables and the output variables. It includes the system parameters. Sometimes the transfer function itself is written in the box.

In a *closed-loop* system, as in Fig. 1, the value of dependent variables can

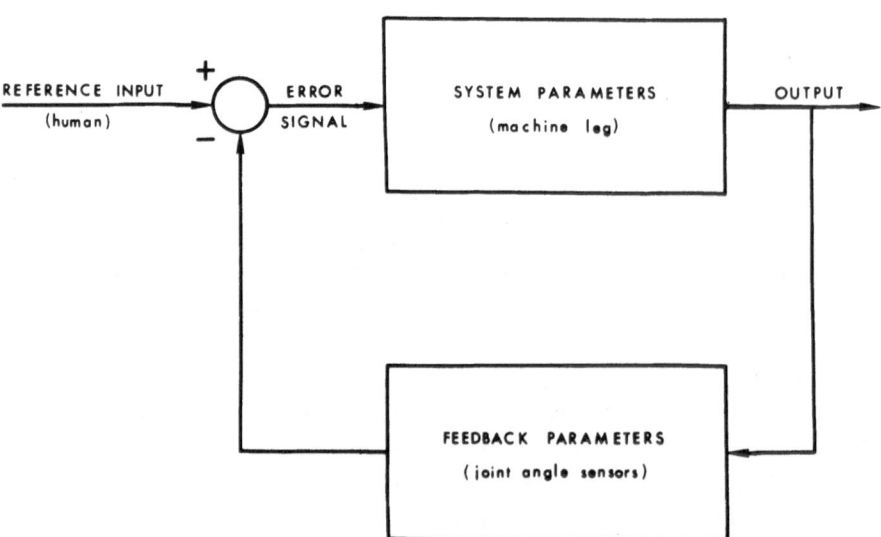

Fig. 1. Block diagram of a servomechanism control system. Arrows represent input and output variables, and boxes enclose system parameters together with actual components (in parentheses) of a locomoting machine. The reference input is "commanded" by the human designer or operator. At the mixer (open circle) the actual feedback output of the system is compared with the standard established by the reference variable, to produce an error signal. Note that feedback parameters are a subset of the total system parameters. Assumptions that are implied by this seemingly simple schematic are given in the text.

influence the value of independent variables or parameters, but in an *open-loop* system there is no such influence. Mathematically, when an output variable affects an input variable, *feedback* exists. The circle in Fig. 1 specifies a *mixer* or "mixing point," a device whose output depicts a particular closed-loop system that is called a *servomechanism,* a term that Hazen invented in 1934. It specifies a feedback control system in which the difference between the reference input and some function of the output is used to supply an *error signal* (physically, it may be a voltage) to the system. The amplified error signal acts to reduce the difference between the output (an actual performance) and the input (the reference) to zero.

We have included this seemingly simple systems diagram, which will look familiar to most readers, for three reasons. First, the concept of state is precise. Arbib (1972) has pointed out that knowledge of the state, as a history or "internal residue of the past" (p. 59) tells us all there is to know about the system in predicting future performance. Since the designer of the system has constructed this history, the performance of the machine will be completely known. A second important concept in systems theory, that of error, has been included to emphasize that it is also precise. It should not be equated with error in any other sense, and especially not as some sort of mistake that the machine makes.

A third reason to study the block diagram of Fig. 1 is that it clarifies the concept of command in systems theory. A reference variable is one type of command. More generally, a *command* is a variable that is established by some means external to the system under consideration. This command is frequently labeled as a "desired value," but the term is unfortunate. Systems theory cannot explain "desire" in any other sense than that the human designer of a system establishes the command value. Having made these important distinctions, we can now turn to descriptions of locomoting machines that have been made to walk and to trot.

The Crawling Gait

In the 1960s McGhee and his collaborators began to study gaits mathematically and to build locomoting machines. In their formulations (McGhee, 1968; McGhee and Jain, 1972; reviewed and summarized by McGhee, 1976), a *leg* is a sequential machine with two possible outputs, up or down, represented by 1 and 0. A *gait* is a periodic sequence of binary k-tuples that represents the successive positions of the k legs. A particular gait can be represented as an *event sequence*. In the case of quadrupeds, an ordering of the integers 1 through 8 suffices for an event sequence because there are 8 (2K) lifting and placing events for the four legs. The number of different arrangements of footfalls can be determined mathematically. There are a large number of theoretical gaits, but observation of animal locomotion (Hildebrand, 1965, 1966; Muybridge, 1887/1957) permits simplifying assumptions. If velocity is not changing, then usually each foot is on the surface about the same length of time as any other foot, such that their stance durations can be considered equal. It is also approximately true that the duty factor, β—the proportion of time that the stance consumes in a complete step cycle (e.g., touchdown to touchdown for a given foot)—is the same for all legs. A gait in which all duty factors are equal is termed *regular*. These assumptions exclude many theoretical gaits and some real footfall patterns of animals. They leave out patterns in which one leg is picked up and set down within the time that another one is also on the ground. Such a gait would have an extra hop by one foot, and while evident in pathological animal locomotion (Wetzel, Atwater, Wait, and Stuart, 1976), need not be included in a simple machine's performance.

The only other important temporal characteristic of gaits is the relative phase, ϕ, between the movements (placings) of any two legs. It is reasonable for slow locomotion to assume that the fraction of a locomotor cycle by which the placing of a forelimb follows the placing of a hind limb is the same for right and left sides of the body. So long as leg mass can be ignored in relation to body mass, then McGhee and Frank (1968) were able to prove a theorem that showed there is one unique gait that maximized a *longitudinal stability margin*. In this gait, there is stability for all successive placing and lifting events, as determined by the position of the vertical projection of the center of gravity on the support pattern that is formed by the feet on the surface. If in this gait the phase, ϕ, is taken between the LH (left hind) and LF (left fore) limbs, it equals the duty factor, β, and $\beta \geq 0.75$. This one unique gait is termed the *quadruped singular crawl* and is depicted in Fig. 2A.

Details of the quadruped singular crawl are available elsewhere (Bessonov and Umnov, 1973; Sun, 1974). If the cycle starts with the LH limb, the footfall order is LH, LF, RH, (right hind), and RF (right fore). This gait is a member of a more general class of *wave gaits* for animals with more than four limbs. In wave gaits, the footfalls on each side of the body are shifted from the rear to the front, and the static stability margin is maximal.

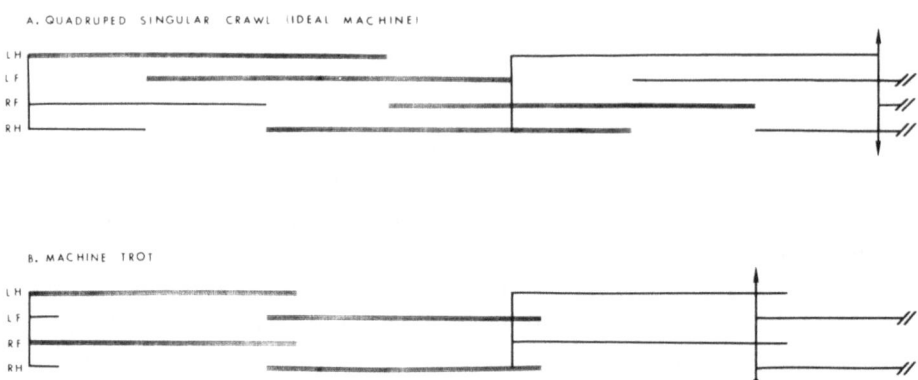

Fig. 2. Gait diagrams of locomotion by a robot quadruped (the "Phoney Pony" reviewed by McGhee, 1976). Left hind, left fore, right fore, and right hind limbs are indicated by LH, LF, RF, and RH, respectively. A complete set of movements in time by the four limbs as the "Pony" moves from left to right, appears as bold horizontal lines and empty spaces. The sequence ends at the long vertical bar and arrow heads. Elapsed time during the *stance* (down period) of a single limb is a horizontal line, and the *swing* (up period) is depicted as empty space, to comprise *one step cycle*. A stride is a four-limb cycle marked by elapsed time between two successive touchdowns of one reference limb (in this figure the LH) and is indicated by a short vertical bar. (A) The unique crawl that maximizes stability. Footfall order is LH, LF, RH, RF. The gait is ideal in the sense that $\beta = \frac{3}{4}$. β is the *duty factor* or stance time/step cycle time. (B) The "Pony" trotting. The gait is very slow, with no period of zero support as would be seen in a rapid animal trot (see Fig. 6).

The quadruped crawl was approximated by a 100 lb machine called the Phoney Pony (McGhee, 1976). Each leg had a knee and a hip powered by electric motors, each joint being driven through a worm gear. The one possible motion at each joint indicated 1 degree of freedom for the hip and 1 degree of freedom for the knee. Three events (forward rotation, backward rotation, and locked) could be achieved by electronic switching, with simple on-off signals sent to the motors. The approach was called *finite state,* in which locomotion was regarded as a discrete sequence of events, rather than consisting of continuously changing motion.

In the Phoney Pony's crawl, discrete feedback signals accomplished the locomotion. A single limb's step cycle can be visualized by imagining that we start to observe its action near the end of the swing (up) phase, when the knee is bent (locked at $-70°$), and the hip is moving forward (flexing). A forward limit sensor detects when the hip is flexed at an angle of $+30°$. Then the hip locks, the knee straightens fully and locks, and the "stiff" leg begins to move backward in the air until a hip angle of $+12°$ is reached. At that point, the foot should be touching the surface so that the power stroke of the stance begins for that leg. The hip continues to move backward during the stance until its angle is $-12°$, and the power stroke ends. The swing begins again, with the knee flexing until the limit of $-70°$ is sensed, and so the cyclic motion proceeds (see diagram in Fig. 10, McGhee, 1976).

It may not be obvious, but it is important to recognize throughout the reading of this chapter that the step cycle of a single limb could not be independent of cycles of the other limbs. Each leg was a "subautomaton" whose control was realized by electronic circuits that were acted on by *intralimb* signals. In addition, overall *interlimb* coordination was under feedback control by way of coupling signals acting between the limbs. The status of each leg (up or down and with joints at particular angles) was conditional, at some key portions of the step cycle, depending on what other legs were doing. McGhee (1976) described the actual solutions for movement in detail, but they can be summarized briefly. The important consideration for stability was to have at least three legs on the ground at all times, in spite of slight differences in power stroke times, or slight slippage on the surface. The self-timing crawl controller achieved coordination through coupling signals between ipsilateral, contralateral, and diagonal legs. The problem of realizing the crawling machine was more complicated than the trotting machine, even though the crawl was dynamically stable in all phases. Nevertheless, only a very small control computer was required. The motion was only in a straight line over a flat floor. An important consideration for animal locomotion is that the Pony did not require continuous ongoing "knowledge" of the leg status. A few "decision points" (i.e., different hip angles in degrees), sufficed.

THE TROTTING GAIT

The trot was implemented for the Phoney Pony before the unique crawl was discovered (McGhee, 1976, traces the history). As might be expected, because it was developed empirically, it bore only slight resemblance to the animal trot. First, it was slow. Second, the "animal" had to have wide tubular feet for stability; and third, it alternately had two or four legs on the surface, as shown in Fig. 2B. An animal trot is a fast gait, with periods of zero support during which no feet are on the surface (reviewed by Wetzel and Stuart, 1976). To begin each step cycle, a *sequence generator* (a central oscillator in the control computer) produced a signal to initiate the power stroke. At the end of the stance phase, the swing began, as in the crawl, when a backward limit sensor detected a hip angle of −12°. In this trotting gait, as in the animal trot, the legs operated in two diagonal pairs. The oscillator produced two start signals, one for each pair, separated by one-half of a complete cycle by the four limbs. There was also an ipsilateral limb-coupling signal to make sure that the same-side leg of the machine was in contact with the ground before the given leg lifted off.

Locomotor machines do not produce much variety. The behavior of an animal that can walk, trot, and gallop is far more challenging to study because of the many additional controlling variables and the different components of its body. In this section, however, we have summarized the minimal requirements that any land-living quadruped must meet. We have also summarized some kinematic terms for the walk and trot that are commonly used for animal locomotion: *intralimb* and *interlimb coordination, stride length, step cycle time, duty factor, stance, swing,* and *gait*. In addition, Fig. 2 shows a preferred schematic for displaying most of these characteristics, the *gait diagram* of Hildebrand (1965, 1966, 1977). The machine or animal is thought of as moving from left to right, with the hind limbs on the outside and the forelimbs on the inside. The diagram is similar to the appearance of actual animal tracks on a surface.

LOCOMOTION BY ANIMALS WITH RESTRICTED NERVOUS SYSTEMS

This section describes stepping (rhythmic limb movements) in spinalized or decerebrated vertebrates and, where present, true locomotion with weight support. These "reduced" preparations should not be thought of as merely the lower portion of an animal whose higher neural regions are eliminated. The reduced preparations are simplified in the sense that they do not respond to the full range of environmental events that control behavior in normal, intact animals. They lack many components that are necessary for some forms of stimulus control, and the remaining components function differently than when the animals were whole.

A living animal contains different structures than does a machine, and its behaviors also differ from those of an automaton, even when the animal has been surgically subdivided. The preparation should be contrasted with a machine system whose designer, as we saw in the section on mechanical locomotion, provided for all of the internal components and the environmental control as well (recall the source of "desired values" in Fig. 1), including the selection of what would be fed back from sensors. A spinal cat has even been observed to show behaviors that appear to obey laws of Pavlovian conditioning (see Patterson, 1976, for review), although it scarcely exhibits advanced learning. Any such effects are beyond the scope of the present chapter. In discussing the reduced preparation here, we assume that all its behaviors are *elicited*. An elicited behavior is drawn from the organism by identifiable preceding *stimuli*.

The reduced preparations are modified internally, and the modifications depend on the site of transection. Sometimes transection is followed by widespread changes for which eliciting stimuli cannot be identified easily. Immediately after decerebration, a cat's extensor muscles become rigid, but not after spinalization. A spinalized dogfish, unlike a spinalized cat, will move continuously until it dies (reviewed by Grillner, 1973b). These behavioral changes must be accounted for, and presumably the necessary and sufficient events will eventually be discovered. Even when eliciting stimuli can be observed, a given environmental event may be followed by a behavior in the decerebrate cat that is different from that of the spinal cat, as we shall see.

The second major section of the chapter also includes experiments on anesthetized, but otherwise intact, mammals. Like the reduced preparation, the anesthetized organism is modified so that stimuli no longer control the same behaviors. Where appropriate, invertebrate data are included for discussion. Occasionally headless animals have been used, especially insects (Pearson and Iles, 1973). In other cases, circuitry of more complete preparations has been examined by recording from single neurons or groups of cells. Unfortunately, it is difficult to separate learned from unlearned control in animals that often are not anesthetized, but which have undergone extensive pinning, excision of portions of the nervous system, and elaborate preparations for recording neural activity.

Two theories traditionally have accounted for behaviors of brain-deprived animals. The first, reflex control, concerns eliciting stimuli. The second theory has prevailed when no apparent eliciting stimuli exist, when the behaviors seem too elaborate to be "simple" reflexes, or when numerous neural pathways are coactive during the behavior. The behavior has been called a "pattern" and an internal "generator" has been sought with the aid of systems and information theory. We will try to show the major characteristics and some restrictions of each theory, and then interpret what is now a large body of facts about stepping by restricted preparations. We will find a rich set of behaviors that often appear surprisingly like those of normal, intact animals. What these locomotor behaviors actually demonstrate is the richness of the phylogenetic endowment. In the last part of this section, we will consider what is inherited.

THE REFLEX CONCEPT AND COORDINATION OF MOVEMENTS

To know what a reflex is, we must study the history of the term. Descartes did not discover the reflex in the seventeenth century, but he is usually credited with the concept of *reflection* of an effect from an external influence or perturbation through the nervous system and back to the environment (Fearing, 1930). The most extensive early twentieth-century treatment was by Sherrington, who devoted his career to reflex control in spinal and decerebrate preparations, usually the domestic cat or the dog. The concept embraced at least an effector organ, a conducting nervous path (presumably interneuronal), and an initiating organ or receptor, where the reaction starts (Sherrington, 1906). What Sherrington failed to emphasize, and what later workers have sometime neglected, is that in his experiments he actually observed behaviors of organisms under particular contingencies.

He and his early collaborators did not observe interneuron activity, nor specify structures, but they noted reflex responses that were *elicited* by a specific preceding stimulus.

In characterizing reflexes for his reduced and usually quiescent animals, Sherrington provided data that have been continuously useful to the neuroscience community. Light and sound are excluded as effective stimuli for these preparations, but a wide array of other kinds of physical energy strike them. Depending on the level of transection of the neuraxis, vestibular inputs may be present or not below the cut. The cerebellum may have been spared. Sherrington also pointed out that "a fraction of the environment is more or less surrounded by the organism itself" (1906, p. 317). In reduced preparations stimuli activate not only skin receptors, but also muscle spindles, tendon organs, and joint receptors.

Reflex behavior is not necessarily of fixed configuration. In the 1906 book, Sherrington described numerous examples of changeable responses, even when many of the same muscles were involved. Upon repetitive stimulation, the flexion reflex elicited by skin stimulation was seen to decrease in strength. Under the same faradic stimulation, the scratch reflex habituated even more rapidly than the flexion reflex. The scratch reflex was and is of considerable interest because of its resemblance to rhythmic movements of stepping. The flexion reflex of a limb could be elicited both by stimulation of the skin—especially of the digits—and by electrical stimulation of the afferent nerve. The scratch reflex, in contrast, required skin stimulation. Rubbing or tickling a saddle-shaped region of skin behind the shoulder elicited hind-limb scratching. Unlike the scratch response, the flexion one could be elicited by single stimuli, and it followed the course of the stimulus closely. Because every reflex was shown to be controlled by more than one applied stimulus and, conversely, any applied stimulus had effects on several different reflex responses, understanding the integration of behavior required that the simple reflex be treated as a "convenient fiction" (Sherrington, 1906, p. 7).

In all the foregoing characterizations of the reflex, the definitions are seen to be precise and useful. Only slight additional comment is required to distinguish reflexes from other kinds of elicited behaviors. A wide variety of behaviors can be elicited by electrical stimulation of central structures in the spinal cord, brain stem, or even the cerebral cortex if it is present. Since a peripheral stimulus is not necessary, we prefer not to call the other elicited behaviors "reflex," and in that way to preserve Sherrington's original usage of the term.

The reflex concept that Sherrington used for reduced preparations was unambiguous. However, sometimes these preparations exhibit complicated sequences of behaviors, including stepping, that seem poorly correlated with stimuli. Sherrington's suggestion that the reflex is "the unit reaction in nervous integration" (1906, p. 7) was questioned more and more frequently, as we shall see in the following section.

THE PATTERN GENERATOR CONCEPT AND COORDINATION OF MOVEMENTS

The concept of reflex emphasized peripheral or external control of behavior. There has been an equally long history in biology of observations that an insect, a spinal frog, or a spinal cat may perform elaborate longlasting behavior in the apparent absence of continuous peripheral stimulation (e.g., Brown-Séquard, 1850; see Weiss, 1941, for early history). Graham Brown (1911, 1914) treated mammalian locomotion, finding that a steady nonrhythmic activation of the spinal cord or brain stem in reduced preparations was followed by rhythmic scratching by one leg, or sometimes by bilateral stepping. Movements often outlasted the stimulus, and an eliciting stimulus could not always be observed. Graham Brown posited paired spinal half-centers to generate the rhythmic movements by a limb, and he rejected an assumption that continuous control by external, reflex stimuli was necessary.

Counterpart reports appeared about complicated motor events for invertebrates. Classical work was performed by Wilson in the 1960s for the locust and was summarized in 1972 just before his untimely death. He isolated from the body the thoracic nerve cord that supplies the flight muscles. Incoming nerves were stimulated electrically with randomly spaced pulses. Activity recorded from the cut motor nerves showed that the rhythmic "output pattern had the basic features of the normal one" (1972, p. 363). Wilson interpreted the data as supporting the hypothesis of a "centrally endogenous pattern generator." This generator was somewhat like one that Jankowska and her collaborators (1967a,b) proposed for the single limb of mammals, in a direct line of descent from Graham Brown's paired half-centers. Alternating flexion and extension in the model of Jankowska *et al.* were provided for by activity in a set of interneurons with inhibitory interconnections. The input to the set did not need to be rhythmic. *This feature of a rhythmic output for a nonrhythmic input seems to be the essential one that has been used to define a "pattern generator" in locomotion.*

Cyclic oscillations of membrane potentials in individual central cells can occur during rhythmic behavior, as in the case of the nonspiking but rhythmic cell in the cockroach, termed *Interneuron I*. Fourtner (1976) gave a historical account and described the locomotor implications of this interesting cell. These cells do not act alone, however. The evidence for cockroaches favors a model in which the flexor motoneurons are excited cyclically and extensor motoneurons are inhibited at the same time (Fourtner, 1976). The models of Wilson (1972), Jankowska, Jukes, Lund, and Lundberg (1967a,b), Jankowska, and others (see Grillner, 1975; Shik and Orlovsky, 1976 for more complicated ring configurations) assume that substantial numbers of cells are required for pattern generation. In no species, to date, has a truly plausible or comprehensive configuration been drawn for the necessary and sufficient cells that are active during a rhythmic episode (see the section on "Components and Circuitry for Rhythm Generation").

A major difficulty is that even for invertebrates it has not been possible to point to cells that unequivocally activate the supposedly relevant interneurons during a rhythmic behavioral act. No one has seriously suggested that a pattern generator would turn itself on, but can we infer an ultimate source of "command" when, as is common, receptors are bypassed and central cells are excited electrically? Excitation of some of these cells elicits activity longer than the stimulus (trigger cells of Davis, 1976; trigger command cells of Stein, 1978), but at other times the response is strictly coincident with continual activation of the "command" cells. Davis (1976) stated that it has become increasingly difficult, however, to distinguish cells with one or the other of these functions. A given cell may do both, depending on conditions—an effect he called "command flexibility." Presumably, cells produce different effects in cells to which they project because they themselves are driven in different ways by inputs from yet other cells. How, then, can we identify the source of relevant nerve cell activity?

The problem of ultimate control has frequently been cast in the language of systems theory and information processing. For example, Davis has said that, "Information theory holds that command flexibility serves the interest of efficiency, and that the property of command should be conferred according to the locus of the 'most important information'" (Davis, 1976, p. 806). A major restriction of the theory is that it does not tell us where to look for the agent that turns on the system. We do not know what is "important" in the clearcut way we knew what turned on a robot. The robot was ultimately activated by a human being.

In place of a systems account, which may be highly useful for describing a system whose elements and states are already specified, we can substitute another explanation. "Important" commands in a surgically subdivided creature have to do with functional relationships between stimuli and responses that were successful for previous individual animals and they were selected during the evolution of that species. In this phylogenetic sense, "command" implies more than a particular configuration of nerve impulses, even in a well-defined set of cells. The essential feature of biological control is not just a set of active neural pathways, nor a given response, no matter how long the behavior lasts or how many muscles are involved. Rather, the "control mechanism" consists of a specific set of conditions (how the animal was subdivided, or what drug it was given) and eliciting stimuli, together with a behavior. Once this history is known, and only then, is it useful to seek the nerve cells that are active at the time. *There is no pattern or generator independent of the prevailing conditions, the eliciting stimuli, and the behavior.*

With the completion of these preliminary reviews of reflex and pattern generator theories, we can now proceed to the facts of locomotion. The interpretations will assume that the ultimate source of control is external to the nervous system; nerve cells do not activate themselves. A living animal must conform to the same external physical exigencies as does a locomotion machine. As it moves about, it must be stable. There must be continuously maintained muscle tension to counter gravity, and also rapid conversion to particular movements in a given direction in the presence of controlling stimuli. We will see that balance is imperfect in reduced preparations, but rhythmic neural and muscular activities occur, sometimes with concomitant stepping upon a surface or in the air. These responses are in all cases dependent on the nature of the transection and the experimental manipulations of drugs or central electrical stimulation, as well as specifiable eliciting stimuli. When we have accounted for all three classes of variables, the functional analysis of stepping by reduced preparations will be complete. As yet, we are still compiling rather simple but necessary descriptive data, often with just one or two variables controlled in any individual experiment.

REDUCED ARTHROPODS, FISH, AND AMPHIBIANS

Insects (cockroaches, Pearson and Iles, 1973; mantids, Roeder, 1937) in which the brain (supraesophageal ganglion) has been removed will walk normally. If the whole head is removed, at least some of the legs will still move for a few cycles (Pearson and Iles, 1973; ten Cate, 1941). Little study has been made of how these creatures balance when they are walking, but a smaller set of behaviors should suffice than for quadrupeds, because a hexapod wave gait may be stable during all of its successive liftings and placings. A wave of placing events moves from the rear to the front along each side of the body, with a constant time interval between the actions of adjacent ipsilateral legs. McGhee (1976) summarized this material and pointed out that both six-legged insects and eight-legged spiders move in wave gaits whose duty factor always is one-half or greater.

In spite of its stable body configuration, an insect is far from a simple locomoting machine. The total set of effective eliciting stimuli has not been listed, even in headless insects. Some stimuli are doubtless attendant on the injury. We would expect abnormal activation of central structures, if only because they are now exposed.

Given that we must assume internal changes in the reduced preparations, a useful experiment to survey the effective stimuli that remain is to eliminate successively, by surgical deafferentation, sources of external activation. When Pearson and Iles (1973) deafferented the limbs, the headless cockroach exhibited a different rhythmic bursting pattern in motor axons than the "normal" pattern. This comparison was made for restrained preparations that were not actually walking, and if the latter data were also available, we could evaluate the stepping by reduced insects with considerable precision. It is encouraging to note that the technical capability to record activity even from central neurons in walking cockroaches now exists (Delcomyn, 1976).

Problems of balance are perhaps greater for flying insects than for walking ones. We have previously noted (in the section "The Pattern Generator Concept and Coordination of Movements) Wilson's (1972) influential work on the flight system of the locust. Equilibrium and flying are lost by deafferentation. "When all the feedbacks are eliminated, the animal cannot really fly, because its wingbeat frequency is too low and it cannot regulate its lift" (Wilson 1972, p. 362). Even the rudimentary cyclic behavior of the preparation seems to have been elicited by a steady stream of wind blowing on its head. In more recent work, Burrows (1976) pointed to numerous sensory receptors that are important in locust flight.

In cockroaches, locusts and scorpions (Root and Bowerman, 1978), the phylogenetic locomotor inheritance is rich, with afferentation influencing central neurons. Extensive documentation of reflex mechanisms has been made for another arthropod, the lobster (Ayers and Davis, 1977a,b; 1978; see also Barnes, 1977). In physically restricted lobsters, a variety of reflexes was described: resistance reflexes that opposed passive movement, "positive feedback" reflexes, distributed reflexes that activated muscles at remote joints, and intersegmental reflexes from one limb to another (Ayers and Davis, 1977b). "Positive feedback" is a term borrowed from systems theory, and its use here was descriptive. These reflexes augmented movement of the stimulated joint. In further work, Ayers and Davis (1978) showed that reflexes elicited by passive joint movements, which seemed by comparison with movements in intact, walking lobsters to be suited for locomotion, also responded in the appropriate "normal" velocity range of joint movement. For example, passive elevation of the coxobasal joint of a limb at high angular velocities elicited retractor discharge. If combined with the "positive feedback" reflex resulting from the subsequent retraction, the retractor discharge that occured during backward walking could be accounted for. The work as a whole supported the importance of reflexes to lobster movement, even to the control of timing. Although the evidence for restricted lobsters was admittedly circumstantial, reflexes to the retractor motor units could specify not only a normal intensity but also the correct phase of discharge relative to the coxobasal joint muscles. Recall that the major requirement of a pattern generator, as a definable set of nerve cells, is that a nonrhythmic input yield a rhythmic output to motoneurons. In arthropods, it is difficult to disentangle afferent from central neurons with respect to rhythm generation.

In at least some fish, a rhythm generator composed of central neurons alone is plausible. Equilibrium control is complex in intact fish, but a spinalized one can be studied readily if it is supported so that the posterior part can swim freely in the water. We have already mentioned that the spinal dogfish will swim continuously (Grillner, 1973b). Intact dogfishes also swim continuously (Alexander, 1977), perhaps for different reasons. Lissman (1946) and later Grillner and Kashin (1976 review) found that stimulation of the cut face of the dogfish's spinal cord could increase the frequency of preexisting movements or elicit swimming movements, should the fish be quiescent. The rhythm (at least the ventral-root activity) appeared after extensive deafferentation, and apparently excitation of any segment of the spinal cord was a sufficient condition. The fish was modified by the transection, such that excitability of cells that produced rhythmic activity of motoneurons was increased. There are species differences in this internal change, because most other species of fish, when spinalized, require a pinch of the tail fin for undulatory waves of the body. In fish, then, there are cross-species distinctions in how susceptible the rhythm generator cells will be to particular sources of activation. Although we do not know why the spinalized dogfish swims continuously, Grillner and Kashin (1976) emphasized that normally the swimming would depend on descending pathways from the brain stem. How these pathways would themselves be activated is another question, and the topic of a later portion of this section.

The built-in susceptibility of rhythm generator cells to incoming neural impulses is apparently less in reptiles than in fish, at least in the spinal turtle (Lennard and Stein, 1977; Stein, 1976). Stimulation of the dorsolateral funiculus (Lennard and Stein, 1977) in turtles with midbody cord section elicited several movements, including lateral movements of the tail and motions suggestive of swimming, backpaddling, withdrawal, tonic extension, and scratching. Swimming and backpaddling did not occur without electrical stimulation. Stimulation restricted to the dorsolateral funiculus was the only one that produced regular swimming movements, and they were restricted to one, usually the contralateral, hind limb. The behaviors continued only so long as electrical stimulation continued. The investigators found a specific internal modification of locomotor pathways in this preparation: a given stimulation frequency and intensity elicited a higher cycling frequency in spinal than intact animals.

Székely and Czéh (1976) found that even with higher, medullary sections, the toad "is extremely inert, moves only in response to strong stimuli, and is incapable of adjusting the movement to the changing external conditions" (p. 767). Spinal reptiles and amphibians seem more like spinal mammals than like fish.

Spinal Mammals

Although a variety of movements are evident in spinal mammals, in no case is there effective stability. The tragedy of paraplegia in man has not been circumvented so far. It is true that if the hindquarters of a spinal kitten become unbalanced, with weight carried mainly by one of the legs, the position of each leg will shift until the posture is more upright (Grillner, 1973a). During forward progression maintained by a moving treadmill belt, the same kitten develops effective and perhaps near-normal muscle forces in the hind limbs, but balance must be provided from the outside. Usually the experimenter supports the base of the tail. Without the brain stem, the spinal behaviors must be viewed as stepping movements only, and not true locomotion.

Do we see any rhythmic behaviors in mammals without peripheral stimulation, as was true for the dogfish? Graham Brown (1913) early described rhythmic limb movements of the cat during "narcosis progression." Similar effects can be observed today with modern anesthetics. Drugs alter animals so that stimuli elicit different behaviors. We must assume that the excitability of a great many neurons is changed from normal. Graham Brown (1914) wrote a speculative account of a "primitive centre" that was discharged rhythmically by asphyxiation and later was replaced in evolution by a "blood-stimulus." The basis for his thinking was probably in large part the observations on narcosis progression.

Graham Brown's (1915) preoccupation with the issue of "spontaneity" also led him to examine behaviors of cat fetuses that were transferred to a warmed saline solution. Although he stated that the resulting stepping movements by the limbs were apparently spontaneous, he seems to have been more attracted to possible asphyxia from pressure on the umbilical cord as an eliciting "blood-stimulus," acting similarly to chemical stimuli that elicit respiration.

Sherrington (1910) gave a remarkably complete account of conditions that produced stepping movements by the limb of a decapitate dog. The skin could be rubbed, squeezed, or pinched with an attached clip, even at remote locations. Effective sites were the perineum, another foot, the neck, back or tail. Intrinsic proprioceptive stimuli were assumed when stepping followed lifting the animal with its hind limbs pendent. Hip extension was a particularly effective stimulus for stepping.

Sherrington maintained a distinction between necessary and sufficient stimuli, and he was fully aware that a movement of given form could be under the control of several variables. He pointed out that although a light touch of the planta elicited a toe movement, the same movement was seen in the suspended leg during stepping when there was no external stimulation. Experiments in which the skin was denervated showed that cutaneous stimuli were not essential for stepping.

Sherrington also documented different forms of flexion or extension, depending again on which controlling stimuli were present. The flexion reflex that was elicited by aversive stimulation differed from the flexion period of the step cycle

(Sherrington, 1910). Under aversive stimulation, the amplitude was greater and passed to extension only with the termination of stimulation; while in the step cycle the flexion was followed by extension even during ongoing stimulation. There were said to be even greater differences between scratching and stepping responses. The character of the flexion phase of the step itself also varied, depending on the source of stimulation. When hind-limb stepping was elicited from a forelimb location or from the neck, there was greater hip flexion than if stepping was elicited by perineal stimulation.

From the preceding paragraphs we see that stepping can sometimes be a reflex behavior, even if it lasts longer than the eliciting stimulus, whereas at other times no eliciting stimuli can be observed. Sherrington and other workers also elicited stepping by exciting central neurons. An effective stimulus was electrical pulses in a train applied to the lateral columns of the spinal cord (Roaf and Sherrington, 1910) or even mechanical stimulation of the cut face of the cord (e.g., Graham Brown, 1911). When stimulus intensity was increased (1) the force and frequency for an individual limb increased, and (2) weak stimulation of the cut end of the cord elicited stepping in one limb, but stronger stimulation elicited bilateral stepping (Sherrington, 1910). These dependencies of stepping on the intensity of stimulation of descending paths were much like those seen in fish, amphibians, and reptiles.

Rhythmic activity has been well documented recently in spinal preparations with modern techniques and drugs. Schomburg, Roesler, and Meinck (1977) studied the high spinal cat. When paralyzed with pancuronium, under intravenous injection of L-dopa (a noradrenergic precursor) and Nialamide (a monoamine oxidase inhibitor that potentiates the effect of dopa), there were rhythmic hind-limb motoneuron discharges. The interaction of reflexes and ongoing rhythmic activity was studied. Reflex effects due to stimulation of forelimb nerves were observed in the hind-limb extensor motoneurons only during the extension phase of the cycle and in flexor motoneurons only during the flexion phase.

Actual stepping has been seen in high spinal animals that were placed on a motor-driven treadmill. It is questionable that it was true locomotion. Halbertsma, Miller, and van der Meché (1976; see also Miller and van der Meché, 1976), like Schomburg et al. (1977), found it necessary to administer L-dopa and Nialamide. Afferent input was necessary in the form of stimulation from the moving treadmill belt. Two patterns of quite normal looking contralateral limb movements, alternating or almost in phase, were seen at different belt speeds. However, the forelimbs were not lifted well during the swing phase, nor did they support the weight of the body. Unfortunately, no footfall patterns have been reported so far for high spinal cats.

Stepping has been studied more extensively in low- than in high-spinal cats (see Grillner, 1975; Wetzel and Stuart, 1976 for reviews). An adult cat that is spinalized at the midthoracic level will step if Clonidine (a stimulator of postsynaptic noradrenergic receptor sites), L-dopa, or another substance is injected systemically (Grillner, 1973a). Hart (1971) reported facilitation of walking in spinal dogs with strychnine. The chemical, as in high-spinal cats, is thought to mimic normal transmitter functions from noradrenergic and perhaps serotonergic axons whose cell bodies lie in the brain stem (reviewed by Grillner, 1975; Shik and Orlovsky, 1976). Much of this valuable work was completed by Grillner and his collaborators (reviewed by Edgerton, Grillner, Sjöstrom, and Zangger, 1976; Grillner, 1976), who pointed out that antidromic activation of normally ascending pathways cannot be excluded (Grillner, 1976).

When drugs are given, neural activity may differ from that when there is no drug present. However, it has not been possible to point to substances that actually elicit specific locomotor behaviors. Rather, the behavioral changes that are due to any drug may be traced to changes in the nerve cells, especially in cases where the same apparent eliciting stimuli are present. It is quite possible that the changes in the nerve cells bring the behavior under the control of some aspect of the apparent eliciting stimulus which was not previously involved in the controlling relationship.

In low-spinal cats, as in higher ones, several conditions must be present in order for stepping to occur (Edgerton *et al.*, 1976; Grillner, 1973a). If the cats are fully grown adults, both chemical activation and external stimulation, as from a treadmill belt, are essential in acute experiments. If a spinal kitten is used, the chemical is not required when the animal survives for weeks or months, but external treadmill movement is still essential. Sometimes a quiescent animal will move if the base of the tail is pinched. As in the high-spinal cat, both alternating and in-phase stepping have been observed, together with forces and timings in the hind limbs that are similar to those in normal cats. As belt speed increases, the alternating cycles of the two hind limbs are replaced by cycles that are almost in phase.

As with "lower" creatures, the presence or absence of afferentation has been manipulated in mammals. In a curarized spinal cat (Edgerton *et al.*, 1976), stimulation of dorsal roots will elicit rhythmic discharges in motoneurons. Rhythmic EMG activity survives local deafferentation in a spinal cat, and remote excitation via a treadmill belt or dorsal root electrical stimulation need not be phasic; that is, patterned to conform to ongoing step cycles by any limb (Grillner and Zangger, 1974). Finally, flexor and extensor motoneuron EPSPs alternated in curarized spinal cats given dopa and Nialamide (Andersson, Grillner, Lindquist, and Zomlefer, 1978).

The issue of what survives deafferentation is still unresolved. Simultaneous records of EMG (or motoneuron activity) together with footfalls during actual stepping are not available for the deafferented, spinal cat. These missing data are important. Engburg and Lundberg (1969) described for intact cats an extensor burst that occurred prior to touchdown of the hind limb. It could not be attributed to a stretch reflex elicited by foot contact, although the authors stated that another proprioceptive reflex was still possible. A similar pre-touchdown burst occurs in the spinal kitten (Grillner, 1973a). This response would ensure adequate extensor force for the early portion of the stance phase. It has been widely construed as evidence for a "central program," or neural pattern generator, so it is unfortunate that we do not know whether or not the "preresponse" depends on either local or remote peripheral stimulation. Is it in any circumstances a reflex?

We have now reviewed a variety of evidence for spinalized mammals. They exhibit complicated stepping movements by all four limbs, elicited by numerous stimuli. Deafferentation has been one attempt to survey residual behaviors and eliciting stimuli, but it has serious shortcomings as a procedure to sort out the multiple controlling variables for stepping. First, it is technically difficult to deafferent surgically after spinalization, to activate the central structures electrically or chemically, and to collect simultaneous records of rhythmic neural or muscle activity plus actual stepping on a surface. There is yet a more serious problem. When a peripheral nerve or dorsal root is sectioned, that specific source of excitation is lost, to be sure, but there may be shifts in excitability in all the cells to which the incoming axons previously projected. Alterations in all remaining cells are impossible to measure with present means.

A stronger technique than deafferentation has been to elicit stepping, by whatever means, and then test the specific effects of one known additional stimulus on a host of behaviors. A number of orderly relationships between eliciting stimuli and responses have been found by this method, as well as some preliminary identification of pathways that are excited or inhibited. Much of this work was done in the 1970s by Grillner and his colleagues (summarized by Forssberg, Grillner, Rossignol, and Wallén, 1976). A light tactile or electrical stimulus to the dorsum of the foot of a low-spinal kitten elicited enhanced flexion if the foot was in the air (swing) but extension if the foot was on the surface (stance). These reflexes are useful in permitting the foot to rise over an object that is encountered in midair, or else to increase force and overcome an obstacle encountered while the foot is placed on the surface. The single peripheral event has produced a flexion reflex ipsilaterally under one condition and an extension reflex in the same limb under another condition. There are also interlimb effects in the "reflex reversal" experiment. When the ipsilateral limb flexed, the contralateral one extended (Forssberg *et al.*, 1976), an effect reminiscent of reflexes that were seen early and described so fully by Sherrington (1906) in quiescent animals. All these behaviors were clearly built in during mammalian evolution, in extensive topographical detail.

Reflexes and concomitant neural activity in low-spinal cats and kittens have been studied in more recent work by Rossignol (1977), who used acute preparations. He concluded from tests with and without rhizotomy at different levels that when the limb was in the extended position, stretch of its flexors elicited reflex inhibition of crossed extensor responses and facilitation of crossed flexor responses. When the limb was flexed, however, the inhibition was removed, and crossed extensor responses predominated. These effects were thought to be appropriate for locomotion, when a limb might be painfully stimulated and flex as the animal moved along (see also Rossignol and Gauthier, 1978). The contralateral limb would extend only if it was in a position to support weight; that is, during the stance but not at the end of it, when the flexors would be fully stretched prior to liftoff.

Experiments that are conceptually similar to those conducted with spinal cats have been done with decerebrate cats as well, including tests of reflex stimulation. In all this work, functional neural "circuitry" is being sought. We therefore turn to preparations with a higher transection, before resuming a general discussion of multiple stimulus control in reduced animals.

CONDITIONS FOR DECEREBRATE STEPPING. Under many circumstances of either rest or motion, and for particular muscle groups (for example, those that elevate the head), a relatively continuous or "tonic" activity in interneurons and motoneurons will stabilize joints. It is not surprising that phylogenetic mechanisms exist to regulate neuronal excitability, and some of these mechanisms have been revealed by transections of the nervous system at different levels. In spinal mammals at rest or stepping, reflexes are of low amplitude, and the behaviors cannot always be elicited. Excitability of cells is quite different in animals with higher transections, and the precise level of the cut can influence the extent of modification. These relatively permanent effects are important because, like drug effects, they constrain all the characteristics of locomotion, when it is present.

Sherrington's (1898) classical transection was at the midbrain level, between the superior and inferior colliculi. In his and all the other preparations discussed in this chapter, we assume that the tissue above the cut is removed or destroyed. This midcollicular animal did not locomote or right itself. If it was not stimulated, the legs were stiff. The rigidity primarily involved physiological extensors (antigravity muscles), and would melt in particular muscle groups if reflexes were elicited. Moreover, the excessive extensor activity was lost after deafferentation, implying an abnormal but continual barrage of afferent impulses. In more recent work with the preparation, Terzuolo and Terzian (1953) found that rigidity could be restored after deafferentation, at least temporarily, with electrical stimulation of the cerebellum. Rigidity of the forelimbs increases, moreover, with midspinal section, as has been known for many years as the Schiff-Sherrington reaction (Ruch and Watts, 1934).

Some progress has been made, then, to identify pathways in which excitability changes occur. These longterm modifications still puzzle us, because for the same transection several combinations of lost or retained excitatory or inhibitory influences, from numerous neural paths, can account for a given effect. Sometimes these multiple variables can be separated. Recent data showed an intimate relationship between labyrinth and neck reflexes in the decerebrate cat (Lindsay, Roberts, and Rosenberg, 1976). Magnus (1924) had thought the neck reflexes acted asymmetrically on the limbs, such that limbs on the side toward which the jaw was rotated became extended and the opposite limbs flexed. He thought that the labyrinth reflexes, in contrast, acted symmetrically on all four limbs. Lindsay et al. (1976) distinguished eliciting stimuli for the neck reflexes from those for the labyrinth by means of selective cervical dorsal root sections. They also could fix or move the head and axis vertebra separately. They confirmed that the neck reflex behaviors were asymmetrical. The labyrinthine reflexes precisely opposed the neck-elicited reflexes and, contrary to the findings of Magnus, were also asymmetrical. Since the labyrinth effects continued after the tilted head was brought to rest, they were assumed to arise from the otoliths instead of from the semicircular canals.

So far no counterpart data exist for neck and labyrinthine reflexes that occur while an animal is locomoting, to complement the important research on quiescent preparations. Another structure, the cerebellum, has also been implicated in widespread excitability influences for many years, and we do know something about its activity during locomotion. First, however, we must review the discovery of how locomotion may be elicited in decerebrate mammals.

Pioneering work began in Russia. Shik, Severin, and Orlovsky (1966a) introduced a postmammillary, precollicular preparation that would step on a motor-driven treadmill so long as a region of the midbrain was stimulated electrically. With a premammillary cut, the cat may step without apparent stimulation (Orlovsky, 1969; Shik and Orlovsky, 1976). Recent work has identified a number of descending pathways from the brain stem whose activation can elicit stepping including the rubrospinal, reticulospinal, vestibulospinal, and pyramidal pathways. Various transmitters have been invoked. The details of this work are available elsewhere (see especially the review by Shik and Orlovsky, 1976) for the cat. Similar study has been made of the mesencephalic fish whose midbrain is stimulated electrically (Kashin, Feldman, and Orlovsky, 1974). There are several effective stimuli for stepping by the brain-stem cat. McCrea (1979) described representative experiments when the section was made from approximately 1 mm rostral to the upper border of the superior colliculus to approximately the caudal border of the optic chiasma. An hour or two after termination of anesthesia, locomotion sometimes occurred "spontaneously." It is not known why these animals step, but other behaviors are also strong. During controlled locomotion (elicited by nonrhythmic trains of electrical stimulation) of a mesencephalic cat, the cardiac output increased by up to 95%, the arterial pressure rose by as much as 35%, and the heart rate increased by as much as 20% above values when the preparation was at rest (Sirota, Sirota, and Shik, 1971).

In other cases brief electrical stimulation of dorsal roots (McCrea, 1979) elicited locomotion that persisted for several seconds after the termination of the stimulus. Once this form of stepping was established, the concomitant activity of an L_7 ventral root filament was recorded, the animal was paralyzed with Flaxedil, and ventral root filament activity was recorded thereafter as *fictive* locomotion (Feldman and Orlovsky, 1975). Pre- and postparalysis records were apparently similar (see also Viala and Vidal, 1978 for comparable experiments on rabbits). Sometimes an additional stimulus was necessary after paralysis to initiate ventral root activity, such as to swing a forelimb rhythmically or gently to stimulate the perianal region manually. These actions seem to be much like the eliciting stimuli that Sherrington (1910) reported for the decapitate cat, except that they are more effective when the transection is higher.

Similar eliciting stimuli have been used by others. Duysens (1976) and Duysens and Pearson (1976; see also Pearson and Duysens, 1976) studied acutely operated premammillary cats that moved on a motor-driven treadmill. Like others before them (including Graham Brown early in the century, whose filmed experiments were described by Lundberg and Phillips, 1973), they found that the afferentation from running on the belt was sufficient to elicit velocity-matched stepping as belt speed was increased. The gaits converted from trotting (diagonal limbs descending together) to galloping (hind limbs descending almost together).

Duysens (1976), like McCrea (1979), found that stepping sometimes continued after the mesencephalic stimulation stopped. Duysens (1976) further noted that a premammillary cat walked with much more "vigor" than a postmammillary one. It sometimes walked on the treadmill spontaneously for hours. In this preparation, it was possible to restrain one limb for recording, while the other legs locomoted on the treadmill.

The presence or absence of the cerebellum in a decerebrate animal changes the locomotion that is seen. If the structure is ablated, extensor rigidity perturbs the step cycle of the forelimbs especially, producing an abnormally brief cycle duration (Orlovsky, 1970). Considerable work on the relevant circuitry has shown that the cerebellum, together with its incoming and outgoing pathways, are active rhythmically with stepping (see Shik and Orlovsky, 1976, for review of Russian work). In a decerebrate cat, activity of the ventral spinocerebellar tract (VSCT) retains its rhythm during elicited locomotion, even after hind-limb deafferentation. In the thalamic preparation, activity of Deiters' neurons (projecting to the ipsilateral lumbosacral cord) and EMGs of ipsilateral hind limb extensor muscles increased before and during the stance, when there was local cooling of the cerebellar cortex (Udo, Oda, Tanaka, and Horikawa, 1976). In premammillary cats, when the medial accessory portion of the olivary nucleus was excited by electrical microstimulation, limb muscle EMGs and, by implication, force were augmented for individual muscles over many step cycles, even after stimulation ceased (Boylls, 1977).

REFLEX FUNCTIONS IN DECEREBRATE PREPARATIONS. The preceding paragraphs showed that many known and unknown stimuli contribute to stepping by the decerebrate cat, for a given level of transection and presence or absence of the cerebellum. Like the spinal cat, the decerebrate one exhibits an extensor burst in the EMG prior to touchdown. In the intact animal an otolith-spinal reflex may contribute to this burst, as occurs when a cat falls suddenly (Watt, 1976), but no such assumption is warranted for the decerebrate animal. We do not know the eliciting stimuli, and there is for the present a more successful way to sort out the complexity of control. The same approach was used with spinal animals. The experimenter identifies an effective stimulus, such as cutaneous stimulation, and then measures its effect on ongoing locomotion. Duysens (1976) and Duysens and Pearson (1976) have tested a number of reflexes. One hind limb was fixed in position, while the other legs of acute premammillary cats walked on a treadmill. The most drastic effects on resetting the locomotor rhythm, as observed in the EMG traces, were seen to follow strong stimulation (e.g., to the sural nerve) that would activate small afferent fibers and presumably would be aversive in intact cats. This stimulation was most effective when it was delivered just outside (before or after) an "immune" period, the transition from extension of the limb to flexion that would normally occur at the end of the stance.

Stimuli of lower intensity were particularly effective in delaying or actually preventing the transition from extension to flexion. Both exteroceptors (by way of large cutaneous afferents that supply the skin of the distal hind limb) and proprioceptors (probably triceps surae Golgi tendon organs) were identified as important. The transition from flexion to extension, during the swing, has also been found susceptible to multiple reflex control. Weak stimulation of the tibial or sural nerve near the end of flexion shortened the flexion and initiated premature extension.

Stronger stimulation exaggerated flexion, so that it looked much like a withdrawal reflex (Duysens, 1976).

CONDITIONS FOR DECORTICATE STEPPING. Perret (1976) observed decorticate cats, with sections above the thalamus. Greater numbers of behaviors were seen than in animals with lower cuts, even when there were no observable eliciting stimuli. Sometimes the locomotor movements were accompanied by tail lashing, pupillary dilation, tachycardia, hypertension, urination, and other signs of what has been called *sham rage* (Bard, 1928). Similar locomotor accompaniments were also reported by Creed, Denny-Brown, Eccles, Liddell, and Sherrington (1932). The significance of these findings is that a host of responses, which may be called emotional ones, occur in high decerebrate or decorticate cats under the same circumstances in which locomotion is seen. To variables that influence skeletal muscle excitability, we must now add those that provide for emotional behaviors, with all the necessary components built in during evolution.

As with spinal animals, the possibility of conditioning cannot be completely excluded (see review by Buchwald and Brown, 1973). No study has determined whether or not locomotor responses can be classically conditioned (respondent conditioning, see the section on "Locomotion by Intact Animals") in decerebrate animals, so we must for the present assume that only the one stimulus function, elicitation, is present. To date, no major differences have been found in locomotion by animals with varying amounts of cortical destruction, so long as the thalamus is intact (Perret, 1976). Unfortunately, quantitative measurements of gaits in decorticate animals have not been made. About all we know is that a decorticate cat or monkey can survive for days or weeks and eventually will locomote with weight support and wander about in a room (Hinsey, Ranson, and McNattin, 1930).

Nor do we know much about how reflexes interact with other elicited behaviors while a decorticate mammal is locomoting. One example will suffice to show that many stimuli control many responses, as we should expect from lower transections. Perret and Cabelguen (1976) recently studied the semitendinosus muscle, a two-joint flexor of the knee and extensor of the hip. Electromyographic activity during locomotion differed, depending on whether the limb was (1) intact; (2) fixed so as to counteract flexion (or a clip was attached to the toes), or (3) deafferented ipsilaterally.

MECHANISMS OF LOCOMOTION BY REDUCED PREPARATIONS

In the three preceding sections, we have documented alterations in neural excitability after particular surgical interventions, as well as concurrent changes in the behaviors elicited by given stimuli. Here we are concerned with what is built into organisms that determines the behaviors of reduced preparations. A single species will be emphasized, the domestic cat, since its locomotion has been studied most widely under a variety of conditions. We will search the controlled behaviors across preparations for rhythmic responses that are seen irrespective of the level of the cut, to see if there are any plausible genetic packages or units. The concept of unit is a difficult one and requires some further explanation. Even when we observe that a behavior of similar form occurs under many conditions of stepping, we do not point to a fixed property that exists in any sense apart from those conditions. Nor does that response, which we label for analysis, exist separately from others that precede, follow, or accompany it. We assume that behavior is continuously variable, even when discrete portions are extracted by scientific rules.

CONDITIONAL NATURE OF THE CONTROLLED BEHAVIORS. We have already pointed out that equilibrium and weight support are present in decorticate cats only if they survive for some time after the surgery. Even then the animals crouch (Hinsey *et al.*, 1930). Perret (1976) pointed out weaknesses and peculiarities in the extension phase of his acutely decorticated cats. They failed to correct misplacements of a foot, and full weight support was not observed when axial fixation of the spine was removed. EMG responses also differed with conditions, as we have seen.

With somewhat lower sections, Graham Brown's premammillary cats (Lundberg and Phillips, 1973) did not place their feet in a successful position on the surface of the belt, and they often stumbled. When the cat was suspended in the air and then lowered to the moving belt, the hind limbs and forelimbs did not always start moving at the same time. Galloping was discoordinated upon even casual inspection, more so than the slower gaits. Recent reports on decerebrate cats have stated that they sometimes limp (Halbertsma *et al.*, 1976) or fail to retract completely the limb during the swing phase, so that the foot grazes the belt (Miller, van der Burg, and van der Merché, 1975).

Nevertheless, the decerebrate cat has an extensive phylogenic repertoire. If it is stepping in the air and is then lowered to the belt, its four-limb cycles match the speed within a few strides (Lundberg and Phillips, 1973), the stance duration of each limb decreases as speed increases, and an extensor burst occurs prior to touchdown (Gambaryan, 1974). Even when the cut is so low that electrical stimulation of the midbrain must be given to elicit stepping, many of the common forms of intact feline locomotion may be seen (Miller *et al.*, 1975). These forms include the slow quadruped wave gait (crawl) depicted in Fig. 2A, trotting (with diagonal footfalls incompletely separated, as in Fig. 2B, or with periods of zero support, as in Fig. 6B), and other forms that will be described briefly here and depicted in the next section (Fig. 6). The *pace* is a gait in which ipsilateral limbs (left hind and fore; then the right hind and fore) touch down and lift off at approximately the same time. Decerebrate cats have galloped at high speeds, with a transverse or rotatory footfall pattern. In the *transverse* gallop, the hind limbs and forelimbs descend in the same right-left order, whereas in the *rotatory* gallop they descend in opposite order (if the right hind limb touches down before the left hind limb, then the left forelimb touches down before the right one). In addition to these complicated interlimb behaviors, decerebrate cats show both single and interlimb reflexes that successfully maintain stepping or terminate it when cutaneous or other stimuli are administered.

Figure 3 summarizes the major single- and interlimb behaviors that we have discussed. The behaviors survive decerebration and in this sense are built in. We cannot at present include detailed descriptions of behaviors of individual muscles, alpha and gamma motoneurons, joint angles or other kinematic events, or force traces. There have been far too few of these records taken simultaneously with footfalls. However, we know enough about some conditions, eliciting stimuli, and behaviors to begin to look for accompanying neuronal activity in this preparation. Figure 3 depicts functional relationships. The rhythmic behaviors are controlled by eliciting stimuli in the presence of other variables termed "conditions."

High-spinal animals, we have already noted, have less successful interlimb coordination than decerebrate ones. We cannot diagram functional relations from the presently available reports. In the chronic low-spinal kitten, however, we see some features that are common to all reduced cat preparations. Functional relations are shown in Fig. 4 between conditions, eliciting stimuli, and behaviors. For the single limb, (1) the pretouchdown extensor burst occurs and (2) stance duration decreases as speed increases. For the two hind limbs, we see (3) conversion of alter-

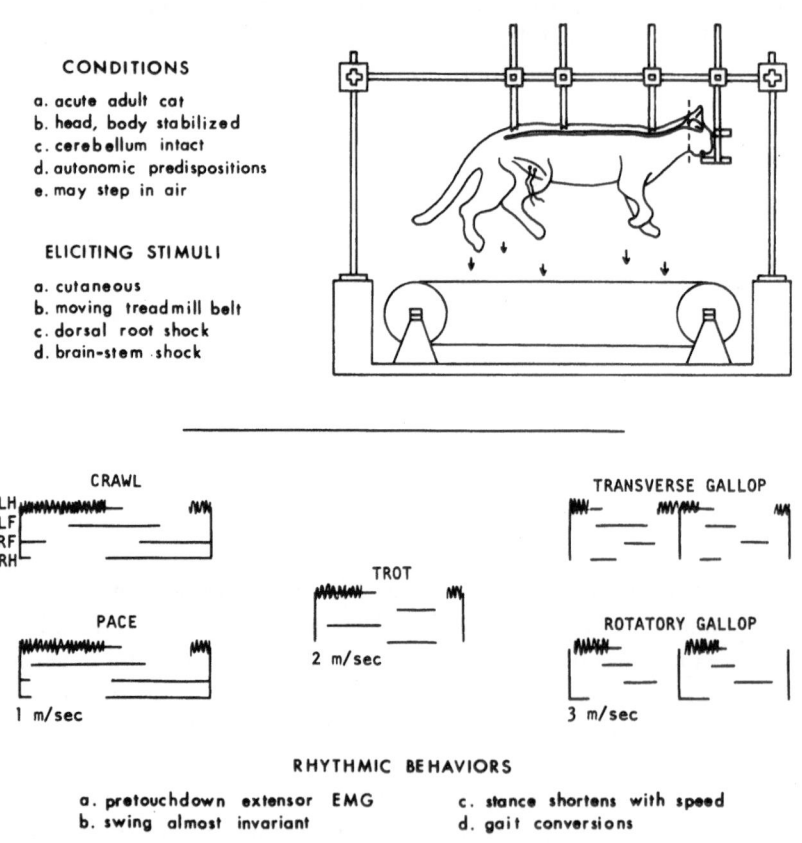

CONDITIONS

a. acute adult cat
b. head, body stabilized
c. cerebellum intact
d. autonomic predispositions
e. may step in air

ELICITING STIMULI

a. cutaneous
b. moving treadmill belt
c. dorsal root shock
d. brain-stem shock

CRAWL

LH
LF
RF
RH

PACE

1 m/sec

TROT

2 m/sec

TRANSVERSE GALLOP

ROTATORY GALLOP

3 m/sec

RHYTHMIC BEHAVIORS

a. pretouchdown extensor EMG
b. swing almost invariant
c. stance shortens with speed
d. gait conversions

Fig. 3. A functional analysis of stepping by an acutely prepared high-decerebrate, premammillary cat. Known conditions under which the stepping is seen, eliciting stimuli, and rhythmic behaviors are specified. The cat is suspended by clamps above a moving treadmill belt. Its limbs are moving, and if lowered to the belt it will step at the velocity imposed by the motor. The lower portion of the figure depicts gait diagrams (same form and notation as in Fig. 2) that have been seen in this preparation (Gambaryan, 1974; Miller *et al.*, 1975). At the top of each diagram, the LH record also shows a representative EMG pattern for a knee extensor (see electrode wires in the cat's vastus lateralis muscle). Note that EMG onset precedes foot touchdown. Numerous stepping behaviors have survived decerebration, including both single- and interlimb sequences.

nating to in-phase movements, and (4) bilateral reflex "reversals" much like those of decerebrate cats (not shown in Fig. 4). Finally, even in spinal animals (5) stepping movements sometimes occur without observable external eliciting stimuli.

There may be other built-in behaviors in spinal animals, and even for those we know about, we have few data. In the future, there will be no substitute for an actual inspection of many instances of stepping by reduced preparations. There is little assurance, moreover, that when we study the remaining components of a surgically subdivided animal, we are revealing functions of the formerly complete animal. A final experiment may be cited to show that extreme care must be taken to distinguish between behaviors of decerebrate and intact cats when we seek the necessary and sufficient circuitry. Jordan and Steeves (1976) treated cats chronically with 6-OHDA (6-hydroxydopamine), to destroy the descending noradrenergic fibers in the spinal cord that are thought to activate spinal cells important for stepping rhythmicity (Grillner, 1976). The hind limbs moved abnormally. There was hyperextension at the end of the stance, which was called "involuntary." Nevertheless, when the same animals were subjected to precollicular decerebration, in most of them the locomotion elicited by brain-stem stimulation appeared to be no different from that of untreated cats. In spite of this uncertainty about the similarity of stepping forms across preparations, we now can say a good deal about which cells are active in reduced preparations from work that we summarize next.

COMPONENTS AND CIRCUITRY FOR RHYTHM GENERATION. If a circuit for stepping control in reduced mammals is to be diagrammed today, numerous components must be included. Usually the elements have been studied in decerebrate cats, so we will be most concerned with those preparations. It is unnecessary to

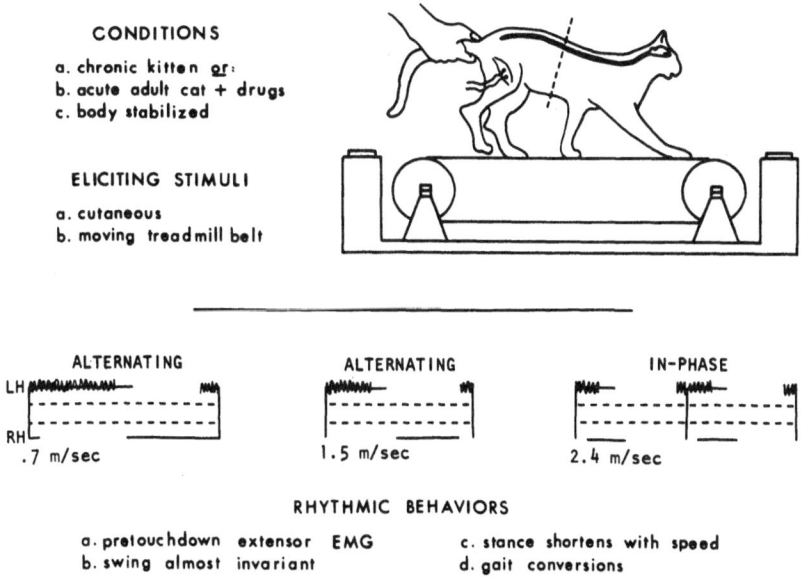

Fig. 4. A functional analysis of stepping by a low-spinal kitten (midthoracic transection). Notation and format are the same as in Fig. 3, except that only hind-limb patterns appear in the gait diagrams (data of Grillner, 1973a). Forelimb step cycles are replaced by dashed horizontal lines.

review here in detail the wealth of data that have been gathered recently (Grillner, 1975; Herman, Grillner, Stein, and Stuart, 1976; Shik and Orlovsky, 1976; Stein, 1977, 1978; Wetzel and Stuart, 1976). The single most significant feature of neural activity during stepping is that so many cells are rhythmically active. Most fire during particular portions of the step cycle of a limb, even if they are already active before the first step is taken.

The afferent fibers that are active during stepping include large group I fibers (Ia from muscle spindles and Ib from Golgi tendon organs), together with smaller group II and III fibers from muscles or elsewhere. A variety of receptors or free nerve endings exists in many locations in the limbs, joints, and skin, with the weight of the evidence being that all the afferent fibers can be active at one time or another during the normal step cycle.

The efferent portion of the nervous system has also been studied in locomoting decerebrate cats. At faster speeds, individual extensor motoneurons maintain a standard frequency of discharge during the stance, and new cells are presumably recruited to deliver increased force (Severin, Shik, and Orlovsky, 1967). Individual trains of impulses in extensor motoneurons become shorter as velocity increases when, as we have noted, the stance duration decreases (early work by Severin *et al.*, 1967; Zajac and Young, 1976). Zajac and Young (1976) found that both flexor motor units, active during the swing, and extensor motor units, in the stance, began each walking step cycle burst with a doublet (an interspike interval $\leq$ 10 msec), an arrangement thought to favor efficient energy usage.

Central to the motoneuron or to the primary sensory afferent, the situation in the spinal cord has not been adequately measured. One of the most puzzling problems is how a muscle might control its own contraction by feedback during locomotion or other behavior (reviewed by Houk, 1972). Many systems accounts incorporate gamma motoneurons that supply the muscle spindle receptor organ. The gamma efferents are coactive with alpha motoneurons during stepping by a decerebrate cat (Severin *et al.*, 1967), a finding that has required new feedback models of "servo-assistance" (reviewed by Matthews, 1972) or "alpha-gamma configuration" (term of Houk, 1972).

Numerous other central cells have been implicated in stepping, or at least in rhythmic alternation of flexor and extensor motoneuron activity. Dorsal horn interneurons are rhythmically active, as are intersegmental neurons, such as those in spinoreticular and propriospinal tracts. Activity in collaterals from these or other ascending paths—perhaps the spino-olivary, VSCT (ventral spinocerebellar), or DSCT (dorsal spinocerebellar)—could drive other spinal cells rhythmically.

We now describe for the cat the progress that has been made to identify lumbar neurons that are important for the rhythmicity of locomotion. The concepts of paired half-centers of Graham Brown (1911, 1914) and later Jankowska *et al.* (1967a,b) have continued to evolve. In the first of the models, that of Graham Brown, there was not even a clear distinction between interneurons and motoneurons that might constitute a pattern generator. In the later model, that of Jankowska and her collaborators, reciprocally related interneurons were posited. Their inhibitory interactions and accumulated refractoriness could produce alternating activity between flexor and extensor motoneurons. In other models, paired half-centers have not been assumed. Pearson and Duysens (1976) have proposed that control of just the flexor motoneuron pool might be sufficient to provide for rhythmic activity in both flexor and extensor motoneurons.

More detailed generators, with specific cells named, have been proposed on the basis of a long history of experimentation on alpha motoneurons, Ia-inhibitory interneurons (IaINs), and Renshaw cells. All three cell types are known to be active when a decerebrate cat steps (reviewed by Wetzel and Stuart, 1976). A recent model of the generator by Miller and Scott (1977) was based upon previously known properties and connections of these cells. The cells' presumed functions were tested in electronic "neurons." For simplicity, the gamma motoneurons were excluded, in spite of the fact that they are rhythmically active during locomotion, because "fictive" locomotion in the decerebrate cat survives deafferentation (Feldman and Orlovsky, 1975). However, a gamma model also was proposed (Miller and Scott, 1977).

Considerable evidence supports the hypothesis that the so-called Ia-inhibitory interneurons are part of a central rhythm generator (Edgerton et al., 1976). These cells produce inhibitory postsynaptic potentials in motoneurons that innervate the antagonists to the muscles from which the Ia afferent fibers arise (reviewed by Hultborn, 1972). They are rhythmically active during stepping by decerebrate preparations (Feldman and Orlovsky, 1975). There are reciprocal inhibitory interconnections between IaINs to antagonists (Hultborn, Jankowska, and Lindström, 1971b); synaptic connections also exist between alpha motoneurons, IaINs, and Renshaw cells. There is a well-known path by which axon collaterals from a motoneuron excite Renshaw cells, that in turn inhibit the parent motoneuron. More recently, Renshaw cells have been shown to inhibit IaINs (thereby providing recurrent facilitation of antagonist motoneurons, Hultborn, Jankowska, and Lindström 1971a), VSCT cells (Lindström and Schomburg, 1973), other Renshaw cells (Ryall, 1970), and some other motoneurons (Jankowska and Smith, 1973).

The proposal that Renshaw cells are part of a spinal rhythm generator was delayed because the cells were thought to be shut down during locomotion (e.g., Severin, Orlovsky, and Shik, 1968). However, McCrea (1979) found for the premammillary cat that Renshaw cell firing often was phase-locked to ventral root discharge. Recurrent IPSPs recorded in the motoneurons showed that this inhibitory pathway could still operate during fictive locomotion. McCrea (1979) then proposed a single-limb generator in which interaction between IaINs and the Renshaw cells could facilitate switching by the spinal generator between flexion and extension phases of locomotion.

A formal circuit diagram, in their case for a single pair of antagonistic "muscles" and one "joint," was drawn by Miller and Scott (1977, see details of their Fig. 1) in conjunction with study of an electronic network composed of six "cells." An "extensor" set composed of an alpha motoneuron, a IaIN, and a Renshaw cell were connected to a corresponding "flexor" set of three cells. Assume that we begin to observe the oscillator while extensor cells are active, and that continual excitatory inputs arrive at IaINs and alpha motoneurons. The "extension" phase is terminated by the inhibitory actions of the "extensor" Renshaw cell on the "extensor" IaIN (leading to disinhibition of the flexor IaIN and inhibition of the extensor motoneuron). Then "flexion" proceeds until terminated by the "flexor" Renshaw cell and IaIN.

The Renshaw cells were postulated to complete a feedback pathway from alpha motoneurons to Ia interneurons. The details of operation of the particular proposed model and cells are less important for present purposes than the investigators' demonstration that the system could account for alternating flexor and extensor activity and even for marked changes in the frequency of oscillation (different stepping speeds) by assuming variations in synaptic gains and the intensity of excitatory signals. The pretouchdown extensor burst in Figs. 3 and 4 *could* be produced by alternating activity in flexor and extensor motoneurons whose rhythm is due to central influences alone. When "interrupt signals" were delivered to the electronic system, some of the changes in rhythm looked similar to behaviors seen in the spinal or the decerebrate cat when reflexes and locomotor movements were combined. Finally, individual electronic oscillators could be coupled to allow one "limb's" activity to entrain the rhythm of another "limb," so that presumably *any* pattern of alternating or in-phase stepping could be duplicated.

The systems work is valuable both to supply practical solutions for artificial locomoting machines and also to suggest how real circuit components—neurons— might be combined during actual locomotion. We should not be particularly concerned at this stage that rhythms at more than one joint cannot be accounted for very well, or by a failure to account for equilibrium of the entire moving body.

Nor is it a major concern that the components of the living cat are so different from machine components, or that there are so many living components. Miller and Scott (1977) indicated numerous inputs to the IaINs, alpha motoneurons, and Renshaw cells from spinal, descending, and afferent neurons (see their Fig. 8). For example, it is well known that at least the following descending tracts can be stimulated to elicit locomotion in decerebrate cats or are active rhythmically during stepping elicited by other means: reticulospinal, rubrospinal, vestibulospinal, and corticospinal (at least indirectly). Moreover, these pathways surely have other functions than in locomotion, as most investigators recognize. The activation from the brain stem "causes much more complex effects than just releasing a certain group of nerve cells that constitute the pattern generator for locomotion" (Grillner, 1976, p. 359). Grillner goes on to say that additional data "can easily be obtained." No theoretical issue surrounds the fact that most of the nervous system is engaged rhythmically when an animal steps. Perhaps we must even propose different generators from moment to moment, but additional experiments will provide answers to all these questions.

However, there are two serious deficiencies in contemporary systems theories, if they purport to explain biological events. The first deficiency is that precision is degraded of both systems and biological concepts when the two sets are equated. A casual survey of just one recent book (Herman, Grillner, Stein, and Stuart, 1976) yielded more than a dozen "circuit" diagrams in which biological labels and systems labels were intermingled. We hear each other substitute "oscillator" for a "set of nerve cells" and "information" for "activity in a particular set of nerve cells." There is no "information" inside the nervous system. Nor is there any "error" or "program" to be "modulated" by afferentation. When we discuss *information,* errors and *noise* in a radio system the terms are precise (Carlson, 1968). Such assertions as that the role of sensory nerve impulses lies in the "correction of errors inherent in genetically determined motor programs" (Wilson, 1972, p. 358) are metaphorical and will mislead us when we try to understand behavior.

The second deficiency in systems theories of animal or human behavior is that they cannot specify the controlling agents. In Fig. 1, the human operator turned on the machine, but how can a "higher center" (presumably a part of the brain) "select the required program" for the human being himself? The true agent or source of control is outside the animal. In reduced preparations, the agents are the eliciting stimuli. In intact animals, as we will see, a far more complete account of behavioral control is required.

LOCOMOTION BY INTACT ANIMALS

The issue of units for locomotor behaviors by intact animals is complicated not only by far larger sets of properties and mechanisms than is true for reduced preparations, but also because of a particularly troublesome set of concepts about volition. Can movements be willed in the sense that a mental agent "commands" a system or activates nerve cells? We specified the controlling agents of decerebrate stepping as external eliciting stimuli, but does not man have some control over himself? Are "volition" or "consciousness" composed of units that warrant scientific accounts and, if so, how should they be studied?

Confusion surrounds all modern motor-control theory, including that for locomotion. Indeed, modern science is threatened by mentalism that is now as widespread as it was in the time of pre-Newtonian mechanics. H. L. Teuber commented on the problem in his remarks to a panel discussion that met at the close of a conference on motor behavior and programming in Marseilles, France, in 1973 (reported in 1974 in a special issue of *Brain Research;* see Paillard and Massion, 1974). "Some participants used the term 'voluntary' with apologies, and in an 'as-if' mode, whereas others insisted on its indispensability, if one were to make sense of the notion of 'levels' of motor organization and of differences in strategies of movement" (p. 536). In speaking of the failure of a monkey that was raised without the sight of its limbs to visually guide them effectively. Teuber asked, "What are the central neural counterparts, or as I like to put it, what are the *physiologic* markers for degrees of voluntariness? Is this way of putting the question really as absurd as it sounds?" (p. 537).

It is absurd to search for physiological markers of voluntariness, and another account of the facts of movement has existed for some years. Its acceptance has been delayed in neurophysiology and neuroanatomy because prescientific concepts of the nature of "mind and body" have accompanied major research discoveries. In the seventeenth century, Descartes proposed that reflex responses could not be controlled directly by the will, yet they could be modified by the action of the soul (Fearing 1930). Generations later, Sherrington (1906) wrote about "body-mind liaison." He thought that it was useful to conceive of two kinds of world stuff, "psychical energy" and "physical" energy. An alternative position, which he seems to have favored less, was that there is really only one substance, either mental or physical. Sherrington wrote that Cajal for a time believed in just a psychical self, but "noticed that to his practical life adherence neither to the one [mental] nor other [physical] seemed to make any difference whatever" (Sherrington, 1906, xxiv, Foreword to the 1948 edition). Assumptions about the relation between mental and physical probably did not make much difference so long as physiologists studied spinal and decerebrate animals. However, a science for intact animals requires a concept of control that is still alien to many students of motor events. In this concept, the brain is not the source of behavior. "As the philosophy of a science of behavior, behaviorism calls for probably the most drastic change ever proposed in our way of thinking about man. It is almost literally a matter of turning the explanation of behavior inside out" (Skinner, 1974, p. 249). The intact nervous system is controlled by what happens to it, and we must discover the outside agents of control.

Like Sherrington, Skinner (1953, 1974) has been concerned with how the term, reflex, has been used. "If we were to assemble all the behavior which falls into the pattern of the simple reflex, we should have only a very small fraction of the total behavior of the organism. This is not what early investigators in the field expected" (1953, p. 49). The Skinnerian theory that is now in use by a small community of behavioral scientists goes far beyond the reflex, but it is compatible with the neurophysiology of locomotion and other behavior. Students of motor control have always been concerned about degradation of rigorous concepts. Matthews (cited in Paillard and Massion, 1974) has deplored the fate of the term "reflex" in recent years. "To transfer the term 'stretch reflex' to any response of the cortex to muscle stretch is, for me, to generalize the term so widely that it can no longer be handled with precision" (p. 545).

In giving the behavioral account of stepping, we will cite Skinner's works many times (1938, 1953, 1957, 1969, 1974). Comprehensive guides to research procedures are available in *Tactics of Scientific Research* (Sidman, 1960) and *Schedules of Reinforcement* (Ferster and Skinner, 1957). As we pointed out in the previous section, systems theories and terms are so widely accepted today as accounting for the control of movement that we will have to reinterpret systems statements as necessary. It should not surprise anyone to learn that theories which can explain machine behavior are inadequate for animal and human behavior.

A Functional Analysis of Complex Behavior

In addition to sensitivity to those portions of the physical world that act on reduced preparations, intact animals share across species, to greater or lesser degree, sensitivity to a broad array of events in the world around them. They see, hear, maintain balance, and move about in relation to those events. Moreover, their responses are a function of previous environmental events, both in the history of the species and in the individual animal. They evolve and learn. The nervous system is necessary for every behavior, not just because some response clusters seem to be built in (see p. 576), but also because the capability to learn is built in. As we develop the theory of external stimulus control, we will also be concerned with other variables that change the organism from time to time. These biological modifications must be accounted for, just as we had to discuss excitability changes of neurons in reduced preparations. The effects are not properly termed stimuli, but they are still dependent on events outside individual cells. Some examples are deprivation, satiation, and emotion.

An intact behaving animal possesses a repertoire of behavior that at any one time is regulated by an intermingling of four known kinds of stimuli. There is some ambiguity of meaning in the literature for the terms that specify these four forms. To avoid this ambiguity the following conventions have been adopted in this chapter. We will list them first and then discuss each. Stimuli *elicit* reflex responses—as described on pages 577–580—*evoke* conditioned responses, *discriminate* operant responses, and *reinforce* the control exerted by evoking and discriminating stimuli. To understand behavior, these stimulus functions must be separated. Only when we know why each stimulus produces a specified behavior is it reasonable to ask questions about what neural circuits might be active at the moment.

THE ELICITING FUNCTION. Behaviors in addition to those that are seen in reduced preparations can be elicited from intact animals. Many of these behaviors are reflexes, of course, but the general term "elicit" implies a controlling function between stimuli and behaviors, so we call this class of behaviors *respondent*. "The kind of behavior that is correlated with specific eliciting stimuli may be called *respondent* behavior and a given correlation *a respondent*. The term is intended to carry the sense of a relation to a prior event" (Skinner, 1938, p. 20).

Some respondent behaviors may be experiences that are commonly called feelings of pain, fear, or pleasure. These experiences exist in addition to autonomic reflexes, such as vasoconstriction to a cold stimulus, but they are treated exactly like any other behavior in a radical behavioristic approach. *Radical behaviorism* "does not insist upon truth by agreement and can therefore consider events taking place in the private world within the skin. It does not call these events unobservable, and it does not dismiss them as subjective. It simply questions the nature of the object observed and the reliability of the observations" (Skinner, 1974, pp. 16–17).

Another special class of elicited behaviors in intact animals has been studied by bypassing the normal afferent routes and stimulating interneurons directly. Muscle contractions or more elaborate movements can be elicited by electrical stimulation of the motor cortex (reviewed by Porter, 1973), and "memories" can be elicited by electrical stimulation of the temporal cortex (Penfield and Roberts, 1959).

EVOKING AND REINFORCING FUNCTIONS. Other behaviors in intact animals are also largely dependent on a preceding stimulus, but we know some of them are "elicited" in the sense of being classically conditioned, so we say they are "evoked." In the classical *stimulus substitution* experiment, a previously neutral stimulus (the conditioned or "conditional" stimulus) acquires the capability to evoke a response that originally was elicited by another stimulus. A tone (CS) comes to evoke conditioned flexion after being paired with a painful foot shock (UCS) which originally elicited limb flexion. In this Pavlovian or respondent conditioning the establishment of control by a new stimulus is termed *reinforcement*. We can distinguish evoked flexion from elicited flexion by manipulating the stimuli. If the *reinforcer* (shock) is omitted, defining *extinction*, control by the conditioned stimulus will decrease. Unfortunately, the procedure has not been performed at all routinely in studying motor control, so that often an observed flexion is controlled by an unknown mixture of eliciting and evoking stimuli. Such a mixture must be assumed for the flexion phase of stepping, and there will also be discriminative control, as we will see in the following paragraphs.

DISCRIMINATING AND REINFORCING FUNCTIONS. The identifiable contingencies between stimuli and responses are not limited to those described above. The most commonly seen behaviors in intact animals are *operants*—Skinner's (1938, 1953) term. The full strength of operant theory has never been exerted to help us understand motor control. Usually, operant conditioning is regarded as a tool with which one can train a monkey to move a lever or change the firing rate of a cortical cell. This is a useful and legitimate application of the theory, but the importance of an experimental analysis lies somewhere other than in training behaviors to arbitrary criteria.

In operant behavior, eliciting stimuli cannot be identified for individual responses, and the empirical probability of an operant, over a given period of observation, ranges continuously between zero and one (Skinner, 1953). "The unit of a predictive science is, therefore, not a response but a class of responses. The word 'operant' will be used to describe this class" (Skinner, 1953, p. 65). Although in Pavlovian (respondent) conditioning a reinforcer is paired with a stimulus, in operant conditioning the reinforcer is contingent on (follows) a response. When instances of a behavior are followed by a particular event (food is usually found to be an effective reinforcer), the frequency of a rare response is seen to increase. A complex act is by its very nature continuous. As was true for respondents (see pp. 577–580, and pp. 590–597) there are no "natural" operant units of locomotion, so there can be no fixed set of neurons to generate them. The unity of particular operant classes is established only when behaviors come under stimulus control, not from *a priori* classification by their topography.

The operant always occurs in an environment, so it is not emitted in the sense of occurring spontaneously. Stimuli that accompany or precede the operant gain a controlling function when reinforcement occurs, but it is not that of eliciting or evoking. It is a discriminative function. The contingency is that a response (such as flexing a limb) is followed by reinforcement (such as presenting food) only on occasions on which the response was preceded by a particular event. That event becomes a discriminative stimulus, or S_D. Stimuli are not defined except in classes that have been specified after a functional relation has been established. Stimuli are continuous, just as was true for responses. Since both are continuous, a search for a complete set of controlling afferent impulses during all "stepping" will be fruitless. It is also unnecessary, because for each behavior, only the existing S_D's (and eliciting or evoking stimuli) need be known.

Operant theory dispenses with the controversy about "voluntary" behavior. "The distinction between voluntary and involuntary behavior is a matter of the kind of control. It corresponds to the distinction between eliciting and discriminative stimuli" (Skinner, 1953, p. 112). The discovery of discriminative control has been tremendously successful in explaining complex behavior. The control is different from that of eliciting or evoking stimuli, yet these stimuli discriminate (cause) behavior just as relentlessly. The probability that our dog will run toward us when we call it can be made as close to 1.00 as we wish.

The occurrence of any particular response is determined by many controlling stimuli and other variables. An example will show the complexity. When a cat eats, the food elicits salivation. If it is fed in a special dish, the empty dish as a conditioned stimulus will come to evoke salivation. The food will also be an operant reinforcer in following, for example, locomotion toward the dish. *Any* prior or concurrent stimuli in the setting may come to discriminate the motion. Besides the food and the dish, other common discriminative stimuli are a call by the owner or the sounds of the food can opener. Finally, "The stimulus which is presented in operant reinforcement may be paired with another in respondent conditioning" (Skinner, 1953, p. 76). From this pairing, the dish will become a *conditioned reinforcer* as well as an S_D and, like the food, will strengthen discriminative control of responses, including locomotion toward the dish.

Of course, many of these behaviors will occur only if it has been a long time since the cat ate. Detailed accounts of variables, such as deprivation and satiation, are beyond the scope of this chapter, but it is useful to list them. In addition to elapsed time after food, water, or other bodily requirements have been given, or consumed, we must note the presence or absence of sleep or other long-term cyclic modifications in the animal, and the circumstances that cause us to call behaviors "emotional." In each case we must account for events that alter organisms and modify stimulus functions, just as we previously noted factors that predisposed reduced preparations to step or not.

Since every stimulus has many effects, and each behavior is controlled by more than one stimulus, the analysis of complex cases is difficult. Some progress has been made for locomotion, however, and we may expect more rapid advances as problems of behavioral control become generally recognized and solved. Students of locomotion have already collected a wealth of observational data for animals living in their home environments and the laboratory, as we shall see in the following sections. We may not always know how control is achieved, but there is ample evidence for discriminated locomotor behavior. First, however, we should review traditional measurements of the structure of locomotor responses.

TRADITIONAL MEASURES OF THE STRUCTURE OF LOCOMOTION

Extensive measurements have been made of stepping at a variety of velocities in many species (reviewed by Gambaryan, 1974; Grillner, 1975; Hildebrand, 1976; Howell, 1944; Shik and Orlovsky, 1976; Stein, 1977, 1978; Wetzel and Stuart, 1976). Kinematics are easy to record, especially the swing and stance of the single limb, and the interlimb sequence of footfalls. Timings are readily taken from cine films. There are also many measurements of joint angles and trajectories of one or more limbs, either from the side (cats, dogs, and people) or the top (lizards or turtles). Sometimes EMG records have been made. Some of the findings are depicted in Figs. 5 and 6. These figures also show controlling variables and should be consulted while the rest of the chapter, especially the section "Terrestrial Mammals Other than Man" is read.

Limited force (kinetic) data and muscle length data have been collected for reduced preparations (Udo, Matsukawa, and Kamei, 1977), with sparse findings also available for intact animals or man (Ariel and Maulucci, 1976; Herman,

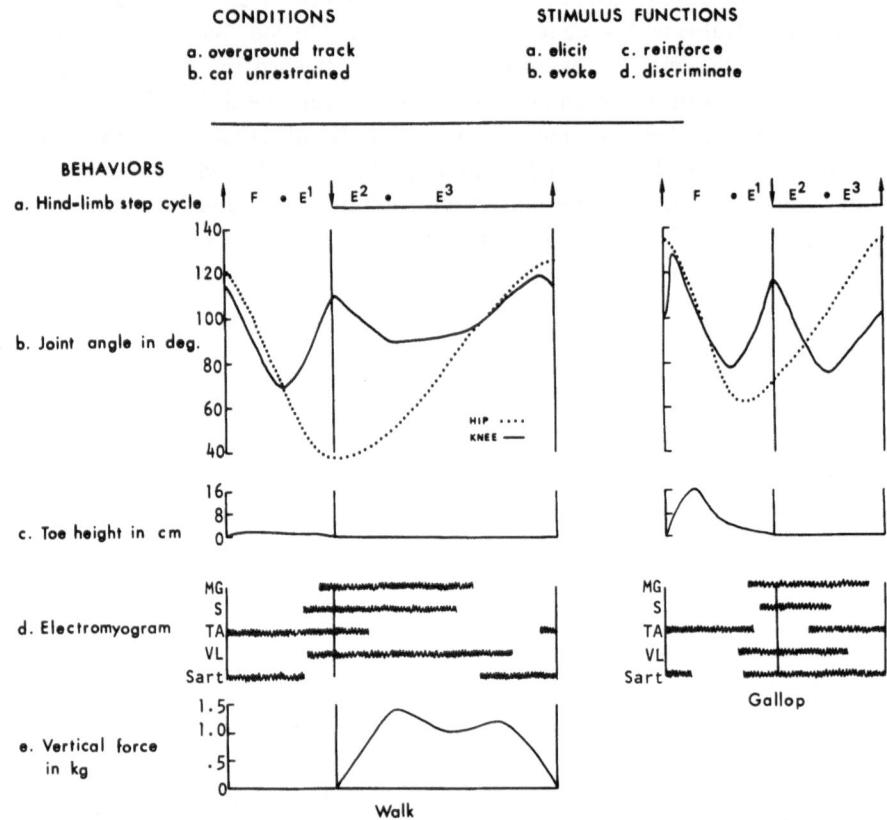

Fig. 5. Functional analysis of locomotion by an intact cat, single hind-limb. All four stimulus functions are assumed to be in force, although few actual relations have been demonstrated by experimentation to date. At the top of the figure are the Philippson step cycle epochs of F, E^1, E^2, and E^3 (see text for detailed description of this well-known diagram). Kinematic and kinetic values for events of the step cycle during walking and running were expressed as percentages of swing and stance from data in a number of experiments by different investigators. Joint angle data for the hip and knee (flexion downward) were taken from Engberg and Lundberg (1969) and Goslow *et al.* (1973). Note the yield in the knee, but not the hip, at the $E^2 \cdot E^3$ junction. The trajectory for the toe was based on data of Wetzel, Atwater, and Stuart (1976). The record shows that during walking, the toe barely clears the surface of the ground, but during galloping, it sweeps high above the ground. EMG durations were estimated from data of Coss *et al.* (1978). MG = medial gastrocnemius (ankle extensor), S = soleus (ankle extensor), TA = tibialis anterior (ankle flexor), VL = vastus lateralis (knee extensor), Sart = sartorius (two joint hip flexor, knee extensor). Timings for the two joint muscle differ with speed (see text). For intact animals vertical force at the foot has been measured only during walking (Manter, 1938). As would be expected, peak force occurs at the peak of the yield when E^2 turns to E^3.

Wirta, Bampton, and Finley, 1976; Loeb, Bak, and Duysens 1977; Manter, 1938; Pedotti, 1977; Prochazka, Westerman, and Ziccone, 1977). There is also considerable information about the energy cost of locomotion (Alexander and Goldspink, 1977; Chassin, Taylor, Heglund, and Seeherman, 1976; Taylor, 1973).

From this wealth of material some topographical features have been noted that seem to occur in small-to-large packages. Here we emphasize terrestrial locomotion, by nonhuman mammals, although most of the material cited could apply to swimming, flying, or even bicycling. The smallest contractile element is a single *motor unit* (one motoneuron together with all the muscle fibers it innervates). Motor units apparently are brought in successively when they are recruited for more forceful movements, but the order is not arbitrary. Small ones and then large ones are recruited (Henneman, Clamann, Gillies, and Skinner, 1974).

Larger "units" of locomotion have been proposed for the step cycle of one leg. A number of distinct behaviors have been observed; one being the sequence of joint movements, which appears to be about the same at all speeds (reviewed in Wetzel and Stuart, 1976). There are, however, small timing changes under different circumstances. When a cat walks or trots overground, the knee and ankle turn from flexion to extension, during the swing, at about the same time; but on a treadmill

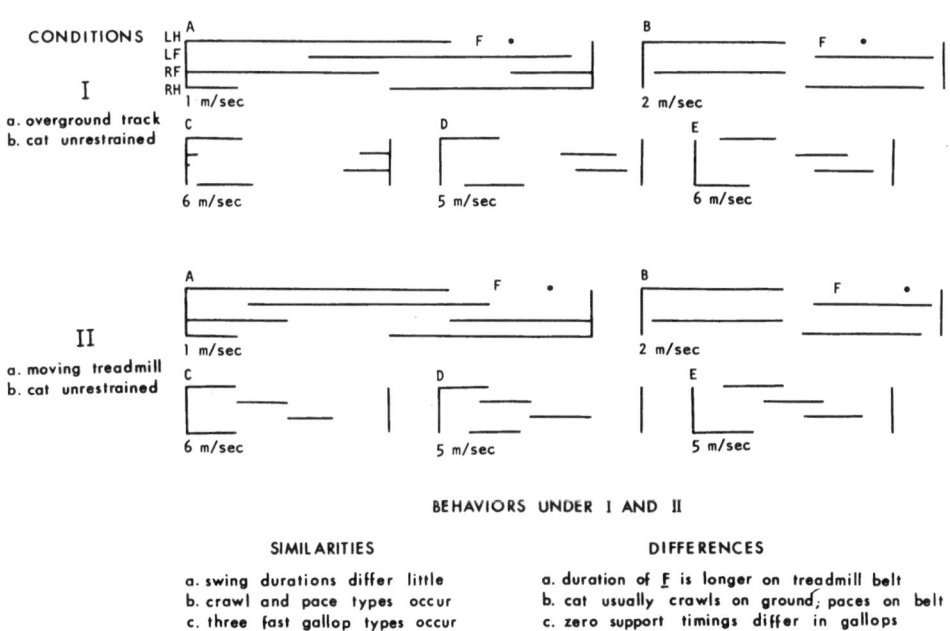

Fig. 6. Functional analysis of locomotion by an intact cat, four-limb sequence. Two conditions have been diagrammed: I, cat moving on a track on open terrain; II, cat moving on a motor-driven treadmill. Behaviors are assumed to be elicited, evoked, reinforced and discriminated by stimuli, as in Fig. 5. Averaged values for timings in the gait diagrams were based on data from Stuart *et al.* (1973); Wetzel *et al.* (1975); Eisenstein *et al.* (1977); Norgren *et al.* (1977). Under each condition, I and II, (A) and (B) show walking and trotting, respectively, with both gaits under positive reinforcement. (C), (D), and (E) show half bounding, transverse galloping, and rotatory galloping, respectively. All gallops were at high speed (5 m/sec or more), under negative reinforcement.

the knee leads the ankle by some 30 msec (Wetzel, Atwater, Wait, and Stuart, 1975).

As can be seen in Fig. 6, unless reflexes are elicited deliberately as by applying a brake to the limb (Orlovsky and Shik, 1965), swing duration for the cat averages 200 msec across velocities from walking through *steady state* galloping (reviewed in Wetzel and Stuart, 1976). Steady state galloping means that the footfall order is invariant from stride to stride. During *mixed galloping,* footfall order changes from stride to stride, and total swing duration can vary widely. The range is from 150–300 msec in the forelimbs and is almost as large in the hind limbs (Wetzel, Anderson, Brady, and Norgren, 1977).

Flexion and extension have appeared to many investigators to be responsible for the swing and stance phases, respectively, of the step cycle. This view could be compatible with that of a central oscillator which drives flexor and extensor moto-neurons alternately (Graham Brown, 1911, 1914, 1915; Jankowska *et al.,* 1967a,b; Miller and Scott, 1977). However, an oscillator account oversimplifies the cycle. First, flexion is longer on the treadmill than overground (Fig. 6), even when swing duration is the same (Wetzel *et al.,* 1975). Second, although Philippson (1905) found one "flexion" or F phase in the hind-limb cycle for dogs, there were three "extension" phases that he termed E^1, E^2, and E^3 (see Fig. 5). E^1 is the beginning of extension, and extensor EMG bursting occurs (as we noted on pp. 585, 589), at the end of the swing prior to touchdown. E^2 is a yield, or active lengthening (with EMG on) of extensors at the knee and ankle, but *not the hip,* upon touchdown. In E^3 the hip, knee, and ankle joints extend together through the remainder of the stance, which, as we previously noted, shortens in duration as velocity increases. A third problem with a simple flexion—extension dichotomy is that two-joint muscles are not simple flexors or extensors. Sartorius, for example, flexes the hip when a cat or dog walks overground, as evidenced by activity at the end of the stance and early in the swing, but when the animal runs there is also a knee extensor action, as shown by activity early in the stance (Fig. 5).

The E^2 "flexion" of the knee and ankle, but not the hip, indicates a division of labor among the extensors at different joints. Detailed knowledge of bursting characteristics of each limb muscle would be highly useful but has been elusive. It has been suggested that there is less "variability" among single-joint extensors than among single-joint flexors or two-joint muscles (Engberg, 1964; Engberg and Lundberg, 1969; Rasmussen, Chan, and Goslow, 1978; Tokuriki, 1974; Wentink, 1976), but it is not always clear what this statement means. It is true that extensors usually are active only once in the step cycle, and all four ankle extensors seem to burst at about the same time in the decerebrate cat (Gambaryan, 1974). However, timings differed within this group for intact cats (Betts, Smith, and Collatos, 1976; Rasmussen *et al.,* 1978). During walking, the soleus muscle became active much earlier than the gastrocnemius muscles, but at faster speeds, both medial and lateral gastrocnemius became active before the soleus (Fig. 5). Even from stride to stride, timings between EMGs of ipsilateral hind-limb and forelimb muscles have been reported to vary a good deal (Coss, Chan, Goslow, and Rasmussen, 1978).

Larger clusters of stepping, flying, or swimming behaviors have been identified by correlations that exist between events at different limbs. A very large number of relations for mammals have been examined, starting many years ago by inspection of animal tracks or footfalls (reviewed by Hildebrand, 1976; Howell, 1944). Different species may have markedly different gaits. Often gaits are tied closely to morphology of the limb. The kangaroo, some rodents, some lizards, and humans have a bipedal gait in which the short forelimbs are not used (Gambaryan, 1974; Sukhanov, 1974). Even among tetrapodal four-limb gaits, many species differences exist. Domestic cats have a duty cycle that is approximately the same for all four limbs in a high-speed gallop (Wetzel *et al.,* 1977). Gaits through pacing and trotting speeds for four-legged locomotion are *symmetrical.* A symmetric pair of legs has the characteristics that the duty factor of either leg is the same as that of the other leg, and the phase shift of one leg relative to the other is exactly equal to one-half (McGhee, 1968). The steps of the two opposite legs alternate, both hind limbs and forelimbs. These relations can be seen in Fig. 6 A, B for both treadmill and overground locomotion. Gaits are asymmetrical if the requirements for symmetry are not met, as in galloping (Fig. 6 C, D, E). Hildebrand (1977) studied asymmetrical gaits in terms of footfall patterns. Five variables expressed timing of events at the ground: two averaged duty cycles (hind and fore); size of two averaged leads (interval between footfalls of one hind limb and the other hind limb, and the same interval for the forelimbs); and the averaged *midtime lag,* that related actions of the forefeet as a pair to those of the hindfeet as a pair. The divisions are appropriate even for mixed galloping (Wetzel *et al.,* 1977). We should note, however, that a simplification of these patterns has been acceptable to several investigators. Relations between contralateral limbs, both fore and hind, have been characterized just as alternating or in-phase (Grillner, 1975; Miller *et al.,* 1975).

Some of the common gait patterns have been displayed in Fig. 6. As was true for events in the single limb (Fig. 5), interlimb relationships vary with conditions. The two common experimental settings are either an overground track or a motor-driven treadmill. They yield rather different behaviors. Under just one set of conditions, on the treadmill, the domestic dog or cat may maintain a forward position on the treadmill belt that varies by only a few cm (Arshavsky, Kots, Orlovsky, Rodionov, and Shik, 1965; Miller *et al.,* 1975; Stuart, Withey, Wetzel, and Goslow, 1973; Wetzel *et al.,* 1977). However, the animal then may stop abruptly or else convert to another gait, within just one stride. At the same slow speed, individual cats in our laboratory have walked with the LH, LF, RH, RF footfall order of the crawl, or they have paced with ipsilateral feet descending together on the treadmill belt. At a single midspeed, they have trotted or galloped. If galloping, the pattern of footfalls is sometimes invariant, or steady state (Eisenstein, Postillion, Norgren, and Wetzel, 1977; Norgren, Seelhorst, and Wetzel, 1977) and sometimes mixed from stride to stride (Wetzel *et al.,* 1977).

The "largest" proposed sets of locomotor behaviors are cyclic movements by much of or all the body, in addition to the limbs. Muscles of the pelvis (e.g., Goslow, Reinking, and Stuart, 1973), head, neck (Tokuriki, 1977; Watt and Wetzel, 1977), tail, and trunk (Tokuriki, 1976) all take part in locomotion. These relations have not been inspected in much detail, but a few statements can be made. The head and neck turn readily from side to side at low or moderate speed. In the absence of turns, the angle of the neck on the body is still extremely variable, but

Watt and Wetzel (1977) saw rhythmic, almost sinusoidal accelerations of the head in the fore-aft and vertical directions. The latter movements were tied to cycling of the forelimbs, but they were independent of cycling of the hind limbs. We should not necessarily conclude that interlimb neurons were inactive, however. Stretch of contralateral dorsal neck muscles in the anesthetized cat is known to excite lumbar alpha and gamma motoneurons (Murthy, Gildenberg, and Marchand, 1978).

After reviewing observations ranging from single motor units to locomotion by the whole animal, we now suggest that it may not be useful to conceptualize such a hierarchical control system. In fact, the data we have summarized reveal nothing about mechanisms of control. Experimenter-controlled behavioral contingencies have been relaxed or lacking altogether (in field situations), so we cannot say whether or not apparent clusters of responses are functional groups. The documented "variability" just means we are recording effects of unknown controlling variables, not that there are "errors" in the animals' locomotion. Separation of control by different stimuli has not been achieved, and the time has come to depart from mere descriptions of locomotion. We are far from knowing enough about intact animals to warrant circuit diagrams of functional pathways.

The following material is organized to show our current limited understanding of how normal and to some extent pathological stepping is multiply controlled. Some data can be interpreted from a functional viewpoint even when they were collected for other reasons.

INVERTEBRATES AND LOWER VERTEBRATES

We do not know whether or not operant or respondent conditioning of locomotor responses occurs for simple arthropods. Tosney and Hoyle (1973) found that the firing frequency of cockroach motoneurons changed when a leg was shocked, and others have found that stepping is highly responsive to external conditions. Adding weights to an insect has been shown to increase the step duration (Bethe, 1930). Gerrids (water striders) have walked if placed on a sloping surface, presumably as an elicited response (Bowdan, 1978). With an increasing angle of slope, Bowdan (1978) saw more walking, measured in steps per unit of elapsed time, a result which was thought to depend on the amount of afferentation from anterior loading of the animal and excitation of hair plates. Rowing patterns on water were shown by statistical analysis to be almost invariant. Walking gaits on a surface, in contrast, included not only a tripodal form but several additional ones. A *tripod* configuration gives a stable base (see pp. 569–571). The front and posterior legs of one side move together with the middle leg of the opposite side, to form a triangle, and then the three remaining legs move together, to form a second triangle (Hoyle, 1976). The gaits other than tripodal in gerrids (Bowdan, 1978) were called "awkward," and a search for the controlling stimuli would be worthwhile, even if they proved just to elicit locomotion.

For the crayfish, crab, or lobster, it is plausible to test for discriminative stimuli, because it has been suggested that "voluntary" movements exist for these creatures (Evoy, 1976). The lobster locomotes in any direction and has walked on a moving treadmill belt (extensive recent work by Davis and Ayers, 1972; Ayers and

Davis, 1977a,b, 1978). If the lobster would not walk with treadmill stimulation alone, mechanical stimulation of the tail appendages or of the antennae sufficed. A group of muscles that contracted together synergistically during movement in one direction were antagonists for movement in the other direction. Specifically, retraction and depression of a limb occurred during forward movement, but protraction accompanied depression during backward walking (Ayers and Davis, 1977a).

Lobster stepping differs from laboratory to laboratory when different contingencies are enforced (cf. Ayers and Davis, 1977a, and MacMillan, 1975), to imply multiple conditioned stimulus control. Ayers and Davis (1977a) sometimes reversed the direction of their treadmill belt and found that animals "frequently resisted belt movement at low velocities" after the reversal (p. 7). That the response was discriminated or evoked was untested, but plausible.

Stimulus control in fish, frogs, and turtles is similarly complicated, with perhaps all four stimulus functions in force. Detailed descriptions are appearing for swimming by fish, in which both kinematic and EMG data have been gathered (Grillner and Kashin, 1976). Frog or other amphibian locomotion (reviewed by Székely and Czéh, 1976) and turtle locomotion (Lennard and Stein, 1977) have also been studied recently. Lennard and Stein (1977) elicited swimming by electrical stimulation of the dorsolateral funiculus in intact turtles and compared the movements with those of spinalized animals, elicited by the same electrical pulses. Also studied were movements of intact turtles in which swimming was produced by tactile stimulation, or in which "spontaneous" (discriminated?) swimming appeared. The shells were restrained. Locomotion of the intact turtle, from whatever source, included forward swimming, backpaddling, withdrawal, tonic extension, and scratching. The behaviors of spinalized turtles were more limited, usually to swimming movements of one hind limb. Intact animals moved the head, neck, and forelimbs, in addition.

A funtional analysis should be performed with this species. Elicited swimming by intact turtles, unlike that of spinal ones, sometimes outlasted the stimulus. Although not tested, the long-term swimming might well have been evoked by conditioned aversive stimuli (almost anything in the water or other surroundings) that were inevitably paired with the shock.

In summary, no real separation has yet taken place between eliciting, evoking, discriminating, and reinforcing stimulus functions for even simple organisms. For the species mentioned in this section, probably the important immediate task is to show that their locomotion is or is not merely elicited. We now turn to terrestrial mammals, including man. Control of their behavior, in contrast to the animals considered so far, is dominated by evoking and discriminative stimuli, with pure respondent behavior more difficult to show.

TERRESTRIAL MAMMALS OTHER THAN MAN

LOCOMOTION AS A COMPLEX REPERTOIRE. We present here some of the important variables that cause locomotion in the laboratory. When an animal such as a domestic cat is the subject, one does not wait for the behavior to emerge by itself. The animal's locomotion is followed after a number of strides by a *positive reinforcer*, such as food at the end of a straight alley. Alternatively, the animal's running can be an escape from aversive stimulation, as by fleeing from an attacking dog, since termination of a *negative reinforcer* strengthens the control of discriminative stimuli. A detailed treatment of the effects of aversive stimulation is available elsewhere (Skinner, 1953).

If a cat or dog walks on a treadmill, rather than overground, the behavior is still under operant control. Perhaps for the lobster there is a strong eliciting function by a moving treadmill belt. However, a cat will not run on the belt without training. Should food be the reinforcer, treadmill training requires a regular schedule of daily feeding (some time after the training session), and *shaping* (Skinner, 1953). Behaviors that are incompatible with running, such as standing still, turning around, or climbing out of the chamber, must be brought under discriminative control. Stimuli other than food, that are aversive and could be made negative reinforcers in escape (for example, loud noises from the belt motor) should not discriminate or reinforce the treadmill running, because these contingencies may interfere with the precision of control that can be achieved with the food reinforcer. Control of fear responses to conditioned aversive stimuli, that were originally neutral but have been paired in the past with running, must be extinguished. Shouting to cats may be unwise because laboratory caretakers often shout when they chase cats back into their cages after cleaning.

The acceptable behavior, approach to food, must be shaped during daily sessions. When discriminative control is established, walking will occur when the belt moves backward. The animal is never fed unless it faces forward, watches the food tray (incompatible turnings of the head and neck are prevented), and has been moving for several successive strides. A host of visual, auditory, and proprioceptive stimuli discriminate responding.

It should be obvious that although an animal can be trained in a couple of weeks to walk, trot, and sometimes gallop on a treadmill, numerous behaviors have been learned. Some of them, perhaps most, are unknown to the experimenter, whose task is to identify functional relationships between stimuli and responses. What, then, are the units of treadmill locomotion? It must be conceded that we do not know how many different responses might be established by different contingencies. We do know that smooth progression at any velocity up to about 6 m/sec is possible for domestic cats. So far, higher speeds have been seen with negative rather than with positive reinforcement contingencies (Eisenstein *et al.*, 1977; Norgren *et al.*, 1977).

Locomotion at a specified speed appears to be functionally separable from locomotion for a specified time. By requiring increasing numbers of strides at low velocity, for example, 1 m/sec, cats have been trained in our laboratory and in others to walk for several minutes. A larger class of behaviors than single strides now is an operant. In order to train faster locomotion, one requires fewer strides at first, but a higher velocity.

In this situation the *primary reinforcer* is food. The external discriminative stimuli and/or conditioned reinforcers include all characteristics of the setting that have been paired with food (its appearance, smell, rattle of the tray, conversation by the experimenter) or the behaviors and stimuli leading up to eating (sound of the treadmill brake, termination of stimulation from the moving belt, touching the food tray with the mouth). The appearance of food is a particularly powerful visual conditioned reinforcer. This stimulus has been paired with food consumption, as well as with every response in the stepping repertoire and all the behaviors leading to food when the treadmill is shut off. To measure the separate contributions of visual, auditory, and cutaneous stimulation from the belt, they must be manipulated systematically, usually one at a time (Sidman, 1960). Reinforcing functions of stimuli can be separated from discriminative functions, since the former follow and the latter precede or accompany behavior.

It is possible to establish a sequence of strides as an operant response and then study such sequences over longer periods of time. As with all behaviors, long-term drifts in response characteristics may be seen over a period of several months, although they do not prevent discernment of stable responding for short-term events up to several weeks (Ferster and Skinner, 1957; Sidman, 1960). A cat that gallops on a treadmill at 2.8 m/sec will probably eventually trot at the same velocity (Eisenstein *et al.*, 1977). This behavior must have been conditioned, and perhaps the reinforcement was escape from aversive stimuli, which were the result of fatigue.

All cats have a long previous history of running toward food or away from aversive stimulation, so functional components of stepping have been built over months or years of the animal's life. They are already assembled into packages that can be brought under the control of new stimuli through food reinforcement. The ease of training perhaps has been unfortunate. Scientists have often assumed that this history can be ignored, although not all workers have been negligent. Roberts (1976) has noted the early learning by herbivores: "These abilities are not inborn" (p. 558), he says and points out the struggles to stand and walk by a newborn foal.

FUNCTIONAL UNITS. We can now examine both the interlimb repertoires and those of the single limb to see whether or not any functional units have been demonstrated. Even when training procedures have not "sharpened" the behaviors very much, some progress has been made. Sharpening means a narrowing of the control exerted by stimuli. When the locomotor pattern is invariant or steady state, at least to arbitrary criteria that may differ from laboratory to laboratory, control can be studied by introducing new stimuli or by altering the animal in other ways. Some experiments are more effective than others. Deafferentation experiments (reviewed by Taub, 1976; Wetzel, Atwater, Wait, and Stuart, 1976) show little more than that locomotion is still possible when all dorsal roots that supply the limbs have been severed. When a single limb is deafferented, cats still exhibit gait stability (see p. 571). If such a cat stumbles, the quadruped crawl may be resumed after a few strides, so long as the other three limbs are moving on a treadmill belt (Wetzel, Atwater, Wait, and Stuart, 1976).

One would expect locomotion to continue in the deafferented preparation, so long as other stimuli were still effective by way of the three intact limbs. We know that there are many eliciting stimuli even in reduced preparations. Proprioceptive or cutaneous stimuli arising from the moving limbs themselves in intact animals have additional functions. They may elicit, evoke, reinforce, or discriminate locomotor behaviors, and they probably also are conditioned reinforcers. After surgical deafferentation, the stimulus array is so changed, as are the mechanics of the moving limbs, that a new pathological repertoire appears almost at once. Probably many of the new operants are negatively reinforced by termination of painful stimulation from falling. The situation may be similar to that of human patients with back or shoulder pain who place their limbs in positions that reduce the pain. The principal difficulty in interpreting behaviors following deafferentation is that so many variables have been altered at once.

More effective laboratory experiments should introduce one new variable at a time. We have some data as to the *grain* of the few behaviors that have been examined. The grain of a repertoire depends on the size of the "smallest response under the functional control of a single variable" (Skinner, 1957, p. 61). The size of the *minimal response* may be reduced by (1) successively requiring responses that correspond more closely to a criterion, such as holding body position within a narrow range of positions on the treadmill belt, and (2) reinforcing only when the movements occur in the presence of a progressively more restricted range of the original stimulus spectrum. One might first give food on trials in which white noise was present, but eventually reinforce only on trials in which a 500 Hz tone was present. Similarly, at first body position might be acceptable within 29 cm, but eventually it would vary by less than 5 cm. When both the behavioral repertoire and the discriminative control are fine grained, we say there is *fine discriminative control*.

Experiments on the grain of treadmill position have not been performed, but we do know something about joint angles during locomotion (Lockard, Traher, and Wetzel, 1976). Cats on a treadmill crouch when running under aversive contingencies, but not under food-reinforcement contingencies, and joint angles can evidently be of extremely fine grain. Within mechanical limits to the excursion, we may assume a continuous range of achievable joint angles at the hip, knee, ankle, and digits (or shoulder, elbow, wrist, and digits). Small extraneous sounds (probably conditioned aversive stimuli) that are made close to the treadmill will cause a cat to crouch slightly, and a louder sound makes it crouch still more. There is functional unity of joint angles across limbs in this instance, because not just one limb, but the whole cat, crouches. In addition, joint angle excursions during locomotion can be functionally separated to large extent from the footfall patterns. When joint angles shift, walking or trotting footfalls need change very little (Lockard *et al.*, 1976).

Other behaviors that are of considerable current interest are electromyographic responses from limb muscles (see recent work by English, 1979). The E^1 extensor burst, for example, has been termed "preprogramming" without there being a program to point to. While the E^1 extensor effect may not be conditionable, it is possible to find out. Twitmyer conditioned the stretch reflex in people in 1902, and at present we should not exclude conditioned control of extension during locomotion, even though the response in spinal animals is elicited. There is a problem of experimental feasibility, however. In the case of the E^1 extensor burst, no controlling stimuli have so far been identified. An easier way to separate elicited from conditioned locomotor responses may be to administer a known stimulus during the step cycle and then establish its effect on different responses.

The "reflex reversal" experiment that has been used with reduced preparations (see p. 586) was studied briefly by Duysens (1976) in intact cats. EMG activity of the triceps surae was recorded during the application of weak stimulation, early in the stance phase, to the sural, posterior tibial, or common peroneal nerve. There was a short latency response of enhanced ankle extensor EMG (increased amplitude and duration) that resembled in form the response in premammillary cats, but far more than simple reflex control was involved in Duysen's (1976) experiments. The stimulation "often caused the normal cat to stop walking, to turn around, or to change the rhythm of stepping so that only short-term changes in the locomotory behavior could be studied" (Duysens, 1976, p. 21–22). Sometimes the cat shook its foot before the swing began. We may assume that the longer latency effects included operants and condititioned respondents that could be brought under separate control. Not much is known about the range of stimuli that control a "reflex reversal," but Prochazka *et al.* (1977) found that excitation of cutaneous afferents is sufficient. They found no increase in the activity of an identified Ia spindle afferent from the ankle extensor group during a "reflex reversal" that was elicited by touching the dorsal skin during locomotion by an intact cat.

The experiments by Duysens typify a high degree of current technical proficiency by many workers who record from identified central or peripheral neurons while an intact, conscious, animal locomotes. The procedures are still difficult, at best, and only scattered data have been taken. Some of the responses look much like those of reduced animals. McElligott (1976) found that some Purkinje cells in intact animals fired cyclically in phase with particular portions of the step cycle, just as was true for decerebrate cats (Orlovsky, 1972). Hoffer and Marks (1976) recorded peripheral, multiunit spike activity from the nerve to the tenuissimus muscle during stepping by the rabbit. Afferent activity to this muscle began sooner, peaked together with, and continued after alpha motoneuron discharge, to imply alpha–gamma coactivation, as had been seen previously in decerebrate cats (Severin, 1970). Prochazka and his collaborators (1977) also found coactivation during stepping by intact cats.

Behaviors of reduced and intact cats may look alike, of course, but be controlled very differently. When the behaviors do not even look alike, analysis has seemed especially difficult. Modern unit recording methods have yielded data from locomoting animals that have been puzzling or even chaotic from a systems engineering point of view. The records seem to be "full of errors," and a few of these findings will be summarized.

Loeb *et al.* (1977) and Loeb and Duysens (1978) recorded from single dorsal root ganglion cells during treadmill stepping by cats. Cells of different conduction velocity were sampled, and the principal findings were that their behaviors could not have been predicted by measures in quiescent animals. A guard hair cell receptor fired at foot touchdown and liftoff, although it was on the dorsum of the foot. Presumably the hairs between the toes rubbed together. Another cell had a receptive field on the plantar surface, yet it fired during the swing and had varying periods of silence during the stance. The firing pattern of a joint afferent fiber could not be explained by examining knee joint angle, joint loading, or velocity of movement alone. The authors emphasized that, "In extrapolating from receptor responses to artificial stimuli applied to immobilized animals, one is handicapped by the inability to predict accurately the occurrence of the effective stimuli during natural movements" (Loeb *et al.*, 1977, p. 1194).

Prochazka *et al.* (1977) found complex Ia activity. They surgically inserted fine wires into a dorsal root of cats and identified Ia afferent cells whose behavior was then followed. They also recorded length of the triceps surae muscle group via an implanted length transducer, and EMGs from biceps femoris and lateral gastrocnemius. When the experimenter pressed down lightly on the animal's back during stepping, a high Ia firing rate followed. This very strong stimulus to fusimotor activity was interpreted as a load compensation response, or else a cortically mediated response involved in "general arousal," "learning," and/or "volition."

What was termed "natural movement" or "voluntary behavior" in the reports cited in the last two paragraphs cannot be accounted for as we saw in an earlier section, by any inner agent in systems theory. Even if an agent, such as the cerebral cortex or a psychic event could be posited, there is a serious problem in working out functional circuits for different locomotor movements on the basis of available data. Only a few cells can be recorded from in any one experiment, so that connections cannot be determined directly. More importantly, the "natural" or "voluntary" movements are evoked or discriminated behaviors. The further course of this work will require stringent control of contingencies of reinforcement and discriminative stimuli. Investigators will also have to abandon the concept of voluntary movement, but it will not disappear without a struggle. Volition has long been held in esteem for man, if for no other creature, and we now turn to the topic of human locomotion.

HUMAN LOCOMOTION

Behaviors of people differ from those of other beings in that human responses come under verbal control, in addition to all the other controlling variables that can be observed in animals. *Verbal behavior* is not just vocal speech or writing, but is all "behavior reinforced through the mediation of other persons" (Skinner, 1957, p. 2). It may include such responses as typing, pointing, clapping, or locomotion. In any normal human, it is safest to assume that all movements come under the control of verbal stimuli produced by other people. The models of feedback or notions of voluntary control that are used to account for locomotion and other behaviors are themselves verbal repertoires, extended by known verbal mechanisms to situations in which the ultimate reinforcers are provided by the external world (see Skinner, 1957, for an account of scientific repertoires).

SPECIAL PROBLEMS IN UNDERSTANDING HUMAN GAITS. Traditional treatments of locomotion have not considered it as a verbal repertoire, even in part, so there has been misunderstanding of its control. Some instances and explanations will be given before we proceed to laboratory descriptions of human locomotion.

Both positive and negative reinforcement contingencies operate when a baby begins to walk. It gains a cookie from a plate when it stands or moves forward, and at the same time it escapes (and later avoids) painful stimulation from falling down and hitting its head on the floor. When we watch an infant who is learning to crawl and walk we should recall especially the aversive stimulation attendant on any behavior other than an erect upright posture. Small children incur minor hurts and injuries all day long. Up to this point, the locomotion is under multiple, but nonverbal control. However, the transition to verbal control is smooth. The cookie is gained either by movement toward the plate or else by movement toward a parent who holds up a cookie and perhaps says "Cookie!" or calls out "Run over here for a cookie!" If the parent is the source of a cookie, then running is a verbal operant whose primary reinforcer is food. The actions and statements of the parent serve as discriminative stimuli and conditioned reinforcers. The three-term contingency of discriminative stimulus, response, and reinforcer accounts for each locomotor response, even though there may be many individual stimuli and responses. The explanation is complete, and no useful addition is made by speaking of learning to walk as "trial and error" with respect to some "correct" behavior (see Skinner, 1953, for full discussion of trial-and-error learning, as was considered important by Thorndike). We only need to say that if a certain consequence is made contingent on behavior, that behavior increases in frequency. The concept of *error* in psychology has been at least as troublesome as has the other concept of *error*, derived from systems theory, in motor control physiology.

A second troublesome concept in locomotion control theory is espceially important for studying humans because of currently popular "biofeedback" training. This concept has a problem term, *feedback*. Feedback in systems theory has a precise meaning, as we have noted in previous sections. The term cannot be used to account for behavior, independently of the prevailing contingencies. Feedback stimulation may elicit, evoke, or discriminate behaviors in the setting, in addition to the presumptive reinforcement function. Alternatively, a given feedback event may not be functionally related at all to the particular behavior that is of interest to the investigator.

A third troublesome concept in locomotion control theory has been that of "consciousness." The role of proprioceptive excitation during progression is particularly unclear, and the vestibular system offers a good example. Balancing behaviors will permit the walker to escape or avoid pain, or just discomfort from holding one's head to one side. Nevertheless, Ayers (1972) pointed out that vestibular functions are often overlooked, and "their importance is taken for granted unless they are either disturbed or have affected the digestive tract. In spite of this lack of conscious awareness of the significance of the vestibular system, the infant likes to be rocked and the toddler enjoys being swung or tossed in the air" (Ayers, 1972, pp. 55–56). From Ayers's statement we suspect that positively reinforcing contingencies may be associated with vestibular function, as well as negative ones, but what does "conscious awareness" mean? Control is evident without awareness, because the child approaches a parent for cookies long before he can say why, if indeed he ever can. The fact of "unconscious" control does not mean that bodily movements cannot be finely discriminated, for they clearly are in the expert guitar player or motorcycle rider. Awareness means more than fine grain, and Skinner (1974) has listed seven things radical behaviorism has to say about consciousness. One of these statements is particularly appropriate here. "A person becomes conscious ... when a verbal community arranges contingencies under which he not only sees an object, but sees that he is seeing it. In this special sense, consciousness or awareness is a social product" (p. 220). We are concerned, then, with more than one behavior. "Walking over rough terrain is one thing; knowing that one is doing so is another" (Skinner, 1974, p. 31). This distinction is important in training children, athletes, and disabled people to locomote more successfully, but these topics are beyond the scope of this chapter (see, however, Goldiamond, 1976, on pathological movements and the effects of stimuli arising from one's own injured body). Our emphasis here is less on the details of which behaviors are observed, than on how the control of all human locomotion is achieved.

LABORATORY FINDINGS AND FUNCTIONAL UNITS. A good many studies have been made of the topography of human stepping, with representative recent studies by Okada, Ishida, and Kimura (1975); Pedotti (1977); and the authors of human reports included in the book edited by Herman, Grillner, Stein, and Stuart (1976). Kinematic facts have also been revealed about gait disorders and treated quantitatively (e.g., Bekey, Chang, Perry, and Hoffer, 1977; Kljacić, Kralj, Bajd, Stanič, and Trnkoczy, 1975). The general procedures of human gait laboratories are very similar from one to another in that subjects are merely told to walk or run, sometimes at a particular speed. Invariably, the performance is regarded as a "choice," without regard to the history of the subject. We must first account, then, for the presence of laboratory locomotion at all.

When an experimenter asks a subject to walk, he *mands* the behavior. A *mand* is a "verbal operant in which the response is reinforced by a characteristic consequence" in a verbal community (Skinner, 1957, p. 35). The experimenter's "reasons" for requesting the locomotion lie in his own reinforcement history and are not important to us here. The subject's "reasons" (the prior and current controlling events) are the important features of a functional analysis. Since early childhood, the subject has been given orders to "Walk to the store for some milk" or "Race me to the corner!" The reinforcer may be positive (the milk) or negative, as in the removal of a threat (of a spanking if the child does not go to the store or of laughter if the race is lost). In a race, of course, there is an added temporal specification that produces a highly complex set of preparatory behaviors which in the trained athlete begin long before the starting signal. The experimental subject's reasons are not all verbal, and perhaps all stimulation that results from successful manipulation of the world is reinforcing (Skinner, 1953). Children and even adults will walk on railroad rails when quite alone.

By the time the subject appears in a human motor control laboratory, the reinforcers for many locomotor behaviors are conditioned and *generalized* (they have been paired with several primary reinforcers, such as food or water). The usual reinforcer is little more than approval by the experimenter, which is sufficient, because reinforcement has resulted in improved performances of many kinds throughout the subject's life. Because of these multiple strengthenings, mands to "walk slowly" or "walk fast" result in extremely reliable and reproducible behaviors in most subjects after only a few practice trials. The walking is quickly discriminated by the few new stimuli present in the laboratory setting. The apparent ease of training has to some extent obscured the powerful mechanisms of control, a problem that sometimes has been unrecognized by the subject, the experimenter, or both. As was true for animal locomotion, the current concern of most students of human locomotion has been kinematic and kinetic descriptions. We will now mention a few of these valuable studies, whose data are also relevant to a functional analysis of locomotor behaviors.

Nelson, Dillman, Lagasse, and Bickett (1972) found a number of kinematic differences between treadmill and overground running by men. On the treadmill, there were longer periods of support, lower vertical velocities, and "less variable" vertical and horizontal velocities than for the same subjects running overground. We can attribute these differences to different discriminative stimuli in the two situations, even without being able to specify them. The experiment also revealed something about velocity control by athletes when they run on open terrain. When requested to move at specified velocities by matching their speed to an automatic pacer that moved alongside them, their matching was poor. The athletes would have had no previous history of such training, so that the resulting "course grain" is not so surprising. We also see from this experiment that it is not necessarily useful to provide "feedback" when its stimulus functions are unknown.

The grain of human locomotor repertoires has received only preliminary experimental attention. Herman, Wirta, Bampton, and Finley (1976) just directed subjects to "select a brisk, natural or comfortable, slow, and extremely slow" (p. 14) rate of walking. Interesting results were obtained, nonetheless (Cook and Cozzens, 1976; Craik, Herman, and Finley, 1976; Herman, Wirta, Bampton and Finley 1976). Subjects "selected" stride frequencies somewhat lower in the laboratory than when studied covertly in shopping centers or other everyday settings (Herman, Wirta, Bampton, and Finley, 1976). Stride frequency is therefore subject to external control, although the effective stimuli have not been identified.

The descriptive data on human locomotor repertoires are orderly and now warrant a full-scale functional analysis. Temporal and force data concerning joint movements showed little variation in the leg from cycle to cycle during walking at a constant frequency (Herman, Wirta, Bampton, and Finley, 1976). In contrast, the amplitude and duration of motor discharges in the EMG were quite variable (Herman, Wirta, Bampton, and Finley, 1976). It is tempting to speculate that a number of different arrangements of muscle activity are functionally interchangeable (equivalent operants) in producing fairly invariant movements of a whole limb, but to do so would require far more extensive sampling of EMGs from all the limb muscles.

Interlimb data (Craik et al., 1976) indicated that when a person walked, there were high correlations between timings of leg and arm movements. The arm muscles, in fact, were cyclically active. This activity was perhaps elicited by stimuli from the moving legs, but it may also have been discriminated by proprioceptive or other stimuli. Although additional observations were not available, several experiments are possible. Merely holding one arm still would reveal to some extent the conditional features of interlimb coordination.

One more study may be cited, in this case to show how data gathered to construct a mathematical model, stated in terms of muscular torques at the joints and the instantaneous length of limb muscles during a stride (Pedotti, 1977), can be given behavioral interpretations. Pedotti's (1977) subjects walked 15 m along a path, "choosing" their own stepping rate, which was then discriminated by a metronome click. A force plate measured ground reactions, kinematics were cinefilmed, and eleven leg EMGs were recorded, three at a time. In many respects, the data for the seven male subjects were similar, but there were some differences that suggested different combinations of muscles could achieve the same torques. Despite the equality of torques at the ankle joint, one subject showed ankle extensor activities in gastrocnemius and soleus that were in phase, while in another subject the two EMGs were much less in phase (cf. Pedotti's Figs. 5c and 6c).

In the reports of both Herman, Wirta, Bampton, and Finley (1976) and Pedotti (1977) there was the fascinating suggestion that EMG responses can be separated functionally to a considerable extent. Perhaps, for example, the E^1 extensor burst could be brought under discriminative control separately in different muscles. Work of this kind has no precedents in locomotion, but there is a relevant literature that concerns other "intended" or "voluntary" movements with humans (for man, see Hammond, 1956, and Hagbarth, 1967; for the monkey, see summary by Evarts and Granit, 1976). In general, investigators have found that the latency and even the presence or absence of EMG bursting in selected arm muscles varied with "intention" or, translated, the discriminative control exerted by instructions to the subjects. These responses were different from short latency bursts that

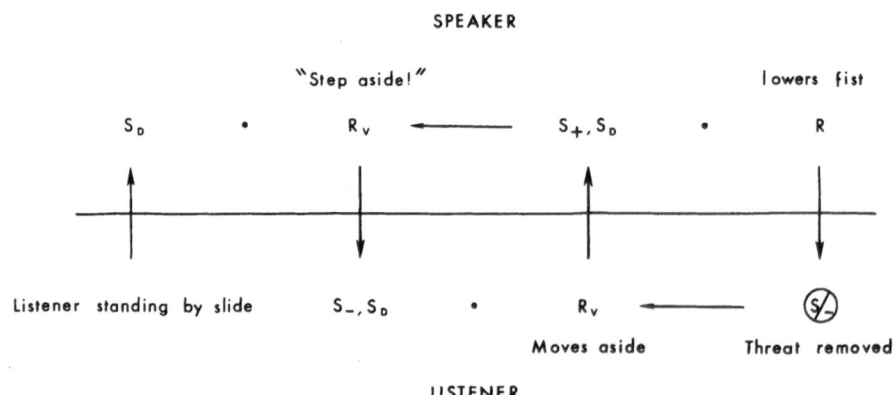

Fig. 7. Human locomotion as a verbal behavior (Fig. 2 modified and reprinted from B. F. Skinner *Verbal Behavior,* ©1957, p. 39, by permission from Prentice Hall, Inc., Englewood Cliffs, N.J.). The horizontal line in the figure separates responses and stimulus functions for the speaker and listener. Vertical arrows cross this line where behaviors of one person have stimulus functions for the other person. Dots between S_Ds (discriminative stimuli) and Rs (responses) indicate that only a short period of time has elapsed between the two events. Horizontal arrows occur between Rs and reinforcers because the strength of the S_D (probability of R in the presence of the S_D) is dependent on its reinforcement history. The sequence of events begins at the left side of the diagram. Below the horizontal line, a *listener* (who is a child standing by a slide) moves aside as a verbal response, R_v. His stepping is discriminated by the mand, "Step aside!" (S_D) of another child, the *speaker* (above the line). For the listener, the speaker's presence is aversive (S_-). When the listener steps aside, the other child lowers his fist and moves away up the ladder. Termination of this threat (circle and diagonal line cancel S_-) reinforces its control over the verbal operant, moving aside (R_v). The behavior of the speaker is also under verbal control. The discriminative control for his mand "Step aside!" is reinforced (S_+) by the listener's compliance.

apparently were elicited. Recall that Duysens (1976) found short latency reflex reversals in intact, stepping cats that were followed by longer latency discriminated operants. We may conclude that evidence for fine discriminative control of human locomotion, as well, could be gained in additional experiments. Such data would place findings of individualized EMG bursting in "biofeedback" experiments on a far more sound scientific basis than is now the case.

We emphasize once again that to understand human locomotion, the reader must be acquainted with the origins and significance of verbal behavior. It is at once the most accessible motor event to us all and the most misunderstood. Figure 7 was modified from a diagram in *Verbal Behavior* (Fig. 2, Skinner, 1957, p. 39). The brief verbal episode in Fig. 7, comprising a few moments of elapsed time, shows how just one stepping incident becomes part of the history of an individual.

The contingencies for a speaker and a listener are separated in the figure by a horizontal line. Other verbal operants are present in the episode, but we are primarily interested here in the "step aside" response. The figure should be scanned in general from left to right for the speaker and also for the listener, but vertical arrows indicate the direction of interactions between the two people. Beginning at the extreme left, we assume that the listener's presence constitutes a discriminative stimulus for the speaker. Perhaps the setting is a playground, and the listener is a child standing at the foot of the ladder on a slide. The speaker is moving toward the slide. The speaker says "Step aside!" which is a verbal response composed of a specification for action and a threat. He may put out his fist also. The operant is, of course, a mand. For the listener, the events of the mand constitute a verbal discriminative stimulus that is also aversive (due to a previous conditioning history), and he steps away from the ladder. The listener's response clears the way for the speaker, who probably lowers his fist, moves up the ladder, and thereby withdraws the threat. There are several reinforcers for the speaker's response, conditioned and unconditioned, but they do not concern us here, and do not appear in Fig. 7. Withdrawal by the speaker reinforces the listener's verbal operant of stepping aside, and the episode ends. Note the complexity of the complete interaction between two people, even when we have not listed every stimulus and response. Each event and each response have several functions for each individual, and the functions are different for the two children.

Conclusions

Numerous suggestions have been made concerning how stimulus functions may be separated and how other variables that control locomotion may be understood. Behavioral control mechanisms have perhaps been given more space than have physiological mechanisms, but the study of psychology has been neglected and is more compelling at this time. We have provided only an introduction to the analysis of locomotion as a complex set of behaviors under multiple control. It should be apparent that no simple set of behavioral techniques can be brought in as a corrective to our present ignorance. Instead, students of motor control must study behavioral theory in as much depth as they now study physiology. There is no real separation, however, between psychology and physiology. We are talking about one science, in which "the behavior of an organism is simply the physiology of an anatomy" (Skinner, 1969, p. 173).

References

Alexander, R. McN. Swimming. In R. McN. Alexander and G. Goldspink (Eds.), *Mechanics and Energetics of Animal Locomotion.* London: Chapman & Hall, 1977.

Alexander, R. McN., and Goldspink, G. (Eds.), *Mechanics and Energetics of Animal Locomotion.* London: Chapman & Hall, 1977.

Andersson, O., Grillner, S., Lindquist, M., and Zomlefer, M. Peripheral control of the spinal pattern generators for locomotion in cat. *Brain Research,* 1978, *150,* 625–630.

Arbib, M. A. *Brain, Machines, and Mathematics.* New York: McGraw-Hill, 1964.

Arbib, M. A. *The Metaphorical Brain. An Introduction to Cybernetics as Artificial Intelligence and Brain Theory.* New York: Wiley-Interscience, 1972.

Ariel, G., and Maulucci, R. A kinetic analysis of the trot in cats. In R. M. Herman, S. Grillner, P. S. G. Stein, and D. G. Stuart (Eds.), *Neural Control of Locomotion.* New York: Plenum, 1976.

Arshavsky, Yu. I., Kots, Ya. M., Orlovsky, G. N., Rodionov, I. M., and Shik, M. L. Investigation of the biomechanics of running by the dog. *Biophysics,* 1965, *10,* 737–746.

Ayers, A. J. *Sensory Integration and Learning Disorders.* Los Angeles: Western Psychological Services, 1972.

Ayers, J. L., Jr., and Davis, W. J. Neuronal control of locomotion in the lobster, *Homarus americanus.* I. Motor programs for forward and backward walking. *Journal of Comparative Physiology,* 1977, *115,* 1–27. (a)

Ayers, J. L., Jr., and Davis, W. J. Neuronal control of locomotion in the lobster, *Homarus americanus.* II. Types of walking leg reflexes. *Journal of Comparative Physiology,* 1977, *115,* 29–46. (b)

Ayers, J. L., Jr., and Davis, W. J. Neuronal control of locomotion in the lobster, *Homarus americanus.* III. Dynamic organization of walking leg reflexes. *Journal of Comparative Physiology,* 1978, *123,* 289–298.

Bard, P. A diencephalic mechanism for the expression of rage with specific reference to the sympathetic nervous system. *American Journal of Physiology,* 1928, *84,* 490–515.

Barnes, W. J. P. Proprioceptive influences on motor output during walking in the crayfish. *Journal of Physiology (Paris),* 1977, *73,* 543–563.

Beer, F. P., and Johnston, E. R., Jr. *Mechanics for Engineers: Statics and Dynamics.* New York: McGraw-Hill, 1962.

Bekey, G. A., Chang, C-W., Perry, J., and Hoffer, M. M. Pattern recognition of multiple EMG signals applied to the description of human gait. *Proceedings of the IEEE,* 1977, *65,* 674–681.

Bernstein, N. A. *On the Construction of Movements.* Moscow: Medgiz, 1947. (Monograph in Russian.)

Bessonov, A. P., and Umnov, N. V. The analysis of gaits in six-legged vehicles according to their static stability. *Proceedings of a Symposium on the Theory of Robots and Manipulators.* Udine, Italy: International Center for Mechanical Science, 1973.

Bethe, A. Studien über die Plastizität des Nervensystems. I. Mitteilung. Arachnoideen und Crustaceen. *Pflügers Archiv für die Gesamte Physiologie des Menschen und der Tiere,* 1930, *224,* 793–820.

Betts, B., Smith, J. L., and Collatos, T. C. Recording fore and hind limb myopotentials during unrestrained movements of cats. *Brain Research,* 1976, *117,* 529–533.

Bowdan, E. Walking and rowing in the water strider, *Gerris remigis.* I. A cinematographic analysis of walking. *Journal of Comparative Physiology,* 1978, *123,* 43–49.

Boylls, C. C., Jr. Olivary unit activity and effect of microstimulation during locomotion. *Society for Neuroscience Abstracts,* 1977, *3,* 55.

Brown-Séquard, C. E. Des rapports qui existent entre la lésion des racines motrices et des racines sensitives. *Comptes Rendus des Séances de la Société de Biologie et de ses Filiales*, 1850, *1*, 15–17.

Buchwald, J. S., and Brown, K. Subcortical mechanisms of behavioral plasticity. In J. D. Maser (Ed.), *Efferent Organization and the Integration of Behavior*. New York: Academic Press, 1973.

Burrows, M. Neural control of flight in the locust. In R. M. Herman, S. Grillner, P. S. G. Stein, and D. G. Stuart (Eds.), *Neural Control of Locomotion*. New York: Plenum Press, 1976.

Carlson, A. B. *Communication Systems: An Introduction to Signals and Noise in Electrical Communication*. New York: McGraw-Hill, 1968.

Chassin, P. S., Taylor, C. R., Heglund, N. C., and Seeherman, H. J. Locomotion in lions: Energetic cost and maximum aerobic capacity. *Zoology*, 1976, *49*, 1–10.

Chestnut, H., and Mayer, R. W. *Servomechanisms and Regulating System Design*. New York: Wiley, 1951.

Cook, T., and Cozzens, B. Human solutions for locomotion. III. The initiation of gait. In R. M. Herman, S. Grillner, P. S. G. Stein, and D. G. Stuart (Eds.), *Neural Control of Locomotion*. New York: Plenum Press, 1976.

Coss, L., Chan, A. K., Goslow, G. E. Jr., and Rasmussen, S. Ipsilateral limb variation in cats during overground locomotion. *Brain Research*, 1978, *15*, 85–93.

Craik, R., Herman, R., and Finley, F. R. Human solutions for locomotion. II. Interlimb coordination. In R. M. Herman, S. Grillner, P. S. G. Stein, and D. G. Stuart (Eds.), *Neural Control of Locomotion*. New York: Plenum Press, 1976.

Creed, R. S., Denny-Brown, D., Eccles, J. C., Liddell, E. G. T., and Sherrington, C. S. *Reflex Activity of the Spinal Cord*. Oxford: Clarendon Press, 1932. (Reprinted in 1972.)

Davis, W. J. Central activation of movements. In R. M. Herman, S. Grillner, P. S. G. Stein, and D. G. Stuart (Eds.), *Neural Control of Locomotion*. New York: Plenum Press, 1976.

Davis, W. J., and Ayers, J. L. Locomotion: Control by positive-feedback optokinetic responses. *Science*, 1972, *177*, 183–185.

Delcomyn, F. An approach to the study of neural activity during behavior in insects. *Journal of Insect Physiology*, 1976, *22*, 1223–1227.

Duysens, J. E. *Reflex control of cat walking*. Doctoral dissertation. University of Alberta, Canada, 1976.

Duysens, J., and Pearson, K. G. Ipsilateral extensor reflexes and cat locomotion. In R. M. Herman, S. Grillner, P. S. G. Stein, and D. G. Stuart (Eds.), *Neural Control of Locomotion*. New York: Plenum Press, 1976.

Edgerton, V. R., Grillner, S., Sjöstrom, A., and Zangger, P. Central generation of locomotion in vertebrates. In R. M. Herman, S. Grillner, P. S. G. Stein, and D. G. Stuart (Eds.), *Neural Control of Locomotion*. New York: Plenum Press, 1976.

Eisenstein, B. L., Postillion, F. G., Norgren, K. S., and Wetzel, M. C. Kinematics of treadmill galloping by cats. II. Steady-state coordination under positive reinforcement control. *Behavioral Biology*, 1977, *21*, 89–106.

Engberg, I. Reflexes to foot muscles in the cat. *Acta Physiologica Scandinavica*, 1964, *62*, Supplementum 235, 1–64.

Engberg, I., and Lundberg, A. An electromyographic analysis of muscular activity in the hindlimb of the cat during unrestrained locomotion. *Acta Physiologica Scandinavica*, 1969, *75*, 614–630.

English, A. W. Interlimb coordination during stepping in the cat: an electromyographic analysis. *Journal of Neurophysiology*, 1979, *42*, 229–243.

Evarts, E. V., and Granit, R. Relations of reflexes and intended movements. In S. Homma (Ed.), *Progress in Brain Research, Vol. 44: Understanding the Stretch Reflex*. Amsterdam: Elsevier/North-Holland Biomedical Press, 1976.

Evoy, W. H. Modulation of proprioceptive information in crustacea. In R. M. Herman, S. Grillner, P. S. G. Stein, and D. G. Stuart (Eds.), *Neural Control of Locomotion*. New York: Plenum Press, 1976.

Fearing, F. *Reflex Action. A Study in the History of Physiological Psychology*. Baltimore: Williams & Wilkins, 1930.

Feldman, A. G., and Orlovsky, G. N. Activity of interneurons mediating reciprocal Ia inhibition during locomotion. *Brain Research,* 1975, *84,* 181–194.

Ferster, C. B., and Skinner, B. F. *Schedules of Reinforcement.* New York: Appleton-Century-Crofts, 1957.

Forssberg, H., Grillner, S., Rossignol, S., and Wallén, P. Phasic control of reflexes during locomotion. In R. M. Herman, S. Grillner, P. S. G. Stein, and D. G. Stuart (Eds.), *Neural Control of Locomotion.* New York: Plenum Press, 1976.

Fourtner, C. R. Central nervous control of cockroach walking. In R. M. Herman, S. Grillner, P. S. G. Stein, and D. G. Stuart (Eds.), *Neural Control of Locomotion.* New York: Plenum Press, 1976.

Gambaryan, P. P. *How Mammals Run.* Translated from the Russian by H. Hardin. New York: Wiley, 1974, pp. 203–259.

Goldiamond, I. Coping and adaptive behaviors of the disabled. In G. L. Albracht (Ed.), *The Sociology of Physical Disability and Rehabilitation.* Pittsburgh: University of Pittsburgh, 1976.

Goslow, G. E., Jr., Reinking, R. M., and Stuart, D. G. The cat step cycle: Hind limb joint angles and muscle lengths during unrestrained locomotion. *Journal of Morphology,* 1973, *141,* 1–41.

Graham Brown, T. The intrinsic factors in the act of progression in the mammal. *Proceedings of the Royal Society (London), Series B,* 1911, *84,* 308–319.

Graham Brown, T. The phenomenon of "narcosis progression" in mammals. *Proceedings of the Royal Society (London), Series B,* 1913, *86,* 140–164.

Graham Brown, T. On the nature of the fundamental activity of the nervous centres; together with an analysis of the conditioning of rhythmic activity in progression, and a theory of the evolution of function in the nervous system. *Journal of Physiology (London),* 1914, *48,* 18–46.

Graham Brown, T. On the activities of the central nervous system of the un-born foetus of the cat; with a discussion of the question whether progression (walking, etc.) is a "learnt" complex. *Journal of Physiology (London),* 1915, *49,* 208–215.

Grillner, S. Locomotion in the spinal cat. In R. B. Stein, K. G. Pearson, R. S. Smith, and J. B. Redford (Eds.), *Control of Posture and Locomotion.* New York: Plenum Press, 1973. (a)

Grillner, S. Locomotion in the spinal dogfish. *Acta Physiologica Scandinavica,* 1973(b), *87,* 31–32A.

Grillner, S. Locomotion in vertebrates: Central mechanisms and reflex interaction. *Physiological Reviews,* 1975, *55,* 247–304.

Grillner, S. Some aspects on the descending control of the spinal circuits generating locomotor movements. In R. M. Herman, S. Grillner, P. S. G. Stein, and D. G. Stuart (Eds.), *Neural Control of Locomotion.* New York: Plenum Press, 1976.

Grillner, S., and Kashin, S. On the generation and performance of swimming in fish. In R. M. Herman, S. Grillner, P. S. G. Stein, and D. G. Stuart (Eds.), *Neural Control of Locomotion.* New York: Plenum Press, 1976.

Grillner, S., and Zangger, P. Locomotor movements generated by the deafferented spinal cord. *Acta Physiologica Scandinavica,* 1974, *91,* 38–39A.

Hagbarth, K.-E. EMG studies of stretch reflexes in man. In L. Widén (Ed.), *Recent Advances in Clinical Neurophysiology. Electroencephalography and Clinical Neurophysiology,* 1967, *Supplement 25,* 74–79.

Halbertsma, J. M., Miller, S., and van der Merché, F. G. A. Basic programs for the phasing of flexion and extension movements of the limbs during locomotion. In R. M. Herman, S. Grillner, P. S. G. Stein, and D. G. Stuart (Eds.), *Neural Control of Locomotion.* New York: Plenum Press, 1976.

Hammond, P. H. The influence of prior instruction to the subject on an apparently involuntary neuro-muscular response. *Journal of Physiology (London),* 1956, *132,* 17P–18P.

Hart, B. L. Facilitation by strychnine of reflex walking in spinal dogs. *Physiology and Behavior,* 1971, *6,* 627–628.

Hazen, H. L. Theory of servomechanisms. *Journal of the Franklin Institute,* 1934, *218,* 279–331.

Hemami, H., Weimer, F. C., and Koozekanini, S. H. Some aspects of the inverted pendulum problem for modeling of locomotion systems. *14th Joint Automatic Control Conference Preprint Volume, Columbus, Ohio,* 1973, pp. 132–137. IEEE Catalog No. 73CHO 750-0 CSS.

Henneman, E., Clamann, H. P., Gillies, J. D., and Skinner, R. D. Rank order of motoneurons within a pool: Law of combination. *Journal of Neurophysiology*, 1974, *37*, 1338-1349.

Herman, R. M., Grillner, S., Stein, P. S. G., and Stuart, D. G. (Eds.). *Neural Control of Locomotion*. New York: Plenum Press, 1976.

Herman, R., Wirta, R., Bampton, S., and Finley, F. R. Human solutions for locomotion. I. Single limb analysis. In R. M. Herman, S. Grillner, P. S. G. Stein, and D. G. Stuart (Eds.), *Neural Control of Locomotion*. New York: Plenum Press, 1976.

Hildebrand, M. Symmetrical gaits of horses. *Science*, 1965, *150*, 701-708.

Hildebrand, M. Analysis of the symmetrical gaits of tetrapods. *Folia Biotheoretica*, 1966, *6*, 1-22.

Hildebrand, M. Analysis of tetrapod gaits: General considerations and symmetrical gaits. In R. M. Herman, S. Grillner, P. S. G. Stein, and D. G. Stuart (Eds.), *Neural Control of Locomotion*. New York: Plenum Press, 1976.

Hildebrand, M. Analysis of asymmetrical gaits. *Journal of Mammalogy*, 1977, *58*, 131-156.

Hinsey, J. C., Ranson, S. W., and McNattin, R. F. The role of the hypothalamus and mesencephalon in locomotion. *Archives of Neurology and Psychiatry*, 1930, *23*, 1-43.

Hoffer, J. A., and Marks, W. B. Long-term peripheral nerve activity during behavior in the rabbit. In R. M. Herman, S. Grillner, P. S. G. Stein, and D. G. Stuart (Eds.), *Neural Control of Locomotion*. New York: Plenum Press, 1976.

Houk, J. C. The phylogeny of muscular control configurations. *Biocybernetics*, 1972, *4*, 125-144.

Howell, A. B. *Speed in Animals*. pp. 217-247. Chicago: University of Chicago Press, 1944.

Hoyle, G. Arthropod walking. In R. M. Herman, S. Grillner, P. S. G. Stein, and D. G. Stuart (Eds.), *Neural Control of Locomotion*. New York: Plenum Press, 1976.

Hultborn, H. Convergence on interneurones in the reciprocal Ia inhibitory pathway to motoneurones. *Acta Physiologica Scandinavics*, 1972, *Supplementum 375*, 1-42.

Hultborn, H., Jankowska, E., and Lindström, S. Recurrent inhibition from motor axon collaterals of transmission in the Ia inhibitory pathway to motoneurones. *Journal of Physiology (London)*, 1971, *215*, 591-612. (a)

Hultborn, H., Jankowska, E., and Lindström, S. Recurrent inhibition of interneurones monosynaptically activated from group Ia afferents. *Journal of Physiology (London)*, 1971, *215*, 613-636. (b)

Jankowska, E., and Smith, D. O. Antidromic activation of Renshaw cells and their axonal projections. *Acta Physiologica Scandinavica*, 1973, *88*, 198-214.

Jankowska, E., Jukes, M. G. M., Lund, S., and Lundberg, A. The effect of DOPA on the spinal cord. 5. Reciprocal organization of pathways transmitting excitatory action to alpha motoneurones of flexors and extensors. *Acta Physiologica Scandinavica*, 1967, *70*, 369-388. (a)

Jankowska, E., Jukes, M. G. M., Lund, S., and Lundberg, A. The effect of DOPA on the spinal cord. 6. Half-centre organization of interneurones transmitting effects from flexor reflex afferents. *Acta Physiologica Scandinavica*, 1967, *70*, 389-402. (b)

Jordan, L. M., and Steeves, J. D. Chemical lesioning of the spinal noradrenaline pathway: Effects on locomotion in the cat. In R. M. Herman, S. Grillner, P. S. G. Stein, and D. G. Stuart (Eds.), *Neural Control of Locomotion*. New York: Plenum Press, 1976.

Kashin, S. M., Feldman, A. G., and Orlovsky, G. N. Locomotion of fish evoked by electrical stimulation of the brain. *Brain Research*, 1974, *82*, 41-47.

Kato, I., and Tsuiki, H. Hydraulically powered biped walking machine with a high carrying capacity. *Proceedings of the Fourth International Symposium on External Control of Human Extremities*. Dubrovnik, Yugoslavia, 1972.

Kljacić, M., Kralj, A., Bajd, T., Stanič, U., and Trnkoczy, A. Some problems of mathematical quantitative gait evaluation. *Proceedings of the Fifth International Symposium on External Control of Human Extremities*. Dubrovnik, Yugoslavia, pp. 279-287, 1975.

Lennard, P. R., and Stein, P. S. G. Swimming movements elicited by electrical stimulation of turtle spinal cord. I. Low spinal and intact preparations. *Journal of Neurophysiology*, 1977, *40*, 768-778.

Lindsay, K. W., Roberts, T. D. M., and Rosenberg, J. R. Asymmetric tonic labyrinth reflexes and their interaction with neck reflexes in the decerebrate cat. *Journal of Physiology (London)*, 1976, *261*, 583-601.

Lindström, S., and Schomburg, E. D. Recurrent inhibition from motor axon collaterals of ventral spinocerebellar tract neurones. *Acta Physiologica Scandinavica*, 1973, *88*, 505–515.

Lissman, H. W. The neurological basis of the locomotory rhythm in the spinal dogfish (*Scyllium canicula, Acanthias vulgaris*). II. The effect of de-afferentation. *Journal of Experimental Biology*, 1946, *23*, 162–176.

Lockard, D. E., Traher, L. M., and Wetzel, M. C. Reinforcement influences upon topography of treadmill locomotion by cats. *Physiology and Behavior*, 1976, *16*, 141–146.

Loeb, G. E., and Duysens, J. The unit activity of primary and secondary afferents from cat hindlimb muscle spindles during normal walking. *Society for Neuroscience Abstracts*, 1978, *4*, p. 300.

Loeb, G. E., Bak, M. J., and Duysens, J. Long-term unit recording from somatosensory neurons in the spinal ganglia of the freely walking cat. *Science*, 1977, *197*, 1192–1194.

Lundberg, A., and Phillips, C. G. T. Graham Brown's film on locomotion in the decerebrate cat. *Journal of Physiology (London)*, 1973, *231*, 90–91P.

Magnus, R. *Die Körperstellung.* Berlin: Springer, 1924.

Manter, J. T. The dynamics of quadrupedal walking. *Journal of Experimental Biology*, 1938, *15*, 522–540.

Matthews, P. B. C. *Mammalian Muscle Receptors and Their Central Actions.* London: Edward Arnold, 1972.

Miller, S., and Scott, P. D. The spinal locomotor generator. *Experimental Brain Research*, 1977, *30*, 387–403.

Miller, S., and van der Meché, F. G. A. Coordinated stepping of all four limbs in the high spinal cat. *Brain Research*, 1976, *109*, 395–398.

Miller, S., van der Burg, J., and van der Meché, F. G. A. Coordination of movements of the hindlimbs and forelimbs in different forms of locomotion in normal and decerebrate cats. *Brain Research*, 1975, *91*, 217–237.

Murthy, K. S. K., Gildenberg, P. L., and Marchand, J. E. Descending long-spinal excitation of lumbar alpha and gamma motoneurons evoked by stretch of dorsal neck muscles. *Brain Research*, 1978, *140*, 165–170.

Muybridge, E. *Animals in Motion.* New York: Dover, 1957. (Originally published, 1887.)

MacMillan, D. L. A physiological analysis of walking in the American lobster (*Homarus americanus*). *Philosophical Transactions of the Royal Society, Series B*, 1975, *270*, 1–59.

McCrea, D. A. *Activity of spinal neurons during controlled locomotion.* Doctoral dissertation. University of Manitoba, Canada, 1979.

McElligott, J. G. Cerebellar neuronal firing patterns in the intact and unrestrained cat during walking. In R. M. Herman, S. Grillner, P. S. G. Stein, and D. G. Stuart (Eds.), *Neural Control of Locomotion.* New York: Plenum Press, 1976.

McFarland, D. J. *Feedback Mechanisms in Animal Behaviour,* New York: Academic Press, 1971.

McGhee, R. B. Some finite state aspects of legged locomotion. *Mathematical Biosciences*, 1968, *2*, 67–84.

McGhee, R. B. Robot locomotion. In R. M. Herman, S. Grillner, P. S. G. Stein, and D. G. Stuart (Eds.), *Neural Control of Locomotion.* New York: Plenum Press, 1976.

McGhee, R. B., and Frank, A. A. On the stability properties of quadruped creeping gaits. *Mathematical Biosciences*, 1968, *3*, 331–351.

McGhee, R. B., and Jain, A. K. Some properties of regularly realizable gait matrices. *Mathematical Biosciences*, 1972, *13*, 179–183.

Nelson, R. C., Dillman, C. J., Lagasse, P., and Bickett, P. Biomechanics of overground versus treadmill running. *Medicine and Science in Sports*, 1972, *4*, 233–240.

Norgren, K. S., Seelhorst, Sr. E., and Wetzel, M. C. Kinematics of treadmill galloping by cats. I. Steady-state coordination under aversive control. *Behavioral Biology*, 1977, *21*, 66–88.

Okada, M., Ishida, H., and Kimura, T. Biomechanical features of bipedal gait in human and nonhuman primates. In P. V. Komi (Ed.), *Biomechanics V-A. Proceedings of the Fifth International Seminar on Biomechanics, Jyväkyla, Finland.* Baltimore: University Park Press, 1975.

Orlovsky, G. N. Spontaneous and induced locomotion of the thalamic cat. *Biophysics*, 1969, *14*, 1154–1162.

Orlovsky, G. N. Influence of the cerebellum on the reticulo-spinal neurones during locomotion. *Biophysics,* 1970, *15,* 928–936.

Orlovsky, G. N. Work of the Purkinje cells during locomotion. *Biophysics,* 1972, *17,* 935–941.

Orlovsky, G. N., and Shik, M. L. Standard elements of cyclic movement. *Biophysics,* 1965, *10,* 935–944.

Paillard, J., and Massion, J. Motor aspects of behaviour and programmed nervous activities. *Brain Research,* 1974, *71, Special Issue,* 189–575.

Patterson, M. M. Mechanisms of classical conditioning and fixation in spinal mammals. In A. H. Riesen, and R. F. Thompson (Eds.), *Advances in Psychobiology,* Volume 3. New York: Wiley, 1976.

Pearson, K. G., and Duysens, J. Function of segmental reflexes ln the control of stepping in cockroaches and cats. In R. M. Herman, S. Grillner, P. S. G. Stein, and D. G. Stuart (Eds.), *Neural Control of Locomotion.* New York: Plenum Press, 1976.

Pearson, K. G., and Iles, J. F. Nervous mechanisms underlying intersegmental co-ordination of leg movements during walking in the cockroach. *Journal of Experimental Biology,* 1973, *58,* 725–744.

Pedotti, A. A study of motor coordination and neuromuscular activities in human locomotion. *Biological Cybernetics,* 1977, *26,* 53–62.

Penfield, W., and Roberts, L. *Speech and Brain-Mechanisms.* Princeton: Princeton University Press, 1959.

Perret, C. Neural control of locomotion in the decorticate cat. In R. M. Herman, S. Grillner, P. S. G. Stein, and D. G. Stuart (Eds.), *Neural Control of Locomotion.* New York: Plenum Press, 1976.

Perret, C., and Cabelguen, J.-M. Central and reflex participation in the timing of locomotor activations of a bifunctional muscle, the semi-tendinosus, in the cat. *Brain Research,* 1976, *106,* 390–395.

Philippson, M. L'autonomie et la centralisation dans le système nerveux des animaux. *Travaux du Laboratoire de Physiologie de l'Institut Solvay (Bruxelles),* 1905, *7,* 1–208.

Porter, R. Functions of the mammalian cerebral cortex in movement. *Progress in Neurobiology,* 1973, *1,* 1–51.

Prochazka, A., Westerman, R. A., and Ziccone, S. P. Ia afferent activity during a variety of voluntary movements in the cat. *Journal of Physiology (London),* 1977, *268,* 423–448.

Ramón y Cajal, S. *Histologie du Système Nerveux de L'Homme et des Vertébrés. Tome II.* Paris: Maloine, 1911.

Rasmussen, S., Chan, A. K., and Goslow, G. E., Jr. The cat step cycle: Electromyographic patterns for hindlimb muscles during posture and unrestrained locomotion. *Journal of Morphology,* 1978, *155,* 253–269.

Roaf, H. E., and Sherrington, C. S. Further remarks on the mammalian spinal preparation. *Quarterly Journal of Experimental Physiology,* 1910, *3,* 209–211.

Roberts, T. D. M. The role of vestibular and neck receptors in locomotion. In R. M. Herman, S. Grillner, P. S. G. Stein, and D. G. Stuart (Eds.), *Neural Control of Locomotion.* New York: Plenum Press, 1976.

Roeder, K. D. The control of tonus and locomotor activity in the praying mantis (*Mantis religiosa* L.). *Journal of Experimental Zoology,* 1937, *76,* 353–374.

Root, T. M., and Bowerman, R. F. Intra-appendage movements during walking in the scorpion *Hadrurus Arizonensis. Comparative Biochemistry and Physiology,* 1978, *59A,* 49–56.

Rossignol, S. The control of crossed extensor and crossed flexor responses. *Society for Neuroscience Abstracts,* 1977, *3,* p. 277.

Rossignol, S., and Gauthier, L. Patterns of contralateral limb responses to nociceptive stimuli during locomotion. *Society for Neuroscience Abstracts,* 1978, *4,* p. 304.

Ruch, T. C., and Watts, J. W. Reciprocal changes to reflex activity of the forelimbs induced by post-brachial "cold-block" of the spinal cord. *Journal of Physiology (London),* 1934, *110,* 362–375.

Ryall, R. W. Renshaw cell mediated inhibition of Renshaw cells: Patterns of excitation and inhibition from impulses in motor axon collaterals. *Journal of Neurophysiology,* 1970, *33,* 257–270.

Schomburg, E. D., Roesler, J., and Meinck, H.-M. Phase-dependent transmission in the excitatory propriospinal reflex pathway from forelimb afferents to lumbar motoneurones during fictive locomotion. *Neuroscience Letters,* 1977, *4,* 249–252.

Severin, F. V. The role of the gamma motor system in the activation of the extensor alpha motor neurones during controlled locomotion. *Biophysics,* 1970, *15,* 1138–1145.

Severin, F. V., Shik, M. L., and Orlovsky, G. N. Work of the muscles and single motor neurones during controlled locomotion. *Biophysics,* 1967, *12,* 762–772.

Severin, F. V., Orlovsky, G. N., and Shik, M. L. Reciprocal influences on work of single motoneurons during controlled locomotion. *Bulletin of Experimental Biology and Medicine,* 1968, *66,* 713–716.

Shannon, C. E. The mathematical theory of communication. *Bell System Technical Journal,* 1948, *27,* 379–423; 623–656.

Sherrington, C. S. Decerebrate rigidity and the reflex co-ordination of movements. *Journal of Physiology (London),* 1898, *22,* 319–322.

Sherrington, C. *The Integrative Action of the Nervous System.* Cambridge: University Press, 1906.

Sherrington, C. S. Flexion-reflex of the limb, crossed extension-reflex, and reflex stepping and standing. *Journal of Physiology (London),* 1910, *40,* 28–121.

Shik, M. L., and Orlovsky, G. N. Neurophysiology of locomotor automatism. *Physiological Reviews,* 1976, *56,* 465–501.

Shik, M. L., Severin, F. V., and Orlovsky, G. N. Control of walking and running by means of electrical stimulation of the mid-brain. *Biophysics,* 1966, *11,* 756–765.

Sidman, M. *Tactics of Scientific Research.* New York: Basic Books, 1960.

Sirota, M. G., Sirota, T. I., and Shik, M. L. Circulation during controlled locomotion in the mesencephalic cat. *Bulletin of Experimental Biology and Medicine,* 1971, *21,* 95–98.

Skinner, B. F. *The Behavior of Organisms. An Experimental Analysis.* New York: Appleton-Century-Crofts, 1938.

Skinner, B. F. *Science and Human Behavior.* New York: Macmillan, 1953.

Skinner, B. F. *Verbal Behavior.* Englewood Cliffs, N.J.: Prentice-Hall, 1957.

Skinner, B. F. *Contingencies of Reinforcement.* Englewood Cliffs, N.J.: Prentice-Hall, 1969.

Skinner, B. F. *About Behaviorism.* New York: Alfred A. Knopf, 1974.

Stein, P. S. G. Mechanisms of interlimb phase control. In R. M. Herman, S. Grillner, P. S. G. Stein, and D. G. Stuart (Eds.), *Neural Control of Locomotion.* New York: Plenum Press, 1976.

Stein, P. S. G. A comparative approach to the neural control of locomotion. In G. Hoyle (Ed.), *Identified Neurons and Behavior of Arthropods.* New York: Plenum Press, 1977.

Stein, P. S. G. Motor systems, with specific reference to the control of locomotion. *Annual Review of Neuroscience,* 1978, *1,* 61–81.

Stuart, D. G., Withey, T. P., Wetzel, M. C., and Goslow, Jr. G. E. Time constraints for interlimb co-ordination in the cat during unrestrained locomotion. In R. B. Stein, K. G. Pearson, R. S. Smith, and J. B. Redford (Eds.), *Control of Posture and Locomotion.* New York: Plenum Press, 1973.

Sukhanov, V. B. *General System of Symmetrical Locomotion of Terrestrial Vertebrates and Some Features of Movement of Lower Tetrapods.* New Delhi: Amerind, 1974.

Sun, S.-S. *A theoretical study of gaits for legged locomotion systems.* Doctoral dissertation. Ohio State University, 1974.

Székely, G., and Cźeh, G. Organization of locomotion. In R. Llinás and W. Precht (Eds.), *Frog Neurobiology.* Berlin: Springer-Verlag, 1976.

Taub, E. Motor behavior following deafferentation in the developing and motorically mature monkey. In R. M. Herman, S. Grillner, P. S. G. Stein, and D. G. Stuart (Eds.), *Neural Control of Locomotion.* New York: Plenum Press, 1976.

Taylor, C. R. Energy cost of locomotion. In J. Bous, K. Schmidt-Nielsen, and S. H. P. Maddrell (Eds.), *Comparative Physiology.* Amsterdam: North-Holland, 1973.

ten Cate, J. Quelques remarques à propos de l'innervation des mouvements locomotaires de la blatte (*Periplaneta americana*). *Archives Néerlandaises de Physiologie,* 1941, *25,* 401–409.

Terzuolo, C., and Terzian, H. Cerebellar increase of postural tonus after de-afferentation and labyrinthectomy. *Journal of Neurophysiology,* 1953, *16,* 551–561.

Tokuriki, M. Electromyographic and joint-mechanical studies in quadrupedal locomotion. III. Gallop. *Japanese Journal of Veterinary Science,* 1974, *36,* 121–132.

Tokuriki, M. Function of the trunk epaxial muscles in the dog's locomotion (walk and trot). *Japanese Journal of Electroencephalography and Electromyography,* 1976, *4,* 109p. (In Japanese.)

Tokuriki, M. Function of neck epaxial muscles in dog's locomotion (walk and trot). *Japanese Journal of Electroencephalography and Electromyography,* 1977, *5,* 53p. (In Japanese.)

Tosney, T., and Hoyle, G. Automatic entrainment for a cellular learning study. *Society for Neuroscience Abstracts,* 1973, p. 238.

Twitmyer, E. B. *A Study of the Knee-Jerk.* Philadelphia: Winston, 1902.

Udo, M., Oda, Y., Tanaka, K., and Horikawa, J. Cerebellar control of locomotion investigated in cats: discharges from Deiters' neurones, EMG and limb movements during local cooling of the cerebellar cortex. In S. Homma (Ed.), *Progress in Brain Research, vol. 44: Understanding the Stretch Reflex.* New York: Elsevier/North-Holland Biomedical press, 1976.

Udo, M., Matsukawa, K., and Kamei, H. Effects of partial cooling of cerebellar cortex at lobules V and IV of the intermediate part in the decerebrate walking cats under monitoring vertical floor reactions. *Brain Research,* 1979, *160,* 559–564.

Viala, D., and Vidal, C. Evidence for distinct spinal locomotion generators supplying respectively fore- and hindlimbs in the rabbit. *Brain Research,* 1978, *155,* 182–186.

Vukobratović, M., Frank, A. A., and Juričić, D. On the stability of biped locomotion. *IEEE Transactions on Biomedical Engineering,* 1970, *BME-17,* 23–35.

Vukobratović, M., Ciric, V., Hristic, D., and Stepanenko, J. Contribution to the study of anthropomorphic robots. *Proceedings of the IFAC V World Congress,* Paris, 1972, paper 18.1.

Watt, D. G. D. Responses of cats to sudden falls: An otolith-originating reflex assisting landing. *Journal of Neurophysiology,* 1976, *39,* 257–265.

Watt, D. G. D., and Wetzel, M. C. Linear head movements of walking and trotting cats. *Society for Neuroscience Abstracts,* 1977, *3,* p. 280.

Weiss, P. Self-differentiation of the basic patterns of coordination. *Comparative Psychological Monographs,* 1941, *17,* 1–96.

Wentink, G. H. The action of the hind limb musculature of the dog in walking. *Acta Anatomica,* 1976, *96,* 70–80.

Wetzel, M. C., and Stuart, D. G. Ensemble characteristics of cat locomotion and its neural control. *Progress in Neurobiology,* 1976, *7,* 1–98.

Wetzel, M. C., Atwater, A. E., Wait, J. V., and Stuart, D. G. Neural implications of different profiles between treadmill and overground timings in cats. *Journal of Neurophysiology,* 1975, *38,* 492–501.

Wetzel, M. C., Atwater, A. E., and Stuart, D. G. Movements of the hindlimb during locomotion of the cat. In R. M. Herman, S. Grillner, P. S. G. Stein, and D. G. Stuart (Eds.), *Neural Control of Locomotion.* New York: Plenum Press, 1976.

Wetzel, M. C., Atwater, A. E., Wait, J. V., and Stuart, D. G. Kinematics of locomotion by cats with a single hindlimb deafferented. *Journal of Neurophysiology,* 1976, *39,* 667–678.

Wetzel, M. C., Anderson, R. C., Brady, T. H., Jr., and Norgren, K. S. Kinematics of treadmill galloping by cats. III. Coordination during gait conversions and implications for neural control. *Behavioral Biology,* 1977, *21,* 107–127.

Wiener, N. *Cybernetics.* New York: Wiley, 1948.

Wilson, D. M. Genetic and sensory mechanisms for locomotion and orientation in animals. *American Scientist,* 1972, *60,* 358–365.

Zajac, F. E., and Young, J. L. Discharge patterns of motor units during cat locomotion and their relation to muscle performance. In R. M. Herman, S. Grillner, P. S. G. Stein, and D. G. Stuart (Eds.), *Neural Control of Locomotion.* New York: Plenum Press, 1976.

Ziwet, A., and Field, P. *Introduction to Analytical Mechanics.* New York: Macmillan, 1912.

Index